Discrete Groups in Space and Uniformization Problems

Mathematics and Its Applications (*Soviet Series*)

Volume 40

Discrete Groups in Space and Uniformization Problems

by

Boris N. Apanasov

Novosibirsk State University,
and Institute of Mathematics,
Siberian Branch of the Academy of Sciences,
Novosibirsk, U.S.S.R.

and

Department of Mathematics,
Oklahoma University,
Norman, Oklahoma, U.S.A.

KLUWER ACADEMIC PUBLISHERS

DORDRECHT / BOSTON / LONDON

Library of Congress Cataloging-in-Publication Data

Apanasov, B. N. (Boris Nikolaevich)
[Diskretnye gruppy preobrazovaniĭ i struktury mnogoobraziĭ. English]
Discrete groups in space and uniformization problems / by Boris N. Apanasov.
p. cm. -- (Mathematics and its applications (Soviet series) ; v. 40)
Rev., enl. translation of: Diskretnye gruppy preobrazovaniĭ i struktury mnogoobraziĭ.
Includes bibliographical references and index.
ISBN 0-7923-0216-8 (alk. paper)
1. Transformation groups. 2. Conformal mapping. I. Title. II. Series: Mathematics and its applications (Kluwer Academic Publishers). Soviet series ; 40.
QA385.A6313 1991
512--dc20 91-18250

ISBN 0-7923-0216-8

Published by Kluwer Academic Publishers,
P.O. Box 17, 3300 AA Dordrecht, The Netherlands.

Kluwer Academic Publishers incorporates
the publishing programmes of
D. Reidel, Martinus Nijhoff, Dr W. Junk and MTP Press.

Sold and distributed in the U.S.A. and Canada
by Kluwer Academic Publishers,
101 Philip Drive, Norwell, MA 02061, U.S.A.

In all other countries, sold and distributed
by Kluwer Academic Publishers Group,
P.O. Box 322, 3300 AH Dordrecht, The Netherlands.

Printed on acid-free paper

Printed in the Netherlands

TO

MY PARENTS,

ALEKSANDRA MICHAILOVNA AND NIKOLAY ALEKSANDROVICH

SERIES EDITOR'S PREFACE

'Et moi, ..., si j'avait su comment en revenir, je n'y serais point allé.'

Jules Verne

The series is divergent; therefore we may be able to do something with it.

O. Heaviside

One service mathematics has rendered the human race. It has put common sense back where it belongs, on the topmost shelf next to the dusty canister labelled 'discarded non-sense'.

Eric T. Bell

Mathematics is a tool for thought. A highly necessary tool in a world where both feedback and non-linearities abound. Similarly, all kinds of parts of mathematics serve as tools for other parts and for other sciences.

Applying a simple rewriting rule to the quote on the right above one finds such statements as: 'One service topology has rendered mathematical physics ...'; 'One service logic has rendered computer science ...'; 'One service category theory has rendered mathematics ...'. All arguably true. And all statements obtainable this way form part of the raison d'être of this series.

This series, *Mathematics and Its Applications*, started in 1977. Now that over one hundred volumes have appeared it seems opportune to reexamine its scope. At the time I wrote

> "Growing specialization and diversification have brought a host of monographs and textbooks on increasingly specialized topics. However, the 'tree' of knowledge of mathematics and related fields does not grow only by putting forth new branches. It also happens, quite often in fact, that branches which were thought to be completely disparate are suddenly seen to be related. Further, the kind and level of sophistication of mathematics applied in various sciences has changed drastically in recent years: measure theory is used (non-trivially) in regional and theoretical economics; algebraic geometry interacts with physics; the Minkowsky lemma, coding theory and the structure of water meet one another in packing and covering theory; quantum fields, crystal defects and mathematical programming profit from homotopy theory; Lie algebras are relevant to filtering; and prediction and electrical engineering can use Stein spaces. And in addition to this there are such new emerging subdisciplines as 'experimental mathematics', 'CFD', 'completely integrable systems', 'chaos, synergetics and large-scale order', which are almost impossible to fit into the existing classification schemes. They draw upon widely different sections of mathematics."

By and large, all this still applies today. It is still true that at first sight mathematics seems rather fragmented and that to find, see, and exploit the deeper underlying interrelations more effort is needed and so are books that can help mathematicians and scientists do so. Accordingly MIA will continue to try to make such books available.

If anything, the description I gave in 1977 is now an understatement. To the examples of interaction areas one should add string theory where Riemann surfaces, algebraic geometry, modular functions, knots, quantum field theory, Kac-Moody algebras, monstrous moonshine (and more) all come together. And to the examples of things which can be usefully applied let me add the topic 'finite geometry'; a combination of words which sounds like it might not even exist, let alone be applicable. And yet it is being applied: to statistics via designs, to radar/sonar detection arrays (via finite projective planes), and to bus connections of VLSI chips (via difference sets). There seems to be no part of (so-called pure) mathematics that is not in immediate danger of being applied. And, accordingly, the applied mathematician needs to be aware of much more. Besides analysis and numerics, the traditional workhorses, he may need all kinds of combinatorics, algebra, probability, and so on.

In addition, the applied scientist needs to cope increasingly with the nonlinear world and the extra mathematical sophistication that this requires. For that is where the rewards are. Linear models are honest and a bit sad and depressing: proportional efforts and results. It is in the non-linear world that infinitesimal inputs may result in macroscopic outputs (or vice versa). To appreciate what I am hinting at: if electronics were linear we would have no fun with transistors and computers; we would have no TV; in fact you would not be reading these lines.

There is also no safety in ignoring such outlandish things as nonstandard analysis, superspace and anticommuting integration, *p*-adic and ultrametric space. All three have applications in both electrical engineering and physics. Once, complex numbers were equally outlandish, but they frequently proved the shortest path between 'real' results. Similarly, the first two topics named have already provided a number of 'wormhole' paths. There is no telling where all this is leading - fortunately.

Thus the original scope of the series, which for various (sound) reasons now comprises five subseries: white (Japan), yellow (China), red (USSR), blue (Eastern Europe), and green (everything else), still applies. It has been enlarged a bit to include books treating of the tools from one subdiscipline which are used in others. Thus the series still aims at books dealing with:

- a central concept which plays an important role in several different mathematical and/or scientific specialization areas;
- new applications of the results and ideas from one area of scientific endeavour into another;
- influences which the results, problems and concepts of one field of enquiry have, and have had, on the development of another.

Take a nice simple space such as the upper half plane or hyperbolic three space (or higher dimensional spaces), take a discrete group of isometries, and take the quotient. The manifolds one obtains inherit a rich strong structure; one speaks about a geometric structure on that manifold. Does one obtain in this way an interestingly large class of manifolds, or maybe all.

The idea has been around for quite some time (and it can take different forms in different contexts; in Riemannian geometry for instance the quotients G/H, G a suitable Lie group, H a suitable closed subgroup, provide the class of symmetric spaces, a class that is rich enough to test many ideas and rich enough in structure to make the testing (i.e. the calculations) possible).

In two-dimensional topology, the Klein-Poincaré uniformization theorem says that closed manifolds are quotients of the upper half plane by a discrete group of isometries and the central — and at the time most startling — idea of Thurston is that a very large class of closed three manifolds should carry such a structure, or , equivalently, admit a metric of constant negative curvature.

This book is devoted to such questions. This is a field in which analysis, geometry and topology interact vigorously and one needs something like a master guide to be able to learn to appreciate and understand the rich and varied landscape. The author of the present volume is such a guide and the present results of his efforts to write the corresponding guide book proves it. It will take the beginner quickly to open problems, it will provide mature scientists with understanding of the interrelations of this field with other specialization areas, and it also provides the experts with new and previously unpublished results.

The shortest path between two truths in the real domain passes through the complex domain.

J. Hadamard

La physique ne nous donne pas seulement l'occasion de résoudre des problèmes ... elle nous fait pressentir la solution.

H. Poincaré

Never lend books, for no one ever returns them; the only books I have in my library are books that other folk have lent me.

Anatole France

The function of an expert is not to be more right than other people, but to be wrong for more sophisticated reasons.

David Butler

Bussum, May 1991

Michiel Hazewinkel

CONTENTS

Series Editor's Preface vii

Preface to the English edition xiii

From the Preface to the Russian edition xv

Acknowledgements xvii

Chapter 1. Conformal mappings 1

§1. Riemannian metric 1
§2. Quasi-conformal mappings and isometry 2
§3. Möbius transformations 4
1. Stereographic projection
2. Reflections with respect to spheres and planes
3. Liouville's theorem
4. Isometric spheres of Möbius transformations
§4. Automorphisms of a ball and half-space 11
1. Half-space automorphisms
2. Automorphisms of the ball
§5. Fixed points of Möbius transformations . . 15
§6. Invariant domains for Möbius subgroups . . 16

Chapter 2. Geometrical structures on manifolds 19

§1. Fundamental groups and coverings 19
1. Homotopy equivalence
2. Fundamental group
3. Coverings
§2. Homology and Mayer-Vietoris sequences 22
1. Singular homology
2. Exact sequences
3. Mayer-Vietoris sequences
§3. Spaces of constant curvature 26
1. Elliptic space
2. Hyperbolic space
3. Hemisphere of imaginary radius
4. The projective model of hyperbolic space
5. Lower hemisphere
6. Poincaré model in a ball

7. Poincaré model in the half-space $\mathbb{R}^n_+$
8. Spheres and horospheres in the hyperbolic space
§4. Representation of the Möbius group in the matrix group . . . 32
§5. $\mathcal{G}$-structures in manifolds 34
§6. Developing map and holonomy 41
§7. Orbifolds and their $\mathcal{G}$-structures . . . 43

Chapter 3. General properties of discontinuous groups . 51

§1. Basic concepts; elementary groups . . . 51
1. Basic definitions
2. Elementary groups
§2. Action of a group on the discontinuity set . . 54
§3. Some classes of Kleinian groups 61
1. Fuchsian groups
2. Quasi-Fuchsian groups
3. Schottky groups
4. Web groups
5. Degenerate groups
6. Function groups
§4. Fundamental domains 64
§5. The limit set: the non-measurability effect and density property 73

Chapter 4. Discrete groups of hyperbolic isometries . . 77

§1. Hyperbolicity and discreteness conditions for a Möbius group 77
§2. Arithmetic construction of co-compact hyperbolic groups 85
§3. Universal bounds of hyperbolic manifold sizes . 87
§4. Margulis lemma and hyperbolic manifold decomposition 93
§5. Precisely invariant horoballs and parabolic Ω-peaks 101
1. Action of a discrete group in horoballs
2. Parabolic Ω-peaks and horoball parametrization
§6. Crystallographic reflection groups and arithmetic 109

Chapter 5. Geometrically finite discrete Möbius groups . 117

§1. Local finiteness 117
§2. The convex Nielsen hull of the limit set . . 122
§3. The limit set of a geometrically finite group . 126
§4. Finiteness theorems for discrete Möbius groups . 141
1. Finiteness in the plane
2. Finiteness in space
§5. Invariant components for geometrically finite Kleinian groups 154
§6. The geometry of the Nielsen hull boundary . . 168

Chapter 6. Geometric and algebraic structures of discrete groups 191

§1. Discrete groups and geometry of spatial tessellations 191
1. A filling of the space X^n
2. Incomplete fillings of the hyperbolic space
3. Locally finite fillings and completeness of hyperbolic structures
4. Regular tessellations in the hyperbolic space
5. Rigidity of tessellations in H^n for $n \geq 4$
§2. Free amalgamated products of Kleinian groups . 204
§3. HNN-extensions of Kleinian groups. Examples . 212
§4. Cayley graph and combinatorial structure of a discrete group 223
1. Visual boundary of the group G
2. Commensurable metrics
3. Group boundary and the limit set
4. Applications to algorithmic problems
§5. Geometric and algebraic convergence of groups . 236
§6. The limit set: Sullivan's ergodic theorems . 241
1. Infinite solid angle of the orbit and ergodicity
2. Conical approximation
3. Horospherical limiit points and conservative action

Chapter 7. Kleinian manifolds 251

§1. Topological methods 251
§2. Topological aspects of combination theorems . 259
1. Free products
2. Free products with amalgamated cyclic subgroups
3. Double of the manifold $M(G)$ with respect to the boundary $\partial M(G)$
§3. Kleinian manifold of a function group . . 263
1. Three-dimensional manifolds
2. Multi-dimensional Kleinian manifolds
3. Non-trivial four-dimensional cobordisms having geometrically finite hyperbolic structures
4. Block-building construction
§4. Isomorphism theorems 283
§5. Geometrically tame ends, Kleinian manifold compactification and the Ahlfors problem on limit-set measure 290

Chapter 8. Uniformization of manifolds 307

§1. The classical approach 307
§2. Gromov's and Thurston's geometric uniformization concepts 310
1. Length spaces and convexity
2. Thurston's structural conjecture

§3. Thurston's hyperbolization theorem and the volumes of hyperbolic manifolds 318
§4. Complexity of 3-manifolds, integrable Hamiltonian systems and hyperbolic volumes 328
1. The Hamiltonian systems and the class $(\mathbb{H})$
2. The complexity of three-dimensional manifolds
3. Computation of volumes of hyperbolic 3-manifolds
§5. Conformal uniformization of manifolds . . . 333
§6. Conformal uniformization of doubles of atoroidal manifolds 344

Chapter 9. Deformation spaces of groups and structures . 355

§1. Deformations of locally homogeneous structures . 355
1. Marked locally homogeneous structures
2. Teichmüller space of a Riemann surface
3. Stable representations and the holonomy theorem
4. Infinitesimal deformations and obstructions
§2. Rigidity theorems for complete hyperbolic structures and isometry groups 365
1. Finite hyperbolic-volume case
2. Infinite-volume hyperbolic structures
3. Rigidity of manifolds with boundary
§3. Deformation space for a Fuchsian group . . 381
1. Teichmüller space $T(G)$ of a Fuchsian group in the plane
2. The space $T_n(G)$ of conjugacy classes of quasi-Fuchsian representations
3. Properties of the space $T_n(G)$: polyhedral point of view
4. Properties of the space $T_n(G)$ defined by the geometry of the orbifold $H^n G$
5. Infinitesimal deformations for bendings and singularities of the spaces $R_n(G)$ and $T_n(G)$
6. Stamping deformations
7. Stamping-with-torsion along a closed geodesic
§4. The boundary of deformation space for a Fuchsian group 409
§5. Deformations of Kleinian groups and conformal structures 415

References 425

Index 467

PREFACE TO THE ENGLISH EDITION

This book is a revised and substantially enlarged edition of the Russian book "Discrete Transformation Groups and Manifold Structures" published by "Nauka" Academic Publishers in Novosibirsk in 1983, which had its origins in 1980, when I started to write a set of notes based on courses I was teaching on Kleinian groups and uniformization at the Novosibirsk State University.

Due to the 1985 D. Reidel Publishing Company contract for English translation and in relation to my course on discrete groups and geometric function theory in the Centre for Mathematics and Mechanics of the Bulgarian Academy of Sciences and "Kl. Ohridski" Sofia University, I wrote a substantial extension of the Nauka book. This work was completed in Spring 1986. This extension includes many results published in Russian, in particular, on geometrical finite Möbius groups, their invariant components of the discontinuity set and their quotient manifolds. These results are not as well known in the English literature as they deserve to be - for example, the remarkable finiteness theorems of Tetenov for spatial Möbius groups - see Section 5 of Chapter 5, Section 3 of Chapter 7 and Section 2 of Chapter 3.

During the publishing process which, unfortunately, has been very protracted, a lot of important results in the field have been obtained and I have tried to note them in several footnotes and special remarks. These results bring reasons for some new points of view in the field, especially for conformal uniformization of manifolds. The reader may find these new approaches in the author's recent book "Geometry of Discrete Groups and Manifolds" which will soon be published in Russian (however, it does not contain some part of the present book). I hope that this sketch of some new results will excite the enthusiasm of readers and make them want to consult the large and rapidly growing literature in this relatively new field. At the present time there seems to be no comprehensive account of this subject except for the case of Kleinian groups in the plane, where recently the well-known books by A.F. Beardon [4] and B. Maskit [10] were published.

Akademgorodok,
Novosibirsk,
August 1990

FROM THE PREFACE TO THE RUSSIAN EDITION

The foundation and development of the theory of discrete transformation groups was accompanied by a lot of stimuli. One (not mentioning the demands of various branches of mathematics such as differential equations, function theory, geometry, number theory, etc.) was the classical uniformization problem for multivalued analytic functions and consequently the uniformization problem for Riemann surfaces which was solved by different methods at the beginning of the 20th century by F. Klein, H. Poincaré and P. Koebe. Studying such groups, H. Poincaré [2] suggested the idea of setting the relation between the theory of Kleinian groups in the plane viewed as discrete isometry groups of the hyperbolic (Lobachevskii) space motions and the topology of manifolds.

Modern development of methods of topology, geometry and the theory of quasi-conformal mappings enabled a beginning to be made in fulfilling the programme of investigation of discrete group actions in space and structures on manifolds, projected by Henri Poincaré and supported in the sixties by Lars. V. Ahlfors. An important part in the progress achieved was played by the basic geometric and topological ideas of L. Greenberg, B. Maskit and A. Marden, and totally revolutionary was the role of the ideas of W. Thurston which resulted in a new approach to the problem of classification for space forms.

This book is devoted to these questions. The emphasis in on the fundamental geometric and topological parts of the subject. Namely, a central place is occupied by the study of geometric and algebraic properties of discrete groups of spatial domain automorphisms, and their action on the boundary of these domains, and also by the development of geometric methods of solving the problems appearing here. The results thus obtained are essential for the presentation of the theory of deformations for discrete groups and related topics connected with uniformization and the topology of manifolds, with the moduli problem for geometric and conformal structures and the deformation rigidity effects appearing here.

The distinctive feature of the book is the systematic use of geometric methods which compensate for the absence of such effective apparatus as analytic function theory on the plane. The use of these methods is especially fruitful in the case of geometrically finite groups. This class (as a result of its crucial role) of discrete groups and structures occupies the central place in the book.

In writing the book I had in mind three classes of readers.

For graduate students approaching the subject for the first time, the book quickly introduces the up-to-date problems. The first introductory chapters, making the book autonomous, are addressed to them, as well as numerous examples continuing the book by S.L. Krushkal', B.N. Apanasov and N.A. Gusevskii [3].

To the second type of reader, i.e., to mature mathematicians working in other fields and to theoretical physicists, the book gives the new knowledge and understanding of the theory of discrete transformation groups and locally homogeneous structures on manifolds.

To the third class, i.e., to the experts, the book addresses the new material included, some of which is published here for the first time.

The book does not pay much attention to the Kleinian group on the plane - this subject is already covered by many classical and recent books.

ACKNOWLEDGEMENTS

I consider it a pleasant debt to express my deep gratitude to Pavel P. Belinskii who rendered to me constant support and from whom I inherited a love of geometric methods in mathematics.

I owe much to my colleagues and friends, with whom I discussed the subject for a long time. It was thanks to Samuil L. Krushkal' that I became acquainted with this fascinating field of mathematics. For a series of valuable remarks I am indebted to N.A. Gusevskii, V.S. Makarov, G.A. Margulis, A.D. Mednykh, V.V. Nikulin, E.B. Vinberg and O.J. Viro. The material for Section 4 of Chapter 8 was kindly presented to be by S.V. Matveev and A.T. Fomenko. Serious assistance was rendered to be by Andrew V. Tetenov, who has read the manuscript and with whom parts of the book were discussed many times. I owe so much to my wife Zara Ramonova for valuable help in the preparation of the manuscript and appreciable assistance. To them all, I express my heartfelt gratitude.

CHAPTER 1

CONFORMAL MAPPINGS

§1. Riemannian metric

By an *n-dimensional topological manifold* one means a topological Hausdorff space M^n with a denumerable basis, in which each point x_i has a neighbourhood U_i homeomorphic to a Euclidean ball $\varphi_i(U_i) = B^n$. If the latter condition is replaced by a homeomorphism to a ball B^n or to a closed ball $\overline{B}^n$, we arrive at the definition of a *manifold with a boundary*; here any point not having a neighbourhood homeomorphic to an open ball forms the boundary ∂M^n of the manifold M^n, which is a manifold of dimension $n-1$. A compact manifold without a boundary is called *closed*. If the changes of charts $\varphi^i \circ \varphi_j^{-1}$ defined in $\mathbb{R}^n$ for $U_i \cap U_j \neq \varnothing$ are smooth, then we say we have a *smooth manifold*.

Now let M^n be a smooth manifold. For each $x \in M^n$, we denote by $T_x(M^n)$ the *tangent space* of the manifold M^n at a point x (an n-dimensional topological vector space). Consider the set

$$T(M^n) = \bigcup_{x \in M^n} T_x(M^n) \tag{1.1}$$

and the natural projection $p : T(M^n) \longrightarrow M^n$ such that $p^{-1}(x) = T_x(M^n)$. The triple $\{T(M^n), M^n, p\}$ is called a *tangent fibre bundle of the manifold* M^n, while the set $p^{-1}(x)$ is called a *fibre* in the fibre bundle *above a point* $x \in M^n$.

The smooth manifold M^n is called a *Riemannian manifold* if the following smooth mapping is defined:

$$g : T(M^n) \longrightarrow \mathbf{R}, \tag{1.2}$$

which is a positive-definite quadratic form in each fibre $T_x(M^n)$. The Riemannian manifold (M^n, g) is a metric space whose metric is generated by those quadratic forms. That is, for points $y_0, y_1 \in M^n$ we put

$$\rho(y_0, y_1) = \inf_{\gamma} \int_{\gamma} g d\gamma(x) = \inf_{\gamma} \int_0^1 g(\gamma(t)) \cdot |\gamma'(t)| dt, \tag{1.3}$$

where the lower bound is taken over all continuously differentiable curves γ in M^n joining points y_0 and y_1, where $\gamma(0) = y_0$ and $\gamma(1) = y_1$. The function $\rho : M^n \times M^n \longrightarrow \mathbb{R}$ has all the properties of a metric. For our purpose, we may use just this property of the Riemannian metric to define it, further setting aside such "refined things" as tangent fibre bundle and the choice of the scalar product in each fibre.

The curve $\gamma \subset M^n$ is described as being the *shortest* if its length is the least amongst all curves with the same ends. It is clear from (1.3) that the curve γ with ends y_0 and y_1 will be the shortest if and only if its length is $\rho(y_0, y_1)$. The curve $\gamma \in M^n$ is called a *geodesic* if any point on it has a neighbourhood such that each arc of the curve γ in this neighbourhood is the shortest one.

§2. Quasi-conformal mappings and isometry

Let M and M' be metric spaces with metrics ρ and ρ', respectively, and let $f : M \longrightarrow M'$ be a homeomorphism preserving the orientation. For each point $y \in M$, we define a linear dilatation $K(f, y)$ of the homeomorphism f:

$$K(f, y) = \overline{\lim_{r \to 0}} \frac{\sup_{\rho(x,y)=r} \rho'(f(x), f(y))}{\inf_{\rho(x,y)=r} \rho'(f(x), f(y))}. \tag{1.4}$$

If $K(f, y) < \infty$, then the homeomorphism f is called *quasi-conformal*.

If $K(f, y)$ is bounded in M by a certain constant $K < \infty$, then f is called *K-quasi-conformal*, and the number

$$K(f) = \operatorname{ess} \sup_{y \in M} K(f, y) \tag{1.5}$$

is called the *maximal quasi-conformality coefficient*. This definition of quasi-conformality is intrinsic, i.e., is in terms of the metric on the space. When we have to deal with homeomorphisms of manifolds or domains in Euclidean space, we can also use many other equivalent definitions of quasi-conformality such as those due to Caraman [1], Reshetnyak [3], and Väisälä [1].

If M and M' are domains in $\mathbb{R}^n$, this definition is equivalent to f being a homeomorphism from the Sobolev class $W_n^1(M)$, and there exists a constant $K < \infty$ such that

$$|\operatorname{grad} f(x)|^n \le K \cdot \det Df(x), \qquad x \in M, \tag{1.6}$$

where $Df(x)$ is the derivative of the mapping f at the point x (Jacobi matrix).

It is evident that $1 \le K(f) < \infty$. If $K(f) = 1$, or, equivalently, if (1.6) becomes an equality in which $K = 1$, then the mapping f is called conformal in M. In that case (see Gehring [1] and Reshetnyak [2]):

1) there exist an orthogonal matrix A and a number (linear distortion) $|f'(y)| > 0$,

$$|f'(y)| = \lim_{x \to y} \rho'(f(x), f(y))/\rho(x, y) \tag{1.7}$$

such that $Df(y) = |f'(y)| \cdot A$;

2) the mapping f preserves the angles between the curves passing through a point y.

The linear dilatation $K(f, x)$ is invariant under the composition of f with a conformal mapping; that is, let D_1, D_2, and D_3 be domains in $\mathbb{R}^n$ and let $f: D_2 \longrightarrow D_3$ and $h: D_1 \longrightarrow D_2$ be a quasi-conformal and a conformal homeomorphism respectively, then almost everywhere

$$K(f \cdot h, x) = K(f, h(x)).$$

As the quasi-conformal homeomorphism $f: D \longrightarrow D'$ is differentiable almost everywhere, we have a matrix-valued function Df defined almost everywhere in D, while almost everywhere

$$\det Df(x) \neq 0.$$

Following Ahlfors [4, 9], we define a symmetrized and normalized Jacobi matrix:

$$M(f, x) = (\det Df(x))^{-2/n} \cdot Df(x)^{\mathrm{T}} \cdot Df(x), \tag{1.8}$$

where A^{T} denotes the transpose of the matrix A. The matrix $M(f, x)$ gives a more exact measure of the quasi-conformality of the mapping f than $K(f, x)$ (see Ahlfors [9]). It follows directly from the definition of quasi-conformality that the equation $M(f, x) = \mathbf{1}$ for almost all $x \in D$ is a necessary and sufficient condition for 1-quasi-conformality, and consequently (Reshetnyak [2]) for the conformality of the homeomorphism f.

Let the conformal homeomorphism f map a domain $D \subset \overline{\mathbb{R}}^n$ into a domain $D' \subset \overline{\mathbb{R}}^n$ and let there be given a Riemannian metric ρ with density $p(x)$ in the domain D. We then have to consider how to define the metric ρ' in D' (the density function p') in order that f preserves distance, i.e., so that we have an isometry $(D, \rho) \longrightarrow (D', \rho')$. For this purpose, we define, for the point $x' = f(x)$, the density $p'(x')$:

$$p'(x') = p'(f(x)) = p(x)/|f'(x)|. \tag{1.9}$$

If here $\gamma \subset D$ is any curve that goes over to a certain curve $\Gamma \subset D'$, $\Gamma(t) = f(\gamma(t))$, then

$$\begin{aligned}\rho'(f(x), f(y)) &= \inf_{\Gamma} \int_{\Gamma} p'(z)|d\Gamma(z)| = \inf_{\Gamma} \int_0^1 p'(\Gamma(t))|\Gamma'(t)|dt = \\ &= \inf_{\gamma} \int_0^1 p'(f(\gamma(t)))\cdot|f'(\gamma(t))|\cdot|\gamma'(t)|dt = \\ &= \inf_{\gamma} \int_0^1 \frac{p(\gamma(t))}{|f'(\gamma(t))|}\cdot|f'(\gamma(t))|\cdot|\gamma'(t)|dt = \\ &= \inf_{\gamma} \int_{\gamma} p(x)|d\gamma(x)| = \rho(x, y).\end{aligned}$$

Then, in particular, it is clear that an automorphism f of the domain $D \subset \overline{\mathbb{R}}^n$ will be an isometry with the given metric $\rho(x, y)$ only if

$$|f'(x)| = p(x)/p(f(x)), \tag{1.10}$$

where the function $p(x)$ is the density of the metric $\rho(x, y)$.

It is evident from (1.10) that a Euclidean isometry f $(p(x) \equiv 1)$ is characterized by the condition $|f'(x)| \equiv 1$. As another example, we consider all conformal automorphisms of the upper half-plane $H^2 = \{z \in \mathbb{C} : \operatorname{Im} z > 0\}$, which are given, as is known, by the linear fractional mappings $w(z) = (az + b)/(cz + d)$ with real coefficients a, b, c, and d, and with $ad - bc = 1$. We see from (1.10) that a metric in H^2 invariant under these automorphisms must be specified with the density $p(z) = 1/\operatorname{Im} z$, since $\operatorname{Im} w(z) = \operatorname{Im} z \cdot |w'(z)|$. This metric converts the upper half-plane into the two-dimensional hyperbolic space (see Chapter 2 for more details).

§3. Möbius transformations

1. *Stereographic projection.* To introduce a metric into the compactified space $\overline{\mathbb{R}}^n = \mathbb{R}^n \cup \{\infty\}$, we consider a construction described as stereographic projection. We assume that $\overline{\mathbb{R}}^n$ is embedded in $\overline{\mathbb{R}}^{n+1}$ as the plane $\{x \in \overline{\mathbb{R}}^{n+1} : x_{n+1} = 0\}$, which can be mapped one-to-one onto the sphere $S^n(0, 1) = \{x \in \mathbb{R}^{n+1} : |x| = 1\}$, by making each point $\tilde{x} = (x, 0)$ of the plane $\overline{\mathbb{R}}^n \subset \overline{\mathbb{R}}^{n+1}$ correspond to the unique point of intersection $\pi(\tilde{x})$ of the straight line passing through $\tilde{x}$ and $e_{n+1} = (0, \ldots, 0, 1)$ with the sphere $S^n(0, 1)$ (see Fig. 1). This mapping $\pi : \overline{\mathbb{R}}^n \longrightarrow S^n(0, 1)$ is called the *stereographic projection.* It can be written in the form $\pi(\tilde{x}) = \tilde{x} + t\cdot(e_{n-1} - \tilde{x})$, where $t \in \mathbb{R}$ is the number such that $|\pi(\tilde{x})| = 1$. Then we have

$$\pi(\tilde{x}) = \left[\frac{2x_1}{|x|^2 + 1}, \ldots, \frac{2x_n}{|x|^2 + 1}, \frac{|x|^2 - 1}{|x|^2 + 1}\right], \tag{1.11}$$

and by definition $\pi(\infty) = e_{n+1}$.

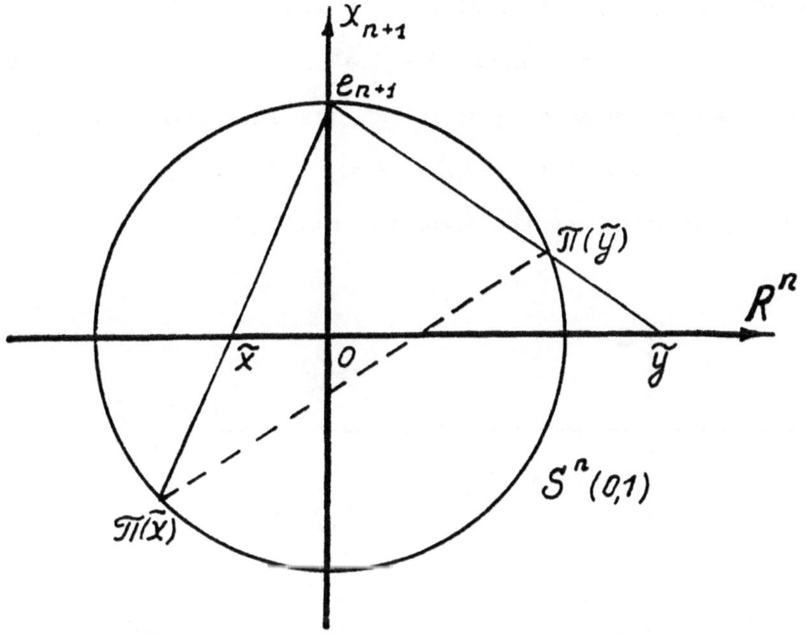

Figure 1

We use the one-to-one mapping in (1.11) to transfer the metric from the sphere $S^n(0, 1)$ (chord length) to $\overline{\mathbb{R}}^n$, where we get a spherical metric $s(x, y)$:

$$s(x, y) = |\pi(\tilde{x}) - \pi(\tilde{y})| = \begin{cases} \dfrac{2|x - y|}{(1 + |x|^2)^{1/2}\cdot(1 + |y|^2)^{1/2}}, & \text{if } x, y \neq \infty \\ \dfrac{2}{(1 + |x|^2)^{1/2}}, & \text{if } y = \infty \end{cases} \tag{1.12}$$

It is readily seen that the topology thereby induced in $\mathbb{R}^n$ coincides with the topology induced there by the Euclidean metric. Therefore, functions continuous in $\mathbb{R}^n$ are continuous also in $\overline{\mathbb{R}}^n$ topology. It is also readily seen that the mapping π is continuous in the Euclidean metric in $\mathbb{R}^n$.

2. *Reflections with respect to spheres and planes.* A reflection (inversion) with respect to a sphere

$S^{n-1}(a, r) = \{x \in \mathbb{R}^n : |x - a| = r\}$ is written in the form

$$\mathcal{T}(x) = a + (r/|x - a|)^2(x - a) = a + r^2(x - a)^*, \qquad (1.13)$$

where x^* denotes inversion in $S^{n-1}(0, 1)$, $x^* = x/|x|^2$.

Reflection with respect to the plane $P(a, r) = \{x \in \mathbb{R}^n : (x, a) = r\}$ (Fig. 2) is written as

$$O(x) = x - 2((x, a) - r) \cdot a^*, \qquad (1.14)$$

where $(*, *)$ is the scalar product in $\mathbb{R}^n$ or, in matrix form

$$O(x) = (\mathbf{1} - 2Q_a) \cdot x + 2ra, \qquad (1.15)$$

where matrix Q_a consists of the elements $a_i a_j/|a|^2$.

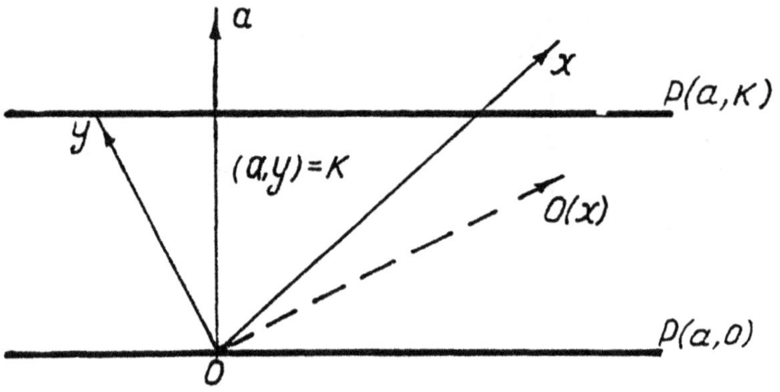

Figure 2

We see from (1.13) - (1.15) that the mappings $\mathcal{T}(x)$ and $O(x)$ are homeomorphisms of $\overline{\mathbb{R}}^n$ (in the spherical metric).

The homeomorphisms of $\overline{\mathbb{R}}^n$ that are superpositions of a finite number of reflections (relative to spheres or planes in $\overline{\mathbb{R}}^n$) are called *Möbius* transformations in $\overline{\mathbb{R}}^n$. The set of all Möbius transformations Möb(n) forms a multiplication group, which is called the *general Möbius group*. A subgroup of this group consisting of orientation preserving automorphisms is called the *Möbius group* and is denoted by $\mathcal{M}_n$.

It is readily seen that a superposition of reflections in planes is a Euclidean isometry. The converse is also true:

PROPOSITION 1.1. *The Euclidean isometries of* $\mathbb{R}^n$ *are exhausted by superpositions of reflections with respect to planes in* $\mathbb{R}^n$.

To prove this, we consider any Euclidean isometry $g(x)$. As the translation $x \to x + a$ is the result of reflections in the planes

$P(a, 0)$ and $P(a, |a|/2)$, we can assume that $g(0) = 0$. The properties of an isometry imply that g preserves vector lengths:

$$|g(x)| = |g(x) - g(0)| = |x|, \tag{1.16}$$

and the angles between vectors:

$$2(g(x), g(y)) = |g(x)|^2 + |g(y)|^2 - |g(x) - g(y)|^2 = 2(x, y). \tag{1.17}$$

It follows from (1.16) and (1.17) that the vectors $g(e_1), \ldots, g(e_n)$ form a basis in $\mathbb{R}^n$, so for each $x \in \mathbb{R}^n$ there exists a vector $p \in \mathbb{R}^n$ such that

$$g(x) = \sum_{i=1}^{n} p_i g(e_i) \tag{1.18}$$

Then in (1.18), $p_i = (g(x), g(e_i)) = x_i$, and so $g(x) = \sum x_i g(e_i)$, i.e., g is a linear transformation that specifies the matrix A, which has rows $g(e_1), \ldots, g(e_n)$. Since the (i, j)-th element of the matrix AA^T is $(g(e_i), g(e_j)) = \delta_{ij}$, we have $AA^T = \mathbf{1}$, i.e., A is an orthogonal matrix.

If $g(e_n) \neq e_n$, we consider the reflection $O_n(x)$ with respect to the plane $P(a, 0)$, where $a = g(e_n) - e_n$. We see from (1.15) that the transformation $O_n \cdot g$ is again specified by an orthogonal matrix, in which the last row coincides with e_n. In fact, the last column in it also coincides with e_n, since the i-th element is $(O_n g(e_i), O_n g(e_n)) = (e_i, e_n) = \delta_{in}$. Then there always exists a transformation O_n (the identity or a reflection with respect to a plane containing 0) such that the matrix for $O_n g$ takes the form

$$\begin{pmatrix} A_1 & 0 \\ 0 & 1 \end{pmatrix}$$

where A_1 is an orthogonal $(n-1) \times (n-1)$-matrix. Then the problem reduces to one with fewer dimensions: what is an isometry of the space $\mathbb{R}^{n-1}$ that leaves 0 fixed? We act as above and get that there exist mappings $O_1, \ldots, O_n$ that are either the identity or are reflections with respect to planes containing 0 such that the mapping $O_1 \circ \ldots \circ O_n \circ g$ is the identity. Then g is a superposition of not more than n reflections with respect to planes containing 0.

COROLLARY 1.2. *Any Euclidean isometry in* $\mathbb{R}^n$ *can be represented as a superposition of not more than* $n + 2$ *reflections.*

3. *Liouville's theorem*:

THEOREM 1.3. *Let* U, V *be open connected subsets of* $\overline{\mathbb{R}}^n$, $n \geq 3$, *and let* $f : U \longrightarrow V$ *be a conformal map. Then* f *is a restriction of a Möbius transformation which is uniquely determined by* f.

To prove this, we need some information from differential geometry.

The values of the normal curvature having stationary values corresponding to the principal directions on the surface are called the *principal curvatures of surface* S at that point. We recall that the principal curvatures at an elliptic point have identical signs, whereas they differ in sign at a hyperbolic point, while at a parabolic point, one or more of the principal curvatures is zero. The lines on S directed at each point along principal directions are called *curvature lines*. On the sphere, and only on it, each line is a curvature one.

If we are given n families of surfaces in a space of dimensions n, each of which is dependent on one real parameter,

$$F_i(x_1, x_2, \ldots, x_n) = C_i, \qquad i = 1, \ldots, n, \tag{1.19}$$

while in the domain of space in which all n functions of (1.19) are defined one surface of each family passes through each point, while the tangent planes to these surfaces at each point in this domain are mutually orthogonal, then we say that the families of surfaces form an n-orthogonal system. Dupin's theorem (see for example Kagan [1]) asserts that the lines on which the surfaces of one of the families in an n-orthogonal system intersect with those in other families are curvature lines. Examples of n-orthogonal families are provided by coordinate planes in Euclidean space and surfaces defined by a spherical coordinate system in $\mathbb{R}^n$.

We note also that any surface $x = x(y_1, \ldots, y_{n-1})$ with normals $n = n(y_1, \ldots, y_{n-1})$ can be included in an n-orthogonal system by supplementing that surface with "parallel surfaces"

$$x(t) = x(y_1, \ldots, y_{n-1}) + n(y_1, \ldots, y_{n-1}) \cdot t$$

and with $n-1$ families of surfaces composed of normals to that surface and dissecting in it coordinate $(n-2)$-dimensional surfaces having $y_k = \text{constant}$ in the coordinate system on the initial surface, whose coordinate lines are curvature ones.

All these facts can be proved at least for objects that are thrice continuously differentiable. Reshetnyak [2] has given the most general proof of Liouville's theorem without any smoothness constraint.

We now consider a proof of the theorem in the smooth case. We first note that by a conformal mapping f, a curvature line of S lying in the domain U becomes a curvature line in the image $f(S)$. In fact, let γ be a curvature line in S, $\gamma_1 = f(\gamma) \subset f(S)$. Then we take γ as one of the coordinate lines $y_i = \text{constant}$ and include the surface S in an n-orthogonal system $\mathcal{T}$. However, as a conformal mapping preserves the angles, the image $f(\mathcal{T})$ of an n-orthogonal system $\mathcal{T}$ is again an n-orthogonal system, so Dupin's theorem gives that the image γ_1 (as the intersection of surfaces in an n-orthogonal system $f(\mathcal{T})$) is a curvature line. This means that the conformal image of a sphere can only be a sphere or a plane, since any line on them is a curvature line.

Now let x_0 be any point in the domain U and $y_0 = f(x_0) \in V$. Applying, if necessary, Möbius transformations, we may assume that

$x_0 = y_0 = \infty$. Consider an n-orthogonal system of planes passing through the point $x_0 = \infty$:

$$x_1 = 0, \ldots, x_n = 0. \tag{1.20}$$

Mapping f converts (1.20) into an n-orthogonal system (of planes) of the form $\sum_{j=1}^{n} c_{ij}(y_j - a_j) = 0$, where (c_{ij}) is an orthogonal matrix, which follows from f being a conformal mapping. Applying a Möbius transformation (Euclidean isometry) again, we may assume that these planes take the form $y_1 = 0, \ldots, y_n = 0$. As straight lines become straight lines in our mapping, i.e., our mapping is an affine one, it follows that it takes the form $y = A \cdot x$, where A is a diagonal matrix. As then we again have that a sphere becomes a sphere, we get directly that the mapping $y = A \cdot x$ is a stretching: $y = r \cdot x$, where r is a scalar constant. This completes the proof of the theorem.

REMARK 1.4. This proof does not apply for $n = 2$, since the family of concentric circles and the family of parallel straight lines on the plane can be included in a 2-orthogonal line system (a distinction from spheres and planes in $\bar{\mathbb{R}}^n$, $n \geq 3$) in a unique fashion: the second family of these systems is represented by the common diameters of the concentric circles or the common perpendiculars for the parallel straight lines. Therefore, there is a difference from the space in that these 2-orthogonal families pass conformally into any 2-orthogonal families, not only into families composed of spheres.

4. *Isometric spheres of Möbius transformations.* Euclidean isometries were described in Theorem 1.1. It is readily seen that the mapping $g \in \text{Möb}(n)$, $|g'(x)| \equiv \text{const}$, can be written in the form $g(x) = r \cdot A(x) + a$, where $A(x)$ is an orthogonal transformation. Such transformations are called similarities.

DEFINITION 1.5. The following sphere is called an *isometric sphere* for the Möbius transformation $g \in \text{Möb}(n)$ different from similarity:

$$I(g) = \{x \in \mathbb{R}^n : |g'(x)| = 1\}. \tag{1.21}$$

This definition is correct. In fact, consider a point $q \in \mathbb{R}^n$ such that $g(q) = \infty$ and let $\mathcal{J}$ be reflection with respect to the sphere $S^{n-1}(q, 1)$, when we can write

$$g \circ \mathcal{J}(x) = r \cdot A(x) + a, \tag{1.22}$$

where $r \in \mathbb{R}$, $a \in \mathbb{R}^n$, and A is an orthogonal transformation. It follows from (1.22) that the following sphere is an isometric sphere for transformation g:

$$\{y \in \mathbb{R}^n : |\det D\mathcal{I}(\mathcal{I}(y))| = r\} = \mathcal{I}(\{x \in \mathbb{R}^n : |\det D\mathcal{I}(x)| = r\}) =$$

$$= \{x \in \mathbb{R}^n : |\det D\mathcal{I}(x)| = 1/r\}.$$

It is readily seen that the image of an isometric sphere $I(g)$ in mapping g is an isometric sphere for the inverse mapping g^{-1}. In fact, as the image $g(I(g))$ is again a certain sphere S near the points of which lengths are not altered by mapping g^{-1} (chain rule), we have by definition that S is the isometric sphere for g^{-1}.

PROPOSITION 1.6. *Any Möbius transformation* $g \in \mathcal{M}_n$ *different from similarity can be represented as a superposition of the inversion* $\mathcal{I}$ *with respect to the isometric sphere* $I(g)$ *and of the reflection* O *with respect to the plane* P_g *such that* $I(g^{-1}) = O(I(g))$ *and possibly the rotation* U *around the centre* q' *of the isometric sphere* $I(g^{-1})$ *for the inverse transformation* g^{-1}.

PROOF. First we note that if a Möbius transformation leaves points on a certain sphere fixed, then either it is an inversion with respect to that sphere or it is the identity mapping. Then if we can show that g coincides with $U \circ O \circ \mathcal{I}$ on the isometric sphere $I(g)$, this will prove the theorem on the basis that $g(q) = U \circ O \circ \mathcal{I}(q) = \infty$.

We consider the points $x^0, x^1, \ldots, x^{n-1}$ on the sphere $I(g)$ such that the vectors $q - x^0, x^1 - x^0, \ldots, x^{n-1} - x^0$ are linearly independent, and we denote their images on the sphere $I(g^{-1})$ by $y^i = g(x^i)$. We have that $g(I(g)) = I(g^{-1})$ and that the interior of $I(g)$ becomes the exterior of $I(g^{-1})$, which applies also for the mapping $O \circ \mathcal{I}(x)$. If $q \neq q'$, we select as the plane P_g the plane passing through the middle of the segment $[q, q']$ and orthogonal to it; if $q = q'$, then we replace here q and q' by x^{i_0} and $g(x^{i_0})$. We then consider the rotation U around the point q' for which $U(O\mathcal{I}(x^{-1})) = y^i$. This can be done because the lengths on the sphere $I(g)$ are unaltered in mappings g and $O \circ \mathcal{I}$. Then it follows that for any point x on $I(g)$ the images $g(x)$ and $U \circ O \circ \mathcal{I}(x)$ coincide, which completes the proof.

COROLLARY 1.7. *A Möbius transformation in* $\overline{\mathbb{R}}^n$ *is dependent on* $(n^2 + 3n + 2)/2$ real parameters.

It is readily seen that these parameters are the following: the coordinates of the points $g(\infty)$ and $g^{-1}(\infty)$ (the centres of the isometric spheres $I(g^{-1})$ and $I(g)$), their common radius, and the parameters of the rotation about $g(\infty)$, i.e., the parameters of an orthogonal transformation in $\mathbb{R}^n$ that preserves the orientation; it is readily seen that there are $n(n-1)/2$ of them.

COROLLARY 1.8. *Any Möbius transformation* $g \in \text{Möb}(n)$ *can be written in the form*

$$g(x) = r^2 U_g V(x) + q' \tag{1.23}$$

where U_g *is an orthogonal transformation,* $r > 0$, $q' \in \mathbb{R}^n$, *while the mapping* $V(x)$ *takes the form*

$$V(x) = \begin{cases} (x-q)/|x-q|^2, & \text{if } \ q' = g(\infty) \neq \infty; \\ x, & \text{if } \ g(\infty) = \infty. \end{cases} \tag{1.24}$$

We note that in the first case $q = g^{-1}(\infty)$ and $q' = g(\infty)$ in (1.24) are the centres of the isometric spheres $I(g)$ and $I(g^{-1})$ having radius r.

PROOF. We note that if $g(\infty) = \infty$, our statement follows from the previous, while if $g(\infty) \neq \infty$, from Proposition 1.6 we write $g(x) = U \circ O \circ \mathcal{I}(x)$. We represent the Euclidean isometry $U \circ O(x)$ as $U_g(x) + a$, where U_g is a certain orthogonal transformation, while the vector $a \in \mathbb{R}^n$ is such that $a = g(\infty) - U_g(q)$. Then

$$\begin{aligned} g(x) &= U \circ O \circ \mathcal{I}(x) = U_g(\mathcal{I}(x)) + g(\infty) - U_g(q) = \\ &= U_g(r^2(x-q)/|x-q|^2 + q) + g(\infty) - U_g(q) = \\ &= r^2 U_g \circ V(x) + g(\infty). \end{aligned} \tag{1.25}$$

Equations (1.25) complete the proof of (1.23).

§4. Automorphisms of a ball and half-space

1. *Half-space automorphisms.* We consider the space $\overline{\mathbb{R}}^n$ as the plane $\{x \in \overline{\mathbb{R}}^{n+1} : x_{n+1} = 0\}$ in the space $\overline{\mathbb{R}}^{n+1}$ and consider the extension of the mappings $g \in \text{Möb}(n)$ into the space. It is clear that such an extension for the similarity $g(x) = r \cdot A \cdot x + a$ will be the similarity $\tilde{g}$ in $\mathbb{R}^{n+1}$, $\tilde{g}(\tilde{x}) = r \cdot \tilde{A} \cdot \tilde{x} + \tilde{a}$, where $\tilde{x} = (x, x_{n+1})$, $\tilde{a} = (a, 0)$, while the orthogonal matrix $\tilde{A} \in O(n+1)$ has the form

$$\tilde{A} = \begin{pmatrix} A & 0 \\ 0 & 1 \end{pmatrix}.$$

For the transformations $g \in \text{Möb}(n)$ different from similarity, we construct the extension as follows. It is evident from Proposition 1.6 that $g(x)$ is a superposition of inversion with respect to the isometric sphere $I(g)$ and a certain Euclidean isometry which is a superposition of not more than $(n+2)$ reflections with respect to $(n-1)$-dimensional planes in $\mathbb{R}^n$. We now consider the sphere S^n in $\mathbb{R}^{n+1}$ orthogonal to the plane $\{x \in \mathbb{R}^{n+1} : x_{n+1} = 0\} = \mathbb{R}^n$ whose intersection with it is the $(n-1)$-dimensional sphere $I(g)$. Similarly, instead of the planes participating in the definition of our isometry, we consider n-planes in $\mathbb{R}^{n+1}$ orthogonal to $\mathbb{R}^n$ whose intersection with it are these $(n-1)$-dimensional planes. The mapping that is a superposition of reflections with respect to this n-dimensional sphere and n-dimensional planes will be the extension of the mapping $g(x)$ in $\overline{\mathbb{R}}^{n+1}$. It is clear that this extension $\tilde{g}$ is unique. It leaves the upper and lower half-spaces $\mathbb{R}^{n+1}_{+} = \{x \in \mathbb{R}^{n+1} : x_{n+1} > 0\}$ and $\mathbb{R}^{n+1}_{-} = \{x \in \mathbb{R}^{n+1} : x_{n+1} < 0\}$ invariant. On the other hand, any conformal automorphism $\tilde{g}$ of the half-space $\mathbb{R}^{n+1}_{+}$ (or $\mathbb{R}^{n+1}_{-}$) induces at its boundary $\partial\mathbb{R}^{n+1}_{+} = \bar{\mathbb{R}}^n$ a conformal mapping g, so the group $\text{Möb}(\mathbb{R}^{n+1}_{+})$ of all conformal automorphisms of $\mathbb{R}^{n+1}_{+}$ is isomorphic to the group $\text{Möb}(n)$.

What should be the metric $\rho(x, y)$ in $\mathbb{R}^{n+1}_{+}$ in which the automorphisms $g \in \text{Möb}(\mathbb{R}^{n+1}_{+})$ are isometries? To answer this question, we find an expression for $|g(x) - g(y)|/|x - y|$ for any points $x, y \in \mathbb{R}^{n+1}_{+}$.

If $\mathcal{T}$ is inversion with respect to a sphere $S^n(a, r)$

$$|\mathcal{T}(x) - \mathcal{T}(y)| = \frac{r^2 \cdot |x - y|}{|a - x| \cdot |a - y|} . \tag{1.26}$$

Equation (1.26) follows readily from considering similar triangles (see Fig.3) with vertices a, x, and y and a, $\mathcal{T}(y)$, and $\mathcal{T}(x)$. At the same time (1.13) shows that the $(n+1)$-th coordinate $[\mathcal{T}(x)]_{n+1}$ of point $\mathcal{T}(x)$ is $r^2 x_{n+1}/|x - a|^2$, so

$$\frac{|\mathcal{T}(x) - \mathcal{T}(y)|^2}{[\mathcal{T}(x)]_{n+1} \cdot [\mathcal{T}(y)]_{n+1}} = \frac{|x - y|^2}{x_{n+1} \cdot y_{n+1}} , \tag{1.27}$$

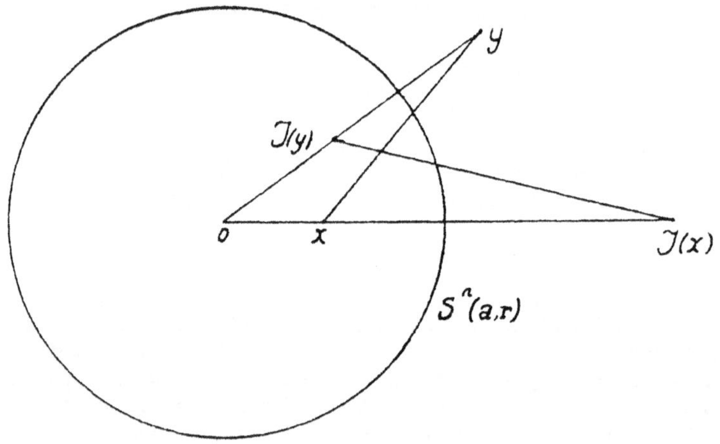

Figure 3

i.e., the ratio in (1.27) is an invariant for inversions with respect to the spheres S^n orthogonal to $\mathbb{R}^n$. It is evident that this ratio is invariant also for reflections with respect to n-dimensional planes orthogonal to $\mathbb{R}^n$. Then Proposition 1.5 gives us that for all $g \in \text{Möb}(\mathbb{R}^{n+1}_{+})$

$$\frac{|g(x) - g(y)|^2}{[g(x)]_{n+1}\cdot[g(y)]_{n+1}} = \frac{|x - y|^2}{x_{n+1}\cdot y_{n+1}}, \tag{1.28}$$

and therefore we have

$$\lim_{y\to x}\frac{|g(x) - g(y)|}{|x - y|} = \frac{[g(x)]_{n+1}}{x_{n+1}}. \tag{1.29}$$

The isometry condition for g follows directly from (1.10) and it follows from (1.29) that the metric $\rho(x, y)$ in $\mathbb{R}^{n+1}_+$ will be an invariant for $\text{Möb}(\mathbb{R}^{n+1}_+)$ if its density $p(x)$ is $1/x_{n+1}$, i.e., the differential of the arc length takes the form $ds = |dx|/x_{n+1}$. The introduction of this metric into that space converts it into a hyperbolic space H^{n+1}, which is dealt with in detail in Chapter 2. Then we have proved the following assertion:

PROPOSITION 1.9. *The Möbius automorphisms of the half-space* $\mathbb{R}^{n+1}_+$ *are hyperbolic isometries.*

2. *Automorphisms of the ball.* Let $\text{Möb}(B^n)$ be the Möbius automorphism group for the ball $B^n(o, 1)$, which consists of superpositions of reflections with respect to spheres (planes) orthogonal to unit sphere $S^{n-1}(0, 1)$. The sphere $S^{n-1}(y, r)$ with centre at a given point $y \in \mathbb{R}^n$, $|y| > 1$, will be orthogonal to the sphere $S^{n-1}(0, 1)$ if its radius is $r = (|y|^2 - 1)^{1/2}$. We see from (1.13) that reflection with respect to $S^{n-1}(y, r)$ is written as

$$\mathcal{T}_y(x) = y + \frac{|y|^2 - 1}{|x - y|^2}(x - y). \tag{1.30}$$

We get from (1.14) and (1.30) that for any point $y \in B^n(0, 1)$ and $y^* = y/|y|^2$ the mapping

$$g_y(x) = (\mathbf{1} - 2Q_y)\mathcal{T}_{y*}(x) \tag{1.31}$$

takes this point y to the origin.

For any other $g \in \text{Möb}(B^n)$; $g(y) = 0$, the transformation $g_y \circ g^{-1}$ leaves the origin fixed, so it is specified by an orthogonal matrix. Then any transformation $g \in \text{Möb}(B^n)$, $g(y) = 0$ is a superposition of the mapping of (1.31) and an orthogonal transformation (rotation): $g = U \circ g_y$, $U \in O(n)$.

We now examine the metrical properties of automorphisms of the ball. A direct calculation gives that

$$D\mathcal{T}_y(x) = ((|y|^2 - 1)/|x - y|^2)(\mathbf{1} - 2Q_{x-y}) \tag{1.32}$$

and the Jacobi matrix for reflection with respect to the plane $P(y, 0)$ is constant and equal to $(\mathbf{1} - 2Q_y)$, so from (1.32) we get for $y \in B^n(0, 1)$ that

$$Dg_y(x) = \frac{1 - |y|^2}{|y|^2 \cdot |x - y^*|^2}(1 - 2Q_y)(1 - 2Q_{x-y*}). \tag{1.33}$$

As the reflection $(1 - 2Q_q)\cdot x$ is a Euclidean isometry, we get from (1.32) that the linear distortion for g_y takes the form

$$|g_y'(x)| = (1 - |y|^2)/|y|^2 \cdot |x - y^*|^2. \tag{1.34}$$

Then in particular this implies:

LEMMA 1.10. *There are isometric spheres* $I(g)$ *and* $I(g^{-1})$ *for the mapping* $g \in \text{Möb}(B^n)$, $g(0) \neq 0$, *which are spheres of radius* $r = \sqrt{1 - |g(0)|^2}/|g(0)|$ *having their centres at* $(g^{-1}(0))^* = g^{-1}(\infty)$ *and* $(g(0))^* = g(\infty)$, correspondingly.

We now find the metric $\rho(x, y)$ in the ball $B^n(0, 1)$ invariant under $\text{Möb}(B^n)$, where we note that the ball $B^n(0, 1)$ is the image of the half-space $\mathbb{R}^n_+$ under the mapping

$$h(x) = \mathcal{T} \circ O(x) = 2(O(x) - e_n)/|O(x) - e_n|^2 + e_n, \tag{1.35}$$

where $O(x) \equiv \bar{x}$ is the reflection with respect to the plane $P(e_n, 0) = \{x \in \mathbb{R}^n : x_n = 0\}$, while $\mathcal{T}(x) = e_n + 2(x - e_n)/|x - e_n|^2$ is an inversion with respect to the sphere $S^{n-1}(e_n, \sqrt{2})$. Then

$$\text{Möb}(B^n) = h\text{Möb}(\mathbb{R}^n_+)h^{-1} \cong \text{Möb}(n-1). \tag{1.36}$$

We use (1.26) to get

$$|h'(x) = \lim_{y \to x} \frac{|h(y) - h(x)|}{|y - x|} = \lim_{y \to x} \frac{|\mathcal{T}(\bar{y}) - \mathcal{T}(\bar{x})|}{|\bar{y} - \bar{x}|} =$$

$$= \lim_{y \to x} \frac{2|\bar{y} - \bar{x}|}{|e_n - \bar{x}| \cdot |e_n - \bar{y}| \cdot |\bar{y} - \bar{x}|} = \frac{2}{|e_n - \bar{x}|^2}. \tag{1.37}$$

Similarly, (1.37) implies

$$1 - |h(x)|^2 = 1 - |\mathcal{T}(\bar{x}) - \mathcal{T}(-e_n)|^2 =$$

$$= 1 - |e_n + \bar{x}|^2 \cdot |\bar{x} - e_n|^{-2} = 4x_n/|\bar{x} - e_n|^2. \tag{1.38}$$

We have directly from (1.37) and (1.38) that

$$|h'(x)| = (1 - |h(x)|^2/2x_n. \tag{1.39}$$

From (1.10) and (1.39) we have an assertion analogous to Proposition 1.9:

PROPOSITION 1.11. *The hyperbolic metric in the ball* $B(0, 1)$ *having the arc differential*

$$ds_1 = 2|dx|/(1 - |x|^2). \qquad (1.40)$$

is invariant for the Möbius group Möb(B^n(0, 1)).

We obtain another proof of the invariance of (1.40) with respect to Möb(B^n) by applying Lemma 1.10 and (1.26) giving the distortion of Euclidean lengths in inversion, with respect to the isometric sphere $I(g)$ in order to calculate the linear distortion

$$|g'(x)| = \lim_{y\to x} |g(x) - g(y)|/|x - y| = (1 - |g(x)|^2)/(1 - |x|^2) \qquad (1.41)$$

together with condition (1.10).

§5. **Fixed points of Möbius transformations**

We have seen above that the action of groups Möb(n) is uniquely extended to the action of the group Möb($\mathbb{R}^{n+1}_+$) in half-space $\mathbb{R}^{n+1}_+$. Moreover, (1.36) and Browder's fixed point theorem show that the following definition is correct.

DEFINITION 1.12. The transformation $g \in$ Möb(n) is called:

1) parabolic if the extended transformation $\tilde{g} \in$ Möb($n+1$) has a unique fixed point in $\overline{\mathbb{R}}^{n+1}_+$ that lies in $\overline{\mathbb{R}}^n = \partial\mathbb{R}^{n+1}_+$;

2) loxodromic if $\tilde{g} \in$ Möb($n+1$) has precisely two fixed points in $\overline{\mathbb{R}}^{n+1}_+$ that lie in $\overline{\mathbb{R}}^n = \partial\mathbb{R}^{n+1}_+$; and

3) elliptic if $\tilde{g} \in$ Möb($n+1$) has a fixed point in $\mathbb{R}^{n+1}_+$.

These classes exhaust group Möb(n). In fact, if $g \in$ Möb(n) has three fixed points p_1, p_2, and p_3 in $\overline{\mathbb{R}}^n$, i.e., is not either loxodromic or parabolic, then all points on the circle $S^1 \subset \overline{\mathbb{R}}^n$ passing through those points are fixed for g, so all points on the two-dimensional sphere $S^2 \subset \overline{\mathbb{R}}^{n+1}$ orthogonal to the plane $\{x \in \mathbb{R}^{n+1} : x_{n+1} = 0\} = \mathbb{R}^n$ and intersecting it on the circle S^1 will be fixed points for the extended transformation $\tilde{g} \in$ Möb($n+1$), so $g \in$ Möb(n) is an elliptic transformation.

It is clear that the Möbius transformation $g \in$ Möb(n) and the conjugate one $\gamma g \gamma^{-1} \in$ Möb(n), where $\gamma \in$ Möb(n), are of the same type, and the parabolic transformation $g \in$ Möb(n) is conjugate to the transformation $\gamma g \gamma^{-1}(x) = A(x) + a$, where the vector $a \in \mathbb{R}^n \setminus \{0\}$ is not orthogonal to the eigen subspace of the orthogonal transformation A. In fact, for this purpose it is sufficient to take as $\gamma \in$ Möb(n) a transformation that converts the unique fixed point p, $g(p) = p$, to ∞. Then the transformation $\gamma g \gamma^{-1}$ is obliged to be a Euclidean isometry, since otherwise the method of contraction mappings can be used to show that there is a fixed point for $\gamma g \gamma^{-1}$ in $\mathbb{R}^n$ different from ∞. If on the other hand the vector $a \in \mathbb{R}^n \setminus \{0\}$ is orthogonal to the eigen subspace of A, it is clear that g is obliged to be an elliptic transformation. If $A = \mathrm{id}$, the parabolic transformation g is called purely parabolic.

The loxodromic transformation $g \in \text{Möb}(n)$ that has precisely two fixed points $p, q \in \overline{\mathbb{R}}^n$ is conjugated by means of the transformation $\gamma \in \text{Möb}(n)$, $\gamma(p) = 0$ and $\gamma(q) = \infty$ with the transformation $\gamma g \gamma^{-1}(x) = r \cdot A(x)$, where $r > 0$, $r \neq 1$, while A is an orthogonal transformation. If $A = \text{id}$, then g is called a hyperbolic transformation.

Finally, an elliptic transformation $g \in \text{Möb}(n)$ having fixed points in $\overline{\mathbb{R}}^n$ is conjugate with a transformation of the form $A(x) + a$, where the vector $a \in \mathbb{R}^n$ is orthogonal to the eigen subspace of the orthogonal transformation A. This follows from the fact that in this case a sphere $S^k \subset \overline{\mathbb{R}}^{n+1}$ of minimal dimension k, $1 \leq k \leq n-1$, that is orthogonal to $\mathbb{R}^n$ and contains a fixed point in $\mathbb{R}^{n+1}_+$ of the extended transformation $\tilde{g}$ and the entire set in $\overline{\mathbb{R}}^n$ of fixed points of g will consist of fixed points for $\tilde{g}$. As $\gamma \in \text{Möb}(n)$ we take a transformation that converts one point from $S^k \cap \overline{\mathbb{R}}^n$ to ∞ and another to 0 to get that $\gamma g \gamma^{-1} \in O(n)$ is an orthogonal transformation.

In dimension $n = 2m + 1$ (and only in that case), there are elliptic transformations $g \in \mathcal{M}_n$ not having fixed points in $\overline{\mathbb{R}}^n$, which are in particular $g(x) = U \circ O \circ \mathcal{I}(x)$ from Theorem 1.6 such that $\dim[I(g) \cap I(g^{-1})] = n - 2$, while $U \neq \text{id}$ is rotation around a straight line passing through the centres of the isometric spheres $I(g)$ and $I(g^{-1})$ (see Apanasov [1]); $(U(x) \neq x$ for $x \in I(g) \cap I(g^{-1}))$. This implies the following assertions:

COROLLARY 1.13. *The fixed points of a transformation* $g \in \mathcal{M}_n$, $g \neq \text{id}$, *lie on a certain* $(n-2)$*-dimensional sphere (plane) in* $\overline{\mathbb{R}}^n$.

COROLLARY 1.14. *Any* $g \in \mathcal{M}_n$ *is uniquely defined by its values at* $n+1$ *points in* $\overline{\mathbb{R}}^n$ *not lying on any* $(n-2)$*-dimensional sphere (plane).*

§6. Invariant domains for Möbius subgroups

We consider a domain $D \subset \overline{\mathbb{R}}^n$, e.g., homeomorphic to a ball. The question arises: what is the group

$$G_D = \{g \in \text{Möb}(n) : g(D) = D\}, \tag{1.42}$$

if D is not a ball or a half-space. We shall see subsequently (Chapter 6) that this group can contain discrete subgroups Γ having a co-compactness property: D is covered by the images $\gamma(F)$, $\gamma \in \Gamma$, where F is a certain compact set in D.

Here another question arises: whether there are D different from balls for which this property of quasi-homogeneity goes over to the stronger property of homogeneity, when F can be taken as a point $x \in D$. The following result due to Kimel'fel'd [1] gives a negative answer to this.

PROPOSITION 1.15. *All conformally homogeneous domains on the sphere* $\overline{\mathbb{R}}^n$, *i.e., domains in which the group of* (1.42) *acts transitively, belong to one of the following types*:

1) $D=\overline{\mathbb{R}}^n$;
2) $D=\mathbb{R}^n$;
3) $D=\overline{\mathbb{R}}^n\setminus S^k$, *where* S^k *is a sphere of dimensions* k, $0\le k\le n-2$;
4) D *is a ball (half-space) in* $\overline{\mathbb{R}}^n$.

One can give a general answer on the domains D for which the group of (1.42) is sufficiently large or, what is the same, on the character of groups of the form of (1.42) by using Greenberg's [1] description of the Lie subgroups of the group Möb(n) on noting that the groups of (1.42) are closed and are consequently Lie subgroups in Möb(n). Also, one can always assume that the action of groups Möb(n) is extended to the half-space $\mathbb{R}^{n+1}_+$, which is converted to the hyperbolic space H^{n+1} by introducing the hyperbolic metric into it (see Proposition 1.9).

If $L\subset H^{n+1}$ is a hyperbolic plane whose dimension does not exceed n, then by $\mathcal{K}(L)$ we denote a subgroup of the Möbius group $\text{Möb}(n)\simeq \text{Isom}\, H^{n+1}$ that leaves plane L pointwise fixed, while by $\mathcal{M}(L)$ we denote a subgroup in that group consisting of isometries of the plane L in the metric induced by the hyperbolic space H^{n+1} and by $\mathcal{M}_+(L)$ we denote the orientation-preserving subgroup. Then the complete stabilizer subgroup for the plane L in that group is obtained as the direct product of the groups $\mathcal{K}(L)\times\mathcal{M}(L)$.

We now formulate the result.

THEOREM 1.16. *For any proper subdomain* $D\subset\overline{\mathbb{R}}^n$, *the group* $G_D\subset\text{Möb}(n)$ *belongs to one of the following types*:

1) G_D *is discrete*;
2) *there exists a plane* $L\subset H^{n+1}$ *such that* $G_D=G\times\mathcal{M}_+(L)$ *for some group* $G\subset\mathcal{K}(L)$;
3) *there exists a straight line* $l\subset H^{n+1}$ *such that* $G_D=G\times\mathcal{M}(l)$, *where* $G\subset\mathcal{K}(l)$, *while* $\mathcal{M}(l)\subset\text{Möb}(n)$ *is a one-parameter loxodromic group with axis* l;
4) G_D *has a common fixed point in* $\overline{\mathbb{R}}^n$ *and is conjugate in* Möb(n) *to a group of Euclidean isometries in* $\mathbb{R}^n$; *and*
5) G_D *has a common fixed point in* H^{n+1} *and is conjugate in* Möb($n+1$) *to a group of orthogonal transformations of the sphere* S^n.

REMARK 1.17. It is clear that the hyperbolic planes L and straight lines l mentioned in the theorem should be spanned by a subset of the boundary ∂D, i.e., one should have

$$\overline{L}\cap\overline{\mathbb{R}}^n\subset\partial D,\qquad \overline{l}\cap\overline{\mathbb{R}}^n\subset\partial D. \tag{1.43}$$

COROLLARY 1.18. *If a domain* $D\subset\overline{\mathbb{R}}^3$ *is bounded by the Jordan surface* $S=\partial D$ *different from a sphere and having an open subset of points in which there exist tangent planes to* S, *then the group* $G_D=\{g\in\text{Möb}(3):g(D)=D\}$ *is conjugate in the group* Möb(3) *either to the orthogonal extension of a one-parameter homothety group in* $\mathbb{R}^3$ *or to some group of Euclidean isometries there.*

The proof of the corollary follows directly from Theorem 1.16, Remark 1.15, and the density properties of loxodromic fixed points (see Lemma 3.22).

To complete the exposition, we formulate here fundamental result used above (Theorem 1 in Greenberg[1]):

THEOREM 1.19. *Let G be a connected Lie subgroup of the group of all Möbius transformations for the sphere $\overline{\mathbb{R}}^n$, which can be identified with the isometry group of the hyperbolic space H^{n+1}, $\partial H^{n+1} = \overline{\mathbb{R}}^n$.*

Then one of the following assertions applies:

1) *the elements of G have a common fixed point in H^{n+1}, while the group G itself is conjugate to a Lie subgroup of the orthogonal group $O(n+1)$;*

2) *the elements of G have a common fixed point in $\overline{\mathbb{R}}^n = \partial H^{n+1}$, while G itself is conjugate to a subgroup of the similarity group in Euclidean space $\mathbb{R}^n$;*

3) *there exists a hyperbolic straight line $l \subset H^{n+1}$ such that $G = \Gamma \times \mathcal{M}(l)$, where $\Gamma \subset \mathcal{K}(l)$, while $\mathcal{M}(l)$ is a one-parameter loxodromic group with axis l;*

4) *there exists a hyperbolic plane $L \subset H^{n+1}$, $2 \leq \dim L \leq n$, such that $G = \Gamma \times \mathcal{M}_+(L)$, where $\Gamma \subset \mathcal{K}(L)$, while $\mathcal{M}_+(L) = \mathcal{M}_n \cap \mathcal{M}(L)$;*

5) $G = \mathcal{M}_n$.

From Theorem 1.19 we get directly:

THEOREM 1.20. *If a Möbius group G in $\overline{\mathbb{R}}^n$ does not have a point in H^{n+1} or a k-dimensional sphere in $\overline{\mathbb{R}}^n = \partial H^{n+1}$, $k \leq n-1$, invariant with respect to G and if there exists a non-empty open subset $U \subset \mathcal{M}_n$ such that $U \cap G = \emptyset$, then G is a discrete group.*

CHAPTER 2

GEOMETRICAL STRUCTURES ON MANIFOLDS

In this chapter I give some necessary preliminary information on constant curvature spaces and covering spaces, and I introduce the important concept of geometrical structure upon a manifold, which is explained in several important examples. I refer the reader to the books by Helgason [1] and Wolf [1] for more detailed acquaintance with the theory of constant curvature spaces. In presenting topics related to geometrical structures on manifolds, I follow Thurston's lectures [1].

§1. Fundamental groups and coverings

1. *Homotopy equivalence.* Let X and Y be topological spaces, with I the segment $[0, 1]$ and let $f: X \longrightarrow Y$ and $g: X \longrightarrow Y$ two continuous mappings. The mappings f and g are called homotopic ($f \simeq g$) if there exists a continuous mapping $F: X \times I \longrightarrow Y$ such that $F(x, 0) \equiv f(x)$ and $F(x, 1) \equiv g(x)$ in X. It is readily shown that the mapping homotopy is an equivalence relation, which splits up the set of all continuous mappings $X \longrightarrow Y$ into pairwise non-intersecting classes, which are called *homotopy classes*. The homotopy class of the mapping f is denoted by $[f]$.

The topological spaces X and Y are called homotopically equivalent if there exist continuous mappings $f: X \longrightarrow Y$ and $g: Y \longrightarrow X$ such that the mappings $f \circ g: Y \longrightarrow Y$ and $g \circ f: X \longrightarrow X$ are homotopic to the identity. If these mappings are themselves identies,then we have a case of homeomorphism. The class of homotopically equivalent spaces is called a homotopy type. The concept of homotopy equivalence is much wider than that of homeomorphism (topological equivalence). For example, a circle and an annulus have the same homotopy type, but they are not homeomorphic.

2. *Fundamental group.* Let X be a topological space and $x_0 \in X$ an arbitrary fixed point. We consider closed curves (loops)

$f_1 : [0, 1] \longrightarrow X$ and $f_2 : [0, 1] \longrightarrow X$, $f_i(0) = f_i(1) = x_0$. By their product we mean the loop

$$f_2 \cdot f_1(t) = \begin{cases} f_1(2t), & 0 \le t \le 1/2; \\ f_2(2t - 1), & 1/2 \le t \le 1; \end{cases} \tag{2.1}$$

This operation preserves homotopy, namely if $f_1 \simeq f_2$, $g_1 \simeq g_2$, then $f_1 \cdot g_1 \simeq f_2 \cdot g_2$.

We now consider the set $\pi_1(X, x_0)$ of homotopy classes of loops at the point x_0 together with the multiplication operation defined by $[f_1] \cdot [f_2] = [f_1 \cdot f_2]$ and (2.1). This set $\pi_1(X, x_0)$ is called the fundamental group of the space X at the point x_0. It is clear that if X is arcwise connected, then for any two points $x_0, y_0 \in X$ the groups $\pi_1(X, x_0)$ and $\pi_1(X, y_0)$ are isomorphic, so we can speak of the abstract fundamental group $\pi_1(X)$ of the arcwise connected space X.

If $\varphi : X \longrightarrow Y$ is a continuous mapping of topological spaces, it induces a homomorphism

$$\varphi_* : \pi_1(X, x_0) \longrightarrow \pi_1(Y, \varphi(x_0)) \tag{2.2}$$

in accordance with the rule $\varphi_*[f] = [\varphi \circ f]$. In particular, if φ is a homeomorphism, then the homomorphism φ_* from (2.2) is an isomorphism.

3. *Coverings.* Let X be a topological space. A *covering space* over X or a *covering* is defined to be a pair $(\tilde{X}, \pi)$ consisting of a topological space $\tilde{X}$ and a continuous mapping $\pi : \tilde{X} \longrightarrow X$, satisfying the condition that each point $x \in X$ has an arcwise connected (open) neighbourhood U such that any connected component of the set $\pi^{-1}(U)$ is mapped homeomorphically onto U under the mapping π. The mapping π is called a *covering mapping* or *projection.*

Covering spaces are closely related to fundamental groups. The following important theorems establish a relation between the fundamental group of a space X and of a space covering it.

LEMMA 2.1 (path lifting theorem). *For any path $f : I \longrightarrow X$ with initial point $x_0 \in X$ there exists a unique covering path $\tilde{f} : I \longrightarrow \tilde{X}$ with initial point $\tilde{x}_0 \in \pi^{-1}(x_0)$ such that $\pi \circ \tilde{f} = f$.*

LEMMA 2.2 (homotopy covering theorem). *If two paths f_1 and f_2 in a space X with origin at a point x_0 and a common end are homotopic, then the paths covering them $\tilde{f}_1$ and $\tilde{f}_2$ with origin at the point $\tilde{x}_0 \in \pi^{-1}(x_0)$ are also homotopic, and further, they have a common end.*

LEMMA 2.3. *The homomorphism $\pi_* : \pi_1(\tilde{X}, \tilde{x}_0) \longrightarrow \pi_1(X, x_0)$, $\tilde{x}_0 \in \pi^{-1}(x_0)$, induced by the projection $\pi : \tilde{X} \longrightarrow X$ is a monomorphism.*

LEMMA 2.4. *Let $\tilde{x}_0$ and $\tilde{x}_1$ be points in $\tilde{X}$ with the same projection $x_0 \in X$. Then the images of the fundamental groups $\pi_1(\tilde{X}, \tilde{x}_0)$ and $\pi_1(\tilde{X}, \tilde{x}_1)$ under the monomorphism π_* are conjugate subgroups of the fundamental group $\pi_1(X, x_0)$; on the other hand, each* *subgroup of the fundamental group* $\pi_1(X, x_0)$ *conjugate to the subgroup* $\pi_*(\pi_1(\tilde{X}, \tilde{x})$ *is obtained as the image* $\pi_*(\pi_1(\tilde{X}, \tilde{x}_1)$ *for some point* $\tilde{x}_1 \in \pi^{-1}(x_0)$.

In other words, the images $\pi_*(\pi_1(\tilde{X}, \tilde{x})$ over all $\tilde{x} \in \pi^{-1}(x_0)$, form a complete class of conjugate subgroups in the group $\pi_1(X, x_0)$.

DEFINITION 2.5. Two coverings $(\tilde{X}, \pi)$ and $(\tilde{X}', \pi')$ over a space X are called equivalent if there exists a homeomorphism $\varphi : \tilde{X} \longrightarrow \tilde{X}'$ such that the diagram

$$\begin{array}{ccc} \tilde{X} & \xrightarrow{\varphi} & \tilde{X}' \\ & {}_{\pi}\searrow \quad \swarrow_{\pi'} & \\ & X & \end{array}$$

commutes.

LEMMA 2.6. *Two coverings* $(\tilde{X}, \pi)$ *and* $(\tilde{X}', \pi')$ *over* X *are equivalent if and only if for any two points* $\tilde{x} \in \tilde{X}$ *and* $\tilde{x}' \in \tilde{X}'$ *for which* $\pi(\tilde{x}) = \pi'(\tilde{x}') = x$ *the subgroups* $\pi_*(\pi_1(\tilde{X}, \tilde{x}))$ *and* $\pi'_*(\pi_1(\tilde{X}', \tilde{x}'_0))$ *belong to the same class of conjugate subgroups in the group* $\pi_1(X, x)$.

DEFINITION 2.7. The name *covering homeomorphism* or *deck transformation* is given to a homeomorphism $f : \tilde{X} \longrightarrow \tilde{X}$ for which the following diagram is commutative:

$$\begin{array}{ccc} \tilde{X} & \xrightarrow{f} & \tilde{X} \\ & {}_{\pi}\searrow \quad \swarrow_{\pi} & \\ & X & \end{array}$$

The set of deck transformations of a covering space $(\tilde{X}, \pi)$ is a group (with respect to the mapping superposition) that we denote by $G(\tilde{X}, X)$. We also denote by $N[\pi_*(\pi_1(\tilde{X}, \tilde{x}))]$ the normalizer of the subgroup $\pi_*(\pi_1(\tilde{X}, \tilde{x}))$ in the fundamental group $\pi_1(X, x))$.

LEMMA 2.8. *For any points* $x \in X$ *and* $\tilde{x} \in \pi^{-1}(x)$ *we have the isomorphism*

$$G(\tilde{X}, \tilde{x}) \cong N[\pi_*(\pi_1(\tilde{X}, \tilde{x}))]/\pi_*(\pi_1(\tilde{X}, \tilde{x})). \tag{2.3}$$

DEFINITION 2.9. A covering $(\tilde{X}, \pi)$ is called *regular* if $\pi_*(\pi_1(\tilde{X}, \tilde{x}))$ is a normal subgroup of the fundamental group $\pi_1(X, x)$ at the point $x = \pi(\tilde{x})$.

LEMMA 2.10. *A covering* $(\tilde{X}, \pi)$ *is regular if and only if all liftings to* $\tilde{X}$ *of any loop at an* $x \in X$ *with initial point at an arbitrary point* $\tilde{x} \in \pi^{-1}(x)$ *are simultaneously either closed or open paths. Then we have the isomorphism*

$$G(\tilde{X}, X) \cong \pi_1(X, x)/\pi_*(\pi_1(\tilde{X}, \tilde{x})). \tag{2.4}$$

DEFINITION 2.11. A covering $(\tilde{X}, \pi)$ of a space X is called *universal* if the space $\tilde{X}$ is simply connected, i.e., it is arcwise connected and has the trivial fundamental group $\pi_1(\tilde{X})$.

It is clear that, if $(\tilde{X}, \pi)$ is a universal covering, then instead of the isomorphisms of (2.3) and (2.4) we have the isomorphism

$$G(\tilde{X}, X) \cong \pi_1(X) \tag{2.5}$$

and the order of the group $\pi_1(X)$ is equal to the number of covering sheets, i.e., the common cardinality of the fibres $\pi^{-1}(x)$, which does not depend on the choice of the point $x \in X$.

LEMMA 2.12. *The group* $G(\tilde{X}, X)$ *acts on the fibres* $\pi^{-1}(x)$, $x \in X$, *transitively if and only if* $(\tilde{X}, \pi)$ *is a regular covering. Then one has a natural homeomorphism of the space* X *onto* $\tilde{X}/G(\tilde{X}, X)$, *while the group* $G(\tilde{X}, X)$ *acts in* $\tilde{X}$ *freely (without fixed points) and discontinuously.*

Here by the discontinuous action of $G(\tilde{X}, X)$ on $\tilde{X}$ we understand a property such that each point $\tilde{x} \in \tilde{X}$ has a neighbourhood U such that $g(U) \cap U = \varnothing$ for all $g \in G(\tilde{X}, X) \setminus \{\mathrm{id}\}$.

The second assertion of Lemma 2.12 is reversed in the following form.

LEMMA 2.13. *If space* X *is arcwise connected and* G *is a discontinuous group of homeomorphisms that acts freely, then* (X, π) *is a regular covering over* X/G, *and* $G = G(X, X/G)$.

LEMMA 2.14. *Let* M *be a topological manifold. Then for any class* $\mathfrak{K}$ *of conjugate subgroups of group* $\pi_1(M, x)$, $x \in M$, *there exists a covering space* $(\tilde{M}, \pi)$ *corresponding to that class* $\mathfrak{K}$, *i.e., such that* $\pi_*(\pi_1(\tilde{M}, \tilde{x})) \in \mathfrak{K}$.

This assertion implies that for any subgroup $H \subset \pi_1(M, x)$ there exists a covering $(\tilde{M}, \pi)$ over M such that $\pi_*(\pi_1(\tilde{M}, \tilde{x}) = H$ for some point $\tilde{x} \in \pi^{-1}(x)$. In that case, we call the subgroup $H \subset \pi_1(M, x)$ the *defining subgroup of the covering* $(\tilde{M}, \pi)$, while as regards the covering itself, we say that it is constructed from the subgroup H.

§2. Homology and Mayer-Vietoris sequences

This section contains definitions and some necessary facts about singular homology and relative homology groups of topological spaces. Here also we introduce exact Mayer-Vietoris sequences, important for what follows (Chapter 5 §5), which enable one to calculate certain homology groups. The proofs of the facts and the detailed exposition of homology theory can be found in standard books on algebraic topology such as Dold [1] and Spanier [1].

1. *Singular homology*. Let X be a topological space, while

Δ_q is a q-dimensional $(q \geq 0)$ standard simplex:

$$\Delta_q = \{X \in \mathbb{R}^{q+1} : \sum_{i=0}^{q} X_i = 1,\ X_0 \geq 0, \ldots, X_q \geq 0\}.$$

A singular q-dimensional simplex of X is a continuous mapping $\delta^q : \Delta_q \longrightarrow X$. We consider the free Abelian group S_qX which is generated by the set of all singular q-simplexes. The elements of this group are called singular q-chains of X, i.e., formal linear combinations of singular q-simplexes with integer coefficients.

We define the boundary homomorphism $\partial_q : S_qX \longrightarrow S_{q-1}X$ by setting it on each singular simplex σ^q from the formula

$$\partial_q(\sigma^q) = \sum_{i=0}^{q} (-1)^i \sigma_i^{q-1},$$

where $\sigma_i^{q-1} = \sigma^q|_{\Delta_{q-1}}$ is a restriction of the mapping σ^q on the i-th face, namely Δ_{q-1}^i of the standard simplex Δ_q. It is readily checked that $\partial_q \partial_{q+1} = 0$, so consequently $\operatorname{Ker} \partial_q \supset \operatorname{Im} \partial_{q+1}$.

The group $H_qX = \operatorname{Ker} \partial_q / \operatorname{Im}\text{-}\partial_{q+1}$ is called the *q-dimensional (singular) homology group.* The elements of the group $Z_qX = \operatorname{Ker} \partial_q = \partial_q^{-1}(0)$ are called *q-dimensional cycles,* while the elements of the subgroup $B_qX = \operatorname{Im} \partial_{q+1} = \partial_{q+1}(S_{q+1}X)$ are called *q-dimensional boundaries.* In this notation $H_qX = Z_qX/B_qX$, while two cycles $z_q, z'_q \in Z_qX$ are homologic (i.e., define one element from H_qX) if their difference is a boundary, $z_q - z'_q \in B_qX$.

The name *singular complex* SX of the space X is given to the sequence

$$0 \longleftarrow \mathbb{Z} \xleftarrow{\varepsilon} S_0X \longleftarrow S_1X \longleftarrow \ldots \longleftarrow S_{q-1}X \xleftarrow{\partial_q} S_qX \xleftarrow{\partial_{q+1}} S_{q+1}X \longleftarrow \ldots$$

Here the epimorphism ε is defined on the zero-dimensional chains $\sum_i k_i \sigma'_0 \in S_0X$ from the formula

$$\varepsilon\left(\sum_i k_i \sigma'_0\right) = \sum_i k_i \in \mathbb{Z}.$$

The above is generalized in a natural fashion to pairs of spaces (X, A). If $i : A \longrightarrow X$ is an inclusion, then the mapping of singular complexes $i : SA \longrightarrow SX$ is monomorphic, and so SA may be considered as a subcomplex of the complex SX. The quotient complex $S(X, A) = SX/SA$ is the (relative) singular complex of the pair (X, A). The resulting homology groups $H_q(X, A)$ of the complex $S(X, A)$ are called relative homology groups of the space $X \bmod A$. We note that the chain $z \in SX$ is a cycle $\bmod A$ if $\partial z \in SA$ and z is a boundary $\bmod A$ if $z = \partial x + y$ for some $x \in SX$, $y \in SA$. It is evident that $H_q(X, A)$ is isomorphic to the quotient group of q-cycles $\bmod A$ with respect to the group of q-boundaries mod A.

This epimorphism ε clearly has the property $\varepsilon\partial_1 = 0$. We use this and determine the reduced complex $\tilde{S}X$ of the complex SX by putting $\tilde{S}_qX = S_qX$ if $q \neq 0$, $\tilde{S}_0X = \operatorname{Ker}\varepsilon$ and $\tilde{\partial}_q = \partial_q$ (we note that $\partial_1(\tilde{S}, X) \subset \tilde{S}_0X$). The homology group for the reduced complex $\tilde{S}X$ is called the reduced homology group of the space X and is denoted by $\tilde{H}X$. If $q \neq 0$, then $\tilde{H}_qX = H_qX$.

2. *Exact sequences.* There is a concise way of summarizing the relations among the homology groups of complexes C', C, and C/C', where C' is a subcomplex of C. This is based on the concept of exact sequences.

A three-term sequence of Abelian groups and homomorphisms

$$G' \xrightarrow{\alpha} G \xrightarrow{\beta} G''$$

is said to be exact at the term G if $\operatorname{Ker}\beta = \operatorname{Im}\alpha$.

A sequence of Abelian groups and homomorphisms

$$\ldots \longrightarrow G_{n+1} \xrightarrow{\alpha_{n+1}} G_n \xrightarrow{\alpha_n} G_{n-1} \longrightarrow \ldots$$

is called an exact sequence if every three-term subsequence of consecutive groups is exact at its middle group.

We put for this sequence that $G'_n = \operatorname{Ker}\alpha_n = \operatorname{Im}\alpha_{n+1}$. Then the given sequence gives rise to short exact sequences

$$0 \longrightarrow G'_n \xrightarrow{\alpha_{n+1}} G_n \xrightarrow{\alpha_n} G'_{n-1} \longrightarrow 0,$$

for every G_n not on one or the other end of the original sequence, and the composite $G_n \longrightarrow G'_{n-1} \longrightarrow G_{n-1}$ equals α_n. Here α_{n+1} is a monomorphism and α_n is an epimorphism, while the kernel of α_n coincides with $\alpha_{n+1}(G'_n)$. Consequently, α_{n+1} is an isomorphism of the group G'_n on the subgroup $\alpha_{n+1}(G'_n) \subset G'_n$, while α_n induces an isomorphism of quotient group $G_n/\alpha_{n+1}(G'_n)$ on the group G'_{n-1}.

An example of an exact sequence is the following (i is the inclusion and j the projection):

$$0 \longrightarrow S_qA \xrightarrow{i} S_qX \xrightarrow{j} S_q(X, A) \longrightarrow 0,$$

which characterizes the above q-dimensional relative homology groups of the space X mod A. If $r : X \longrightarrow A$ is a retraction, then $r_* : S_qX \longrightarrow S_qA$ splits it for all q:

$$S_qX = S_qA \oplus A_q(X, A).$$

3. *Mayer-Vietoris sequences.* There exists an exact sequence linking the homology of the union of two spaces with the homology of each of them and the homology of the intersection. This sequence enables one to employ induction to compute the homology of spaces glued together from pieces whose homology is known. The exactness of this

sequence is closely related to the following property.

Let X_1 and X_2 be subspaces of the topological space X (denoted by the triad $(X; X_1, X_2)$) and let $i_\nu : X_\nu \to X$ be inclusions, $\nu = 1, 2$.

The triad $(X; X_1, X_2)$ is called excisive if one of the following equivalent conditions holds:

a) $i_{1*} : H(X_1, X_1 \cap X_2) \cong H(X_1 \cup X_2, X_2)$;
b) $i_{2*} : H(X_2, X_1 \cap X_2) \cong H(X_1 \cup X_2, X_1)$;
c) $(i_{1*}, i_{2*}) : H(X_1, X_1 \cap X_2) \oplus H(X_2, X_1 \cap X_2) \cong H(X_1 \cup X_2, X_1 \cap X_2)$;
d) $i_* : HS\{X_1, X_2\} \cong HS(X_1 \cup X_2) = H(X_1 \cup X_2)$;
e) $\bar{i}_* : H[S\{X_1, X_2\}/S(X_1 \cap X_2)] \cong H[S(X_1 \cup X_2)/S(X_1 \cap X_2)] = H(X_1 \cup X_2, X_1 \cap X_2)$;
f) $p_* : H(X, X_1 \cup X_2) \cong H[SX/S(X_1 \cup X_2)] \cong H[SX/S\{X_1, X_2\}]$,

where $S\{X_1, X_2\}$ is the least subcomplex of the complex $S(X_1 \cup X_2)$, containing the complexes SX_1 and SX_2, with i the inclusion and p the projection.

CW-spaces and their CW-subspaces give examples of excisive triads, as well as do triads for which X_1 and X_2 are open in $X_1 \cup X_2$. An example of a non-excisive triad is $X = \mathbb{R}$, $X_1 = (-\infty, 0)$, $X_2 = (0, +\infty)$.

For every triad $(X; X_1, X_2)$ the sequence:

$$0 \to S(X_1 \cap X_2) \xrightarrow{(j_1, -j_2)} SX_1 \oplus SX_2 \xrightarrow{(i_1, i_2)} S\{X_1, X_2\} \to 0,$$

is exact where i_ν and j_ν are inclusions. If the triad is excisive, the homology sequence of this exact sequence has the form

$$\begin{aligned} \ldots \to H_{q+1}(X_1 \cup X_2) &\xrightarrow{d_*} H_q(X_1 \cap X_2) \xrightarrow{(j_{1*}, -j_{2*})} \\ &\to H_q X_1 \oplus H_q X_2 \xrightarrow{(i_{1*}, i_{2*})} H_q(X_1 \cup X_2) \xrightarrow{d_*} \ldots \end{aligned} \tag{2.6}$$

This exact sequence is called the (absolute) Mayer-Vietoris sequence of the triad $(X; X_1, X_2)$.

We also have, for any triad $(X; X_1, X_2)$, an exact sequence

$$0 \to SX/S(X_1 \cap X_2) \xrightarrow{(j_1, -j_2)} SX/SX_1 \oplus SX/SX_2 \xrightarrow{(i_1, i_2)} SX/S\{X_1, X_2\} \to 0,$$

where i_ν and j_ν are projections. If the triad is excisive, the corresponding homology sequence takes the form

$$\begin{aligned} \ldots \to H_{q+1}(X, X_1 \cup X_2) &\xrightarrow{d_*} H_q(X, X_1 \cap X_2) \xrightarrow{(j_{1*}, -j_{2*})} \\ &\to H_q(X, X_1) \oplus H_q(X, X_2) \xrightarrow{(i_{1*}, i_{2*})} H_q(X, X_1 \cup X_2) \xrightarrow{d_*} \ldots \end{aligned} \tag{2.7}$$

This exact sequence is called the relative Mayer-Vietoris sequence of the triad $(X; X_1, X_2)$.

The boundary operator d_* of the Mayer-Vietoris sequence (2.6) ((2.7) respectively) coincides with the following composition:

$$H_{q+1}(X_1 \cup X_2) \longrightarrow H_{q+1}(X_1 \cup X_2, X_2) \cong H_{q+1}(X_1, X_1 \cap X_2) \xrightarrow{\partial_*} H_q(X_1 \cap X_2),$$

and correspondingly:

$$H_{q+1}(X_1 \cup X_2) \xrightarrow{\partial_*} H_q(X_1 \cup X_2, X_1) \cong H_q(X_2, X_1 \cap X_2) \longrightarrow H_q(X, X_1 \cap X_2),$$

where all the maps other than ∂_*, are induced by inclusions.

The reduced homology gives us from (2.6) for the excisive triad $(X; X_1, X_2)$ having $X_1 \cap X_2 \neq \emptyset$ that we have an exact sequence:

$$\begin{aligned} \ldots \longrightarrow \tilde{H}_{q+1}(X_1 \cup X_2) &\xrightarrow{d_*} \tilde{H}_q(X_1 \cap X_2) \xrightarrow{(j_{1*}, -j_{2*})} \\ &\longrightarrow \tilde{H}_q X_1 \oplus \tilde{H}_q X_2 \xrightarrow{(i_{1*}, i_{2*})} \tilde{H}_q(X_1 \cup X_2) \xrightarrow{d_*} \ldots \end{aligned} \tag{2.8}$$

It is called the reduced Mayer-Vietoris sequence of $(X; X_1, X_2)$.

§3. Spaces of constant curvature

There are three **kinds** of geometry which possess a notion of distance and which look the same from any viewpoint with the observer's head turned in an orientation, i.e., which have constant sectional curvature. For positive sectional curvature, we get elliptic or spherical geometry, and for zero curvature we get Euclidean geometry, and for negative curvature, we have hyperbolic or Lobachevskii geometry.

1. *Elliptic space.* A simple model for an n-dimensional space of constant positive curvature k is the sphere $S^n(0, 1/k)$ in Euclidean space $\mathbb{R}^{n+1}$. The distance between points on the sphere is defined as the length of the arc on the great circle joining those points. The great circles act as straight lines in this geometry. However, the analogy with Euclidean straight lines is disrupted by the fact that the great circles intersect at two diametrically opposite points. It is convenient to identify those points and define the distance between the points by means of the shortest arc. The resulting new space of constant positive curvature k is called the elliptic space or Riemannian space. If we consider straight lines passing through the points instead of the diametrically opposite points on the sphere $S^n(0, 1/k)$, the elliptic space will be considered as the space of all straight lines in $\mathbb{R}^{n+1}$ passing through the origin, and we arrive at the concept of the projective space $\mathbb{RP}^n$.

The projective space $\mathbb{RP}^n$ can also be obtained from the n-dimensional ball having its centre at the origin by identifying diametrically opposite points on its boundary, the sphere S^{n-1}. In the space $\mathbb{RP}^n$ we can introduce homogeneous coordinates by noting that the coordinates $(x_0, x_1, \ldots, x_n)$ of different points on a straight line through the origin of $\mathbb{R}^{n+1}$ differ only by a common propertionality factor. The distance ρ in $\mathbb{RP}^n$ between two straight lines is given by the following formula if one of them passes through a point $x \in \mathbb{R}^{n+1}\setminus\{0\}$ and the other through a point $y \in \mathbb{R}^{n+1}\setminus\{0\}$:

$$\cos^2 k\rho = \frac{(x, y)^2}{(x, x)\cdot(y, y)}, \quad 0 \le k\rho \le \pi/2, \tag{2.9}$$

where (x, y) denotes the scalar product in $\mathbb{R}^{n+1}$.

The isometry of the spherical space is defined by orthogonal transformation in $\mathbb{R}^{n+1}$, and the same applies to the elliptic space and the projective space $\mathbb{RP}^n$.

2. *Hyperbolic space* H^n. To define a space of constant negative curvature k, we consider the following bilinear form in the space $\mathbb{R}^{n+1}$:

$$[x, y] = -x_0y_0 + x_1y_1 + \ldots + x_ny_n. \tag{2.10}$$

We also consider the set V_n of straight lines in $\mathbb{R}^{n+1}$ passing through the origin and lying within the cone $[x, x] = 0$, i.e., lines for the points on which $[x, x] < 0$. For two elements in V_n, i.e., for two straight lines, one passing through a point $y \in \mathbb{R}^{n+1}$ and the other through a point $x \in \mathbb{R}^{n+1}$, we define the distance ρ induced by (2.10) as

$$\cosh^2 k\rho = \frac{[x, y]^2}{[x, x]\cdot[y, y]}, \tag{2.11}$$

which is analogous to (2.9). Then, for points lying within the cone $[x, x] = 0$ we have $[x, y]^2/[x, x]\cdot[y, y] \ge 1$, which implies that the distance ρ defined by (2.11) is a real non-negative number that becomes zero only if the lines coincide. It is readily seen that this satisfies the usual axioms for distance, i.e., the symmetry and triangle axioms.

The resulting metrical space (V_n, ρ) is called an *n-dimensional hyperbolic space* or *Lobachevskii space*. The quantity k is the curvature of this space. The straight lines lying in the cone $[x, x] = 0$ are points at infinity of the Lobachevskii space. The isometries of the Lobachevskii space (V_n, ρ) are defined by the linear transformations of the space $\mathbb{R}^{n+1}$ with determinant equal to one keeping the form $[x, x]$ invariant and takes to itself each half-cone $[x, x] = 0$. Such transformations are called Lorentz transformations. They form the group $SO(n, 1)$.

3. *Hemisphere of imaginary radius*. As a second model for the n-dimensional hyperbolic space H^n we consider the upper half of the following hyperboloid in $\mathbb{R}^{n+1}$:

$${}^1H^n = \{x \in \mathbb{R}^{n+1} : [x, x] = -x_0^2 + x_1^2 + \ldots + x_n^2 = -1,\ x_0 \ge 1\}, \tag{2.12}$$

i.e., a hemisphere of radius i in the space having the pseudometric defined by (2.10). The surface (2.12) intersects each straight line passing through the origin and lying within the cone $[x, x] = 0$ at one and only one point. We put this straight line into correspondence with that point of intersection to get the mapping of V_n into the hemisphere of (2.12) of imaginary radius i, and we thus transform it to a model of the n-dimensional hyperbolic space. The distance ρ in the space ${}^1H^n$ is given by

$$\cosh k\rho = [x, y] \tag{2.13}$$

and is induced by the arc differential

$$k^2 ds^2 = -dx_0^2 + dx_1^2 + \ldots + dx_n^2. \tag{2.14}$$

Then (2.14) is a positive-definite form in ${}^1H^n$, because there we have a relation for the differential dx_i:

$$x_0 dx \quad - x_1 dx_1 - \ldots - x_n dx_n = 0. \tag{2.15}$$

The space ${}^1H^n$ has constant negative curvature k in the metric of (2.13) and its isometries are again Lorentz transformations in $\mathbb{R}^{n+1}$ that preserve ${}^1H^n$. From (2.15) one obtains that straight lines passing through the origin intersect ${}^1H^n$ at right angles in the pseudometric of (2.14), and the same consequently applies to planes having this property. This means that the intersections of ${}^1H^n$ with planes passing through the origin are geodesics in this hyperbolic space model.

4. *The projective model of hyperbolic space.* Consider the set of points in the plane $\{x \in \mathbb{R}^{n+1} : x_0 = 1\}$, lying within the cone $[x, x] = 0$. It is clear that this is an n-dimensional unit ball $B^n(0, 1)$ to which ${}^1H^n$ is projected along straight lines passing through the origin in accordance with

$$(x_0, x_1, \ldots, x_n) \longrightarrow (x_1/x_0, x_2/x_0, \ldots, x_n/x_0). \tag{2.16}$$

As linear transformations of straight lines in the space V_n correspond to projective transformations of the plane $\{x : x_0 = 1\} \subset \mathbb{R}^{n+1}$, the isometries of this hyperbolic space model are projective transformations of an n-dimensional space that preserve the interior of the ball $B^n(0, 1)$. As the mapping (2.16) projects the geodesics in ${}^1H^n$ to chords, so the chords of the ball $B^n(0, 1)$ are geodesics in this projective model. From (2.11) we get that the distance $\rho(x, y)$ in this realization of H^n between the points $x, y \in B^n(0, 1)$ is defined by

$$\rho(x, y) = \frac{1}{2k} \ln \left(\frac{|x - \alpha| \cdot |y - \beta|}{|x - \beta| \cdot |y - \alpha|} \right), \tag{2.17}$$

where α and β are intersection points of the sphere $S^{n-1}(0, -1/k)$

and the chord passing through the points x and y, with the point x lying within the segment $[\alpha, y]$ of this chord. We see from (2.17) that $\rho(x, y)$ tends to infinity if either of the points x or y tends to the sphere $S^{n-1}(0, 1)$ at infinity.

As regards the angles between geodesic planes in this model, we note the following. If $L \subset B^n(0, 1)$ is a geodesic plane intersecting $S^{n-1}(0, 1)$ on some sphere S^{n-2}, then there exists a point $x \in \overline{\mathbb{R}}^n$ such that all tangent planes to the ball $B^n(0, 1)$ at points on the sphere S^{n-2} contain it. Then all the geodesics in the hyperbolic space (chords in $B^n(0, 1)$) being the segments of straight lines passing through this point x intersect the geodesic plane L at right angles.

5. *Lower hemisphere.* $S^n_-(0, 1) = \{x \in \mathbb{R}^{n+1} : |x| = 1, x_{n+1} < 0\}$ gives us a fourth model for the hyperbolic space. To introduce a hyperbolic metric into it, we consider the mapping $p : B^n(0, 1) \longrightarrow S^n_-(0, 1)$, which is the orthogonal projection with respect to the plane $\{x_{n+1} = 0\} = \mathbb{R}^n$. The inverse mapping p^{-1} is put as

$$p^{-1}(x_1, \dots, x_n, x_{n+1}) = (x_1, \dots, x_n); \quad \sum_{i=1}^{n+1} x_i^2 = 1. \tag{2.18}$$

Providing $S^n_-(0, 1)$ with the metric induced from $B^n(0, 1)$ by this projection, we transform the hemisphere into a hyperbolic space model in which the geodesics are circles in $S^n_-(0, 1)$ orthogonal to the boundary sphere $\partial S^n_- = S^{n-1}(0, 1) \subset \mathbb{R}^n$ (the chord projections are geodesics from $B^n(0, 1)$). The sphere $S^{n-1}(0, 1)$ consists of points at infinity in this model. The isometries in this model are specified by the Möbius transformations of space $\overline{\mathbb{R}}^{n+1}$ that preserve $S^n_-(0, 1)$ because they should convert the k-dimensional hemispheres orthogonal to the sphere $S^{n-1}(0, 1)$ into the hemispheres having the same property.

The angle conservatism in Möbius mappings implies that the angles in this model coincide with those measured in the Euclidean metric.

6. *Poincaré model in a ball.* The steographic projection π introduced in Chapter 1 mapped the space $\overline{\mathbb{R}}^n$ into the sphere $S^n(0, 1)$. We consider the action of the inverse mapping π^{-1} of the hemisphere $S^n_-(0, 1)$. From (1.10) we get that

$$\pi^{-1}(x) = (x_1/(1 - x_{n+1}), \dots, x_n/(1 - x_{n+1})), \tag{2.19}$$

which means that $\pi^{-1}(S^n_-(0, 1)) = B^n(0, 1) \subset \mathbb{R}^n$. The stereographic projection converts Euclidean spheres (planes) into spheres and preserves the angles. So the images of the geodesics in the model $S^n_-(0, 1)$ with the mapping of (2.18) are arcs of Euclidean circles (straight lines) orthogonal to the boundary $S^{n-1}(0, 1)$ of the ball $B^n(0, 1)$, which thus become geodesics for the metric in $B^n(0, 1)$ induced by the mapping $\pi^{-1}(S^n_-(0, 1)) \longrightarrow B^n(0, 1)$. It can be shown that this metric $\rho(x, y)$ in $B^n(0, 1)$ has the following arc differential:

$$k^2ds^2 = \frac{4|dx|^2}{(1 - |x^2|)^2} \tag{2.20}$$

and is defined by

$$\rho(x, y) = \frac{1}{k} \ln [(1-h)/(1+h)], \tag{2.21}$$

where $h = |y|$ if $x = 0$ and $h = |x-y| \cdot |x|/|x - |x|^2 y|$ if $x \neq 0$. The unit ball having the metric of (2.21) is converted to a space of constant negative sectional curvature k having the above geodesics. The isometries of it are again specified by Möbius automorphisms, as in the case of the hemisphere $S^n(0, 1)$. This follows also from Theorem 1.10.

7. *Poincaré model in the half-space* $\mathbb{R}^n_+$. In the half-space $\mathbb{R}^n_+ = \{x \in \mathbb{R}^n : x_n > 0\}$, the metric $\rho(x, y)$ that converts $\mathbb{R}^n_+$ into the n-dimensional hyperbolic space H^n can be induced by various model mappings. One such mapping is obtained by the superposition of the rotation U of the space $\mathbb{R}^{n+1}$ specified by the $(n+1) \times (n+1)$ matrix

$$U = \begin{pmatrix} 1 & & 0 \\ 0 & 0 & -1 \\ & 1 & 0 \end{pmatrix},$$

and the mapping $\pi^{-1} : S^n(0, 1) \longrightarrow \overline{\mathbb{R}}^n$, inverse to the stereographic projection π:

$$\pi^{-1} \circ U : S^n_-(0, 1) \longrightarrow \mathbb{R}^n_+ \tag{2.22}$$

Another such mapping is inverse to the Möbius mapping in $\overline{\mathbb{R}}^n$ and maps unit ball $B^n(0, 1)$ to $\mathbb{R}^n_+$. It is readily seen that this mapping and the mapping of (2.22) induce the same metric $\rho(x, y)$ in half-space $\mathbb{R}^n_+$, in which the arc differential is

$$k^2ds^2 = |dx|^2/x_n^2. \tag{2.23}$$

The hyperbolic space $H^n \cong \mathbb{R}^n_+$ having the metric of (2.23) has curvature $k < 0$, and its geodesics are semicircles (rays) in $\mathbb{R}^n_+$, orthogonal to the boundary plane $\mathbb{R}^{n-1} = \{x \in \mathbb{R}^n : x_n = 0\}$, lying at infinity. The isometries of H^n are Möbius automorphisms of the half-space (see Theorem 1.8) whose set forms a group isomorphic to the group Möb$(n-1)$.

8. *Spheres and horospheres in the hyperbolic space.* By a sphere $S(y, r)$ with centre $y \in H^n$ and radius $r > 0$ we mean, as usual, a set

$$S(y, r) = \{x \in H^n : \rho(x, y) = r\}.$$

The sphere $S(y, r)$ is an $(n-)$-dimensional orthogonal manifold for the family of geodesics passing through the point y.

We obtain the notion of a horosphere, as a limit case for a sphere in H^n, when changing the radius correspondingly we drive its centre to a point at infinity. Namely, a *horosphere* $S_y \subset H^n$ with a centre $y \in \partial H^n$ is an $(n-1)$-manifold orthogonal to the family of geodesics $\gamma : \mathbb{R} \longrightarrow H^n$ in H^n, having y as their endpoint:

$$\lim_{t \to +\infty} \gamma(t) = y.$$

We call such geodesics *asymptotic*: there exists a constant $c > 0$ such that $\rho(\gamma_1(t), \gamma_2(t)) \le c$ for $t \ge 0$ and for such geodesics γ_1, γ_2.

Another (equivalent to the previous) definition of a horosphere is obtained by replacing the Euclidean distance function $x \longrightarrow \rho(x, y)$ for a sphere $S(y, r) \subset H^n$ by the *Busemann function* for the horosphere $S_y \subset H^n$.

Namely, for each point at infinity $y \in \partial H^n$ and each geodesic $\gamma \subset H^n$, $\gamma(+\infty) = y$, we may define a function $f_\gamma : H^n \longrightarrow \mathbb{R}$ such that for all $x \in H^n$

$$f_\gamma(x) = \lim_{t \to +\infty} (\rho(x, \gamma(t)) - t).$$

The function f_γ has the following qualities (cf. Busemann [1], Eberlein-O'Neil [1]):

1. f_γ is uniformly continuous and convex; particularly,

$$|f_\gamma(x_1) - f_\gamma(x_2)| \le \rho(x_1, x_2).$$

2. If the geodesic β is asymptotic to the geodesic γ, then

$$f_\gamma(\beta(s)) - f_\gamma(\beta(t)) = t - s;$$

in particular, f_γ has no minimum.

3. If the geodesic β is asymptotic to the geodesic γ, then $f_\gamma - f_\beta = \text{const}$.

These properties show the correctness of the definition of the Busemann function $f_y(x)$ for a point $y \in \partial H^n$ as the function $f_\gamma(x)$, where γ is some geodesic ending at y.

Now we can define a *horosphere* $S_y \subset H^n$ *with a centre at* $y \in \partial H^n$ and passing through a fixed point $s \in H^n$ as a set

$$S_{y,z} = \{x \in H^n : f_y(x) = f_y(z)\}.$$

This horosphere bounds the *horoball*

$$B_{y,z} = \{x \in H^n : f_y(x) < f_y(z)\}.$$

It is essential to note here that in Poincaré models of the hyperbolic space H^n (in the ball $B^n(0, 1)$ or in the half-space $\mathbb{R}^n_+$) the horospheres S_y are represented by Euclidean spheres tangent to the sphere at infinity ∂H^n at the point y. In particular, the horospheres $S_\infty \subset \mathbb{R}^n_+ \cong H^n$ centred at ∞ are represented by Euclidean planes

$$H_t = \{x \in \mathbb{R}^n_+ : x_n = t\}, \qquad t > 0.$$

One can see straight away from the latter that the restriction of the hyperbolic metric to the horosphere $S_y \subset H^n$ converts the horosphere to the Euclidean space $\mathbb{R}^{n-1}$, as far as for all $x_1, x_2 \in H_t$

$$\rho(x_1, x_2) = |x_1 - x_2|/t.$$

§4. Representation of the Möbius group in the matrix group

If we take the upper half-space $\mathbb{R}^n_+$ with the metric in which the arc differential takes the form $ds = |dx|/x_n$ as a model for the n-dimensional hyperbolic space, then we have, as noted, that each isometry of it is an extension to $\mathbb{R}^n_+$ of a Möbius mapping of $\bar{\mathbb{R}}^{n-1}$ and vice versa. At the same time, in the model ${}^1H^n$ having the metric (2.13) with the curvature -1, the isometries are linear transformations of the space $\mathbb{R}^{n+1}$ that preserve the quadratic form $[x, x]$ and the upper half of the hyperboloid ${}^1H^n$. In other words, the isometries are the transformations specified by the $(n+1)\times(n+1)$ Lorentz matrices with determinant one, which form the orthogonal group $SO(n, 1)$.

We consider in sequence the isometries (2.16), (2.18), and (2.22) of the spaces ${}^1H^n$, $B^n(0, 1)$, $S^n_-(0, 1)$ and H^n correspondingly. Taking a superposition of these we finally get the isometry

$$\varphi : {}^1H^n \longrightarrow H^n,$$

which acts in accordance with

$$\varphi(x_0, x_1, \ldots, x_n) = (x_1/(x_0 - x_n), \ldots, x_{n-1}/(x_0 - x_n), 1/(x_0 - x_n)).$$

The isometry $\varphi^{-1} : H^n \longrightarrow {}^1H^n$, inverse to φ, acts in accordance with

$$\varphi^{-1}(x_1, \ldots, x_n) = (|x|^2 + 1, 2x_1, \ldots, 2x_{n-1}, |x|^2 - 1)/2x_n.$$

This isometry φ for the hyperbolic space models specifies an isomorphism for their isometry groups, which extends to the isomorphism

$$\varphi_* : \text{Möb}(n-1) \longrightarrow SO(n,1)/\{\pm 1\}, \tag{2.24}$$

which acts in accordance with the rule that each transformation $g \in \text{Möb}(n-1)$ is put into correspondence with an $(n+1)\times(n+1)$-matrix $\tilde{T} = (t_{ij})_{i,j=0,1,\dots,n}$, such that

$$\tilde{T}\cdot x \equiv \varphi^{-1}\circ \tilde{g}\circ\varphi(x), \qquad x \in {}^1H^n,$$

where $\tilde{g} \in \text{Möb}(\mathbb{R}^n_+) = \text{Isom}\, H^n$, $\tilde{g}|_{\mathbb{R}^{n+1}} = g$.

By virtue of Corollary 1.8, any Möbius transformation is a superposition of translations, stretchings, the inversion $\mathcal{J}(x) = x/|x|^2$, and orthogonal transformations. Therefore it is sufficient to know which matrices from $SO(n,1)$ correspond to transformations of the form

$$\begin{aligned} &1.\ \ g_v(x) = x + v, \quad \text{where } v \in \mathbb{R}^{n-1};\\ &2.\ \ U(x) = U\cdot x, \quad \text{where } U \in O(n-1);\\ &3.\ \ g_r(x) = r\cdot x, \quad \text{where } r > 0,\ \ r \neq 1;\\ &4.\ \ \mathcal{J}(x) = x/|x|^2. \end{aligned} \tag{2.25}$$

Using the forms of the isometries φ, φ^{-1} and (2.24), we get after some computation that

$$\tilde{U} = \begin{pmatrix} 1 & 0 & 0\\ 0 & U & 0\\ 0 & 0 & 1\end{pmatrix}, \qquad \tilde{\mathcal{J}} = \begin{pmatrix} 1 & & 0\\ & 1 & \\ 0 & & -1\end{pmatrix},$$

$$\tilde{T}_v = \begin{pmatrix} (|v|^2+2)/2 & v_1 \ \dots\ v_{n-1} & -|v|^2/2\\ v_1 & 1 \ \dots\ 0 & -v_1\\ \vdots & \vdots \quad\ \ \vdots & \vdots\\ v_{n-1} & 0 \ \dots\ 1 & -v_{n-1}\\ |v|^2/2 & v_1 \ \dots\ v_{n-1} & (2-|v|^2)/2 \end{pmatrix}, \tag{2.26}$$

$$\tilde{T}_r = \begin{pmatrix} (r^2+1)/2r & 0 \ \dots\ 0 & (r^2-1)/2r\\ 0 & 1 \ \dots\ 0 & 0\\ \vdots & \vdots \quad\ \ \vdots & \vdots\\ 0 & 0 \ \dots\ 1 & 0\\ (r^2-1)/2r & 0 \ \dots\ 0 & (r^2+1)/2r \end{pmatrix}.$$

We use the representation of Corollary 1.8 for the Möbius transformation $g \in \text{Möb}(n-1)$, $g(\infty) \neq \infty : g(x) = rU\mathcal{J}(x-q) + q'$, where q and q' are the centres of the isometric spheres $I(g)$ and $I(g^{-1})$ of radius $\sqrt{r} > 0$ to get that

$$\tilde{T} = (t_{ij})_{0\le i,j\le n} = \tilde{T}_{q'}\cdot\tilde{T}_r\cdot\tilde{U}\cdot\tilde{\mathcal{J}}\cdot\tilde{T}_{-q'} \tag{2.27}$$

In particular, we have

$$t_{00} = [(|q|^2+1)(|q'|^2+1) + r^2 - 2r(U(q'), q)]/2r;$$

$$t_{i0} = (U(q'), e_i) - (q, e_i)(|q'|^2+1)/r, \qquad 1 \le i \le n-1; \tag{2.28}$$

$$t_{n0} = [(1-|q|^2)(|q'|^2+1) - r^2 + 2r(U(q'), q)]/2r,$$

where $(*, *)$ is the scalar product in $\mathbb{R}^n$ and e_i is the i-th coordinate vector.

Conversely, if we know the matrix $\tilde{T} = (t_{ij}) \in SO(n, 1)$, then the corresponding Möbius transformation $\tilde{g} \in \text{Möb}(\mathbb{R}^n_+)$ (the isometry H^n) acts in accordance with the formulae: $y = \tilde{g}(x)$,

$$y_i = \frac{(t_{i0} + t_{in}) \sum_{m=1}^{n} x_m^2 + 2 \sum_{m=1}^{n-1} t_{im} x_m + t_{i0} + t_{in}}{t_0 \sum_{m=1}^{n} x_m^2 + 2 \sum_{m=1}^{n-1} (t_{0m} - t_{nm}) x_m + t_1}, \qquad 1 \le i \le n-1, \tag{2.29}$$

$$y_n = \frac{2x_n}{t_0 \sum_{m=1}^{n} x_m^2 + 2 \sum_{m=1}^{n-1} (t_{0m} - t_{nm}) x_m + t_1},$$

where

$$t_0 = t_{00} + t_{0n} - t_{n0} - t_{nn}, \qquad t_1 = t_{00} + t_{nn} - t_{0n} - t_{n0}.$$

REMARK. Kobayashi-Nomizu [1] have given a somewhat different approach to the representation $\text{Möb}(n-1)$ in the form of the Lorentz group $SO(n, 1)/\{\pm 1\}$. See Apanasov [3] for the above representation.

As any finitely generated matrix group contains a normal torsion-free subgroup of finite index without torsion (not containing elements of finite order) (Mal'tsev [1], Selberg [1]), we get the following important fact from this representation.

THEOREM 2.15. *In any finitely generated group* $G \subset \text{Möb}(n)$, *there exists a normal torsion-free subgroup* $G_0 \subset G$ *having finite index* $\mu(G_0, G)$ *in the group* G.

Here and subsequently, the index $\mu(G_0, G)$ for a subgroup G_0 in a group G means the cardinality of the quotient group G/G_0.

§5. $\mathcal{G}$-structures in manifolds

The topological manifold M^n specified by its atlas $\{(U_i, \varphi_i)\}_{i \in \mathcal{I}}$ can also be represented as composed of pieces of the

Euclidean space $\mathbb{R}^n$, which are glued together by homeomorphisms of $\mathbb{R}^n$ taken from a set $\mathcal{G}$. This set $\mathcal{G}$ should satisfy the following obvious conditions, which transform it to a pseudogroup of local homeomorphisms between open sets in $\mathbb{R}^n$:

1) the restriction g_0 of any element $g \in \mathcal{G}$ to an open set in its domain is an element of $\mathcal{G}$;

2) the superposition $g \circ h$ of any elements $g, h \in \mathcal{G}$ (if defined) is an element of $\mathcal{G}$;

3) an inverse element for $g \in \mathcal{G}$ is an element of $\mathcal{G}$; and

4) if $U = \bigcup_i U_i \subset \mathbb{R}^n$ and $g : U \longrightarrow V$ is a local homeomorphism such that $g_i = g|_{U_i} \in \mathcal{G}$ for all i, then $g \in \mathcal{G}$.

DEFINITION 2.16. The manifold M^n derived by gluing together pieces of $\mathbb{R}^n$ by means of homeomorphisms from a pseudogroup $\mathcal{G}$ with the properties 1) - 4) is called a $\mathcal{G}$-manifold.

We note that, if in this definition we replace $\mathbb{R}^n$ by some topological manifold X, then, if the action of $\mathcal{G}$ on X is transitive, we do not obtain new manifold types.

If now we are given a manifold X and a homeomorphism group G acting in it, the restrictions of its elements to open sets in X form the pseudogroup $\mathcal{G}$. Then by the (G, X)-manifold we understand the manifold obtained by gluing together the pieces of X by means of elements of this pseudogroup.

If $\mathcal{G}$ is a pseudogroup of local C^r-diffeomorphisms in $\mathbb{R}^n$, the $\mathcal{G}$-manifold is a C^r-manifold (a smooth manifold for $r \geq 1$).

If $\mathcal{G}$ is a pseudogroup of local piecewise-linear homeomorphisms, then the $\mathcal{G}$-manifold is a PL-manifold.

If G is the group of affine transformations in $\mathbb{R}^n$, the $(G, \mathbb{R}^n)$-manifold is affine.

If G is the Möbius group Möb(n) acting on the n-dimensional sphere $S^n = \overline{\mathbb{R}}^n$, $n \geq 3$, the $(S^n, \text{Möb}(n))$-manifold is called conformal. For $n = 2$ similarly, the structure on the oriented surface M^2 coincides with the definition of a projective structure in M^2 because $\mathcal{M}_2$ is naturally identified with the group PSL(2, $\mathbb{C}$), as is evident from Kra [1], Kra and Maskit [1], and Hejhal [1].

If G is the isometry group correspondingly for the Euclidean space $X = \mathbb{R}^n$, the elliptic space $X = \mathbb{RP}^n$ (or S^n), or the hyperbolic space $X = H^n$, then the (G, X)-manifold is correspondingly a Euclidean, elliptic, or hyperbolic one.

EXAMPLE 2.17 (*Figure-eight knot complement*) (Thurston [1]). There is a complete hyperbolic structure on the complement of the figure-eight knot L - see Fig. 4. This three-dimensional manifold $M = \overline{\mathbb{R}}^3 \backslash L$ has the fundamental group

$$\pi_1(M) = \{x, y : xyx^{-1}yx = yxy^{-1}xy\}. \tag{2.30}$$

This knot may be stretched onto a tetrahedron with faces A, B, C, and D. Fig. 5 shows the form of this stretching from above. It is

Figure 4

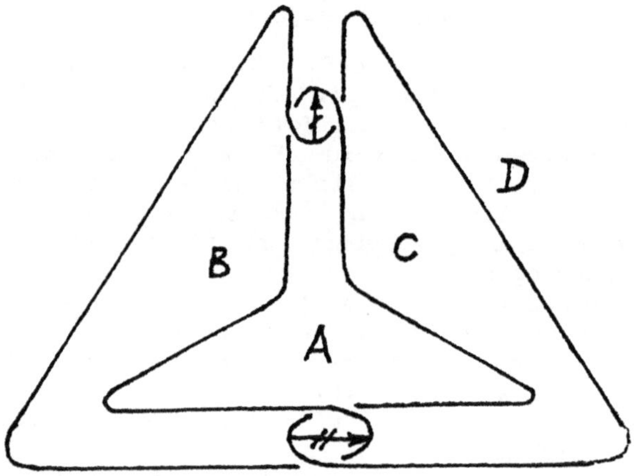

Figure 5

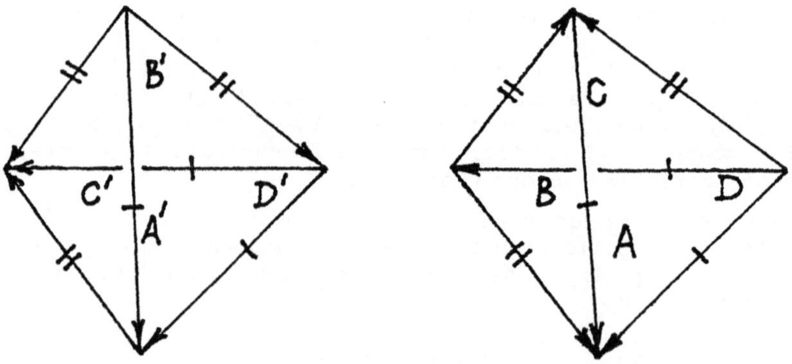

Figure 6

obtained by joining the links of the knot by means of oriented segments denoted by $-|\!\to$ and $-\|\!\to$ and letting the resulting one-dimensional complex span four membranes A, B, C, and D. The boundary of each membrane consists of the indicated segments joined by the parts of the knot L. If one contracts these parts of the membrane boundaries into points, the resulting two-dimensional complex will be the boundary of a tetrahedron. The original two-dimensional complex splits up $\bar{\mathbb{R}}^3$ into two domains homeomorphic to a ball. These two domains in our contraction become two tetrahedrons as shown in Fig. 6, while the manifold $M = \bar{\mathbb{R}}^3 \backslash L$ is obtained by identifying the corresponding tetrahedron faces with the previously removed vertices. Since six dihedral angles meet at each edge of the complex, tetrahedra with

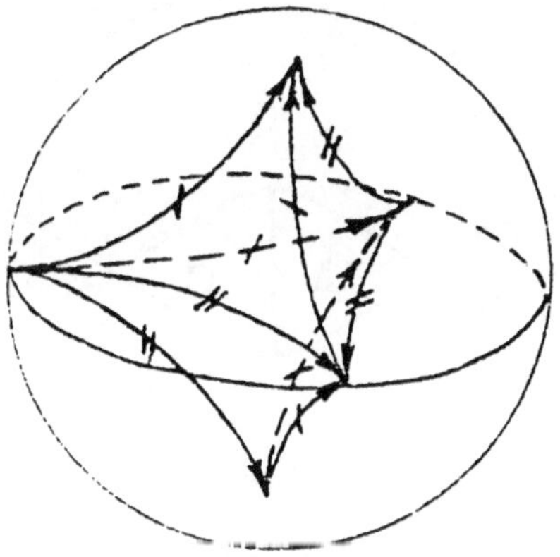

Figure 7

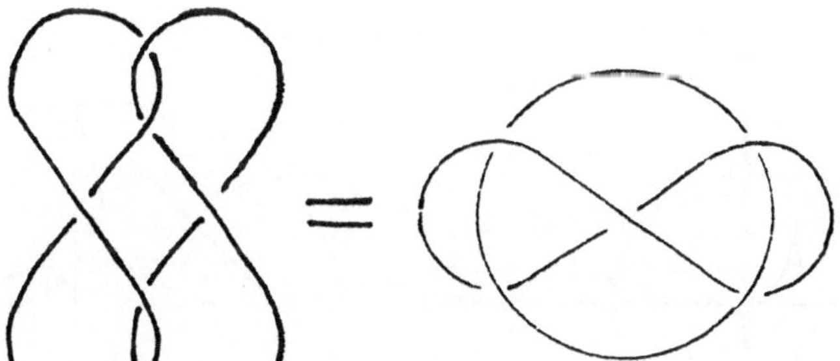

Figure 8

dihedral angles equal to $\pi/3$ are needed for a conformal gluing in this way. Such tetrahedra can be taken to be regular hyperbolic tetrahedra (simplexes) with their vertices at infinity (Fig. 7). It is easy to construct hyperbolic motions realizing the needed gluing, and thus a certain hyperbolic structure is defined on the manifold $M = \bar{\mathbb{R}}^3 \setminus L$.

EXAMPLE 2.18 (*Whitehead link*) (Thurston [1]). There is a hyperbolic structure on the complement of the Whitehead link as shown in Fig. 8. The fundamental group of this manifold $M = \bar{\mathbb{R}}^3 \setminus L$ has the following representation:

$$\pi_1(M) = \{x, y : (y^{-1}xyx^{-1})(y^{-1}x^{-1}yx)(yx^{-1}y^{-1}x)(yxy^{-1}x^{-1}) = 1\} \tag{2.31}$$

We represent the link L in the form shown in Fig. 9 and join the components of the link by oriented segments $-|\!\rightarrow$, $-\|\!\rightarrow$, and $\longrightarrow$, as shown in Fig. 9. We let the resulting one-dimensional complex,

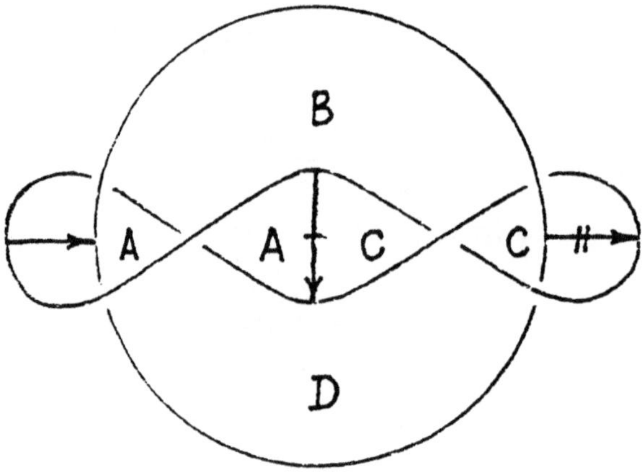

Figure 9

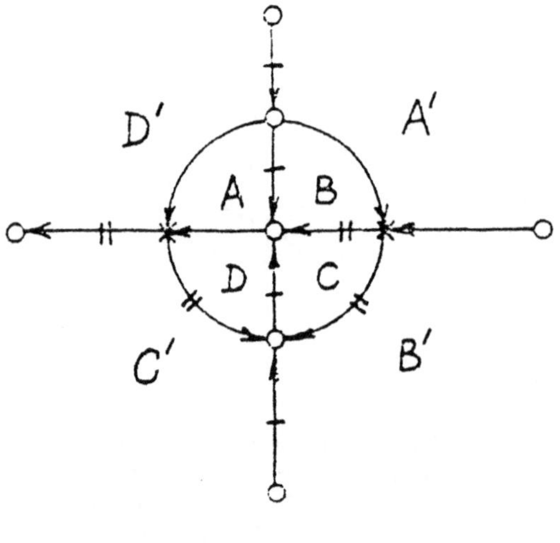

Figure 10

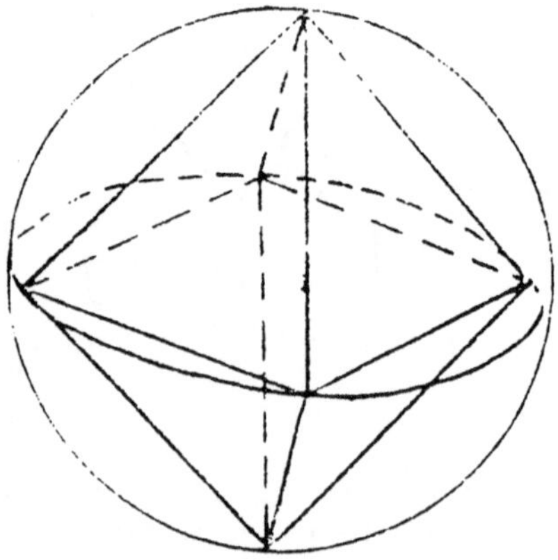

Figure 11

span the membranes A, B, C, D, and E, whose boundaries consist of these oriented segments joined by the link L parts. If these link parts contract to points, the two-dimensional complex becomes the common boundary of two domains B_1 and B_2 homeomorphic to a ball, where $\partial B_1 = A \cup B \cup C \cup D \cup E$, and $\partial B_2 = A' \cup B' \cup C' \cup D' \cup E'$; here the corresponding domains in the boundaries differ in orientation. We glue E and E' by the identity map and may assume that the two-dimensional complex containing the membranes $A, A', \ldots, D, D'$, obtained on contracting the parts of L, is the boundary of an octahedron whose spherical development is shown in Fig. 10; the vertices of the octahedron correspond to the link parts. Then our manifold $\bar{\mathbb{R}}^3 \backslash L$ is obtained by gluing together the corresponding sides of the octahedron with removed vertices and is a complex consisting of one octahedron, four triangles, and three edges. Since four dihedral angles meet at each edge of the complex, an octrahedron with dihedral angles equal to $\pi/2$ is needed for a conformal gluing in this way (the dihedral angles in the Euclidean octahedron are about 109°). We consider the projective model for the hyperbolic space in $B^3(0, 1)$. Then one can show that the octahedron, shown in Fig. 11, is the desired one: it has vertices at infinity and right dihedral angles. It is easy to construct hyperbolic motions realizing the necessary gluing, and thus it shows that the manifold $\bar{\mathbb{R}}^3 \backslash L$ has a hyperbolic structure.

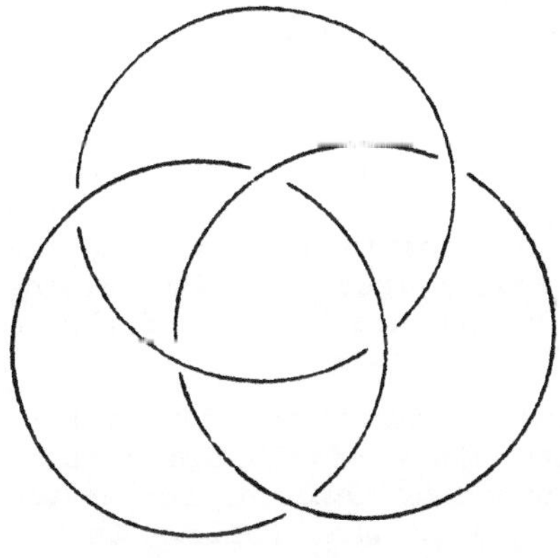

Figure 12

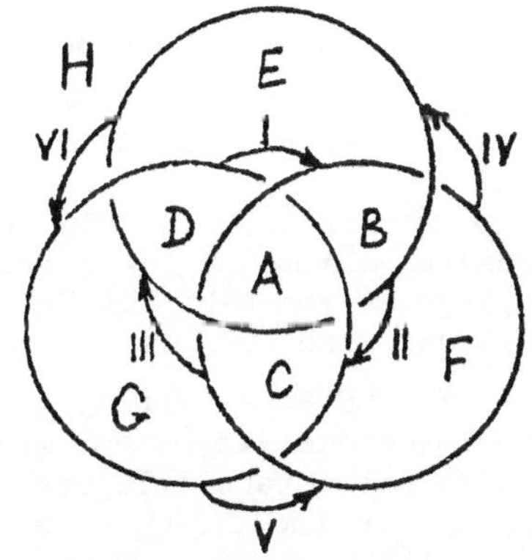

Figure 13

EXAMPLE 2.19 (*Borromean rings*) (Thurston [1]). There is a hyperbolic structure on the complement of Borromean rings (Fig. 12). This manifold $\bar{\mathbb{R}}^3 \backslash L$ is obtained by gluing the faces of two octahedra with the removed vertices. It is analogous to the previous manifold. Namely, from L we first get a one-dimensional complex and then a two-dimensional one (Fig. 13) by spanning membranes on it. This complex becomes the boundary of the octahedra whose spherical developments are shown in Fig. 14, when the parts of L contract to points. In that case, after gluing the faces of the octahedra with the

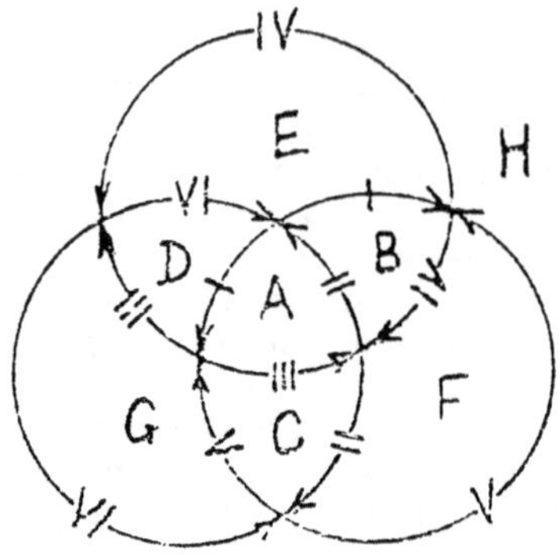

Figure 14

removed vertices in this way, we have a complex consisting of two octahedra, eight triangles, and six edges. The gluing conditions imply that the dihedral angles in the octahedra must be right angles, therefore these octahedra can be taken as regular octahedra in the projective model of H^3. However, gluing mappings can then serve as hyperbolic isometries in that model. This shows that the complement to Borromean rings has a hyperbolic structure.

EXAMPLE 2.20 (*Euclidean torus*). There is a Euclidean structure on an n-dimensional torus defined as the Cartesian product of n circles: $T^n = S^1 \times \ldots \times S^1$. The fundamental group $\pi_1(T^n)$ is a free Abelian group of rank n. The torus T^n may be obtained by gluing the opposite sides of an n-dimensional parallelipiped. The gluing mappings may be taken as Euclidean isometries (translations), and thus a Euclidean structure is introduced into the torus T^n.

EXAMPLE 2.21 (*Affine torus*). An affine structure is induced in a two-dimensional torus by gluing the opposite sides of a quadrilateral that is not a parallelogram. This is performed by means of two affine mappings in the plane. However, we shall see in what follows that a torus with that structure is not a complete $(G, \mathbb{R}^2)$-manifold.

EXAMPLE 2.22 (*Torus layer*). In the interior of a torus layer $M = T^2 \times [0, 1]$ it is possible to introduce a Euclidean structure. It follows from the fact that $\operatorname{int} M$ is homeomorphic to the quotient space $\mathbb{R}^3/G$, where the discrete group G is generated by the Euclidean isometries

$$g_1 : x \longmapsto x + e_1, \qquad g_2 : x \longmapsto x + e_2.$$

EXAMPLE 2.23. We take the manifold $(T^2 \times [0, 1])/\mathbb{Z}_2$ as the compact three-manifold M^3, where the group $\mathbb{Z}_2$ acts on the torus T^2 as a covering transformation group upon a Klein bottle, flipping the segment [0, 1]. In the interior of this manifold M^3, we introduce a Euclidean structure, since $\operatorname{int} M$ is homeomorphic to the orbit space $\mathbb{R}^3/G$ of the discrete Euclidean isometry group G, whose generators are the elements $g_1 : x \longmapsto x + e_1$ and $g_2 : x \longmapsto (-x_1, x_2 + 1, -x_3)$.

§6. Developing map and holonomy

Let M be a certain (G, X)-manifold having atlas $\{(U_i, \varphi_i)\}_{i\in\mathcal{J}}$, where the homeomorphisms $\varphi_i : U_i \longrightarrow X$ are related by the changes of charts γ_{ij} on the intersections $U_i \cap U_j$:

$$\gamma_{ij} : x \longmapsto \varphi_j \circ \varphi_i^{-1}(x), \qquad x \in \varphi_i(U_i \cap U_j). \tag{2.32}$$

We further assume that the manifold X from pieces of which M is glued is a real analytic manifold, while the group G acting there is the group of real analytic diffeomorphisms X. Then each element $g \in G$ is completely defined by its values on any open set from X, so we can assume that (2.32) defines a locally constant mapping of the manifold X to the group G (constant on any intersection $\varphi_i(U_i \cap U_j)$).

We now consider a certain chart (U_1, φ_1) and any path α in manifold M with origin $x_0 \in U_1$. From (2.32) we can consider the direct analytic extensions of φ_1 to the charts U_j, $U_j \cap U_1 \neq \varnothing$. By induction on the chain of charts covering the path α, we get that the analytic extension of φ_1 to the component $\alpha \cap U_i$ will take the form $g \circ \varphi_i$, where $g \in G$. If we now take the universal covering manifold $\tilde{M}$ as the set of pairs (x, α), where $x \in M$, while the path $\alpha \subset M$ joins the points x_0 and x and this set is factorized by homotopy equivalence, we get from what has gone before that there exists a global analytic extension for φ_1 defined on the universal covering manifold $\tilde{M}$. This is the mapping

$$d : \tilde{M} \longrightarrow X \tag{2.33}$$

and is called a *development*. It follows from (2.32) that the mapping d is a local (G, X)-diffeomorphism defined uniquely modulo superposition with an element of the group G Then (2.33) induces a (G, X)-structure in the universal covering manifold $\tilde{M}$.

This uniqueness of the development d gives us that each deck transformation $T_\alpha \in G(\tilde{M}, M)$ (see (2.5)) corresponds to a unique element g_α of a group G such that

$$g_\alpha d = dT_\alpha. \tag{2.34}$$

The mapping $d_* : \pi_1(M) \cong G(\tilde{M}, M) \longrightarrow G$, acting in accordance with the rule of (2.34) is called the *holonomy* for the manifold M. It is readily seen that it is a homomorphism.

In general, the holonomy of the manifold M does not define a

(G, X)-structure in M. For example, one can show that the holonomies corresponding to an affine structure in the two-dimensional torus constructed in Example 2.21 can correspond to other affine structures constructed in a similar fashion. However, there is an important special case where the holonomy defines a (G, X)-structure.

DEFINITION 2.24. The manifold M is called a complete (G, X)-manifold if $(\tilde{M}, d)$ is a covering over X. Here $\tilde{M}$ is the universal covering manifold for M and d is the development (2.33).

We now prove the sufficient completeness condition:

THEOREM 2.25. *Let G be a group of analytic diffeomorphisms acting transitively upon the manifold X and such that for any point $x \in X$ the isotropy group $G_x = \{g \in G : g(x) = x\}$ is compact. Then any closed (G, X)-manifold is complete.*

PROOF. Let $T_x(X)$ be a tangent space of the manifold X at a point x and let Q be a positive-definite quadratic form in $T_x(X)$. We consider the set $\{g(Q),\ g \in G_x\}$ and use Haar's measure in the compact group G_x to average the quadratic form Q in such a way that it becomes invariant with respect to G_x. We define the quadratic form as the image of the form Q in the diffeomorphism $g \in G$, $g(x) = y$ in the tangent space $T_y(X)$ at any point $y \in X$, namely we put

$$(ds^2)_y = g(Q). \tag{2.35}$$

It is clear that the definition of (2.35) is independent of the choice of the transformation $g \in G$ that transfers the point x to the point y. Then on the manifold X we have defined a G-invariant Riemannian metric. It follows from this that it induces (G-gluing) a Riemannian metric in any (G, X)-manifold that is invariant with respect to any (G, X)-mapping.

If now M is a closed (G, X)-manifold, then there exists a certain number $\varepsilon > 0$ such that any ball $B(y, \varepsilon) \subseteq M$ in the Riemannian metric is convex and contracts to a point. If we now consider the preimage in the developing map d of any ball $B(x, \varepsilon/2) \subset X$, then by virtue of the choice $\varepsilon > 0$, this will be the union of pairwise non-intersecting images in $\tilde{M}$ that are homeomorphic to the ball $B(x, \varepsilon/2)$. Then the pair $(\tilde{M}, d)$ is a cover over X, i.e., M is a complete (G, X)-manifold.

If X is a simply connected manifold, then any complete (G, X)-manifold M is readily obtained from the image of the fundamental group $d_*(\pi_1(M)) = \Gamma$ with respect to the holonomy in the form of the quotient X/Γ.

For example, any closed elliptic manifold (and correspondingly a hyperbolic or Euclidean manifold) has the sphere (correspondingly, hyperbolic or Euclidean space) as its own universal covering space. Such manifolds are defined by their holonomies d_* and are obtained in the form of the quotient X/Γ, where $\Gamma = d_*(\pi_1(M))$.

For non-compact manifolds, i.e., for manifolds glued from non-compact pieces of X, the question of completeness becomes much more complicated, and we examine it subsequently in the hyperbolic case (see Theorem 6.6 in Chapter 6 §1).

We complete this section with a theorem (Thurston [1]) on the relationship between the completeness of a (G, X)-manifold and the completeness of it as a metric space (in a certain metric in this (G, X)-manifold).

PROPOSITION 2.26. *Let the group* G *of analytical diffeomorphisms act transitively on the manifold* X *and let any isotropy subgroup* G_x *of it be compact. Then the following assertions are equivalent*:

(1) M *is a complete* (G, X)*-manifold*;

(2) M *is complete as a metric space*;

(3) *there exists an* $\varepsilon > 0$ *such that any closed ball* $B(x, \varepsilon) \subset M$ *is compact*;

(4) *for any* $k > 0$, *any closed ball* $B(x, k)$ *is compact*; *and*

(5) *there exists a family* $\{M_\alpha\}_{\alpha\in\mathbb{R}}$ *of compact sets that exhausts* M *and is such that* $M_{\alpha+\beta}$ *contains a neighbourhood of radius* β *about* M_α.

PROOF. (2) $\Rightarrow$ (1). To show that the developing map $d : \tilde{M} \longrightarrow X$, which is a local homeomorphism, is a cover, it is sufficient to show that any path $\gamma : [0, 1] \longrightarrow X$ can be lifted to a path in $\tilde{M}$. We assume that $I_1 \subset [0, 1]$ is a maximally connected set such that the path $\gamma : I_1 \longrightarrow X$ can be lifted to $\tilde{M}$. However, $\tilde{M}$ together with the manifold M is a complete metric space, and the developing map is a local homeomorphism, so I_1 is simultaneously a closed and open subset in the interval $[0, 1]$, i.e., $I_1 = [0, 1]$. This proves the completeness of the manifold M.

The proofs of the assertions (2) $\Leftarrow$ (3) $\Leftrightarrow$ (4) $\Leftrightarrow$ (5) are obtained by means of usual elementary arguments. We now show that (1) $\Rightarrow$ (3). For this we note that the universal covering space $\tilde{X}$ of the manifold X satisfies (2), (3), (4), and (5). In fact, we have seen in the proof of Theorem 2.15 that in X we can introduce a G-invariant Riemannian metric, which follows from the transitiveness of G and the compactness of G_x, $x \in X$. But as $\tilde{G}$ is a group of (G, X)-diffeomorphisms of the space $\tilde{X}$ and is also transitive, while the groups $\tilde{G}_y$ are compact for all $y \in \tilde{X}$, there exists a $\tilde{G}$-invariant metric in $\tilde{X}$. Then we take $\varepsilon > 0$ such that the closed ball $B(y, \epsilon) \subset \tilde{X}$ centred at a certain point $y \in \tilde{X}$ is compact, and the transitiveness of group $\tilde{G}$ gives us that this is so at any point $\tilde{x} \in \tilde{X}$. We now assume that M is a complete (G, X)-manifold. Then it is covered by the space $\tilde{X}$. Then from what has preceded we have that M satisfies (3), which completes the proof of the theorem.

§7. Orbifords and their $\mathcal{G}$-structures

Following Satake [1] and Thurston [1, Chapter 13], we introduce an object that extends the concepts of a manifold and a manifold with a boundary.

The orbifold $\mathcal{O}$ is a topological space locally appearing as the orbit space for a properly discontinuous group acting on a certain manifold, or, what is equivalent, it is a topological space appearing locally as $\mathbb{R}^n$ modulo the action of a finite group. We give a formal definition (Thurston [1]):

DEFINITION 2.27. The orbiford $\mathcal{O}$ consists of a Hausdorff space $X = X_{\mathcal{O}}$ with a certain additional structure. X has the covering $\{U_i\}$ by a collection of open sets, which is closed under a finite intersection. We put each U_i into correspondence with a finite group G_i and the action of G_i on the open set $\tilde{U}_i \subset \mathbb{R}^n$ and the homeomorphism

$$\varphi_i : U_i \longrightarrow \tilde{U}_i/G_i. \tag{2.36}$$

As $U_i \subset U_j$, there exists the injective homomorphism

$$f_{ij} \; ; \; G_i \hookrightarrow G_j \tag{2.37}$$

and the embedding

$$\tilde{\varphi}_{ij} : \tilde{U}_i \hookrightarrow \tilde{U}_j \tag{2.38}$$

equivarient with respect to f_{ij} (i.e., for $g \in G_i$ and $x \in \tilde{U}_i$ we have $\tilde{\varphi}_{ij}(g(x)) = f_{ij}(g)\tilde{\varphi}_{ij}(x))$, and which are such that the diagram below is commutative:

$$\begin{array}{ccc} \tilde{U}_i & \xrightarrow{\quad \tilde{\varphi}_{ij} \quad} & \tilde{U}_j \\ \downarrow & & \downarrow \\ \tilde{U}_i/G_i & \xrightarrow{\quad \varphi_{ij} = \tilde{\varphi}_{ij}/G_i \quad} & \tilde{U}_j/f_{ij}G_i \\ \uparrow \varphi_i & & \downarrow \\ U_i \subset U_j & \xrightarrow{\quad \varphi_j \quad} & \tilde{U}_j/G_j \end{array} \tag{2.39}$$

We note that the $\tilde{\varphi}_{ij}$ are defined modulo composition with elements of the group G_j, and correspondingly, the f_{ij} are defined modulo conjugation by elements of G_j. In general, it is not true that $\tilde{\varphi}_{ik} = \tilde{\varphi}_{jk} \circ \tilde{\varphi}_{ij}$ when $U_i \subset U_j \subset U_k$; however, there exists an element $g \in G_k$ such that

$$g \circ \tilde{\varphi}_{ik} = \tilde{\varphi}_{jk} \circ \tilde{\varphi}_{ij} \tag{2.40}$$

and

$$g \circ f_{ik}(g_i) \cdot g^{-1} = f_{jk} \circ f_{ij}(g_i), \qquad g_i \in G_i, \tag{2.41}$$

It is clear that the orbifold structure should not be dependent on the covering $\{U_i\}$: two coverings give the same orbifold structure if their union satisfies the definition.

By a $\mathcal{G}$-orbifold, where $\mathcal{G}$ is a pseudogroup, we mean an orbifold in whose definition all the mappings and actions of the groups relate to $\mathcal{G}$ (see §5).

Similarly, if we are given a manifold X and a group G of homeomorphisms acting in it, then by a (G, X)-orbifold we mean the $\mathcal{G}$-orbifold obtained by gluing pieces of X by means of elements from the pseudogroup $\mathcal{G}$ formed by restrictions of the elements of the group G to open sets in X. In that sense one can speak of the following orbifolds: differentiable, conformal, Euclidean, elliptic, and hyperbolic.

In this situation, we say that $X = X_{\mathcal{O}}$ is a modelling space for $\mathcal{O}$ or $\mathcal{O}$ is topologically $X_{\mathcal{O}}$.

In what follows, we shall operate in the smooth category, so we have to assume that the action of the group G_i in the tangent space at its fixed point is defined by some subgroup of $GL(n, \mathbb{R})$. Due to the finiteness of G_i, this action should be determined by the action of a subgroup of the orthogonal group $O(n)$.

It follows from the definition of the orbifold $\mathcal{O}$ that if we make the neighbourhood of a point $x \in \mathcal{O}$ arbitrarily small, the groups put into correspondence with it stabilize to the group G_x, which is the local group at point x. The singular set $\Sigma_{\mathcal{O}}$ of the orbifold $\mathcal{O}$ consists of those points $x \in X$ for which $G_x \neq \{\mathrm{Id}\}$. It is evident that $\mathcal{O}$ is a manifold if $\Sigma_{\mathcal{O}} = \varnothing$. The singular set $\Sigma_{\mathcal{O}}$ is closed, since its intersection with any coordinate chart is closed. $\Sigma_{\mathcal{O}}$ is nowhere dense, as follows from the fact that a non-trivial homeomorphism in a manifold that leaves an open set fixed cannot have finite order.

We consider some orbifolds in order to illustrate these concepts (Thurston [1]).

EXAMPLE 2.28 (*a barber's shop*). Consider the group G generated by reflections in the planes $P(e_i, 0)$ and $P(e_i, 1)$ of the space $\mathbb{R}^3$. Group G is an infinite dihedral group $D_\infty = \mathbb{Z}_2 * \mathbb{Z}_2$. The quotient $\mathbb{R}^3/G$ is the layer $\{x \in \mathbb{R}^3 : 0 \leq x_1 \leq 1\}$. Physically, this corresponds to two mirrors on parallel walls.

EXAMPLE 2.29 (*a rectangular pillow*). Let G be the subgroup of index 2 of the group H of Euclidean plane isometries consisting of elements that preserve orientation and which is generated by reflections with respect to the four sides of the rectangle R. The group H is isomorphic to $D_\infty \times D_\infty$, while the quotient $\mathbb{R}^2/H$ (billiard table) is the rectangle R. The factor space $\mathbb{R}^2/G = \mathcal{O}$ is obtained by identifying the sides of the two rectangles R and R' by relections in their common edge. Topologically, the orbifold $\mathcal{O}$ is a sphere having four singular points corresponding to points in $\mathbb{R}^2$ having non-trivial isotropy groups ($\mathbb{Z}_2$). This sphere has a Riemannian metric

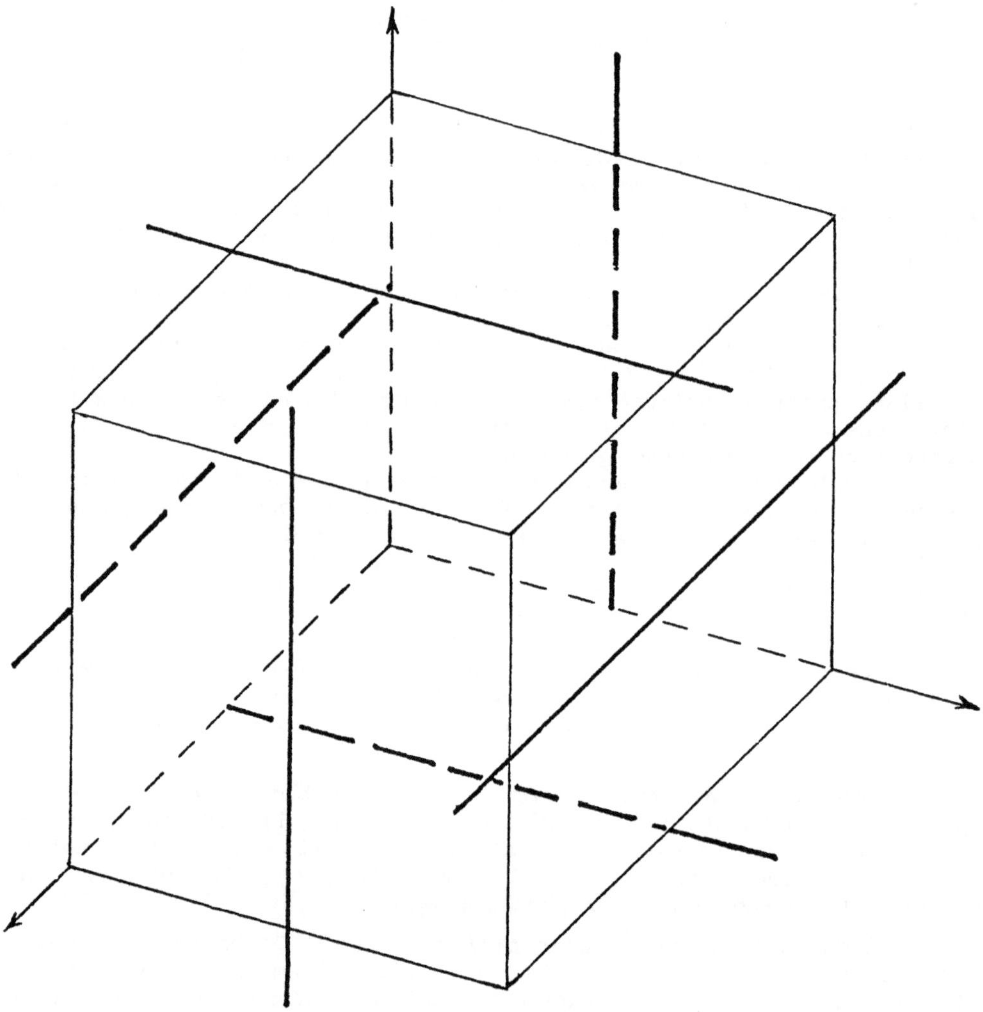

Figure 15

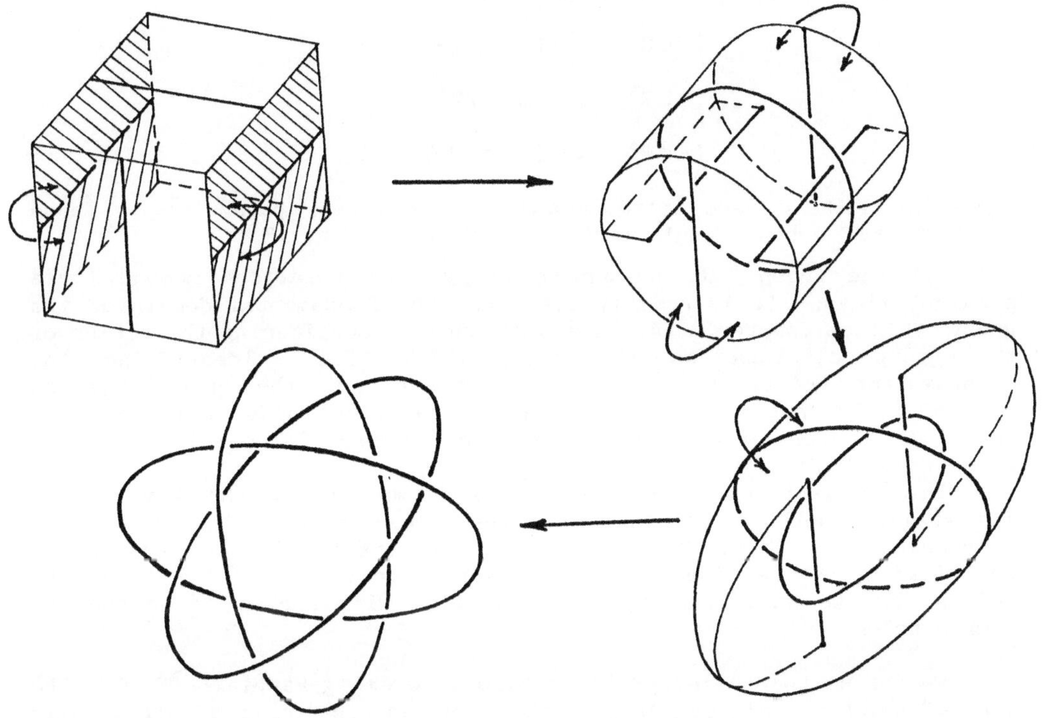

Figure 16

having zero curvature in the complement to these four points p_i and has curvature $K_{p_i} = \pi$, concentrated at each of the four points p_i. The neighbourhood of each point p_i is a cone, whose vertex angle (at the surface) is $\pi = 2\pi - K_{p_i}$.

EXAMPLE 2.30 (*a singular set here is the Borromean rings*). We consider three families of straight lines in $\mathbb{R}^3$ of the form

$$\{x \in \mathbb{R}^3 \ : \ x = (t,\ n,\ m + 1/2)\},$$

$$\{x \in \mathbb{R}^3 \ : \ x = (m + 1/2,\ t,\ n)\},$$

$$\{x \in \mathbb{R}^3 \ : \ x = (n,\ m + 1/2,\ t)\},$$

where n and m are integers and t is a real parameter. These lines cut off cubes of unit lattice as shown in Fig. 15.

Let the group G be generated by 180° rotations around these straight lines: it is readily seen that the fundamental domain of the group G (see Chapter 3 §4) is a unit cube. Considering the action of the group elements on the cube sides, i.e., identifying the G-equivalent points on the cube sides, we get the quotient space $\mathcal{O} = \mathbb{R}^3/G$. In that identification, each cube side folds up twice as an open book. Here we have to examine the images of the rotation axes, which form the singular set $\Sigma_{\mathcal{O}}$ (Fig. 16).

The quotient space $\mathcal{O} = \mathbb{R}^3/G$ is homeomorphic to the sphere S^3. Its singular set $\Sigma_{\mathcal{O}}$ consists of three linked circles in the form of Borromean rings (Fig. 12). In the complement $S^3 \setminus \Sigma_{\mathcal{O}}$, we have a Euclidean structure, i.e., a metric having curvature 0, while at the points $\Sigma_{\mathcal{O}}$ we have cone-type singularities with angles π along the link circles.

We define the notion of the orbifold covering mapping $\tilde{\mathcal{O}} \to \mathcal{O}$, the idea of which roughly speaking is that its restriction to a sufficiently small neighbourhood appears as $\tilde{U}/G' \to \tilde{U}/G$, where $\tilde{U}$ is an open ball in $\mathbb{R}^n$ and G' is a subgroup of G. More precisely, a covering orbifold $\tilde{\mathcal{O}}$ for the orbifold $\mathcal{O}$ is one with the projection $p : X_{\tilde{\mathcal{O}}} \to X_{\mathcal{O}}$ for modelling spaces for which at each point $x \in X_{\mathcal{O}}$ there exists a neighbourhood $U = \tilde{U}/G$ (where $\tilde{U} \subset \mathbb{R}^n$ is open) such that each connectivity component V_i of the set $p^{-1}(U)$ is isomorphic to $\tilde{U}/G_i$, where $G_i \subset G$ is a certain subgroup. Here the isomorphism should be compatible with the projection.

The following situation is a basic example of this covering. Let the group G act on a manifold M. Then M is a covering orbifold for the orbifold M/G; moreover, for any subgroup $G' \subset G$, the orbifold M/G' is also a covering one for the orbifold M/G. For example, in Example 2.29, the "rectangular pillow" is a double covering for the "billiard table".

It can be shown that any orbifold $\mathcal{O}$ has the universal covering orbifold $\tilde{\mathcal{O}}$ defined in the following sense. If $x \in X_{\mathcal{O}} \setminus \Sigma_{\mathcal{O}}$ is a basis (fixed) point in the orbifold $\mathcal{O}$, then the universal covering orbifold $\tilde{\mathcal{O}}$, $p: \tilde{\mathcal{O}} \to \mathcal{O}$, is a connected orbifold covering $\mathcal{O}$ having basis point at $\tilde{x}$, which is projected in x and is such that for any other covering orbifold $\mathcal{O}$, $p' : \tilde{\mathcal{O}}' \to \mathcal{O}$, having a basis point x', $p'(x') = x$, there exists a lifting $q : \tilde{\mathcal{O}} \to \mathcal{O}$ of the mapping p that is a covering mapping for orbifolds (see Thurston [1], Proposition 13.2.4, and see also Scott [4]).

If $\mathcal{O}$ is a good orbifold (i.e., $\mathcal{O}$ has a covering orbifold $\mathcal{O}'$, $\Sigma_{\mathcal{O}'} = \emptyset$), then $\mathcal{O}$ is covered by the simply connected manifold M having a trivial structure in the orbifold sense. Then one can show directly that M is a universal covering orbifold by using the fact that each orbifold covering $\mathcal{O}$ should be isomorphic to M/G' for some subgroup G' of deck transformations group G for manifold M over $\mathcal{O}$.

DEFINITION 2.31. The deck transformation group for the universal covering orbifold $\tilde{\mathcal{O}}$ is called the fundamental group $\pi_1(\mathcal{O}) = \pi_1^{orb}(\mathcal{O})$ of the orbifold $\mathcal{O}$.

DEFINITION 2.32. A topological space that is locally either the space $\mathbb{R}^n$ or the half-space $\{x \in \mathbb{R}^n : x_n \geq 0\}$ modulo the action of a finite group (see Definition 2.30) is called an orbifold with boundary.

DEFINITION 2.33. The subspace $X_{\mathcal{O}_1} \subset X_{\mathcal{O}}$ that is locally a space $\mathbb{R}^k \subset \mathbb{R}^n$ $(k \leq n)$ modulo finite-group action is called a suborbifold $\mathcal{O}_1$ of the orbifold $\mathcal{O}$.

If $X_{\mathcal{O}}$ is a topological manifold, one should not confuse $\partial X_{\mathcal{O}}$ with $\partial \mathcal{O}$ or with $X_{\partial \mathcal{O}}$, or the fundamental group $\pi_1(\mathcal{O})$ with the fundamental group $\pi_1(X_{\mathcal{O}})$.

We finish this survey of orbifold properties with the following theorems.

THEOREM 2.34 (Dunbar [1]). *Let N be a subgroup of the fundamental group $\pi_1(\mathcal{O})$ of a good orbifold $\mathcal{O}$ having modelling space $X_{\mathcal{O}}$ and generated by deck transformations having fixed points (we note that N is a normal subgroup).*

Then

$$\pi_1(X_{\mathcal{O}}) \cong \pi_1(\mathcal{O})/N.$$

THEOREM 2.35 (Thurston [1], 13.3.2). *If G is the group of real analytic diffeomorphisms for the real analytic manifold X, then any (G, X)-orbifold $\mathcal{O}$ is a good one. Then the developing map is defined:*

$$d : \tilde{\mathcal{O}} \to X$$

and the holonomy homomorphism

$$d_* : \pi_1(\mathcal{O}) \to G$$

THEOREM 2.36. *Under the conditions of Theorem 2.35, let G be an isometry group acting transitively in X. Then if the orbifold $\mathcal{O}$ is closed or metrically complete, $\mathcal{O}$ is a complete (G, X)-orbifold, i.e., the developing map is a cover. In particular, if X is simply connected, then $\tilde{\mathcal{O}} = X$ and $\pi_1(\mathcal{O})$ acts as a discrete subgroup of the group G.*

We note finally that the orbifold concept gives us a mechanism for examining the discrete actions of groups in manifolds, which generalizes the free action to action with fixed points. With such action, most of the concepts defined for manifolds are transferred to orbifolds: in particular, Euler characteristics, orientability, and the concept of a Haken manifold. A more detailed exposition can be found in Thurston [1], Dunbar [1], [2], and Scott [4], where there are proofs of the theorems in this section.

CHAPTER 3

GENERAL PROPERTIES OF DISCONTINUOUS GROUPS

§1. Basic concepts; elementary groups

1. *Basic definitions.* Here we shall examine the subgroups G of the group $\text{Möb}(n)$ of the Möbius transformations for the n-dimensional conformal sphere $S^n = \overline{\mathbb{R}}^n$. In the previous chapter, we established the natural isomorphism

$$\text{Möb}(n) \cong SO(n+1, 1)/\{\pm 1\}, \tag{3.1}$$

induced by the isometry of (2.22) for two models of the $(n+1)$-dimensional hyperbolic space. Then the topology in $\text{Möb}(n)$ can be induced by the topology of $SO(n+1, 1)$.

We have seen in Chapter 1 §4.2 that for the ball $B^{n+1}(0, 1)$, any element $g \in \text{Möb}(B^{n+1})$ takes the form $U \circ g_y$, where $U \in O(n+1)$, $y \in B^{n+1}$, $g_y(y) = 0$, as (1.31) shows. This indicates that in $\text{Möb}(B^{n+1})$ we may introduce the product topology $O(n+1) \times B^{n+1}$. The topology in $\text{Möb}(n)$ induced by this topology by means of the isomorphism $\text{Möb}(B^{n+1}) \cong \text{Möb}(n)$, see (1.36), coincides with that introduced by means of (3.1).

A subgroup $G \subset \text{Möb}(n)$ is called a *discrete* group if it is a discrete subset of the topological group $\text{Möb}(n)$. All such groups are countable.

Let a group $G \subset \text{Möb}(n)$ act in an open set $D \subset \overline{\mathbb{R}}^n$. One says that it acts discontinuously in D if for each compact subset $K \subset D$ there exists only a finite number of elements $g \in G$ such that $g(K) \cap K \neq \emptyset$. A group $G \subset \text{Möb}(n)$ is called a *discontinuous* or *Kleinian* group in $\overline{\mathbb{R}}^n$ if there exists a non-empty open set $D \subset \overline{\mathbb{R}}^n$ in which it act discontinuously.

It is clear that the discontinuity of a group G involves discreteness (the absence of elements close to the identity). The converse is incorrect. A classical example of this is provided by the Picard group, which consists of all linear fraction transformations $w(z) = (az+b)/(cz+d)$ having integer complex coefficients. It is clear that this group is discrete, but it does not act discontinuously in any open subset of the plane: the images of any point in the plane are dense in it.

For a group $G \subset \text{Möb}(n)$ and for any point $x \in \overline{\mathbb{R}}^n$, we consider the G-orbit of the point x, i.e., the set $G(x) = \{g(x) : g \in G\}$. The set of converging points for the orbits $G(x)$, $x \in \overline{\mathbb{R}}^n$ of the group G is called the *limit* set of the group G and is denoted by $\Lambda(G)$. The complement of this $\Omega(G) = \overline{\mathbb{R}}^n \setminus \Lambda(G)$ is called the discontinuity set of the group G. The following theorem can be proved by elementary methods:

PROPOSITION 3.1. *Let a group $G \subset \text{Möb}(n)$. Then the following assertions are equivalent:*

a) *G is a discontinuous group in $\overline{\mathbb{R}}^n$;*

b) $\Omega(G) \neq \varnothing$;

c) *there exist a point $x \in \text{Möb}(n)$ and an open neighbourhood of it $U(x)$ such that the set $\{g \in G : g(U) \cap U \neq \varnothing\}$ is finite.*

From the definition of the limit set for any group $G \subset \text{Möb}(n)$ (including a discontinuous or discrete one) it follows that is closed and G-invariant, and, which is the same, that the discontinuity set $\Omega(G)$ is open and G-invariant. The converse is also readily shown: any closed set $L \subset \overline{\mathbb{R}}^n$ consisting of more than two points and that is invariant with respect to a discrete group G contains the limit set $\Lambda(G)$. This, in particular, implies that the group $\tilde{G} \subset \text{Möb}(n+1)$ obtained by extension of the action of the group G to $\overline{\mathbb{R}}^{n+1}$ (see Chapter 1 §4) has the same limit set: $\Lambda(\tilde{G}) = \Lambda(G) \subset \overline{\mathbb{R}}^n$, and consequently acts discontinuously in the half-space $\overline{\mathbb{R}}^{n+1}_+$.

2. *Elementary groups.* The limit set of a discrete group $G \subset \text{Möb}(n)$ is either infinite (and in that case perfect) or empty (in that case G is a finite group), or consists of one or two points. The groups corresponding to the last three cases are called *elementary*. The following assertion is proved to describe such groups.

LEMMA 3.2. *Let $g_1, g_2 \in G \subset \text{Möb}(n)$ with g_1 a loxodromic transformation. If g_1 and g_2 have precisely one common fixed point, then the group G is not discrete.*

PROOF. Without loss of generality we assume that $g_1(0) = 0$, $g_2(0) \neq 0$ and $g_1(\infty) = g_2(\infty) = \infty$. Then we apply Corollary 1.8 to get

$$g_1(x) = \alpha \cdot U_1(x), \quad g_2(x) = \beta U_2(x) + b, \tag{3.2}$$

where $\alpha, \beta > 0$, $\alpha \neq 1$, $b \in \mathbb{R}^n \setminus \{0\}$ and U_1, U_2 are orthogonal transformations. We now consider elements of the following form in the group G:

$$h_n(x) = g_2 g_1^n g_2^{-1} g_1^{-n}(x) = U_2 U_1^n U_2^{-1} U_1^{-n}(x) - \alpha^n U_2 U_1^n U_2^{-1}(b) + b. \tag{3.3}$$

If here the transformation g_1 is such that $U_1^m = 1$ in (3.2) for some $m \in \mathbb{Z} \setminus \{0\}$, we consider a number n in (3.3) that is a multiple of m. If on the other hand the rotation U_1 has infinite order, then for any $\varepsilon > 0$ there exists a number n_ε such that $|U_1^{n_\varepsilon}(x) - x| < \varepsilon$ for all x from the bounded ball $B_n(0, r)$. Letting ε tend to zero we get that, as in the first case, there is an infinite sequence of distinct elements h_n in the group G (because $0 < \alpha < 1$) that converges to a non-degenerate Möbius transformation $h : x \longmapsto x + b$, which contradicts the discreteness of the group G.

We now describe all the elementary groups:

PROPOSITION 3.3. *Group* $G \subset \text{Möb}(n)$ *is elementary if and only if* G *has a finite-index Abelian subgroup.*

PROOF. In the first place, $L(G)$ is empty if and only if G is a finite group of elliptic elements. Namely, if G is infinite, then there would exist points x_0, y_0 and a sequence $\{g_i\} \subset G$ which converges to y_0 c-uniformly in $\bar{\mathbb{R}}^n \setminus \{x_0\}$ (cf. Belinskii [1]); thus we would obtain $y_0 \in \Lambda(G) \neq \emptyset$.

In the second place, the limit set $\Lambda(G) = \{x_0\}$ if and only if the group G is an infinite group which contains only elliptic and parabolic elements which fix x_0. Namely, if $\Lambda(G) = \{x_0\}$, then $\{g(x_0)\} = g(\Lambda(G)) = \{x_0\}$ for each $g \in G$, since $\Lambda(G)$ is G-invariant. Thus, by Lemma 3.2, each element of G is either of finite order or infinite order with one fixed point. Conversely, if G contains no loxodromic elements and $G(x_0) = x_0$, then in the case $|G| = \infty$ we can choose a sequence of distinct elements $\{g_j\} \subset G$ which converges to some y_0 c-uniformly in $\bar{\mathbb{R}}^n \setminus \{x_0\}$. Since G contains no loxodromic elements, then $\{y_0\} = \{x_0\} = \Lambda(G)$.

In the last case, $\Lambda(G)$ contains exactly two points x_0 and y_0 if and only if G is an infinite group which contains only loxodromic elements which either fix or interchange the points x_0 and y_0. Namely, if $\Lambda(G) = \{x_0, y_0\}$, each $g \in G$ satisfies one of the following conditions:

$$g(x_0) = x_0, \quad g(y_0) = y_0$$

or

$$g(x_0) = y_0, \quad g(y_0) = x_0$$

since $\Lambda(G)$ is G-invariant. We see that G does not contain parabolic elements since no such element g could satisfy the first condition. Also each g which satisfies the last condition is an elliptic element of even order. Conversely, G contains (as above) a loxodromic element which fixes x_0 and y_0. Hence $\{x_0, y_0\} \subset \Lambda(G)$. If $y^* \in \Lambda(G)$, then we can choose point x^* and a sequence $\{g_i\}$ in G which converge to y^* in $\bar{\mathbb{R}}^n \setminus \{x^*\}$. By hypothesis there exists a subsequence (g_j) whose elements either all fix or all interchange the two points x_0, y_0 and, therefore, either $x_0 \neq x^*$ and $y^* = \lim g_j(x_0)$ or $y_0 \neq x^*$ and $y^* = \lim g_j(y_0)$. Thus $\Lambda(G) = \{x_0, y_0\}$ and the proof is complete.

The sharpening of this theorem is the following theorem of Bieberbach.

THEOREM 3.4. *Let* G *be a discrete group of Euclidean isometries. Then*:

1) *if* $\overline{\mathbb{R}}^n/G$ *is compact, the normal subgroup* G^* *of translations is a free Abelian group of rank* n *having finite index* $\mu(G^*, G)$ *in the group* G;

2) *if* $\overline{\mathbb{R}}^n/G$ *is not compact, there exists a normal subgroup* G^* *having finite index* $\mu(G^*, G)$ *that is a free Abelian group of rank* k, $0 \le k \le n-1$. *Then up to conjugation of the group* G *by an isometry of* $\mathbb{R}^n$, *the groups* G *and* G^* *have the following structure.*

We write $\mathbb{R}^n = \mathbb{R}^k \times \mathbb{R}^{n-k}$ *and let* $g : x \longmapsto U \cdot x + a$ $(U \in O(n),\ a \in \mathbb{R}^n)$ *be any element from* G; *then* $U(\mathbb{R}^k) = \mathbb{R}^k$, $U(\mathbb{R}^{n-k}) = \mathbb{R}^{n-k}$, $a \in \mathbb{R}^k \times \{(0, \ldots, 0)\}$. *If* g *is an element in the group* G^*, *then* $U|_{\mathbb{R}^k} = \mathrm{id}$, *i.e.,* $g|_{\mathbb{R}^k}$ *is a translation.*

See for example Vinberg [3] and Wolf [1] for modern proofs of this fact. See Remark 4.18 for another approach to Bieberbach's theorem.

§2. Action of a group on the discontinuity set

The definition of the discontinuity set $\Omega(G)$ for a discontinuous group $G \subset \text{Möb}(n)$ implies discontinuity in the action of the group G there. We consider the space of the orbits $G(x)$, $x \in \Omega(G)$, i.e., the quotient space $\Omega(G)/G$. Let $\pi : \Omega(G) \longrightarrow \Omega(G)/G$ be the natural projection.

PROPOSITION 3.5. *Let* $G \subset \text{Möb}(n)$ *be a Kleinian group. Then* $\Omega(G)/G$ *is an* n*-dimensional orbifold. If here the group* G *acts freely on* $\Omega(G)$, *then* $\pi : \Omega(G) \longrightarrow \Omega(G)/G$ *is a covering and* $\Omega(G)/G$ *is a manifold.*

PROOF. As it is obvious that the structural conditions for an orbifold are met, it is sufficient to prove separability for the topological space $\Omega(G)/G$. We consider two distinct orbits $G(x_1)$ and $G(x_2)$, $G(x_1) \cap G(x_2) = \varnothing$. Due to the discontinuity of the group G, any compact neighbourhood of a point $x_1 \in \Omega(G)$ contains only a finite number of images $g(x_2)$ of a point $x_2 \in \Omega(G)$. Therefore, there exists a neighbourhood $U_1 \subset \Omega(G)$ of the point x_1 such that $U_1 \cap G(x_2) = \varnothing$. However, then $x_2 \cap G(U_1) = \varnothing$ and, as above, there exists a neighbourhood $U_2 \subset \Omega(G)$ of the point x_2 such that $G(U_1) \cap G(U_2) = \varnothing$. This proves the separability of $\Omega(G)/G$.

Similarly, if G acts in $\Omega(G)$ without fixed points, for any point $x \in \Omega(G)$ one can find a neighbourhood U that does not intersect with the images $g(x)$, $g \in G\setminus\{\mathrm{id}\}$. In that neighbourhood, the projection π is a homeomorphism, so π is a covering and $\Omega(G)/G$ is a manifold.

We note that for $G \subset \text{Möb}(2)$ and $G \subset \mathcal{M}_3$ we know that topologically $\Omega(G)/G$ is always a manifold (without allowing for the orbifold structure); for $n \ge 4$ or for $G \subset \mathcal{M}\text{öb}(3)$ with orientation reversing involutions fixing two points, it may also be not a manifold but only a pseudo-manifold (consider for example the quotient space of the 4-dimensional ball by a group $\mathbb{Z}_k$ acting freely on its boundary with a lens space as the quotient 3-space).

The natural projection π defines in the orbifold $\Omega(G)/G$ the structure of a conformal orbifold (see Chapter 2 §7). It branches over the projections of the subdomains $S^k \cap \Omega(G)$ of the k-dimensional spheres S^k, $0 \le k \le n-2$, which have the non-trivial stabilizers

$$G_{S_k} = \{g \in G : g(x) = x \quad \text{for all} \quad x \in S^k\},$$

which act as structural groups for the orbifold $\Omega(G)/G$.

The discontinuity set $\Omega(G)$ itself splits up into connected components Ω_i, of which there may be a countable set (they are also called *components of the group G*). The stabilizers

$$G_{\Omega_i} = \{g \in G : g(\Omega_i) = \Omega_i\} \tag{3.4}$$

of these components are subgroups in G and are Kleinian groups themselves. If the stabilizer of (3.4) coincides with the group G itself, the component Ω_i is called *invariant*. If $\Omega_1, \Omega_2, \ldots$ is a complete set of components not equivalent one to another, the orbifold $\Omega(G)/G$ is represented in the form

$$\Omega(G)/G = \Omega_1/G_{\Omega_1} \cup \Omega_2/G_{\Omega_2} \cup \ldots \tag{3.5}$$

as the union of connected n-dimensional orbifolds.

One therefore has to consider how many components the Kleinian group G can have, how many of these may be invariant, and what is the homotopy type of these components.

Answers to these questions have long been known for planar Kleinian groups. The number of connected components for a Kleinian group $G \subset \text{Möb}(2)$ is one, two, or ∞; not more than two of them may be invariant, and if there are two invariant components, the other components of the group are atoms, namely simply connected domains having trivial stabilizers (Accola [1]). All finitely generated groups having two invariant components are quasi-Fuchsian groups (Ahlfors [2] and Accola [1]), i.e., they do not have other components of the discontinuity set.

The situation alters completely as the dimension increases.

Firstly, for any dimension $n \ge 3$, there exist finitely generated Kleinian groups $G \subset \mathcal{M}_n$ whose discontinuity sets each consist of two invariant components Ω_0 and Ω_1, one of which is contractible and the other is not even simply connected (has non-trivial homotopy groups $\pi_k(\Omega_1)$, $1 \le k \le n-2$).

Secondly, for any $n \ge 3$, there exist analogous groups $G \subset \mathcal{M}_n$, $\Omega(G) = \Omega_0 \cup \Omega_1$, having two invariant components with non-trivial homotopy groups $\pi_k(\Omega_i)$, $i = 0, 1$, $k = 1, \ldots, n-2$.

Thirdly, for $n \ge 3$, there exist finitely generated Kleinian groups $G \subset \mathcal{M}_n$, each having two invariant components $\Omega_0, \Omega_1 \subset \Omega(G)$, and, in addition, each having an infinite number of non-invariant components:

$$\Omega(G)\setminus(\Omega_0 \cup \Omega_1) = \bigcup_{j=2}^{\infty} \Omega_j.$$

Here the non-invariant (contractible) components Ω_j, $j \geq 2$, are not atoms, i.e., they have non-trivial stabilizers $G_j \subset G$.

The three types of finitely generated (and even geometrically finite) Kleinian groups $G \subset \mathcal{M}_n$, $n \geq 3$, containing two invariant components each are clearly not quasi-Fuchsian ones (see above for $n = 2$). The first two classes will be constructed in Chapter 9 §4 as the boundary groups for the deformation space of the Fuchsian group $\Gamma \subset \mathcal{M}_n$. An example of the third type will be constructed in Chapter 5 §5, see Example 5.66.

Finally, the result for the number of invariant connected components for the discontinuity set $\Omega(G)$ of a group $G \subset \mathcal{M}_n$, $n \geq 3$, is determined quite differently from $n = 2$, as the following theorem shows constructively (see Chapter 5 §5 for geometrically finite groups, where the situation is different):

THEOREM 3.6 (Tetenov [1], [2]). *For any natural numbers* k, k_0, $n : k \geq 1$, $k \geq k_0 \geq 0$, $n \geq 3$, *there exist Kleinian groups* $G \subset \mathcal{M}_n$ *whose discontinuity sets split up into* k *invariant components each*:

$$\Omega(G) = \Omega_1 \cup \ldots \cup \Omega_k, \tag{3.6}$$

$$\pi_j(\Omega_i) = 0, \quad 1 \leq j \leq n-1, \quad 0 \leq i \leq k_0. \tag{3.7}$$

To prove the theorem, we need the following simple criterion of invariantness for a component of the discontinuity set, see Accola [1]:

LEMMA 3.7. *Let* $G \subset \text{Möb}(n)$ *be a Kleinian group, and let* F_0 *be a component of its fundamental domain, which contains a certain point* x_0. *If here for each generator* g *of the group* G, *the points* $g(x_0)$ *and* x_0 *can be joined by a path lying in* $\Omega(G)$, *then the component* Ω_0 *of the set* $\Omega(G)$ *containing* F_0 *is* G*-invariant.*

We also need the following lemma (Brown [1]):

LEMMA 3.8. *Let a topological space* X *be the union of the increasing sequence* $V_1 \subset V_2 \subset \ldots \subset V_i \subset \ldots$ *of open subsets* V_i, *each of which is homeomorphic to* $\mathbb{R}^n$. *Then* X *is homeomorphic to* $\mathbb{R}^n$.

PROOF OF THEOREM 3.6. Given numbers $n \geq 3$, $k \geq 1$, $k \geq k_0 \geq 0$, we construct the group $G = G(n, k, k_0) \subset \text{Möb}(n)$ with the properties of (3.6) and (3.7). For this we construct a system Σ of $(n-1)$-dimensional spheres in $\overline{\mathbb{R}}^n$ that satisfy the following conditions.

C1. For all distinct $\sigma_i, \sigma_j \in \Sigma$, we have

$$\text{int}\,\sigma_i \cap \text{int}\,\sigma_j = \varnothing.$$

C2. The set

$$F_{\Sigma} = \overline{\mathbb{R}}^n \setminus \overline{\bigcup_{\sigma\in\Sigma} \operatorname{int} \sigma} \tag{3.8}$$

is non-empty and splits up into k connected components (the spherical polyhedra) $P_1, \ldots, P_k$.

C3. If P is a connected component of the set F_{Σ}, then for any $\sigma\in\Sigma$ the set $\sigma\cap\overline{P}$ has a non-empty interior in the sphere σ.

C4. The components P_m, $m\le k_0$, of the set F_{Σ} are topological balls and for any $\sigma\in\Sigma$ the domain $\operatorname{int}_{\sigma}(\sigma\cap\overline{P})$ is contractible in σ.

If a family Σ having these properties has been constructed, we can take as a Kleinian group $G=G(n, k, k_0)$ satisfying (3.6) and (3.7) the group $G\subset \mathcal{M}_n$, generated by the compositions of an even number of inversions with respect to spheres from the family Σ.

The discontinuity of the group G follows from condition C1 and from the non-emptiness of the set (3.8), see Corollary 3.20. Here F_{Σ} will be a fundamental set of the group G, i.e., $\Omega(G)\subset G(\overline{F}_{\Sigma})$.

From condition C3 and Lemma 3.7 we see that for all $m=1, \ldots, k$, the sets

$$\Omega_m = \operatorname{int}(G(\overline{P}_m)) \tag{3.9}$$

are invariant components of the discontinuity set $\Omega(G)$. Also C2 implies that F_{Σ} is the union of the polyhedra $P_1, \ldots, P_k$, so $\Omega(G)=\Omega_1\cup\ldots\cup\Omega_k$.

Finally, from C4 we get that the components Ω_m, $m\le k_0$, are homeomorphic to a ball. This is obtained by applying Lemma 3.8 to the increasing sequence of topological balls

$$B_N = \operatorname{int}\left[\bigcup_{g\in G(N)} g(\overline{P}_m)\right],$$

where $G(N)$ is a subset of the group G consisting of all words (on the free generators of the group G) having length less than or equal to N, i.e., words of the form

$$\mathcal{I}_{k_1}\circ\ldots\circ\mathcal{I}_{k_i},$$

where $\mathcal{I}_{k_j}$ are inversions with respect to the spheres $\sigma_{k_j}\in\Sigma$, while $i\le N$. The proof of homeomorphism to a ball for the sets B_N is obtained by induction on N with the use of C4 on the contractibility of the domain $\operatorname{int}_{\sigma}(\sigma\cap\overline{P}_m)$ in a sphere $\sigma\in\Sigma$.

As the assertion of Theorem 3.6 is obvious for $k=1, 2$ (for $k=1$, the groups are elementary, while for $k=2$ they are the Fuchsian groups $G\subset\mathcal{M}_n$), it remains to prove this for $k\ge 3$. We perform the constructions when the dimension n of the space and the number k of components in the discontinuity set are three; the constructions are analogous for larger k and n.

We first construct a family Σ satisfying C1 - C3, and we then show how to ensure C4 is satisfied, or rather, the topological ball condition for the components P_m of the set (3.8).

We start by taking two closed non-intersecting surfaces S_1^0, S_2^0 in the Euclidean space and "cover" each of them by a family of spheres such that firstly the union of the balls bounded by these spheres contains the given surface and lies in some neighbourhood of it and secondly the points of tangency of the spheres in this family lie on that surface. In this way, we get a family of spheres Σ_0 satisfying conditions C1 and C2. The set $\bigcap_{\sigma\in\Sigma_0} \text{ext}\,\sigma$ will split up into three connected components: the spherical polyhedra P_0^0, P_1^0, P_2^0. Each of the spheres $\sigma\in\Sigma_0$ adjoins only two polyhedra, the unbounded one P_0^0 and one of the bounded ones. For condition C3 to hold, it is necessary to "join up" each such sphere with the polyhedron with which it is not yet adjacent.

To do this we remove from Σ_0 one sphere adjoining P_1^0 and one adjoining P_2^0. This gives a family Σ_0'. The set $\bigcap_{\sigma\in\Sigma_0'} \text{ext}\,\sigma$ will be connected and there are holes at the boundaries of P_1^0 and P_0. On each of the spheres $\sigma_i\in\Sigma_0'$ we take a domain δ_i, bounded by a loop γ_i, such that $\delta_i\subset\sigma_i\cap P_0^0$. We now construct two non-intersecting surfaces S_1^1 and S_2^1 that lie in P_0^0 and are such that each surface S_j^1, $j=1, 2$, is a branched tube joining the hole in the boundary of P_j^0 to the loops γ_i in those spheres σ_i which bound the polyhedron P_{3-j}^0. Then these two surfaces S_1^1 and S_2^1 will divide the set $\bigcap_{\sigma\in\Sigma_0'} \text{ext}\,\sigma$ into three components, and each of the spheres $\sigma_i\in\Sigma_0'$ will adjoin each of them.

As previously for S_1^0 and S_2^0, we cover each of the surfaces S_1^1 and S_2^1 by a certain sphere family Σ_1 such that firstly

$$\sigma_i\cap\sigma_j = \varnothing \quad \text{for all} \quad \sigma_i\neq\sigma_j, \quad \sigma_i, \sigma_j\in\Sigma_1\cup\Sigma_0' \tag{3.10}$$

and secondly

$$\bigcap\{\text{ext}\,\sigma_i : \sigma_i\in\Sigma_1\cup\Sigma_0'\} = P_0^1\cup P_1^1\cup P_2^1. \tag{3.11}$$

Satisfying (3.10) and (3.11) leads to the family $\Sigma_1\cup\Sigma_0'$ satisfying C1 and C2; C3 will be met only for the spheres $\sigma\in\Sigma_0'$. Each of these spheres $\sigma\in\Sigma_1$ will adjoin only two polyhedra: the unbounded one P_0^1 and some bounded one P_i^1, $i=1, 2$. We remove one sphere each from the family Σ_1 from the boundary of P_1^1 and P_2^1 to get a new family Σ_1' (Fig. 17).

We repeat this process with conditions analogous to (3.10) and (3.11) an unlimited number of times to get the sequence of sphere families Σ_m', $m=0, 1, 2, \ldots$, such that

$$\bigcap_i \text{ext}\,\sigma_i = P_0^m\cup P_1^m\cup P_2^m. \tag{3.12}$$

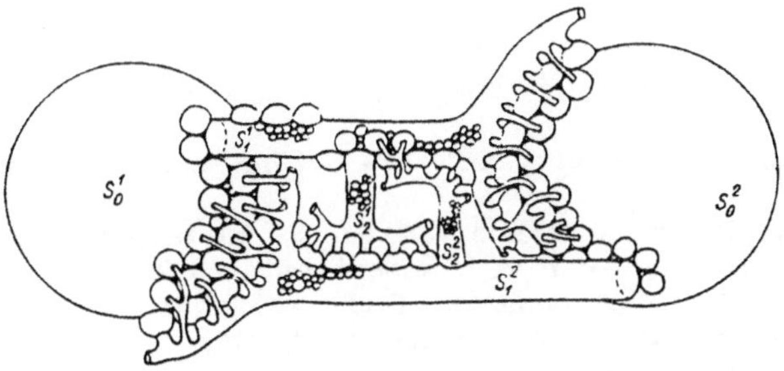

Figure 17

The intersection in (3.12) is taken on all the spheres

$$\sigma_i \in \bigcup_{j=0}^{m-1} \Sigma'_j \cup \Sigma_m. \tag{3.13}$$

Each of the families of (3.13) satisfies C1 and C2, and C3 fails only for the spheres $\sigma_i \in \Sigma_m$. We put

$$\Sigma = \bigcup_{m=0}^{\infty} \Sigma'_m, \quad P_j = \bigcup_{m=0}^{\infty} \bigcap_{i=m}^{\infty} P^i_m \quad \text{for} \quad j=0,\ 1,2. \tag{3.14}$$

The connectivity of P_1 and P_2 follows from the fact that these polyhedra are constructed in (3.14) as the union of an increasing sequence of connected polyhedra P^i_1 and P^i_2 in each case. The connectivity of the polyhedron $P_0 = \bigcap_{i=m}^{\infty} P^i_0$ can be obtained as follows: at each step in (3.12), before we construct the surfaces (branched tubes) S^m_1 and S^m_2, we take in P^{m-1}_0 a connected neighbourhood V_{m-1} of its boundary and require that S^m_1 and S^m_2 and the spheres of the family Σ_m are contained in V_{m-1}. Here the neighbourhoods V_i will be chosen such that

$$V_i \supset V_{i+1}, \quad \bigcap_{i=0}^{\infty} V_i = \varnothing, \tag{3.15}$$

and the sets $P^i_0 \backslash V_i$ are connected ones.

This construction with the additional conditions of (3.15) gives us the following sequences of inclusions:

$$\begin{array}{ccccc} V_0 & \supset & V_1 & \supset \ldots \supset \ldots & = \varnothing \\ \cap & & \cap & & \cap \\ P_0^0 & \supset & P_0^1 & \supset \ldots \supset \ldots & = P_0 \\ \cup & & \cup & & \parallel \\ P_0^0 \backslash V_0 & \subset & P_0^1 \backslash V_1 & \subset \ldots \subset \ldots & = P_0 \end{array}$$

This implies the connectivity of P_0 (the union of connected sets $P_0^m \backslash V_m$).

Then we finally get a sphere family Σ satisfying C1 - C3.

As we wish to derive assertion (3.7) of the theorem (contractibility of the components Ω_i), we require that the family Σ must obey condition C4. To ensure that this is so, we make some changes in our construction.

Firstly, as S_1^0 and S_2^0 we take two spheres touching one another at one point. This enables us to consider a family Σ_0 such that P_0^0, P_1^0 and P_2^0 will be homeomorphic to a ball.

Secondly, for each of the branched tubes S_1^m and S_2^m to split up P_0^{m-1} into contractible parts, we construct S_1^m and S_2^m as follows. We join the hole in the boundary of P_j^{m-1}, $j = 1, 2$, and the domains δ_i on the spheres $\sigma_i \in \Sigma'_{m-1}$ bounding P_{3-j}^{m-1} by means of a branched contractible curve ξ_m^j lying at the boundary of P_0^{m-1}. "Pushing" this curve onto the boundary of each of the sets $\sigma_i \cap P_o^{m-1}$, we make sure that the curve ξ_m^j does not split up any such set into parts. We then glue the surface S_j^m to ∂P_0^{m-1} along the curve ξ_m^j (Fig. 18). This refinement of our construction ensures that C4 is met, so we have proved the last assertion (3.7) in the theorem.

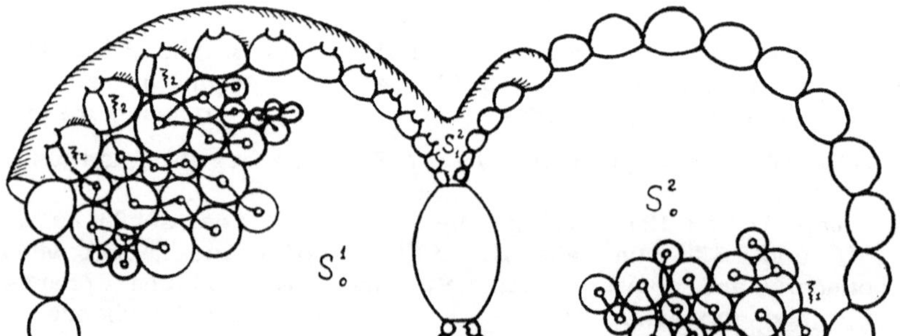

Figure 18

§3. Some classes of Kleinian groups

1. *Fuchsian groups.* A Kleinian group $G \subset \text{Möb}(n)$ is called a *Fuchsian* group if it leaves invariant a certain ball (half-space) $B^n \subset \overline{\mathbb{R}}^n$. The limit set $\Lambda(G)$ of the Fuchsian group G lies on the invariant sphere $S^{n-1} = \partial B^n$. If it coincides with the sphere, G is called a *Fuchsian group of the first kind.* If, on the other hand, $S^{n-1} \backslash \Lambda(G)$ is not empty, G is a *group of the second kind.* In that case, $\Lambda(G)$ is a perfect set, nowhere dense in the sphere S^{n-1}. A Fuchsian group of the first kind has two invariant components $\text{int}\, S^{n-1} = B^n$ and $\text{ext}\, S^{n-1} = \overline{\mathbb{R}}^n \backslash \overline{B}^n$; there is one invariant component for a group of the second kind, namely $\Omega(G)$ itself.

2. *Quasi-Fuchsian groups.* These groups represent a direct generalization of Fuchsian ones, namely a group G whose $\Omega(G)$ is the union of two G-invariant contractible domains Ω_0 and Ω_1 $(\pi_i(\Omega_j) = 0,\ i = 1, \dots, n-1,\ j = 0, 1)$, with the limit set $\Lambda(G)$ as the common boundary. We note that for $n \geq 3$, it is insufficient to specify this property from the closure of Ω_0 alone because there exist groups G for which the Jordan surfaces $\Lambda(G)$ (even a topological $(n-1)$-sphere) do not split up $\overline{\mathbb{R}}^n$ into two contractible components (Example 7.22). In the space of dimension $n \geq 3$ even the existence of non-trivial quasi-Fuchsian groups obtained by quasi-conformal conjugation from Fuchsian ones (i.e., ones distinct from the Fuchsian groups) was unknown until recently. These were first constructed by Apanasov and Tetenov [1] and Apanasov [11], [12] in connection with the study of Fuchsian group deformation space. Examples of quasi-Fuchsian groups will be constructed in Chapter 9 §3. This is the type of the group G_1 in §3.4.

3. *Schottky groups.* Each such group is a Kleinian group $G \subset \mathcal{M}_n$ with generators $g_1, \dots, g_m$ for which there are $2m$ non-intersecting closed topological balls $B_1, B_1', \dots, B_m, B_m'$ in the space $\overline{\mathbb{R}}^n$ such that $g_i(\text{ext}\, B_i) = \text{int}\, B_i'$, $i = 1, \dots, m$. It is readily seen that G is free and that all its elements are loxodromic transformations. Following Myrberg [1], one can show that any non-elementary Kleinian group $G \subset \mathcal{M}_n$ contains non-cyclic Schottky subgroups.

4. *Web groups.* A Kleinian group $G \subset \text{Möb}(n)$ is called a *web group* if each component Ω_i of $\Omega(G)$ has a quasi-Fuchsian group G_i of the first kind as its stabilizer in G. Often one adds here the requirement that the group G is finitely generated. The web groups were first considered by Abikoff [1] for $n = 2$. They have the property that their $\Lambda(G)$ contain points not lying on the boundary of any component $\Omega_i \subset \Omega(G)$:

$$\Lambda(G) \neq \bigcup_i \partial\Omega_i.$$

For $n \geq 3$, the non-trivial web groups (distinct from the quasi-Fuchsian ones) were first constructed by the author; see Krushkal', Apanasov, and Gusevskii [1], [3]. I give one of the possible constructions on the assumption that the space is of dimension three. See Chapter 5 §6 (Example 5.86) for other examples of web groups based on other ideas.

EXAMPLE 3.9. Let Q be the unit coordinate cube. From its vertices we describe the spheres S_i, $1 \le i \le 8$, having radius $\sqrt{3}/3$. As the spheres S_i, $g \le i \le 14$, we take spheres that do not intersect one with another and have their centres coincident with the centres of the faces of Q and having radii $(1/2 - \sqrt{3}/6)$. All these spheres S_i, $1 \le i \le 14$, intersect one with another only at angles of $\pi/3$, and their interiors cover the boundary of the cube Q. We consider a group $G_1 \subset \mathcal{M}_3$, consisting of compositions of an even number of inversions $\mathcal{T}_i$ with respect to the spheres S_i, $i = 1, \ldots, 14$. From Corollary 3.20, it follows that this group is a Kleinian one, and the fundamental set for it can be taken in the form

$$F(G_1) = \left[\bigcap_{i=1}^{14} \text{ext } S_i\right] \cup \mathcal{T}_1 \left[\bigcap_{i=1}^{14} \text{ext } S_i\right].$$

Then $F(G_1)$ splits up into two components: the bounded $F_0 \subset Q$ and the unbounded $F_1 \subset \overline{\mathbb{R}}^3 \setminus \overline{Q}$.

We show that the discontinuity set $\Omega(G_1)$ splits up into two invariant components Ω_0 and Ω_1, $\Omega_0 \cup \Omega_1 = \Omega(G_1)$. In fact, we assume the converse, and from the openness of $\Omega(G_1)$ we obtain its arcwise connectedness. Let $\gamma \subset \Omega(G_1)$ be a curve joining the points $0 \in F_0$ and $\infty \in F_1$. One can assume that it intersects the polyhedra $g(F_0)$ and $g(F_1)$, $g \in G_1$, while avoiding the edges and is such that each intersection consists of not more than one component and is also such that the number of intersected polyhedra is minimal (due to the compactness of $\gamma \subset \Omega(G_1)$, this number is always finite). However, then the intersected polyhedra contain some polyhedra $g_1(F_0)$ and $g_2(F_1)$, $g_i \in G_1$, having common faces, which is clearly impossible. As $F_0 \cup F_1 = F(G_1)$ is the fundamental set of G_1, we apply Lemma 3.7 to get that $\Omega(G_1)$ splits up into two invariant components:

$$\Omega_0 = G_1(\overline{F}_0) \subset Q, \quad \Omega_1 = G_1(\overline{F}_1). \tag{3.16}$$

The construction of G_1 shows that Ω_0 and Ω_1 are homeomorphic to a ball, i.e., that G_1 is a quasi-Fuchsian group. Also, the methods of Chapter 9 enable one to show that G_1 is obtained by a non-trivial quasi-conformal deformation from a certain cocompact Fuchsian group in $\overline{\mathbb{R}}^3$.

We now define G as the product of G_1 and the free Abelian rank three group Γ, generated by translations by the coordinate vectors e_1, e_2, and e_3. The group $G = \langle G_1, \Gamma\rangle$ is Kleinian, since there are no G-equivalent points within F_0. The stabilizer G_∞ coincides with Γ, while each element $\gamma \in \Gamma$ translates the isometric sphere of an element $g \in G \setminus \Gamma$ onto the isometric sphere of the element $\gamma g \gamma^{-1} \in G \setminus \Gamma$. Then Lemma 3.11 and the fact that the sides of Q (the fundamental polyhedron for Γ) are pairwise Γ-equivalent imply that

F_0 is a fundamental domain for G. Then the discontinuity set of the group G

$$\Omega(G) = \bigcup_{g \in G} g(\overline{F}_0) \tag{3.17}$$

splits up into a countable set of non-invariant components Ω_i, $i = 0, 1, 2, \ldots$, with the first component Ω_0 defined by (3.16), while the others are images of it in the mappings $\gamma_i \in G \setminus G_1$:

$$\Omega_i = \gamma_i(\Omega_0), \qquad i \geq 1. \tag{3.18}$$

It follows directly from (3.17) and (3.18) that the stabilizer for each component Ω_i will be the subgroup $\Gamma_i \subset G$ having the form $\Gamma_i = \gamma_i G_1 \gamma_i^{-1}$, $i = 1, 2, \ldots, \Gamma_0 = G_1$. It is clear that all these stabilizers $G_{\Omega_i} = \Gamma_i = \gamma_i G_1 \gamma_i^{-1}$ are quasi-Fuchsian groups. Then $G = \langle G_1, \Gamma \rangle$ is a web group.

5. *Degenerate groups.* This name is given to non-elementary finitely generated Kleinian groups in $\overline{\mathbb{R}}^n$, $n \geq 2$, whose discontinuity sets are contractible domains, i.e., have trivial homotopic groups. So far, there have been no constructive examples given of such groups, and this remains one of the unsolved problems in Kleinian group theory. In the planar case $(n = 2)$, Bers [4] has given a proof for the existence of such groups, which is based on the properties of Teichmüller spaces and dimension arguments.

Extending the action of a degenerate group $G \subset \text{Möb}(2)$ from the plane into the space $\overline{\mathbb{R}}^n$, $n \geq 3$, we, of course, obtain a degenerate group there. However, it is not clear whether there are specifically spatial degenerate groups, such as, for example, ones for which the dimension of the limit set is $n - 1$.

Groups with analogous properties, but infinitely generated, are readily constructed in any dimension. For this purpose, it is sufficient to take a smooth $(n-1)$-disc $D \subset \mathbb{R}^n$ and consider a countable set Σ of non-intersecting spheres σ_i touching the disc D from one side in such a way that the points of contact are dense in $\overline{D}$ and form a family Σ', reflecting Σ with respect to D. Then the group $G \subset \mathcal{M}_n$ with the contractible discontinuity set $\Omega(G)$ and the limit set $\Lambda(G)$ containing the closed disc $\overline{D}$ will be generated by parabolic transformations g_i with the isometric spheres $\sigma_i \in \Sigma$ (the spheres $\sigma_i' \in \Sigma'$ are isometric for g_i^{-1}). These properties are readily proved by means of Lemma 3.7, by analogy with Accola's example [1].

6. *Function groups.* These are Kleinian groups in $\overline{\mathbb{R}}^n$, each of which has an invariant component in its discontinuity set. Such groups, in particular, are Fuchsian and quasi-Fuchsian groups, as well as Schottky groups and degenerate groups; see also Examples 6.16 and 6.18 and Theorem 7.61 in Chapter 7 §5. A particular place is taken in

the class of function groups by finitely generated ones, each having an invariant contractible component. We call these b-groups (we retain the name for such groups in the plane, which arose because the first such groups (distinct from quasi-Fuchsian ones) were obtained as boundary groups in Teichmüller spaces of Fuchsian groups, see Bers [4] and Maskit [2]). Spatial b-groups will be considered in Chapter 8 §5.

§4. **Fundamental domains**

A fundamental domain for a discontinuous (Kleinian) group $G \subset \text{Möb}(n)$ is an open subset $F \subset \Omega(G)$ of the discontinuity set not containing G-equivalent points and such that $\Omega(G) \subset G(\overline{F})$. One adds some of its boundary points to F to get the fundamental set of the group G, i.e., a set containing one point from each orbit $G(x)$, $x \in \Omega(G)$. The orbifold $\Omega(G)/G$ is obtained from $\overline{F}$ by identifying the G-equivalent points of the boundary ∂F.

The concept of a fundamental domain plays a special part in the theory of discontinuous transformation groups. It is employed as a combinatorial-geometrical method of examining groups going back to studies by Poincaré [1], [2] on Fuchsian and Kleinian groups. This method on the one hand enables one to study the structure of a Kleinian group as an abstract group (i.e., to derive the generators and the defining relations) and thus to establish a relation between the theory of Kleinian groups and combinatorial group theory (see Lyndon and Schupp [1], Magnus, Karrass and Solitar [1], and Cannon [1]), while on the other hand it enables one to prove discontinuity and to find the fundamental domain for a transformation group having given generators. An important part is played here by fundamental domains having a special form, namely fundamental polyhedra.

THEOREM 3.10. *Let* $G \subset \text{Möb}(n)$ *be a discontinuous group such that* $\infty \in \Omega(G)$ *and* $G_\infty = \{\text{id}\}$. *Then the set*

$$P(G) = \{x \in \overline{\mathbb{R}}^n : \sup_{g \in G\setminus\{\text{id}\}} |g'(x)| < 1\} = \overline{\mathbb{R}}^n \setminus \overline{\bigcup_{g \in G\setminus\{\text{id}\}} (\text{int}\, I(g))} \tag{3.19}$$

is a fundamental domain of the group G.

PROOF. As any element $g \in G\setminus\{\text{id}\}$ maps $\text{ext}\, I(g)$ into $\text{int}\, I(g^{-1})$, it is clear that $P(G)$ does not contain G-equivalent points. It remains to show that $\Omega(G) \subset G(\overline{P(G)})$. For this, we first show that for any $k > 0$, the radii r_g of the spheres $I(g)$, $g \in G$, have the property:

$$\text{card}\{g \in G : r_g \geq k\} < \infty. \tag{3.20}$$

It follows from the conditions in the theorem that there exists a sufficiently large ball B containing the orbit $G(\infty)\setminus\{\infty\}$, i.e., the centres of all isometric spheres $I(g)$. We extend the group G to the group $\tilde{G} \subset \text{Möb}(n+1)$ and assume that $\tilde{G}$ (or, more precisely, $\gamma\tilde{G}\gamma^{-1}$) acts in a certain ball, from which we see that the isometric spheres $I(g)$ are uniquely defined by their centres. Then if there existed elements g_1 and $g_2 \in G$ having arbitrarily close centres for the isometric spheres and radii greater than k, the image $g_1 g_2^{-1}(\infty)$ would not lie in the ball B. This proves (3.20). From this in turn it follows that there exists a sufficiently large number R such that all isometric spheres $I(g)$, $g \in G\setminus\{\text{id}\}$ lie in a ball having radius R.

We now assume that there exists a point x_0 and its neighbourhood $B_0 = B^n(x_0, \rho)$ such that $B_0 \cap G(\overline{P(G)}) = \emptyset$. It is clear that then there exists $g_0 \in G$ such that $x_0 \in \overline{\text{int}\, I(g_0)}$. It follows from our assumption that B_0 does not contain the centre q_0 of the isometric sphere $I(g_0)$ having radius r_0. This implies that $\rho < r_0 < R$. The image $B_1 = g_0(B_0)$ is a ball having radius ρ_1. We use (1.26) to get that

$$\rho_1 \ge \frac{r_0^2 \cdot \rho}{r_0^2 - \rho^2} > k \cdot \rho, \qquad (3\ 21)$$

with equality attained if $x_0 \in I(g_0)$, while the constant $k = (1 - \rho^2/R^2)^{-1} > 1$. We replace B_0 by the ball B_1, whose centre $x_1 \in \overline{\text{int}\, I(g_1)}$ for a certain $g_1 \in G$, with radius r_1 of $I(g_1)$, and consider its image $B_2 = g_1(B_1)$, which is a ball of radius $\rho_2 > 0$. Due to $\rho_1 > \rho$, we have by analogy with (3.21) that

$$\rho_2 \ge \frac{r_1^2 \cdot \rho_1}{r_1^2 - \rho_1^2} > \frac{\rho_1}{1 - \rho_1^2/R^2} > k \cdot \rho_1, \qquad (3.22)$$

We continue this process and obtain a ball, namely the image $g(B_0)$ of the ball $B_0 = B(x_0, \rho)$, which has the arbitrarily large radius of (3.22) and by assumption does not intersect with $G(\overline{P(G)})$. This contradiction proves the theorem.

A fundamental domain $P(G)$ having the form of (3.19) is called the *isometric fundamental polyhedron*. The meaning of this name is evident from the following. Let the group G be a Fuchsian one, i.e., one that leaves invariant a ball (half-space) B^n, which by the introduction of the Poincaré metric analogous to (2.18) or (2.20) is converted to the hyperbolic space. Then the isometric spheres $I(g)$, $g \in G$, will be planes in that space, so $P(G) \cap B^n$ is a hyperbolically convex polyhedron.

The conditions $G_\infty = \{\text{id}\}$ and $\infty \in \Omega(G)$ do not restrict the generality of Theorem 3.10, since the transition to the conjugate group hGh^{-1}, $h \in \mathcal{M}_n$, can always ensure that they are obeyed. However, this sometimes leads to a complication. Following Ford [1] we give below a direct construction for a fundamental domain for a Kleinian group $G \subset \text{Möb}(n)$ having a non-trivial stabilizer G_∞ for the points at infinity.

LEMMA 3.11. *Let P_∞ be a certain fundamental polyhedron (with pairwise equivalent sides) for the subgroup $G_\infty = \{g \in G : g(\infty) = \infty\}$ of* *a Kleinian group* $G \subset \text{Möb}(n)$. *Then*

$$P_\infty \cap \{x \in \overline{\mathbb{R}}^n : \sup_{g \in G \setminus G_\infty} |g'(x)| < 1\}$$

is a fundamental domain for group G.

The proof is elementary (see Ford [1], §35) and follows from the fact that the elements $h \in G_\infty$ translate the isometric spheres $I(g) = \{x \in \mathbb{R}^n : |g'(x)| = 1\}$ of the elements $g \in G \setminus G_\infty$ to isometric spheres $I(hgh^{-1})$.

A disadvantage of an isometric fundamental polyhedron is the special role that is played by the point at infinity: it is a form of centre for $P(G)$. This deficiency can be eliminated. We now give a construction of the fundamental polyhedron for a discontinuous group $G \subset \text{Möb}(n)$ having its centre at any point $x \in \Omega(G)$, $G_x = \{\text{id}\}$.

We introduce a function analogous to the Jacobian $\det Dg(x)$ of a Möbius transformation g. Given $y \in \overline{\mathbb{R}}^n$ we consider a transformation $\gamma \in \text{Möb}(n)$ that transfers it to infinity and define for a transformation $g \in \text{Möb}(n)$ that

$$\alpha_y(g, x) = \det[D\gamma g\gamma^{-1}(\gamma(x))]. \tag{3.23}$$

We show that this definition is independent of the choice of the transformation $\gamma \subset \text{Möb}(n)$, $\gamma(y) = \infty$. Let $\gamma_1 \in \text{Möb}(n)$, $\gamma_1(y) = \infty$ and $\gamma_1 \neq \gamma$. Then $\gamma\gamma_1^{-1}$ is a similarity transformation: $r = \text{const} = \det D\gamma\gamma_1^{-1}(x)$, $\det D\gamma_1\gamma^{-1}(x) = 1/r$, and since

$$\det[D\gamma g\gamma^{-1}(\gamma(x))] = \det[D\gamma\gamma_1^{-1}\gamma_1 g\gamma_1^{-1}\gamma_1\gamma^{-1}(\gamma\gamma_1^{-1}\gamma_1(x))]$$

$$= \det[D\gamma_1 g\gamma_1^{-1}(\gamma_1(x))]. \tag{3.24}$$

The definition of $\alpha_y(g, x)$ implies directly that:

A1. $\alpha_\infty(g, x) = \det Dg(x)$;
A2. $\alpha_y(g, x) = \alpha_{\gamma(y)}(\gamma g\gamma^{-1}, \gamma(x))$ for all $\gamma \in \text{Möb}(n)$;
A3. $\alpha_y(\gamma g, x) = \alpha_y(\gamma, g(x)) \cdot \alpha_y(g, x)$ for all $g, \gamma \in \text{Möb}(n)$.

This function also has two important properties:

A.4. $\alpha_y(g, x) = (|x - y|/|g(x) - y\ )^{2n} \cdot \det Dg(x)$;
A.5. $\alpha_y(g, x) = \alpha_x(g^{-1}, y)$, except the case $x = y = g(y)$.

To prove property A4, we consider the inversion with respect to the sphere $S^{n-1}(y, 1)$ as the mapping γ in property A2. Then $\gamma^{-1} = \gamma$, and the Jacobian matrix takes the form

$$D\gamma(x) = (\mathbf{1} - 2Q_{x-y})/|x - y|^2, \tag{3.25}$$

where the determinant of the matrix $(\mathbf{1} - 2Q_{x-y})$ is -1. One can therefore write the following chain of equations, which demonstrates A4:

$$\alpha_y(g, x) = \alpha_\infty(\gamma g\gamma^{-1}, \gamma(x) = \det D\gamma g\gamma^{-1}(\gamma(x)) =$$

$$= (|x-y|/|g(x)-y|)^{2n}\det[(1-2Q_{g(x)-y})\cdot Dg(x)\cdot(1-2Q_{x-y})] =$$

$$= (|x-y|/|g(x)-y|)^{2n}\det Dg(x).$$

We now prove property A5. Let $x = y$. Then for any transformation $\gamma \in \text{Möb}(n)$, $\gamma(y) = \infty$, we have

$$\alpha_y(g, y) = \det D\gamma g\gamma^{-1}(\infty)$$

and (3.26)

$$\alpha_y(g^{-1}, y) = \det D\gamma g^{-1}\gamma^{-1}(\infty)$$

These two Jacobians (3.26) become zero (which proves A5) if the mapping $\gamma g\gamma^{-1}$ is not a similarity, i.e., if $g(y) \neq g$.

Now let $x \neq y$. From A2 we see that it is sufficient to prove A5 for $x = 0$ and $y = \infty$. Then the left-hand side of A5 is $\det Dg(0)$, while the right-hand side takes the form $\det[D\mathcal{J}g^{-1}\mathcal{J}(0)]$, where $\mathcal{J}(x) = x/|x|^2$. We use the matrix representation derived in Chapter 2 §4 with (3.25), where a direct calculation shows that this Jacobian is precisely $\det Dg(0)$, which completes the proof of A5.

We now fix a point $y \in \overline{\mathbb{R}}^n$ and define an $(n-1)$-dimensional sphere $I_y(g)$ for a transformation $g \in \text{Möb}(n)$, $g(y) \neq y$:

$$I_y(g) = \{x \in \overline{\mathbb{R}}^n : \alpha_y(g, x) = 1\}, \tag{3.27}$$

which is analogous to the isometric sphere $I(g)$ (compare (1.21)) of the transformation g:

$$I_\infty(g) = I(g). \tag{3.28}$$

It can be shown that the transformation g takes $I_y(g)$ to $I_y(g^{-1})$, but then their radii are mutually inverse. By analogy with Theorem 3.10, one can prove:

THEOREM 3.12. *Let $G \subset \text{Möb}(n)$ be a discontinuous group such that $y \in \Omega(G)$ and $G_y = \{\text{id}\}$. The set*

$$P_y(G) = \{x \in \overline{\mathbb{R}}^n : \sup_{g\in G\setminus\{\text{id}\}} \alpha_y(g, x) < 1\} = \overline{\mathbb{R}}^n \setminus \overline{\bigcup_{g\in G\setminus\{\text{id}\}} \text{int } I_y(g)}, \tag{3.29}$$

is a fundamental polyhedron for the group G.

The polyhedron $P_y(G)$ has the point y as its "centre" and for $y = \infty$ coincides, due to (3.28), with the isometric fundamental polyhedron of (3.19). If we assume, as above, that the group G is a hyperbolic isometry group in the ball $B^n = B^n(0, 1)$ (or half-space R^n_+), we can characterize the polyhedra (3.29) in terms of the metric $d(x, y)$ in the hyperbolic space.

THEOREM 3.13. *Let a discrete isometry group $G \subset \text{Möb}(B^n)$ for the hyperbolic space B^n and a point $y \in B^n$ be such that $G_y = \{\text{id}\}$.*

Then the fundamental polyhedron $P_y(G)$ *coincides with the normal fundamental Dirichlet polyhedron having its centre at the point* y:

$$B^n \cap P_y(G) = D_y(G) = \bigcap_{g \in G \setminus \{id\}} \{x \in B^n : d(x, y) < d(g(x), y)\}. \tag{3.30}$$

To prove the theorem, we note that $I_y(g)$ is the sphere orthogonal to the boundary sphere $S^{n-1}(0, 1)$, with respect to which the points y and $g^{-1}(y)$ are symmetrical, so it constitutes a set of points equidistant from y and $g^{-1}(y)$. Then

$$I_y(g) \cap B^n = \{x \in B^n : d(x, y) = d(x, g^{-1}(y))\}. \tag{3.31}$$

As the sides of the polyhedron $P_y(G)$ lie on the spheres of (3.27), this proves (3.30).

We note that here $P_y(G)$ is unaltered if the point y is replaced by the point $y^* = y/|y|^2$. From this it follows that $B^n \cap P(G) = D_0(G)$. In the formulation of Theorem 3.13, we replace the ball model $B^n(0, 1)$ for the hyperbolic space by the half-space model $H^n \equiv R^n_+$, which gives us an assertion analogous to (3.31):

$$P_y(G) \cap H^n = D_y(G) = \bigcap_{g \in G \setminus \{id\}} \{x \in H^n : d(x, y) < d(g(x), y)\}, \tag{3.32}$$

where $y \in H^n$, $G_y = \{id\}$.

We can also introduce a further type of fundamental polyhedron, which is new and which has not previously been considered; we call this a pseudo-Dirichlet polyhedron for a discrete isometry group G in H^n, and consequently for its action on the sphere $\partial H^n = \overline{\mathbb{R}}^{n-1}$, where G is Kleinian if $\Omega(G) \neq \emptyset$.

The pseudo-Dirichlet polyhedron $D^*_y(G)$ is different from the Dirichlet polyhedron $D_y(G)$ only for the dimensions $n \geq 4$ of the space H^n and, besides, as it will be shown in Chapter 5 §3, it possesses some kind of stability of the number of sides with respect to the choice of its centre $y \in H^n$ (which is analogous to the stability of the Dirichlet polyhedron $D_y(G)$ in dimension $n \leq 3$ - cf. Macbeath [3], Jorgensen-Marden [2]).

That is, for a discrete group $G \subset \mathrm{Isom}\, H^n$, we fix a point $y \in H^n$, having the trivial stabilizer $G_y \subset G$, and for each element $g \in G$, we consider some "pseudo-Dirichlet" half-space $H^+_y(g)$ in the hyperbolic space H^n, $y \in H^+_y(g)$. Below we define the pseudo-Dirichlet hyperplanes $H_y(g)$ bounding the half-spaces $H^+_y(g)$ according to types of elements $g \in G$ instead of the Dirichlet ones $S_y(g)$,

$$S_y(g) = \{x \in H^n : d(x, y) = d(x, g(y))\}.$$

Let $p \in \partial H^n$ be a parabolic fixed point of the group G and let L_p be the k-dimensional G_p-invariant horocycle on the horosphere $S_p \subset H^n$ with centre at p, $y \in S_p$, corresponding to the Bieberbach decomposition (Theorem 3.4) of the Euclidean space

$\mathbb{R}^{n-1} = \mathbb{R}^k \times \mathbb{R}^{n-k-1} \cong S_p$, in accordance with the action of the stabilizer G_p on S_p as a group Γ of Euclidean isometries. Namely, the action of any element $T \in \Gamma$ on $\mathbb{R}^{n-1} \cong S_p$ has the form $T(x) = U(x) + a$, $a \in \mathbb{R}^k \times \{0\} \subset \mathbb{R}^{n-1}$, where the orthogonal transformation $U \in O(n-1)$ has the following form:

$$U = \begin{bmatrix} U_1 & 0 \\ 0 & U_2 \end{bmatrix}, \quad U_2 \in O(n-k-1), \quad U_1 \in O(k).$$

The orthogonal transformation U_1 here, since the lattice $G_p(y^*) \subset L_p \cong \mathbb{R}^k$ is G_p-invariant, must be of finite order, while the integer $k = \text{rank } G_p$ is such that $1 \leq k \leq n-1$. Besides in the k-dimensional horocycle L_p, the group G_p acts as a Euclidean crystalographic group with the compact quotient space L_p/G_p. Here we denote by y^* a point in L_p which is the orthogonal projection of the centre y. It is the centre of the k-dimensional (compact) Dirichlet polyhedron for G_p on the horocycle L_p:

$$L_p \cap D_{y^*}(G_p) = L_p \cap D_y(G_p).$$

DEFINITION 3.14. A parabolic element $T \in \Gamma$, $T(x) = U(x) + a$, is called an *irrational screw translation* if its orthogonal part U has infinite order. If $G_p \cong \Gamma$ has such elements, then we say that the parabolic point p and its stabilizer $G_p \subset G$ have irrational type (in the oppposite case - rational type).

At this moment we are ready to define our pseudo-Dirichlet planes $H_y(g)$ as follows.

DEFINITION 3.15. Let p be an irrational parabolic fixed point of the group G. We define now the pseudo-Dirichlet hyperplane $H_y(g) \subset H^n$ for any element g from the stabilizer-subgroup G_p of the point p as a hyperplane orthogonal to G_p-invariant horocycle L_p, which intersects L_p in the horocycle

$$S_y(g) \cap L_p = S_{y^*}(g) \cap L_p,$$

i.e., $H_y(g) = S_{y^*}(g)$.

If an element $g \in G$ does not belong to any parabolic subgroup of irrational type, then we define its pseudo-Dirichlet plane $H_y(g)$ as the Dirichlet one $S_y(g)$.

Now we shall define the pseudo-Dirichlet polyhedron $D^*_y(G)$ by means of some reconstruction of the Dirichlet polyhedron $D_y(G)$.

For it we take a subset $Y_D \subset H^n$ such that $H^n \setminus Y_D$ is of measure zero (it is the union of a countable set of $(n-1)$-planes) and such that any Dirichlet polyhedron $D_y(G)$, $y \in Y_D$, is a simple polyhedron: no two of its boundary parabolic vertices are G-equivalent (see Lemma 3.17). This is equivalent to the fact that for any $g \in G \setminus \{1\}$ the Dirichlet planes $S_y(g)$ and $S_y(g^{-1})$ do not contain distinct parabolic points p and $g(p) \neq p$. The pseudo-Dirichlet planes $H_y(g) = S_{y^*}(g)$ for the elements g from the irrational parabolic subgroups G_p

contain parabolic points which are distinct from p only in the case of the centres $y \in Y_D$ lying in a countable set of planes which are orthogonal to the G_p-invariant horocycles L_p. Therefore, after removing from the set Y_D this countable set of planes (the set of parabolic points is countable), we have a set $Y(G) \subset H^n$ such that:

1. its complement in H^n (of measure zero) is a countable set of planes;
2. for any $y \in Y(G)$ neither of the Dirichlet planes $S_y(g)$ and $S_y(G^{-1})$ nor the pseudo-Dirichlet planes $H_y(g)$ and $H_y(g^{-1})$ contain distinct parabolic points p and $g(p) \neq p$.

Now we enumerate the conjugacy classes of the irrational parabolic subgroups in the group G by their fixed points $p_1 = \infty$, $p_2, \ldots$ on the boundary of a simple Dirichlet polyhedron $D_y(G)$ with a centre $y \in Y(G)$, and denote $D_y^0(G) = D_y(G)$. Here we define the algorithmical process of our reconstruction of polyhedra by the following rule:

$$D_y^0(G) \longrightarrow D_y^1(G) \longrightarrow, \ldots, \longrightarrow D_y^i(G) \longrightarrow, \ldots, \longrightarrow D_y^*(G)$$

$$D_y^i(G) = R_i \cap \{H_y^+(g) : g \in G_{p_i} \setminus \{1\}\}, \tag{3.33}$$

$$R_i = \operatorname{int}(\bigcup\{g(\overline{D_y^{i-1}(G)}) : g \in G_{p_i}\}), \qquad i \geq 1.$$

Here the polyhedra $D_y^i(G)$ are fundamental polyhedra for the group G with pairwise G-equivalent sides due to the following two facts.

Firstly, if it is so for $(i-1)$, then $\bigcup G \setminus G_{p_i}(R_i) \cup R_i$ is a tessellation of the hyperbolic space H^n with $G \setminus G_{p_i}$-equivalent sides.

Secondly, $D_y^*(G_{p_i}) = \bigcap\{H_y^+(g) : g \in G_{p_i} \setminus \{1\}\}$ is such a fundamental polyhedron for the subgroup G_{p_i} (see Lemma 3.11).

Now we note an important fact: all new polyhedra $D_y^i(G)$ are simple, i.e., no two boundary parabolic vertices of every such polyhedron (and of the polyhedron's G-images) are G-equivalent due to the definition of the set $Y(G)$ of centres.

As a result of this process we obtain finally (in some non-degenerate cases, particularly for a finite set of irrational parabolic points p_i see Chapter 5 §3) for any centre $y \in Y(G)$ a simple fundamental polyhedron which is called the *pseudo-Dirichlet polyhedron* $D_y^*(G)$.

In this reconstruction process we replace all Dirichlet planes the closures of which contain irrational parabolic points by pseudo-Dirichlet ones. This fact allows us to formulate the following conjecture.

CONJECTURE 3.16. The following polyhedron

$$\bigcap\{H_y^+(g) : g \in G \setminus \{1\}\}, \quad y \in Y(G) \subset H^n,$$

gives a likely equivalent definition of the pseudo-Dirichlet polyhedron $D_y^*(G)$ for a geometrical finite group $G \subset \operatorname{Isom} H^n$. Therefore the pseudo-Dirichlet polyhedron $D^*_y(G)$ is a convex one.

We now examine the behaviour of points lying on the boundary ∂P of some fundamental polyhedron P of a discrete group $G \subset \text{Möb}(n)$, namely the polyhedra $P_y(G)$, $D_y(G)$ and $D^*_y(G)$. Such points x belong to one of the following categories:

α) $x \in \Lambda(G)$;
β) $x \in \Omega(G)$, x lies on one face of polyhedron P; and
γ) $x \in \Omega(G)$, x lies on two or more faces of P.

It is readily shown that for a point $x \in I_y(g) \cap \overline{P_y(G)}$ (correspondingly, $x \in S_y(g) \cap \overline{D^*_y(G)}$) of the category β), the image $g(x)$ is again a boundary point of $P_y(G)$ (correspondingly, $D^*_y(G)$) and $g(x) \in I_y(g^{-1})$ ($g(x) \in S_y(g^{-1})$), while the set of such points forms a set of pairwise G-equivalent 1-codimensional sides of $P_y(G)$ or $D^*_y(G)$. The points $x \in \partial P_y(G)$ (or $\partial D^*_y(G)$) of the category γ) form the k-dimensional sides of these polyhedra, $0 \le k \le n-1$.

Amongst the boundary points x of the fundamental polyhedra $P_y(G)$, $D_y(G)$ and $D^*_y(G)$), we distinguish those fixed for the parabolic or elliptic elements $g \in G$. These may form cycles, each consisting of a set of G-equivalent k-dimensional sides, $0 \le k \le n-2$. However, the following assertion can be proved (compare Apanasov [8]).

LEMMA 3.17. *The set of points* $y \in \Omega(G)$, *for which the polyhedra* $P_y(G)$ *of a Kleinian group* $G \subset \text{Möb}(n)$ *have equivalent parabolic or elliptic points at the boundaries, lies in the union of not more than a countable set of* $(n-1)$*-dimensional spheres.*

PROOF. Let a point $y \in \Omega(G)$ be such that in $\partial P_y(G)$ we have equivalent parabolic or elliptic points p and $g(p)$, where $g \in G \setminus G_p$. By S_p we denote the maximal pointwise fixed side (of dimension $\le n-1$) of $P_y(G)$ in the elliptic case such that $p \in S_p$ and $g(S_p) \subset \partial P_y(G)$; in the parabolic case, we put $S_p = \{p\}$. It is evident that for any point $q \in S_p$ we have $\alpha_y(g, q) = 1$, and then, due to the property A5, we get for all $q \in S_p$ that

$$\alpha_q(g^{-1}, y) = 1.$$

However, the set of points $y \in \overline{\mathbb{R}}^n$ that satisfy this for $q \in S_p$ is a sphere of dimension m, $m \le n-1$:

$$S^m(p) = \bigcap_{q \in S_p} I_q(g^{-1}).$$

Thus the set of points $y \in \Omega(G)$ for which $P_y(G)$ has G-equivalent elliptic or parabolic points at the boundary is contained in the union of spheres S^m, $m \le n-1$, which has the same cardinality as a set whose elements are either parabolic fixed points or

k-dimensional spheres, pointwise fixed by some elements $g \in G$. This completes the proof.

REMARK. The analogue of Lemma 3.17 applies also for pseudo-Dirichlet polyhedra and for Dirichlet polyhedra, which follows from Theorem 3.13 (cf. our arguments about the properties of the set $Y(G) \subset H^n$ in the definition of $D^*_y(G)$.

Let us now have a certain convex fundamental polyhedron P for a discrete isometry group G in a space X of constant section curvature. By $O_G = \{g \in G : \dim(\overline{P} \cap g(\overline{P})) = n-1\}$ we denote the set of elements of the group G that identify the $(n-1)$-dimensional sides of the polyhedron P.

THEOREM 3.18. *The set O_G generates the group G. The defining relations for G can be taken as all edge relations of the following two forms*:

$$g_1 g_2 = \mathbf{1}, \quad \textit{where} \quad g_1, g_2 \in O_G; \quad g_1 g_2 \cdot \ldots \cdot g_k = \mathbf{1}, \tag{3.34}$$

$$\textit{where} \quad g_i \in O_G, \quad g_i g_{i+1} \neq \mathbf{1}, \quad g_1 g_2 \cdot \ldots \cdot g_m \neq \mathbf{1},$$

for $m < k$, *and* $\dim\{\overline{P} \cap g_1(\overline{P})) \cap g_1 g_2 \overline{P} \cap \ldots \cap g_1 \cdot \ldots \cdot g_{k-1}(\overline{P})\} = n-2$.

To prove the first assertion, for a fixed point $x_0 \in P$, we consider any element $g \in G \setminus \{\mathrm{id}\}$, and let $x = g(x_0)$ be joined to x_0 by a path $\alpha \subset X$. Then there exists a closed domain $D \subset X$ such that $\alpha \subset \operatorname{int} D$. Due to the discontinuity of the group $G_0 \subset G$ generated by the set O_G, the compact D is covered by a finite number of images $\gamma(\overline{P})$, $\gamma \in G_0$. In particular, one can find $\gamma_0 \in G_0$ such that $\gamma_0(P)$ covers a neighbourhood of the point $x = g(x_0)$. It then follows that $g = \gamma_0 \in G_0$, i.e., $G_0 = G$. The second assertion is obtained by considering all the cycles of the boundary edges of P, i.e., the sets of G-equivalent edges of P and from the fact that a discontinuous group cannot contain infinite-order elliptic elements. For the details see Gernstenhaber [1] and Macbeath [1].

A theorem that establishes the reverse relationship also applies.

THEOREM 3.19. *Let P be a convex polyhedron in n-dimensional Euclidean, elliptic, or hyperbolic space X, and let each $(n-1)$-dimensional side A_i of P be given an isometry g_i of X such that*:

1) *for each A_i there exists an $(n-1)$-side A_j such that* $g_i g_j = \mathrm{id}$;

2) *for each $(n-2)$-dimensional side E of P there exists a sequence $A_1, A_1, \ldots, A_k$ of its $(n-1)$-dimensional sides such that* $g_1 g_2 \cdot \ldots \cdot g_k = \mathrm{id}$, *and*

$$\overline{P} \cap g_1(\overline{P}) \cap g_1 g_2(\overline{P}) \cap \ldots \cap g_1 g_2 \cdot \ldots \cdot g_{k-1}(\overline{P})) = E \tag{3.35}$$

and the polyhedra (3.35) do not have pairwise common interior points. Then the group $G \subset \operatorname{Isom} X$, generated by the transformations $\{g_i\}$, is *discrete. If also* P *is bounded in* X, *it is a fundamental polyhedron of* G.

This theorem follows from Aleksandrov's [1] more general result on space filling (see also Abels [1]). In the case of an unbounded polyhedron $P \subset X = H^n$, these conditions are insufficient to fill H^n with images $g(\overline{P})$, $g \in G$, i.e., P may also be not a fundamental polyhedron for G. Chapter 6 §1 deals with this topic at the same time as giving a proof of a generalization of Aleksandrov's theorem to the non-compact (unbounded) case.

The following facts immediately follow from Theorem 3.19:

COROLLARY 3.20. *An isometry group* G *in the constant curvature space* X *generated by reflections in the* $(n-1)$*-sides of a convex polyhedron* $P \subset X$ *is discrete if and only if*:

1) *all dihedral angles of* P *are integer parts of* 2π, *and*

2) *if a dihedral angle at an* $(n-2)$*-dimensional side* E *of* P *is an odd part of* 2π, *then* P *has a symmetry plane passing through* E.

COROLLARY 3.21. *Let a group* $G \subset \text{Möb}(n)$ *be generated by a finite number of involutions* $g_i = O_i\mathcal{J}_i$, *where* $\mathcal{J}_i$ *is the inversion with respect to a sphere* $S^{n-1}(x^i, r_i)$, $x^i_n = 0$, *and* O_i *is the reflection in the hyperplane* $L_i \subset \mathbb{R}^n$ *passing through the point* x^i *and orthogonal to the subspace* $\mathbb{R}^{n-1} \times \{0\} \subset \mathbb{R}^n$ *and which is also a symmetry plane for the polyhedron* $P \subset \mathbb{R}^n_+$ *bounded by the spheres* $S^{n-1}(x^i, r_i)$ *and possibly by the plane* $\{x \in \overline{\mathbb{R}}^n : x_n = 0\} = \overline{\mathbb{R}}^{n-1}$.

Then G *is discrete if and only if all the dihedral angles of* P *are integer parts of* π.

The application of these results is not restricted only to a Kleinian group $G \subset \text{Möb}(n)$ as a group of spherical, Euclidean, or hyperbolic isometries, i.e., the elementary or Fuchsian groups G. This is because the extension of any discontinuous group $G \subset \text{Möb}(n)$ into the half-space $\mathbb{R}^{n+1}_+$ is a discrete isometry group in the hyperbolic space H^{n+1}.

§5. The limit set: the non-measurability effect and density property

If $G \subset \text{Möb}(n)$ is a non-elementary group, its limit set is obtained by closing the orbit $G(x)$ of any point $x \in \overline{\mathbb{R}}^n$. As the transformations $\gamma g \gamma^{-1}$ are of the same type for all $\gamma \in \text{Möb}(n)$, it follows from this that the limit set for a non-elementary Kleinian group $G \subset \text{Möb}(n)$ is the closure of any of the following sets:

1) the set of loxodromic fixed points;

2) the set of parabolic fixed points (if it is not empty);

3) the set of irrational parabolic fixed points (if it is not empty);

4) the set of hyperbolic fixed points (if it is not empty).

The first assertion can be strengthened (Greenberg [1]).

LEMMA 3.22. *If* g *is a loxodromic element of a non-elementary Kleinian group* $G \subset \text{Möb}(n)$, *then for each open neighbourhood* U *of any limit point* $x \in \Lambda(G)$ *there exists a* $g_1 \in G$ *such that the fixed points of the loxodromic element* $g_1 g g_1^{-1}$ *lie in* U.

We say that the fixed points of loxodromic elements of a discrete group $G \subset \text{Möb}(n)$ are dense in $\Lambda(G) \times \Lambda(G)$ if for any $z_1, z_2 \in \Lambda(G)$ and any of their neighbourhoods $U_1, U_2 \subset \overline{\mathbb{R}}^n$ there exists a loxodromic element in G having one fixed point in U_1 and another in U_2 (see Gottschalk and Hedlund [1], pp. 121-123, and Greenberg [1]).

THEOREM 3.23. *The fixed points of the loxodromic elements of a non-elementary Kleinian group* $G \subset \text{Möb}(n)$ *are dense in* $\Lambda(G) \times \Lambda(G)$.

PROOF. We first show that there are loxodromic elements in G. As G is non-elementary, it must contain a non-elliptic element γ. If γ is parabolic (i.e., is not a loxodromic element), then we can assume that $\gamma(\infty) = \infty$, i.e., γ is a Euclidean isometry. Theorem 3.3 demonstrates the existence of $g \in G$ such that $g(\infty) \neq \infty$. Direct calculation shows that

$$I(\gamma^m g) = I(g), \quad I(g^{-1}\gamma^{-m}) = \gamma^m(I(g^{-1})).$$

It is clear that these isometric spheres do not intersect for sufficiently large m, so the element $\gamma^m g$ is loxodromic.

We consider any distinct points $z_1, z_2 \in \Lambda(G)$ and let $U_1, U_2 \subset \overline{\mathbb{R}}^n$ be their disjoint neighbourhoods. Then following Greenberg's proof [1], we use Lemma 3.22: there exists a loxodromic element $g_1 \in G$ having fixed points in U_1 and a certain element $\gamma \in G$ such that $g_2 = \gamma g_1 \gamma^{-1}$ has fixed points in U_2. Considering G as a subgroup in $\text{Möb}(n+1)$, we can assume that G acts in the ball B^{n+1} (with the hyperbolic metric introduced there) as an isometry group (see Chapter 1 §4).

Let l_1 and l_2 be the axes of the hyperbolic translations g_1 and g_2, and let L be a hyperbolic $(n-1)$-dimensional plane which is orthogonal to l_1 and which intersects $\partial B^{n+1} \cong \overline{\mathbb{R}}^n$ inside the neighbourhood U_1. L separates B^{n+1} into two half-spaces. Let H_1 be the one of these whose closure contains the point z_2. The plane $\gamma(L)$ is orthogonal to the axis $l_2 = \gamma(l_1)$ and is translated along this axis by the translations g_2^n. If n is large enough, we can assume that $H_2 = g_2^n\gamma(H_1)$ is contained in H_1 and $V_2 = \overline{H}_2 \cap \partial B^{n+1}$ is contained in U_2. Then for $V_1 = \overline{H}_1 \cap \partial B^{n+1}$ we have that $V_2 \subset V_1$ and $g_2^n\gamma(V_1) = V_2$. This fact and Brouwer's fixed point theorem show that $g_2^n\gamma$ has a fixed point in $V_2 \subset U_2$. Similarly, since for $V_i' = \overline{\partial B^{n+1} \setminus V_i}$, $i = 1, 2$, we have $V_1' \subset V_2'$ and $(g_2^n\gamma)^{-1}(V_2') = V_1'$, we have that $V_1' \subset U_1$ also has a fixed point for $g_2^n\gamma$. Since $g_2^n\gamma$ maps H_1 onto H_2, it moves every point on the hyperbolic line between the fixed points, and therefore it cannot be elliptic. Thus $g_2^n\gamma$ is the loxodromic element in G having the necessary property and the theorem is established.

Theorem 3.23 shows already (see Schmidt [1], §4) that the group G (by some natural conditions) acts conservatively on the limit set $\Lambda(G)$ (see Chapter 6 §6), i.e., has the property that, for any set A of positive measure, $g(A)\cap A$ has positive measure for infinitely many elements $g\in G$. Another approach to this, namely the effect of non-measurability for fundamental sets in $\Lambda(G)$, has been observed by Krushkal' [2] (compare Metzger [1]). Now we describe it more precisely.

From the limit set $\Lambda(G)$ for a discrete group $G\subset \text{Möb}(n)$, we delete the fixed points of the elements of G (the dimension of this set not more than $n-1$). The remaining set $L\subset\Lambda(G)$ does not contain the points x for which $g_1(x)=g_2(x)$ for distinct elements $g_1, g_2\in G$. We can therefore form a fundamental limit set F_Λ for the group G in L (using the axiom of choice), which will contain only one point from each orbit $G(x)$, $x\in L$. We denote again by μ_n the spherical n-dimensional Lebesgue measure, while by $P(G)\subset H^{n+1}$, $\partial H^n=\overline{\mathbb{R}}^n$, we denote a fundamental Dirichlet polyhedron for G, which has any centre.

THEOREM 3.24. *Any fundamental limit set* F_Λ *for a discrete group* $G\subset\text{Möb}(n)$, $\mu_n(\Lambda(G))>0$, *satisfying the condition*

$$\mu_n(\overline{P}(G)\cap\Lambda(G)) = 0, \tag{3.36}$$

is non-measurable.

To prove the theorem, it is more convenient to assume that G acts in the sphere S^n and in the ball B^{n+1}, $\partial B^{n+1}=S^n$. We consider in S^n the spherical metric

$$ds^2 = 4|ds|^2/(1+|x|^2)^2, \tag{3.37}$$

and denote by

$$[g'(x)] = ds(g(x))/ds(x) = |g'(x)|(1+|x|^2)/(1+|g(x)|^2) \tag{3.38}$$

the linear distortion of the mappings $g\in\text{Möb}(n)$ in the spherical metric of (3.37).

We now assume that some fundamental set F_Λ is measurable. Then, due to the fact that the images $g(F_\Lambda)$ for distinct elements $g\in G$ do not intersect, we have

$$\mu_n(\Lambda(G)) = \int_{F_\Lambda}\sum_{g\in G}[g'(x)]^n d\mu_n(x), \tag{3.39}$$

where $[g'(x)]$ from (3.38). It follows from (3.39) that the following series should converge almost everywhere in L:

$$\sum_{g\in G}[g'(x)]^n. \tag{3.40}$$

Further, if necessary, we remove from L the countable set of $(n-1)$-dimensional spheres for which $[g_1'(x)] = [g_2'(x)]$ for distinct $g_1, g_2 \in G$, and $x \in S^n$, which gives us that there is a largest term $[g_0'(x)]^n$ at each of the remaining points $x_0 \in F_\Lambda$, where the series (3.40) converges, i.e., such that $[g'(x_0)] < [g_0'(x_0)]$ for all $g \neq g_0$. Then for $g \in G\setminus\{\mathrm{id}\}$ we have

$$[g'(g_0(x_0))] = [(gg_0)'(x_0)]/[g_0'(x_0)] < 1. \tag{3.41}$$

But the series (3.40) converges also at the points $g_0(x_0)$, so it follows from (3.41) that

$$\sup_{g\in G\setminus\{\mathrm{id}\}}[g'(g_0(x_0))] < 1. \tag{3.42}$$

From condition (3.36) and from the property of (3.42) it now follows directly that the point $g_0(x_0)$ should belong to the set $\overline{P}(\overline{G}) \cap \Omega(G)$. The resulting contradiction completes the proof.

COROLLARY 3.25. *Let a group $G \subset \text{Möb}(n)$ satisfy the conditions of the theorem. Then any subset $F \subset F_\Lambda$ having positive exterior measure is non-measurable.*

REMARK 3.26. The above relationship between this effect and the theory of ergodic action for a discrete group of hyperbolic isometries on the sphere at infinity, as well as with the rigidity of deformations of such discrete groups and with the theory of harmonic functions, will be considered further in Chapter 6 §6. Here we illustrate this relationship in the following simple case.

Let G be a Fuchsian torsion-free group in a disc B^2 having a compact fundamental polygon (and consequently $\Lambda(G) = \partial B^2 = S^1$). Assuming that G has a measurable fundamental limit set $F_\Lambda \subset S^1$, we can take a function $h(z) \neq \text{const}$ bounded in F_Λ and extend it compatible with G to the circle S^1 by means of $h(g(z)) = h(z)$, $g \in G$. Then this function should define a bounded harmonic function $u(z) \neq \text{const}$ in the disc B^2 that coincides in S^1 with $h(z)$ and is also automorphic for G, which is impossible.

REMARK 3.27. Condition (3.36) of Theorem 3.24 is also necessary, since for any discrete group $G \subset \text{Möb}(n)$ with $P(G) \subset H^{n+1}$, $\mu_n(\overline{P}(G) \cap \Lambda(G)) > 0$, the set $\overline{P}(G) \cap \Lambda(G)$ should lie in some fundamental limit set F of the group G. Also condition (3.36) cannot be replaced by the analogous condition on the boundary of an n-dimensional fundamental polyhedron for the group G in the n-sphere $\overline{\mathbb{R}}^n$ because the polyhedron $P(G)$ may have a set of limit vertices having positive measure not attainable from a spheric polyhedron. For examples of these groups, see the book by Krushkal', Apanasov, and Gusevskii [3] and Remark 9.20.

CHAPTER 4

DISCRETE GROUPS OF HYPERBOLIC ISOMETRIES

§1. Hyperbolicity and discreteness conditions for a Möbius group

In this section, we are concerned with what properties of a group $G \subset \text{Möb}(n)$ are sufficient for, firstly, this group to act in a certain ball $B^n \subset \overline{\mathbb{R}}^n$, or, if we consider B^n as the hyperbolic space model, for G to be an isometry group for hyperbolic space, and secondly, for G to be a discrete group. We begin with a simple property of groups $G \subset \mathcal{M}_n$ (compare van Vleck [1] and Apanasov [4]), which are not necessarily discrete.

THEOREM 4.1. *If group $G \subset \mathcal{M}_n$ $(n \geq 2)$ does not have a common fixed point and consists of hyperbolic, parabolic without torsion, and type $n-2$ elliptic elements (the dimension of the fixed-point set is equal to $n-2$), then G leaves invariant a certain circle $S^1 \in \overline{\mathbb{R}}^n$ and any spherical disc $B^2 \subset \overline{\mathbb{R}}^n$, $\partial B^2 = S^1$. In other words, there exists $h \in \mathcal{M}_n$ such that $hGh^{-1} \subset \text{Isom}\, H^2 = \text{Möb}(1)$.*

PROOF. As G lacks a common fixed point, this implies that G contains a non-elliptic element γ. If γ is not hyperbolic, one can assume that γ is a Euclidean translation and that there exists $g \in G$, $g(\infty) \neq \infty$. Direct calculation shows that the mapping $\gamma^m g \in G \backslash G_\infty$ and the inverse to it have the following isometric spheres:

$$I(\gamma^m g) = I(g), \quad I(g^{-1}\gamma^{-m}) = \gamma^m(I(g^{-1})).$$

It is clear that these spheres do not intersect for large m, so one can always assume that G contains hyperbolic elements. Let one of them be a homothety $\gamma : x \longmapsto rx$, $0 < r < 1$. The condition for the absence of a common fixed point in the group and Lemma 3.21 imply that G contains another hyperbolic element g having fixed points p, q, $0 < |p| \leq |q| < \infty$. The subgroup $\langle\gamma, g\rangle \subset G$ leaves the plane $\mathbb{R}^2$ invariant, which passes through the origin and the points p and q, since $\mathbb{R}^2$ contains the centres of the isometric spheres $I(g)$ and $I(g^{-1})$. Moreover, plane $\overline{\mathbb{R}}^2$ contains a straight line $\overline{\mathbb{R}}^1$ containing all finite fixed points for the hyperbolic elements in group $\langle\gamma, g\rangle$. Namely, let us introduce a complex structure $\mathbb{C}$ in the plane R^2 so for $z \in \mathbb{C} = R^2$ we have $\gamma(z) = rz$, $g(z) = (az+b)/(cz+d)$, where $a, d, b, c \in \mathbb{C}$, $ad - bc = 1$, $c \neq 0$ $(g(\infty) \neq \infty)$. Because γ, g and γg are not loxodromic, traces $\operatorname{tr}\gamma = \sqrt{r} + 1/\sqrt{r}$, $\operatorname{tr} g = a + d$ and $\operatorname{tr}\gamma g = \sqrt{r}a + d/\sqrt{r}$ are real, as a and d, too. Since the fixed points p and q of g have the form

$$((a-d) \pm \sqrt{(a+d)^2 - 4})/2c,$$

their ratio p/q is real. Therefore, the points p and q lie in a subspace $R^1 \subset R^2$, $0 \in R^1$. The arbitarity in choosing the hyperbolic elements γ and g from G shows that $\overline{\mathbb{R}}^1$ contains finite fixed points for all hyperbolic elements in G. Then, as conjugation in $\mathcal{M}_n$ does not change the type of transformations, we at once have $G(\overline{\mathbb{R}}^1) = \overline{\mathbb{R}}^1$. The condition that G lacks loxodromic elements gives us that $\overline{\mathbb{R}}^1$ cannot contain the fixed points for elliptic mappings. Therefore, G leaves any two-dimensional plane N containing the straight line $\overline{\mathbb{R}}^1$ invariant, and the same applies to the half-planes N_+ and N_1 bounded by this straight line. This completes the proof.

We note that the Theorem 4.1 conditions that are necessary for $n=2$ are certainly not so for $n \geq 3$. In even dimensions, $n = 2m$, one can have a criterion of a similar type. To formulate this, we introduce the following notion.

DEFINITION 4.2. Given a set $X \subset \overline{\mathbb{R}}^n$, the least dimension for the spheres (planes) in the space $\overline{\mathbb{R}}^n$ containing the set X is called the visible dimension $\dim_{vis} X$ of the set $X \subset \overline{\mathbb{R}}^n$. In other words, $\dim_{vis} X = \dim V(X)$, where $V(X) \subset \overline{\mathbb{R}}^n$ is a spherical convex hull for set X.

THEOREM 4.3 (Volynets [1]). *Let $G \subset \mathcal{M}_{2m}$, $m \geq 1$, be a Kleinian group whose limit set has visible dimension $\dim_{vis} \Lambda(G) \geq 2m-2$. Then G has an invariant ball (half-space) $B = B^{2m} \subset \mathbb{R}^{2m}$ if and only if for any element $g \in G$ there exists a ball (half-space) B_g invariant with respect to g.*

We need some preliminary assertions to prove the theorem.

Let g be a Möbius transformation in $\overline{\mathbb{R}}^n$ that perhaps is orientation reversing and is such that $g(\infty) \neq \infty$. Then $g = U \circ O \circ \mathcal{J}$ (see Theorem 1.6), where $\mathcal{J}$ is inversion with respect to the isometric sphere $I(g)$ and O is reflection in the hyperplane P_g orthogonal at the point $(q+q')/2$ to the segment $[q, q']$, where q and q' are the centres of $I(g)$ and $I(g^{-1})$, while U is rotation or rotation with reflection with respect to q'. Then with the exception of the case $q = q'$ (when g is elliptic, $g^2 = \mathrm{id}$), we can put each transformation g into one-to-one correspondence with an orthogonal transformation $U_g \in O(n)$ (which corresponds to U).

Moreover, if g is a loxodromic transformation or $n = 2m$, the necessary and sufficient condition for it to have an invariant ball is that U_g contains an eigenvalue of one. This follows from the fact that, firstly, the latter condition is equivalent to

$$P_g^* = \overline{P}_g \cap \operatorname{ext} I(g) \cap \operatorname{ext} I(g^{-1}) \cap \{x \in \overline{\mathbb{R}}^n : U(x) = x\} \neq \varnothing,$$

and, secondly, that any sphere having its centre at a point $x \in P_g^*$ and orthogonal to $I(g)$ and $I(g^{-1})$ clearly bounds a g-invariant ball.

Direct calculation shows that the Jacobi matrix $Dg(x)$ takes the form

$$Dg(x) = \lambda(x) \cdot U_g \circ W_x \equiv \lambda(x) \cdot V_g(x),$$

where $\lambda(x) > 0$ and the orthogonal transformation $W_x \in O(n)$ corresponds to rotation through an angle 2α in the direction from the vector $(q' - q)$ to the vector $(x - q)$ in the two-dimensional plane passing through the points q, q', and x. Here α denotes the angle between those vectors. In particular, if α is 0 or π, i.e., x lies on the straight line $L(q, q')$ passing through q and q', we have $W_x = \mathrm{id}$. It proves the following assertion.

LEMMA 4.4. *Let g, $g(\infty) \neq \infty$, be a loxodromic transformation in $\overline{\mathbb{R}}^n$, or, if $g \in \mathcal{M}_n$ is not loxodromic, let $n = 2m$. Then g has an invariant ball if and only if the orthogonal transformation $V_g(x)$ has an eigenvalue of one at point x lying on the straight line $L(q, q')$.*

In what follows, by multiplicity $n_1(g)$ of the eigenvalue one of orthogonal transformation $U_g \in O(n)$ we understand the dimension of a subspace in $\overline{\mathbb{R}}^n$ whose basis is provided by the vectors v having the property $U_g(v) = v$.

LEMMA 4.5. *Let G be a discontinuous group of Möbius transformations in $\overline{\mathbb{R}}^n$, $n \geq 2$ (maybe orientation reversing) and whose limit set has visible dimension n.*

Then

$$\min_{g \in \mathcal{H}_+(G)} n_1(g) \leq 1, \qquad \min_{g \in \mathcal{H}_-(G)} n_1(g) \leq 1,$$

where $\mathcal{H}_-(G)$ and $\mathcal{H}_+(G)$ are sets of loxodromic elements from G that are orientation reversing or preserving, correspondingly.

PROOF. We prove the first inequality (the second is proved entirely analogously). It follows that the sets $\mathcal{H}_+(G)$ and $\mathcal{H}_-(G)$ are non-empty because group G is not elementary (see Lemma 3.22).

We assume that our assertion is incorrect; let g_0 be a loxodromic element in G such that

$$n_1(g_0) = \min_{g \in \mathcal{H}_+(G)} n_1(g) \geq 2.$$

We conjugate G and assume that $g_0(x) = \lambda \cdot U_0(x)$, where $0 < \lambda < 1$, while the orthogonal transformation $U_0 \in O_+(n)$ has a pointwise fixed subspace

$$E = \{x \in \mathbb{R}^n : U_0(x) = x\}, \quad \dim E \geq 2.$$

We select a loxodromic element $g_1 \in \mathcal{H}_+(G)$ such that the orthogonal projection on the subspace E of the two-dimensional plane L passing through the origin and the points $z_1, z_2 \in \mathbb{R}^n$, fixed for g_1, is two-dimensional; such a choice is possible since $\dim_{vis}\Lambda(G) = n$ and the fixed points for the loxodromic elements in G are dense in $\Lambda(G) \times \Lambda(G)$ (see Theorem 3.23).

We now consider elements from $\mathcal{H}_+(G)$ having the form

$$g_{i,j} = g_0^{2m_i+1} \cdot g_1^{n_j}.$$

Here the sequence (m_i) and (n_j) of integers tending to $+\infty$ are such that

$$\lim_{i\to\infty} U_0^{m_i} = \mathrm{id}, \quad \lim_{j\to\infty} U_{g_1}^{n_j} = \mathrm{id}.$$

It is readily seen that, when (i, j) tends to (∞, ∞), the fixed points for the loxodromic elements $g_{i,j}$ tend to the origin and to the repulsive fixed point z_1 of the transformation g_1. Also, simple calculations show that then the radii of the isometric spheres $I(g_{i,j})$ tend to zero.

The chain rule gives

$$V_{g_{i,j}}(x) = V_{g_0^{2m_i+1}}(g_1^{n_j}(x)) \cdot V_{g_1^{n_j}}(x).$$

At the same time we see that

$$\lim_{i\to\infty} V_{g_0^{2m_i+1}} = \lim_{i\to\infty} U_0^{2m_i+1} = U_0$$

and for points $x \in [0, z_1]$

$$\lim_{j\to\infty} V_{g_1^{n_j}}(x) = V,$$

where $V \in O(n)$ corresponds to a non-trivial rotation in a two-dimensional plane L through an angle twice the angle between the segments $[z_1, z_2]$ and $[z_1, 0]$.

It then follows directly that for points $x \in [0, z_1]$

$$\lim_{i,j\to\infty} V_{i,j}(x) = U_0 \cdot V.$$

We now assume that $x \in \mathbb{R}^n$ is an eigenvector for the orthogonal transformation $U_0V \in O_+(n)$, i.e., $U_0V(x) = x$, and let $y = V(x)$. Two cases are possible: $y \neq x$ and $y = x$.

If $y \neq x$, then $y - x = y - V^{-1}(y)$ and therefore the vector $(y - x)$ is parallel to the plane L (invariant but not fixed for V). On the other hand, $y - x = y - U_0V(x) = y - U_0(y)$ so $(y - x)$ is orthogonal to the subspace E (U_0 is a rotation around E). This contradicts the choice of g_1 related to the non-orthogonality of L to subspace E.

If $y = x$, then $U_0(y) = y$ and consequently $y \in E$. Also, $V(y) = y$, i.e., vector y is orthogonal to plane L, which shows that the subspace of such vectors y has the dimension of the subspace

$$\{x \in \mathbb{R}^n : U_0V(x) = x\}$$

and these do not exceed $(\dim E - 2)$. We get directly from this for sufficiently large i and j that

$$n_1(g_{i,j}) = \dim\{x \in \mathbb{R}^n : V_{g_{i,j}}(x) = x\} \le \dim E - 2.$$

The contradiction of this fact to the minimality

$$n_1(g_0) = \dim E$$

completes the proof of Lemma 4.5.

PROOF OF THEOREM 4.3. As the necessity of the conditions in the theorem is obvious, we prove the sufficiency. We assume that for any ball $B \subset \overline{\mathbb{R}}^n$ (by ball we also understand a half-space) $G(B) \neq B$, and we show that then there exists an element $g \in G$ also not having any invariant ball.

We consider separately the cases where $\dim_{vis} \Lambda(G)$ is $2m$, $2m-1$, and $2m-2$.

1. Let $\dim_{vis} \Lambda(G) = 2m$. Then Lemma 4.5 gives

$$\min_{g \in \mathcal{H}(G)} n_1(g) \le 1.$$

However, the space has even dimension, $n = 2m$, and the group G elements preserve orientation, so there exists a loxodromic element $g_* \in G$ such that $n_1(g_*) = \dim\{x \in \mathbb{R}^n : V_{g_*}(x) = x\} = 0$. Lemma 4.4 shows that the transformation g_* does not have an invariant ball.

2. Let $\dim_{vis} \Lambda(G) = 2m-1$. Then there exists a G-invariant sphere S^{2m-1} containing $\Lambda(G)$ and there is a transformation $g_0 \in G$, $g_0(\operatorname{int} S^{2m-1}) = \operatorname{ext} S^{2m-1}$. We conjugate G and assume that S^{2m-1} is the hyperplane $\overline{\mathbb{R}}^{2m-1} \subset \overline{\mathbb{R}}^{2m}$. Considering the action of G on $\overline{\mathbb{R}}^{2m-1}$, we get from $\dim_{vis} \Lambda(G) = 2m-1$ and Lemma 4.5 that the minimal multiplicity of the eigenvalue one is zero for the orthogonal transformations $U_g \in O(2m)$, corresponding to loxodromic elements $g \in G$ that reverse the orientation of $\overline{\mathbb{R}}^{2m-1}$. Application of Lemma 4.4 again gives the existence of an element $g_* \in G$ that does not have any invariant ball.

3. $\dim_{vis} \Lambda(G) = 2m-2$. Then, as above, we can assume that the unbounded limit set $\Lambda(G)$ is contained in the G-invariant subspace $\overline{\mathbb{R}}^{2m-2}$ having the two-dimensional subspace P as orthogonal complement. It is obvious that $\overline{\mathbb{R}}^{2m-2}$ and P are invariant with respect to any orthogonal transformation V_g, $g \in G$.

Our assumption that group G does not have any invariant ball implies that either there exists a $g \in G$ such that U_g does not have any eigenvector in P corresponding to an eigenvalue one, or that there exist two loxodromic elements $g_1, g_2 \in G$ such that the eigenvectors $v_1, v_2 \in P$,

$$U_{g_1}(v_1) = v_1, \quad U_{g_2}(v_2) = v_2$$

are not parallel.

In the second case, it is readily checked that the orthogonal transformation $U_{g_1g_2}$ for element g_1g_2 does not have any eigenvector in P corresponding to eigenvalue one. Then we again have the first case.

In the first (main) case, we consider the restriction of the action of group G on $\overline{\mathbb{R}}^{2m-2}$ and repeat the proof of Lemma 4.5 for those elements $g \in G$ for which the orthogonal transformations $U_g \in O(n)$ do not have vectors $v \in P$ such that $U_g(v) = v$. Then, as above, we use Lemma 4.4 to demonstrate that there is an element $g_* \in G$ not having any invariant ball.

Theorem 4.2 is thus completely proved.

If now group $G \subset \text{Möb}(n)$ has an invariant ball, it can be considered as a subgroup of group $\text{Isom}\, H^n$. The following discreteness conditions apply for such groups.

THEOREM 4.6. *Let G be an isometry group for hyperbolic space H^n for which there are no fixed point from $\overline{H}^n$ and no invariant k-dimensional plane from H^n, $k \le n-1$. Then either G is dense in* $\text{Isom}\, H^n$ *or G is discrete.*

The proof of this theorem, which extends results due to Nielsen [2], Siegel [1], and Fenchel and Nielsen [2] for planar groups, is obtained directly from Greenberg's description [1] of the closed subgroups in $\text{Möb}(n)$ (compare Theorem 1.19).

More detailed discreteness conditions depend on the space dimension and on the behaviour of the elliptic elements in G. We denote by $H(G)$ the minimal plane (line) in H^n invariant with respect to the group $G \subset \text{Isom}\, H^n$. If $\dim_{\text{vis}} \Lambda(G)$ on a sphere is k, then $\dim H(G) = k+1$ and $H(G)$ is the hyperbolic plane having dimensions $k+1$, whose boundary at infinity $\partial H(G) \subset \partial H^n$ is a k-sphere containing $\Lambda(G)$.

THEOREM 4.7. *Let G be a non-elementary isometry subgroup of H^n and* $\dim H(G) = 2m$. *If the identity is not a limit for elliptic elements of G, then G is discrete.*

COROLLARY 4.8. *Let G be a loxodromic isometry group in H^n such that* $\dim H(G) = 2m$. *Then G is discrete.*

Greenberg [1] has given proofs for these theorems and also gives the following example, which shows that the dimension of $H(G)$ must necessary be even.

EXAMPLE 4.9. To construct a non-discrete loxodromic group G of isometries in H^3 having $H(G) = H^3$, we note that the extension of any orientation preserving isometry g in H^3 to the plane $\overline{\mathbb{C}} = \partial H^3$ gives a unique linear fractional transformation $\gamma : z \longmapsto (az + b)/(cz + d)$, $ad - bc = 1$. This mapping γ, and consequently also g, will be loxodromic if and only if the trace $\operatorname{tr}\gamma = a + d$ is not real.

We denote by $\{U_m\}_{m\in\mathbb{N}}$ a basis of the open sets in $\mathbb{R}^6$ and consider the sequence $\{P_m\}$ of points $P_m = (a'_m, a''_m, b'_m, b''_m, c'_m, c''_m)$ from $\mathbb{R}^6$ such that $P_m \in U_m$ for all m and the coordinates of all P_i, $i \le m$, are independent transcendentals (over the rational field). We then put

$$a_m = a'_m + ia''_m, \quad b_m = b'_m + ib''_m, \quad c_m = c'_m + ic''_m, \quad d_m = (1 + b_m c_m)/a_m \tag{4.1}$$

and consider the group $\Gamma = \langle\gamma_1, \gamma_2, \ldots\rangle$, generated by the transformations $\gamma_m : z \longmapsto (a_m z + b_m)/(c_m z + d_m)$ with the coefficients (4.1). The construction of the group shows that it is dense in $\mathcal{M}_2$ and is therefore non-discrete. We show that Γ is purely loxodromic. For this purpose it is sufficient to show that either $\operatorname{tr}\gamma \notin \mathbb{R}$, or $\gamma = \mathrm{id}$ for any $\gamma \in \Gamma$.

In fact, γ is a word $W(\gamma_{m_1}, \ldots, \gamma_{m_r})$ on the group Γ generators and

$$\operatorname{tr}\gamma_1 = a + d = \mathcal{R}(a_{m_1}, b_{m_1}, c_{m_1}, \ldots, c_{m_r}), \tag{4.2}$$

where $\mathcal{R} = U + iV$, with U and V real, rational functions. If $V = 0$ for γ, then because of transcendental independence, V is identically zero. However, then the fact that $\mathcal{R}$ is analytic gives us that the function $\mathcal{R}$, defined by (4.2), is identically equal to a real constant R. This means that for any $\gamma_1, \ldots, \gamma_r \in \mathcal{M}_2$ and $\gamma = W(\gamma_1, \ldots, \gamma_r)$ we have $\operatorname{tr}\gamma = R$. We put here $\gamma_1 = \gamma_2 = \ldots = \gamma = \mathrm{id}$ and get that $R = 2$. If now $\gamma_1, \ldots, \gamma_r$ are free generators of a free, hyperbolic group, then $\operatorname{tr}\gamma = 2$ implies that $\gamma = \mathrm{id}$, i.e., that $W(\gamma_1, \ldots, \gamma_r) = \mathrm{id}$.

This means that the word W can be freely reduced to the identity, so $\gamma = W(\gamma_{m_1}, \ldots, \gamma_{m_r}) = \mathrm{id}$. We thus get that the hyperbolic isometry group G in H^3, obtained by extending group Γ from the plane $\overline{\mathbb{C}}$ is a free, non-discrete group having a countable set of distinct axes, which means that it is not commutative and has $H(G) = H^3$.

The above discreteness conditions for Möbius groups in general are difficult to examine even for two-generated groups. It would be desirable to have effectively examinable discreteness conditions for Möbius groups in each dimension formulated as conditions on generators of groups G. Discreteness condition of that type is Jørgensen necessary condition [2] on the trace function $\tau(X) = \operatorname{tr} X = a + d$, where $X \in \mathcal{M}_2$, $X : z \longmapsto (az + b)/(cz + d)$, $ad - bc = 1$.

THEOREM 4.10. *Let the Möbius transformations* X *and* Y *generate a discrete group* $G \subset \mathcal{M}_2$. *Then we have*

$$|\tau^2(X) - 4| + |\tau(XYX^{-1}Y^{-1}) - 2| \geq 1,$$

with the exception of one of the following cases, in which group G is elementary:

1. *G is cyclic or is a finite extension of a cyclic group and* $|\tau(X) - 4| < 1$;

2. *X is elliptic or loxodromic and* $|\tau^2(X) - 4| < 1/2$. *Here two fixed points for X are permutated by element Y, so Y is elliptic,* $Y^2 = \mathrm{id}$;

3. *X is parabolic; its fixed point is fixed for Y also. Then either Y is elliptic and has order 2, 3, 4, or 6 or Y is parabolic and G is a free Abelian group.*

There is some prospect of obtaining the multi-dimensional analogue of the Jørgensen[1] conditions in connection with Gilman's [3], [4] proposed geometric approach to the Jørgensen inequality and with a precise geometric meaning of necessary conditions for discreteness of hyperbolic torsion groups in H^n, $n \leq 3$ (in terms of distances d between fixed points of the group), that is given by recently deep results of Derevnin [1], [2] and Derevnin-Mednykh [1]. In particular, the last paper contains the following result.

Given a discrete group $G \subset \mathrm{Isom}\, H^3$ with torsion, a point $x \in H^3$ is called an icosahedral point of G if its stablizer subgroup $G_x \subset G$ is isomorphic to the icosahedral orthogonal group $A_5 \subset O(3)$. This icosahedral point x is the intersection of axes of elliptic elements in G_x of order 3 and 5.

THEOREM 4.11. *Let G be a discrete group of isometries of H^3, and let P and Q be distinct icosahedral points of G. Then:*

(i) $\cosh d(P, Q) \geq \cosh d_1 = (2 + \sqrt{5})/2$;

(ii) *if P and Q do not lie on a common elliptic axis of fifth order of G, then*

$$\cosh d(P, Q) \geq \cosh d_2 = (5 + 3\sqrt{5})/4;$$

(iii) *if P and Q do not lie on a common elliptic axis of fifth and third order of G, then*

$$\cosh d(P, Q) \geq \cosh d_3 = 2 + \sqrt{5}.$$

---

[1] Note added in proof. Recently G. Martin [2] presented such an analogue where the main concept is the calculation of the Zassenhaus neighbourhood shape in $SO(1, n+1)$ which, in the Hilbert-Schmidt norm, has the form $\{A : \|A - E\| < 1 - \sqrt{3}\}$:

THEOREM. *Let $f, g \in \mathrm{M\ddot{o}b}(n)$ generate a discrete non-elementary group $\langle f, g\rangle$, then* $\max\{\|g^i f g^{-i}\| : i = 0, 1, \ldots, n\} \geq 2 - \sqrt{3}$, $\max\{\|f - \mathrm{Id}\|, \|[f, g^i] - \mathrm{Id}\| : i = 1, \ldots, n\} \geq 2 - \sqrt{3}$. *Moreover, if f is non-elliptic, then it suffices to consider only those terms with $i = 0$ or $i = 1$.*

All the estimates are sharp and are obtained for groups generated by reflections in the sides of a tetrahedron.

If the two generators X and Y of group $G \subset \operatorname{Isom} H^3 \cong \text{Möb}(2)$ are elliptic transformations, one can have discreteness conditions for a group $G \subset \operatorname{Isom} H^3$ that are simultaneously necessary and sufficient. We say that an elliptic element is primitive if it provides rotation through an angle $2\pi/n$, where n is a natural number.

THEOREM 4.12. *Let a group $G \subset \operatorname{Isom} H^3$ be generated by two primitive elliptic elements X and Y having orders n and m correspondingly $(n \le m)$ and with crossed mutually orthogonal axes in H^3. Then the group $G = \langle X, Y \rangle$ is discrete if and only if one of the following conditions is satisfied:*

1. *the commutator $[X, Y]$ is a loxodromic element;*
2. *$[X, Y]$ is a parabolic element;*
3. *$[X, Y]$ is rotation through an angle $4\pi/l$ and one of the following is realized:*

a) *the triple (n, m, l) belongs to the set*

$$\begin{Bmatrix} (3,3,5), (4,5,3), (5,5,3), (3,3,6), (3,4,4), \\ (3,6,3), (4,4,4), (4,6,3), (5,6,3), (6,6,3), \end{Bmatrix}$$

b) *l is a positive integer, $1/l + 1/m < 1/2$;*

c) *$n = 2$ and the integer $m = 2l$ is greater than seven;* and

d) *the integer $n = m = 2l$ is greater than seven.*

The proof of this assertion (see Klimenko [1]) is obtained geometrically (compare Beardon [1], Theorem 11.5.2) by examining the polyhedra arising on the intersection of dihedral angles in H^3 whose edges are crossed axes of the elliptic generators of the group. Here an important part is played by the well-known description of all finite subgroups for the orthogonal group $O(3)$.

§2. **Arithmetic construction of co-compact hyperbolic groups**

The purpose here is to give a construction for discrete isometry groups in hyperbolic space H^n having any dimensions $n \ge 2$ and having compact fundamental polyhedra in H^n. Such groups are called *co-compact*. The construction is based on a certain arithmetic construction arising for $n = 2$ in works by Fricke [1] and Bianchi [1] (see also Fricke and Klein [1]), and for $n \ge 3$ in works by B.A. Venkov [1], [2], and it is based on considering automorphism groups for indefinite quadratic forms. For the further development of it, see Mostow and Tamagawa [1] and Borel [1], and also compare §6 of this chapter.

Then in $\mathbb{R}^{n+1}$ we consider the quadratic form

$$f_d(x) = -\sqrt{d}x_0^2 + x_1^2 + \dots + x_n^2, \tag{4.3}$$

where d is a natural number, free from squares and different from one. Then we take the upper half of the hyperboloid $\{x \in \mathbb{R}^{n+1} : f_d(x) = -1,\ x_0 > 0\}$ and, as in §2.3, obtain a model for the hyperbolic space H^n (a hemisphere having an imaginary radius) whose isometry group is the index two subgroup $O'(f_d, \mathbb{R})$ in the group $O(f_d, \mathbb{R}) \subset GL_{n+1}(\mathbb{R})$, consisting of linear transformations of space $\mathbb{R}^{n+1}$ that preserve the quadratic form f_d and map the upper half of the hyperboloid into itself.

We now define a group $G = G(n, d)$ for any $n \geq 2$ as a group $O'(f_d, A_d)$, where A_d is the ring of integers in the quadratic field $Q(\sqrt{d})$; in other words,

$$G(n, d) = O'(f_d, A_d) = \{M = (a_{ij}) \in GL_{n+1}(\mathbb{R}) : \\ a_{ij} = n_{ij} + m_{ij}\sqrt{d},\quad f_d(M \cdot x) = f_d(x);\quad n_{ij}, m_{ij} \in \mathbb{Z}\}. \tag{4.4}$$

THEOREM 4.13. *For all* $n \geq 2$ *and for all natural numbers* d *distinct from squares and from one,* $G = G(n, d)$ *is a discrete co-compact isometry group in the hyperbolic space* H^n.

PROOF (Sullivan [5]). We first give the proof of co-compactness for the analogous group $G = O(f, \mathbb{Z}) \subset GL_4(\mathbb{R})$, which preserves the quadratic form $f(x) = -7x_0^2 + x_1^2 + x_2^2 + x_3^2$ and consists of integer matrices. The discreteness here obviously follows from that of $GL_n(\mathbb{Z})$ in $GL_n(\mathbb{R})$. To prove compactness for the orbifold H^3/G, due to the transitive action on H^3 from $O(f, \mathbb{R}) \cong \operatorname{Isom} H^3$, it is sufficient to prove compactness for the factor group $O(f, \mathbb{R})/O(f, \mathbb{Z})$, which can be identified with the $O(f, \mathbb{R})$-orbit $L_0 = O(f, \mathbb{R})(\mathbb{Z}^4) \subset L$ of the integer lattice $\mathbb{Z}^4 \subset \mathbb{R}^4$.

Here by L we denote the set of lattices in $\mathbb{R}^4$, i.e., of discrete subgroups isomorphic to $\mathbb{Z}^4$. The lattices in $\mathbb{R}^n$ are characterized (up to a compact set of possibilities) by two invariants: the volume V of the fundamental domain $V = \det M$, where matrix M is composed of basis vectors, and the minimal distance S of points in the lattice from the origin. We introduce in L the topology for the basis vectors or an equivalent topology induced by the quotient topology $GL_n(\mathbb{R})/GL_n(\mathbb{Z}) \cong L$ and prove the compactness of $O(f, \mathbb{R})$-orbit L_0 by the following compactness criterion:

THEOREM 4.14 (Hermite and Mahler [1]). *A subset of lattices* L_0 *has a compact closure in* L *if* $\inf_{L_0} S > 0$ *and* $\sup_{L_0} V < \infty$.

We now show that $\inf S > 0$ on the set $L = O(f, \mathbb{R})$-orbit of the integer lattice $\mathbb{Z}^4 \subset \mathbb{R}^4$, namely on reducing form f by $\bmod 8$, we see that the restriction of form f to $\mathbb{Z}^4 \setminus \{0\} \subset \mathbb{R}^4$ takes only non-zero integer values. Let

$$U = \{x \in \mathbb{R}^4 : f(x) \leq 1/2\}$$

be the neighbourhood of the origin. As $f(0) = 0$ and f is continuous, U does not contain points from $\mathbb{Z}^4 - \{0\}$, so due to the form f being preserved by the group $O(f, \mathbb{R})$, we get that no point from an arbitrary lattice from $O(f, \mathbb{R})$-orbit L_0 of integer lattice $\mathbb{Z}^4$ is contained in U. This implies that $\inf_{L_0} S$ is positive.

We have $\sup_{L_0} V \leq 1$. In fact, $V(\mathbb{Z}^4) = 1$, while all elements M of group $O(f, \mathbb{R})$ leave form f invariant, so $\det M = 1$.

Finally, L_0 is closed in L. If the sequence $\{L_j\} \subset L_0$ converges to a lattice $L_* \in L$, then the restriction to a point in L_* for form f (continuous and integer on points of lattices L_j) takes integer values. Let $\{V_i\}$ and $\{U_i\}$ be correspondingly the bases for L_* and L_j with sufficiently large number j. We transfer to subsequences and can assume that U_i are close to V_i, $f(U_i) = f(V_i)$ and $f(U_i \pm U_k) = f(V_i \pm V_k)$, which implies that $f(U_i, U_k) = f(V_i, V_k)$ for $1 \leq i, k \leq 4$. Therefore, a linear transformation that transforms L_j into L_* leaves form f invariant, i.e., is an element of group $O(f, \mathbb{R})$. Then lattice L_* lies in the set L_0.

These properties of the $O(f, \mathbb{R})$-orbit L_0 and Theorem 4.14 complete the proof of compactness for H^3/G.

In the general case $G = O(f_d, A_d)$, this proof is modified in a standard fashion. It is necessary to consider separately only the proof of discreteness for group G because now the integer ring A_d in field $Q(\sqrt{d})$ is not discrete in $\mathbb{R}$.

We note firstly that with the embedding $(m + n\sqrt{d}) \longmapsto (m + n\sqrt{d}, m - n\sqrt{d})$, $m, n \in \mathbb{Z}$, the image of ring A_d forms a discrete subring in the product $\mathbb{R} \times \mathbb{R}$. This embedding induces an embedding of the group $G = O'(f_d, A_d)$ in the group

$$GL_{n+1}(\mathbb{R} \times \mathbb{R}) = GL_{n+1}(\mathbb{R}) \times GL_{n+1}(\mathbb{R}), \qquad (4.5)$$

and forms a discrete subgroup there.

As the group $G = G(n, d)$, specified in (4.4), leaves the form f_d of (4.3) invariant, its image in the second cofactor $GL_{n+1}(\mathbb{R})$ in (4.5) leaves the form $+\sqrt{d}x_0^2 + x_1^2 + \ldots + x_n^2$ invariant, i.e., it is a subgroup of a compact orthogonal group. This immediately implies discreteness of the image of G in the first cofactor $GL_{n+1}(\mathbb{R})$ in (4.5), which coincides with group G itself. This completes the proof of Theorem 4.13.

§3. Universal bounds of hyperbolic manifold sizes

The purpose of this section is to prove the following theorem:

THEOREM 4.15. *There exist constants $r = r(n) > 0$, dependent only on the dimensions n, $n \geq 2$, and such that any complete hyperbolic n-dimensional manifold allows isometric embedding of the ball $B(r) \subset H^n$ having radius r.*

The following is an obvious equivalent reformulation for discrete groups of hyperbolic isometries:

THEOREM 4.16. *Any fundamental Dirichlet polyhedron* $P \subset H^n$ *for any discrete torsion-free group* $G \subset \mathrm{Isom}\, H^n$, $n \geq 2$, *contains a ball having radius* $r = r(n)$ *dependent only on* n.

Before we prove Theorem 4.16 we give two assertions (Apanasov [5]) on convergent transformation sequences.

LEMMA 4.17. *Let a sequence* $\{g_i\}$ *of hyperbolic isometries in* H^n *converge to the identity. Then there exist its subsequence* $\{g_{i_j}\}$ *and a sequence* $\{\gamma_j\} \subset \mathrm{Isom}\, H^n$ *such that* $g = \lim_{j\to\infty} \gamma_j g_{i_j} \gamma_j^{-1}$ *is a parabolic transformation.*

The proof is divided into three parts in accordance with whether g_i have:

1) each one fixed point $p_i \in \overline{\mathbb{R}}^{n-1} = \partial H^n$,
2) at least one fixed point each $p_i \in H^n$,
3) each two different fixed points p_i, $q_i \in \overline{\mathbb{R}}^{n-1} = \partial H^n$.

It is readily seen that in case 1) the $\{\gamma_j\}$ elements should be taken as the transformations that conjugate g_{i_j} with screw translations $x \longmapsto U_{i_j}(x) + a_j$, where $a_j \in \mathbb{R}^{n-1}$, $U_{ij} \in O(n-1)$, $U_{ij}(a_j) = a_j$, $|a_j| = 1$. Here the subsequence $\{g_{i_j}\}$ is chosen such that the limits $\lim_{j\to\infty} U_{ij} = \mathrm{id}$, $\lim a_j = a \in S^{n-1}(0, 1)$ exist.

The proof in case 2) is given by assuming that we have the ball Poincaré model in $B^n(0, 1)$ for the hyperbolic space and consider its isometries w_i such that $w_i(p_i) = 0$. Then the compactness of the orthogonal group $O(n)$ implies that there exist $w \in O(n)$ and a subsequence $\{g_{i_j}\}$ such that

$$\lim_{j\to\infty} w w_{i_j} g_{i_j} w_{i_j}^{-1} w^{-1} = \mathrm{id}, \tag{4.6}$$

where the (4.6) transformations are orthogonal.

We denote the distortions of the coordinate vector e_1 in the transformation (4.6) by r_j and let $x_j = \sqrt{1 - r_j} \cdot e_1$. Then the desired conjugating sequence $\{\gamma_j\}$ consists of transformations

$$\gamma_j = v_j w w_{i_j} \tag{4.7}$$

where $v_j \in \text{Möb}(B^n)$ are hyperbolic translations having the properties

$$v_j(-x_j) = 0, \quad v_j(\infty) = x_j/(1 - r_j). \tag{4.8}$$

With conditions (4.8) on γ_j, the sequence $\gamma_j g_{i_j} \gamma_j^{-1}$ converges to the parabolic transformation $g \in \text{Möb}(B^n)$, in which the isometric spheres $I(g)$ and $I(g^{-1})$ have radius one and correspondingly centres

$(1, q_2, \ldots, q_n)$ and $(1, -q_2, \ldots, -q_n)$, $\sum_{i=2}^{n} q_i^2 = 1$.

In case 3) we can assume, without loss of generality, that $\lim p_i = p$ and $\lim q_i = q$. Then there exists a sequence $w_i \in \operatorname{Isom} H^n$ such that

$$w_i(p_i) = \infty, \quad w_i(q_i) = 0, \quad w = \lim w_i \in \operatorname{Isom} H^n. \tag{4.9}$$

We use (4.9) to get the respresentation

$$w_i g_i w_i^{-1}(x) = r_i U_i(x) + a_i, \tag{4.10}$$

in which the orthogonal transformations U_i converge to the identity, while the positive numbers r_i converge to one, and the lengths of the vectors $a_i \in \mathbb{R}^{n-1}\setminus\{0\}$ tend to zero.

It is readily seen from the representation (4.10) that it is sufficient to take a subsequence g_{i_j} such that there exists $\lim_{j\to\infty} a_{i_j}/|a_{i_j}|$ and where one takes the elements of the sequence $\{\gamma_j\} \subset \operatorname{Isom} H^n$ as the transformations $\gamma_j = w_{i_j}/|a_{i_j}|$.

LEMMA 4.18. *Let a sequence* $\{g_i\}$ *of hyperbolic isometries in* H^n *be such that on a certain point* $x_0 \in H^n$, *we have* $\lim g_i(x_0) = x_0$. *Then for a certain sequence* $\{m_i\}$ *of integers*

$$\lim_{i\to\infty} g_i^{m_i} = \mathrm{id}. \tag{4.11}$$

PROOF. We perform constructions for the hyperboloid model of the hyperbolic space (Chapter 2 §3.3) on the assumption that the point x_0 is $e_0 = (1, 0, \ldots, 0)$. We write each isometry g_i in the form $g_i = U_i V_i$, where $U_i \in O(n)$, while V_i are hyperbolic rotations, $\lim V_i(e_0) = e_0$. But then the limit of the norm $\|V_i\|$ is one for the corresponding matrices V_i. We now define a sequence $\{\varepsilon_j\}$, $\varepsilon_j > 0$, $\lim \varepsilon_j = 0$. Then there exists a $K_j > 0$ such that for any $U \in O(n)$ one can find $m \in \mathbb{Z}$, $|m| \le K_j$, for which $\|U^m - 1\| < \varepsilon_j$. Now let the number i_j be such that

$$|\,\|V_i\|^{K_j} - 1| < \varepsilon_j \quad \text{for all} \quad i \ge i_j. \tag{4.12}$$

Then for number i, $i_j \le i < i_{j+1}$ we choose $m_i \in \mathbb{Z}$ such that

$$|m_i| \le K_j, \quad \|U_i^{m_i} - 1\| < \varepsilon_j. \tag{4.13}$$

Assertion (4.11) follows directly from conditions (4.12) and (4.13).

PROOF OF THEOREM 4.16. By $d(x, y)$ we denote hyperbolic distance, while by $D(x, r)$ we denote the hyperbolic ball with its centre at point x and radius r, $r > 0$. Let $P(G)$ be the fundamental Dirichlet polyhedron with its centre at the origin for a discrete group $G \subset \text{Möb}(B^n)$ of hyperbolic isometries (in the ball Poincaré model $B^n(0, 1)$).

The proof is given by assuming the converse. Namely we assume that for any number m there exists a discrete torsion-free group (not containing elliptic elements) $G_m \subset \text{Möb}(B^n)$ such that for all $w \in \text{Möb}(B^n)$

$$D(0, 1/m) \not\subset P(wG_mw^{-1}). \tag{4.14}$$

Then the groups G_m contain elements g_m for which $d(0, g_m(0)) \leq 2/m$, and we have a sequence $\{g_m\}$ of isometries in B^n having the property $\lim_{m\to\infty} g_m(0) = 0$. On the basis of Lemma 4.18, we can assume that

$$\lim_{m\to\infty} g_m = \text{id}. \tag{4.15}$$

Without changing the symbols, we transfer from groups acting in $B^n(0, 1)$ to conjugate groups acting in the half-space H^n. The conjugating mapping is chosen such that the origin transfers to the point $e_n = (0, \dots, 0, 1)$. Then after this we consider the sequence of transformations $w_m \in \text{Möb}(H^n) = \text{Möb}(n-1)$, whose isometric spheres $I(w_m)$ and $I(w_m^{-1})$ pass through the point e_n, while the centre of $I(w_m)$ remains fixed in transformation g_m. Then

$$w_m g_m w_m^{-1}(x) = r_m U_m(x) + a_m, \tag{4.16}$$

where $U_m \in O(n-1)$, $r_m > 0$, $a_m \in \mathbb{R}^{n-1}$.

It follows from (4.15) that $\lim U_m = \text{id}$, $\lim r_m = 1$, $\lim a_m = 0$.

Now let the sequence $\{\gamma_j\} \subset \text{Möb}(n-1)$ and the sequence m_j be chosen in accordance with Lemma 4.17, i.e., the transformation

$$g = \lim_{j\to\infty} \gamma_j w_{m_j} g_{m_j} w_{m_j}^{-1} \gamma_i^{-1} \tag{4.17}$$

is parabolic. We use the initial assumption and put

$$G'_{m_i} = \gamma_i w_{m_i} G_{m_i} w_{m_i}^{-1} \gamma_i^{-1}, \tag{4.18}$$

to get again that there exist elements $v_i \in G'_{m_i}$ that converge to the identity and do not have fixed points in common with the transformations of (4.17) (we exhaust, as above, all elements of the (4.16) form from the stabilizer at the point ∞ in group $w_{m_i} G_{m_i} w_{m_i}^{-1}$). We denote the elements from (4.17) in group G_{m_j} by h_j and thus get a sequence of doubly generated subgroups $\langle v_i, h_i \rangle$ of the groups G'_{m_j} whose generators v_i and h_i converge correspondingly to identity and parabolic transformation (from (4.17)), are not elliptic and do not have common fixed points. Here one encounters a contradiction with the discreteness in these groups, since the h_i differ little from translation by a unit vector, while the isometric spheres $I(v_i)$ have arbitrarily large radii (compare Theorem 4.27).

We can give a detailed proof that $\langle v_i, h_i \rangle$ are not discrete groups for sufficiently large i (here we simultaneously give the concept behind the proof of Theorem 6.46).

We conjugate the group $\langle v_i, h_i \rangle$ by means of $\gamma_i \in \text{Möb}(n-1)$, which converge to a certain Möbius transformation γ and which leave the point $e_n \in H^n$ fixed. As the v_i converge to the identity, the mappings γ_i can be chosen such that

$$\gamma_i v_i \gamma_i^{-1}(x) = r_i U_i(x) + a_i, \tag{4.19}$$

where $U_i \in O(n-1)$, $a_i \in \mathbb{R}^{n-1}$, $r_i > 0$ and $\lim U_i = \text{id}$, $\lim a_i = 0$, $\lim r_i = 1$. These groups and their generators will be denoted as before (we omit the subscript i, which is taken as sufficiently large). As the mappings v and h do not have common fixed points, h are not similarities due to (4.19), so we apply Corollary 1.8 to get

$$h(x) = \rho \cdot A \cdot \mathcal{J}(x-q) + q', \tag{4.20}$$

where $A \in O(n-1)$, $\mathcal{J}(x) = x/|x|^2$, with q and q' the centres of the isometric spheres $I(h)$ and $I(h^{-1})$ having radius $\sqrt{\rho}$.

If q is a fixed point for v, we consider the element h_1 in group $\langle v, h \rangle$,

$$h_1(x) = hvh^{-1}(x) = r^{-1}AUA^{-1}(x-q') + q', \tag{4.21}$$

where U and r are from (4.19). Then Lemma 3.2 implies that h_i should leave q fixed: $h_1(q) = q$. That and (4.21) give $q = q'$, and therefore h should be an elliptic transformation, which is impossible. Therefore q is not a fixed point for v.

We construct the sequence $\{h_i\} \subset \langle v, h \rangle$ having the elements

$$h_1 = hvh^{-1}, \ \ldots, \ h_i = h_{i-1}vh_{i-1}^{-1}, \ \ldots \tag{4.22}$$

As before (4.19) is applied to $\langle v, h \rangle$, the second generators converge to a parabolic transformation, there exists a constant $k > 0$, independent of h, such that

$$d(h^{-1}(e_n), e_n) \le k. \tag{4.23}$$

At the same time, the generators v are sufficiently close to the identity so one can assume that the constant $k > 0$ from (4.23) has the property that $d(v(x), x) \le k$ for all x such that $d(x, e_n) \le k$. In particular,

$$d(vh^{-1}(e_n), h^{-1}(e_n)) \le k.$$

This involves the bound

$$d(h_1(e_n),\ e_n) = d(hvh^{-1}(e_n),\ e_n) \leq k. \tag{4.24}$$

Induction by i gives us that the following applies for all elements of $\{h_i\}$:

$$d(h_i(e_n),\ e_n) \leq k, \tag{4.25}$$

which are analogous to (4.24).

Then to complete the proof of the non-discreteness for $\langle v, h\rangle$ it is enough to show distinctness for elements in $\{h_i\}$. For this purpose, we substitute representations (4.19) and (4.20) for v and h into (4.22) and for $i = 1$ we have

$$\begin{aligned} \rho_1 A_1 \mathcal{J}(x - q_1) + q'_1 &= h_1(x) = hvh^{-1}(x) = \\ &= \rho\mathcal{J}[r\rho AUA^{-1}\mathcal{J}(x - q') + A(v(q) - q)] + q', \end{aligned} \tag{4.26}$$

where $v(q) - q \neq 0$, as has been shown. From (4.26) we get by simple calculation that

$$\rho_1 = r\rho^2/|v(q) - q|^2.$$

The uniform bound of (4.23) for h implies that the numbers ρ are uniformly bounded from below by a certain number $\rho_0 > 0$, while q lies in a certain ball $B^{n-1}(0, K) \subset \mathbb{R}^{n-1}$. Then as the transformations v are close to the identity, we have $\rho_1 \geq 2\rho$.

We write formulae analogous to (4.26) for h_i, $i \geq 2$, and get by induction on i that all the elements in $\{h_i\}$ have isometric spheres $I(h_i)$ having radius $\sqrt{\rho_i}$, where

$$\rho_{i+1} \geq 2\rho_i.$$

It follows directly from the bound of (4.25) and this property that $h_i \neq h_j$ for $i \neq j$, and therefore the groups $\langle v, h\rangle$ are non-discrete. This contradiction completes the proof of Theorem 4.16.

REMARK 4.19. The universal property of discrete hyperbolic groups in Theorem 4.16 (it is related to the profound Kazhdan-Margulis theorem [1]) was proved independently for $n = 2$ by different but essentially planar methods in several papers simultaneously: Harvey [1], Marden [2], and Sturm and Shinnar [1]; Apanasov [5] derived Theorem 4.16 for $n \geq 3$. Wielenberg [1] later obtained another proof of Theorem 4.16, which for $n = 3$ allows extension to groups with torsion. Recently, Waterman [1] established numerical values for the constant $r = r(3)$ for planar Kleinian groups, i.e., for groups of isometries in the three-dimensional hyperbolic space H^3. He showed that $r(3) \geq 1/25$; in the case of discrete torsion groups, the corresponding radius for the hyperbolic ball is not less than 1/300. See also Chapter 8 §3.

§4. Margulis lemma and hyperbolic manifold decomposition

We define a decomposition of a hyperbolic manifold M into thin and thick submanifolds; for a sufficiently small $\varepsilon > 0$, we consider the manifold $M_{(0,\varepsilon]} \subset M$, consisting of points $x \in M$ through which pass non-trivial loops of length not exceeding ε (uncontractible in M). The complement in M of the interior $M_{(0,\varepsilon)}$, i.e., the manifold $M_{[\varepsilon,\infty)} = M \setminus M_{(0,\varepsilon)}$ consists, by definition, of those points through which pass only loops (non-trivial in M) having lengths not less than ε.

To examine this decomposition, we naturally transfer to the universal covering, which in the case of complete M is the hyperbolic space H^n, and take in the discrete group $G \subset \operatorname{Isom} H^n$ (the image of $\pi_1(M)$ by holonomy homomorphism) for each point $x \in H^n$ the subgroup $G_\varepsilon(x) \subset G$, generated by all elements $g \in G$ for which the hyperbolic distance $d(x, g(x) \le \varepsilon$. By $G'_\varepsilon(x)$ we denote its subgroup generated by the elements whose derivatives also differ by ε from the identity.

THEOREM 4.20 (Margulis's lemma). *For each n there exists $\varepsilon = \varepsilon(n) > 0$ such that for any discrete group $G \subset \operatorname{Isom} H^n$ and for each point $x \in H^n$ the subgroup $G'_\varepsilon(x)$ is Abelian, while the group $G_\varepsilon(x)$ has an Abelian subgroup of finite index.*

PROOF. In any Lie group H, the commutator mapping $[*\ ,\ *] : H \times H \longrightarrow H$ has a zero derivative at the point (id, id). Therefore there is an upper bound to the size of the commutator for two small elements set by a certain constant (the product of their sizes), and for any discrete subgroup $G'_\varepsilon \subset G$, generated by small elements, one has finiteness for the lower central series

$$G'_\varepsilon \supset [G'_\varepsilon, G'_\varepsilon] \supset [G'_\varepsilon, [G'_\varepsilon, G'_\varepsilon]] \supset \ldots \supset \mathrm{id}.$$

This follows from a lower bound to the size of elements from G'_ε. Then the group G'_ε is nilpotent. It is familiar that nilpotency of a hyperbolic isometry group implies that it is Abelian (this can be demonstrated by means of the classifications for H^n isometries), so G'_ε is an Abelian group.

The proof of the second statement is based, firstly, on finding $\varepsilon_1 > 0$ such that the groups $G'_{\varepsilon_1}(x)$ are always Abelian, and secondly, on the choice of an $\varepsilon < \varepsilon_1$ such that the product of the generators of $G_\varepsilon(x)$ will lie in $G'_\varepsilon(x)$.

More precisely, let there be a sufficiently large $N > 0$ such that any set, consisting of at least N elements from the orthogonal group $O(n)$, contains at least one pair of elements whose distance apart is not more that $\varepsilon_1/3$.

We take ε_2, $0 < \varepsilon_2 \le \varepsilon_1/3$, such that, for any isometries $f_1, f_2 \in \operatorname{Isom} H^n$ that translate a point x by a distance $\le \varepsilon_2$, the derivative at point x of mapping $f_1 \circ f_2$ is estimated within $\varepsilon_1/6$ by the product of the derivatives at the point x of f_1 and f_2.

Now let $\varepsilon = \varepsilon_2/2N$. Then the product of $2N$ isometries that each translate x a distance not more than ε will translate x a distance not exceeding ε_2. Let $g_1, \dots, g_k \in G$ be the generators of the group $G_\varepsilon(x)$. We consider the cosets $\gamma G'_{\varepsilon_1}(x)$, where $\gamma \in G_\varepsilon(x)$, and show that they have representatives whose words in the generators $g_1, \dots, g_k$ have lengths less than N. In fact, if $\gamma = g_{i_1}, \dots, g_{i_m}$ is any word having length $m \geq N$, then one can write $\gamma = \alpha \cdot \delta \cdot \beta$, $\alpha, \delta, \beta \neq \mathrm{id}$, where the word $\delta\beta$ has length not greater than N, and the derivative of δ is within $\varepsilon_1/3$ of the identity. This implies that $(\alpha\beta)^{-1} \cdot (\alpha\delta\beta) = \beta^{-1}\delta\beta$ lies in $G'_{\varepsilon_1}(x)$, so we have that the coset $\gamma G'_{\varepsilon_1}(x) = \alpha\beta \cdot G'_{\varepsilon_1}(x)$. This fact and induction prove the property of cosets stated above.

We thus see that the Abelian group $G'_{\varepsilon_1}(x)$ has a finite index in the group generated by the groups $G_\varepsilon(x)$ and $G'_{\varepsilon_1}(x)$, so $G'_{\varepsilon_1}(x) \cap G_\varepsilon(x)$ has a finite index in $G_\varepsilon(x)$.

The Margulis lemma is correct in the more general case of discrete isometry groups for a Riemannian manifold having bounded curvature if one replaces Abelian groups by nilpotent ones in the formulation.

EXAMPLE 4.21. For $n \geq 3$, a discrete Abelian group Γ of isometries of the hyperbolic space H^n can be either of the following:

a cyclic group $\mathbb{Z}$ (having loxodromic or parabolic generator) or

the free Abelian group $\mathbb{Z}^k$ having rank k, $k \leq n-1$ (conjugating to a Bieberbach group, whose generators are Euclidean translations, which may be screw ones), or

the group $\mathbb{Z} \times \mathbb{Z}_m$ (conjugate to the hyperbolic translation and rotation group along a certain geodesic in H^n), or

a finite, order m, Abelian group conjugate to the subgroup of the orthogonal group $O(n)$ (for $n=3$, $\mathbb{Z}_2 \times \mathbb{Z}_2$).

It is clear that, if $\varepsilon > 0$ is sufficiently small (not exceeding $1/m$), Γ cannot be the group $G'_\varepsilon(x)$ in the latter two cases. If we reduce ε further (make ε less than the lengths of the hyperbolic translations in G), we see that only parabolic subgroups Γ of group G correspond to arbitrarily thin components $M_{(0,\varepsilon]}$ of the manifold $M = H^n/G$, $G \subset \mathrm{Isom}\, H^n$, i.e., subgroups conjugate to a Euclidean isometry group of $\mathbb{R}^{n-1}$.

REMARK 4.22. Let Γ be a discrete torsion-free Euclidean isometry group in $\mathbb{R}^{n-1}$. Then Γ acts in the half-space $\mathbb{R}^n_+$ as the isometry group in the Poincaré model of the hyperbolic space H^n. Then for each point $'x \in \mathbb{R}^{n-1}$ there exists $t_\varepsilon = t_\varepsilon('x)$ such that for each point x on the geodesic ray

$$\{('x,\, x_n) \in \mathbb{R}^{n-1} \times \mathbb{R}_+ = \mathbb{R}^n_+ \cong H^n :\ x_n \geq t_\varepsilon('x)\}$$

we have the obvious equality $\Gamma_\varepsilon(x) = \Gamma$.

This fact on the one hand shows that the statement of Bieberbach's theorem (Theorem 3.4) on the existence of a finite-index Abelian subgroup in Γ is a special case of Margulis's lemma.

On the other hand, it is useful to obtain bounds for the numbers $t_\varepsilon('x)$.

Here there are three different cases:

a) Γ consists of Euclidean translations;

b) for any $g \in \Gamma\backslash\{\mathrm{id}\}$, there exists $m \in \mathbb{Z}$ such that g^m is a Euclidean translation; and

c) there exist $g \in \Gamma$ for which no degrees g^m, $m \in \mathbb{Z}$, are Euclidean translations (g is a screw translation with irrational rotation).

In the first case (the only one for the orientable case in dimension $n = 3$), one conjugates Γ by homothety and assumes that the minimal length of translation vectors in Γ is one. Then elementary trigonometry of the hyperbolic plane H^2 (see Chapter 2 §3 or Beardon [4], §7.2) shows that for the hyperbolic distance $d(g, g(x))$ (where $g : x \longmapsto x + a$, $|a| = 1$), we have the formula

$$d(x, g(x)) = 2\,\mathrm{Arc\,sinh}(|x - g(x)|/2\sqrt{x_n \cdot g(x)_n}) = 2\,\mathrm{Arc\,sinh}(1/2x_n).$$

Then we see at once that the ε-thin part of $M_{(0,\varepsilon]}$ in manifold $M = H^n/\Gamma$ in the case a) is the quotient

$$(H^n/\Gamma)_{(0,\varepsilon]} = \{x \in H^n : x_n \geq t_\varepsilon(\Gamma)\}/\Gamma, \quad t_\varepsilon(\Gamma) = |a|/2\sinh(\varepsilon/2), \tag{4.27}$$

where $t_\varepsilon(\Gamma)$ is independent of the point $'x \in \mathbb{R}^{n-1}$, while the number $|a|$ is the minimum length for the translation vectors from Γ.

In the case b), there is a number m_Γ such that for any $g \in \Gamma$ there exists an $m \leq m_\Gamma$ for which g^m is a Euclidean translation. Then, as above, we have that for any $'x \in \mathbb{R}^{n-1}$ we certainly have

$$t_\varepsilon('x) \leq (|a| \cdot m_\Gamma)/2\sinh(\varepsilon/2) = t_\varepsilon(\Gamma),$$

where $|a|$ is the minimum length for the (screw) translation vectors from Γ.

To refine this bound, we can assume that the translation vectors from Γ form the basis of the subspace $\mathbb{R}^k \subset \mathbb{R}^{n-1}$, $1 \leq k \leq n-3$, $\mathbb{R}^{n-1} = R^k \times \mathbb{R}^{n-k-1}$ (see Theorem 3.4), and denote $'x = (x^0, x^1) \in \mathbb{R}^k \times \mathbb{R}^{n-k-1}$. Then any element $g \in \Gamma$ is written as

$$g : x \longrightarrow g('x, x_n) = g(x^0, x^1, x_n) = (x^0 + a, U(x^1), x_n),$$

where $a \in \mathbb{R}^k\backslash\{0\}$, $U \in O(n-k-1) \subset O(n-1)$.

Then we assume that $|a| = 1$ to get

$$d(x, g(x)) \leq 2\,\mathrm{Arc\,sinh}(\sqrt{1 + 4|U(x^1)|^2}/2x_n).$$

Then we get at once inequalities that refine the above:

$$|a|/(2\sinh(\varepsilon/2)) \le t_\varepsilon('x) \le \inf \begin{cases} |a|\sqrt{1+4(|x^1|/|a|)^2}/(2\sinh(\varepsilon/2)), \\ (|a|\cdot|m_\Gamma|)/(2\sinh(\varepsilon/2)) \end{cases} \qquad (4.28)$$

Inequalities (4.28) show that in the case b) the ε-thin part $M_{(0,\varepsilon]}$ of the manifold $M = H^n/\Gamma$ is obtained by the union of the parabolic quotient of a horoball i.e., of $\{x \in H^n : x_n \ge t_\varepsilon(\Gamma)\}/\Gamma$ and a compact manifold

$$\{x \in H^n : t_\varepsilon('x) \le x_n \le t_\varepsilon(\Gamma)\}/\Gamma.$$

Compactness in the latter follows from the fact that, firstly, the translation vectors in Γ form a basis in $\mathbb{R}^k$ and, secondly, that for the points $'x = (x^0, x^1) \in \mathbb{R}^k \times \mathbb{R}^{n-k-1}$, for which $|x^1|$ is large enough (namely $(|x^1|/|a|)^2 \ge (m_\Gamma^4 - 1)/4$), we have

$$t_\varepsilon('x) = t_\varepsilon(\Gamma).$$

In the case c), Γ has irrational screw Euclidean translations g:

$$g : x \longmapsto U(x) + a, \quad U \in O(n-1), \quad a \in \mathbb{R}^{n-1}\backslash\{0\},$$

$$U(a) = a, \quad U^m \ne \mathrm{id} \quad \text{for all} \quad m \in \mathbb{Z}\backslash\{0\}.$$

We denote by $\mathbb{R}^k$, $1 \le k \le \operatorname{rank}\Gamma \le n-3$, a subspace of $\mathbb{R}^{n-1}$ whose basis is formed by the translation vectors for the irrational screw translations from the group Γ, while $|a|$ denotes their minimum length. We write a point $x \in H^n$ as $x = (x^0, x^1, x_n)$, where $x^0 \in \mathbb{R}^k$, $x^1 \in \mathbb{R}^{n-k-1}$, $x_n > 0$, to get inequalities analogous to (4.28):

$$|a|/(2\sinh(\varepsilon/2)) \le t_\varepsilon('x) \le |a|\sqrt{1+4(|x^1|/|a|)^2}/(2\sinh(\varepsilon/2)). \qquad (4.29)$$

It follows from (4.27) - (4.29) that in all cases the ε-thin submanifold in H^n/Γ has the property

$$\{x = (x^0, x^1, x_n) \in H^n : x_n \ge |a|\sqrt{1+4(|x^1|/|a|)^2}/(2\sinh(\varepsilon/2))\}/\Gamma \subseteq$$

$$\subseteq V_\infty(\Gamma, \varepsilon)/\Gamma = (H^n/\Gamma)_{(0,\varepsilon]} \subset \{x \in H^n : x_n \ge |a|/(2\sinh(\varepsilon/2))\}/\Gamma.$$

Here $V_\infty(\Gamma, \varepsilon)$ denotes the parabolic ε-neighbourhood of the parabolic vertex ∞, which in the general case replaces the ε-horoball appearing on the right of this relation, i.e.,

$$V_\infty(\Gamma, \varepsilon) = \{('x, x_n) \in H^n : x_n \ge t_\varepsilon('x)\}.$$

In the case $n = 3$, which we have seen occupies an exceptional place in our descriptions of ε-thin parts of complete hyperbolic manifolds H^n/G, one can give a complete description. Here one has either the quotient space of a horoball, having its centre at the parabolic vertex p by the action of its stabilizer $G_p \subset G$ (cyclic or free Abelian group of rank two for orientable manifolds) or

the quotient space of a hyperbolic cylinder

$$D_r(l) = \{x \in H^3 : d(x, l) \le r\}$$

with the geodesic axis $l \subset H^3$ and radius $r > 0$ by the action of the cyclic loxodromic subgroup having axis l.

However, in fact $\varepsilon > 0$ with the above property can be taken as the same ε_0 for all the discrete groups $G \subset \operatorname{Isom} H^3$. We formulate this as follows.

COROLLARY 4.23 (Thurston [1]). *There exists a constant* $\varepsilon > 0$ *such that each component of* $M_{(0,\varepsilon]}$ *for any complete orientable hyperbolic 3-manifold* M *is either*:

1) *the quotient of a horoball modulo the action of groups* $\mathbb{Z}$ *or* $\mathbb{Z} \times \mathbb{Z}$, *or*

2) *the quotient of a cylinder* $D_r(l)$, $r > 0$, *modulo the action of the group* $\mathbb{Z}$, *where* l *is a geodesic.*

PROOF. Let $x \in M_{(0,\varepsilon]}$, where $\varepsilon > 0$ satisfies Theorem 4.20. let $\tilde{x} \in H^3$ be a lifting of point x in $\tilde{M} = H^3$. Then a certain deck transformation $g \in G \subset \operatorname{Isom} H^3$ moves $\tilde{x}$ through a distance not more than ε. If g is loxodromic, we denote its axis by l. All rotations around l, hyperbolic translations along l and uniform contractions of H^3 along geodesics, orthogonal to l, commute with g. This shows that $\tilde{M}_{(0,\varepsilon]}$ contains a cylinder $D_r(l)$, where $r = d(l, \tilde{x})$, since $d(y, g(y)) \le \varepsilon$ for all $y \subset D_r(l)$. If transformation g is parabolic with fixed point $p \in \partial H^3$, then $M_{(0,\varepsilon]}$ contains a horoball centred at p bounded by a horosphere passing through $\tilde{x}$. Then $\tilde{M}_{(0,\varepsilon]}$ is the union of horoballs and solid cylinders $D_r(l)$. Whenever two of these are not disjoint, they correspond to two deck transformations $g_1, g_2 \in G \subset \operatorname{Isom} H^3$, which move some point $\tilde{x}$ by a distance $\le \varepsilon$. Theorem 4.20 implies that g_1 and g_2 must commute, so the corresponding horoballs or solid cylinders must be concentric. The precise invariance of these horoballs and solid cylinders with respect to their stabilizers completes the proof.

REMARK 4.24. For a given hyperbolic manifold M of finite topological type, one can consider $\varepsilon = \varepsilon(M)$ such that the ε-thin part $M_{(0,\varepsilon]}$ will consist only of parabolic components of the first form. However, a universal constant ε having that property (i.e., a constant that strengthens the Margulis constant from Theorem 4.20 or Corollary 4.23) does not exist even for hyperbolic manifolds of finite topological type, since Bonahon and Otal [1] have constructed such manifolds having closed geodesics of arbitrarily small length.

In proving the corollary, we have made substantial use of the fact that any discrete group $G \subset \operatorname{Isom} H^3$ has precisely G_p-invariant horoballs $B_p \subset H^3$ centred at its parabolic points $p \in \partial H^3$ (see Definition 4.2 in §5), i.e., such that $\gamma(B_p) = B_p$ for all $\gamma \in G_p$ and

$g(B_p) \cap B_p = \emptyset$ for all $g \in G \setminus G_p$.

It is pointed out (Apanasov [21]) that this property in general does not occur for discrete isometry groups of the hyperbolic space having dimension greater than three, as is evident from the example of a Kleinian group $G \subset \mathcal{M}_n$, $n \geq 3$, that will be constructed in the proof of Theorem 4.25. Then for any complete hyperbolic manifold M having dimension greater than three, it is impossible to give a universal description of $M_{(0,\varepsilon]}$ as in Corollary 4.23 (for the geometrically finite case, see Corollary 5.45 and Chapter 7 §1).

THEOREM 4.25. *There exist Kleinian groups* $G \subset \mathcal{M}_n$, $n \geq 3$, *whose parabolic points* p *do not have any precisely* G_p*-invariant horoballs* $B_p \subset H^{n+1}$.

The proof of the theorem consists of constructing an infinitely generated group $G \subset \mathcal{M}_n$, $n \geq 3$, having that property.

Let (m_i) be some monotone natural number sequence:

$$\lim_{i \to \infty} m_i = \infty \tag{4.30}$$

while $U \in O(n)$ is an infinite-order rotation in $\mathbb{R}^n$ that leaves the axis $R \cdot e_1$ fixed. For $n = 3$,

$$U = \begin{bmatrix} 1 & 0 & 0 \\ 0 & \cos q & \sin q \\ 0 & -\sin q & \cos q \end{bmatrix},$$

where $0 < q < \pi$ and q is not commensurate with 2π. For simplicity, we consider in what follows that $n = 3$ (the general case is analogous, see Remark 4.26). The Kleinian group G takes the form

$$G = \langle G_0, h \rangle, \tag{4.31}$$

where h is an irrational screw translation in $\mathbb{R}^n$,

$$h : x \longmapsto U \cdot x + e_1,$$

and the group G_0 is a free group (infinitely generated):

$$G_0 = \langle g_1, g_2, \ldots, g_k, \ldots \rangle \tag{4.32}$$

with hyperbolic generators g_i having the isometric spheres

$$I(g_i) = \{x \in \mathbb{R}^n : |g_i'(x)| = 1\} = S^{n-1}(z_i, m_i) \tag{4.33}$$

with radii m_i. The parabolic points in the group form the G-orbit:

$$G(\infty) = \{g(\infty) : g \in G\},$$

so their properties are defined by those of ∞. The horoballs centred at ∞ are

$$B_\infty = H_r = \{x \in H^{n+1} : x_{n+1} > r\}$$

and for any $r > 0$ cannot be precisely G_∞-invariant ones, since they intersect with an infinite set of isometric spheres $I(g)$, $g \in G\setminus G_\infty$ (in particular for generators g_i such that $m_i > r$).

To construct the (4.32) generators for G_0 having the properties of (4.30) and (4.33), we denote the integer part of a number $a \in \mathbb{R}$ by $[a]$, where

$$a_i = -[1/2 - 2qm_i/\pi], \quad b_i = [a_i/2]$$

denote correspondingly the number of turns in π for vector e_2 on rotation U^{2m_i} and the number of complete rotations of e_2 on rotation U^{2m_i}, i.e., through the angle $2qm_i$ (see Polya-Szegö [1], §8). Here let K_i and n_i be integers such that

$$n_i \geq 2m_i, \quad q(n_i - 1) < 2\pi(b_i + 1),$$

$$k_i \geq 2m_i, \quad q(k_i - 1) < \pi(a_i + 1),$$

$$t_i^1 = qn_i - 2\pi(b_i + 1) > 0,$$

$$t_i^2 = qk_i - \pi(a_i + 1) > 0.$$

We put $t_i = \min\{t_i^1, t_i^2\}$ to get the minimum angle formed by the vectors $\{U^k(e_2), \ 0 < k \leq 2m_i\}$ with the vectors e_2 and $-e_2$. In other words, $t_i > 0$,

$$t_i = \min_{0<k\leq 2m_i} \{|\mathrm{arc\,Cos}(e_2, U^k(e_2))|, |\mathrm{arc\,Cos}(-e_2, U^k(e_2))|\}. \tag{4.34}$$

As m_i increases without bound, the vectors $U^{k_i}(e_2)$ and $U^{n_i}(e_2)$ tend correspondingly to $-e_2$ and e_2 (due to the choice of n_i and k_i), i.e.,

$$\lim_{i\to\infty} t_i = 0 \tag{4.35}$$

Now let the points $z_i = (0, l_i, 0)$, $z_i' = (0, -l_i, 0)$ be such that

$$l_i = m_i/\mathrm{Sin}(t_i/2) \tag{4.36}$$

By virtue of (4.30), (4.35) and (4.36), we can assume that the following inequalities apply for all $i \geq 1$, where we transfer to subsequences if necessary:

$$1/2 < l_i - m_i < l_i + m_i < l_{i+1} - m_{i+1}. \tag{4.37}$$

We consider the disjoint balls in $\mathbb{R}_n$:

$$B_0 = B^n(0, 1/2)$$
$$B_i = B^n(z_i, m_i), \quad B_i' = B^n(z_i', m_i), \quad i \geq 1, \tag{4.38}$$

bounded by the spheres S_0, S_i and S_i'.

Then one can define the generators of (4.32) as hyperbolic mappings g_i such that

$$I(g_i) = S_i, \quad I(g_i') = S_i',$$

$$g_i(\text{ext } S_i) = \text{int } S_i' = B_i'.$$

It follows from (4.37) that G_o from (4.32) is a free, purely hyperbolic group, and for any $g \in G_o$, $g \neq \text{id}$, there exists an i_0 such that

$$g(\bigcap_i (\text{ext } S_i \cap \text{ext } S_i') \subset B_{i_0} \cup B_{i_0}'. \tag{4.39}$$

To complete the proof it remains to prove that group (4.31) is discrete. We have proved something more, namely that G is a Kleinian group in $\overline{\mathbb{R}}^n$, i.e., its discontinuity set $\Omega(G) \neq \emptyset$. This follows from the fact that for all $g^* \in G$, $g^* \neq \text{id}$, we have

$$g^*(B_0) \cap B_0 = \emptyset, \tag{4.40}$$

i.e., $B_o \subset \Omega(G)$.

We have for integer $m \neq 0$ and for all i

$$(B_i \cup B_i') \cap h^m(B_i \cup B_i') = \emptyset, \quad B_0 \cap h^m(B_0) = \emptyset. \tag{4.41}$$

In fact, (4.41) follows for B_0 from the fact that h is a translation on the unit-length vector. For $i \geq 1$ and for all integers m, $|m| \geq 2m_i$, (4.41) follows from the fact that B_i and B_i' have diameters $2m_i$, while the translation h^m is on the vector having length $|m| \geq 2m_i$. For the other integers m, $m \neq 0$, $|m| < 2m_i$, (4.41) follows from definition (4.34) for the numbers t_i and the related (4.36) choice of the numbers l_i defining the centres z_i and z_i' of balls B_i and B_i'. In fact, this choice ensures that the sectors

$$V_i = \{t \cdot x \in \mathbb{R}^n : x \in B_i, \ t > 0\}$$

$$V_i' = \{t \cdot x \in \mathbb{R}^n : x \in B_i', \ t > 0\}$$

do not intersect the set

$$\bigcup \{U^m(V_i) : m \in \mathbb{Z}\setminus\{0\}, \quad |m| < 2m_i\}.$$

We put

$$D = \bigcup\{h^m(\bigcup_i B_i \cup B'_i \cup B_0) : m \in \mathbb{Z}\setminus\{0\}\}.$$

From (4.41) we immediately get for all i and $m \neq 0$:

$$h^m(B_i \cup B'_i) \subset D, \quad h^m(B_0) \subset D$$
$$(B_i \cup B'_i) \cap D = \emptyset. \tag{4.42}$$

At the same time, (4.39) and (4.41) imply that for any $g \in G_0$, $g \neq \mathrm{id}$, there exists an i_0 such that

$$g(B_0 \cup D) \subset B_{i_0} \cup B'_{i_0}. \tag{4.43}$$

We now write any element $g^* \in G$, $g^* \neq \mathrm{id}$, as a word having the form

$$g^* = g_1 h^{n_1} \cdot \ldots \cdot h^{n_k} g_{k+1}, \tag{4.44}$$

where $g_1, \ldots, g_{k+1}$ are elements in G_0 in general different from its generators; $g_i \neq \mathrm{id}$ for $i \neq 1$, $i \neq k+1$, with g_1 and g_{k+1} possibly identities, and $n_i \neq 0$ for all i. Then induction on the word length in (4.44) with (4.42) and (4.43) serve to prove property (4.40), which completes the proof of Theorem 4.25.

REMARK 4.26. If we do not require an analytic form for the angles t_i but take them directly from (4.34), the construction is in no way dependent on the dimension $n \geq 3$. Our above construction of discrete groups $G \subset \mathrm{Isom}\, H^n$, $n \geq 4$, also shows, as pointed out by Epstein and Penner [1], that the G-orbit of any parabolic horoball $B_p \subset H^n$ is not in general discrete for $n \geq 4$. The situation for finitely generated groups remains open. We note also that a similar approach to proof of Theorem 4.25 is contained in an after paper by Ohtake [1].

§5. Precisely invariant horoballs and parabolic Ω-peaks

1. *Action of a discrete group in horoballs.* Let $G \subset \mathrm{M\ddot{o}b}(n)$ be a discrete group having parabolic elements with the fixed point $p \in \overline{\mathbb{R}}^n$. The group G also acts as a subgroup of group $\mathrm{Isom}\, H^{n+1}$ in half-space $\mathbb{R}^{n+1}_+ \cong H^{n+1}$ having the Poincaré hyperbolic metric. As we have seen in the previous section, it is important to know the action of G in the small horoballs $B_p \subset H^{n+1}$ having their centres at parabolic points $p \in \partial H^{n+1} = \overline{\mathbb{R}}^n$ (in the Poincaré model B_p is a Euclidean $(n+1)$-dimensional ball tangent to $\overline{\mathbb{R}}^n$ at point p). We conjugate G and may assume that $p = \infty$. Then the horoballs $B_\infty \subset H^{n+1}$, having their centres at ∞, are the half-spaces

$$H_\alpha = \{x \in \mathbb{R}^{n+1} : x_{n+1} > \alpha\}, \quad \alpha > 0.$$

DEFINITION 4.27. Domain D is called precisely invariant with respect to subgroup H of discrete group G (= precisely H-invariant) if

$$\gamma(D) = D \quad \text{for} \quad \gamma \in H,$$

$$g(D) \cap D = \varnothing \quad \text{for} \quad g \in G \setminus H.$$

It is clear that the action of G in precisely H-invariant domains is completely defined by H.

THEOREM 4.28. *If a discrete group* $G \subset \text{Möb}(n)$, $n \geq 2$, *either contains a Euclidean translation or, for a certain Euclidean t-neighbourhood* $U_t(E)$ *of a subspace* $E \subset \mathbb{R}^n$ *which is invariant with respect to the parabolic stabilizer* $G_\infty \subset G$, *we have the inclusion*

$$G(\infty) = \{g(\infty) : g \in G\} \subset U_t(E), \tag{4.45}$$

then there exists $\alpha_0 > 0$ *such that the horoballs* $H_\alpha \subset H^{n+1}$ *are precisely* G_∞*-invariant for all* $\alpha \geq \alpha_0$. *If here in group* G_∞ *there is a Euclidean translation on a vector* $a \in \mathbb{R}^n$, *then* $\alpha_0 \leq |a|$.

PROOF. Let initially G_∞ contain a Euclidean translation h by a vector $a \in \mathbb{R}^n$. Instead of group G we consider the conjugate group $\psi G \psi^{-1}$, where $\psi : x \longmapsto x/|\alpha|$, which gives us that there is a translation on a unit-length vector in that group, while $|a|^2$ is the ratio of the radii for the isometric spheres of transformations $g \in G \setminus G_\infty$ and $\psi g \psi^{-1}$. It is then sufficient for $|a| = 1$ to show that the radii of the isometric spheres $I(g)$ for $g \in G \setminus G_\infty$ do not exceed one.

We assume that there exists an element g in $G \setminus G_\infty$ such that the radius r of $I(g)$ is greater than one. We conjugate the group by a Euclidean translation and assume that $I(g^{-1}) = S^{n-1}(0, r)$. We consider a sequence of elements from $G \setminus G_\infty$

$$g_0 = g; \quad g_i = g_{i-1} h g_{i-1}^{-1}, \quad i \geq 1. \tag{4.46}$$

By q_i and q_i' we denote the centres of the isometric spheres $I(g_i)$ and $I(g_i^{-1})$ having radii r_i, while by U_i we denote orthogonal transformations, $\mathcal{J}(x) = x/|x|^2$. We have the following representations (Corollary 1.8):

$$g_i(x) = r_i^2 U_i \mathcal{J}(x - q_i) + q_i'. \tag{4.47}$$

From (4.46) and (4.47) we get a representation for g_1 in terms of $r = r_0$, $U = U_0$ and $O_a(x)$ (the latter being the reflection with respect to the subspace $P(a, 0) = \{x \in \mathbb{R}^n : (x, a) = 0\}$):

$$\begin{aligned} g_1(x) &= g h g^{-1}(x) = r^2 U \mathcal{J}[r^2 U^{-1} \mathcal{J}(x) + a] = \\ &= r^2 U(r^2 O_a \mathcal{J}(U^{-1}(x) + a r^2) + a) = r^4 U O_a U^{-1} \mathcal{J}(x + r^2 U(a)) + r^2 U(a). \end{aligned} \tag{4.48}$$

Then

$$q_1 = -r^2U(a), \quad q'_1 = r^2U(a), \quad r_1 = r^2, \quad U_1 = UO_aU^{-1}. \tag{4.49}$$

By analogy with (4.48), we get for $i \geq 2$ that

$$r_i^2U_i\mathcal{J}(x-q_i) + q'_i = g_{i-1}hg_{i-1}^{-1}(x) =$$

$$= r_{i-1}^4U_{i-1}O_aU_{i-1}^{-1}\mathcal{J}(x - q'_{i-1} + r_{i-1}^2U_{i-1}(a)) + q'_{i-1} + r_{i-1}^2U_{i-1}(a) = \dots$$

$$= r^{2^{i+1}}U_i\mathcal{J}[x - \sum_{j=0}^{i-2} r^{2^{j+1}}\cdot U_j(a) + r^{2^i}U_{i-1}(a)] + \sum_{j=0}^{i-1} r^{2^{j+1}}\cdot U_j(a). \tag{4.50}$$

Then formulae analogous to (4.49) follow (compare Theorem 4.30):

$$q_i = \sum_{j=0}^{i-2} r^{2^{j+1}}U_j(a) - r^{2^i}U_{i-1}(a), \quad r_i = r^{2^i},$$
$$q'_i = \sum_{j=0}^{i-1} r^{2^{j+1}}U_j(a), \quad U_i = \left[\prod_{j=0}^{2^i-1} (U^{(-1)^j}O_a)\right]\cdot O_a. \tag{4.51}$$

Here U_i is the reflection with respect to a subspace $P(U_{i-1}(a), 0) \subset \mathbb{R}^n$. If now the transformation $U \in O(n)$ is such that $UO_aU^{-1}O_a$ has finite order $m \in \mathbb{Z}$, then we shall consider numbers i that are multiples of m and such that $U_i = O_a$. If on the other hand $UO_aU^{-1}O_a$ has infinite order, as the orthogonal group $O(n)$ is compact, we can choose the i such that the U_i become ε-close to the reflection O_a. This follows from (4.51), so we can assume that

$$\lim_{i\to\infty} U_i = O_a. \tag{4.52}$$

We now consider the matrices $T_m = (t_{ij}^m) \in SO(1, n+1)$, corresponding to the (4.46) elements in the isomorphism $\varphi^*: \text{Möb}(n) \longrightarrow SO(1, n+1)/\pm 1$, induced by the isometry $\varphi: H^{n+1} \longrightarrow {}'H^{n+1}$ of hyperbolic space models, see Chapter 2 §4. The Lorentz transformations $T_m\cdot x$ transform the vector $e_0 \in {}'H^{n+1}$ into the vector $T_m\cdot e = (t_{00}^m, t_{10}^m, \dots, t_{n+1,0}^m)$, whose coordinates satisfy

$$(t_{00}^m)^2 - \sum_{i=1}^{n+1} (t_{i0}^m)^2 = 1. \tag{4.53}$$

The isometry φ is such that $\varphi(e_{n+1}) = e_0$. Therefore, taking into account the bound $|t_{i0}^m| \leq |t_{00}^m|$, $1 \leq i \leq n+1$, following from (4.53), from the uniform boundedness of t_{00}^m we obtain uniform boundedness for the hyperbolic distances:

$$d(e_{n+1}, g_m(e_{n+1})) \leq C < \infty. \tag{4.54}$$

To bound the term t_{00}^m in matrix T_m we consider its dependence on r and U governed by (2.28) and (4.51). As $r > 1$, for $m \longrightarrow \infty$

we get the following equivalences:

$$\begin{aligned} &(1+|q'_m|^2)(1+|q_m|^2) \approx r_m^4 - 2r_m^3 = r^{2^{m+2}} - 2r^{3\cdot 2^m} \\ &2r_m^2(q_m, U_m(q'_m)) \approx 2r_m^2|q_m|\cdot|q'_m| \approx 2r_m^4 - 2r_m^3 = 2r^{2^{m+2}} - 2r^{3\cdot 2^m}, \end{aligned} \tag{4.55}$$

where $(*, *)$ is the scalar product in $\mathbb{R}^n$, while $\approx$ denotes equality within quantities of orders not exceeding the order of $r_{m-1}^4 = r^{2^{m+1}}$, where $r > 1$.

From (2.28) and (4.55) we get uniform boundedness for term t_{00}^m in $T_m \in SO(1, n+1)$, and thus the (4.54) bound. By virtue of our assumption $(r > 1)$, we have $r_{m+1} > r_m$, so all the elements in (4.46) are pairwise distinct. Therefore, boundedness in the orbit $\{g_m(e_{n+1})\}$ contradicts the discreteness of group G. We have thus shown that, if there is a Euclidean translation by a vector $a \in \mathbb{R}^n$ in G_∞, the radii of the spheres $I(g)$ for $g \in G\backslash G_\infty$ do not exceed $|a|$.

Now let G_∞ contain only (irrational) screw translations $h(x) = V(x) + a$, $V \in O(n)$, $a \in \mathbb{R}^n\backslash\{0\}$, $V(a) = a$. We show that here the radii of the isometric spheres $I(g)$, $g \in G\backslash G_\infty$, are bounded by a certain constant dependent only on G.

We assume the converse. Then from (4.45) we can assume that the norm of the orthogonal transformation V is less than ε for small $\varepsilon > 0$ and that there exists an element $g \in G - G_\infty$ such that

$$g(x) = r^2 U\mathcal{J}(x-q) + q',$$

where $\mathcal{J}(x) = x/|x|^2$; $q, q' \in \mathbb{R}^n$, $U \in O(n)$, $r > 0$, satisfy the conditions:

1) the centres q and q' of the isometric spheres $I(g)$ and $I(g^{-1})$, having radius r, are at Euclidean distance less than $\varepsilon/2$ from the subspace $E \subset \mathbb{R}^n$ of translation vectors from G_∞; and
2) the radius r is greater than $1 + \varepsilon$.

In fact, to meet these conditions it is sufficient to replace h by a sufficiently high degree of it h^{m_ε} and to conjugate G by means of the homothety $x \longmapsto x/(m_\varepsilon \cdot |a|)$.

Then, without changing the symbols, we assume that these conditions are met and consider $\{g_i\}$ in G, where

$$g_0 = g, \quad g_{i+1} = g_i h g_i^{-1}, \quad i \ge 0. \tag{4.56}$$

A direct calculation analogous to (4.48) - (4.51) shows that for the elements g_i,

$$g_i(x) = r_i^2 U_i \mathcal{J}(x - q_i) + q'_i$$

we have

$$\begin{aligned} q_i &= q'_{i-1} - r^2_{i-1}U_{i-1}V^{-1}\mathcal{J}(a_i), \\ q'_i &= q'_{i-1} + r^2_{i-1}U_{i-1}\mathcal{J}(a_i), \\ U_i &= U_{i-1}O_{a_i}VU^{-1}_{i-1}, \\ r_i &= r^2_{i-1}/|a_i|. \end{aligned} \tag{4.57}$$

Here O_y denotes the reflection with respect to the subspace $P(y, 0) \subset \mathbb{R}^n$, orthogonal to a vector

$$y \in \mathbb{R}; \quad U_i \in O(n), \quad a_i = V(q_{i-1}) - q_{i-1} + a.$$

Our assumptions then imply that $r_0 = r > 1 + \varepsilon > |a_i| > 1$, and so from (4.57) we get that the radii $r_i > r^{(3/2)^i}$ tend to infinity as i increases, while the U_i for sufficiently large i are close to the transformation O_aV (the corresponding i-th reflection is defined by a vector close to q'_i). These facts and (4.57) demonstrate the following equivalence (compare (4.55)):

$$\begin{aligned} (1 + |q'_i|^2)(1 + |q_i|^2) &\approx r_i^4 - 2(r_i \cdot r_{i-1})^2 \\ 2r_i^2(U_i(q'_i), q_i) \approx 2r_i^2|q'_i||q_i| &\approx 2r_i^4 - 2(r_i \cdot r_{i-1})^2: \end{aligned} \tag{4.58}$$

where $(*, *)$ denotes the scalar product in $\mathbb{R}^n$ and $\approx$ denotes equality within quantities of orders not exceeding the order of r^4_{i-1} ($r^4_{i-1} > r^{(3/2)^i}$ tends to infinity as i increases).

By analogy with the previous, we apply the (4.58) asymptote to the formulae for the term t^i_{00}, see (2.28), and again have uniform boundedness for term t^i_{00} of $T_i \in SO(1, n+1)$, corresponding to the element g_i in (4.56), and consequently we get the bound of (4.54) for that sequence.

By virtue of our assumption, we have $r > 1$, so $r_{i+1} > r_i$ for all i, which means that the $\{g_i\}$ elements are pairwise distinct, and that with the boundedness of the orbit $\{g_i(e_{n+1})\}$ proves that G is not discrete. That contradiction completes the proof of boundedness for the radii of the isometric spheres $I(g)$, $g \in G\backslash G_\infty$ in the screw-translation case.

This radius boundedness for $I(g)$, $g \in G\backslash G_\infty$ implies precisely G_∞-invariance for any horoball H_α not intersecting the isometric spheres, since

$$g(\operatorname{ext} I(g)) = \operatorname{int} I(g^{-1}), \quad g \in G\backslash G_\infty,$$

implies

$$g(H_\alpha) \cap H_\alpha = \varnothing, \quad g \in G\backslash G_\infty.$$

Theorem 4.23 is completely proved.

REMARK 4.29. The statement in the theorem for $n=2$ is well known (see, for example, Lehner [1] or Shimizu [1], Lemma 4) and has a simple proof (here G_∞ always consists of Euclidean translations). For $n \geq 3$, the first proof for this statement was given by Apanasov [3] for the case where G_∞ contains Euclidean translations (see also Apanasov [13]). Apanasov [14], [19] has studied the case where G_∞ consists of screw translations. The resulting condition (4.45) is relevant. This is demonstrated by the examples of the groups $G \subset \mathcal{M}_n$, $n \geq 3$, constructed in the proof of Theorem 4.25.

We note that, as Theorem 4.28 shows, the condition (4.45) is equivalent that ∞ is the parabolic vertex with rank $G_\infty < n$.

G.A. Margulis pointed out to me that this proof gives a stronger theorem, which I formulate for Euclidean translations:

THEOREM 4.30. *Let group* $G \subset \text{Möb}(n)$, $n \geq 2$, *contain the Euclidean translation* h *by a vector* $a \in \mathbb{R}^n$. *If here* $G \setminus G_\infty$ *has an element* g *for which the radius of the isometric sphere* $I(g)$ *is greater than* $|a|$, *then sequence* $\{g_i\} \subset G$ *from* (4.46) *converges to* h.

PROOF. We will not continue our technical expressions but instead give a proof for this due to G.A. Margulis. We transform (4.46) into the commutator sequence $\tilde{g}_i$:

$$\tilde{g}_1 = ghg^{-1}h^{-1}, \ \ldots, \ \tilde{g}_{i+1} = \tilde{g}_i h \tilde{g}_i^{-1} h^{-1}.$$

It is clear that $\tilde{g}_i = g_i h^{-1}$, so the following equations are equivalent:

$$\lim g_i = h \iff \lim \tilde{g}_i = \text{id}. \tag{4.59}$$

By H we denote the parallel translation group, and let $h_1, h_2 \in H$. We put $f = h_1 g h_2$, where f and g have isometric spheres with equal radius and consider the sequences

$$f_0 = f, \quad f_1 = fhf^{-1}, \ \ldots, \ f_{i+1} = f_i h f_i^{-1}, \ \ldots$$

$$\tilde{f}_1 = fhf^{-1}h^{-1}, \ \ldots, \ \tilde{f}_{i+1} = \tilde{f}_i h \tilde{f}_i^{-1} h^{-1}, \ \ldots$$

Then for $i \geq 1$ we have $f_i = h_1 g_i h_1^{-1}$ and $\tilde{f}_i = h_1 \tilde{g}_i h_1^{-1}$, so to prove (4.59) it is sufficient to show that

$$\lim_{i\to\infty} \tilde{f}_i = \text{id}.$$

For transformations γ sufficiently close to the identity, the commutator sequence $\gamma_1 = [\gamma, h] = \gamma h \gamma^{-1} h^{-1}, \ \ldots, \ \gamma_{i+1} = [\gamma_i, h], \ \ldots$ converges to the identity. This assertion follows from the following facts:

1. for any two elements c and d sufficiently close to the identity, the sequence $c_1 = [c, d], \ldots c_{i+1} = [c_i, d]$ converges to an identity; and

2. element h can be made sufficiently close to the identity because the translation h is unipotent. In fact we replace h by $\delta h\delta^{-1}$ for a suitable element δ and the element γ by $\delta\gamma\delta^{-1}$.

We now assume that the radius of the isometric sphere $I(g)$ for $g\in G\backslash G_\infty$ is sufficiently large. Then, due to the above assertion, we have $h_1, h_2\in H$, together with the Euclidean motion w and the Möbius transformation $v\in$ Möb(n) sufficiently close to the identity such that $h_1gh_2 = vw$. As above, we replace g by h_1gh_2 and get that $\tilde{g}_2$ is sufficiently close to the identity, i.e., (4.59) applies.

To complete the proof, it remains to use $r_{i+1} = r_i^2$ (see (4.51)), which provides for the existence of an element $g\in G$ having a sufficiently large radius for the isometric sphere $I(g)$.

2. *Parabolic* Ω*-peaks and horoball parametrization.* We define the action of the Möbius group on the space of linear elements in $\overline{\mathbb{R}}^n$, i.e., on the unit tangent bundle $T_1(\overline{\mathbb{R}}^n)$ of the sphere $\overline{\mathbb{R}}^n$, which coincide for odd n with the product $\overline{\mathbb{R}}^n\times S^{n-1}(0,1)$ in the following way (Apanasov [8]). For the pair $c=(p,\zeta)\in\overline{\mathbb{R}}^n\times S^{n-1}$, we consider a transformation $w_c\in\mathcal{M}_n$ that transfers the point $p\in\overline{\mathbb{R}}^n$ to ∞, while the direction $\xi\in S^{n-1}$ at p is transferred to the direction of the n-coordinate vector $e_n\in S^{n-1}$. Now let $(*,*)$ denote the scalar product in $\mathbb{R}^n$. We define for any pair $c=(p,\xi)\in\overline{\mathbb{R}}^n\times S^{n-1}$ the family $\mathcal{B}(c)$ consisting of the balls

$$B_r = \{x\in\overline{\mathbb{R}}^n\backslash\{p\} : (w_c(x), e_n) > r\}.$$

It is evident that the distinct pairs $c, c_1\neq c$ correspond to distinct families $\mathcal{B}(c)\neq\mathcal{B}(c_1)$ and vice versa.

Now let $g\in$ Möb(n). For each pair $c=(p,\zeta)\in\overline{\mathbb{R}}^n\times S^{n-1}$ there is a uniquely defined family of tangent balls $B_r^* = g(B_r)$, which defines the pair $c^*=(p^*,\zeta^*)\in\overline{\mathbb{R}}^n\times S^{n-1}$ uniquely. Therefore, we can define the action of g on $T_1(\overline{\mathbb{R}}^n)$ by putting $g(c)=c^*$.

DEFINITION 4.31. Let $G\in$ Möb(n) be a Kleinian group, with $\Omega\subset\Omega(G)$ an open set invariant with respect to G (possibly $\Omega=\Omega(G)$) and $p\in\partial\Omega$ be a fixed parabolic point for G. We call the pair $c=(p,\zeta)\in\overline{\mathbb{R}}^n\times S^{n-1}$ a parabolic Ω-peak for G if it is fixed for the stabilizer $G_p\subset G$ and if there exists a ball B_r from the family $\mathcal{B}(c)$ contained in the domain Ω and precisely invariant with respect to G_p.

The definition shows that the Fuchsian group $G\in$ Möb(n) that leaves a ball $B=B^n(a,1)$ invariant has each fixed parabolic point satisfying the condition of Theorem 4.23 (i.e., a parabolic vertex) coresponding to two parabolic Ω-peaks: a B-peak $(p, p-a)$ and a $(\overline{\mathbb{R}}^n\backslash\overline{B})$-peak $(p, a-p)$. One can show that a similar property occurs in any group derived from Fuchsian quasi-conformal deformations. However, a group at the boundary of the Fuchsian group deformation space (see Chapter 9 §4) can have parabolic points not corresponding to any parabolic Ω-peaks in $\overline{\mathbb{R}}^n$ but having Ω-peaks in $\overline{\mathbb{R}}^{n+1}$. That property occurs in a fixed point for so-called accidental (screw) parabolic elements in the boundary group, i.e., elements corresponding to loxodromic elements in the Fuchsian group for group isomorphism.

Now let $c = (p, \zeta)$ be a certain parabolic Ω-peak for a Kleinian group G in $\overline{\mathbb{R}}^n$, while $B \in \mathcal{B}(c)$ is the maximal ball contained in $\Omega(G)$. We consider a Möbius transformation w such that

$$w(p) = \infty, \quad w(B) = \mathbb{R}^n_+, \tag{4.60}$$

and that the maximal precisely G_p-invariant ball from $\mathcal{B}(c)$ transfers into the half-space $H^n_1 = \{x \in \mathbb{R}^n_+ : x_n > 1\}$. That w is defined uniquely within Euclidean isometries of half-space $H^n = \mathbb{R}^n_+$

DEFINITION 4.32. The function

$$\rho_c(x) = \rho_{c,G}(x) = \exp\{-(w(x), e_n)\}$$

is called a parametrizing one.

It is evident that this definition is correct and that for any $\gamma \in \text{Möb}(n)$ we have

$$\rho_{c,G}(x) = \rho_{\gamma(c),\gamma G \gamma^{-1}}(\gamma(x)). \tag{4.61}$$

Also $\rho_c(x)$ parametrizes balls in the $\mathcal{B}(c)$ family for the Ω-peak $c = (p, \zeta)$, since that family consists of the balls

$$B_r(c) = \{x \in \overline{\mathbb{R}}^n \setminus \{p\} : \rho_c(x) < r\}. \tag{4.62}$$

We call the ball $B_r(c)$ a horoball, having radius r and centred at the Ω-peak c. The maximal ball from $\mathcal{B}(c)$ lying in $\Omega(G)$ corresponds to the horoball having radius $r = 1$. Also, the maximal precisely G_c-invariant ball has radius $r = 1/e$.

REMARK 4.33. The property (4.61) of $\rho_c(x)$ shows that elements of G transpose horoballs of equal radii with centres at parabolic Ω-peaks belonging to a G-orbit. Moreover, for Kleinian group $G \in \text{Möb}(n)$, $n \geq 2$, with parabolic Ω-peaks satisfy the Theorem 4.28 conditions, there exists a constant $K = K(G)$ such that horoballs $B_k(c)$ and $B_k(c')$ do not intersect for any Ω-peaks $c = (p, \zeta)$ and $c' = (p', \zeta')$ with $p \neq p'$. This is evident for the case of finitely many distinct G-orbits of Ω-peaks.

§6. Crystallographic reflection groups and arithmetic

Let P be a convex polyhedron bounded by the hyperplanes H_i, $i \in I$, in an n-dimensional Euclidean, elliptic, or hyperbolic space X^n. Corollary 3.19 gives the necessary and sufficient conditions for $G \subset \text{Isom}\, X^n$, generated by the reflections g_i with respect to the planes H_i, to be a discrete group, where P is its fundamental polyhedron. These consist in that P is a Coxeter polyhedron, i.e., for any $i, j \in I$, $i \neq j$, the hyperplanes H_i and H_j either do not intersect or else form a dihedral angle containing P equal to π/n_{ij}, where $n_{ij} \in \mathbb{Z}$, $n_{ij} \geq 2$. Here the relations

$$g_i^2 = \text{id}, \quad (g_i g_j)^{n_{ij}} = \text{id}$$

are the defining relations for G, which we call a *reflection group*.

Weyl [1], Cartan [1], [2], and Coxeter [1], [2] have studied discrete reflection groups in Euclidean space and thus in the sphere; in particular, a complete classification has been obtained in terms of symmetric Gram matrices. In the hyperbolic space H^n, $n \geq 3$, (the case $n = 2$ was described by Poincaré [1]), such a description of co-finite reflection groups, i.e., groups generated by reflections in sides in a finite-volume convex polyhedron P, has been given by Vinberg [1], [3], [4] (we subsequently call them *crystallographic reflection groups*). His results are formulated in terms of Gram matrices, and the proofs are based on the Perron-Frobenius theorem on matrices containing non-negative elements. We formulate an existence theorem here.

We consider V_n, a model for the n-dimensional hyperbolic space, where the points in it are rays in $\mathbb{R}^{n+1}$ lying in the cone $[x, x] = -x_0^2 + x_1^2 + \ldots + x_n^2 < 0$, see Chapter 2 §3. Let $P \subset V_n$ be a convex polyhedron, while the vectors $e_i \in \mathbb{R}^{n+1}$ are perpendicular to the i-th hyperplanes H_i containing n-dimensional i-th face of polyhedron P (in the metric defined by the bilinear form $[x, y]$, see (2.7)). The e_i are directed away from P and are normalized by $[e_i, e_i] = 1$. There is no linear relationship between the $\{e_i\}$ containing non-negative coefficients. If now the hyperplanes H_i and H_j for two faces of P form an angle α_{ij}, we put

$$a_{ij} = [e_i, e_j] = -\cos \alpha_{ij};$$

and if they do not intersect, then

$$a_{ij} = [e_i, e_j] \leq -1.$$

Matrix $A = (a_{ij})$ is called the *Gram matrix* for hyperbolic polyhedron P and has the following properties:

1) the rank of A is $n + 1$;

2) $a_{ii} = 1$, $a_{ij} \leq 0$ $(i \neq j)$;

3) A is indecomposable, i.e., it is not a direct sum of two matrices.

THEOREM 4.34 (Vinberg [1]). *If the symmetric matrix $A = (a_{ij})$ is to be a Gram matrix for a certain hyperbolic polyhedron, it is necessary and sufficient for it to satisfy conditions* 1) - 3) *and the following conditions*:

4) *A has at least one principal submatrix that is either a positive matrix having order n (C^+-matrix) and rank $n - 1$ or is a non-negative matrix having order $n + 1$ (C^0-matrix) and rank $n - 1$*

(*such submatrices are called nodal ones*);

5) *for any nodal submatrix* B_1 *and any* C^+*-submatrix* B *in it, having rank* $n-1$*, there exists a nodal submatrix* B_2 *containing* B *and different from* B_1.

Symmetrical matrices satisfying conditions 1) - 5) are called C^--matrices.

This theorem in particular implies a complete description for the groups generated by reflections in faces of a hyperbolic simplex. For that purpose it is only necessary to enumerate the corresponding C^--matrices for which the non-diagonal elements have the form

$$a_{ij} = -\cos \pi/m_{ij} \quad \text{or} \quad a_{ij} < -1.$$

In that description, it is convenient to use Coxeter diagrams, which are constructed as follows. We take vertices to a number equal to the matrix order; vertices v_i and v_j are joined by $(m_{ij}-2)$-fold line or by a simple line having index m_{ij}, where $a_{ij} = -\cos \pi/m_{ij}$; if $a_{ij} < -1$, the vertices are joined by a dashed line, with statement of the number $-a_{ij}$.

Then for the necessary description, one merely needs to sketch any connected diagram having the following properties:

1) the number of vertices is $n+1$;

2) the diagram is not contained in the diagram tables for the Coxeter C^+-matrices and the C^0-matrices for the Euclidean reflection groups (see Coxeter and Moser [1]); and

3) any subdiagram composed of n vertices either is the union of some diagrams from the Coxeter C^+-matrix tables or is contained in Coxeter C^0-matrix table.

Subdiagrams of the first type in condition 3) correspond to the ordinary vertices of simplex P, while subdiagrams of the second type correspond to vertices at infinity.

The groups for which simplex P is bounded were first described from a combinatorial point of view by Lanner [1]; they exist for $n \leq 4$. There are nine such for $n=3$, and only five for $n=4$. Figure 19 shows diagrams for them. It has been shown by computer methods that the corresponding groups for an unbounded simplex exist only for $n \leq 9$.

Andreev [1], [2] has derived the crystallographic reflection group classification for three-dimensional hyperbolic space (see Kaplinskaya [1] for its generalization in the case of simplicial prisms). This classification amounts to an exhaustive description of the Coxeter polyhedra P contained in H^3, whose most important property from the geometrical viewpoint is that they are acute-angled: none of the dihedral angles in the polyhedron exceeds $\pi/2$.

An acute-angled spherical polyhedron not containing opposing points is a simplex (Cartan [2] and Coxeter [1]), so the polyhedral angles at the vertices in a bounded acute-angled polyhedron in H^n or $\mathbb{R}^n$ are simplicial. A convex polyhedron having that property is called *simple*, and the combinatorial type for it is called the simple combinatorial one.

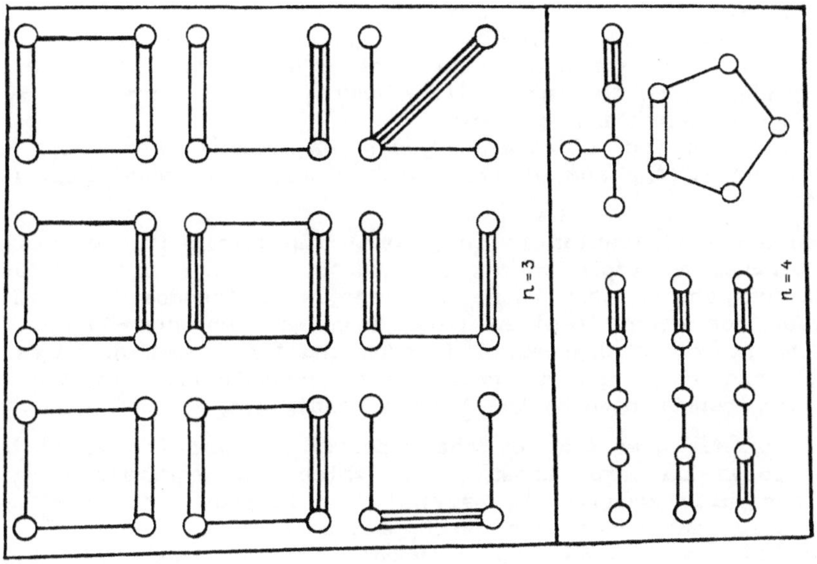

Figure 19

THEOREM 4.35 (Andreev [1]). *Given a simple combinatorial type of the polyhedron* P *different from a simplex or a triangular prism and given dihedral angles not exceeding* $\pi/2$, *the three-dimensional hyperbolic space* H^3 *contains a bounded acute-angled polyhedron* P *of prescribed type if and only if the following conditions are satisfied*:

1) *if some three sides of* P *meet at one vertex, the sum of the dihedral angles between them is greater than* π;

2) *if some three sides of* P *are pairwise adjacent but do not meet at one vertex, the sum of the dihedral angles between them is less than* π;

3) *if some four sides of* P *are adjacent on a circle (as the side faces of a four-sided prism), the sum of the dihedral angles between them is less than* 2π.

An unbounded acute-angled polyhedron having finite volume in H^3 may have four sides meeting at certain vertex in ∂H^3. The following criterion applies for such a polyhedron.

THEOREM 4.36 (Andreev [2]). *If an acute-angled polyhedron* P *having, first, a finite volume; second, a given combinatorial type different from a simplex and triangular prism; and third, given dihedral angles, is to exist in* H^3, *it is necessary and sufficient to meet the following conditions*:

1) *if three sides meet at some vertex in* P, *the sum of the dihedral angles between them is not less than* π;

2) *if four sides meet at some vertex in* P, *all the dihedral*

angles between them are $\pi/2$;

3) *if some three sides are pairwise adjacent but do not meet at one vertex, the sum of the dihedral angles between them is less than* π;

4) *if some side* $S_1 \subset \partial P$ *is adjacent to the sides* S_2 *and* $S_3 \subset \partial P$, *while* S_2 *and* S_3 *are not adjacent but have the common vertex* $p \notin S_1$, *then the sum of the dihedral angles formed by* S_1 *with* S_2 *and* S_3 *is less than* π; *and*

5) *if some four sides are adjacent on a circle but do not meet at a single vertex, the sum of the dihedral angles between them is less than* 2π.

Such a classification is in general impossible in the space H^n, $n \geq 4$, because a simple convex polyhedron in H^n, $n \geq 4$, can have more dihedral angles than there are degrees of freedom (the number of the latter for each m-sided polyhedron is $mn - n(n+1)/2$). For $n = 3$, the number of degrees of freedom and the number of edges in P coincide, and this is the reason for compatibility in the linear inequalities represented by the Andreev conditions.

The maximum dimension of the hyperbolic space for which bounded Coxeter polyhedra are known (or, which is equivalent, compact crystallographic groups) is seven. Makarov [4] has derived geometrically some examples of such polyhedra in H^4 and H^5, while Bugaenko [1] has derived such polyhedra in H^6 and H^7 by their Coxeter diagrams. They correspond to linear transformation groups over the ring of integers in the field $\mathbb{Q}(\sqrt{5})$ that preserve quadratic forms (see §2):

$$-\frac{\sqrt{5}+1}{2}x_0^2 + x_1^2 + \ldots + x_n^2; \quad n = 6, 7.$$

Figure 20 shows the Coxeter diagrams for these polyhedra.

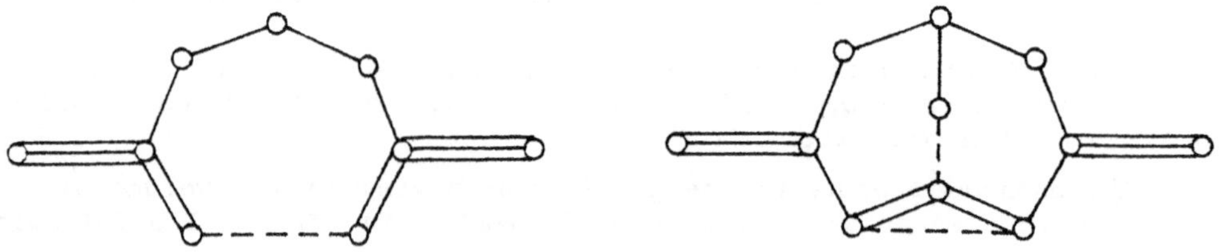

Figure 20

The maximum dimension of the hyperbolic space H^n for which any crystallographic reflection groups are known, i.e., for which Coxeter polyhedra have been constructed (unbounded but of finite volume) is nineteen. Geometrical constructions for these examples for $n \leq 10$ are to be found in papers by Makarov [1] - [7] and for $n \leq 19$ Coxeter diagrams for such groups are to be found in papers by Vinberg [1], [2],

[4], [6] and by Vinberg and Kaplinskaya [1], while for $n=18$ see Meyer [1].

The crystallographic reflection groups in H^n and the commensurable ones constitute a very particular case of co-compact and co-finite discrete groups $G \subset \operatorname{Isom} H^n$ (P with bounded volume). For example, the arithmetic discrete groups in H^{2n+1} related to non-commutative algebraic extensions over the field $\mathbb{Q}$ are never commensurable with the crystallographic reflection groups (Vinberg [1]). Also, for any n, there are infinitely many incommensurate arithmetic discrete co compact groups in H^n, see, for example, the construction in §2, while recent results for the arithmetic case, due to Nikulin [3], [4], and for the general case to Vinberg [7], [8], show that there are no discrete co-compact reflection groups at all for sufficiently large $n \geq 30$.

We give some definitions in order to formulate these results and ones on arithmetic crystallographic reflection groups in H^n.

Let K be a totally real algebraic number field, with A its ring of integers. The quadratic form

$$f(x) = \sum_{i,j=0}^{n} a_{ij}x_ix_j \qquad (a_{ij} = a_{ij} \in A)$$

with signature $(n, 1)$ is called *admissible* if the following quadratic form is positive-defined for any non-identical embedding $\sigma : K \longrightarrow \mathbb{R}$:

$$f^{\sigma}(x) = \sum_{i,j=0}^{n} a^{\sigma}_{ij}x_ix_j.$$

Mostow and Tamagawa [1] have shown (for the case $K=\mathbb{Q}$ for the first time see Venkov [1], see also the proof of Theorem 4.10) that here the group $O'(f, A)$ of integer linear transformations that preserve the form f and each connected component of $\{x \in \mathbb{R}^{n+1} : f(x) = 1\}$ is a discrete co-finite subgroup in $\operatorname{Isom} H^n$. Here $O'(f, A)$ is also co-compact if, additionally, $K \neq \mathbb{Q}$ or $K=\mathbb{Q}$ but f does not represent zero in $\mathbb{Q}$ (Venkov [1]).

By $O_r(f, A)$ we denote the subgroup of $O'(f, A)$, generated by all reflections there; if this subgroup has a finite index, form f is called *reflective*.

The crystallographic reflection group $G \subset \operatorname{Isom} H^n$ is called *arithmetic* if it is a finite-index subgroup in a group of form $O'(f, A)$, where f is an admissible quadratic form (automatically reflective) over real field K, the base field for group G. This definition agrees (Vinberg [1]) with the general definition of arithmetic discrete subgroups in semisimple Lie groups (see for example Ragunatan [1]); Vinberg [1] has given a criterion for crystallographic reflection groups in H^n to be arithmetic in terms of Gram matrices. The first examples for non-arithmetic crystallographic reflection groups in H^3 (it disproves a Selberg conjecture) were given by V.S. Makarov [1], [2].

Numerous finiteness theorems have been proved recently for the crystallographic and arithmetic subgroups of $\operatorname{Isom} H^n$, mainly by Nikulin and Vinberg.

The first such result is a theorem (Nikulin [1], [2]) on the finite number of hyperbolic reflective lattices over $\mathbb{Z}$ having rank ≥ 5 (for fixed dimension of H^n). Nikulin extended that theorem to any hyperbolic lattice with any base field and then proved in general that the number of reflective forms is finite within proportionality and integer equivalence for any fixed n and any given degree for the base field K. A corollary from this is the finiteness (within commensurability) of a number of arithmetic discrete reflection groups for a given n and degree of the base field.

Moreover, Nikulin [1], [2] proved that there are no hyperbolic lattices S over $\mathbb{Z}$ of rank $S \geq 20$ for which the automorphism group of S is generated by 2-reflections to within finite index. The following Nikulin finiteness theorem is a sharpening and generalization of this result.

THEOREM 4.37 (Nikulin [4]). *If the degree N of base field K is sufficiently large ($N \geq N_0$, where N_0 is some calculated constant), then there are no arithmetical groups generated by reflections in the hyperbolic space H^n of dimension $n \geq 10$, respectively, there are no hyperbolic lattices S, rank $S \geq 11$, which the automorphism group is generated by reflections modulo finite index.*

On proving this result it is of great importance (along with some interesting arithmetical properties of Gram matrix elements which are close to the properties of Pissot-Vijayaraghavan numbers - see, for example Cassels [1, Ch.8]) some recent results on combinatorics for convex polyhedra having simplicial sides (Stanley [1] and Danilov [1]), which by virtue of the duality can be reformulated for simple polyhedra. That is, Nikulin [4] used these results to obtain an upper bound for the mean number of l-dimensional sides in a k-dimensional side of a simple polyhedron P for $l < k \leq \frac{n+1}{2}$:

THEOREM 4.38. *Let P be the convex closed polyhedron in n-space, $n \geq 2k-1$, $k \geq 2$, with simplicial angles. Then the mean number of vertices of its k-dimensional sides does not exceed*

$$2C_n^k/(C_{[n/2]}^k + C_{n-[n/2]}^k).$$

This bound implies that the mean complexity for a side of a given dimension k in a simple polyhedron $P \subset H^n$ tends to that for the k-dimensional cube as n increases, and this observation for three-dimensional sides plays a key part in proving the above Nikulin's finiteness theorems.

We now give the following particular case of Nikulin's bound.

THEOREM 4.39. *The mean number of vertices (sides) of a two-dimensional side for a simple convex polyhedron $P \subset H^n$ is less than*

$$\begin{cases} 4(n-1)/(n-2) & \textit{for even } n, \\ 4n/(n-1) & \textit{for odd } n. \end{cases}$$

This, in other words, means that a simple polyhedron of high dimensions should have a great number (in a certain sense) of quadrangular and triangular two-dimensional sides.

Vinberg [7], [8] used Theorem 4.39 to obtain the following finiteness theorems (cf. Theorem 4.37) that, with Nikulin's results, lead one to hope for a classification of all reflective forms.

THEOREM 4.40. *There are no reflective forms in the hyperbolic space H^n having dimensions $n \geq 30$. In other words, in H^n there are no co-compact or arithmetic reflection groups for $n \geq 30$.*

THEOREM 4.41. *There are no arithmetic co-compact reflection groups in H^n for $n \geq 22$ having base fields different from $\mathbb{Q}(\sqrt{5})$, $\mathbb{Q}(\sqrt{2})$ and $\mathbb{Q}(\operatorname{Cos} 2\pi/7)$.*

THEOREM 4.42. *For $n \geq 14$, there are no arithmetic co-compact reflection groups in H^n having base fields different from*

$$\mathbb{Q}(\sqrt{2}),\ \mathbb{Q}(\sqrt{3}),\ \mathbb{Q}(\sqrt{5}),\ \mathbb{Q}(\sqrt{6}),\ \mathbb{Q}(\sqrt{2}, \sqrt{3}),\ \mathbb{Q}(\sqrt{2}, \sqrt{5}) \ \textit{and} \ \mathbb{Q}(\operatorname{Cos} 2\pi/m),$$

where $m = 7, 9, 11, 15, 16$, or 20.

REMARK 4.43. Recently A.G. Khovanskii [1] obtained an important generalization of Nikulin's estimates of sides complexity for simple polyhedra (Theorem 4.38 and Corollary 4.39) in the case of almost simple polyhedra. This result allowed M.N. Prohorov [1] to extend Vinberg's finiteness theorem on the case of crystallographic groups with unbounded fundamental polyhedra. More exactly one can formulate this result in such form:

THEOREM 4.44. *There are no crystallographic reflection groups in the hyperbolic space H^n of dimension $n > 995$.*

Research on crystallographic and arithmetic groups is important and interesting not only in itself but also for algebraic geometry. We describe probably the first and certainly the most interesting example of that.

This is related to type K3 surfaces, which are non-singular simply connected projective algebraic surfaces over $\mathbb{C}$ whose canonical class is zero. Examples are:

1) a two-sheet cover over the projective plane $\mathbb{C}P^2$ branched over a non-singular curve of degree six;

2) a quartic in $\mathbb{C}P^3$;

3) the intersection of a quadric and a cubic in $\mathbb{C}P^4$; and

4) the intersection of three quadrics in $\mathbb{C}P^5$.

For each surface X having type K3, there is an intersection form in the lattice of its algebraic cycles. This is an even integer quadratic form having signature $(1, n)$, where $n + 1 \leq 20$ are the dimension of the algebraic cycle lattice. Let f_X be the opposite form having signature $(n, 1)$.

In 1971, Pyatetskii-Shapiro and Shafarevich [1] proved Torelli's global theorem for type K3 surfaces X and as a corollary obtained a description in terms of form f_X for the automorphism group of an X surface within a finite central extension. In particular, Aut X is finite if and only if group $O'(f_X, \mathbb{Z})$ contains a finite-index subgroup generated by reflections associated with integer vectors e for which $f_X(e) = 2$ (see also Nikulin [1], [2]). Such reflections are called 2-reflections, and the integer quadratic form of the signature $(n, 1)$ whose automorphism group contains a finite-index subgroup generated by 2-reflections is called a 2-reflective form.

In classical studies by the Italian algebraic-geometry school in the early years of this century, algebraic surfaces were classified (see, for example, Shafarevich et al. [1]), which left open the description of birational automorphism groups for algebraic surfaces (in particular, finiteness) only for two classes of algebraic surface: K3 type and Enriques surfaces, the quotients of K3-type surfaces by involutions without fixed points. For this problem, Nikulin [1], [2] described all 2-reflective even integer quadratic forms having $(n, 1)$ signature for $n \geq 4$. The number of them is finite and they are all related in the above fashion to K3 surfaces. The maximum possible n for them is 18. The even forms cover all possibilities, since an odd form is 2-reflective if and only if the canonically related even form is 2-reflective. This description was completed by the proofs of finiteness for 2-reflective forms for $n = 3$ (Vinberg) and $n = 2$ (Nikulin [6], see Theorems 1.3 and 2.5).

These results with the finiteness theorems for arithmetic groups complete the description of type K3 surfaces X having finite groups Aut X. Nikulin's description [7] of Enriques surfaces X having finite groups Aut X completes the description of algebraic surfaces having finite birational automorphism groups.

CHAPTER 5

GEOMETRICALLY FINITE DISCRETE MÖBIUS GROUPS

§1. Local finiteness

So far, we have considered a special type of polyhedra as fundamental polyhedra for discrete Möbius groups $G \subset \text{Möb}(n)$ on n-dimensional sphere $\overline{\mathbb{R}}^n$ or in the half-space $\mathbb{R}^{n+1}_+$ having hyperbolic Poincaré metric, i.e., in H^{n+1}: either an isometric polyhedron $P_y(G)$ centred at a point $y \in \Omega(G)$, $G_y = \{\text{id}\}$, or a Dirichlet polyhedron $D_y(G) \subset H^{n+1}$ $(y \in H^{n+1}$, $G_y = \{\text{id}\})$, which may also be obtained from an isometric polyhedron $P_y(G)$ (Theorem 3.13), or else the special polyhedron of (3.33), which we have called a *pseudo-Dirichlet* one. Before we describe their characteristic properties (in hyperbolic space H^n, $n \geq 2$), we give a definition for a polyhedron that does not use the convexity, namely by a hyperbolic polyhedron we understand a subset P in H^n with the properties:

a) $P = \text{cl}(\text{int}\, P)$ (int and cl are taken in H^n);

b) the boundary ∂P is contained in a locally finite family of hyperbolic hyperplanes in H^n, and if that family is finite, P is said to be finitely-sided;

c) a side (or $(n-1)$-side) of P is a set $S \subset \partial P$ such that S lies in a certain hyperbolic hyperplane $H \subset H^n$; $\text{int}_H S$ is a component of the set $\text{int}_H (H \cap \partial P)$ and $S = \text{cl}_H(\text{int}_H S)$. Then S is a polyhedron in H and we can define inductively an i-dimensional side of P as a side of a polyhedron which itself is an $(i+1)$-side of the polyhedron P (where we understand by n-side of P, which may also be not connected, the closure of a connected component for $\text{int}\, P$).

We now distinguish the main properties of a fundamental polyhedra P for a discrete isometry group G in hyperbolic space H^n (we do not assume connectedness or convexity for P):

F1) P is a polyhedron in H^n.

F2) The boundary ∂P in H^n consists only of sides, i.e., any compact $K \subset H^n$ intersects only a finite number of sides $S_i \subset \partial P$.

F3) P is a fundamental domain for G.

F4) The sides of P are pairwise identified by the elements of G.

F5) Each point in H^n has a neighbourhood that intersects only a finite number of distinct images $g(P)$, $g \in G$.

All these properties clearly apply for the polyhedra $P_y(G)$ (F5 by virtue of (3.20)), and thus for the Dirichlet polyhedra $D_y(G)$ and the pseudo-Dirichlet ones $D^*_y(G)$. At first sight, F5 is redundant, but in fact it is not a consequence of F1 - F4, as Beardon [2] pointed out, who examined F5 for Kleinian groups in the plane.

DEFINITION 5.1. If a fundamental domain P has property F5, we say that it has local finiteness.

For $n=2$, the necessary and sufficient condition for a fundamental domain F in a discrete group $G \subset \mathrm{Isom}\, H^2$ to be locally finite is homeomorphism for the surfaces H^2/G and $\overline{F}/G$, where $\overline{F}$ is the closure of F in H^2 (Beardon [2]). The sufficient condition for local finiteness for a fundamental domain $F \subset H^2$ having an attainable boundary ∂F is that for any point $x \in \partial F$ its multiplicity $n(x)$, i.e., the number

$$n(x) = \mathrm{card}\{g \in G : g(x) \in \partial F\} \qquad (5.1)$$

is not less than two. This implies that for convex polyhedra $F \subset H^2$, local finiteness follows from F1 - F4 (compare Greenberg [4]).

The situation alters for the case $n \geq 3$: Tetenov's results [3] (see below) show that for local finiteness it is not sufficient to have the homeomorphism $\overline{F}/G \longrightarrow H^n/G$, nor side-pairing a fundamental polyhedron F, nor convexity of polyhedron F.

DEFINITION 5.2. In dimension n, $n \geq 3$, we call a regular decomposition of a ball its decomposition into multihedral angles that any of the sides is a common side for two such angles, and any two such angles adjoin either a full $(n-1)$-side or a k-dimensional side, $1 \leq k \leq n-2$, or have a common vertex, and any point in the space belongs either to the interior or to a k-dimensional side $(1 \leq k \leq n-1)$ of one of the decomposition angles.

LEMMA 5.3. *Let* $\Sigma = \{\sigma_1, \sigma_2, \ldots\}$ *be a regular decomposition for a ball* $B^n(Q, r) \subset H^n$. *Then there exists a discrete isometry group* G_Σ *in* H^n *together with a fundamental polyhedron* $P_\Sigma \subset H^n$, *which are such that the* $(n-1)$*-sides of* P_Σ *are pairwise equivalent, while the vertices* P_Σ *form a set of* G_Σ*-equivalent points* $\{Q_i\}$, *the multihedral angles of which are equal to the angles* σ_i *of the decomposition* Σ.

PROOF. To construct G_Σ with these properties, we take a point Q in H^n and construct the decomposition of the solid angle having vertex Q. For each multihedral angle σ_i in Σ we take a pair of hyperbolic hyperplanes, S_i and S'_i, which lie entirely within the angles σ_1 and σ_i correspondingly; we select also a hyperbolic translation g_i that transforms S_i into S'_i. We specify also that S_i and S_j do not intersect for distinct i and j. When these conditions are met, G_Σ as generated by the hyperbolic translation g_i, will be discontinuous, and, as a fundamental polyhedron P, we can take the component $H^n \setminus \bigcup_i (S_i \cup S_i')$, containing the point Q. Let $P_i = P \cap \sigma_i$. As $H^n = \bigcup_i \bar{\sigma}_i$, P is contained in the union of the closures for its subpolyhedra P_i. Therefore, the polyhedron

$$P_\Sigma = \bigcup_i [g_i^{-1}(P_i) \cup P_i] \tag{5.2}$$

is a fundamental one for G_Σ and satisfies the requirement of the lemma (see Fig. 21).

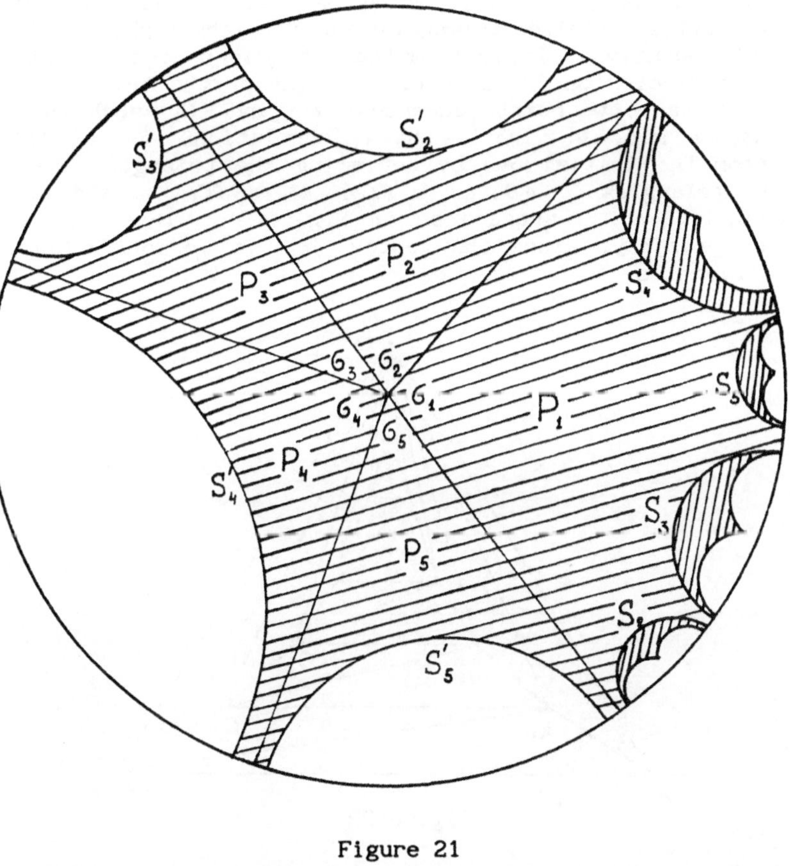

Figure 21

REMARK 5.4. This group G_Σ is a Fuchsian group in an n-sphere S^n of the second kind, but we can use a Kleinian combination to extend it to a group of the first kind having the same properties.

REMARK 5.5. If the decomposition Σ in Lemma 5.1 consists of an infinite number of angles σ_i, it is evident that property F5 does not apply for P_Σ in the neighbourhood of any vertex Q_i in it. The following example gives a construction that realizes this.

EXAMPLE 5.6. We use Lemma 5.3 and Remark 5.5 to construct a discrete hyperbolic isometry group $G_\Sigma \subset \operatorname{Isom} H^n$ $(n \geq 3)$ together with the convex fundamental polyhedron P_Σ for it that is not locally finite.

For that purpose it is sufficient to find a regular decomposition for the solid angle in which there is an edge common to an infinite number of decomposition angles, or what is the same, to indicate in a sphere S^{n-1} a regular cellular decomposition in which there would be a vertex common to an infinite number of cells.

Without loss of generality, we assume for simplicity that $n = 3$. We take a certain finite triangulation for the sphere $S^{n-1}(=S^2)$. Here and subsequently it is implied that the sides of the triangles are arcs of great circles on the sphere. In one of the triangles (denoted by ABC), we take two point sequences A_i and C_i such that $A_0 = A$, $C_0 = C$, $\lim A_i = \lim C_i = B$, while the triangulation Σ of the sphere obtained from the initial one by replacing the triangle ABC by the system of triangles A_iBA_{i+1}, $A_iC_iA_{i+1}$, C_iBC_{i+1} and $C_iA_{i+1}C_{i+1}$, where $i = 0, 1, 2, \ldots$, appears as in Fig. 22. This Σ will have the desired property, since the vertex B is common to an infinite number of triangles.

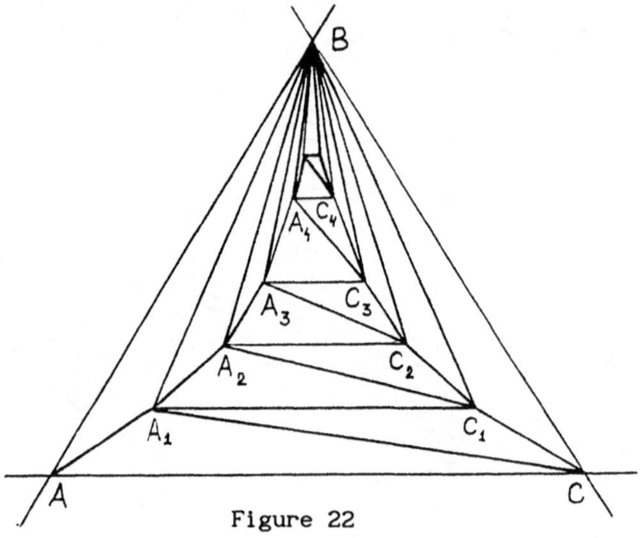

Figure 22

This G_Σ, a discrete group of H^3 isometries, constructed on this triangulation of S^2 by means of Lemma 5.3, (it may be assumed to be such that $\Lambda(G_\Sigma) = S^2$), and this with its fundamental polyhedron P_Σ (see (5.2)) are such that:

1) P_Σ is not locally finite (in the sense of F5);
2) $\operatorname{cl} P_\Sigma/G_\Sigma \longrightarrow H^n/G_\Sigma$ is a homeomorphism; and

3) the sides of P_Σ are pairwise equivalent.

REMARK 5.7. In H^n, $n \geq 3$, it is possible to construct fundamental polyhedron without side-pairing and not locally finite in which the multiplicity (5.1) for the boundary points is bounded, namely takes only the values 2 and 3 (Tetenov [3]).

Although there are such pathological examples, one can indicate the sufficient conditions for local finiteness for $n \geq 3$. Examples show that these conditions are exact.

THEOREM 5.8 (Tetenov [3]). *Let $P \subset H^n$ be a convex fundamental polyhedron for isometry group G in H^n, $n \geq 2$, which satisfies the conditions* F1 - F4 *and for which all points $x \in \partial P$ on the boundary have the multiplicity $n(x) < \infty$. Then P is locally finite.*

As a finite-sided polyhedron is always the union of a finite number of convex polyhedra, Theorem 5.5 gives us directly that:

COROLLARY 5.9. *Local finiteness applies for any fundamental polyhedron for a discrete isometry group in hyperbolic space H^n, $n \geq 2$, having a finite number of pairwise equivalent sides.*

The proof of Theorem 5.8 is independent of the dimension $n \geq 3$ and for simplicity is given for $n = 3$ by a contradiction.

Let x be a point in H^3 such that any neighbourhood of that point intersects an infinite number of different images $g_i(P)$, $g_i \in G$, of the polyhedron P. If $x_i \in g_i(P)$ is a point sequence tending to x, while y is a certain point within P, then we can join each of the points x_i to the point $g_i(y)$ via a certain curve γ_i lying within the polyhedron $g_i(P)$. As for $i \to \infty$ the distance between the points x and $g_i(y)$ also tends to infinity, any sphere $S^2(x, r)$ centred at x with radius $r > 0$ will intersect an infinite number of curves γ_i, and therefore on any such sphere there are points limiting for $\gamma_i \cap S^2(x, r)$. The topological arguments imply that there is a curve γ limiting for the γ_i curves that passes through x and intersects each of the spheres $S^2(x, r)$. Consequently, the neighbourhoods of the points on γ will intersect an infinite number of different images $g_i(P)$, $g_i \in G$, of P.

The points on γ cannot belong to the interior of any of the polyhedra $g(P)$, $g \in G$, and equally cannot be interior points in the sides of the polyhedra $g(P)$, since in that case they would lie in the interior of the union of two (closed) polyhedra $g'(\overline{P})$ and $g''(\overline{P})$. Then γ can consist only of segments from the edges of the polyhedra $g(P)$ or of their vertices. We now take an arc w in a certain polyhedron $g_{i_0}(P)$, which has a subinterval in common with the curve. Arc w is the intersection of two sides of the polyhedron $g_{i_0}(P)$, and two other polyhedra $g_{i_1}(P)$ and $g_{i_2}(P)$ adjoin $g_{i_0}(P)$ on those sides, which have the same common arc w. We continue this process to get that w is a common edge for a certain sequence of polyhedra $g_{i_m}(P)$, $m = 0, 1, 2, \dots$. As $n(z) < \infty$ for $z \in w$, only a finite number of $g_{i_m}(P)$ can adjoin w, and w, excluding the ends, must lie in the interior of the union of the closure of these polyhedra. Therefore the points on γ cannot be interior points for the edges of the polyhedra $g(P)$, $g \in G$.

The line γ cannot consist entirely of vertices from the polyhedra $g(P)$, but it also may not contain points not belonging to any closed polyhedron $g(\overline{P})$ because P is a fundamental polyhedron. That contradiction completes the proof.

In what follows, it is everywhere assumed, if the converse is not stated, that fundamental polyhedra $P \subset H^{n+1}$ for discrete groups $G \subset \text{Möb}(n)$ are selected to have properties F1 - F5.

§2. The convex Nielsen hull of the limit set

Let $K \subset \overline{H}^n$ be a certain set. We say that it is convex (in hyperbolic geometry) if for any two points $x_1, x_2 \in K$ the geodesic interval (x_1, x_2) also belongs to K. The convex hull $CH(K)$ of set K is defined as the least convex set in $\overline{H}^n$ containing K.

DEFINITION 5.10. The hyperbolic convex hull H_G for the limit set $\Lambda(G)$ of a group G of isometries in H^n (or Nielsen hull) is the convex hull

$$H_G = CH(\Lambda(G)) = \bigcap\{T \subset H^n : T \supset \Lambda(G) \text{ and } T \text{ is convex in } H^n\}. \qquad (5.3)$$

The Nielsen hull for a discrete group $G \subset \text{Möb}(n)$ is the Nielsen hull for the isometry group in the hyperbolic space H^{n+1} that is the extension of G in $\mathbb{R}^{n+1}_+$.

As an example, we consider the isometry group G in H^n that is the extension of a certain Fuchsian group $\Gamma \subset \text{Möb}(n-1)$ in the sphere at infinity $\partial H^n = \overline{\mathbb{R}}^{n-1}$. Then $\Lambda(G) = \Lambda(\Gamma)$ is contained in a certain sphere S^{n-2} having dimension $n-2$, and therefore $H_G = H_\Gamma$ lies in the hyperbolic $(n-1)$-plane spanned on the sphere S^{n-2}.

If we take as G a discrete isometry group in H^n which does not act discontinuously on the sphere at infinity $\partial H^n = \overline{\mathbb{R}}^{n-1}$, i.e., having $\overline{\mathbb{R}}^{n-1}$ as its limit set, then $H_G = \overline{H}^n$.

In the general case, when the limit set $\Lambda(G)$ does not lie on any $(n-2)$-sphere in $\overline{\mathbb{R}}^{n-1}$, H_G has a non-empty interior, since it contains n-simplexes having their vertices at the points $x^0, \ldots, x^n \in \Lambda(G)$ and lying in general positions. Then $H_G \backslash \Lambda(G) = \text{cl}_H(\text{int}_H H_G)$, and the boundary $\partial_H H_G$ is a pleated hypersurface in H^n bending along $(n-2)$-dimensional hyperbolic convex polyhedra having complete geodesics as one-dimensional skeleton that intersect only on the sphere at infinity ∂H^n (at extreme points in the limit set $\Lambda(G)$).

In terms of Riemannian geometry, the exterior curvature of the boundary $\partial_H H_G$ is 0, while the interior curvature is the sectional curvature of the surrounding space H^n, which is -1. §6 deals in more detail with the geometry of the pleated surface $\partial_H H_G$, where it is shown that the properties are dependent on dimension n.

The discrete group G acts discontinuously in $H_G \setminus \Lambda(G)$ and is an isometry group there. We consider the orbit space for G in the usual way in $H_G \setminus \Lambda(G)$ to get a convex hyperbolic orbifold (see Chapter 9 §1), which is denoted subsequently by $\mathcal{O}_G$:

$$\mathcal{O}_G = [H_G \setminus \Lambda(G)]/G. \tag{5.4}$$

If G is a torsion free group, then the quotient $\mathcal{O}_G$ is a manifold (we denote it also by M_G). If G acts discontinuously in $\partial H^n = \overline{\mathbb{R}}^{n-1}$, $\mathcal{O}_G$ has the boundary that has a hyperbolic singular structure with path metric induced by the hyperbolic metric in H^n. There exists a unique hyperbolic retraction $\rho : \mathcal{O}(G) \longrightarrow \mathcal{O}_G$ for the orbifold with boundary

$$\mathcal{O}(G) = [\overline{H}^n \setminus \Lambda(G)]/G \tag{5.5}$$

onto the convex suborbifold $\mathcal{O}_G$ induced by the retraction

$$\tilde{\rho} : \overline{H}^n \longrightarrow H_G. \tag{5.6}$$

To define the (5.6) retraction, we put

$$\tilde{\rho}(x) = \begin{cases} x, & \text{if } x \in H_G; \\ x_0, & \text{if } x \in \overline{H}^n \setminus H_G, \end{cases} \tag{5.7}$$

where x_0 is a point from H_G closest to point x if $x \in H^n \setminus H_G$; if on the other hand $x \in \partial H^n \setminus \Lambda(G)$, we take as nearest point x_0 the first point of tangency with H_G for a horosphere in H^n centred at x. This definition in (5.7) is correct because H_G is a convex set, while a ball and a horoball are strictly convex in H^n.

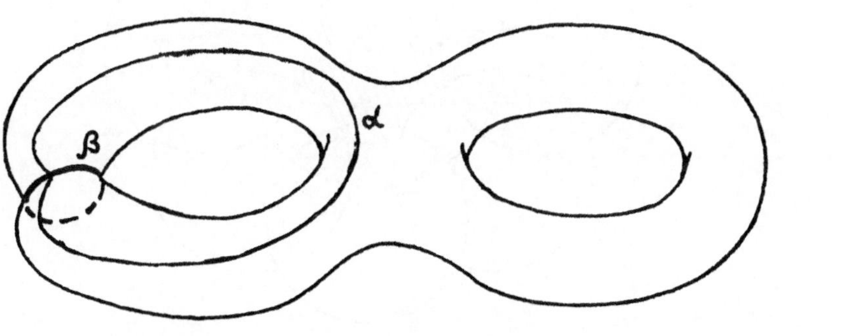

Figure 23

EXAMPLE 5.11 (*Mickey Mouse example*) (see Thurston [1]). Let $\Gamma \subset \mathrm{Isom}\, H^3$ be obtained by extending from the plane $\overline{\mathbb{C}} \subset \partial H^3$ a Fuchsian group (also denoted by Γ) having an invariant disc $U \subset \mathbb{C}$

and uniformizing a closed Riemann surface $S = U/\Gamma$ of genus two. The 3-manifold $M(\Gamma)$ is the surface layer $S \times [-1, 1]$ (see Chapter 7 §3), whose retract is the convex surface $S \cong S \times \{0\} = M_\Gamma$.

We consider a closed geodesic α on the surface S (Fig. 23); in disc U, this corresponds to the axis $\tilde{\alpha}$ of the hyperbolic generator γ for group Γ with respect to which the sides a and a' of the fundamental polygon are equivalent (they lie on the isometric circles $I(\gamma)$ and $I(\gamma^{-1})$). We move these circles apart (increasing their radii) in such a way that the angles in the circle arc polygon remain as before, while the circle ∂U will intersect the swallowed circles at angles less than $\pi/2$ (see Fig. 24). That deformation of the fundamental polygon generates a quasi-conformal deformation f for Γ (see Chapter 9 §2 for more details), and converts it to the quasi-Fuchsian group $G = f\Gamma f^{-1}$. For group G, the convex hull H_G has dimension three, while the boundary consists of two pleated hyperbolic surfaces S_0 and S_1. One of them, S_0, is the image of disc U under the superposition $\tilde{\rho} \circ f$ of the retraction $\tilde{\rho} : \overline{H}^3 \to H_G$ and the quasi-conformal deformation f, i.e., $S_0 = \tilde{\rho}(f(U))$ can be represented as follows.

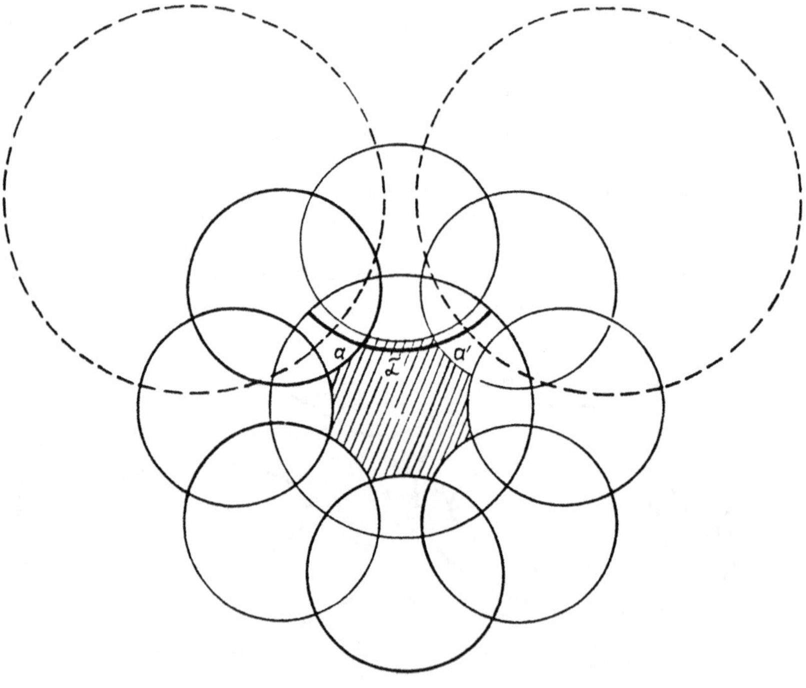

Figure 24

A simple loop, homotopic in the surface S, to a unique geodesic β, corresponds to the sides a and a' of the fundamental polygon for Γ. Then the hyperbolic plane $H \subset H^3$, spanned on ∂U, intersects its image $f\gamma^{-1}f^{-1}(H)$ on a geodesic $\tilde{\beta} \subset H^3$, being the lifting of the geodesic $\beta \subset S$ in the universal covering $\tilde{M}_\Gamma = H$ of S (see Fig. 25). In result, all images $g(H)$ with respect to the elements g of the group G cut off a certain domain from H^3 (bounded in $\mathbb{R}^3_+$ in the Euclidean sense), whose boundary in H^3 is the ruled surface S_0, which is everywhere flat, apart from the geodesics, which are the images of the geodesic $\tilde{\beta}$. The projection of that domain on the plane $\partial H^3 = \overline{\mathbb{R}}^2$ is shown in Fig. 25.

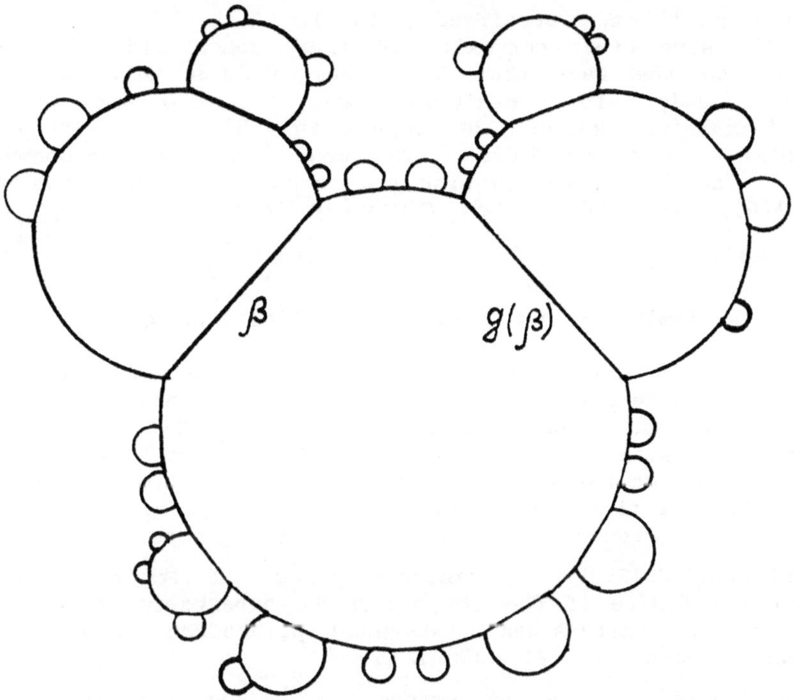

Figure 25

The boundary of manifold M_G (topologically equivalent to $S \times [-1, 1]$) consists of two surfaces, one of which is S_0/G. This surface is obtained from the surface $S = U/\Gamma$ by bending it along the geodesic β through a certain angle. The second surface S_1/G has a more complicated structure.

In what follows, the following simple assertion is useful, as it establishes a relation between the geometry of the Nielsen hull and that of the hyperbolic orbifold H^n/G. Namely, one can define a convex core $C(\mathcal{O})$ in the hyperbolic orbifold $\mathcal{O} = H^n/G$ (or a minimal convex retract) as the minimal of the convex complexes $X \subset \mathcal{O}$, for which embedding induces the isomorphism $\pi_1(X) \longrightarrow \pi_1(\mathcal{O})$ of the fundamental groups (in the orbifold sense). It turns out that

$C(\mathcal{O}) = [H_G \setminus \Lambda(G)]/G = \mathcal{O}_G$, i.e. we have:

THEOREM 5.12. *For a non-elementary discrete group* $G \subset \operatorname{Isom} H^n$ *the orbifold* $\mathcal{O}_G = [H_G \setminus \Lambda(G)]/G$ *is the minimal convex retract for the orbifold* $\mathcal{O}(G) = [H^n \cup \Omega(G)]/G$ *(convex core of the orbifold* H^n/G*).*

PROOF. The convexity of $\mathcal{O}_G \subset H^n/G \subset \mathcal{O}(G)$ is evident from the convexity of the Nielsen hull H_G. The definition of H_G implies the isomorphism $G \cong \pi_1(\mathcal{O}_G) \longrightarrow \pi_1(\mathcal{O}(G)) \cong G$, induced by the embedding $\mathcal{O}_G \subset \mathcal{O}(G)$. The property of the retraction $\mathcal{O}(G) \longrightarrow \mathcal{O}_G$ is obtained from the construction of the retraction ρ that is induced by the retraction $\tilde{\rho} : \overline{H}^n \longrightarrow H_G$, specified in (5.7).

At the same time, convexity of some suborbifold $\mathcal{O}' \subset \mathcal{O}(G)$ is equivalent to the fact that the image of its universal covering $\tilde{\mathcal{O}}' \subset \tilde{\mathcal{O}}(G)$ under the developing map $d : \tilde{\mathcal{O}}(G) \longrightarrow H^n \cup \Omega(G)$ is $d_*(\pi_1(\mathcal{O}'))$-invariant subset D convex in H^n. By virtue of the isomorphism $\pi_1(\mathcal{O}') \cong \pi_1(\mathcal{O}(G))$ we have that the holonomy group $d_*(\pi_1(\mathcal{O}')) \subset \operatorname{Isom} H^3$ coincides with the group G. This implies that $\overline{D} \cap \partial H^n \supset \Lambda(G)$, so $\overline{D} \supset H_G$. Then $C(\mathcal{O}) = [H_G \setminus \Lambda(G)]/G$.

§3. The limit set of a geometrically finite group

Theorem 3.18 shows that if the Kleinian group $G \subset \text{Möb}(n)$ has a finite-sided fundamental polyhedron $P \subset S^n$, then that group is finitely generated. For Fuchsian groups $G \subset \text{Möb}(2)$, the converse is also true: finite set of generators of a group implies that its fundamental polygon has a finite number of sides. In what follows, we shall see the important role this geometrical property of a fundamental domain plays for research on discrete Möbius groups.

DEFINITION 5.13. A discrete group $G \subset \text{Möb}(n)$ is called *geometrically finite* if for its action in hyperbolic space $H^{n+1} \cong \mathbb{R}^{n+1}_+$ there exists a finite-sided fundamental polyhedron $P(G) \subset H^{n+1}$ that satisfies the conditions F1 - F5 in §1.

REMARK 5.14. The half-space $\mathbb{R}^{n+1}_+$ in Definition 5.13 is necessary. There are examples of Kleinian groups $G \subset \text{Möb}(n)$, $n \geq 2$, having finite-sided fundamental polyhedra in $\overline{\mathbb{R}}^n$ but not having such in $\mathbb{R}^{n+1}_+$. This property for a finitely generated Kleinian group in the plane is a consequence of Ahlfors's finiteness theorem, see Beardon and Jørgensen [1]; for similar examples for $n \geq 3$, see Krushkal'-Apanasov-Gusevskii [3] (Example 52) and Apanasov [11].

Degenerate groups in the plane were the first examples of discontinuous finitely generated groups that were not geometrically finite (i.e., geometrically infinite); Greenberg [2] pointed out this. Till now, no constructive examples of degenerate groups have been obtained, but there are notable examples of doubly generated discrete groups $G \subset \mathcal{M}_2$ (acting non-discontinuously in the plane, i.e., Fuchsian groups $G \subset \mathcal{M}_3$ of the first kind) that are geometrically infinite but which have fairly simply constructed manifolds $\mathbb{R}^3_+/G$, see Jørgensen [3] and Jørgensen and Marden [1].

Parabolic cusps. In the description of geometrically finite groups given below, a particular part is played by fixed points for parabolic elements in such a group, or more precisely, parabolic cusps, which are points that correspond to "thin ends" of the $\mathbb{R}_+^{n+1}/G$ manifold.

More precisely (see Apanasov [13]), let $G \subset \text{Möb}(n)$ be a discrete group and $z \in \overline{\mathbb{R}}^n$ a fixed point for a parabolic element $\gamma \in G$. By G_z^* we denote the maximal free Abelian subgroup of a finite-index in the stabilizer $G_z \subset G$ for point z (see Theorem 3.4) and by k, $1 \le k \le n$, we denote the rank of G_z^*.

DEFINITION 5.15. A cusp domain of a discrete group $G \subset \text{Möb}(n)$ at a parabolic fixed point $z \in \overline{\mathbb{R}}^n$ is an open subset $U_z \subset \overline{\mathbb{R}}^n$ such that:

C1) U_z is precisely invariant with respect to $G_z \subset G$, i.e. for all $\gamma \in G_z$ and $g \in G \setminus G_z$ we have

$$\gamma(U_z) = U_z, \quad g(U_z) \cap U_z = \emptyset;$$

C2) there exist a $t > 0$ and a transformation $h \in \text{Möb}(n)$, $h(z) = \infty$, such that

$$\{x \in \mathbb{R}^n : \sum_{i=k+1}^{n} x_i^2 > t\} \subset h(U_z),$$

where $k = \text{rank}\, G_z^*$, $1 \le k \le n-1$.

DEFINITION 5.16. A fixed point $z \in \overline{\mathbb{R}}^n$ for a parabolic element in the discrete group $G \subset \text{Möb}(n)$ is called a *parabolic cusp* for the group G if one of the following conditions is satisfied:

1) the subgroup $G_z^* \subset G_z$ has rank n; or

2) there exists a cusp domain U_z at point z.

Then $\text{rank}\, G_z^*$ is called the *rank* of the parabolic cusp z.

REMARK 5.17. It follows directly from the definitions that,

firstly, conditions 1) and 2) defining a parabolic cusp are mutually exclusive (a cusp domain U_z exists only if $\text{rank}\, G_z^* \le n-1$);

secondly, the cusp domains U_z do not contain limit points of the group G; and

thirdly, our definition coincides with the standard definition for a cusp point in a Kleinian group $G \subset \text{Möb}(2)$ on the plane (see Beardon and Maskit [1]).

REMARK 5.18. Not all parabolic fixed points for a discrete group $G \subset \text{Möb}(n)$ and even not all the parabolic Ω-peaks there are parabolic cusps; examples of such discrete groups G can be obtained as the free products of a free Abelian group Γ having rank k, $1 \le k \le n-1$, and a countable family of cyclic loxodromic groups Γ_i (Fig. 26 shows the projection of the fundamental polyhedron for such a group on $\partial H^{n+1} = \overline{\mathbb{R}}^n$). The same property occurs in the groups constructed in the proof of Theorem 4.26. As an example of Ω-peaks that are not parabolic cusps for a discrete group $G \subset \text{Möb}(3)$ one can take as the extensions to $\overline{\mathbb{R}}^3$ of the actions of discrete groups in the plane

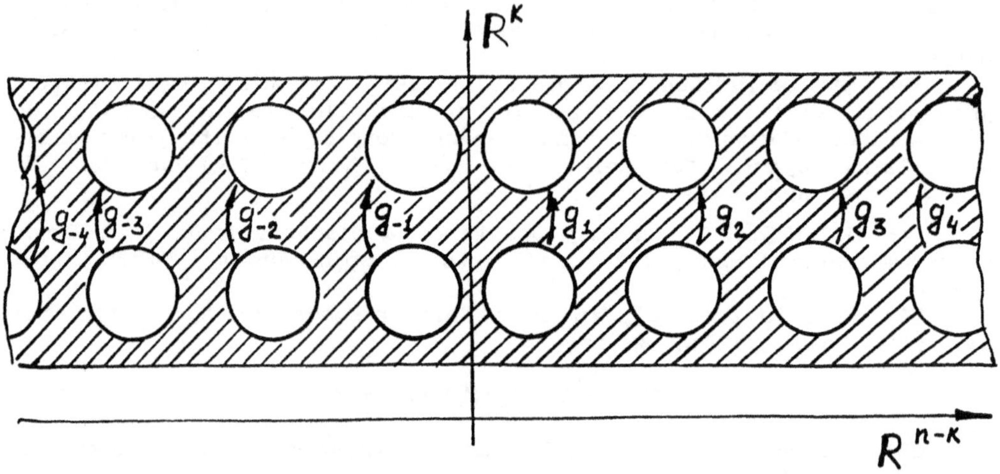

Figure 26

constructed by Jørgensen and Marden [1]. In such groups, the stabilizers G_∞ for the fixed parabolic point ∞ have ranks of one, but at the same time the parabolic point ∞ does not have a cusp domain.

On the other hand, it follows directly from the definition of a parabolic cusp $z \in \overline{\mathbb{R}}^n$ for the discrete group $G \subset \text{Möb}(n)$ that there exists an open set (hyperbolic cusp domain) $\tilde{U}_z \subset \mathbb{R}^{n+1}_+ \cong H^{n+1}$ such that for a number $t > 0$ and a transformation $h \in \text{Möb}(n)$, $h(z) = \infty$, we have from condition C2 in Definition 5.15 that:

1) $\tilde{U}_z$ is precisely G_z-invariant; and

2) $h(\tilde{U}_z) \supset \{x \in \mathbb{R}^{n+1}_+ : \sum_{i=k+1}^{n+1} x_i^2 > t\}$.

These properties imply that the horoball $h^{-1}(\{x \in \mathbb{R}^{n+1}_+ : x_{n+1}^2 > t\}) \subset H^{n+1}$ is precisely invariant with respect to G_z, so (∞, e_n) is a Ω-peak (more precisely, a $\mathbb{R}^{n+1}_+$-peak) for the group G.

Approximation points. We specify a point $x \in \overline{\mathbb{R}}^n$, and from the definition of the limit set $\Lambda(G)$ for a group $G \subset \text{Möb}(n)$ we get that its points are approximated by points in the orbit $G(x)$. However, that approximation is not uniform. Amongst the limit points, one can distinguish points having good approximation. Hedlund [1] first introduced these.

DEFINITION 5.19. Let $G \subset \text{Möb}(n)$, $G_\infty = \{\text{id}\}$ be a discrete group acting in the ball $B^n(0, 1)$; a limit point $z \in \Lambda(G) \subseteq S^{n-1}(0, 1)$ is called an *approximation point of the group* G if there exist a point $x \in \Omega(G) \subset \overline{\mathbb{R}}^n$, a constant $k > 0$, and a sequence $\{g_m\}$ of distinct elements in G such that

$$|z - g_m(x)| < k \cdot r_m^2, \tag{5.8}$$

where r_m is the radius of the isometric sphere $I(g_m)$.

From formula (1.26) for the distortion for the inversion, it follows that for any $g \in \text{Möb}(n)$, $g(\infty) \neq \infty$, and for all $x, y \notin \{\infty, g^{-1}(\infty)\}$ we have

$$|g(x) - g(y)| = \frac{r^2 \cdot |x - y|}{|x - g^{-1}(\infty)| \cdot |y - g^{-1}(\infty)|}. \tag{5.9}$$

Then we see that for any fixed $x, y \in \Omega(G)$ there exist $k_1 > 0$ and $k_2 > 0$ such that for all numbers m, up to a finite set of them,

$$k_1 \cdot r_m^2 \leq |g_m(x) - g_m(y)) \leq k_2 \cdot r_m^2. \tag{5.10}$$

On comparing (5.8) and (5.10) we get that the point x in Definition 5.19 can be taken as any point from the discontinuity set $\Omega(G)$.

We put $y = z$ in (5.9) to get that for an approximation point z and for any point x, $x \neq z$, there exist a $k > 0$ and a sequence of distinct elements $\{g_m\} \subset G$ such that

$$|g_m(x) - g_m(z)| \geq k. \tag{5.11}$$

We note that the speed of approximation to the point z by G-images of the points $x \in \Omega(G)$, given in (5.8), is exact. In fact, we put $g = g_m^{-1}$, $x = z$, and $y = 0$ in (5.9) to get that

$$|z - g_m(\omega)| \geq k_3 \cdot r_m^2,$$

where the constant $k_3 = (2\sup\{|g(\infty)| : g \in G\setminus\{\text{id}\}\})^{-1} > 0$ is dependent only on G.

It is readily seen that fulfilment of (5.11) at some point x, $x \neq z$, is equivalent to that it is fulfilled uniformly on compact subsets from $\overline{\mathbb{R}}^n\setminus\{z\}$.

The above properties of approximation points are metrical ones. We now prove two topological properties of such points.

A Stolz domain at a point $z \in S^{n-1}$ is a cone in the ball $B^n(0, 1)$ having the form

$$S_z = \{x \in B^n : |z - x| \leq k(1 - |x|)\}. \tag{5.12}$$

We consider the hyperbolic geodesic σ in B^n joining the points x and z; then from (5.11) follows the existence of a compact $C_0 \subset B^n$ such that for the same sequence $\{g_m\}$ we have

$$g_m(\sigma) \cap C_0 \neq \varnothing. \tag{5.13}$$

Conversely, let there exist a compact $C_0 \subset B^n$ and a sequence of distinct elements $\{g_m\} \subset G$ such that (5.13) is satisfied. Then for the same $\{g_m\}$ and point x, (5.11) is satisfied, since otherwise the images $g_m(\sigma)$ would converge to $\Lambda(G)$, which contradicts (5.13).

We now take the compact C_0 as the set $\{x \in B^n : d(x, 0) \leq d_0\}$, where $d(x, y)$ is the hyperbolic distance, $d_0 > 0$, and where K denotes the d_0-neighbourhood of the geodesic σ:

$$K = \{x \in B^n : d(x, \sigma) \leq d_0\}. \tag{5.14}$$

Then we see that (5.13) holds if and only if $g_m(x) \longrightarrow z$ in domain (5.14) for all points x from C_0, i.e., the images of point x tend to z in the Stolz domain S_z.

The above may be summarized as:

THEOREM 5.20 (Beardon and Maskit [1]). *Let $G \subset \text{Möb}(n)$ be a discrete group in the ball B^n and z its limit point. Then the following statements are equivalent:*

A1. *z is an approximation point.*

A2. *For some (or all) point $x \in \Omega(G)$ there exists a sequence $\{g_m\} \subset G$ of distinct elements together with a constant $k > 0$ such that*

$$|g_m(x) - z| < kr_m^2.$$

A3. *For some (or all) point x, $x \neq z$, there exists a sequence $\{g_m\} \subset G$ of distinct elements together with a constant $k > 0$ such that*

$$|g_m(x) - g_m(z)| \geq k.$$

A4. *There exists a sequence $\{g_m\} \subset G$ of distinct elements such that $|g_m(x) - g_m(z)|$ is bounded away from zero uniformly on compact subsets of $\overline{\mathbb{R}}^n \setminus \{z\}$.*

A5. *For any hyperbolic line $\sigma \subset B^n$ ending at z, there exists a sequence $\{g_m\} \subset G$ of distint elements and a relatively compact subset $C \subset \mathbb{B}^n$ such that*

$$g_m(\sigma) \cap C \neq \varnothing.$$

A6. *For some (or all) point $x \in B^n$ there exist a sequence $\{g_m\} \subset G$ of distinct elements and a Stolz domain S_z at point z such that $g_m(x)$ converges to z in S_z.*

We note that A5 is independent of the conjugation to hyperbolic isometry, so if we consider a Möbius transformation γ, $\gamma(B^n) = \mathbb{R}^n_+$, the group $\Gamma = \gamma G \gamma^{-1}$ is a group in the half-space $\mathbb{R}^n_+ \cong H^n$, and the approximation points for G pass into those for $\Gamma = \gamma G \gamma^{-1}$. In the special case when $\Lambda(G)$ is a proper subset of S^{n-1} we can choose γ

so that $\infty \notin \Lambda(\gamma G \gamma^{-1})$. In this case let σ be the vertical line through z and we conclude that z is an approximation point if and only if there exist a sequence of distinct elements $\{g_m\} \subset \Gamma$ and $k > 0$ such that

$$|g(\infty) - g(z)| \geq k. \tag{5.15}$$

REMARK 5.21. If $g \in \text{Möb}(n)$ is a loxodromic transformation, its fixed points are approximation ones, and as the sequences $\{g_m\}$ in A5 one needs to take $\{g^m\}$ or $\{g^{-m}\}$, while one takes the axis of element g (an invariant geodesic) as σ. If g is a parabolic transformation from group $G \subset \text{Möb}(n)$, and its fixed point z is a Ω-peak, then that point z is not an approximation one. This clearly follows from A5 and from the fact that no orbit can approach z in a Stolz domain at z, since the action of G in a precisely G_z-invariant horoball is determined by the parabolic stabilizer G_z of z.

From A5 in Theorem 5.20, we have directly:

COROLLARY 5.22. *Approximation points (in particular loxodromic fixed points) cannot lie on the boundary of a fundamental polyhedron* $P(G) \subset H^n$ *for a discrete group* $G \subset \text{Isom}\, H^n$ *having the properties* F1 - F5 *from* §1.

LEMMA 5.23. *Let* $G \subset \text{Möb}(n)$ *be a discrete group having a finite-sided fundamental polyhedron* $P \subset H^{n+1}$. *Then the set* $\overline{P} \cap \Lambda(G) = P^*$ *of its limit vertices either is empty or else consists of a finite number of parabolic cusps for the group* G.

PROOF. We note firstly that no point from set P^* can be a fixed point for a loxodromic element from G due to Corollary 5.22.

Now let $x \in P^*$ (we may assume that $x = \infty$). As the number of pairwise equivalent sides of polyhedron P is finite, at the boundary ∂P, there exists only a finite set of distinct points G-equivalent to the point $\infty = x$. Let these be the points $g_i(\infty) \in P^*$, $0 \leq i \leq l$, where $g_i \in G$, $g_0 = \text{id}$. We denote by G^* the maximal free Abelian subgroup in the stabilizer G_∞, which may be trivial. Bieberbach's theorem (Theorem 3.4) shows that G^* has a finite index in G_∞, and therefore there exists a finite number of elements $\gamma_1, \ldots, \gamma_m \in G_\infty$ such that

$$G_\infty = G^* \cup G^*\gamma_1 \cup \ldots \cup G^*\gamma_m. \tag{5.16}$$

If now for some element $g \in G$ the point ∞ lies in $g(P^*)$, we consider a certain geodesic ray $\sigma \subset g(P)$ running to the point ∞ on the boundary $\partial(g(P))$. The existence of such a ray σ follows from the finite-sidedness of P. For ray σ there exists an element $g_i \in G$, $0 \leq i \leq l$, such that the geodesic ray $g_i g^{-1}(\sigma)$ is ending at ∞, i.e., the element $g_i g^{-1}$ lies in the stabilizer G_∞. Then (5.16) implies that the element $g \in G$ lies in one of the finite number of sets $G^* h_j$, $h_j \in G$, $0 \leq j \leq ml$. It shows that one of the following possibilities occurs:

1) $G^* = \{\text{id}\}$. Then a neighbourhood of ∞ in $\overline{H}^{n+1}$ intersects

only a finite number of images of the polyhedron P and therefore $\infty \in \Omega(G)$;

2) the free Abelian group G^* has rank n, and then, by definition, ∞ is a parabolic cusp; and

3) G^* has rank k, $1 \le k \le n-1$.

In the last case, G^* is generated by k (screw) Euclidean translations by linearly independent vectors $a_i \in \mathbb{R}^n$. Moreover, Theorem 3.4 shows that up to the conjugation of group G by a Euclidean isometry, any element $\gamma \in G^*$ has the following form (we assume that $\mathbb{R}^n = \mathbb{R}^k \times \mathbb{R}^{n-k}$, where the translation vectors $\{a_i\}$, $1 \le i \le k$, form the basis of $\mathbb{R}^k$):

$$\gamma(x) = U(x) + a, \quad U \in O(n), \quad U|_{\mathbb{R}^k} = \text{id}, \quad U(\mathbb{R}^{n-k}) = \mathbb{R}^{n-k}, \quad a \in \mathbb{R}^k. \tag{5.17}$$

Then it follows that any image $g(P)$, $g \in G\setminus G_\infty$ of the polyhedron P lies within one of a finite number of surfaces $S_j \subset \mathbb{R}^{n+1}$, $0 \le j \le ml$, bounded in the Euclidean sense, or within their images $\gamma(S_j)$ for translations $\gamma \in G^*$. Property (5.17) for the elements γ of the group G^* shows that all the surfaces $G^*(S_j)$, for a certain $t > 0$, lie within the following set:

$$\{x \in \mathbb{R}^n : \sum_{i=k+1}^{n} x_i^2 \le t\}. \tag{5.18}$$

The exterior U_∞ of the set (5.18) satisfies the conditions C1 and C2 in Definition 5.15 for a cusp domain, i.e., in that case the point $x = \infty \in P^*$ is a parabolic cusp for G, which completes the proof.

We now pass on to the main result in this section, which provides a criterion for geometrical finiteness of a discrete group $G \subset \text{Möb}(n)$ in terms of its limit set.

THEOREM 5.24. *A discrete group* $G \subset \text{Möb}(n)$, $n \ge 2$, *is geometrically finite if and only if its limit set* $\Lambda(G)$ *consists of approximation points and parabolic cusps.*

Before we prove this, we make some observations and comments.

REMARK 5.25. This result (Apanasov [13]) is a generalization of the Beardon-Maskit [1] theorem for discrete groups in the plane ($n = 2$), which for $n = 2$ demonstrates equivalence between the condition for Theorem 5.24 on the limit set $\Lambda(G)$ and the condition for finite-sidedness in any convex fundamental polyhedron $P \subset H^3$ for the group G. A similar assertion (i.e., one that uses the convexity condition) is incorrect for $n \ge 3$, as the author first observed in 1973 (Apanasov [2], [4]), see Theorem 5.26 below. The exception for $n = 2$ occurs because there the parabolic elements of the group $G \subset \mathcal{M}_2$ are conjugated to Euclidean translations (without torsion) and so all parabolic points in the group are parabolic Ω-peaks in $\overline{\mathbb{R}}^3$. Theorem 4.26 shows that the situation is different for $n \ge 3$.

THEOREM 5.26. *Let* $G \subset \text{Isom}\,\mathbb{R}^n$, $n \ge 3$, *be a discrete Euclidean isometry group having a maximal free Abelian subgroup* $G^* \subset G$ *with rank* k, $1 \le k \le n-2$, *having translation subspace* $\mathbb{R}^k$. *If in* G^* *there exist elements* g, $g(x) = U(x) + b$, $b \in \mathbb{R}^k \times \{0\} \subset \mathbb{R}^n$, $U|_{\mathbb{R}^k} = \mathrm{id}$, $U \in O(n)$, *having infinite-order orthogonal components* U, *then any Dirichlet polyhedra* $D_y(G)$ *centred at points* $y \in \mathbb{R}^n \setminus \mathbb{R}^k$ *are infinite sided.*

We get directly from Theorems 5.26 and 3.13:

COROLLARY 5.27. *Let the group* G *in Theorem* 5.26 *act in the hyperbolic space* H^{n+1}, $n \ge 3$, *and let the hyperbolic plane* $H^{k+1} \subset H^{n+1}$ *be such that* $\partial H^{k+1} \supset \mathbb{R}^k$. *Then for any point* $y \in H^{n+1} \setminus H^{k+1}$ *the Dirichlet polyhedron* $D_y(G) \subset H^{n+1}$, *centred at* y *has an infinite set of faces.*

PROOF. The proof of Theorem 5.25 is given for $n = 3$ and $k = 1$. The general case differs only in formal details.

Let the generator for the cyclic group $G \subset \text{Möb}(3)$ be the screw translation T, $T(x) = U(x) + e_1$, where $U \in O(3)$ is an irrational rotation in $\mathbb{R}^3$ around the coordinate axis $\tau = \mathbb{R} \cdot e_1$ by angle α incommensurable with 2π, i.e., $U^k \ne \mathrm{id}$, for $k \in \mathbb{Z} \setminus \{0\}$ and

$$U = \begin{bmatrix} 1 & 0 & 0 \\ 0 & \cos\alpha & \sin\alpha \\ 0 & -\sin\alpha & \cos\alpha \end{bmatrix}. \tag{5.19}$$

We consider a point $y \in \mathbb{R}^3 \setminus \tau$. Then the Dirichlet polyhedron in the Euclidean metric ρ, having its centre at y,

$$D_y(G) = \{x \in \mathbb{R}^3 : \rho(x, y) \le \rho(g(x), y),\ g \in G\} \tag{5.20}$$

goes over to an isometric polyhedron P_y for group $G_y = h_y^{-1} G h_y$, where h_y is inversion in $\overline{\mathbb{R}}^3$ with respect to the unit sphere with the centre at the point y, $h_y(y) = \infty$ (compare the proof of Theorem 3.13). We note that the isometric spheres $I(\gamma)$ and $I(\gamma^{-1})$ with centres at q and q' always contain the point y. Also, $I(\gamma) \cap I(\gamma^{-1}) = \{y\}$ only if $y \in \tau$. If on the other hand $y \notin \tau$, then $I(\gamma)$ and $I(\gamma^{-1})$ intersect on a circle containing y. The generator γ is the superposition of inversion with respect to $I(\gamma)$, reflection in the plane containing the circle $I(\gamma) \cap I(\gamma^{-1})$, and rotation around the straight line passing through the points y and q' (see Theorem 1.6).

As the centres of the isometric spheres $I(g)$ of the elements $g = \gamma^n$, $n \in \mathbb{Z} \setminus \{0\}$, in the group G_y form a G_y-orbit for the point ∞, the set Q of such centres can be written as

$$Q = \{g_n = \gamma^n(\infty) : n \in \mathbb{Z} \setminus \{0\}\} = \{h_y T^n(y) : n \in \mathbb{Z} \setminus \{0\}\}. \tag{5.21}$$

The image of Q in the transformation h_y is the set $\{T^n(y) : n \in \mathbb{Z} \setminus \{0\}\}$, lying on a spiral around the first coordinate axis τ, so the points q_n in (5.21) lie on a spiral converging from two sides (at zero angles) to the point y (Fig. 27).

At the same time, all the isometric spheres $I(g)$, $g \in G_y$ contain point y, so the vectors

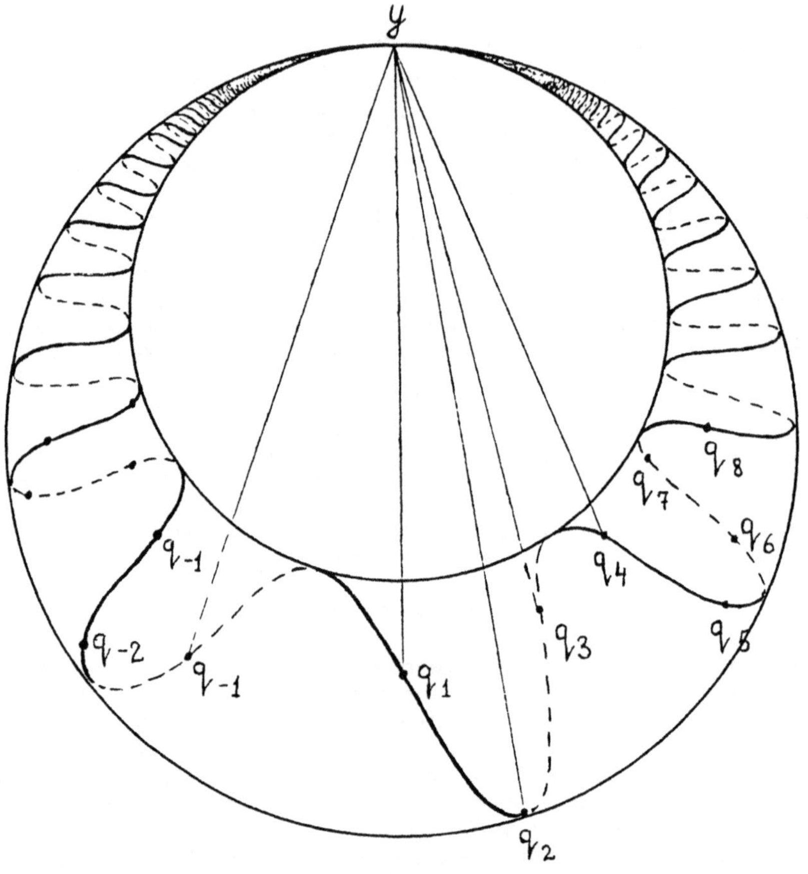

Figure 27

$$V_n = (y - \gamma^{-n}(\infty))/|y - \gamma^{-n}(\infty)|$$

are unit normals for the isometric spheres $I(\gamma^n)$. This and the irrationality condition for the rotation angle α and the fact that the radii r_n for the spheres $I(\gamma^n)$ tend to 0 for $|n| \to \infty$ show that, for any integer n different from zero, the isometric spheres for elements γ^n and γ^{-n} have segments not contained in the union of the interiors of the spheres $I(\gamma^i)$ and $I(\gamma^{-i})$, where $|i| < n$. Then we get directly that the isometric fundamental polyhedron for the group G_y has an infinite set of sides, and the same applies to (5.20), which completes the proof of Theorem 5.26.[1]

COROLLARY 5.28. *For* $n \geq 3$, *there exist geometrically finite groups* $G \subset \text{Möb}(n)$ *for which any Dirichlet polyhedron* $D_y \subset H^{n+1}$ *centred at any point* $y \in H^{n+1}$ *is infinite-sided.*

To prove this, it is sufficient to consider as G the group obtained by Klein combination from several (two) groups $G_1, \dots, G_m \subset \text{Möb}(n)$ conjugate to the group constructed in the proof of Theorem 5.26. These groups G_i should be far enough apart in the sense that the (k_i+1)-dimensional planes $H^{k_i+1} \subset H^{n+1}$, $1 \leq k_i \leq n-2$, $1 \leq i \leq m$, based on the translation subspace for the G_i should not intersect $\bigcap_{i=1}^{m} H^{k_i+1} = \emptyset$. Then any Dirichlet polyhedron $D_y(G)$, $y \in H^{n+1}$, will have, due to Corollary 5.27, an infinite set of sides in the neighbourhood of at least one of the parabolic cusps $p_1, \dots, p_m$ of the group G.

REMARK 5.29. The above groups $G \subset \text{Isom}\, H^{n+1}$ are geometrically finite, i.e., although the Dirichlet polyhedron $D_y(G) \subset H^{n+1}$ is infinite-sided (near the parabolic cusps of G), the pseudo-Dirichlet polyhedron $D_y^*(G)$, constructed in (3.33), at the same time has a finite number of sides. This clearly follows for the above groups from the construction and the definition of $D_y^*(G)$ - see also Theorem 6.15. This observation plays an important part in the general case.

PROOF OF THEOREM 5.24 (necessity). We assume that the group G is a discrete isometry group in the Poincaré ball model (in $B^{n+1}(0, 1)$) of the $(n+1)$-dimensional hyperbolic space. Let $P \subset B^{n+1}$ be a finite-sided fundamental polyhedron for the group G. Lemma 5.23 implies that the set $P_0 = \overline{P} \cap S^n(0, 1)$ consists of parabolic cusps and points from the discontinuity set $\Omega(G)$. We may assume here that P contains the origin.

We consider any point $z \in S^n$ and the straight line σ joining the origin 0 to the point z. There are two possibilities: σ intersects either a finite number of images of sides of P or an infinite one.

---

[1] NOTE ADDED IN PROOF. The recently Bowditch [1] approach to the proof of this example is as follows. Let l be the ray through y, perpendicular to τ. Suppose $D_y(G)$ is finite sided. So we have $D_y(G) = \bigcap\{S_y^+(g) : g \in G_0\}$ where $G_0 \subset G$ is finite and $S_y^+(g) = \{x : d(x, y) < d(x, g(y))\}$, are half-spaces with the boundaries $S_y(g) = \partial S_y^+(g)$. Note that l is never parallel to $S_y(g)$ for $g \neq \text{id}$. It follows that $l \subset \text{int}\, F_y$, where $F_y \subseteq F$ is the set of (parallel classes of) rays lying in $D_y(G)$ and F is the $(n-1)$-sphere of infinitely extended rays emanating from y. G acts on F as a non-discrete rotation group fixing τ. Thus, for some $g \in G$, $\text{int}\, F_y \cap g(\text{int}\, F_y) \neq \emptyset$, so that $D_y(G) \cap g(D_y(G)) \neq \emptyset$, contradicting the fundamental property of $D_y(G)$.

In the first case, there exists $g \in G$ such that $g(z) \in P_0$ and therefore either $z \in \Omega(G)$ or z is a parabolic cusp for G. In the second case, there exists a side S of P infinitely many of whose images intersect σ. Then z is a limit point. We consider a sequence of distinct elements $\{g_m\} \subset G$ and a sequence $\{y^m\} \subset S$ of points such that

$$g_m(\sigma) \cap S = \{y^m\}, \tag{5.22}$$

We consider, if necessary, subsequences to get a point $y \in S$, to which the points y^m converge in the Euclidean sense. If here $y \in B^{n+1}$, its compact neighbourhood intersects an infinite set of images of the geodesic σ, so z is an approximation point (A5 in Theorem 5.20). If y lies in S^n, it is a limit point, and then from Lemma 5.23 we have that y is a parabolic cusp. Then we conjugate G to arrive at its action in the half-space H^{n+1} (we may assume that the geodesic σ joins z and $0 \in \mathbb{R}^n \subset \partial H^{n+1}$, while $y = \infty$).

If the stabilizer G_∞ has a free Abelian subgroup of rank n, then the projection on $\mathbb{R}^n$ of the polyhedron P is a bounded set, so the sequence $\{y^m\}$ of points on the side S, $\infty \in \overline{S}$, is such that $\lim_{m\to\infty} y^m_{n+1} = \infty$, and the Euclidean distance $|g_m(0) - g_m(z)|$ between the ends of the geodesics $g_m(\sigma)$ passing through the points $y^m \in S$ will tend to ∞. This and (5.15) for the approximation points shows that z is an approximation point.

If G_∞ is a finite extension of a free Abelian group having rank k, $1 \le k \le n-1$, then there exists a cusp domain

$$U_\infty = \{x \in \mathbb{R}^n : \sum_{i=k+1}^{n} x_i^2 > t > 0\}, \tag{5.23}$$

not containing limit points, including images of the point z. As above, from

$$\lim_{m\to\infty} y^m_{n+1} = \infty \tag{5.24}$$

we get that there exists a sequence $\{\gamma_m\} \subset G^*_\infty \subset G_\infty$ such that for a certain $R > 0$

$$|\gamma_m g_m(z)| \le R \tag{5.25}$$

and moreover

$$\lim_{m\to\infty} |\gamma_m g_m(0)| = \infty. \tag{5.26}$$

Property (5.26) of sequence $\gamma_m g_m$ follows from (5.25) and the fact that the geodesics joining $\gamma_m g_m(0)$ to $\gamma_m g_m(z)$ have points with arbitrarily large $(n+1)$-coordinates, see (5.24).

(5.26) shows on the one hand that $\{\gamma_m g_m\}$ has infinitely many distinct elements, while on the other, it shows, along with property (5.15) that z is an approximation point.

<u>Sufficiency</u>. Let $P \subset H^{n+1}$ be any fundamental Dirichlet polyhedron for a discrete group $G \subset \text{Möb}(n)$ (whose limit set consists of approximation points and parabolic cusps). We note that, if there are two pairs of sides S_1, S'_1 and S_2, S'_2 that are identified by the elements g_1 and g_2, we have $g_1 = g_2$ only if $S_1 = S_2$. In fact, let Q_1 and Q'_1 be half-spaces in H^{n+1} corresponding to S_1 and S'_1 and containing P. If S_2 does not lie in the n-plane containing S_1, then $S_2 \subset Q_1$ and therefore $g_1(S_2) \cap Q'_1 = \emptyset$. This means that $g_1(S_2)$ is a side of P only if S_1 and S_2 lie in the same n-plane, i.e., $S_1 = S_2$.

We now assume that P has an infinite set of sides. Then we see that there exists a countable set of distinct G-images of P converging at a certain point $z \in \partial P$. Then z is obliged to be limit point and, moreover, Corollary 5.22 shows that z is a parabolic cusp of G.

We consider the G-invariant set of cusp domains $\tilde{U}_z \subset H^{n+1} \cup \Omega(G)$ at all parabolic cusps z of G. Then we see that the subset P_0 of the polyhedron P, that is a complement to the cusp domains $\tilde{U}_z$, is compact and therefore $\partial P \cap \overline{P}_0$ consists of a finite number of sides (if this were not so, we would have a limit point in ∂P that is not a parabolic cusp). This observation implies directly that the number of parabolic cusps lying at the boundary of P is finite, which is equivalent (Corollary 5.22) to the set $P^* = \partial P \cap \Lambda(G)$ being finite. Using the finiteness of the set P^* (= the finiteness of the set of conjugacy classes for parabolic subgroups in G) we have the finite reconstruction process (3.33) defining the pseudo-Dirichlet polyhedron $D^*_y(G)$ with a centre $y \in Y(G) \subset H^{n+1}$. It has only a finite set of parabolic cusps. We denote these parabolic cusps by $z_1, \ldots, z_m$, while their cusp domains in $H^{n+1} \cup \Omega(G)$ are denoted by $\tilde{U}_1, \ldots, \tilde{U}_m$.

The structure of the pseudo-Dirichlet polyhedron $D^*_y(G)$ in such cusp domains is described by the following.

LEMMA 5.30. *Let $D^*_y(G)$ be a fundamental pseudo-Dirichlet polyhedron for the discrete group $G \subset \text{Isom}\, H^n$, $n \geq 4$, $y \in Y(G) \subset H^n$. Then its sides are pairwise G-equivalent and, for any parabolic cusp $p \in \partial D^*_y(G)$ with cusp domain $\tilde{U}_p$, the set of sides for $\tilde{U}_p \cap D^*_y(G)$ is finite.*

PROOF. The first statement follows directly from the construction (3.33) of $D^*_y(G)$. The second statement follows from compactness property of the quotient space L_p/G_p for the action of G_p (by Euclidean isometry) on G_p-invariant k-dimensional horocycle $L_p \subset H^{n+1}$, $k = \text{rank}\, G_p$ (see Theorem 3.4) and from precisely G_p-invariantness of cusp domain U_p. Namely, the last fact gives:

$$D^*_y(G) \cap U_p = D^*_y(G_p) = D_{y*}(G_p)$$

where y^* is the orthogonal projection of y to horocycle L_p.

Now, Lemma 5.30 completes the proof of Theorem 5.24, since the polyhedron $D^*_y(G)\setminus(\tilde{U}_1\cup\ldots\cup\tilde{U}_m)$ is relatively compact in $H^{n+1}\cup\Omega(G)$, i.e., it has only a finite set of sides.

COROLLARY 5.31. *Let* $G\subset \text{Möb}(n)$, $n\geq 2$, *be a discrete group. Then an alternative arises: either all fundamental pseudo-Dirichlet polyhedra* $D^*_y(G)\subset H^{n+1}$, $y\in Y(G)\subset H^{n+1}$, *have a finite number of sides or all* $D^*_y(G)$ *have countable sets of sides.*

Note that for $n=2$ such an alternative is valid for Dirichlet polyhedra as well (Marden [1]); however, for $n\geq 3$, this is not true (see Theorem 5.27).

COROLLARY 5.32. *The limit set* $\Lambda(G)$ *for a geometrically finite group* $G\subset \text{Möb}(n)$, $n\geq 2$, *either has a zero* n*-dimensional Lebesgue measure or* $\Lambda(G)=\overline{\mathbb{R}}^n$.

The proof amounts to the approximation points being uniformly approximated for a geometrically finite group, i.e., there exists a constant $k>0$ such that for each approximation point $z\in\Lambda(G)$ (we assume that G acts in the ball $B^n(0,1)$):

$$|z-g_m(\infty)| \leq k\cdot r_m^2 \tag{5.27}$$

where r_m is the radius of the isometric sphere $I(g_m)$.

In the proof of necessity for Theorem 5.24, we saw that, if $\{y^m\}$ is a point sequence converging to one of a finite number of parabolic cusps of the polyhedron $\overline{P}$, then there exist elements γ_m from its stabilizer such that the points $\gamma_m g_m(0)$ lie outside a certain neighbourhood of the set $\{\gamma_m g_m(z)\}$, so the bound of (5.27) follows from the finiteness of the set $\overline{P}\cap\Lambda(G)$, formula (5.9), and property (5.10) of the approximation points.

We now assume that $\Lambda(G)\neq S^{n-1}(0,1)$. We consider for any m coverings of the approximation point set by balls having radii $|z-g_m(\infty)|$ and allow m to tend to ∞. Then from (5.27) and Lemma 5.20 we get that $\Lambda(G)$ has zero measure, since the parabolic set is countable. The proof is completed.

REMARK 5.33. Ahlfors [3] proved this theorem for $n=2$, where his proof employs the properties of harmonic functions in hyperbolic space, which was extended by Su-Shing Chen [1] to the case of groups $G\subset \text{Möb}(n)$, where the approximation point set has a positive measure. Our proof follows the Beardon-Maskit [1] scheme for Kleinian groups in the plane.

Corollary 5.32 can be strengthened considerably if one shows that the Hausdorff dimensions $\dim_H\Lambda(G)<n$ for geometrically finite Kleinian groups $G\subset \text{Möb}(n)$.

DEFINITION 5.34. Let $X\subset\overline{\mathbb{R}}^n$ be a set from the Borel σ-algebra in the space $\overline{\mathbb{R}}^n$ and let, for a certain $\alpha>0$ and for each $\varepsilon>0$,

$$h_{\alpha,\varepsilon}(X) = \inf\sum_i d(E_i)^\alpha,$$

where the lower bound is taken over all finite or countable coverings

$\{E_i\}$ of the set X by balls (or cubes) E_i whose diameters $d(E_i)$ do not exceed ε. Then

$$\lim_{\varepsilon\to 0} h_{\alpha,\varepsilon}(X) = h_\alpha(X)$$

is called the *Hausdorff* *α-measure* for the set X. The Hausdorff dimension of X is called a number

$$\dim_H X = \sup\{\alpha : h_\alpha(X) \neq 0\}.$$

THEOREM 5.35. *The Hausdorff dimension of the limit set of any geometrically finite Kleinian group* $G \subset \text{Möb}(n)$, $n \geq 1$, *is less than* n.

This result is familiar for $n=1$, i.e., for finitely generated Fuchsian groups of the second kind on the plane (Beardon [5] and Patterson [1]). For $n=2$, it was first proved by Sullivan [6], whose proof can be extended to any n. Previously, Sullivan ([1], §3) proved Theorem 5.35 for compact-type Kleinian groups $G \subset \text{Möb}(n)$, subject to the additional condition that the n-dimensional Lebesgue measure $\mu_n(\Lambda(G)) = 0$ (this constraint is removed by Corollary 5.32). Tukia [3] has given the proof of Theorem 5.35 for the general case.

The condition for geometrical finiteness of group $G \subset \text{Möb}(n)$ in Theorem 5.35 is important, since Sullivan [2] constructed an example of a finitely generated Kleinian group $G \subset \mathcal{M}_2$ (geometrically infinite) for which $\dim_H \Lambda(G) = 2$, but the two-dimensional Lebesgue measure was zero.

We demonstrate the proof of Theorem 5.35 for the case of the group $G \subset \text{Möb}(n)$ of the compact-type, i.e., one having the compact orbifold

$$O(G) = [H^{n+1} \cup \Omega(G)]/G.$$

We show firstly that there exists an integer q such that for decomposing any n-dimensional cube $Q \subset \mathbb{R}^n$ into q^n equal subcubes (forming the family $K(Q, q)$), one can find at least one subcube $Q' \in K(Q, q)$ that does not intersect $\Lambda(G)$. The proof of this is based on a compactness property.

We assume that $r = d(Q \cap \Lambda(G))/d(Q) \geq 1/2$ (if $r < 1/2$, the existence of q $(q=4)$ is obvious). By z_Q we denote the centre of Q, while by s_Q we denote the edge length of the cube Q, and by $\bar{z}_Q = (z_Q, s_Q)$ we denote a point in hyperbolic space $H^{n+1} \cong \mathbb{R}^{n+1}_+$. Let $x, y \in Q \cap \Lambda(G)$ be such that $|x-y| \geq d(Q)/2$, and let the geodesic $\gamma \subset H^{n+1}$ join x and y. Then for the hyperbolic metric d_h:

$$d_h(\gamma, \bar{z}_Q) < C_0 = \log 2 + \log n + \sqrt{n}.$$

This inequality shows that

$$d_h(H_G, \bar{z}_Q) < C_0$$

where H_G is the hyperbolic convex hull of $\Lambda(G)$, see §2. In our case, H_G/G is compact and therefore there exists a compact set $\mathbb{K} \subset H^{n+1}$ such that for any cube $Q \subset \mathbb{R}^n$, $Q \cap \Lambda(G) \neq \emptyset$, we have $\bar{z}_Q \in G(\mathbb{K})$.

Now let $Q_0 \subset \mathbb{R}^n$ be a fixed "standard cube" and β be a similarity in $\mathbb{R}_+^{n+1}$, $\beta(Q) = Q_0$. Then we can put

$$\beta = hg, \tag{5.28}$$

where $g \in G$, $g(\bar{z}_Q) \in \mathbb{K}$, while h is a certain element from $\text{Möb}(n)$. As $h^{-1}(\bar{z}_{Q_0}) \in \mathbb{K}$, the set of elements h from (5.28) is a compact set in $\text{Möb}(n)$. Also

$$\beta(Q \cap \Lambda(G)) = Q_0 \cap hg(\Lambda(G)) = Q_0 \cap h(\Lambda(G)), \tag{5.29}$$

where $h(\Lambda(G))$ is nowhere dense in $\mathbb{R}^n$, while h runs through a compact set in $\text{Möb}(n)$. This demonstrates the existence of the necessa number q.

Now let $\mathcal{L}(Q) = \{Q' \in K(Q, q) : Q' \cap \Lambda(G) \neq \emptyset\}$. The following inequality derives from the above property of q:

$$\sum_{Q' \in \mathcal{L}(Q)} d(Q')^{\alpha} \le C \cdot d(Q)^{\alpha}, \tag{5.30}$$

where $\alpha = n$ and $C = 1 - 1/q^n$. The inequality (5.30) remains true for $\alpha < n$ (slightly) and slightly larger C, $1 - 1/q^n < C < 1$. Therefore, if we transfer from $\mathcal{L}(Q)$ to the family $\mathcal{L}(Q')$, where $Q' \in \mathcal{L}(Q)$, (5.30) gives us the bound $\dim_H \Lambda(G) < n$ via inductive arguments.

This proof can be generalized to geometrically finite groups $G \subset \text{Möb}(n)$, not having fixed parabolic points of rank n (Tukia [6]). Here the sets $h(\Lambda(G))$ in (5.29) form a family of subsets in $\overline{\mathbb{R}}^n$ compact in the Hausdorff topology induced by the following metric in the family $\mathcal{C}$ of closed subsets in $\overline{\mathbb{R}}^n$:

$$\rho(X, Y) = \sup\{q(x, Y),\quad q(y, X) : x \in X,\quad y \in Y\}. \tag{5.31}$$

Here $X, Y \in \mathcal{C}$ and q is the spherical metric in $\overline{\mathbb{R}}^n$ defined by stereographic projection (1.11) on the sphere S^n.

The proof of Theorem 5.35 for the general case (Tukia [3]) is obtained by improving the above compactness method, namely as above, we define a family $\mathcal{L}(Q)$ of subcubes for a given cube Q that cover the set

$$(Q \cap \Lambda(G)) \setminus \{\text{parabolic points of } G\}.$$

Here the family $\mathcal{L}(Q)$ satisfies (5.30) for certain $\alpha < n$ and $C < 1$, and also a certain additional condition near the parabolic points of the group G (which also holds for families $\mathcal{L}(Q')$ for all $Q' \in \mathcal{L}(G)$).

The construction of the $\mathcal{L}(Q)$ families enables one to employ the inductive method again and get the required inequality:

$$\sup\{\alpha : \lim_{\varepsilon\to 0} h_{\alpha,\varepsilon}(\Lambda(G) \neq 0\} < n.$$

§4. Finiteness theorems for discrete Möbius groups

1. *Finiteness in the plane.* A basic fact for Fuchsian groups in the plane is that such a group Γ is finitely generated if and only if there is a fundamental (hyperbolic) polygon bounded by a finite number of sides, i.e., it is geometrically finite. That fact has many geometrical, algebraic and metrical consequences. For example, if $\Gamma \subset \mathcal{M}_2$ is a finitely generated Fuchsian group in a disc U, then:

1) the surface $S = U/\Gamma$ has a finite topologic type;

2) the covering $\pi : U \longrightarrow S$ is branched at not more than over a finite number of points;

3) Γ has only a finite number of equivalence classes for elliptic elements;

4) there exists only a finite number of disjoint Γ-orbits for parabolic fixed points;

5) the lengths of the hyperbolic translations in U (of the hyperbolic elements in Γ) form a discrete subset in $\mathbb{R}$;

6) the intersection of two finitely generated subgroups is again a finitely generated group; and

7) the limit set $\Lambda(G)$ consists only of approximation points and parabolic cusps.

This basic fact does not apply for Kleinian groups in the plane (= Fuchsian groups in $\overline{\mathbb{R}}^3$), as there exist finitely generated Kleinian groups each of which does not have a finite-sided fundamental polyhedron in H^3. Such groups have been called geometrically infinite (see previous section). The situation here is rectified in part by Ahlfors's finiteness theorem, the most profound theorem in the theory of Kleinian groups in the plane (Ahlfors [2], Kulkarni and Shalen [1]):

THEOREM 5.36. *Let $G \subset \mathcal{M}_2$ be a finitely generated Kleinian group and $\Omega \subset \overline{\mathbb{R}}^2$ its discontinuity set. Then*

1) *the quotient Ω/G has a finite number of components;*

2) *each component in Ω/G is obtained from a closed Riemann surface by a finite number of punctures, and*

3) *there exists only a finite number of points in Ω/G over which the covering $\pi : \Omega \longrightarrow \Omega(G)$ is branched.*

Another formulation for Ahlfors's theorem can be given as (see Beardon and Jørgensen [1]):

COROLLARY 5.37. *A finitely generated Kleinian group $G \subset \mathcal{M}_2$ has a finite-sided fundamental polygon in $\overline{\mathbb{R}}^2$ (such a polygon in particular is the intersection of the sphere $\overline{\mathbb{R}}^2$ and the closure of a Dirichlet polyhedron in $\overline{H}^3$, which is not always connected).*

This formulation of Ahlfors's theorem shows that the infinite set of sides for the fundamental polyhedron $P(G) \subset H^3$ of the finitely generated discrete group $G \subset \mathcal{M}_2$ accumulates to points from $P^* = \overline{P} \cap \Lambda(G)$ not lying at the boundary of the planar fundamental domain for the group G. Ahlfors and Sullivan showed that the spherical measure for that set P^* is zero. Also, there is a conjecture that the set P^* is finite. A partial confirmation of that conjecture is Sullivan's [3] theorem on the finiteness of the parabolic-point subset in P^*:

THEOREM 5.38. *Let* $G \subset \mathcal{M}_2$ *be a discrete group having* N *generators. Then the number* N_C *of conjugation classes of non-trivial maximal parabolic subgroups of* G *is finite and satisfies* $N_C \le 5N - 4$.

The proof of this is based on examining constructions proposed by Ahlfors [2] in the proof of his finiteness theorem, and in particular on the construction of Borel series $\sum a_n(z - b_n)^{-1}$, where b_n is the centre of an n-th horoball in the sphere at infinity $\partial H^3 = \overline{\mathbb{C}}$, while a_n is a complex number related to the horoball's Euclidean volume. These Borel series are used to define crossed homomorphisms for the group G in the space of fourth degree polynomials and in the space of G-invariant automorphic forms having degree -2. Here Ahlfors's arguments [2] are interpreted in terms of holomorphic square vector fields.

Another important part of Sullivan's proof is his ergodic theorem on the non-existence of a G-invariant vector field (Sullivan [1], see also Theorem 9.24) together with bounds for the volumes of disjoint horoballs whose centres lie in a certain concentric ring.

We note that for this 3-dimensional case, the set of parabolic horoballs in H^3 is "discrete". The situation is different in higher dimensions: for $n \ge 4$ there exist non-discrete orbits of parabolic horoballs for some discrete groups $G \subset \mathrm{Isom}\, H^n$. This was proved recently by Epstein-Penner [1] for groups constructed in the proof of Theorem 4.25 (Apanasov [21]).

At this point, we finish the survey of finiteness properties for $n = 2$. A notable exposition (with proofs) is to be found in Greenberg's [4] lecture.

We further show that, if we restrict consideration to discrete geometrically finite subgroups in $\text{Möb}(n)$, $n \ge 3$, then most of the finiteness properties for Fuchsian groups in that case are present (although some unpleasant properties remain: for example, a geometrically finite group may contain geometrically infinite subgroups, see examples given by Jørgensen [3] and Jørgensen and Marden [1], as well as Corollary 5.52) - see also Remark 7.25.

2. *Finiteness in space.* Let $G \subset \text{Möb}(n)$ be a discrete torsion free group whose action is considered also in the half-space $\mathbb{R}^{n+1}_+$, having the hyperbolic Poincaré metric ($\mathbb{R}^{n+1}_+ \cong H^{n+1}$). As previously in (5.4) and (5.5), we define the manifold

$$M(G) = [H^{n+1} \cup \Omega(G)]/G$$

and its minimal convex retract $C(M) = M_G \subset M(G)$, which can be put in the following form (see Theorem 5.12):

$$C(M) = M_G = [H_G \backslash \Lambda(G)]/G,$$

where H_G is the hyperbolic convex hull of $\Lambda(G)$.

As we have seen previously, it is possible that

$$\dim M_G < \dim M(G) = n+1.$$

This property of the retract M_G requires us to distinguish its ε-neighbourhood for an $\varepsilon > 0$, which we denote by $U_\varepsilon(M_G) \equiv C_\varepsilon(M)$.

For any manifold M and for an $\varepsilon > 0$, by $M_{(0,\varepsilon]} \cup M_{[\varepsilon,\infty)} = M$ we denote the Margulis decomposition (see Chapter 4 §4), which cuts off ε-thin parts of manifold M.

The main result of this section is the following finiteness theorem.

THEOREM 5.39 (Apanasov [14], [19]). *For a discrete torsion-free group* $G \subset \text{Möb}(n)$, $n \geq 2$, *the following statements are equivalent*:

GF1) *the group* G *is geometrically finite*: *there is some finite sided fundamental polyhedron* $P \subset H^{n+1}$ *for* G;

GF2) *for some (any)* $\varepsilon > 0$, *the manifold* $U_\varepsilon(M_G)$ *has a finite hyperbolic volume*;

GF3) *for some (any)* $\varepsilon > 0$, *the* ε*-thick manifold* $(M_G)_{[\varepsilon,\infty)}$ *is compact*;

GF4) *the limit set* $\Lambda(G)$ *consists of approximation points and parabolic cusps*;

GF5) *the manifold* $M(G)$ *has finitely many ends, and each such end is conformally equivalent to a standard cusp end, i.e., the end of* $(\overline{H}^{n+1} \setminus \{\infty\})/\Gamma$ *for a discrete torsion free parabolic group* Γ, $\Gamma(\infty) = \infty$ *(i.e.,* $\Gamma \subset \text{Isom}\,\mathbb{R}^n$*)*.

REMARK 5.40. The condition that G is a torsion-free group is not essential for all properties of finiteness, except GF2) (where, to provide that the order of elliptic elements of G is bounded, some additional conditions are required - either $\text{vol}\, M(G) < \infty$ or finite generation of G). For GF1) and GF4) it follows from Theorem 5.24. For the cases GF3) and GF5) there are finite-sheeted coverings of the orbifolds $O(G) = [H^{n+1} \cup \Omega(G)]/G$ and its retract $O_G \subset O(G)$ by manifolds M_0 and $C(M_0)$ with a subgroup $G_0 \subset G$ of finite index associated to them. Since the covering is finite-sheeted and $\mu(G_0, G) < \infty$, properties of the limit sets $\Lambda(G)$ and $\Lambda(G_0)$ and of the orbifolds $O(G)$, O_G and the manifolds $M_0 = M(G_0)$, $C(M_0) = M_{G_0}$ are analogous.[1]

---

[1] NOTE ADDED IN PROOF. A recent work by Bowditch [1] contains some different approach to the proof of Theorem 5.39 for hyperbolic torsion groups (cf. Corollary 5.42). He replaces the notion of the fundamental polyhedron in GF1) by a similar notion of a convex cell complex K with

Theorem 5.39 and the above Remark 5.40 demonstrate correctness for the following definition, which is equivalent to Thurston's definition [1] for three-dimensional manifolds:

DEFINITION 5.41. The structure of the complete hyperbolic orbifold $\mathcal{O}$, $\dim \mathcal{O} = n \geq 2$, having holonomy d_*, is geometrically finite if the discrete group $G = d_*(\pi_1(\mathcal{O})) \subset \mathrm{Isom}\, H^n$ is geometrically finite.

We get directly from Theorem 5.39 and Remark 5.40:

COROLLARY 5.42. *The hyperbolic orbifold* $\mathcal{O}$, $\dim \mathcal{O} = n \geq 2$, $\mathrm{rank}\, \pi_1(\mathcal{O}) < \infty$, *having holonomy* d_*, *has a geometrically finite structure if and only if one of the following equivalent conditions is satisfied*:

I) *for a certain (any)* $\varepsilon > 0$, *the minimal convex retract* $C(\mathcal{O})$ *of the orbifold* $\mathcal{O}$ *has an* ε*-neighbourhood* $C_\varepsilon(\mathcal{O}) \subset \mathcal{O}$, *having a finite volume*;

II) *for a certain (any)* $\varepsilon > 0$, *the* ε*-thick suborbifold* $C(\mathcal{O})_{[\varepsilon,\infty)}$ *is compact; and*

III) the limit set for the discrete group $d_*(\pi_1(\mathcal{O}))$, *where* $d_* : \pi_1(\mathcal{O}) \longrightarrow \mathrm{Isom}\, H^n$ *is the holonomy, consists of approximation points and parabolic cusps.*

REMARK 5.43. For torsion free discrete groups $G \subset \text{Möb}(2)$ in the plane $\overline{\mathbb{C}}$, Thurston (Chapter 8.4 in [1]) proposed a different scheme for proving the equivalence of GF1) - GF3), without using the properties of approximation points, but requiring Ahlfors's finiteness theorem for planar Kleinian groups. Greenberg [4] has described the topological situation for the equivalence GF1) $\Longleftrightarrow$ GF5) for the case $n \leq 2$ subject to additional constraints: all parabolic points of the group are parabolic cusps. He also described the history of these topics.

DEFINITION 5.44. A fundamental polyhedron $P \subset H^n$ for the discrete group $G \subset \mathrm{Isom}\, H^n$ is called *convex* at its ends if for each point $z \in P^* = \overline{P} \cap \Lambda(G)$ there exists a horoball $B_z \subset H^n$ centred at the point z (a Euclidean ball in $\mathbb{R}^n_+ \cong H^n$, tangent at z to the boundary plane $\overline{\mathbb{R}}^{n-1}$) such that $P \cap B_z$ contains the geodesic ray $[x, z)$ for any $x \in P \cap B_z$.

To prove Theorem 5.39, we require the following property of polyhedra, convex at their ends (in particular, pseudo-Dirichlet polyhedra (3.33)), compare Apanasov [14], [19]:

LEMMA 5.45. *Let* $G \subset \mathrm{Isom}\, H^n$, $n \geq 2$, *be a discrete torsion free group and let the manifold* $[M_G]_{[\varepsilon,\infty)}$ *be compact. Then the limit-vertex set* $P^* = \overline{P} \cap \Lambda(G)$ *for any fundamental polyhedron* $P = P(G) \subset H^n$, *convex at its ends, is either empty or consists of parabolic fixed points of the group* G.

---

polyhedron in GF1) by a similar notion of a convex cell complex K with a finite set K/G of orbits under G (cf. Aleksandrov's complex in Chapter VI §1).

PROOF. For the closure $\overline{P}$ of the polyhedron P in $\overline{H}^n = H^n \cup \partial H^n$ we denote by P_H the polyhedron in the Nielsen hull H_G:

$$P_H = \overline{P} \cap H_G \setminus \Lambda(G),$$

where one gets the manifold M_G by identifying its G-equivalent sides.

We assume that $P^* \neq \emptyset$. Then we can assume that $\infty \in P^*$ and that P_H contains the geodesic ray σ:

$$\sigma = \{x \in H^n : x = t \cdot e_n, \quad t \geq 1\} \subset \operatorname{int} P_H.$$

We consider the point sequence $y^i = e_n \cdot \exp(i)$, $i \geq 0$, on the ray σ, where $d(y^i, y^{i+1}) = 1$ is the distance between them.

From the compactness condition for the ε-thick submanifold $[M_G]_{[\varepsilon,\infty)}$ of $M_G = P_H/G$, it follows that G has a sequence $\{g_i\}$ of elements such that

$$\lim_{i\to\infty} \gamma_i g_i \gamma_i^{-1} = \mathrm{id}, \tag{5.32}$$

where $\gamma_i : x \longrightarrow x \cdot \exp(-i)$.

The element sequence $g_i \in G$ is obtained as follows (see the proof of Theorem 4.20 and Kazhdan and Margulis [1]). For any $\varepsilon > 0$, we find a number i_ε such that for all $i \geq i_\varepsilon$ there are elements $g_{i,\varepsilon} \in G$ that translate the points $\gamma_i^{-1}(e_n) \in \sigma$ into ε-close ones (the elements $g_{i,\varepsilon}$ correspond to non-trivial loops in M_G having lengths less than ε). We now consider a sufficiently small $\varepsilon > 0$, and for any i we can obtain an arbitrarily large number of such elements $g_{i,\delta} \in G$ that translate the point $\gamma_i^{-1}(e_n)$ into points ε-close to it. We consider the corresponding frame in the tangent space to H^n and select ε-near ones amongst them. Then from the elements $g_{i,\delta} \in G$ we select an element $g_i \in G$, which at the point $\gamma_i^{-1}(e_n)$ has the derivative differing by not more than ε from the identity, while the point $\gamma_i^{-1}(e_n)$ goes over to an ε-close one. This demonstrates the existence of the sequence $\{g_i\} \subset G$ having the property (5.32).

At the same time, we have

$$d(e_n, \gamma_i g_{i+1} \gamma_i^{-1}(e_n)) = d(e_n/e, \gamma_{i+1} g_{i+1} \gamma_{i+1}^{-1}(e_n/e)$$

which with (5.32) shows that

$$\lim_{i\to\infty} d(e_n, \gamma_i g_{i+1} \gamma_i^{-1}(e_n)) = 0.$$

This relation shows that the sequence $\{\gamma_i g_{i+1} \gamma_i^{-1}\}$ converges to a non-degenerate transformation.

Then we apply Theorem 6.46 to the sequence of discrete torsion-free groups $\gamma_i G_i \gamma_i^{-1}$ to get that the transformations $\gamma_i g_i \gamma_i^{-1}$ and $\gamma_i g_{i+1} \gamma_i^{-1}$ for sufficiently large $i \geq i_0$ have a common fixed point. It is clear that the same property occurs in the

We write the obvious equations

$$d(y^i, g_i(y^i)) = d(\gamma_i^{-1}(e_n), g_i(\gamma_i^{-1}(e_n)) = d(e_n, \gamma_i g_i \gamma_i^{-1}(e_n))$$

and see from (5.32) that

$$\lim_{i\to\infty} d(y^i, g_i(y^i)) = 0$$

and consequently the images $g_i(\sigma)$ of the geodesic σ have a convergence point.

Images of a geodesic in a discrete group having a common fixed point and consisting either of parabolic elements or of loxodromic ones converge either to that point or to the common axis of the loxodromic elements. At the same time, fixed points of loxodromic elements in the discrete group G cannot lie at the boundary of the fundamental polyhedron P (Corollary 5.22), so all the elements g_i for $i \geq i_0$ are parabolic transformations, and the point $\infty \in P^* = \overline{P} \cap \Lambda(G)$ is their common fixed point. This completes the proof of Lemma 5.45.

COROLLARY 5.46. *Under the conditions of Lemma* 5.45, *the set* P^* *of limit vertices of the fundamental polyhedron* $P \subset H^n$ *is finite and consists (if it is not empty) of parabolic cusps for the group* $G \subset \operatorname{Isom} H^n$.

PROOF. We can assume that the point $z \in P^*$ is ∞; then its stabilizer $G_\infty \subset G$, due to Lemma 5.45, is a parabolic group, i.e., a Euclidean translation group (possibly screw translations) in $\mathbb{R}^{n-1} = \partial H^n \setminus \{\infty\}$, $\operatorname{rank} G_\infty = k$, $1 \leq k \leq n-1$, with the space $E = \mathbb{R}^k \subset \partial H^n$ of translation vectors (see Theorem 3.4).

Suppose that $z = \infty$ is not a parabolic cusp. Now, if $\Lambda(G)$ is contained in some Euclidean t-neighbourhood $U_t(E) \subset \mathbb{R}^n$ of E, then Theorem 4.33 gives a constant $K > 0$ which bounds the radii of the isometric spheres $I(g)$ for $g \in G \setminus G_\infty$. So, $(t+K)$-neighbourhood U_{t+k} of E has the complement $U_\infty \subset \mathbb{R}^{n-1}$, which is a cusp domain for ∞, i.e., $z = \infty$ is a parabolic cusp for G.

Suppose now that $k < n-1$, $E = \mathbb{R}^k$, and there is a sequence $\{y^m\} \subset \Lambda(G)$ such that

$$\lim_{m\to\infty} \sum_{i=k+1}^{n} (y_i^m)^2 = \infty. \tag{5.33}$$

We may regard that the sequence $\{y^m\}$ lies in the projection of the polyhedron P to ∂H^n, i.e., in the fundamental polyhedron $P(G_\infty) \cap \mathbb{R}^{n-1}$. Joining the points y^m to ∞ by geodesics contained in the Nielsen hull H_G we see that they intersect the polyhedron $P_H = P \cap H_G$. Then the Margulis lemma (Theorem 4.20) and Remark 4.22 about ε-thin components show that the property (5.33) contradicts the fact that the part of P_H obtained by cutting off ε-thin pieces (for small $\varepsilon > 0$) is compact. This contradiction proves that $z = \infty$ is a parabolic cusp in this case, too.

Simultaneously at the expence of compactness of the ε-cut-off

Simultaneously at the expence of compactness of the ε-cut-off polyhdron $[P_H]_{[\varepsilon,\infty)}$ we obtain the finiteness of the set P^*. The proof is complete.

THE PROOF OF THEOREM 5.39 consists of the proof of the assertions

$$GF1) \Rightarrow GF2) \Rightarrow GF3) \Rightarrow GF4)$$

since Theorem 5.24 closes that chain; we will prove the last statement GF5) in Chapter 7 §1.

GF1) $\Rightarrow$ GF2). Let $P \subset H^{n+1} \cong \mathbb{R}^{n+1}_+$ be a certain finite-sided fundamental polyhedron for the group $G \subset \text{Möb}(n)$ having properties F1 - F5; then Lemma 5.23 shows that the set $P^* = \overline{P} \cap \Lambda(G)$ is either empty or consists of a finite number of parabolic cusps of the group G. Let ∞ be one of those cusps, which means (Definition 5.16) that either the limit set $\Lambda(G)$ lies in the set $L \subset \overline{\mathbb{R}}^n = \partial H^{n+1}$ that is Euclidean isometric to the set

$$\{x \in \mathbb{R}^n : x_{k+1}^2 + \ldots + x_n^2 < t\}, \quad t > 0, \quad k = \text{rank}\, G_\infty < n,$$

or that the stabilizer G_∞ has a normal finite-index subgroup generated by translations on n linearly independent vectors from the space $\mathbb{R}^n$.

In the first case, the convex Nielsen hull H_G is contained in $L \times [0, \infty)$. As then the set L has bounded sizes in $n-k$ linearly independent directions, there is a horoball B_∞ (i.e., a half-space $H_\alpha = \{x \in H^{n+1} : x_{n+1} > \alpha\}$) such that $H_G \cap B_\infty = [L \times [0, \infty)] \cap B_\infty$. At the same time, the projection of the polyhedron P on $\partial H^{n+1} = \overline{\mathbb{R}}^n$ in the remaining k linearly independent directions also has bounded sizes, since translations from the stabilizer G_∞ act in those directions (in general, screw ones). Then we have the inequalities

$$\begin{aligned} &\text{vol}\, U_\varepsilon[M_G \cap (B_\infty/G_\infty)] = \\ &\quad = \text{vol}[U_\varepsilon(H_G) \cap P \cap B_\infty] \leq \text{vol}[U_\varepsilon(L \times [0, \infty)) \cap B_\infty \cap P] < \infty, \end{aligned} \tag{5.34}$$

which shows that the hyperbolic volume of an ε-neighbourhood $U_\varepsilon(M_G)$ of the end of the manifold M_G corresponding to the parabolic cusp ∞ is finite.

To consider the second case, we note that if the group G_∞ contains a free Abelian subgroup of rank n, while z is any limit point from $\Lambda(G)$, then the convex hull $H(G_\infty(z)) \subset H_G$ of the lattice $G_\infty(z)$ contains a certain half-space $H_\alpha = \{x \in H^{n+1} : x_{n+1} > \alpha\}$. Therefore in the second case, there exists a horoball $B_\infty = H_\alpha$ centred at ∞ such that $B_\infty = H_G \cap B_\infty$. This implies an inequality for the volume $\text{vol}\, U_\varepsilon[M_G \cap (B_\infty/G_\infty)]$, analagous to (5.34), i.e., it shows that in this case also the volume of an ε-neighbourhood of the end of the manifold M_G corresponding to ∞ is finite.

Therefore, the sum of volumes of ε-neighbourhoods of all ends of the manifold M_G is finite. The compactness for the ε-thick submanifold of the manifold $U_\varepsilon(M_G)$ completes the proof that $\text{vol}\, U_\varepsilon(M_G) < \infty$.

GF2) ⇒ GF3). The Margulis decomposition definition for the manifold $M = M_{(0,\varepsilon]} \cup M_{[\varepsilon,\infty)}$ implies that each point from $(M_G)_{[\varepsilon,\infty)}$ has an embedded $\varepsilon/2$-ball neighbourhood in the $\varepsilon/2$-neighbourhood $U_{\varepsilon/2}(M_G)$, which has finite volume. It is clear that there can be only finitely many such disjoint balls, which demonstrates compactness for the manifold $(M_G)_{[\varepsilon,\infty)}$.

GF3) ⇒ GF4). Let the manifold $(M_G)_{[\varepsilon,\infty)}$ be compact and let $P \subset H^{n+1}$ be a certain fundamental polyhedron of the group G, convex at its ends (see Definition 5.41) and having properties F1 - F5 from §1. Examples in particular being all Dirichlet polyhedra $D_y(G)$ as well as the pseudo-Dirichlet polyhedra $D_y^*(G)$ constructged in (3.33) whose intersections with horoball neighbourhoods $B_p \subset H^{n+1}$ for the parabolic cusps p of the group G are finite-sided fundamental polyhedra for the parabolic stabilizers $G_p \subset G$ - see Lemma 5.30.

Lemma 5.45 and Corollary 5.46 show that the set $P^* = \overline{P} \cap \Lambda(G)$ is either empty or consists of finitely many parabolic cusps p for the group G. Then the Margulis lemma (Theorem 4.20) and Remark 4.22 show that, for sufficiently small ε, the compact manifold $(M_G)_{[\varepsilon,\infty)}$ has the following properties.

First, $(M_G)_{[\varepsilon,\infty)}$ coincides with the manifold M_G if the polyhedron P does not have limit vertices $(P^* = \varnothing)$.

Second, if $P^* \neq \varnothing$, then $(M_G)_{[\varepsilon,\infty)}$ is obtained (up to a compact set, set Remark 4.22) as a result of the following construction.

Due to the finiteness of P^*, we consider the horoballs $B_\delta(p) = B_\delta(c) \subset H^{n+1}$ at all parabolic fixed points p of the group G specified by (4.62) and having sufficiently small equal radii $\delta = \delta(\varepsilon)$ at G-equivalent points, which are specified by the parametrizing G-invariant function $\rho_c(x)$ (Definition 4.32), i.e. G-invariant family of disjoint horoballs $B_\delta(p)$ which are precisely invariant with respect to their stabilizers $G_p \subset G$. Also, as the union of all these horoballs $B_\delta(p)$ is G-invariant, we get that the domain

$$H_G \backslash \bigcup_p B_\delta(p), \tag{5.35}$$

is invariant with respect to the group G (where the union is taken over all parabolic fixed points in G). Factorizing (5.35) by the action of the group G gives (up to a certain compact submanifold $M_c \subset M_G$) the compact manifold $(M_G)_{[\varepsilon,\infty)}$:

$$(M_G)_{[\varepsilon,\infty)} \cong [H_G \backslash \bigcup_p B_\delta(p)]/G. \tag{5.36}$$

That compact manifold M_c arises by virtue of a possible difference between the parabolic ε-neighbourhoods $V_p(G, \varepsilon)$ and the horoballs $B_\delta(p)$: see (4.27) - (4.29).

Lemma 3.16 and the pseudo-Dirichlet construction (see (3.33)) show that the Dirichlet polyhedron $D_y(G)$ and the pseudo-Dirichlet polyhedron $D_y^*(G)$ centred at $y \in Y(G) \subset H^{n+1}$ will not have equivalent parabolic points at their boundaries. We put $P = D_y^*(G)$ and let

$P^* = \{p_1, \ldots, p_l\}$, where $G(p_i) \cap G(p_j) = \emptyset$ for $i \neq j$. The finiteness of this set follows from Corollary 5.46. Then (5.36) is rewritten as

$$(M_G)_{[\varepsilon,\infty)} \cong [H_G \cap (\overline{P} \setminus \bigcup_{i=1}^{l} B_\delta(p_i))]/G. \tag{5.37}$$

Repetition of the arguments in the sufficiency proof of Theorem 5.24 then shows that G has a finite-sided fundamental polyhedron in H^{n+1}. This and Theorem 5.24 show finally that $\Lambda(G)$ for G consists of approximation points and parabolic cusps, i.e., that GF4) holds. The proof is complete.

I complete this section with some simple consequences from the finiteness Theorem 5.39.

COROLLARY 5.47. *If* $G \subset \text{Möb}(n)$, $n \geq 2$, *is a geometrically finite group, then*:

1) G *has only a finite number of equivalence classes of elliptic elements*;

2) G *has only a finite number of orbits of parabolic fixed points*;

3) *the lengths of the loxodromic elements in* G *(translations in the hyperbolic space* H^{n+1}*) form a discrete subset in* $\mathbb{R}$; *and*

4) $\Omega(G)/G$ *consists of a finite number of connected components.*

COROLLARY 5.48. *Let* $G \subset \text{Möb}(n)$, $n \geq 2$, *be a discrete non-discontinuous group in* $\overline{\mathbb{R}}^n$; *then the following assertions are equivalent*:

1) *the group* G *is geometrically finite*;

2) *the hyperbolic orbifold* H^{n+1}/G *has a finite volume*;

3) *the orbifold*

$$[H^{n+1} \setminus \bigcup_{p} B_\delta(p)]/G,$$

is compact, where the union is taken over all the parabolic points p *in* G, *the horoballs* $B_\delta(p) \subset H^{n+1}$ *with centres at the points* p *all having the same sufficiently small radius* $\varepsilon > 0$; *and*

4) $\overline{\mathbb{R}}^n$ *consists of approximation points and parabolic cusps of* G;

5) H^{n+1}/G *has finitely many ends, and each such end is equivalent to* $E^n \times [0, \infty)$, *where* E^n *is some Euclidean* n*-space form (finite covered by* n*-torus* T^n*).*

COROLLARY 5.49. *If* G_1 *and* $G_2 \subset \text{Möb}(n)$, $n \geq 2$, *are commensurable discrete groups, then* G_1 *and* G_2 *simultaneously are or are not geometrically finite.*

The proofs of these corollaries follow simply from Theorem 5.39. Selberg [2] and Garland and Raghanathan [1] (see also Wielenberg [1]) first derived the proof of assertion 1) $\Leftrightarrow$ 2) in Corollary 5.48; for $n = 2$, there is a topological proof for Corollary 5.49 without the use of GF4) in Theorem 5.39 (see Greenberg [4]).

There is a further consequence concerned with the action of the geometrically finite Kleinian group $G \subset \text{Möb}(n)$, $n \geq 2$, near a fixed point $p \in \overline{\mathbb{R}}^n$ for a parabolic element whose stabilizer has a free Abelian subgroup of rank $n-1$.

COROLLARY 5.50. *In a geometrically finite Kleinian group* $G \subset \text{Möb}(n)$, $n \geq 2$, *there is a normal subgroup* $G^* \subset G$ *of finite index having the following property: for any parabolic point* $p \in \overline{\mathbb{R}}^n$ *of* G *with stabilizer of rank* $n-1$ *there exist* $\varepsilon = \varepsilon(p) > 0$ *and* $\zeta = \zeta(p) \in S^{n-1}(0, 1)$ *such that the action of* G^* *in the horoballs*

$$B_1 = B^n(p + \varepsilon\zeta, \varepsilon), \quad B_2 = B^n(p - \varepsilon\zeta, \varepsilon) \tag{5.38}$$

is completely determined by the stabilizer G^*_p, *i.e., for all* $\gamma \in G^*_p$ *and* $g \in G^* \setminus G^*_p$ *we have*

$$\gamma(B_i) = B_i, \quad g(B_i) \cap (B_1 \cup B_2) = \varnothing, \quad i = 1, 2.$$

PROOF. We see directly from Theorem 5.39 and Corollary 5.47 that there are only finitely many distinct orbits $G(p_1), \ldots, G(p_m)$ for the parabolic points of the group G. Here each of the points p_i should be a parabolic cusp. It follows from $\operatorname{rank} G_{p_i} = n-1$ that the cusp domain U_{p_i} is the union of two balls B_1 and B_2 as in (5.38) tangent at p_i, where $\varepsilon > 0$ is sufficiently small.

The stabilizer G_{p_i} may contain elements that permute B_1 and B_2. We select the corresponding subgroups (indices not exceeding $2(n-1)$) free from those elements to obtain the subgroup $G^* \subset G$ of finite index (in particular, as G^* we can take a torsion-free subgroup having finite index in G). When we pass to G^*, each of the components B_1 and B_2 of the cusp domain U_{p_i} becomes precisely $G^*_{p_i}$-invariant.

REMARK 5.51. Bieberbach's theorem (Theorem 3.4) implies that each point p from Corollary 5.50 in the orbifold $\Omega(G^*)/G^*$ corresponds to two isolated ends whose neighbourhoods are homeomorphic to "the punctured solid Euclidean n-space forms", i.e., there are finite-sheeted coverings of them by "the punctured solid n-torus":

$$\mathcal{T}^n = T^{n-1} \times [0, 1) = S^1 \times \ldots \times S^1 \times [0, 1), \tag{5.39}$$

which is the quotient of the half-space $\{x \in \mathbb{R}^n : x_n \geq 1\}$ with respect to the free Abelian group generated by Euclidean translations by the vectors $e_1, \ldots, e_{n-1}$.

COROLLARY 5.52. *Let* $G \subset \text{Möb}(n)$, $n \geq 2$, *be a geometrically finite Kleinian group and let* Ω_0 *be a component of its discontinuity set. Then the stabilizer* $G_0 = \{g \in G : g(\Omega_0) = \Omega_0\}$, *is a geometrically finite group.*

PROOF. We can assume that G is a torsion-free group (Theorem

2.15) and that Ω_0 is a non-invariant component and that the complement $\Omega(G)\setminus\Omega_0$ has not less than two connected components. If that is not so, the stabilizer G_0 of Ω_0 is a subgroup of G of index 2, and by Theorem 5.39 is geometrically finite (the limit set $\Lambda(G_0)=\Lambda(G)$ consists of approximation points and parabolic cusps).

First we note that $\Lambda(G_0)=\partial\Omega_0$. Really, $\partial\Omega_0$ is contained in the set of accumulation points of the orbit $G(x_0)$ of any $x_0\in\Omega_0$, i.e., in $\Lambda(G)$. Since the number of classes of G-equivalent components of $\Omega(G)$ is finite (Corollary 5.47), the set $\partial\Omega_0$ is contained in the set of accumulation points of the suborbit $G_0(x_0)\subset G(x_0)$, i.e., in $\Lambda(G_0)$.

Two cases are possible: either $\operatorname{int} H_{G_0}=\varnothing$ or $\operatorname{int} H_{G_0}\neq\varnothing$.

The first case occurs (see §2) only when the component Ω_0 is a ball $B^n\subset\overline{\mathbb{R}}^n$. Here $\Lambda(G_0)=\partial B^n=S^{n-1}$ and $H_0=H_{G_0}\setminus\Lambda(G_0)$ is an n-dimensional plane in H^{n+1}, whose boundary ∂H_0 is the $(n-1)$-sphere $\Lambda(G_0)$ (H_0 is a boundary totally geodesic component of the hyperbolic convex hull $H_G\cap H^{n+1}$). We show that H_0 is precisely invariant with respect to G_0. It is evident that $G_0(H_0)=H_0$. If, on the other hand, for certain points $x', y'\in H_0$ and for $g\in G$ we have $y'=g(x')$, then $g\in G_0$. In fact, in our case the restriction ρ_0 of the retraction $\tilde{\rho}:\overline{H}^{n+1}\longrightarrow H_G$, specified in (5.7) to the component $\Omega_0=B^n$ is a conformal homeomorphism for the space $\overline{\mathbb{R}}^{n+1}$, or more precisely, ρ_0 is a $\pi/2$ rotation around the $(n-1)$-dimensional sphere $\Lambda(G_0)=S^{n-1}$. At the same time, $\tilde{\rho}$ is compatible with the action of the group G and so

$$\rho_0^{-1}(y') = \tilde{\rho}^{-1}(y') = \tilde{\rho}^{-1}(g(x')) = g(\tilde{\rho}^{-1}(x')) = g(\rho_0^{-1}(g(x'))).$$

In other words, in Ω_0 there exist points $x=\rho_0^{-1}(x')$ and $y=\rho_0^{-1}(y')$ such that $y=g(x)$. As Ω_0 is precisely invariant with respect to G_0, we have $g\in G_0$.

From the precisely G_0-invariant property of H_{G_0} we have the relation

$$H_{G_0}/G_0 = G(H_{G_0})/G\subset H_G/G.$$

This fact and the statement GF2) of Theorem 5.39 show the finiteness of the hyperbolic volume $\operatorname{vol} U_\varepsilon(H_{G_0}/G_0)$ of an ε-neighbourhood of the manifold $H_{G_0}/G_0\subset H^{n+1}/G_0$. Now the geometrical finiteness for the subgroup G_0 follows from GF2) of Theorem 5.39.

Now suppose that $\operatorname{int}^{*} H_{G_0}\neq\varnothing$. Hence this convex domain is precisely G_0-invariant modulo some finite G-orbit. In other words, if

$$G = G_0\cup g_1G_0\cup g_2G_0\cup\dots$$

is the decomposition of the group G on distinct factor-classes by the subgroup G_0, then the following property

$$\operatorname{int} H_{G_0} \cap g_i(\operatorname{int} H_{G_0}) \neq \emptyset \tag{5.40}$$

holds only for a finite number of elements $g_1, \ldots, g_k$ representing distinct classes.

Namely, Corollary 5.47 shows that among the components $\Omega_i = g_i(\Omega_0)$ (which are distinct from each other) there exist only a finite number of non-G-equivalent ones: $\Omega_1, \ldots, \Omega_k$. Moreover, if for some elements g and g' of the group G, $g(\Omega_0) = \Omega_i$ and $g'(\Omega_0) = \Omega_i$ hold, then $g' = gg_0$ for some element $g_0 \in G_0$, i.e., g and g' are in the same i-th class of our decomposition.

By using the above-mentioned property (5.40) of the almost precisely G_0-invariantness of $\operatorname{int} H_{G_0}$, and the statement GF2) of Theorem 5.39 for G, we have, as in the case $\operatorname{int} H_{G_0} = \emptyset$, that

$$\operatorname{vol}(H_{G_0}/G_0) = \operatorname{vol}(\operatorname{int} H_{G_0}/G_0) \le k \cdot \operatorname{vol}(\operatorname{int} H_{G_0}/G) < k \cdot \operatorname{vol}(H_G/G) < \infty.$$

The volume finiteness of the $(n+1)$-dimensional manifold H_{G_0}/G_0 results in the volume finiteness of the ε-neighbourhood of H_{G_0}/G_0. This fact and GF2) of Theorem 5.39 give us the geometrical finiteness of the subgroup G_0. The proof is complete.

REMARK 5.53. For $n=2$, the assertion in Corollary 5.52 was proved by Marden [1], and his proof empolyed specific facts for Kleinian groups in the plane, namely Ahlfors's finiteness theorem and the description of the manifold $M(G)$ for a B-group $G \subset \mathcal{M}_2$.

COROLLARY 5.54. *Let a Kleinian group* $G \subset \text{Möb}(n)$, $n \ge 3$, *be a finite extension of a torsion free group, and its limit set* $\Lambda(G)$ *be the sphere* $S^k \subset \overline{\mathbb{R}}^n$, $1 \le k \le n-2$.

Then G *is geometrically finite if and only if one of the following conditions is satisfied*:

a) *the orbifold* $\mathcal{O} = (\overline{\mathbb{R}}^n \setminus S^k)/G$ *is compact*; *and*

b) *any end of the orbifold* $\mathcal{O}$ *corresponds to a maximal parabolic subgroup in* G *of rank* k.

PROOF. We consider the convex hull $H_G \subset \overline{H}^{n+1}$ of the limit set $\Lambda(G)$; the conditions of the theorem imply that $H_G = \overline{H}^{k+1}$, where $\partial H^{k+1} = \Lambda(G) = S^k$ is a k-dimensional sphere. G acts on $\overline{H}_G = \overline{H}^{k+1}$ as a discrete subgroup of the group $\operatorname{Isom} H^{k+1}$, so Corollary 5.48 can be applied to its action on H^{k+1}. To elucidate the relation between its assertions and conditions a) and b), we consider the retraction

$$\rho : \mathcal{O}(G) = [\overline{H}^{n+1} \setminus \Lambda(G)]/G \longrightarrow \mathcal{O}_G = [\overline{H}^{k+1} \setminus \Lambda(G)]/G,$$

defined by the retraction $\tilde{\rho} : \overline{H}^{n+1} \longrightarrow H_G = \overline{H}^{k+1}$ from (5.7).

As $k \le n-2$, the action of G on H^{k+1} is completely determined by its action on the discontinuity set $\Omega(G) = \overline{\mathbb{R}}^n \setminus S^k$. In particular, due to the continuity of ρ, condition a) for the compactness of the orbifold $\Omega(G)/G = \mathcal{O}$ implies the compactness of the minimal convex retract $\mathcal{O}_G$, which is equivalent to the geometrical finiteness of G.

If the orbifold $\mathcal{O} = \Omega(G)/G$ is non-compact, the same applies for $\mathcal{O}(G)$. Each end of it corresponds to a limit point at the boundary of the fundamental polyhedraon $P(G) \subset H^{n+1}$, which is a fixed point for a parabolic element of the group G. This follows from one-to-one correspondence of the ends of $\mathcal{O}(G)$ to the ends of the minimal convex retract $\mathcal{O}_G$, which in our case $(k \le n-2)$ are defined by the ends of $\mathcal{O}$, and from condition b) they correspond to the parabolic points of G.

We show that all parabolic points p in G correspond to parabolic H^{n+1}-peaks, i.e., there are precisely G_p-invariant horoballs $B_p^{n+1} \subset H^{n+1}$. In fact, the stabilizer G_p has rank k by virtue of condition b). We assume that $p = \infty$ and $S^k = \overline{\mathbb{R}}^k$ and see that the subspace $\mathbb{R}^k$, generated by translation vectors of the stabilizer G_∞, coincides with $\Lambda(G)$. Then the condition in Theorem 4.28 is satisfied, which shows that ∞ is a parabolic H^{n+1}-peak for G. The intersections between the precisely G_p-invariant horoballs $B_p^{n+1} \subset H^{n+1}$ and the $(k+1)$-plane $H^{k+1} = H_G \setminus \Lambda(G)$ are the precisely G_p-invariant horoballs $B_p^{k+1} \subset H^{k+1}$.

It follows from our condition b) that what has been said above on the ends of $\mathcal{O}(G)$ that the orbifold

$$[H^{k+1} \setminus \bigcup_p B_p^{k+1}]/G, \tag{5.41}$$

is compact, where the union is taken over all parabolic points p in G, while the disjoint horoballs B_p^{k+1} have the same sufficiently small radius $r \le r(G)$, i.e., form a G-invariant family. In fact, in our construction we have eliminated all the parabolic ends in $\mathcal{O}_G$, and there are no other ends by virtue of the condition.

The compactness of the orbifold (5.41) and assertion 3) in Corollary 5.48 prove that G is geometrically finite.

To prove that conditions a) and b) are necessary, we assume that the orbifold $\mathcal{O} = \Omega(G)/G$ is non-compact. The geometrical finiteness of G and assertion 2) in Corollary 5.48 imply that the $(k+1)$-dimensional volume for the orbifold $\mathcal{O}_G = H^{k+1}/G$ is finite. This means that all the ends in $\mathcal{O}_G$ correspond to parabolic cusps $p \in S^k$, whose stabilizers have rank k (since $S^k = \partial H^{k+1} = \Lambda(G)$). Then all the ends of $\mathcal{O}$ correspond to parabolic points having rank k stabilizers, i.e., condition b) is satisfied. This completes the proof of Corollary 5.54.

REMARK 5.55. If we take discrete geometrically infinite doubly generated groups G_1 and G_2, given by Jørgensen and Marden [1], and extend their action from $\mathbb{R}^2$ to $\mathbb{R}^4$, we obtain examples of groups for Corollary 5.54. In these cases, the parabolic fixed point $p = \infty$ in each of the groups $G_i \subset \mathcal{M}_4$, $i = 1, 2$, corresponds to a unique end of the orbifold

$$\mathcal{O}_i = (\mathbb{R}^4 \backslash \mathbb{R}^2)/G,$$

$\Lambda(G_i) = \overline{\mathbb{R}}^2$, $i = 1, 2$, but $\operatorname{rank}(G_i)_\infty = 1$ does not coincide with the dimension 2 of the limit sphere $\overline{\mathbb{R}}^2$. Corollary 5.54 proves that these groups are geometrically infinite.

§5. Invariant components for geometrically finite Kleinian groups

Chapter 3 §2 shows that the properties of the connected components of the discontinuity set $\Omega(G)$ for a Kleinian group $G \subset \text{Möb}(n)$ for $n \geq 3$ may differ considerably from those for Kleinian groups in the plane, which relates not only to the homotopy properties but also to the number of them, including the number of G-invariant components. In this section, we show that the properties of the discontinuity set of a geometrically finite group $G \subset \text{Möb}(n)$, are in the main (compare Example 5.65) analogous to those for Kleinian groups in the plane, i.e., we have the following two facts proved by Tetenov [3], [4]:

THEOREM 5.56. *Let $G \subset \text{Möb}(n)$, $n \geq 3$, be a Kleinian group whose discontinuity set $\Omega(G)$ contains two invariant contractible components Ω_0 and Ω_1 with compact quotients Ω_i/G, $i = 1, 2$.* Then

$$\mathcal{O}(G) = [H^{n+1} \cup \Omega(G)]/G$$

is compact, the group G is geometrically finite, and $\Omega(G) = \Omega_0 \cup \Omega_1$.

THEOREM 5.57. *Let $G \subset \text{Möb}(n)$, $n \geq 3$, be a non-elementary geometrically finite Kleinian group having an invariant contractible component $\Omega_0 \subset \Omega(G)$. Then we have alternatives*:

a) *either $\Omega(G) = \Omega_0 \cup \Omega_1$ and Ω_1 is also G-invariant*; *or*

b) *$\Omega(G) \backslash \Omega_0$ consists of an infinite set of non-invariant components (finite modulo G).*

We need some auxiliary lemmas to prove these statements. The limit set is unchanged by passing to a finite-index subgroup or a normal subgroup, so everywhere we assume, if necessary, that the group does not have elliptic elements and orientation reversing ones.

LEMMA 5.58. *Let $G \subset \text{Möb}(n)$, be a Kleinian group having an invariant component $\Omega_0 \subset \Omega(G)$. Then for any connected component Ω_i $(i \geq 1)$ of the set $\Omega(G) \backslash \Omega_0$ and for any point $A \in \overline{\mathbb{R}}^n$, the following sets*

$$\overline{\mathbb{R}}^n \backslash \Omega_i, \quad \partial\Omega_i, \quad \overline{\mathbb{R}}^n \backslash (\Omega_i \cup \{A\}), \quad \partial\Omega_i \backslash \{A\}$$

are connected.

PROOF. As Ω_0 is G-invariant, the set $\Omega_0 \cup \Lambda(G) = \overline{\Omega}_0$ is connected, and the same applies to the union of it with any number of connected components in the complement to it in $\overline{\mathbb{R}}^n$ (i.e., with the components Ω_i). In particular, the sets

$$\overline{\mathbb{R}}^n\backslash\Omega_i = \overline{\Omega}_0 \cup \{\Omega_j : j \neq i\}$$

are connected for $i \geq 1$.

To prove that the set $\partial\Omega_i$, $i \geq 1$, is connected, we consider any open neighbourhood V of it, for which each connected component has a non-empty intersection with $\partial\Omega_j$. The sets $X = \Omega_i \cup V$ and $Y = (\overline{\mathbb{R}}^n\backslash\Omega_i) \cup V$ are connected and open and in their union give the entire space $\overline{\mathbb{R}}^n$, while in their intersection they give the neighbourhood V. Therefore, the triad $(\overline{\mathbb{R}}^n; X, Y)$ is excisive (see Chapter 2 §2), and for it we have an exact (reduced) Mayer-Vietoris sequence, as in (2.8), of which the first terms are

$$0 = H_1(\overline{\mathbb{R}}^n) \xrightarrow{d_*} \widetilde{H}_0(V) \xrightarrow{(j_{1*},\, j_{2*})} \widetilde{H}_0(X) \oplus \widetilde{H}_0(Y) \xrightarrow{(i_{1*},\, i_{2*})} \widetilde{H}_0(\overline{\mathbb{R}}^n) = 0. \quad (5.42)$$

Here i_{k*} and j_{k*} are homomorphisms induced by the inclusions i_k and j_k, $k = 1, 2$, while d_* is a boundary operator. Here the terms $\widetilde{H}_0(X)$ and $\widetilde{H}_0(Y)$ in (5.42) are zero, since the sets X and Y are connected. It shows that $\widetilde{H}_0(V) = 0$, and thus that the neighbourhood V is connected. As the choice of V is arbitrary, $\partial\Omega_i$ is connected too.

We replace each of the sets in the previous proof by the intersection with $\overline{\mathbb{R}}^n\backslash\{A\}$ which gives us a proof that the set $\partial\Omega_i\backslash\{A\}$ is also connected, and as $\partial\Omega_i\backslash\{A\}$ is the boundary set $\overline{\mathbb{R}}^n\backslash(\Omega_i \cup \{A\})$ in the space $\overline{\mathbb{R}}^n\backslash\{A\}$, this means that the set $\overline{\mathbb{R}}^n\backslash(\Omega_i \cup \{A\})$ is connected, which completes the proof of Lemma 5.58.

LEMMA 5.59. *Let Ω_0 be an invariant contractible component for the Kleinian group $G \subset$ Möb(n), while p is a parabolic cusp for G, rank $G_p = k < n$. Then, if Ω_0 contains the cusp domain U_p, then $\Omega_0 = \overline{\mathbb{R}}^n\backslash\{p\}$, and G is elementary.*

PROOF. Conjugating G by a Möbius transformation h, $h(p) = \infty$ and considering a finite-index subgroup of G, we may assume that in $\mathbb{R}^n$ there acts a free Abelian group G_∞ of Euclidean isometries (rank $G_\infty = k < n$) that leaves invariant the subspace $\mathbb{R}^k$, the t-neighbourhood $\mathbb{R}^n\backslash\{U_\infty\}$ of it, and the set $\Omega_0\backslash\{U_\infty\}$.

As $\Omega_0\backslash\{U_\infty\}$ is a strong deformation retract for the contractible component Ω_0, $\Omega_0\backslash U_\infty$ is contractible, as is $\mathbb{R}^n\backslash U_\infty$. Also, $\Omega_0\backslash U_\infty$ is an open set in $\mathbb{R}^n\backslash U_\infty$.

We put

$$M = (\mathbb{R}^n\backslash U_\infty)/G_\infty, \quad N = (\Omega_0\backslash U_\infty)/G_\infty.$$

The manifold M is the product of the $(n-k)$-dimensional ball B^{n-k} and the k-dimensional torus T^k, while its boundary ∂M is $T^k \times S^{n-k-1}$, and that boundary lies in N by virtue of the lemma conditions.

As N and M are manifolds of $K(G_\infty, 1)$ type, their homology groups coincide, so we consider exact homology sequences for the pairs $(N, \partial M)$ and $(M, \partial M)$ to get (on the basis that $\partial M \subset N$) that $H_*(N, \partial M) = H_*(M, \partial M)$. Then, in particular, we have

$$H_n(N, \partial M) = H_n(M, \partial M) = \mathbb{Z}.$$

This means that N is compact and coincides with M, so $\Omega_0 \backslash U_\infty = \mathbb{R}^n \backslash U_\infty$ and $\Omega_0 = \mathbb{R}^n$, which is what we set out to prove.

REMARK 5.60. This proof, due to Tetenov, is a substantial simplification in proving the analogous statement from Tetenov's works (Lemma 3 in [4] and Lemma 3.2.3 in [5]), where a more cumbersome technique was used, and the same applies to the proof of the following lemma (see Tetenov, [4, Lemma 4] and [5, Lemma 3.2.4]).

LEMMA 5.61. *Let* Ω_0 *be an invariant contractible component for the Kleinian group* $G \subset \text{Möb}(n)$, *while* p *is a parabolic cusp of rank* $n-1$. *If the cusp domain* U_p *for the cusp* p *does not intersect* Ω_0, *then* $\overline{\mathbb{R}}^n \backslash \Omega_0$ *is not connected (therefore all the other components are not invariant).*

PROOF. We assume, as in the previous lemma, that $p = \infty$ and $G_\infty \cong \mathbb{Z}^{n-1}$ and put for the cusp domain $U = U_\infty$ that

$$M = (\mathbb{R}^n \backslash U)/G_\infty \supset N = \Omega_0/G_\infty.$$

We have also that the subset N of the manifold M is open in M and has an empty intersection with the boundary ∂M, while the embedding $N \subset M$ is a homotopy equivalence. As $H_{n-1}(N) = \mathbb{Z}$, there exists in N a compact subset K such that the mapping $H_{n-1}(K) \longrightarrow H_{n-1}(N)$, induced by the embedding, is surjective.

We show that the two components S_1 and S_2 of ∂M in M lie in distinct components of the complement $M \backslash N$. If this is not so, $\partial M = S_1 \cup S_2$ is contained in a certain open connected component for the set $M \backslash K$, and therefore there exists a simple path $\gamma \subset M \backslash K$, joining S_1 and S_2. For this path γ we readily show that $H_{n-1}(M \backslash \{\gamma\}) = 0$. In fact, we can embed the manifold $M \approx T^{n-1} \times [-1, 1]$ in S^n (as a torus layer). Its complement in S^n consists of two components and therefore the set $(S^n \backslash M) \cup \gamma$ is connected. This implies that $\tilde{H}^0((S^n \backslash M) \cup \gamma) = 0$, and then by the Alexander duality theorem (see, for example, Dold [1], 8.15) we have that $H_{n-1}(M \backslash \gamma) = 0$.

We consider the commutative embedding diagram

$$\begin{array}{ccccc} & & M\backslash\gamma & & \\ & \nearrow & & \searrow & \\ K & & & & M \\ & \searrow & & \nearrow & \\ & & N & & \end{array}$$

The homomorphisms of the $(n-1)$-dimensional homologies induced by these embeddings also form a commutative diagram:

$$\begin{array}{ccccc} & & H_{n-1}(M\setminus\gamma) & & \\ & \overset{\varphi}{\nearrow} & & \overset{\psi}{\searrow} & \\ H_{n-1}(K) & & & & H_{n-1}(M) \\ & \underset{\chi}{\searrow} & & \underset{\theta}{\nearrow} & \\ & & H_{n-1}(N) & & \end{array}$$

where χ is epimorphic, θ an isomorphism, and φ a zero mapping.

Then we have a contradiction at once, since $\psi\circ\varphi(H_{n-1}(K))=0$, while $\theta\circ\chi(H_{n-1}(K))=H_{n-1}(M)=\mathbb{Z}$, which shows that $M\setminus N$ is non-connected and therefore that the set $\overline{\mathbb{R}}^n\setminus\Omega_0$ is the same.

Lemma 5.58 shows (from the converse) that here all other components $\Omega_i\subset\Omega(G)$ distinct from Ω_0 cannot be invariant, which completes the proof of the lemma.

PROOF OF THEOREM 5.56. Considering a subgroup of finite index, we may assume that the group G acts freely in the components Ω_0 and Ω_1. We denote by $N_i=\Omega_i/G$, $i=0, 1$, $M=[H^{n+1}\cup\Omega(G)]/G$, the corresponding manifolds, which are $K(G, 1)$-type spaces, since $\pi_k(N_i)$ and $\pi_k(M)$ are trivial for $k\geq 2$, while $\pi_1(N_0)=\pi_1(N_1)=\pi_1(M)=G$. The embeddings $i_\nu: N_\nu\longrightarrow M$, $\nu=0, 1$, induce homotopy-group isomorphisms, and the Whitehead [1] theorem shows that they are homotopy equivalences that induce isomorphisms of homology groups for the manifolds N_0, N_1, and M.

We consider the manifold $\hat{M}$ obtained by gluing two copies of the manifold M (M and $M'=M$) along the components N_0 and N_1 of the boundary; the triad $(\hat{M}; M, M')$ is excisive (see Chapter 2 §2) and for it there is an exact Mayer-Vietoris sequence (2.6) enabling one to calculate the groups $H_k(\hat{M})$:

$$0\longrightarrow H_{n+1}(\hat{M})\xrightarrow{d_*}H_n(N_0)\oplus H_n(N_1)\xrightarrow{j_*}H_n(M)\oplus H_n(M')\xrightarrow{i_*}H_n(\hat{M})\longrightarrow\dots$$

In this sequence, d_* is a boundary homomorphism, $H_n(N_0)=H_n(N_1)=H_n(M)=\mathbb{Z}$, and the homomorphism j_* translates the element (k, l) of the direct sum $H_n(N_0)\oplus H_n(N_1)=\mathbb{Z}\oplus\mathbb{Z}$ into the element $((k+l), \ -(k+l))$ in the group $H_n(M)\oplus H_n(M')=\mathbb{Z}\oplus\mathbb{Z}$. Here the kernel $\operatorname{Ker} j_*$ is a cyclic group consisting of all elements of the group $H_n(N_0)\oplus H_n(N_1)$ having the form $(-k, k)$. This kernel coincides with the group $H_{n+1}(\hat{M})$ and therefore $H_{n+1}(\hat{M})=\mathbb{Z}$, so the $(n+1)$-dimensional manifold $\hat{M}$ is closed. This means that the boundary ∂M of M is the union of N_0 and N_1, while M itself is compact. This and GF3) in 5.39 at once demonstrate the geometric finiteness of G. Also, from these equations:

$$\Omega(G)/G=\partial M=N_0\cup N_1=(\Omega_0/G)\cup(\Omega_1/G)$$

we get directly that $\Omega(G)=\Omega_0\cup\Omega_1$. Theorem 5.56 is proved.

PROOF OF THEOREM 5.57. We have seen in §3 and §4 that a geometrically finite group $G\subset\text{Möb}(n)$, can have only parabolic cusps (finitely many) as limit points at the boundary $\partial P \cap \overline{\mathbb{R}}^n$ of the fundamental polyhedron $P \subset H^{n+1}$. Let $p_1, \ldots, p_m$ be the maximal set of such non-equivalent vertices. Definition 5.16 for a parabolic cusp p implies that there exists an open set $\widetilde{U}_p \subset \overline{\mathbb{R}}^{n+1}_+$ (a cusp domain in $\overline{\mathbb{R}}^{n+1}_+$) having the following properties (see §3 for details):

1) $\widetilde{U}_p$ is precisely G_p-invariant, i.e. $\gamma(\widetilde{U}_p) = \widetilde{U}_p$ and $g(\widetilde{U}_p) \cap \widetilde{U}_p \neq \varnothing$ for $\gamma \in G_p$, $g \in G\backslash G_p$;

2) there exist an integer k, $1 \leq k \leq n$ (the rank of the parabolic cusp p), an element $h \in \text{Möb}(n)$, $h(p) = \infty$, and a number $t > 0$ such that

$$h(\widetilde{U}_p) \supset \{x \in \overline{\mathbb{R}}^{n+1}_+ : \sum_{i=k+1}^{n+1} x_i^2 > t\}; \tag{5.43}$$

3) for $k < n$, the set $\widetilde{U}_p \cap \overline{\mathbb{R}}^n$ is the cusp domain U_p at the point p.

Moreover, due to the finiteness for a number of the cusps p_i on ∂P, the parameters $t_i = t(p_i)$ in (5.43), which define the sizes of the cusp domains $\widetilde{U}_{p_i} = \widetilde{U}_i$, can be chosen sufficiently large such that, for $i \neq j$,

$$G(\widetilde{U}_i) \cap G(\widetilde{U}_j) = \varnothing.$$

We transfer to a finite-index subgroup and see that each orbit $G(\widetilde{U}_i)$ in the manifold $M = M(G) = [\mathbb{R}^{n+1}_+ \cup \Omega(G)]/G$ corresponds to a submanifold $W_i = \widetilde{U}_i/G_{p_i}$, homeomorphic to

$$B^{n-k} \times (0,\ 1) \times T^k, \tag{5.44}$$

if $\operatorname{rank} p_i = k < n$, and correspondingly $T^n \times (0,\ 1)$ if $\operatorname{rank} p_i = n$. Here $T^k = S^1 \times \ldots \times S^1$ is a k-dimensional torus. When we have excised all the submanifolds W_i, $i = 1, \ldots, m$, from $M(G)$, we are left with the compact submanifold M_C (see GF5) in Theorem 5.39 and Chapter 7 §1), which we write as

$$M_C = [(\mathbb{R}^{n+1}_+ \cup \Omega(G)) \setminus \bigcup_{i=1}^{m} G(\widetilde{U}_i)]/G. \tag{5.45}$$

We order the points p_i, $1 \leq i \leq m$, in such a way that

$$\operatorname{rank} G_{p_i} \equiv \operatorname{rank} p_i = \begin{cases} < n-1 & \text{for } i \leq r, \\ = n-1 & \text{for } r < i \leq s, \\ = n & \text{for } i > s, \end{cases}$$

where $r \leq s \leq m$ are some integers.

By $N_0 = \Omega_0/G$ we denote a component of the boundary ∂M of $M = M(G)$, and by $N_{0C} = M_C \cap N_0$ the intersection of it with the boundary of the (5.45) manifold; as above, contractibility and G-invariance for Ω_0 imply that the embedding $N_0 \longrightarrow M(G)$ induces isomorphisms for all homotopy groups, and it is a homotopy equivalence due to Whitehead's [1] theorem.

The complement $M \setminus M_C$ consists of m conected components W_i having (5.44) form; the closures $\overline{W}_i$ for those components in M have the form $T^{k_i} \times [0, 1]^{n-k_i} \times (0, 1]$, where $k_i = \operatorname{rank} G_{p_i}$, and the intersection of $\overline{W}_i$ with M_C is $T^{k_i} \times [0, 1]^{n-k_i} \times \{1\}$, while the complement in ∂W_i, i.e., the subset $T^{k_i} \times \partial([0, 1]^{n-k_i}) \times (0, 1]$ is the intersection of W_i and ∂M.

Let $k_i < n-1$, i.e., $i \le r$. Then $\overline{W}_i \cap \partial M$ are connected and $\overline{W}_i \cap N_0 = \varnothing$, which is a direct consequence of Lemma 5.59 and the group G being non-elementary, since $\overline{W}_i \cap \partial M$ is then contained in N_0 if $\tilde{U}_i \cap \overline{\mathbb{R}}^n = U_i \subset \Omega_0$.

If $k_i = n$, the cusp domain for the point p_i in $\overline{\mathbb{R}}^{n+1}$ is the union $\tilde{U}_i$ and its reflection with respect to $\mathbb{R}^n$ and is contained in the open G-invariant contractible set

$$\mathbb{R}^{n+1}_+ \cup \Omega_0 \cup \mathbb{R}^{n+1}_-.$$

(The contractibility follows from that of the sets $\mathbb{R}^{n+1}_+ \cup \Omega_0$, $\mathbb{R}^{n+1}_- \cup \Omega_0$ and of the intersection of them, Ω_0.) We apply Lemma 5.59 to get this case that

$$\mathbb{R}^{n+1}_+ \cup \Omega_0 \cup \mathbb{R}^{n+1}_- = \overline{\mathbb{R}}^{n+1} \setminus \{p_i\}$$

and consequently $\Omega_0 = \overline{\mathbb{R}}^n \setminus \{p_i\}$. The contradiction of this and the condition that the group G is not elementary shows that $k_i < n$ always.

Finally, let $k_i = n-1$, i.e., $i > r$; then $\overline{W}_i \cap \partial M$ are non-connected and are of the form $T^{n-1} \times \{0, 1\} \times (0, 1)$; as for $k_i < n-1$, Lemma 5.59 gives, from the fact that G is not elementary, that $\overline{W}_i \cap \partial M$ are not contained in N_0. If $\overline{W}_i \cap N_0 \ne \varnothing$, then $\overline{W}_i \cap (\partial M \setminus N_0) \ne \varnothing$, and each of the sets $\overline{W}_i \cap N_0$ (with $i = r+1, \ldots, q$; $q \le s$) is $T^{n-1} \times \{0\} \times (0, 1]$. If $\overline{W}_i \cap N_0 = \varnothing$, (for $i = q+1, \ldots, s$), then the set $\Omega_0 \cap U_i$. is empty. This with Lemma 5.61 shows that $\overline{\mathbb{R}}^n \setminus (\Omega_0 \cup \{p_i\})$ is non-connected, which, in accordance with Lemma 5.58, means that the group G does not have invariant components distinct from Ω_0. It corresponds to point b) in the theorem.

As each of the sets $\overline{W}_i \cap M_C$ is a strong deformation retract of $\overline{W}_i$, M_C is also a strong deformation retract of the manifold M, and this applies also for the pairs (M_C, N_{0C}) and (M, N_0), so

$$H_*(M_C, N_{0C}) = H_*(M, N_0) = 0. \tag{5.46}$$

We consider $q - r$ copies of the product $\hat{T}^n \times [0, 1]$ ($q = m$ if b) does not apply), where $\hat{T}^n$ is a solid n-dimensional torus, i.e., a set

$$\hat{T}^n \approx [0, 1]^2 \times T^{n-2} \approx \bar{B}^2(0, 1) \times T^{n-2}.$$

We identify the subset $T^{n-1} \times [0, 1]$ of the boundary of each of such product with one of the sets $\bar{W}_i \cap M_C$, $r+1 \le i \le q$, where we assume that each of the sets $T^{n-1} \cap N_{0C}$ is identified with one of $\bar{W}_i \cap N_{0C}$. Then from M_C we get the compact $(n+1)$-dimensional manifold $\hat{M}$, whose boundary contains the closed component $\hat{N}_0$, obtained from the n-manifold N_{0C} by gluing to its boundary $q-r$ copies of the solid n-dimensional tori $\hat{T}^n$.

Each of the pairs

$$(\hat{T}^n \times [0, 1], \hat{T}^n \times \{0\}), \quad (T^{n-1} \times [0, 1], T^{n-1} \times \{0\}), \quad (M_C, N_{0C}) \tag{5.47}$$

is homotopically trivial. The manifold pair $(\hat{M}, \hat{N}_0)$ is the union of the pair (M_c, N_{0C}) and the pair

$$(\bigcup_{j=1}^{q-r} (\hat{T}^n \times [0, 1])_j, \quad \bigcup_{j=1}^{q-r} (\hat{T}^n \times \{0\})_j),$$

whose intersection is the pair

$$(\bigcup_{j=1}^{q-r} (T^{n-1} \times [0, 1])_j, \quad \bigcup_{j=1}^{q-r} (T^{n-1} \times \{0\})_j).$$

We consider the (2.7) exact Mayer-Vietoris sequence for the pairs of triads

$$(\hat{N}_0; \ N_{0C}, \ (\bigcup_{j=1}^{q-r} (\hat{T}^n \times \{0\})_j) \subset (\hat{M}; \ M_{C}, \ \bigcup_{j=1}^{q-r} (\hat{T}^n \times [0, 1])_j),$$

and get from (5.46) and the homotopy triviality of the (5.47) pairs that

$$H_*(\hat{M}, \hat{N}_0) = 0. \tag{5.48}$$

The manifold $\hat{N}_0$ is closed and so $H_n(\hat{N}_0) = \mathbb{Z}$, and (5.48) then implies that $H_n(\hat{M}) = \mathbb{Z}$.

We now consider an exact sequence for the pair $(\hat{M}, \partial\hat{M})$:

$$\begin{gathered} 0 \longrightarrow H_{n+1}(\hat{M}, \partial\hat{M}) \longrightarrow H_n(\partial\hat{M}) \longrightarrow H_n(\hat{M}) \longrightarrow \ldots \\ \ldots \longrightarrow H_1(\partial\hat{M}) \longrightarrow H_1(\hat{M}) \longrightarrow H_1(\hat{M}, \partial\hat{M}) \longrightarrow H_0(\partial\hat{M}) \longrightarrow H_0(\hat{M}) \end{gathered} \tag{5.49}$$

Due to (5.48), here the homomorphism

$$H_1(\partial\hat{M}) \longrightarrow H_1(\hat{M})$$

is surjective because $\partial\hat{M} \supset \hat{N}_0$, while $H_1(\hat{N}_0) = H_1(\hat{M})$, and then the kernel of the homomorphism

$$H_1(\hat{M}, \partial\hat{M}) \longrightarrow H_0(\partial\hat{M})$$

is trivial and it is an embedding. At the same time, $H_0(\partial\hat{M}) = \mathbb{Z}^k$, where k is the number of connected components of $\partial\hat{M}$, and therefore $H_1(\hat{M}, \partial\hat{M})$ does not contain finite-order elements.

As $\hat{M}$ is compact, Poincaré's duality theorem (see Dold [2], Chapter 8 §7) gives us that the groups $H_n(\hat{M})$ and $H^1(\hat{M}, \partial\hat{M})$ are isomorphic. At the same time, as $H_0(\hat{M}, \partial\hat{M})$ is trivial, the theorem on universal coefficients (see Dold [1], Chapter 6 §4) gives $H^1(\hat{M}, \partial\hat{M}) = H_1(\hat{M}, \partial\hat{M})$, so $H_1(\hat{M}, \partial\hat{M}) = H_n(\hat{M}) = \mathbb{Z}$.

As the homomorphism $H_0(\partial\hat{M}) \longrightarrow H_0(\hat{M})$ is surjective and the subsequence in (5.49) is exact:

$$H_1(\hat{M}, \partial\hat{M}) \longrightarrow H_0(\partial\hat{M}) \longrightarrow H_0(\hat{M})$$

it follows that $H_0(\partial\hat{M}) = \mathbb{Z} \oplus \mathbb{Z}$, which shows that the boundary of $\hat{M}$ consists of two connected components:

$$\partial\hat{M} = \hat{N}_0 \cup \hat{N}_1. \tag{5.50}$$

By K and L we denote disjoint closed regular neighbourhoods for the components $\hat{N}_0$ and $\hat{N}_1$ of the boundary $\partial\hat{M}$ of the $(n+1)$-dimensional manifold $\hat{M}$. As K and L are Euclidean neighbourhood retracts, the Čech cohomology for the pair $(\hat{M}\setminus\mathring{K}, L)$ coincide with the usual ones, and Poincaré's duality theorem gives us that $H^k(\hat{M}\setminus\mathring{K}, L)$ and $H_{n+1-k}(\hat{M}\setminus L, \mathring{K})$ are isomorphic (Dold [1], Chapter 8).

On the other hand, the pairs $(\hat{M}\setminus\mathring{K}, L)$ and $(\hat{M}\setminus L, \mathring{K})$ are homotopically equivalent to the pairs $(\hat{M}, \hat{N}_0)$ and $(\hat{M}, \hat{N}_1)$ correspondingly, so the groups $H^k(\hat{M}, \hat{N}_1)$ and $H_{n+1-k}(\hat{M}, \hat{N}_0)$ are isomorphic for any k, and thus

$$H_*(\hat{M}, \hat{N}_1) = 0. \tag{5.51}$$

This means that the embedding $\hat{N}_1 \longrightarrow \hat{M}$ induces an isomorphism of homology groups.

We now assume that assertion b) is not obeyed. Then each of the submanifolds $\overline{W}_i$ in M, $i = r+1, \ldots, m$, intersects with N_0 on the follow subset of its boundary:

$$T^{n-1} \times \{0\} \times (0, 1] = \overline{W}_i \cap N_0,$$

while the subset

$$T^{n-1} \times \{1\} \times (0, 1]$$

is the intersection of $\overline{W}_i$ with one of the other components N_j of ∂M. We show that there is only one such component. in fact, for $i = 1, \ldots, r$, the sets $\overline{W}_i \cap \partial M$ and ∂W_i are connected and therefore, when one deletes the submanifold W_i from M, the number of connected components in ∂M is unaltered. There is also no change in that number when the submanifolds W_i, $i = r+1, \ldots, m$, are replaced by $\hat{T}^n \times [0, 1]$. Hence the number of components in the boundary $\partial\hat{M}$ is equal to that in ∂M, and our statement about the number of components N_j follows from (5.50).

We consider the component N_1 of the boundary ∂M of M. Its strong deformation retract is the manifold $M_C \cap \hat{N}_1$. Each loop in the manifold $\hat{N}_1$ is homotopic in $\hat{N}_1$ to a certain loop in the manifold $M_C \cap \hat{N}_1$, so the embedding $N_1 \longrightarrow M$ induces the epimorphism $\pi_1(N_1) \longrightarrow \pi_1(M)$. This shows (see Lemma 7.12) that the inverse image of the component $N_1 = \partial M \setminus N_0$ for the natural projection $\Omega(G) \longrightarrow \partial M$ is the connected component $\Omega_1 = \Omega(G) \setminus \Omega_0$, which is clearly invariant with respect to G. It shows that assertion a) is obeyed and completes the proof of Theorem 5.57.

These results can be sharpened for $n = 3$.

COROLLARY 5.62. *Let* $G \subset \text{Möb}(3)$ *be a Kleinian group whose discontinuity set* $\Omega(G)$ *contains two invariant simply connected components* Ω_0 *and* Ω_1, *and let* $N_i = \Omega_i/G$, $i = 0, 1$, *be compact. Then*

$$[H^4 \cup \Omega(G)]/G$$

is compact, G *is geometrically finite, and* $\Omega(G) = \Omega_0 \cup \Omega_i$.

PROOF. This follows directly from Theorem 5.56 if we prove that the components Ω_0 and Ω_1 are contractible, and this is a consequence of the following statement.

LEMMA 5.63. *Let a Kleinian group* $G \subset \text{Möb}(3)$ *have an invariant component* $\Omega_0 \subset \Omega(G)$. *Then for any other connected component* $\Omega_i \subset (\Omega(G) \setminus \Omega_0)$ *we have*

$$H_2(\Omega_i) = 0, \quad \pi_2(\Omega_i) = 0.$$

PROOF. The first equation follows directly from Alexander's duality theorem (see, for example, Dold [1], 8.15), according to which $\tilde{H}^0(\overline{\mathbb{R}}^3 \setminus \Omega_i) = H_2(\Omega_i)$. The necessary for it equation $\tilde{H}^0(\overline{\mathbb{R}}^3 \setminus \Omega_i) = 0$, i.e., conectedness of a set $\overline{\mathbb{R}}^3 \setminus \Omega_i$, is obtained from Lemma 5.58.

If the group $\pi_2(\Omega_i)$ is non-trivial, the sphere theorem (Theorem 7.9) implies that there would be a two-dimensional sphere S embedded in Ω_i that is non-contractible there. Such a sphere decomposes $\overline{\mathbb{R}}^3$ into two balls, one of which contains the set $\overline{\mathbb{R}}^3 \setminus \Omega_i$, while on the other S contracts to a point. This contradiction completes the proof of Lemma 5.63 and Corollary 5.62.

We apply Lemma 5.63 to get the following from Theorem 5.42:

COROLLARY 5.64. *Let* G *be a geometrically finite group in* $\overline{\mathbb{R}}^3$ *containing an invariant simply connected component* Ω_0. *Then one of the following occurs*:

1) $\Omega(G) = \Omega_0$;

2) $\Omega(G) = \Omega_0 \cup \Omega_1$, *the component* Ω_0 *is contractible, and* Ω_1 *is invariant with respect to* G *and aspheric in dimension* 2 *(i.e.,* $\pi_2(\Omega_1) = 0$);

3) *the set* $\Omega(G)\setminus\Omega_0$ *consists of an infinite set (finite modulo* G*) of non-invariant aspheric in dimension 2 components.*

The simply connectedness condition in Corollary 5.64 is important, as is shown by the example below.

EXAMPLE 5.65. *The geometrically finite Kleinian group* G *in the space* $\overline{\mathbb{R}}^3$ *having two invariant (not simply connected) components and an infinite set of non-invariant ones* (Tetenov [4]).

To construct a group $G \subset \text{Möb}(3)$ having those properties, it is sufficient to construct a finite system Σ of spheres σ_i in $\overline{\mathbb{R}}^3$ such that, for any $\sigma_i \neq \sigma_j$, either $\text{int}\,\sigma_i \cap \text{int}\,\sigma_j = \varnothing$ or σ_i and σ_j intersect at an angle π/m, $m \in \mathbb{Z}$, and the set

$$F_\Sigma = \overline{\mathbb{R}}^3 \setminus (\overline{\bigcup_{\sigma_i \in \Sigma} \text{int}\,\sigma_i}) \tag{5.52}$$

consists of three components F_0, F_1, F_2, where the last two adjoin (on an open set) any sphere $\sigma \in \Sigma$. Then the group $G = G_\Sigma$, consisting of compositions of an even number of reflections in the sphere $\sigma \in \Sigma$, will be the desired one. Its discontinuity follows from Corollary 3.20 applied to the extension of G into $H^4 \cong \mathbb{R}^4_+$, where $G \subset \text{Isom}\,H^4$ (compare Example 3.9).

The announced properties of components in the discontinuity set, i.e., the invariance of the components

$$\Omega_1 = \text{int}\,G(\overline{F_1 \cup \mathcal{I}_1(F_1)}), \quad \Omega_2 = \text{int}\,G(\overline{F_2 \cup \mathcal{I}_1(F_2)}) \tag{5.53}$$

and the non-invariance of the components (infinitely many) in the complement

$$\Omega_1(G)\setminus(\Omega_1 \cup \Omega_2) = \text{int}\,G(\overline{F_0 \cup \mathcal{I}_1(F_0)}), \tag{5.54}$$

follows from Lemma 3.7 and the existence of the spheres $\sigma \in \Sigma$ that do not intersect $\overline{F}_0$ (here $\mathcal{I}_1$ is a reflection in σ_1).

We form Σ from the following nine spheres in $\overline{\mathbb{R}}^3$:
$\sigma_1 = \{x \in \overline{\mathbb{R}}^3 : x_2 = 0\}$, $\sigma_2 = \{x \in \overline{\mathbb{R}}^3 : x_1 = x_2\}$, $\sigma_3 = \{x \in \overline{\mathbb{R}}^3 : x_1 + x_2 = 3\}$;

σ_4 has radius 3 and its centre at $(0, 0, -\sqrt{\sqrt{6} + 26/3})$;
σ_5, σ_6, and σ_7 have radius $\sqrt{6}/3$ and their centres correspondingly at $(1, 0, 0)$, $(3, 0, 0)$, and $(2, 1, 0)$; and
σ_8 and σ_9 have radius $\sqrt{3}/3$ and centres correspondingly at $(2, 0, 0)$ and $(1, 1, \sqrt{2}/3)$.

As $\text{int}\,\sigma_i$ we take the components of $\overline{\mathbb{R}}^3\setminus\sigma_i$ that do not contain the point $A_1 = (1, 1/2, 10)$. Then the set F_Σ in (5.52) consists of F_1 which contains A_1, F_0 which contains $A_0 = (1, 1/2, -10)$, and F_2 which contains $A_2 = (3/2, 1/2, -1/2)$. A direct calculation shows that the dihedral angles in the spherical polyhedra F_j are $\pi/2$, $\pi/3$, $\pi/4$ or $\pi/6$. Also, all the spheres $\sigma_i \in \Sigma$ intersect $\overline{F}_1$ and $\overline{F}_2$ on certains discs, and $\overline{F}_0$ does not intersect the σ_i for $i = 5, \ldots, 9$.

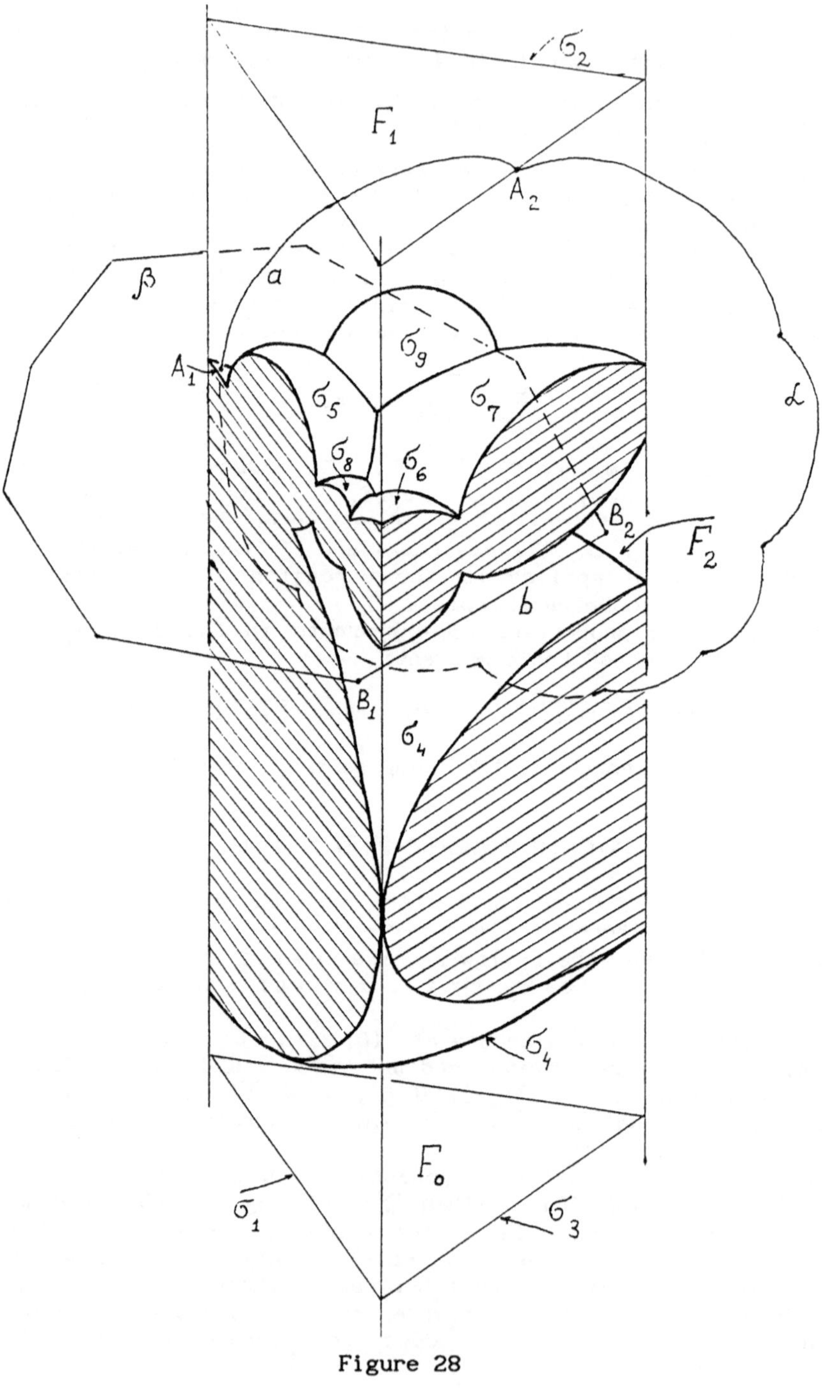

σ_2
F_1
A_2
β
a
σ_9
A_1
σ_5
σ_7
α
σ_8
σ_6
B_2
F_2
b
B_1
σ_4
σ_4
F_0
σ_1
σ_3

Figure 28

Figure 28 shows the resulting spherical polyhedra F_i, where the planes σ_1, σ_2, σ_3 are taken as transparent, while their intersections with $\operatorname{int}\sigma_i$, $i \ge 4$, are hatched. Our constructions imply that Ω_1 and Ω_2 from (5.53) are G-invariant, while the non-invariant components in (5.54) form a G-orbit for the half-space

$$\Omega_0 = \{x \in \mathbb{R}^3 : x_3 < -\sqrt{\sqrt{6} + 26/3}\} = \operatorname{int} G_0(\overline{F_0 \cup \mathcal{T}_1(F_0)}).$$

The stabilizer $G_0 \subset G$ of this component is a Fuchsian group of the first kind in $\overline{\mathbb{R}}^3$, which is generated by an even number of reflections in the sides of F_0, i.e., reflections with respect to σ_1, σ_2, σ_3, and σ_4.

One can prove that the invariant components Ω_1 and Ω_2 are non-connected in a direct fashion without using Corollary 5.64. In fact, let $a \subset \overline{F}_1$ be a topological segment joining the points A_1 and A_2, which lie in $\operatorname{int}(\overline{F}_1 \cap \sigma_4)$ and $\operatorname{int}(\overline{F}_1 \cap \sigma_3)$ correspondingly, and let $b \subset \overline{F}_2$ be a topological segment joining the points B_1 and B_2, which lie in $\operatorname{int}(\overline{F}_2 \cap \sigma_1)$ and $\operatorname{int}(\overline{F}_2 \cap \sigma_2)$ correspondingly (Fig. 28). The closed curves α and β consisting of the union of a finite number of images of the segments a and b on the reflection in the sides of F_i containing their ends are evidently linked in $\overline{\mathbb{R}}^3$. Here $\alpha \subset \Omega_1$ and $\beta \subset \Omega_2$, and consequently these curves are non-contractible in Ω_i. We consider the images of α and β by the translations g_m belonging to the group G,

$$g_m : (x_1, x_2, x_3) = x \longmapsto (x_1 + 3m, x_2 + 3m, x_3),$$

which gives an infinite set of alternately linked loops contained in the distinct components Ω_1 and Ω_2 of the discontinuity set $\Omega(G)$. This means that $\operatorname{rank} \pi_1(\Omega_i) = \infty$, $i = 1, 2$. It completes the construction.

These properties of the connected components of $\Omega(G)$ for the geometrically finite Kleinian group $G \subset \text{Möb}(n)$ are of topological character. Moreover, they can be characterized from the metrical viewpoint also, i.e., one can prove the following theorem (Tukia [6]), which shows that under certain circumstances these components Ω cannot be very thin in comparison with their Euclidean diameters $\operatorname{diam}_E \Omega$ (however, we transform it with Möbius transformations).

THEOREM 5.66. *Let $\Omega_0 \subset \Omega(G)$ be an invariant component for a geometrically finite Kleinian group $G \subset \text{Möb}(n)$, $n \ge 2$, not containing parabolic elements. Then there exists a constant $C > 0$ such that for any $h \subset \text{Möb}(n)$ and any ball $B = B^n(x, t)$ centred at $x \in \overline{h(\Omega_0)} \cap \mathbb{R}^n$ with radius $t \le \operatorname{diam}_E h(\Omega_0)/2$ there exists a ball $B' \subset B \cap h(\Omega_0)$ such that*

$$C \cdot \operatorname{diam}_E B \le \operatorname{diam}_E B'.$$

PROOF. This is obtained by applying the compactness method used in the proof of Theorem 5.35 about the Hausdorff dimension of the limit set for a geometrically finite Kleinian group $G \subset \text{Möb}(n)$.

We consider the following families for any point $x \in \mathbb{R}^{n+1}_+$ and any set $X \subset \mathbb{R}^{n+1}_+$ (we assume that Möb(n) acts also in $\mathbb{R}^{n+1}_+$):

$$M_x = \{h \in \text{Möb}(n) : h(x) = e_{n+1}\}, \quad M_X = \bigcup_{x \in X} M_x.$$

As $M_{e_{n+1}}$ is a subgroup in Möb(n) that is isomorphic as a topological group to the orthogonal group $O(n+1)$, we have that M_x and M_X are compact for any $x \in \mathbb{R}^{n+1}_+$ and any compact set $X \subset \mathbb{R}^{n+1}_+$.

We now define the following family for a non-elementary Kleinian group $G \subset \text{Möb}(n)$ and for $X \subset \mathbb{R}^{n+1}_+$:

$$\mathcal{L}(X) = \mathcal{L}(X, G) = \{h(\Lambda(G)) \; ; \; h \in M_X\}.$$

As we have the following obvious relation for any $h \in \text{Möb}(n)$

$$M_{h(X)} = M_X \circ h^{-1},$$

we see readily that the family $\mathcal{L}(X)$ is compact (in Hausdorff topology, see (5.31)) for any compact set $X \subset \mathbb{R}^{n+1}_+$, and we have

$$\mathcal{L}(G(X)) = \mathcal{L}(X). \tag{5.55}$$

In other words, the family M_X for the compact $X \subset \mathbb{R}^{n+1}_+$ is also a compact, and for $h \in M_{G(X)}$ there exist $g \in G$ and $h' \in M_X$ such that $h = h'g$ and

$$h(\Lambda(G)) = h'(\Lambda(G)). \tag{5.56}$$

We now take in place of $X \subset \mathbb{R}^{n+1}_+$ the fundamental polyhedron $P_H(G)$ for the Kleinian group G in the hyperbolic convex hull H_G for the limit set $\Lambda(G)$ (see §2), i.e., $P_H(G) = P(G) \cap H_G$, where $P(G) \subset \mathbb{R}^{n+1}_+$ is a fundamental polyhedron of the group G. Then, due to GF3) in Theorem 5.39, P_H is a compact in $\mathbb{R}^{n+1}_+$, so from (5.55) we have that

$$\mathcal{L}_G = \mathcal{L}(H_G, G) \tag{5.57}$$

is a compact family of closed subsets in $\overline{\mathbb{R}}^n$ (in the topology induced by the (5.31) Hausdorff metric).

We now consider extremal cases in the theorem: when x lies in $\partial h(\Omega_0) \cap \mathbb{R}^n = h(\Lambda(G)) \cap \mathbb{R}^n$. Let α be a similarity in $\mathbb{R}^{n+1}_+$ such that $\alpha(x, t) = e_{n+1}$, where $t \le \text{diam}_E\, h(\Omega_0)/2 = \text{diam}_E h(\Lambda(G))/2$. Then (5.56) and the compactness of $\mathcal{L}_G$ in (5.57) imply that

$$\alpha(\Omega_0) = h(\Omega_0),$$

where h belongs to a compact subset $M \subset \text{Möb}(n)$ dependent only on G. Compactness of M completes the proof.

The assertion in Theorem 5.66 remains true also if a geometrically finite group G contains parabolic elements, provided that its components are not thin near the parabolic cusps. This observation allowed Tukia [6] to prove the following theorem by means of a minor modification in the above proof:

THEOREM 5.67. *Let* $\Omega_0 \subset \Omega(G)$ *be an invariant component for a geometrically finite Kleinian group* $G \subset \mathrm{M\ddot{o}b}(n)$, $n \geq 2$, *not containing parabolic subgroups of rank* n. *Also, let* Ω_0 *contain a connected component of the cusp domain* U_p *for any parabolic cusp* p *for* G. *Then there exists a constant* $C > 0$ *such that, for any* $h \subset \mathrm{M\ddot{o}b}(n)$ *and for any ball* $B = B^n(x, t)$ *with centre at* $x \in \overline{h(\Omega_0)} \cap \mathbb{R}^n$ *and radius* t, $t \leq \mathrm{diam}_E\, h(\Omega_0)/2$, *there exists a ball* $B' \subset B \cap h(\Omega_0)$ *having the following property:*

$$C \cdot \mathrm{diam}_E\, B \leq \mathrm{diam}_E\, B'.$$

The condition that G does not contain parabolic subgroups of rank n is very important for Theorem 5.67, as is evident from the example of a group $G = G_0 * \Gamma$, where Γ is the group generated by the translations $x \longmapsto x + e_i$, $i = 1, \ldots, n$, while G_0 is a Kleinian group, where the complement to its fundamental polyhedron is contained in the unit cube $Q = \{x \in \mathbb{R}^n : |x_i| < 1/2\}$.

The importance of the second condition in Theorem 5.67 is less obvious for Kleinian groups in space; it is evident from the following construction based on considering a boundary group of the deformation space for a Fuchsian group in $\overline{\mathbb{R}}^n$ (compare Apanasov [11], [12]).

EXAMPLE 5.68. Again let Q be a unit cube in $\mathbb{R}^3$. We take the spheres S_i, $1 \leq i \leq 8$, centred at the vertices of Q with radii $\sqrt{3}/3$. As four other spheres S_i, $9 \leq i \leq 12$, we take disjoint spheres having radii $(1 - \sqrt{3}/3)/2$ and whose centres are the face centres for Q:

$$(0, +1/2, 0), \quad (0, 0, \pm 1/2).$$

These spheres do not intersect the spheres

$$S_{13} = \{x \in \overline{\mathbb{R}}^3 ;\, x_1 = -1/2\}, \quad S_{14} = \{x \in \overline{\mathbb{R}}^3 ;\, x_1 = 1/2\}.$$

The spheres S_i and S_j in the constructed family

$$\Sigma = \{S_i, \quad 1 \leq i \leq 14\}$$

intersect one another (apart from the pair $(i, j) = (13, 14)$ only at the angles of $\pi/3$ and $\pi/2$; S_{13} and S_{14} are tangent at infinity.

The group $G = G_\Sigma \subset \mathcal{M}_3$, consisting of superpositions of an even number of reflections $\mathcal{T}_i$ in the spheres S_i in Σ is a Kleinian group and has two invariant components:

$$\Omega_0 = \mathrm{int}\, G(\overline{F}_0), \quad \Omega_1 = \mathrm{int}\, G(\overline{F}_1),$$

where $F_0 \subset Q$ and F_1 are components of the fundamental polyhedron for the group G:

$$F_0 \cup F_1 = \operatorname{int}(\overline{F_\Sigma \cup \mathcal{I}_1(F_\Sigma)}), \quad F_\Sigma = \mathbb{R}^3 \setminus \bigcup_{i=1}^{14} \operatorname{int} S_i.$$

The infinity is a parabolic cusp with the stabilizer G_∞:

$$G_\infty = \langle T \rangle, \quad T : x \longmapsto x + 2e_1.$$

Then the domain $\{x \in \mathbb{R}^3 : x_2^2 + x_3^2 > 4\}$ is the cusp domain U_∞, and its complement contains the excessively thin component Ω_0,

$$\{x \in \mathbb{R}^3 : x_2^2 + x_3^2 < 1\} \subset \Omega_0 \subset \{x \in \mathbb{R}^3 : x_2^2 + x_3^2 < 4\},$$

for which the statement in Theorem 5.67 is not true.

It can be shown that this group G lies at the boundary of the deformation space for a certain co-compact Fuchsian group $G_1 \subset \mathcal{M}_3$ analogous to the group G_1 in Example 3.9, while the element T is an accidental parabolic element (i.e., it corresponds to a loxodromic element in G_1 in the isomorphism $G_\Sigma \longrightarrow G_1$).

Groups having similar properties are discussed in more detail in Chapter 9 §4.

The next theorem shows that all components of the discontinuity set $\Omega(G)$ are sufficiently thick if G does not contain parabolic elements (see Tukia [6] for $n=2$).

THEOREM 5.69. *Let $G \subset \text{Möb}(n)$, $n \geq 2$, be a geometrically finite Kleinian group not containing parabolic elements. Then there exists a $C > 0$ such that, for any component $\Omega_i \subset \Omega(G)$, any $h \in \text{Möb}(n)$, and any ball $B = B^n(x, t)$ having centre $x \in \overline{h(\Omega_i)} \cap \mathbb{R}^n$ and radius t, $t \leq \operatorname{diam}_E h(\Omega_i)/2$ there exists a ball $B' \subset B \cap h(\Omega_i)$ having the following property*:

$$C \cdot \operatorname{diam}_E B \leq \operatorname{diam}_E B'.$$

PROOF. Let Ω_i be some connected component of $\Omega(G)$; Corollary 5.52 shows that its stabilizer $G_i = \{g \in G : g(\Omega_i) = \Omega_i\}$ is a geometrically finite group, so Theorem 5.66 gives us a constant $C_i > 0$ having the required property (for the component Ω_i). The proof is completed by the number of orbits $G(\Omega_i)$ for the components of $\Omega(G)$ being finite, which follows from Corollary 5.47.

§6. **The geometry of the Nielsen hull boundary**

In §2 we introduced the convex Nielsen hull $H_G \subset H^{n+1} \cup \Lambda(G)$ for the limit set $\Lambda(G)$ of a discrete group $G \subset \text{Möb}(n)$. Here we study some properties of its boundary $H^* = \partial_H H_G$, which in general are dependent on n (see Thurston [1], §8.5 for $n=2$).

Here we shall assume that the limit set $\Lambda(G) \subset \overline{\mathbb{R}}^n$ of the Kleinian group $G \subset \text{Möb}(n)$ does not lie on any $(n-1)$-dimensional sphere $S^{n-1} \subset \overline{\mathbb{R}}^n$. This does not restrict the generality because, if $\Lambda(G)$ lies in a k-dimensional sphere $S^k \subset \overline{\mathbb{R}}^n$, having minimal dimension k, $0 \le k \le n-1$, one can consider G as a subgroup in $\text{Möb}(k)$ whose Nielsen hull lies in $\overline{H}^{k+1}$ and already satisfies our assumption.

With that assumption we have that $\text{int}\, H_G \neq \emptyset$, while the boundary $H_G^* = \partial H_G \cap H^{n+1}$, being the boundary of a convex body, is an n-dimensional pleated surface in H^{n+1} in which one can introduce the path metric d_*, induced by the hyperbolic metric d. Namely, for any two points x and y in the same component of the surface H_G^*, the distance $d_*(x, y)$ is defined as the infimum of the hyperbolic lengths of rectifiable paths in H^* joining x to y.

REMARK 5.70. The set of rectifiable paths in H^* joining x to y is non-empty - see Epstein-Marden [1]. In fact, let $p_0 \in \text{int}\, H_G$. Then $H^n \backslash H_G$ is homeomorphic to $H^* \times (0, \infty)$, where the first coordinate is where a ray from p_0 crosses H^*, and the second coordinate is the distance along the ray from H^*. Let C be the component of H_G^* containing x and y. The corresponding component of $H^n \backslash H_G$ is homeomorphic to $C \times (0, \infty)$ and is clearly connected by rectifiable paths. In order to connect x and y by a rectifiable path, we push them out a little by a radial push from p_0 to points x' and y'. Under the retraction $\rho : \overline{H}^n \longrightarrow H_G$ (see (5.7)), a rectifiable path in $H^n \backslash H_G$ joining x' to y' is mapped to a rectifiable path in H_G^*, because ρ is distance decreasing:

$$d(\rho(x), \rho(y)) \le d(x, y)$$

for any $x, y \in H^n$. For the proof of this, let L be the line containing $\rho(x)$ and $\rho(y) \neq \rho(x)$, oriented from $\rho(x)$ to $\rho(y)$, and let $A = [\rho(x), \rho(y)]_H$, $A \subset H_G$. We must have $\pi(x) \le \rho(x) < \rho(y) \le \pi(y)$ for the orthogonal projection $\pi : H^n \longrightarrow L$, since $\rho(x)$ is the nearest point of H_G to x, and is therefore the nearest point of A to x, and similarly for y. The result follows, since π is distance decreasing.

Our study of the boundary H_G^* of the Nielsen hull is based on considering a certain special ball cover $\mathcal{B}(G)$ for the discontinuity set $\Omega(G)$ of the group $G \subset \text{Möb}(n)$, $n \ge 2$ (see Apanasov [28], [29]).

DEFINITION 5.71. An n-dimensional open ball $B \subset \Omega(G) \subset \overline{\mathbb{R}}^n$ is called a *(strictly) maximal ball* for the Kleinian group $G \subset \text{Möb}(n)$ if its boundary sphere ∂B contains at least $n+1$ limit points $x^0, \ldots, x^n$ from $\Lambda(G)$ not lying in any $(n-2)$-dimensional sphere, i.e., if the sphere of lowest dimension containing the limit subset $\Lambda(G) \cap \partial B$ is the sphere ∂B itself. By $\mathcal{B}(G)$ we denote the set of all (strictly) maximal balls for the Kleinian group G.

It is not a trivial fact that maximal balls exist for any G, i.e., that the family $\mathcal{B}(G)$ is non-trivial. It is demonstrated by an ellipse in the plane (or ellipsoid in space): any circle inscribed in it touches the ellipse only at two points. As for an infinitely generated Kleinian group $G \subset \text{Möb}(n)$, $n \geq 2$, the interior of an ellipsoid (an ellipse for $n=2$) may be a component of the discontinuity set, the example implies that $\mathcal{B}(G)$ can be trivial for an infinitely generated group $G \subset \text{Möb}(n)$.

The situation is simplest for $n=2$, where we have:

THEOREM 5.72. *Let* $G \subset \text{Möb}(2)$ *be a finitely generated Kleinian group whose limit set does not lie on a circle. Then* $\Omega(G)$ *is covered by maximal discs from the family* $\mathcal{B}(G)$, *finite modulo action of* G.

PROOF. As the limit set is unaltered by transfer to a finite-index subgroup, one can assume that $G \subset \mathcal{M}_2$ and does not have torsion (Theorem 2.15). Then G acts freely on the surface $H^* = H^*_G$ and, due to Ahlfors's finiteness theorem (Theorem 5.36), H^*/G consists of a finite number of surfaces of finite type. In other words, the hyperbolic area $\text{Area}\, H^*/G$ is bounded (cf. Epstein-Marden [1]).

The n-dimensional convex hull for some set X in n-dimensional Euclidean space is bounded by an $(n-1)$-dimensional surface consisting of $(n-1)$-dimensional polyhedra having their vertices at the extreme points of X (the Krein-Milman theorem). As the extreme points for the Nielsen hull are points from $\Lambda(G) \subset \partial H^3$, (we may consider the projective model of H^n in the ball $B^n \subset \mathbb{R}^n$), the boundary $H^* \subset H^3$ consists of polygons lying in planes of support for H_G in H^3 and bounded by disjoint complete geodesics having their ends at extreme limit points. Below, the geodesics $\gamma = (x, y)_H \subset H^*$, $x, y \in \Lambda(G)$, not lying inside flat pieces of the boundary H^*, will be called the *bending geodesics* (limiting or not).

To the set of bending geodesics in H^* (which obviously is closed) there corresponds a geodesic lamination (Definition 7.58) in H^*/G consisting of the union of closed or infinite geodesics. We consider the double for each component of $(H^*/G)\setminus L$ (the complete hyperbolic surface), which gives (see Thurston [1], §8.5) that the Euler characteristic $\chi(H^*/G)$ is equal to half of the Euler characteristic for the double for $(H^*/G)\setminus L$. Then this and the Gauss-Bonnet formula implies that $\text{Area}((H^*/G)\setminus L) = \text{Area}\, H^*/G$, and therefore the lamination L has zero measure. This shows that $(H^*/G)\setminus L$ consists of totally geodesic domains D_i, bounded either by closed geodesics or by infinite ones, and therefore having areas $\geq \pi$ (π is the hyperbolic area of the hyperbolic triangle with vertices at infinity). Then from $\text{Area}\, H^*/G < \infty$ we get that the number of domains D_i is finite. Each of them corresponds to a G-orbit consisting of disjoint totally geodesic maximal subdomains in H^* forming the inverse image $\pi^{-1}(D_i)$ for the natural projection $\pi : H^* \longrightarrow H^*/G$. Let $\tilde{D}_i$ be one of these domains lying in a certain plane of support H_i for the convex hull H_G. As $\tilde{D}_i$ is bounded by at least three distinct complete infinite geodesics, its set of points at infinity $\partial\tilde{D}_i \cap \partial H^3$ contains at least three distinct points from $\partial H_i \cap \Lambda(G)$. Thus the domains $\tilde{D}_i \subset H^*$ correspond to maximal discs $B_i \in \mathcal{B}(G)$, amongst which only finitely many are non-equivalent because the set of domains $D_i \subset (H^*/G)\setminus L$ is finite.

Now let $z \in \Omega(G)$. Then for a certain i we have, due to

Area $L = 0$, that $\pi \circ \rho(z) \subset \overline{D}_i$. If then $\pi \circ \rho(z) \in \operatorname{int} D_i$, we have $\rho(z) \in \operatorname{int} \widetilde{D}_i$ for a certain $\widetilde{D}_i \subset \pi^{-1}(D_i)$, and as the mapping $\rho^{-1}|_{\operatorname{int}\widetilde{D}_i}$ is a homeomorphism, the point z lies inside the maximal disc $B_i \in \mathcal{B}(G) : \partial B_i = \partial H_i$, $\widetilde{D}_i \subset H_i$.

If $\pi \circ \rho(z) \in \partial D_i$, $\rho(z)$ lies in a certain bending geodesic $\gamma = (x, y)_H \subset H^*$ and there are two possible cases: either γ is the common boundary geodesic for two domains $\widetilde{D}_i, \widetilde{D}_j \subset H^*$, $\operatorname{cl} \widetilde{D}_i \cap \operatorname{cl} \widetilde{D}_j = \gamma$ or γ is the limit for a certain orbit $g_m(\widetilde{D}_i)$, $g_m \in G$, $\widetilde{D}_i \subset H^*$.

In the first case, if z does not lie inside one of the maximal discs $B_i, B_j \in \mathcal{B}(G)$, defined by the domains $\widetilde{D}_i$ and $\widetilde{D}_j$, then z is a common boundary point for the discs B_i and B_j. As $x, y \in \partial B_i \cap \partial B_j$, this gives the inclusion $\Lambda(G) \subset \partial B_i = \partial B_j$, which is impossible from the theorem assumption.

In the second (limit) case, we have alternatives: either there exists $\widetilde{D}_i \subset H^*$, such that $\gamma \subset \partial\widetilde{D}_i$, or there exist $\widetilde{D}_i, \widetilde{D}_j \subset H^*$ such that their orbits $g_m(\widetilde{D}_i)$ and $g_k(\widetilde{D}_j)$ converge to γ from different directions. We replace $\widetilde{D}_i$ and $\widetilde{D}_j$ be the corresponding discs $B_i, B_j \in \mathcal{B}(G)$ to get that either $z \in \operatorname{int} B_i$ or there exist bending geodesics $\gamma_i = (x_i, y_i)_H \subset \partial\widetilde{D}_i$, $\gamma_j = (x_j, y_j)_H \subset \partial\widetilde{D}_j$, such that $x = \lim_{m\to\infty} g_m(x_i) = \lim_{k\to\infty} g_k(x_j)$ and $y = \lim_{m\to\infty} g_m(y_i) = \lim_{k\to\infty} g_k(y_j)$. But then either z lies in $g_{m_0}(B_i)$ or in $g_{k_0}(B_j)$, or else $\lim_{m\to\infty} g_m(B_i) = B_i^* \subset \Omega(G)$, $\lim_{k\to\infty} g_k(B_j) = B_j^* \subset \Omega(G)$, $B_i^* \cap B_j^* = \varnothing$ and $z \in \overline{B}_i^* \cap \overline{B}_j^* = \partial B_i^* = \partial B_j^*$. As in the last case $\Lambda(G) \subset \partial B_i^* = \partial B_j^*$, this contradicts the theorem assumption, which completes the proof.

The situation becomes more complicated for $n \geq 3$ because, firstly, there is no spatial analogue of Ahlfors's finiteness theorem, which forces us to consider only geometrically finite groups.

Another difficulty is related to the infinite-order orthogonal components of the group G elements. Namely, let ∞ be an irrational parabolic cusp for G whose stabilizer G_∞ ($\operatorname{rank} G_\infty = k \leq n - 2$) has the element g, $g : x \longmapsto U(x) + a$, where the rotation $U \in O(n)$ has infinite order and $U(a) = a$. Then the G_∞-orbit of any half-space $B_\infty \subset \Omega(G)$, having limit points in the boundary plane forms the domain

$$W_\infty = G(B_\infty) = \{x \in \mathbb{R}^n : \sum_{i=k+1}^{n} x_i^2 > t\} \tag{5.58}$$

where t is the Euclidean distance of the plane ∂B_∞ to the subspace $\mathbb{R}^k \subset \mathbb{R}^n$, whose basis is provided by translation vectors from G_∞.

We call the domain $W_\infty \subset \Omega(G)$ the *maximal cylinder domain* for the irrational parabolic cusp ∞. The boundary cylinder ∂W_∞ is densely covered (in the direction sense) by limit points for G as the rotation U is irrational, so each half-space $B \subset W_\infty$, $\partial B \cap \partial W_\infty \neq \varnothing$, cannot lie in $\mathcal{B}(G)$. It is important that the domain W_∞ contains the cusp domain U_∞ in which the retraction ρ is a homeomorphism. That property enables us subsequently to operate with cylinder domains as maximal balls.

Another type of maximal (ball) domain is given by maximal cone domains $C(x, y)$, $(x, y \in \Lambda(G))$ which are obtained by an irrational rotation U of a certain ball (a half-space for $x=0$, $y=\infty$) $B_0 \subset \Omega(G)$, $\partial B_0 \cap \Lambda(G) \setminus \{x, y\} \neq \varnothing$ around a k-dimensional sphere containing x and y. So U can be either an orthogonal component of a loxodromic element $g \in G$, $(g(x) = x,$ $g(y) = y)$ or a limit one for the orthogonal components of a certain element sequence $g_m \in G$, for which x and y are repulsive and attractive points.

EXAMPLE 5.73. A maximal cone domain $C(0, \infty)$ arises for a Kleinian group $G \subset \text{Möb}(3)$, $G = \langle G_0, h \rangle$, where $G_0 \subset \text{Möb}(3)$ and $h \in \text{Möb}(3)$ have the following characteristics:

1) The transformation h is the homothety $h : x \longmapsto r \cdot U(x)$, where $r > 1$, while the rotation U around the axis $\mathbb{R} \cdot e_3$ has infinite order.

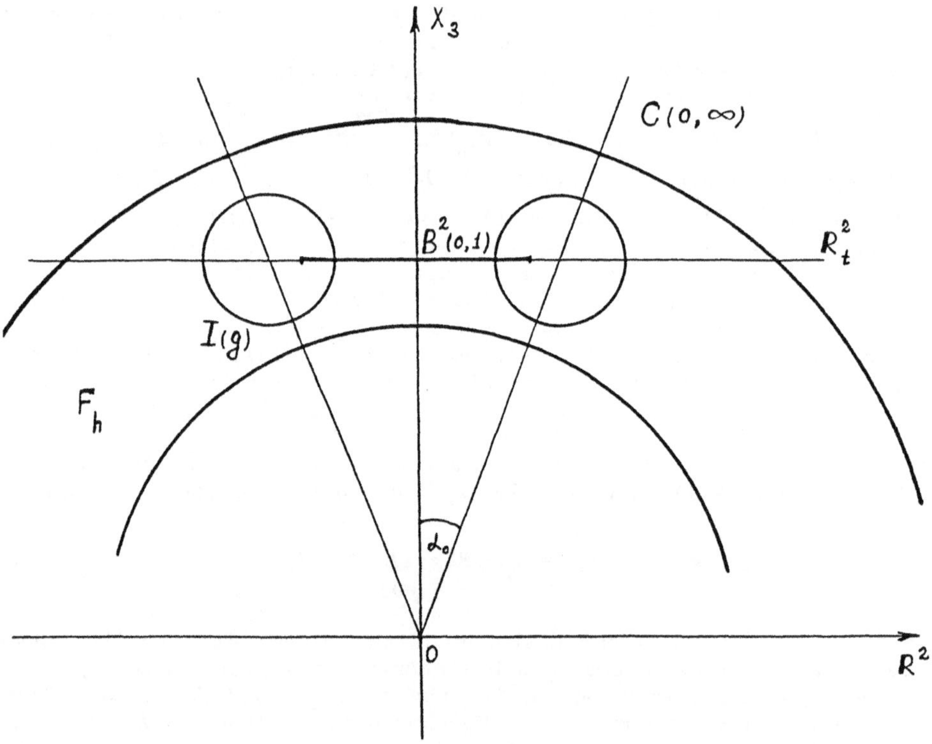

Figure 29

2) The Fuchsian group G_0 acts in the disc $B^2(0, 1)$ in the plane $\mathbb{R}^2_t = \{x \in \mathbb{R}^3 : x_3 = t\}$, where $t > 0$ is such that all the

isometric spheres $I(g)$ for the elements $g \in G_0$ lie inside a certain spherical layer F_h, fundamental for the cyclic group $\langle h \rangle$ (see Fig. 29).

Then $C(0, \infty)$ has the following cone at its boundary:

$$\partial C(0, \infty) = \{x \in \overline{\mathbb{R}}^3 : t^2(x_1^2 + x_2^2) = r_0^2 x_3^2, \quad x_3 \geq 0\}, \tag{5.59}$$

where $r_0^2 = \sup\{x_1^2 + x_2^2 : x = (x_1, x_2, t) = g(\infty), \ g \in G_0 \backslash \mathrm{id}\}$, while the angle of the cone is $\alpha_0 = \arctan(r_0/t)$.

REMARK 5.74. The G constructed above having maximal cone domain $C(0, \infty)$ in its discontinuity set $\Omega(G)$ has connected $\Omega(G)$, but that property can occur also for quasi-Fuchsian groups obtained from Fuchsian co-compact groups $\overline{\mathbb{R}}^3$ by quasi-conformal deformation, namely see Chapter 9 §3.5, Remark 9.47.

We refer the reader to the paper Apanasov [29] where he can find some approach for the inductive proof of the following theorem.

THEOREM 5.75. *Let $G \subset \mathrm{Möb}(n)$, $n \geq 2$, be a geometrically finite Kleinian group whose limit set is not contained in any $(n-1)$-dimensional sphere. Then the complement in $\Omega(G)$ to the maximal cylinder and cone domains for G is covered by the maximal balls $B \in \mathcal{B}(G)$.*

The approach used in the said proof of this theorem (which we don't use there) is based on a quite cumbersome technique (which needs further perfection) using a variation of non-maximal balls in the discontinuity set $\Omega(G)$.

REMARK 5.76. The covering $\mathcal{B}(G)$ is invariant with respect to G because, for any $B \in \mathcal{B}(G)$ and for $g \in G$, the ball $g(B)$ has limit points $g(\Lambda(G) \cap \partial B)$ on its boundary sphere. This also applies to maximal cylinder-domain and cone-domain families in $\Omega(G)$ (they are the union of "weakly" maximal balls).

LEMMA 5.78. *Let $G \subset \mathrm{Möb}(n)$ be a finitely generated Kleinian group for $n = 2$ and a geometrically finite one for $n \geq 3$ whose limit set $\Lambda(G)$ is not contained in an $(n-1)$-dimensional sphere. Also for each maximal ball $B_i \in \mathcal{B}(G)$ let the subdomain $B_i' \subset B_i$ signify the interior of the Nielsen hull of the limit subset $\partial B_i \cap \Lambda(G)$ in the Poincaré hyperbolic metric of the ball B_i.*

Then the set

$$\Omega' = \bigcup \{B_i' : B_i \in \mathcal{B}(G)\}$$

is the open set of bijectivity of the retraction ρ for the set $\Omega_b = \bigcup\{B_i : B_i \in \mathcal{B}(G)\} \subseteq \Omega(G)$. The restriction of the retraction ρ to Ω' is a conformal homeomorphism: given a connected component $B_i' \subset \Omega'$, $\rho|_{B_i'}$ is the rotation in $\mathbb{R}^{n+1}$ through the angle $\pi/2$ around the fixed sphere ∂B_i.

REMARK 5.79. One can also define maximal bijective domains $W'(p)$ and $C'(x, y)$ for the retraction ρ in the maximal cylinder- and cone-domains $W(p)$ and $C(x, y)$ by analogy with the case of maximal balls, which we now consider; $W'(p)$ and $C'(x, y)$ are "minimal convex hulls" for the limit sets $\partial W(p) \cap \Lambda(G)$ and $\partial C(x, y) \cap \Lambda(G)$, where the convexity is understood in the sense that the geodesic planes in the ball $B \in \mathcal{B}(G)$ are replaced by the spheres in $W(p)$ and $C(x, y)$ orthogonal to rulings of the cylinders and cones, respectively.

PROOF OF LEMMA 5.78. Let $B_0 \in \mathcal{B}(G)$ be any maximal ball from $\Omega(G)$. If B_0 coincides with a connected component of $\Omega(G)$, then the restriction $\rho|_{B_0}$ of ρ to the n-ball B_0 is the rotation in $\overline{\mathbb{R}}^{n+1}$ around the fixed $(n-1)$-dimensional sphere ∂B_0 through the angle $\pi/2$. Therefore, ρ conformally maps B_0 into the n-dimensional hyperbolic plane $H_0 \subset H^{n+1}$ (a hemisphere in $\mathbb{R}^{n+1}_+$) that is a connected component of $H^* = \partial_H H_G$. It is clear that then $B_0' = B_0$.

Now let B_0 not coincide with any connected component of $\Omega(G)$. Then non-emptiness of $B_0' \subset B_0$ is a direct consequence of the Definition 5.71 of the maximal ball.

We now describe B_0' more constructively. Let B_j, $j \in J = J(B_0)$ be all the balls from $\mathcal{B}(G)\backslash B_0$ or the balls from the maximal cylinder- or cone-domains intersecting B_0. For each $j \in J$, we consider the $(n-1)$-dimensional spheres $s_j = s_{0j}$ and $\sigma_j = \sigma_{0j}$, which intersect $S_0 = \partial B_0$ in the $(n-2)$-dimensional sphere $S_0 \cap \partial B_j$. Here s_j is orthogonal to S_0, while σ_{0j} divides the dihedral angle $B_0 \cap B_j$ into halves (we call s_j a plane in B_0 and σ_{0j} a bisector sphere). For $s_j = s_{0j}$ and σ_{0j}, we define in B_0 the $\text{ext}_0\, s_{0j}$ and $\text{ext}_0\, \sigma_{0j}$ as those of the components $B_0\backslash s_{0j}$ and $B_0\backslash\sigma_{0j}$ that contain B_0' (i.e., they adjoin the spherical $(n-)$-ball $\partial B_0\backslash B_j$). Then as the $(n-1)$-dimensional balls $B_j \cap S_0$, $j \in J$, cover $\Omega(G) \cap S_0$, we have

$$B_0' = \bigcap_{j \in J} \text{ext}_0\, s_{0j} \subset B_0'' = \bigcap_{j \in J} \text{ext}_0\, \sigma_{0j}. \tag{5.64}$$

It follows from the G-invariance of $\mathcal{B}(G)$ (see Remark 5.75) and (5.64) that for different balls $B_i, B_j \in \mathcal{B}(G)$ one always has

$$B_i' \cap B_j' = \varnothing, \quad B_i'' \cap B_j'' = \varnothing. \tag{5.65}$$

This immediately gives precise invariance for the domains B_i' and B_i'' with respect to the stabilizer $G_i \subset G$ for the ball $B_i \in \mathcal{B}(G)$ (which may be trivial), i.e., for all $h \in G_i$ and $g \in G\backslash G_i$ we have

$$h(B_i') = B_i', \quad h(B_i'') = B_i''; \quad g(B_i') \cap B_i' = g(B_i'') \cap B_i'' = \varnothing. \tag{5.66}$$

This property combines with the fact that the retraction ρ in the domain $B_i' \subset B_i \in \mathcal{B}(G)$ is the restriction of the rotation in $\overline{\mathbb{R}}^{n+1}$ around the $(n-1)$-dimensional spheres ∂B_i through the angle $\pi/2$ to show that ρ is a conformal homeomorphism in Ω'. At the same time, for all $j \in J(B_0)$ such that the sphere of lowest dimension containing the limit subset $\Lambda(G) \cap \partial B_0 \cap \partial B_j$ is the $(n-2)$-sphere $\partial B_0 \cap \partial B_j$ itself, the closed domain B_{0j} bounded in the balls B_0 and B_j correspondingly by the hyperbolic planes s_{0j} and s_{j0} (see Fig. 30) is contracted by the retraction ρ to the $(n-1)$-dimensional plane $H_0 \cap H_j \subset H^{n+1}$, where the n-planes H_0 and H_j have as their spheres at infinity the spheres ∂B_0 and ∂B_j. That is, for any circle δ orthogonal simultaneously to the planes s_{0j} and s_{j0} and to the spheres ∂B_0, ∂B_j, σ_{0j} (Fig. 30) we have

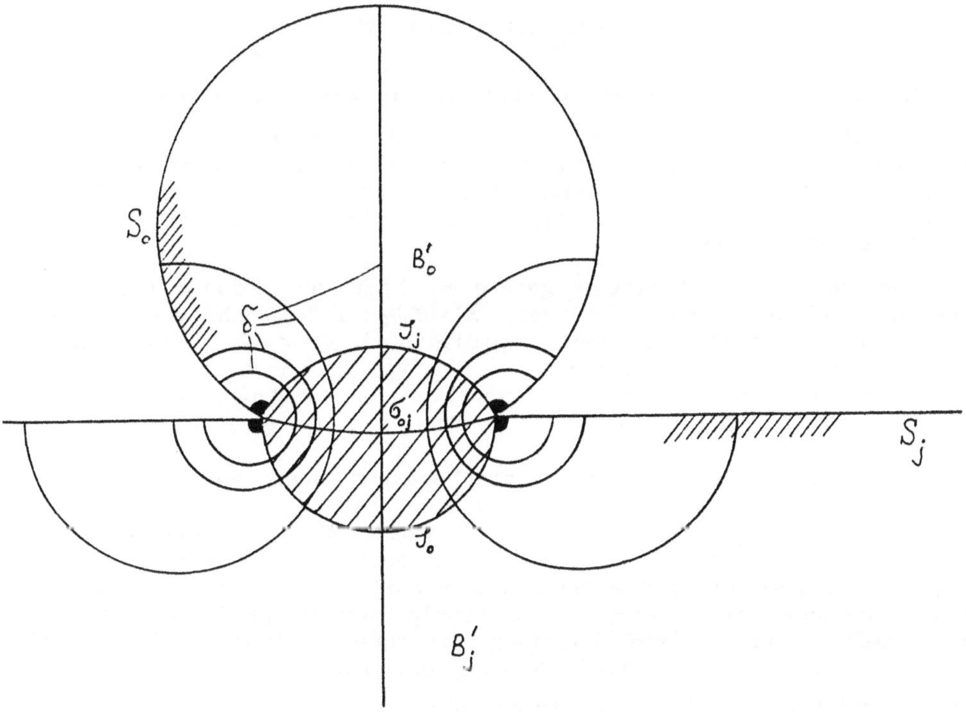

Figure 30

$$\rho(\delta \cap B_{0j}) = \rho(\delta \cap \sigma_{0j}). \tag{5.67}$$

Therefore the closure of $\rho'(\Omega')$ coincides with $\rho(\Omega_b)$. This completes the proof of the lemma.

LEMMA 5.80. *Let $B_0 \in \mathcal{B}(G)$ be a maximal ball for a finitely generated (for $n=2$) or a geometrically finite (for $n \geq 3$) Kleinian group $G \subset$ Möb(n). Then its stabilizer $G_0 \subset G$ is either trivial or geometrically finite.*

PROOF. Due to the precisely G_0-invariance for the bijectivity domain $B_0' \subset B_0$, there corresponds to the orbit $G(B_0')$ a connected submanifold $M_0 = G(B_0')/G \subset \Omega(G)/G$ (we assume that G does not have torsion), and which goes over on the retraction $M(G) \longrightarrow M_G$ to a connected and totally geodesic submanifold $M_0^* = G(\rho(B_0'))/G \subset H^*/G$ having a bounded n-dimensional hyperbolic volume. For $n=2$, this follows from Ahlfors's finiteness theorem (Theorem 5.36, compare Theorem 5.72), while for $n \geq 3$, it follows from the geometrical finiteness of G (Theorem 5.39). Then we get that $\operatorname{vol} U_\varepsilon(M_0^*) < \infty$ is bounded, where $U_\varepsilon(M_0^*) \subset H^{n+1}/G$ is an ε-neighbourhood of M_0^*. At the same time, $\Lambda(G_0) \subset \partial B_0 \cap \Lambda(G)$ and therefore $H_{G_0} \subset \rho(B_0')$. This implies

$$\operatorname{vol} U_\varepsilon(H_{G_0}/G_0) \leq \operatorname{vol} U_\varepsilon(M_0^*) < \infty,$$

which gives (with Theorem 5.39) that G_0 is geometrically finite.

COROLLARY 5.81. *Under the conditions of Lemma 5.80, the intersection P_0 of the fundamental pseudo-Dirichlet polyhedron for the stabilizer G_0 for the maximal ball $B_0 \in \mathcal{B}(G)$ and the bijectivity domain B_0' has finite hyperbolic volume, and, in particular, $\operatorname{vol} B_0' < \infty$ if $G_0 = \{\mathrm{id}\}$.*

We see for a finitely generated Kleinian group $G \subset \mathcal{M}_2$ from Theorem 5.72 and Lemma 5.78 that $H^* = \partial_H H_G \subset H^3$ is the closure of the union of hyperbolic polygons $g(\mathrm{cl}_H \rho(B_i'))$, $g \in G$, adjoining one another on bending geodesics $\gamma_i = (x, y)_H$, where $x, y \in \partial B_i \cap \Lambda(G)$, such that the geodesic γ_i^0 in the disc B_i, having ends x and y, lies in $\partial B_i'$ (here $B_i, \ldots, B_m$ form the fundamental set for the action of G in $\mathcal{B}(G)$).

The following statement describes an analogous situation in dimensions $n \geq 3$, where by a bending geodesic of the $(n-)$-dimensional pleated surface $H^* = \partial_H H_G \subset H^{n+1}$ we mean a geodesic $\gamma = (x, y)_H \subset H^*$, with $x, y \in \Lambda(G)$ such that γ does not lie inside any totally geodesic k-dimensional piece of H^*, $k \geq 2$. We call such a bending geodesic $\gamma = (x, y)_H$ a simple bending geodesic if its ends $x, y \in \Lambda(G)$ lie on the boundary of some maximal ball $B \in \mathcal{B}(G)$; otherwise, we call γ a limit bending geodesic.

LEMMA 5.82. *Let a Kleinian group $G \subset \mathrm{Möb}(n)$, $n \geq 2$, be such that:*

1) *$\Lambda(G)$ does not lie in any $(n-1)$-dimensional sphere;*
2) *G is finitely generated for $n=2$; and*
3) *for $n \geq 3$, G is geometrically finite and $\Omega(G)$ is covered by the maximal-ball family $\mathcal{B}(G)$.*

Then there exists a finite family $\mathcal{B}_f = \{B_1, \ldots, B_m\} \subset \mathcal{B}(G)$ such that:

a) $G(\bigcup_{i=1}^m B_i) = \Omega(G)$ and $G(B_i') \cap G(B_j') = \varnothing$ for $i \neq j$, $1 \leq i, j \leq m$;
b) *for each component $\Omega_0 \subset \Omega(G)$ there exists a subfamily $\mathcal{B}_0 \subset \mathcal{B}_f$ of joint balls $B_i \in \mathcal{B}(G)$ for which the hyperbolic polyhedra $\rho(B_i') \subset \rho(\Omega_0) \subset H^*$ are joined to each other by common sides of some finite subset of their non-degenerate polyhedral angles $W_1, \ldots, W_k$;*

c) *the angles* W_j *have for* $n=2$ *the value* π, *while for* $n \geq 3$ *they have the combinatorial type of angle*

$$\Pi_j = \{(tx, x_n) \in \mathbb{R}^{n-1} \times \mathbb{R} : x_n > 0, \quad t > 0, \quad x \in Q_j\}, \tag{5.68}$$

where Q_j *is a convex bounded Euclidean polyhedron lying in an* $(n-2)$*-dimensional plane in* $\mathbb{R}^{n-1}$ *which does not pass through the origin; here the edge* $\{(0, x_n) \in \mathbb{R}^{n-1} \times \mathbb{R}^n\} \subset \overline{\Pi}_j$ *corresponds to the geodesic edge of* W_j, *which is a simple bending geodesic* $\gamma = (x, y)_H \subset H^*$;

d) *to every simple bending geodesic* $\gamma = (x, y)_H \subset H^*$ *(G-equivalent to an edge of some angle* W_j*) there corresponds a finite subset* $\mathcal{B}(\gamma)$ *of maximal balls such that an arc* $\gamma' \subset \Omega(G)$ *of the circle passing through* x *and* y *is the central "axis" of the spherical polyhedral angle*

$$B(\gamma) = \bigcap \{B : B \in \mathcal{B}(\gamma)\} \tag{5.69}$$

and such that $\rho(\gamma') = \gamma$.

PROOF. The case $n=2$ has already been considered above, so we assume that $n \geq 3$. Let $P^{n+1} \subset H^{n+1} \cup \Omega(G)$ be a fundamental finite-sided polyhedron for G not having equivalent parabolic points in ∂P^{n+1} (the Dirichlet polyhedron or the pseudo-Dirichlet one - see (3.33) and Lemma 3.16). Let $p_1, \ldots, p_m$ be all the parabolic cusps for G in ∂P^{n+1} and let $\tilde{U}_i \subset H^{n+1} \cup \Omega(G)$ be precisely G_{p_i}-invariant cusp domains for those cusps. Then the polyhedron

$$P_0^{n+1} = P^{n+1} \setminus \bigcup_{i=1}^{m} \tilde{U}_i \subset H^{n+1} \cup \Omega(G) \tag{5.70}$$

is compact (Theorem 5.39), as is the polyhedron $P_0^n = P_0^{n+1} \cap \Omega(G)$.

As the retraction ρ is a conformal homeomorphism in the cusp domains $U_i = \tilde{U}_i \cap \Omega(G)$ (rotation in $\overline{\mathbb{R}}^{n+1}$ with respect to the central spheres of the complements $\overline{\mathbb{R}}^n \setminus U_i$ through the angles $\pi/2$), the bending geodesics $\gamma \subset H^*$ cannot end in the parabolic cusps of G. Also, the condition 3) implies that all cusp domains for G are covered by a G-orbit for a finite number of maximal balls. This condition 3) and compactness of P_0^n gives us finally the finite subfamily $\mathcal{B}_f \subset \mathcal{B}(G)$, which satisfies property a): the elements $B_i \in \mathcal{B}_f$ are defined by the condition $B_i' \cap P^n \neq \emptyset$.

The following modification of $\mathcal{B}_f$ is made by adding obedience to b). For every class of G-equivalent components, let us consider as its representative Ω_0, $\Omega_0 \cap P^n \neq \emptyset$. Besides, in place of the balls $B_i \in \mathcal{B}_f$ lying in $G(\Omega) \setminus \Omega_0$, we consider their images $B_j = g_j(B_i) \subset \Omega_0$, $g_j \in G$, which form the family $\mathcal{B}_0$ satisfying property b). The finiteness of classes of G-equivalent components (Theorem 5.39) implies that we thus obtain ultimately a finite family $\mathcal{B}_f$ satisfying b).

The compactness of the polyhedron (5.70), properties a) and b), and Corollary 5.81 together show that the polyhedral angles of the finite-sides polyhedra $P_0^n \cap B_i'$, $B_i \in \mathcal{B}_f$, are non-degenerate and that they include only finitely many non-equivalent ones. Then for any component $\Omega_0 \subset \Omega(G)$, the hyperbolic polyhedra $\rho(B'_i) \subset H^*$, $B_i \in \mathcal{B}_0$, will have only finitely many non-equivalent polyhedral angles each, $W_1, \ldots, W_k$, all of which are non-degenerate. This gives final obedience to b).

To prove c) and d), it is sufficient to show that for any simple bending geodesic $\gamma = (x, y)_H \subset H^*$ there exists only a finite number of balls $B \in \mathcal{B}(G)$ containing the points x and y in ∂B (we denote that ball family by $\mathcal{B}(\gamma)$). This follows from the fact that otherwise the polyhedra $B' \subset B \in \mathcal{B}(\gamma)$, having x and y as vertices, should each have an infinite number of various polyhedral (non-degenerate) angles - a contradiction with b). Therefore, the family $\mathcal{B}(\gamma)$ is finite and the angle $B(\gamma)$, defined in (5.69), whose axis is the circle arc $\gamma' \subset \Omega(G)$, having ends x and y. This arc is the intersection of the corresponding bisectrix spheres σ_{ij} defined by the balls $B_i, B_j \in \mathcal{B}(\gamma)$ (Fig. 31).

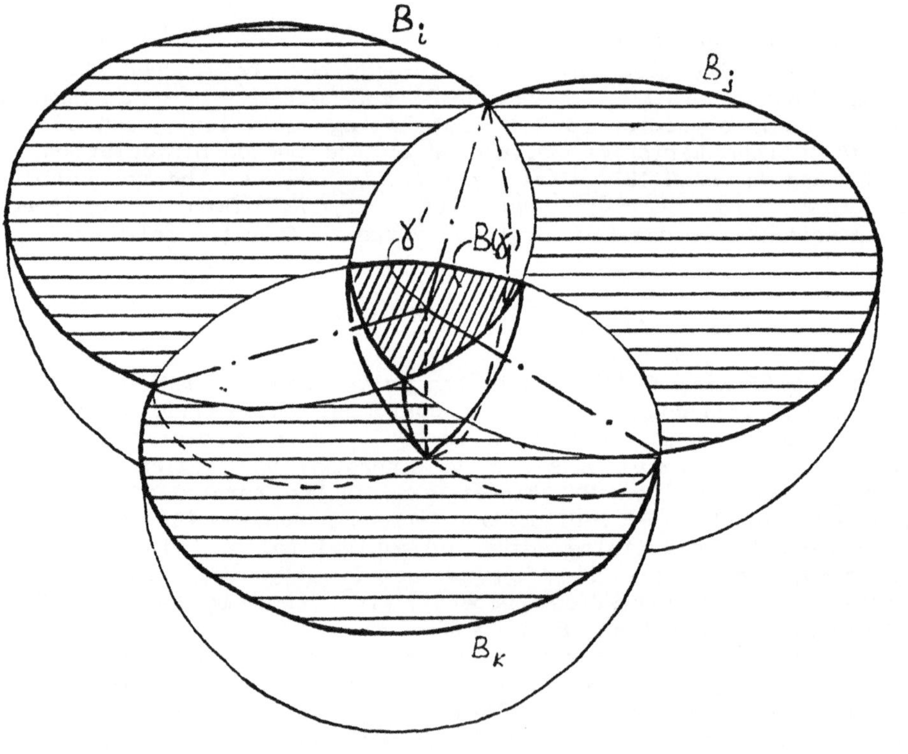

Figure 31

We note further that, not only do we have $\rho(\gamma') = \gamma \subset H^*$, but also that for any geodesic $\gamma_i^0 = (x, y)_i$ in the i-th ball $B_i \in B(\gamma)$ we have $\rho(\gamma_i^0) = \gamma$. Here the geodesic γ_i^0 in the ball B_i lies in the intersection of the hyperbolic planes $s_{ij} \subset B_i$ defined by the maximal balls joined to B_i, namely $B_j \subset \mathcal{B}(\gamma)\setminus B_i$. Here γ_i^0 is a geodesic edge for a polyhedral angle of the polyhedron $B_i' \subset B_i$ that goes over by the conformal mapping $\rho|_{B_i'}$ to a certain polyhedral angle W_j in the hyperbolic n-dimensional polyhedron $\rho(B_i') \subset H^*$. Due to the conformality of this map, W_j is non-degenerate and is of combinatorial type (5.68). The lemma is proved.

REMARK 5.83. The angle $\rho^{-1}(W_j)$ of the polyhedron $B_i' \subset B_i \in \mathcal{B}(\gamma)$, having a geodesic edge $\gamma_i^0 \subset \bigcap_k s_{ik}$ (where the planes $s_{ik} \subset B_i$ correspond to the balls $B_k \in \mathcal{B}(\gamma)\setminus B_i$), is contained in a spherical polyhedral angle V_j of the same combinatorial type. That angle V_j has as edge the arc $\gamma' \subset B(\gamma)$ (see (5.69)) and is formed by the intersection of balls bounded by the bisectrix spheres σ_{ij}. We note that

$$\bigcup\{V_j : V_j \subset B_j \in \mathcal{B}(\gamma)\} \tag{5.71}$$

is a full neighbourhood of the arc γ' having ends $x, y \in \Lambda(G)$, $\rho(\gamma') = \gamma = (x, y)_H \subset H^*$. In other words, the polyhedral angles V_j form a decomposition of a full neighbourhood of the arc γ' (into dihedral angles for $n = 3$, see Fig. 31).

Note here that limit bending geodesics in H^* are contained in the closure of the orbits of simple bending geodesics. Such limit bending geodesics for the retraction ρ correspond to the circular arcs in $\Omega(G)$ to which the orbits $G(\gamma_i^0)$ converge, where γ_i^0 are the geodesic edges of the polyhedra $B_i' \subset B_i$, $B_i \in \mathcal{B}(G)$.

A good illustration of the maximal ball covering method is given by quasi-Fuchsian groups in the class constructed below. The groups in that class may be called "pea-pod-groups" because their limit sets resemble a pod of some plant (Fig. 32).

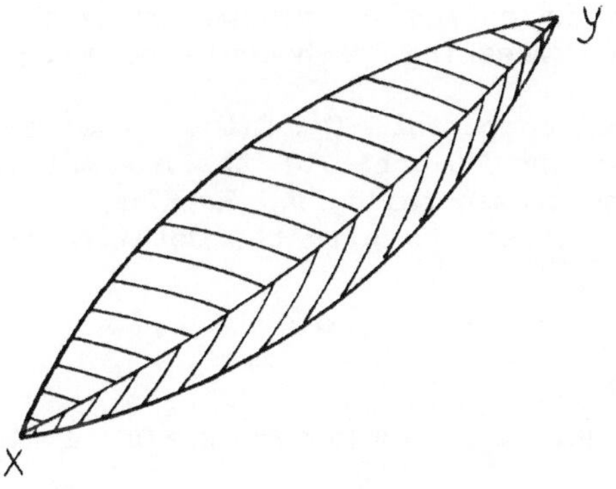

Figure 32

We consider balls $D_1, \ldots, D_k \subset \overline{\mathbb{R}}^n$, $n \geq 3$, $k \geq n$, in general positions, whose intersection

$$D(x, y) = \bigcap_{i=1}^{k} D_i$$

is a k-hedral spherical angle with vertices x and y. For $n = 3$ and $k = 3$, $D(x, y)$ is shown in Fig. 32. Let $h \in \text{Möb}(n)$ be a loxodromic transformation having fixed points x and y, $h(D(x, y)) = D(x, y)$. As a fundamental domain F_h for the cyclic group $\langle h \rangle$ one can take the exterior of two non-intersecting spheres I, I', $x \in \text{int}\, I$, $y \in \text{int}\, I'$, which are orthogonal to the sides of $D(x, y)$.

Suppose now that there is a finite family $\mathcal{S}$ of spheres $S_i \subset \mathbb{R}^n$, $i \in \mathcal{I}$, decomposed into subfamilies:

$$\mathcal{S} = \bigcup_{i,j=1}^{k} (\mathcal{S}_{ij} \cup \mathcal{S}_i), \tag{5.72}$$

where the spheres $S \in \mathcal{S}_{ij}$ are orthogonal to the sphere $\partial D_i \cap \partial D_j$ containing an $(n-2)$-dimensional side of $D(x, y)$; the spheres $S \in \mathcal{S}_i$ are orthogonal to the $(n-1)$-dimensional side of $D(x, y)$ lying in ∂D_i. In addition spheres from the different subfamilies $\mathcal{S}_{ij}$ and $\mathcal{S}_{ml}$, or correspondingly $\mathcal{S}_i$ and $\mathcal{S}_j$, do not intersect each other and intersection angles of these spheres along themselves and with I and I' are integer parts of π.

If the interiors of the spheres in this family $\mathcal{S}$ cover the surface $F_h \cap \partial D(x, y)$, their properties and Theorem 6.9 imply that the group

$$G = \langle G_0, h \rangle,$$

where G_0 is the group generated by the reflections in the spheres $S_i \in \mathcal{S}$, is a discrete quasi-Fuchsian group (cf. proof in Chapter 3 §3.4). This group is called a "pea-pod-group"; its limit set is shaped like a finely breaking k-hedral pod with given vertices x and y.

For the unbounded component $\Omega_0 \subset \Omega(G)$, the maximal ball covering $\mathcal{B}_0(G)$ is obtained as the G-orbit of the maximal balls B_i that are the complements to the closed balls $\overline{D}_i$, $B_i = \overline{\mathbb{R}}^n \setminus \overline{D}_i$.

To explain this better, we give the detailed construction for one such pea-pod-group in $\overline{\mathbb{R}}^3$.

EXAMPLE 5.84. Let $x = 0$, $y = \infty$, $D_i = \{x \in \mathbb{R}^3 : x_i > 0\}$, $i = 1, 2, 3$,

$$D(x, y) = D(0, \infty) = D_i = \{x \in \mathbb{R}^3 : x_i > 0, \quad i = 1, 2, 3\}.$$

We take the disjoint spheres $S_{23} = S^2(x^1, r)$, $S_{13} = S^2(x^2, r)$, $S_{12} = S^2(x^3, r)$ having radius $r > 0$ and their centres at the points $x^i = \rho \cdot e_i$, $i = 1, 2, 3$, where $2\rho \cdot \text{Sin}\, \pi/8 = r \cdot \sqrt{3}$, i.e.,

$$r = \rho\cdot\sqrt{\frac{2-\sqrt{2}}{3}}$$

We consider also the spheres

$$S_1 = S^2\left[\frac{\rho}{\sqrt{2}}(e_2+e_3),\ r\right],\ S_2 = S^2\left[\frac{\rho}{\sqrt{2}}(e_1+e_3),\ r\right],\ S_3 = S^2\left[\frac{\rho}{\sqrt{2}}(e_1+e_2),\ r\right],$$

and the spheres

$$I = S^2(0,\ R),\quad I' = S^2(0,\ R'),$$

where

$$R = \frac{-r + \sqrt{4\rho^2 - 3r^2}}{2},\quad R' = (\rho^2 - r^2)/R.$$

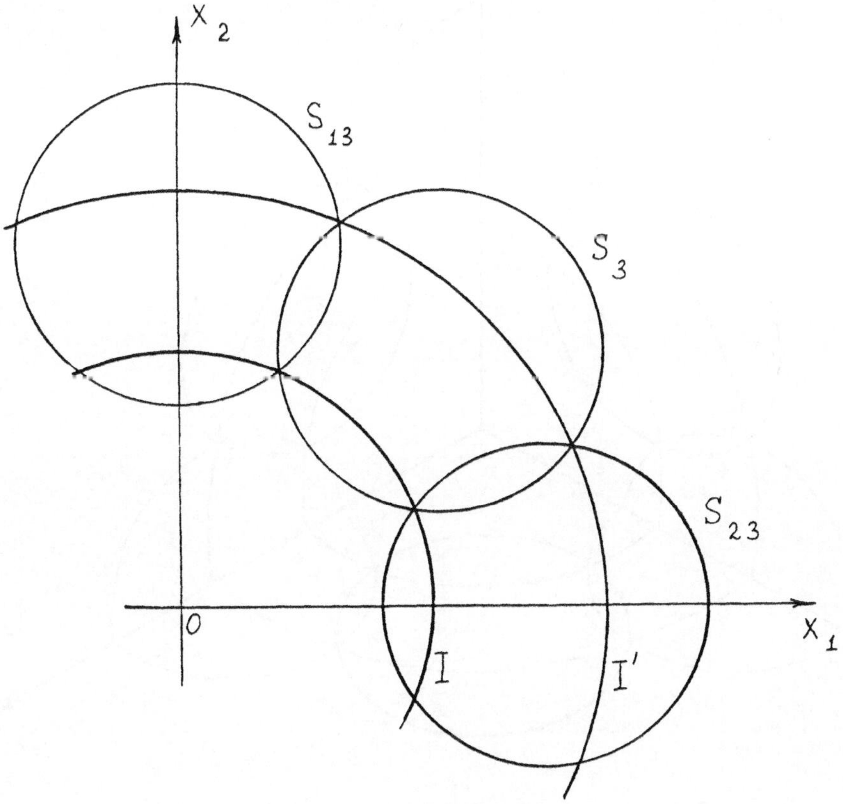

Figure 33

Then the spheres I and I' bound the spherical layer F_h which is a fundamental polyhedron for the cyclic group generated by $h \in \text{Möb}(3)$,

$$h : x \longmapsto \lambda x, \quad \lambda = 2(\rho^2 - r^2)/(2\rho^2 - r^2 - r\sqrt{4\rho^2 - 3r^2}).$$

The spheres S_i, $i = 1, 2, 3$, are disjoint because the distance between the centres of any two of them is ρ. Figure 33 shows the intersection of these spheres with the plane $\{x : x_3 = 0\}$, while Fig. 34 shows the intersections of I, I', and the spheres of the family $\mathcal{S}$ with boundary of the trihedral angle $D(0, \infty)$.

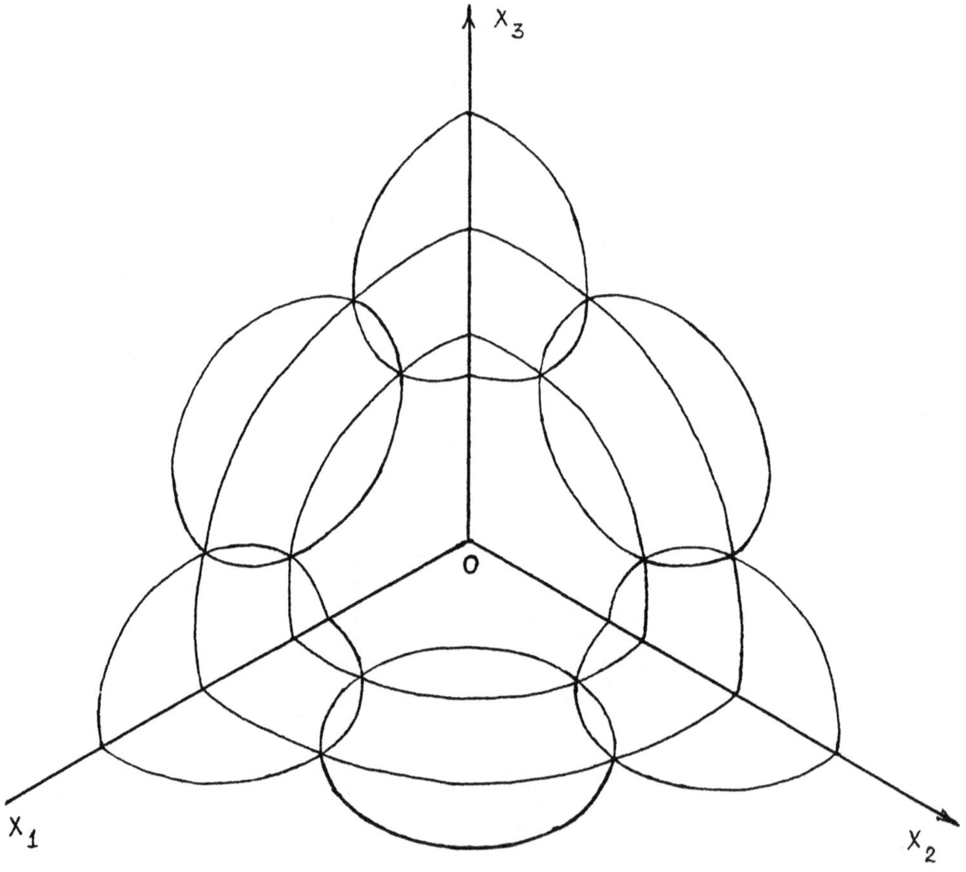

Figure 34

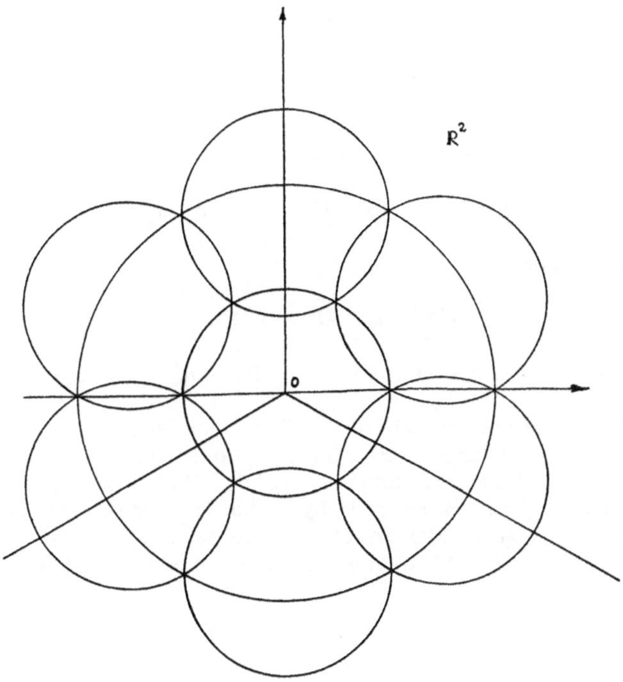

Figure 35

It is readily seen that all dihedral angles of the polyhedron

$$P = \text{ext}\, I \cap \text{int}\, I' \bigcap_i \text{ext}\, S_i \bigcap_{i,j} \text{ext}\, S_{ij}$$

are $\pi/3$ and therefore the group $G \subset \text{Möb}(3)$ generated by reflections in the faces of P is Kleinian. Moreover, one can show by the deformation method in Chapter 9 §3 that G is obtained by means of a certain quasi-conformal mapping $f : \overline{\mathbb{R}}^3 \longrightarrow \overline{\mathbb{R}}^3$ compatible with G from a Fuchsian group $G_0 = fGf^{-1} \subset \text{Möb}(3)$, acting in H^3. Figure 35 shows the intersections with $\partial H^3 \backslash \overline{\mathbb{R}}^2$ for the spheres bounding the fundamental polyhedron $P(G_0) \subset H^3$, $\text{vol}\, P(G_0) < \infty$. Moreover, that conjugation f of the group G_0 may be included in a continuous family of such quasi-conformal conjugations of G_0 by $f_\alpha : \overline{\mathbb{R}}^3 \longrightarrow \overline{\mathbb{R}}^3$, where the parameter α (the planar angle in the vertex of the pod) runs through values from $2\pi/3$ to $\pi/2$. Here $f_{2\pi/3} = \text{id}$, $f_{\pi/2} = f$ (see Chapter 9 §3.5).

It is readily seen that the unbounded component $\Omega_0 \subset \Omega(G)$ for the group G is the G-orbit of the complement $\mathbb{R}^3 \backslash D(0, \infty)$, i.e., the maximal-ball family $\mathcal{B}_0(G)$ constructed in Lemma 5.77, in our case (for Ω_0) is the union of the G-orbits for three maximal balls B_i, $i = 1, 2, 3$:

$$B_i = \{x \in \mathbb{R}^3 : x_i < 0\}.$$

In each of these maximal balls B_i, the bijectivity domains for the retraction $\rho : \Omega_0 \longrightarrow H_0^* \subset H^* \subset H^4$ are such that, for sufficiently large $R > 0$, their intersections with balls $B_i(R) = \{x \in B_i : x_i < -R\}$ have the form:

$$B_i'(R) = B_i(R) \cap B_i' = \{x \in \mathbb{R}^3 : x_i < -R, \quad x_j > 0 \quad \text{for} \quad j \neq i\}$$

whose geodesic edges

$$\gamma_i^0 = \{x \in \mathbb{R}^3 : x_i < -R, \quad x_j > 0 \quad \text{for} \quad j \neq i\}$$

are identified by the retraction $\rho : \Omega_0 \longrightarrow H_0^*$ to one ray γ_R in the bending geodesic $\gamma = (0, \infty)_H \subset H^4$.

The full neighbourhood for the geodesic ray γ_R in the three-dimensional surface $H_0^* \subset \partial_H H_G$ consists of the union of the closures of three dihedral angles $\rho(B_i'(R))$, $i = 1, 2, 3$, but it is clear that three right dihedral hyperbolic angles cannot give a full neighbourhood for a geodesic in the space H^3.

This observation illustrates the reasons for the following theorem.

THEOREM 5.85. *Let* $G \subset \text{Möb}(n)$ $n \geq 3$, *be a Kleinian group having at least three maximal balls* $B_i \subset \Omega(G)$, *with two common limit points* $x, y \in \Lambda(G)$ *on their boundaries and with* $\text{int}(\cap B_i) \neq \emptyset$. *Then the boundary* $H_G^* \subset H^{n+1}$ *of the convex Nielsen hull* H_G *is not isometrically developable in the* n*-dimensional hyperbolic plane* H^n.

PROOF. The statement follows directly from Lemmas 5.78 and 5.82. Indeed, for an isometric development of H^* in the n-plane H^n to exist, it is necessary that the polyhedral angles W_j of all the n-dimensional hyperbolic convex polyhedra $\rho(B_j')$, $B_j' \subset B_j \in \mathcal{B}(\gamma)$, having as their common edge the simple bending geodesic $\gamma \subset H^*$, $\gamma = (x, y)_H$, should form a full neighbourhood of some geodesic in H^n. However, to these polyhedral angles there correspond the strictly larger polyhedral angles V_j of the domains $B_j'' \subset B_j$ (see (5.64) and (5.71)) having as their common edge the arc $\gamma' \subset \Omega(G)$ (= the axis of $B(\gamma)$ from (5.71)) and forming its full neighbourhood. Using the conformality for ρ in bijectivity domains B_j', we see that the above necessary condition for an isometric development is not satisfied if the number $\#\mathcal{B}(\gamma)$ of mutually intersecting maximal balls B, $\partial B \supset \{x, y\}$, is greater than 2. The proof is complete.

REMARK 5.86. The quasi-conformal conjugation between the group G from Example 5.84 and a Fuchsian group $G_0 \subset \text{Möb}(3)$ shows that this development $H^* \subset H^4$ in the plane H^3 can be performed via a quasi-isometry that expands the dihedral angles $\pi/2$ to $2\pi/3$.

From quasi-Fuchsian pea-pod-groups $G \subset \text{Möb}(n)$ one can readily construct web groups $G' \subset \text{Möb}(n)$ in the space $\overline{\mathbb{R}}^n$, $n \geq 3$ (see the definition and an example in Chapter 3 §3).

EXAMPLE 5.87. The web groups $G' \subset \text{Möb}(n)$, $n \geq 3$, are obtained by combining some pea-pod-group $G = G(x, y)$ constructed from the angle $D(x, y)$ and the family (5.72) together with a finite group $\Gamma(x, y) \subset \text{Möb}(n)$, having fixed points x and y and isomorphic to a certain subgroup of the orthogonal group $O(n)$. Here the balls $D_1, \ldots, D_k$, defining the angle $D(x, y)$, should satisfy the condition that $D(x, y)$ is a fundamental polyhedron for the group $\Gamma(x, y)$.

For $n = k = 3$, there is a complete list of such finite groups in Möb(3) defined by the angles between three edges of the pod $D(x, y)$.

In particular, in our example the finite group

$$\Gamma(0, \infty) \subset O(3)$$

has order eight and is generated by rotations through $\pi/2$ around the coordinate axes.

From Lemma 3.11 one can readily check that this group $G' = \langle G(x, y), \Gamma(x, y)\rangle$ is a Kleinian group whose fundamental polyhedron is provided by a component of the fundamental polyhedron for the group $G(x, y)$, lying inside $\Omega_1 \subset D(x, y)$, $\Omega_1 \subset \Omega(G)$. Here all the connected components $\Omega_i \subset \Omega(G')$ will be G'-equivalent to the component Ω_1, so their stabilizers are conjugate in G' to G'_{Ω_1}, i.e., to the quasi-Fuchsian group $G = G(x, y)$. This shows that G' is a web group.

We shall show how one can apply the method of covering $\Omega(G)$ with maximal ball domains to prove the possibility of replacing the retraction $\rho : \Omega(G) \longrightarrow H^* = \partial_H H_G$, which is not a bijective mapping, by a K-quasi-conformal homeomorphism F that commutates with the action of the group $G : FgF^{-1} = g$ for all $g \in G$ (G-equivariance), i.e., more precisely (compare Apanasov [28], [29]):

THEOREM 5.88. *Let* $G \subset \text{Möb}(n)$ *be a finitely generated for* $n = 2$ *and geometrically finite for* $n \geq 3$ *Kleinian group whose limit set does not lie in any* $(n-1)$*-dimensional sphere. In the latter case* $n \geq 3$ *assume additionally that the discontinuity set admits a maximal ball covering. Then there exists a G-equivariant K-quasi-conformal mapping*

$$F : \Omega(G) \longrightarrow \partial_H G_G$$

that can be extended to the identity in $\Lambda(G)$.

REMARK 5.89. The quasi-conformality constant K of the map F in the theorem is dependent on G. Moreover, there are families of Kleinian groups $G_i \subset \text{Möb}(n)$ having non-contractible components of $\Omega(G_i)$ (even for $n = 2$ - see Epstein-Marden [1], 2.16) such that their constants $K_i = K(G_i)$ are uniformly unbounded. But in the case of quasi-Fuchsian groups in the plane $(n = 2)$ such a universal constant exists. This follows from a remarkable theorem of D. Sullivan [8] playing an important part in Thurston's hyperbolization theorem (see Chapter 8 §3) - for the proof see D. Epstein and A. Marden [1]. Our Theorem 5.88 is the spatial analogy of Sullivan's theorem. In the case

$n=2$ our constructions give a new (more simple) approach to the proof of this theorem.

PROOF OF THEOREM 5.88. In accordance with Lemma 5.82, we consider the finite maximal-ball family $\mathcal{B}_f \subset \mathcal{B}(G)$, whose G-images cover all $\Omega(G)$, and we fix a certain ball $B_0 \in \mathcal{B}_f$. Let us specify a certain decomposition of G with respect to the stabilizer $G_0 \subset G$ (perhaps $G_0 = \{\mathrm{id}\}$) of the ball B_0:

$$G = g_0 G_0 \cup g_1 G_0 \cup \ldots, \quad g_0 = \mathrm{id}. \tag{5.73}$$

We first construct the G-equivariant quasi-conformal mapping

$$\varphi_0 : \bigcup_{i=0}^{\infty} g_i(B_0') = O_0' \longrightarrow O_0'' = \bigcup_{i=0}^{\infty} g_i(B_0''). \tag{5.74}$$

For this purpose we consider some fundamental polyhedron $P^{n+1} \subset H^{n+1} \cup \Omega(G)$ for G such that its intersection with $\Omega(G)$ has a finite number of sides: for $n \geq 3$, this follows from G being geometrically finite (Theorem 5.24), and for $n=2$ from Theorem 5.37 (the Beardon-Jørgensen [1] form of Ahlfors's finiteness theorem). Without loss of generality we assume that $P^{n+1} \cap B_0' = P_0' \neq \varnothing$.

We define quasi-conformal mappings ψ_i, $i \geq 0$, in P_0', and together with it in all polyhedra $P_i' = P^{n+1} \cap g_i(B_0')$, if they are non-empty, in such a way. If the sphere $S_i = \partial B_i$ has parabolic points p for the group G, then rank $p < n$ and there exist cusp domains $U_p \subset \Omega(G)$ forming finitely many disjoint G-orbits and intersecting the ball B_i. We define the mappings ψ_i as the identity on the intersections $P_i' \cap U_p$, while the complements to these domains in the polyhedra P_i' are denoted by $P_i = P_i' \setminus \bigcup_p U_p$. Then Theorems 5.39 and 5.36 show that these polyhedra P_i, $i \geq 0$, are compact.

Assuming that the Poincaré metric has been introduced into B_0, we denote by $s_1, \ldots, s_N$ the set of all $(n-1)$-dimensional planes in B_0 that contain $(n-1)$-dimensional sides of the polyhedron P_0 lying in $\partial B_0'$ ($s_j = s_{0j}$ spanned on the $(n-2)$-dimensional sphere $\partial B_0 \cap \partial B_j$, $B_j \cap B_0 \neq \varnothing$, $B_j \in \mathcal{B}(G)$). We consider an $\varepsilon > 0$ such that for any two disjoint sides (having any dimension) of P_0, their ε-neighbourhoods do not intersect. We also denote the $\varepsilon \cdot 2^{k-n}$-neighbourhood of a k-dimensional side $s \subset \partial P_0$ as $U_\varepsilon(s)$.

Now let a geodesic $\gamma^0 \subset B_0$ having ends $x, y \in S_0 \cap \Lambda(G)$ be transformed under the retraction ρ into the simple bending geodesic $\gamma = (x, y)_H \subset H^*$ (i.e., γ^0 lies in the intersection of the planes $s_j \subset B_0$ corresponding to the maximal balls $B_j \subset \mathcal{B}(\gamma) \subset \mathcal{B}(G)$, see Lemma 5.82) and contains an edge of the polyhedral angle $P_0(\gamma^0) \subset P_0$. We define

$$P_0(\gamma^0, \varepsilon) = P_0 \cap U_\varepsilon(\gamma^0). \tag{5.75}$$

Together with the angle $P_0(\gamma^0, \varepsilon)$ running along γ^0, we consider the angle $V_0(\gamma') \subset B_0''$ containing it (having the same combinatorial type), which is bounded, instead of by the planes $s_j = s_{0j}$, by the bisector spheres σ_{0j}, $(B_j \in \mathcal{B}(\gamma)\setminus B_0)$ and having as edge the arc γ' with ends x and y (= the axis of the angle $B(\gamma)$ from (5.69)). In the angle $V_0(\gamma')$, we consider the neighbourhood $V_0(\gamma', \varepsilon)$ of the arc γ', which is the union of $P_0(\gamma^0, \varepsilon)$ and the following domain adjoining it. This domain is cut out from $B_0''\setminus B_0'$ by the geodesics in the ball B_0, which are simultaneously orthogonal to the planes s_{0j} and the spheres σ_{0j} (where $B_j \in \mathcal{B}(\gamma)$) and which start from the points of the set

$$\partial_{P_0} U_\varepsilon(\gamma^0) \cap (\bigcup\{s_{0j} : B_j \in \mathcal{B}(\gamma)\}). \tag{5.76}$$

Now let $\psi_{\gamma^0} : P_0(\gamma^0, \varepsilon) \longrightarrow V_0(\gamma', \varepsilon)$ be a quasi-conformal mapping for non-degenerate polyhedral angles that leaves invariant the geodesics starting from the set (5.76) as well as the $(n-1)$-dimensional planes in B_0 intersecting these angles orthogonal to their edges γ^0 and γ' correspondingly. Writing the angles $P_0(\gamma^0, \varepsilon)$ and $V_0(\gamma', \varepsilon)$ as the products $\gamma^0 \times A^0$ and $\gamma' \times A'$, where A^0 and A' are cones over $(n-2)$-dimensional polyhedra $Q^0 \subset Q'$ having the same combinatorial type and assuming that the vertices of the cones A^0 and A' are common, we may say that the mapping ψ_{γ^0} "expands" A^0 to A' along spheres orthogonal to the common axis of these cones and to their sides (cf. Gehring and Väisälä [1] and Väisälä [1]). The quasi-conformality coefficient for the mapping ψ_{γ^0} is depending on the polyhedral angles $P_0(\gamma^0, \varepsilon)$ and $V_0(\gamma', \varepsilon)$ (dihedral for $n=3$, see Fig. 36).

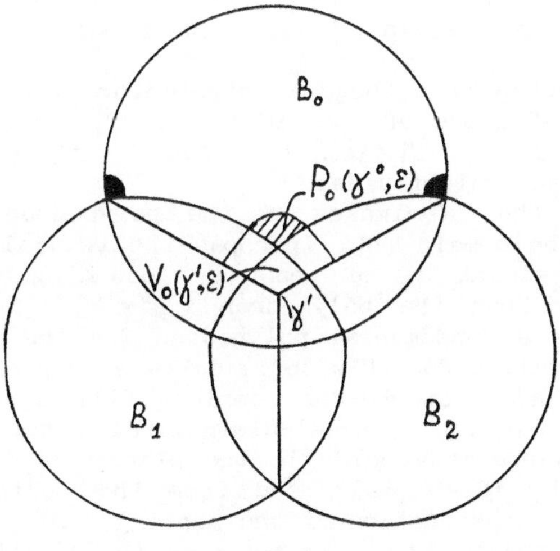

Figure 36

At the same time as ψ_{γ^0}, we similarly define the quasi-conformal mappings ψ_s in the neighbourhoods $U_\varepsilon(s) \subset P_0$ of the k-dimensional sides s for all angles $P_0(\gamma^0)$, $2 \le k \le n-2$, which coincide with ψ_{γ^0} at the boundaries $\partial U_\varepsilon(\gamma^0)$ and which leave invariant the geodesics and $(n-1)$-planes in B_0 orthogonal to the sides s.

Finally, the K_0-quasi-conformal mapping ψ_0 in $P'_0 \subset B'_0$ is defined as follows:

1) $\psi_0 = \mathrm{id}$ in $P'_0 \cap U_p$ and in the complement in P_0 to $\bigcup\{U_\varepsilon(s) : s \subset \partial P_0 \cap \partial B'_0\}$.

2) $\psi_0 = \psi_{\gamma^0}$ on $P_0(\gamma^0, \varepsilon)$ for any geodesic $\gamma^0 \subset \partial B'_0$ which is an edge of the polyhedron $P'_0 \subset B'_0$.

3) $\psi_0 = \psi_s$ in $U_\varepsilon(s)$ for any k-dimensional $(2 \le k \le n-2)$ side $s \subset \partial P_0 \cap \partial B'_0$.

4) ψ_0 is a stretching along the family of geodesics in B_0 which are simultaneously orthogonal to the mutually corresponding planes s_{0j} and the spheres σ_{0j}, $1 \le j \le N$; ψ_0 transfers every $(n-1)$-dimensional j-th side $(\subset \partial B'_0)$ of the complement

$$P_0 \backslash (\bigcup\{U_\varepsilon(s) : s \subset \partial P'_0 \cap \partial B'_0, \quad 0 \le \dim s \le n-2\})$$

into a domain in the sphere σ_{0j} and coincides on the boundary of that complement with the mappings ψ_{γ^0} and ψ_s.

The similar mappings for the polyhedra P'_i are denoted by ψ_i, and we define for all $i \ge 0$ the polyhedra $P''_i = \psi_i(P'_i) \subset g_i(B''_0) = B''_i$. Here one may assume that all polyhedra P'_i, which have pairwise G-equivalent sides, where gathered beforehand by means of the elements $g_i^{-1} \in G$ (before the construction of ψ_i) into one connected polyhedron $P'_* \subset B'_0 \subset B_0$, and their images were then returned together with the ψ_i via $g_i \in G$ to the initial balls $B_i = g_i(B_0)$. It gives us the following fact:

The construction of orthogonal stretchings for ψ_i preserves the G-identification of sides of the polyhedra P'_i and P'_j in the sense that the polyhedra $P''_i = \psi_i(P'_i)$ and $P''_j = \psi_j(P'_j)$ have the corresponding G-equivalent sides.

Also, since the G_0-images of the polyhedron $P'_* \subset B'_0$ gave a tessellation of the domain B'_0 (perhpas the trivial one: $P'_* = B'_0$ for $G_0 = \{\mathrm{id}\}$) and tesselations of the B'_0 bounding $(n-1)$-dimensional planes in B_0, then the polyhedron $P''_* \subset B''_0$ G-gathered from the polyhedra P''_i is a fundamental polyhedron for the group G_0 in B''_0. The property sufficient for this to be so (see Chapter 6 §1) is that the images $G_0(P''_*)$ form a one-to-one complete filling (i.e., tesselation) of neighbourhoods of the $(n-2)$-dimensional sides in $B''_0 \backslash B'_0$. It follows that this property holds in the planes $s_j \subset B_0$ containing the $(n-1)$-dimensional sides $\partial B'_0$ and from the "orthogonality" of the stretchings in the construction of the mappings ψ_i.

Now for each point $x \in O'_0$ we put (see (5.74)) that

$$\varphi_0(x) = g \circ \psi_i \circ g^{-1}(x),$$

where $g \in G$ is such that $g(P^{n+1}) \ni x$, and the number i is such that $g^{-1}(x) \in P_i'$. The mapping $\varphi_0 : O_0' \longrightarrow O_0''$ is a compatible with G extension (by symmetry) of the mappings ψ_i, $i \geq 0$, and therefore is a G-equivariant K_0-quasi-conformal mapping onto O_0''. The construction of the mappings ψ_i implies the identity of the continuous extension of φ_0 to the limit vertices of B_1', i.e., to the set $G(\partial B_0) \cap \Lambda(G)$.

If one considers the similar K_i-quasi-conformal mappings φ_i in the G-orbits of other maximal balls from $\mathcal{B}_f = \{B_0, \dots, B_m\}$ (see Lemma 5.81), we get a G-equivariant K-quasi-conformal homeomorphism

$$f : \bigcup\{B' : B \in \mathcal{B}(G)\} = \Omega' \longrightarrow \Omega'' = \bigcup\{B'' : B \in \mathcal{B}(G)\}, \tag{5.77}$$

extending identically to the limit set $\Lambda(G)$. The quasi-conformality coefficient of f here is $K = \max_{0 \leq i \leq m} K_i$.

Lemma 5.77 shows that $\rho|_{\Omega'}$ is a conformal homeomorphism; also, Lemma 5.81 and Remark 5.82 show that $\Omega(G)\backslash\Omega''$ is the closure of a union of $(n-1)$-dimensional spherical polyhedra lying on the bisector spheres σ_{ij} $(B_i, B_j \in \mathcal{B}(G)$, $B_i \cap B_j \neq \varnothing)$ and having, as their one-dimensional skeleton, the circular arcs γ' joining in $\Omega(G)$ the "extremal" limit points $x, y \in \Lambda(G)$ (the ends of the simple bending geodesics $\gamma = (x, y)_H \subset H^*$). Each such arc γ' is an axis of the spherical angle $B(\gamma)$ (see (5.69)), which is decomposed by the spheres σ_{ij} into a finite number of polyhedral angles V_k with the common edge γ' (see (5.71)). Here V_k is the f-image of a corresponding polyhedral angle from $B_k' \subset B_k \in \mathcal{B}(\gamma)$ lying in it.

We use the fact that the mappings φ_k in the balls B_k are stretchings along geodesics and planes in the balls B_k orthogonal simultaneously to corresponding hyperbolic planes and bisectrix spheres, and thus the same applies for the mapping f, to get the analogue of (5.67) (see Fig. 30). This gives compatibility of the mapping f and the retraction ρ, which, in particular, implies that all the points z_k on the geodesics $\gamma_k^0 \subset \partial B_k'$ having ends x, y (where $(x, y)_H = \gamma \subset H^*$) which are projected orthogonally to a common point $z \in \gamma' \subset B(\gamma)$ (and therefore go over on the retraction ρ to the common point $\rho(z)$), are mapped by the mapping f into the same point equal to z. This compatibility of f and ρ shows that the mappings $\rho \circ f^{-1}$ of joint polyhedra B_i'' and B_j'' are glued by the retraction ρ into a common G-equivarient homeomorphism F,

$$F(x) = \begin{cases} \rho \circ f^{-1}(x), & x \in \Omega'', \\ \rho(x), & x \in \Omega(G)\backslash\Omega'', \end{cases} \tag{5.78}$$

which is the desired K-quasi-conformal G-equivariant homeomorphism of $\Omega(G)$ onto H^*. It is evident that F can be extended to the identity in $\Lambda(G)$. The theorem is proved.

Note that the proof scheme for the theorem is not changed in the general case (without the maximal ball covering condition), and the only changes are related to considering our constructions for bijectivity domains in maximal conical and cylindrical domains $W(p)$ and $C(x, y)$ - see Remark 5.79.

COROLLARY 5.89. *Let* $G \subset \text{Möb}(n)$, $n \geq 3$, *be a geometrically finite group whose limit set does not lie in any* $(n-1)$*-dimensional sphere; then the function* $\rho_G(x, y) = d_*(F(x), F(y))$ *is defined for points* x, y *from one (any) component of* $\Omega(G)$ *and defines a* G*-invariant distance in* $\Omega(G) \subset \overline{\mathbb{R}}^n$.

We note that the construction (5.78) of the mapping $F : \Omega(G) \longrightarrow H^*$, defining the matric ρ_G in $\Omega(G)$ by means of the path metric d_* in H^*, readily shows that the metric ρ_G is commensurable with the well-known quasi-hyperbolic metric k_Ω introduced by Gehring and Palka [1]. The metric k_Ω is defined in the domain $\Omega \subset \mathbb{R}^n$ by the length differential $|dx|/\rho(x, \partial\Omega)$, where $\rho(x, \partial\Omega)$ is the Euclidean distance of the point x to the boundary $\partial\Omega$. For Möbius transformations $h \in \text{Möb}(n)$, the metric k_Ω is quasi-invariant:

$$k_\Omega(x, y)/2 \leq k_{\Omega'}(h(x), h(y)) \leq 2k_\Omega(x, y),$$

where $\Omega' = h(\Omega) \subset \mathbb{R}^n$.

CHAPTER 6

GEOMETRIC AND ALGEBRAIC STRUCTURES OF DISCRETE GROUPS

§1. Discrete groups and geometry of spatial tessellations

There is a well-known close relation between the discrete groups $G \subset \operatorname{Isom} X$ of motions in the space X with constant sectional curvature and regular tessellations of that space. Namely, each discrete group G can be put into correspondence with regular tessellations of the space X into the congruent polyhedra $g(P)$, $g \in G$, where $P \subset X$ is some fundamental polyhedron of the group G (see Theorem 3.18). Conversely, each regular tessellation of X into congruent polyhedra $P_i \cong P$ can be put into correspondence (not uniquely and subject to certain conditions) with a discrete group $G \subset \operatorname{Isom} X$, for which P is a fundamental polyhedron (see Theorem 3.19). The relation is given concrete form in classical theorems by Poincaré [1], [2] on the fundamental polygon and polyhedron and their various generalizations due to Aleksandrov [1], Maskit [6], de Rham [1], Seifert [2], Morokuma [1], and Mostow [3]. The purpose of this section is to give a criterion for the regular tessellations of the space X free from any constraints on the character of the tessellation group or on the volumes of the polyhedra of a tessellation, the number of them, and so on, including the cases discussed previously (see above). We shall see that, on going from dimension $n=3$ to $n \geq 4$, we get marked changed in the tessellation properties, comparable with those in the transition from $n=2$ to $n=3$ (see Apanasov [17], [23], [26]).

Here we have set aside the question of choosing the tessellations $G(P)$ of X having the simplest structure around the edges and vertices, which should be stable under a change in the "centres" of the polyhedra P. For the groups $G \subset \operatorname{Isom} H^n$, this topic has been examined for Dirichlet polyhedra, for $n=2$ by Macbeath [3] and for $n=3$ in a recent study by Jørgensen and Marden [3]. Dirichlet polyhedra do not have such properties of simplicity and stability for $n \geq 4$ (at least if there are parabolic elements, see Theorem 5.26 and Remark 5.29). However, it follows from Lemma 5.30 and Corollary 5.31 that the pseudo-Dirichlet polyhedra introduced in (3.33) are good candidates for having these properties: see Apanasov [30].

1. *A filling of the space* X^n. For this section (compare Chapter 3 §4 and Chapter 5 §1), we do not require that the polyhedra are convex or have any homotopic properties. Namely, by a polyhedron P in n-dimensional space X^n we mean a connected set such that $P = \mathrm{cl}(\mathrm{int}\, P)$ (the interior and closure in X^n), while the boundary ∂P belongs to the locally finite union of $(n-1)$-dimensional planes in X^n. If that union is finite, P is called *finite-sided*. A set $S \subset P$ is called a *side* or an $(n-1)$-*side* if the set S is contained in a certain $(n-1)$-dimensional plane $X \subset X^n$, while $\mathrm{int}_X S$ is a connected component for the set $\mathrm{int}_X(X \cap \partial P)$ and $S = \mathrm{cl}_X(\mathrm{int}_X S)$. Then the $(n-1)$-dimensional side S is a polyhedron in X and thus we have an inductive definition for the i-sides of the polyhedron P as the sides for the $(i+1)$-dimensional sides of P.

By a filling of the space X^n with polyhedra, we understand the following construction (Aleksandrov [1]).

We specify a locally finite cell complex K^n formed by polyhedra from X^n. The complex may be infinite and considered abstractly, i.e., it forms a space $|K^n|$ without embedding in X^n. In addition to the complex, we specify a continuous projection $\pi : |K^n| \to X^n$, which together with K^n satisfies the following conditions.

C1. Amongst the polyhedra in the complex, there is only a finite number of them $P_1, \ldots, P_m$ that are geometrically unequal (in X^n), which are called model polyhedra.

C2. Each $(n-1)$-dimensional side of the polyhedron $P \in K^n$ is the side of precisely one other polyhedron $P' \in K^n$.

C3. Any two polyhedra $P, P' \in K^n$ join a polyhedron chain, i.e., a finite sequence of polyhedra in the complex, where each two of the neighbouring polyhedra are adjacent on an $(n-1)$-dimensional side.

C4. The 1-dimensional side Q^i, $0 \le i \le n-2$, of the polyhedron $P \in K^n$ is a side of another $P' \in K^n$ if and only if these polyhedra are joined by a chain in which each two adjacent polyhedra are adjacent on an $(n-1)$-dimensional side Q, $Q^i \subset Q$.

C5. The restriction of the projection π to any $P \in K^n$ is an isometry.

C6. If $P, P' \in K^n$ are adjacent on a side Q^{n-1}, the images $\pi(P)$ and $\pi(P')$ lie (in a neighbourhood of the image $\pi(Q^{n-1})$) on different sides of the latter. (If P and P' are convex, then those images in general lie on different sides of the plane of the side $\pi(Q^{n-1})$; in the opposite case, this is required only in a neighbourhood of that side.)

THEOREM 6.1 (Aleksandrov [1]). *Let conditions* C1 - C6 *be satisfied and the model polyhedra* $P_1, \ldots, P_m \subset X^n$ *be compact. Then* $\pi(|K^n|) = X^n$, *and further, the projection* π *here is a one-to-one projection if and only if it is of one-to-one type around each* $(n-2)$-*dimensional side of the complex* K^n.

In the case of an n-dimensional simply connected Riemannian manifold, there is a similar result due to Mostow [3] for a single model polyhedron P whose interior is a cell.

In what follows, we consider only periodic (regular) fillings, i.e., ones for which the identifications of the model polyhedron sides generate a group acting transitively on K^n (modulo a finite union of model polyhedra), and K^n is called regular.

The filling of the space X^n is called a *tessellation* if any point in the space belongs to at least one polyhedron and distinct polyhedra do not have common interior points. Theorem 6.1 gives the necessary and sufficient conditions for which a filling of the space X^n by compact polyhedra is a tessellation of it.

Note that there are tessellations of Euclidean and hyperbolic spaces by compact polyhedra that are not regular, i.e., that do not have any group periodicity. Examples of such tessellations answer the question posed by Hilbert [1, §18]. We note amongst them the tessellation of the plane $\mathbb{R}^2$, known as "Penrose's two chickens" (see, for example Berge [1], 1.7.1 and 1.9.16); see also Robinson [1]; one also has a tessellation of $\mathbb{R}^5$ by congruent star-shaped polyhedra (Stein [1]) or tessellations of the hyperbolic plane by congruent convex polygons (Robinson [2]).

If X^n is regular filled in accordance with the rules C1 - C6, the space $|K^n|$ for a regular complex K^n covers the orbifold $\mathcal{O}$ glued from model polyhedra by means of a given side identification. At the same time, any complete $(X^n, \operatorname{Isom} X^n)$-orbifold $\mathcal{O}$ is assigned in the form of the quotient space X^n/G by the action of the discrete holonomy group $G = d_*(\pi_1(\mathcal{O})) \subset \operatorname{Isom} X^n$ (see Chapter 2 §7), and thus it can be obtained by identifying the $(n-1)$-dimensional sides of a fundamental polyhedron $P(G) \subset X^n$. It is evident that the orbifold $\mathcal{O} = X^n/G$ corresponds to a regular complex having the natural projection and satisfying conditions C1 - C6 with a single model (fundamental) polyhedron $P = P(G)$, so we have:

LEMMA 6.2. *Let the regular complex* K^n *having a projection* π *and conditions* C1 - C6 *correspond to the orbifold* $\mathcal{O}$ *obtained from the model polyhedra* $P_1, \ldots, P_m \subset X^n$ *by means of a given isometric identification of their sides. Then* $\pi(|K^n|) = X^n$ *if and only if* $\mathcal{O}$ *is a complete orbifold.*

2. *Incomplete fillings of the hyperbolic space.* The criterion for the X^n tessellation (Theorem 6.1) also gives completeness conditions for the compact polyhedron filling. However, the theorem conditions are insufficient for filling completeness in the case of the hyperbolic space H^n, $n \geq 2$, if the model polyhedra P_i include unbounded ones.

This effect was first discovered (evidently by Hadamard) for ideal simplexes filling of the hyperbolic plane H^2, in which the usual parabolic identification of the parallel sides in the simplex was replaced by their identification by means of a hyperbolic translation having the vertex at infinity of the simplex as a fixed point (see also Maskit [6] and Beardon [2]). Thurston [1, Chapter 4] observed an analogous situation on H^3 filling by ideal simplexes with vertices at infinity, when he examined hyperbolic structures in the figure-eight knot complement in the sphere S^3 (see Example 2.17). All these incomplete fillings have the following property of "local topological similarity".

For any polyhedron $P \subset H^n$, we consider an $(n-1)$-dimensional abstract unoriented cell complex isomorphic to the boundary of the polyhedron ∂P in the category of abstract complexes. Any complex obtained in this way is called an n-dimensional abstract polyhedron. We say that the polyhedra P and P' have the same combinatorial type ($P \sim P'$) if the corresponding abstract polyhedra coincide, i.e., there exists a one-to-one correspondence between the sides of P and P' that preserves the dimensions and the incidence relation.

Now let the two model polyhedron sets $P_1^1, \dots, P_m^1$ and $P_1^2, \dots, P_m^2$, where $P_i^1 \sim P_i^2$ for $i = 1, \dots, m$, specify the infinite complexes K_1^n and K_2^n having projections $\pi_1 : |K_1^n| \longrightarrow H^n$ and $\pi_2 : |K_2^n| \longrightarrow H^n$ and which satisfy the above conditions C1 - C6. We consider their connected minimal subcomplexes $K_1 \subset K_1^n$ and $K_2 \subset K_2^n$, which are formed by polyhedra such that their projections adjoin (by certain k-dimensional sides, $0 \le k \le n-1$) the sets

$$\bigcup_{i=1}^{m} \pi_1(P_i^1) \qquad \text{and} \qquad \bigcup_{i=1}^{m} \pi_2(P_i^2)$$

respectively. We say that the pairs (K_1^n, π_1) and (K_2^n, π_2) (and correspondingly, the fillings defined by them) are locally topologically similar if all the complexes K_1 and K_2 have similar adjacency relations for their polyhedra and this similarity is induced by a homeomorphism:

$$\pi_1(|K_1|) \longrightarrow \pi_2(|K_2|). \tag{6.1}$$

In other words, K_1^n and K_2^n, having the projections π_1 and π_2, are locally similar if they have identical schemes for identifying the sides of the model polyhedra (they have identical combinatorial types) and if the sums of the dihedral angles at the sides in the projections of the polyhedra coincide for any isomorphic cycles of $(n-2)$-dimensional sides in those complexes.

One can introduce a natural topology on the set of locally topologically similar fillings of the space H^n factorized with respect to $\operatorname{Isom} H^n$. Namely, the fillings (K_1^n, π_1) and (K_2^n, π_2) are close if and only if there exists a mapping $H^n \longrightarrow H^n$ close to the identity and such that it is locally the (6.1) homeomorphism.

In Chapter 2 §5, we constructed a complete hyperbolic structure on the Whitehead link complement M_W (Example 2.18), which corresponded to a regular tessellation of the hyperbolic space H^3 into congruent regular ideal octahedra with vertices at infinity and right dihedral angles (see Figs. 8 - 11), which we denote by (K_W, π_W). We note that the Whitehead link is universal (see Chapter 7 §1 and Thurston [5] as well as Hilden, Losano and Montesinos [2]).

In the next theorem, we describe for all incomplete regular fillings (K_α, π_α) of H^3 by ideal octahedra that are locally topologically similar to (K_W, π_W). (See Thurston [1] for a more general result.)

THEOREM 6.3. *The set* $\mathcal{H}_W$ *of (incomplete) hyperbolic structures on the Whitehead link complement* M_W *not isometric one with another and corresponding to regular octahedral fillings of* H^3 *specified by the complexes* (K_α, π_α) *and locally topologically similar to the tessellation* (K_W, π_W) *can be embedded as an open subset into* $\mathbb{C}^2$.

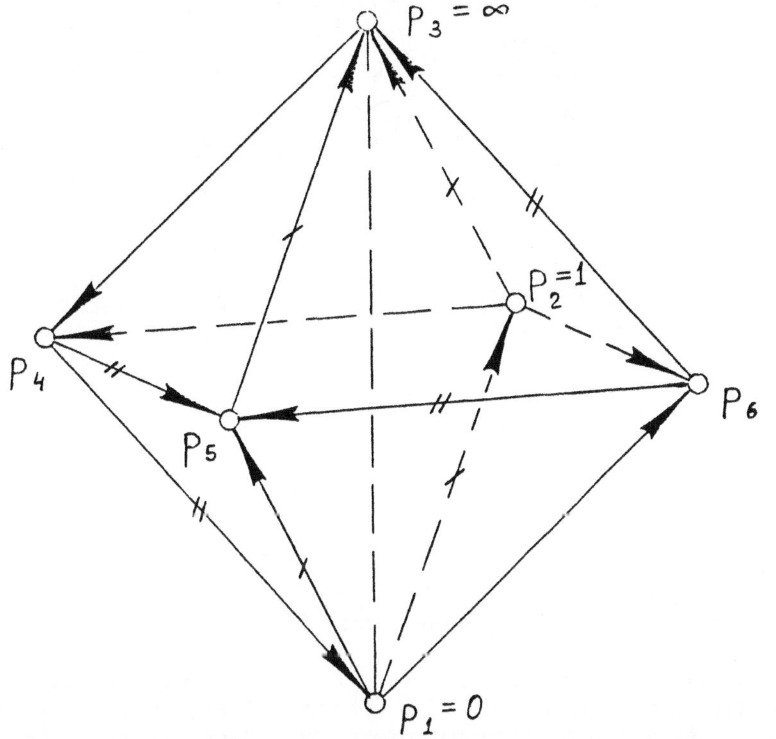

Figure 37

PROOF. We describe the set of octahedra $P_\alpha \subset H^3$ having vertices at infinity $\partial H^3 = \overline{\mathbb{C}}$ $(H^3 \cong \mathbb{R}^3_+)$ and the identifications of the sides similar to the identifications in Fig. 37, in which the sum of the dihedral angles at equivalent edges is 2π (the one-to-one property of π_α around the edges of the complex K_α).

For this we split up the octahedron P_α into four ideal simplexes having the common edge (p_1, p_3). Such simplexes in H^3 are characterized modulo $\operatorname{Isom} H^3$ by their three dihedral angles at any vertex, whose sum is π, or, what is equivalent, by the Euclidean triangle on which the simplex intersects a horosphere having its centre at some vertex. One can assume that this triangle lies in the complex plane $\mathbb{C}$ in such a way that its vertices are 0, 1, and a point z,

$\operatorname{Im} z > 0$. Then any ideal simplex in H^3 having vertices at infinity $\partial H^3 = \overline{\mathbb{C}}$ can be specified by the following parameters (see Thurston [1, Chapter 4]).

We assign the complex numbers z, v, and w from $\mathbb{C}_+ = \{z \in \mathbb{C} : \operatorname{Im} z > 0\}$ to each of the three pairs of opposite edges in the simplex, which are obtained by considering the vertex Euclidean triangles as above. Then these numbers have the obvious relations

$$v = 1/(1 - z), \quad w = (z - 1)/z. \tag{6.2}$$

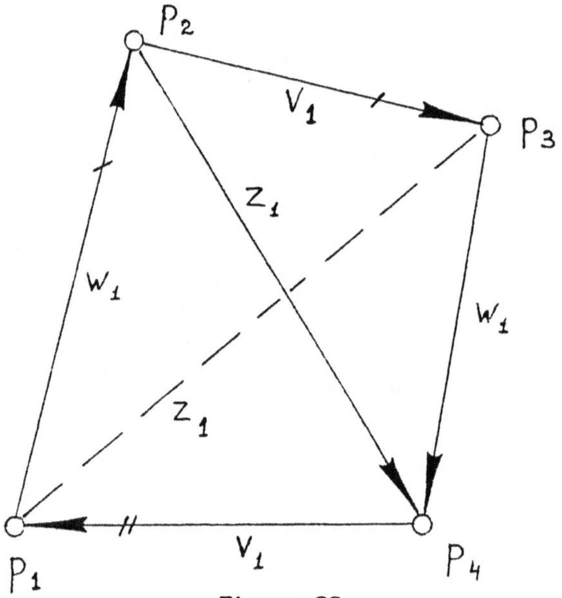

Figure 38

Then we have the simplexes σ_i, $1 \le i \le 4$, that are described by the parameter set $(z_i, v_i, w_i) \in \mathbb{C}_+^3$, which satisfy eight relations of (6.2) type. Also, in all the simplexes we assign the parameter z_i to the common edge (p_1, p_3) (Fig. 38 shows σ_i). The conditions for the sum of the dihedral angles at each of the four edge cycles (p_1, p_3), $\longrightarrow$, $-\!/\!\!\rightarrow$, $-\!/\!/\!\!\rightarrow$ to be equal to 2π give as follows (see Fig. 38):

$$\begin{gathered} z_1 z_2 z_3 z_4 = 1, \quad w_1 v_4 z_1 z_2 w_2 v_3 = 1 \\ v_1 w_2 v_3 w_4 w_3 v_4 w_1 v_2 = 1, \quad v_2 w_3 z_3 z_4 v_1 w_4 = 1. \end{gathered} \tag{6.3}$$

Relations (6.2) and (6.3) are equivalent to one relation on $z_1, z_2, z_3 \in \mathbb{C}_+$:

$$\frac{(z_1 - 1)(z_2 - 1)}{z_3 - 1} = \frac{1 - z_1 z_2 z_3}{z_1 z_2 z_3}. \tag{6.4}$$

Then the unique (due to Mostow's rigidity Theorem 9.12) parameters

$z_1 = z_2 = z_3 = i$ correspond to the complete octahedral filling (tessellation) of H^3, i.e., octahedra with right angles. Then the mapping

$$\varphi : \mathcal{H}_w \cong \{(K_\alpha, \pi_\alpha)\} \longrightarrow \mathbb{C}^2$$

defined by the formula $\varphi(\alpha) = (z_1(\alpha), z_2(\alpha))$ is the desired embedding with an open image $\varphi(\mathcal{H}_w) \subset \mathbb{C}^2$, where $z_1(\alpha)$ and $z_2(\alpha)$ are such that $z_1(\alpha)$, $z_2(\alpha)$ and $z_3(\alpha)$ lie in $\mathbb{C}^+$ and are related by (6.4).

REMARK 6.4. To describe the incomplete fillings of H^3 by the octahedra themselves, mentioned in Theorem 6.3, we specify the vertices in the model octahedron defining them (with side identification everywhere as in Fig. 37). For this purpose we specify three vertices: $p_1 = 0$, $p_2 = 1$, $p_3 = \infty$, and we thus exclude isometric cases. Then (6.4) gives us that the other vertices p_4, p_5, p_6 in P_α are any points in the complex plane $\mathbb{C}_* \subset \partial H^3$ satisfying the conditions:

$$p_5/p_4 \in \mathbb{C}_+, \quad p_6/p_5 \in \mathbb{C}_+,$$

$$\frac{(p_4 - 1)(p_4 - p_5)}{(p_6 - 1)(p_6 - p_5)} = \frac{p_4}{p_5 \cdot p_6}.$$

The set of incomplete octahedral fillings for H^3 described here has real dimensions 4 and contains a one-parameter family:

$$P_1 = 0, \quad p_2 = 1, \quad p_3 = \infty, \quad p_4 = it, \quad p_5 = -1, \quad p_6 = -it; \quad t > 0.$$

Fillings from this family have been considered by Makarov and Gutsul [2].

3. *Locally finite fillings and completeness of hyperbolic structures.* In the previous section, we have shown that conditions C1 - C6 for the filling (K^n, π) of H^n do not provide filling completeness if the model polyhedra include unbounded ones. We shall see below that complete fillings should be characterized by the following property.

DEFINITION 6.5. The filling (K^n, π) of the space X defined by the rules C1 - C6 is called *locally finite* if the following condition is satisfied:

C7. Any compactum from X can intersect only a finite number of images $\pi(P)$ of the polyhedra in K^n.

Bieberbach's theorem (Theorem 3.4) implies that a filling of $\mathbb{R}^n$ by unbounded polyhedra induces a filling in a certain subspace $\mathbb{R}^k \subset \mathbb{R}^n$ by bounded polyhedra. This shows that a filling being locally finite is meaningful only in the hyperbolic case, since in the case of the bounded model polyhedra $P \subset X$ in K^n, it follows from K^n being locally finite.

We note that the condition C7 is not dependent on the local finiteness of K^n. If a discrete group G is generated by isometries of H^n that identify the sides of the model polyhedra in K^n, then the condition C7 is equivalent to the local finiteness of the fundamental set for the group G (possibly partially fundamental), which is equal to the union $\bigcup P_i$ (see Chapter 5 §1).

THEOREM 6.6. *A hyperbolic structure on the orbifold* $\mathcal{O}$, $\dim \mathcal{O} \geq 2$, *having holonomy* d_* *is complete if and only if the polyhedra* $P_1, \ldots, P_m \subset H^n$, *having a finite number of sides whose identification by the generators of the group* $G = d_*(\pi_1(\mathcal{O}))$ *gives the orbifold* $\mathcal{O}$, *specify a complex* K^n *having property* C7.

REMARK 6.7. The local isometry property for the projection π readily shows that π is a proper mapping (i.e., the inverse image $\pi^{-1}(C)$ of any compactum $C \subset H^n$ is a compact set) if and only if condition C7 is met. This means that the sufficiency of condition C7 (local finiteness of a filling) in the theorem can be obtained from the general properties of proper mappings in topological space. However, we give a constructive proof (although it is longer) for the sufficiency, in which the reasons for the possible incompleteness of $\mathcal{O}$ are demonstrated (this will subsequently be used substantially).

PROOF OF THEOREM 6.6. The theorem for compact model polyhedra obviously follows from Aleksandrov [1] Theorem 6.1, so in what follows, we consider the orbifold as non-compact. Also, if the model polyhedra include non-convex ones, we split them up by means of $(n-1)$-dimensional planes into a finite number of convex model polyhedra, and thus we pass to a new set of convex model polyhedra, since that rearrangement does not alter the group G: all the resulting additional gluing mappings are identity ones.

Necessity. Let us assume that condition C7 is not satisfied for the model polyhedra $P_1, \ldots, P_m$ but the orbifold $\mathcal{O} = (\bigcup P_i)/G$ is complete. Then there exist a compactum $C \subset H^n$, a polyhedron $P_0 = P_{i_0}$, $1 \leq i_0 \leq m$ and a sequence of distinct elements $\{g_m\}$ in the group $G = d_*(\pi_1(\mathcal{O}))$ such that the sequence of points $\{y^m\}$, $y^m \in C \cap g_m(P_0)$, converges to a certain point $y \in C$. As G is discrete, the image sequence $g_m(y_0)$ for some point $y_0 \in P_0$ converges to a limit point $z \in \Lambda(G)$ (if necessary, we take a subsequence). We use the condition for the polyhedra P_i to be convex (see above) and get that for all j the geodesic segment, joining the points y^j and $g_j(y_0)$, lies itself in the polyhedron $g_j(P_0)$. As j increases, those segments converge to a geodesic ray $\gamma = [y, z)$. It is clear that any neighbourhood of any point on that ray γ intersects all the images $g_j(P_0)$ for sufficiently large j. At the same time, the completeness condition for $\mathcal{O}$ implies that $\bigcup P_i$ contains a fundamental set of the group G, and therefore the G-images of that set cover all the space H^n. The multiplicity of this covering is one at points covered by the interior of some polyhedron $g(P_i)$, $g \in G$, while it is two at points covered by the interior of some $(n-1)$-dimensional side of $g(P_i)$, and it is finite at the other points. This follows from the fact that the finite-sided polyhedra P_i each have only a finite number of unequal dihedral and polyhedral angles. For any compacta in H^n, we thus get finiteness of its polyhedra cover, which contradicts the above γ property.

Sufficiency. Let us assume that the orbifold $O = (\bigcup P_i)/G$ is not complete. We take a Poincaré model for the hyperbolic space H^n in a Euclidean ball and denote by $\overline{P}$ the closure in $\overline{\mathbb{R}}^n$ of the polyhedron $P \subset H^n \cong \mathbb{R}^n_+$. The group $G = d_*(\pi_1(O))$ leaves the set H_p invariant:

$$H_p = G(\bigcup_{i=1}^{m} \overline{P}_i) \subset \overline{H}^n.$$

The convex hyperbolic hull $H_G \subset \overline{H}^n$ of the limit set $\Lambda(G)$ of the group G (see Chapter 5 §2) is also invariant with respect to G, so the set

$$H_G^0 = (H_p \cap H_G) \setminus \Lambda(G)$$

is invariant with respect to G. One factorizes H_G^0 with respect to G to get a hyperbolic orbifold with boundary O_G^0 lying in the orbifold O and which is a deformation retract (see (5.6)) of an orbifold $\overline{O}$ containing O:

$$O_G^0 = H_g^0/G \subset O \subset \overline{O} = \left[\bigcup_{i=1}^{m} \overline{P}_i \setminus \Lambda(G)\right]/G.$$

The condition that the orbifold O is incomplete implies the incompleteness of the retract O_G^0. In fact, the origin of the geodesic ray running into the incomplete end of the orbifold O and having a finite length (see Helgason [1], Chapter 1 §10) can be taken in the orbifold O_G^0, which will then contain all of that ray, due to the convexity of the Nielsen hull. Also O_G^0 contains a finite number of isolated ends, and each of them corresponds to a point in the set

$$P^* = (\bigcup_{i-1}^{m} \overline{P}_i) \cap \Lambda(G).$$

In fact, the points of the set $\partial H^n \cap (\bigcup \overline{P}_i \setminus \Lambda(G))$ (the complement to P^*) by the retraction (5.6) go over to finite points of H^n and cannot correspond to ends of the orbifold O_G^0. At the same time, the finiteness of P^* follows readily from the finite-sidedness of the polyhedra P_i, which shows that the number of ends in O_G^0 is finite, and these have a one-to-one correspondence with the equivalence classes of the limit vertices of the polyhedra P_i, i.e. with points of the set P^*/G.

Let a point $p \in P^*$ have a stabilizer G_p in G; we consider the set of geodesic rays in H^n having their ends at the point p, which can be identified with a horosphere centred at p (a sphere in $\mathbb{R}^n_+$ tangent to $\partial\mathbb{R}^n_+$ at p). As G_p for p can be a finite extension either of a parabolic group (a free Abelian group of rank $\leq n-1$) or of a cyclic loxodromic group (Theorem 3.3), there exists a minimum distance $d(p)$ between that horosphere and its images by the transformations $g \in G_p$, and therefore each class $p^* \in P^*/G$ can be put into correspondence with the number $d(p^*) \geq 0$, equal to $d(p)$, where p is any point from the class p^*.

If for all $p^* \in P^*/G$ we have $d(p^*) = 0$, the geometric finiteness of G (the finiteness of P^*) enables us to construct a G-invariant system of distinct horospheres in H^n having their centres at points in P^*. We cut out the horoballs from the P_i bounded by these horospheres and get, on gluing together the remaining parts, an orbifold (with boundary) having a compact deformation retract $\mathcal{O}_0$. In other words, this horoball cutting corresponds in to the orbifold $\mathcal{O}^0_G = H^0_G/G$ to cutting out all the ends; consequently, one gets a compact suborbifold $\mathcal{O}_0 \subset \mathcal{O}^0_G$. We now recede from this horosphere system a distance $t > 0$ and cut out the new horoballs from the polyhedra P_i. The compact suborbifold $\mathcal{O}_t \supset \mathcal{O}_0$ obtrained in $\mathcal{O}^0_G$ is a t-neighbourhood of $\mathcal{O}_0$. We continue this process to get the exhaustion of the orbifold $\mathcal{O}^0_G$ by the compact suborbifolds $\mathcal{O}_t$, where always $\mathcal{O}_{t+s}$ is an s-neighbourhood of the orbifold $\mathcal{O}_t$. This contradicts the incompleteness of $\mathcal{O}^0_t$ and shows that there is a point $p \in P^*$ such that $d(p) > 0$, i.e. its stabilizer G_p contains loxodromic elements. However, the fixed points of these loxodromic elements cannot lie at the boundary of a locally finite fundamental set (perhaps partial fundamental) that is the union of a finite number of convex polyhedra (see Corollary 5.22). The contradiction with condition C7 (local finiteness of a filling) completes the proof of Theorem 6.6.

This proof contains information on the limit vertices of the model polyhedra, which gives the proof for the following theorem:

COROLLARY 6.8. *A hyperbolic structure with holonomy* d_* *in the orbifold* $\mathcal{O}$, $\dim \mathcal{O} \geq 2$, *obtained by isometric side-pairing for the finite-sided polyhedra* $P_1, \ldots, P_m \subset H^n$, *is complete if and only if* $\bigcup_{i=1}^{m} \partial P_i$ *does not contain fixed points of loxodromic elements in the holonomy group* $d_*(\pi_1(\mathcal{O})) \subset \mathrm{Isom}\, H^n$.

4. *Regular tessellations in the hyperbolic space.* Here we describe conditions under which a filling of the space H^n, determined by the complex (K^n, π) (i.e., by a process of periodic gluing of hyperbolic polyhedra on entire sides) is a regular tessellation in H^n.

THEOREM 6.9. *Let a filling of the hyperbolic space* H^n, $n \geq 2$, *be determined by the regular complex* (K^n, π) *having finite-sided model polyhedra* $P_1, \ldots, P_m \subset H^n$; *then that filling is a regular tessellation in* H^n *if and only if it satisfies the local finiteness condition* C7, *while the projection* π *is a one-to-one map around each* $(n-2)$*-dimensional side of* K^n.

PROOF. The assertion in the theorem on the equivalence of the completeness of the filling $(\pi(|K^n|) = H^n)$ to condition C7 is a direct consequence of Theorem 6.6 and Lemma 6.2; to prove the second property of a tessellation (π is a homeomorphism) it is sufficient to show that

π is a covering by virtue of H^n being simply connected. We shall prove this in a similar way to Aleksandrov [1] in the compact case.

For this purpose, we define special (ball) neighbourhoods in the space $|K^n|$. Namely, if a point $y \in |K^n|$ lies within some polyhedron P, then as its neighbourhood we take the ball $B(y, r_y)$ contained in P and having centre y and a certain radius $r_y > 0$. Now let y lie within a certain k-dimensional side Q^k of the complex K^n $(k \neq 0)$. We consider all the polyhedra in the complex that contain Q^k and construct in each of them ball sectors, having their centres at y that do not intersect any other sides apart from Q^k and those sides of the polyhedron that contain Q^k. By virtue of the finite-sidedness condition, one can obtain an $r_y > 0$ such that the ball sectors in the polyhedra adjoining Q^k can take a common radius $r_y > 0$. The union of all these ball sectors gives the ball neighbourhood $B(y, r_y)$ of the point y. Similarly, one defines the ball neighbourhood for a vertex y in the complex as the union of ball sectors in all the polyhedra containing y. These ball sectors have radius r_y such that they intersect only those sides of the polyhedra that contain the vertex y. The $r_y > 0$ is the same for all polyhedra, which is obtained again from the finite-sidedness condition for P_i, $1 \leq i \leq m$.

We now take some point y in one of the model polyhedra and consider all the points $y' \in |K^n|$ corresponding to it. The radii r_y for the ball neighbourhoods of the points y' can be taken as a common value $R_y > 0$.

This is obvious if y lies within some polyhedron P_i, since then $R_y = r_y$. For a point y lying on a k-side Q^k, $0 \leq k \leq n-1$, there will be only a finite number of corresponding points y' in each of the model polyhedra. It follows from the finite sidedness condition and the fact that the stabilizer of the side Q^k in the group G is finite, and by virtue of the condition C7 it cannot contain loxodromic elements - see Corollary 6.8. Therefore, with $R_y = \min r_y$, the ball neighbourhoods $B(y', R_y)$ of the points y' corresponding to the point y will be the desired ones, since any of the polyhedra in the complex is equal to one of the model polyhedra.

We now show that we can define the covering for the projection π by taking the covering neighbourhood $U(x)$ of any point $x \in H^n$ as the ball $B^n(x, R_y) = \pi(B(y, R_y))$, where a certain point y is taken from $\pi^{-1}(x)$. In fact, apart from the trivial case $(y \in \operatorname{int} P_i)$, the boundary surface of the ball neighbourhood $B(y, R_y)$ consists of $(n-1)$-dimensional spherical polyhedra cut out from an $(n-1)$-dimensional sphere by ball sectors. These polyhedra form a complex K^{n-1} having properties C1 - C4 and the space $|K^{n-1}|$, which is the boundary surface of the ball neighbourhood $B(y, R_y)$. The projection π naturally defines a continuous projection for the space $|K^{n-1}|$ on the $(n-1)$-dimensional sphere having radius R_y and its centre at the point $x \in H^n$, where $x = \pi(y)$, which follows from π being isometric in the polyhedra $P \in K^n$. Therefore, the $(n-1)$-dimensional projection has property C4 and C5 and by virtue of Theorem 6.1, $|K^{n-1}|$ is projected onto the whole sphere. Each $(n-3)$-dimensional side of K^{n-1} is an intersection of a certain

$(n-2)$-dimensional side of K^n that contains y and a sphere of radius R_y with centre at the point y. Therefore we have the result that the one-to-one property for π around each $(n-2)$-dimensional side of K^n implies a one-to-one property for the $(n-1)$-dimensional projection around the $(n-3)$-dimensional sides of K^{n-1}. Therefore, the space $|K^{n-1}|$ is homeomorphically projected onto the sphere $S^{n-1}(x, R_y)$ (Theorem 6.1). Then this and the isometric property of π in the ball sectors give us a homeomorphism for the ball neighbourhood $B(y, R_y)$ onto the ball $B^n(x, R_y)$ provided by π. There is freedom in choosing y from $\pi^{-1}(x)$, while the radii R_y of the ball neighbourhoods coincide for all such points. It completes the proof of the covering property for π. Then as H^n is simply connected, we have that π is a homeomorphism, which completes the proof of Theorem 6.9.

5. *Rigidity of tessellations in* H^n *for* $n \geq 4$. We have seen in part 2 of this section in the example of the octahedral regular tessellation (K_W, π_W) in the hyperbolic space H^3 that there exist continuous locally homeomorphic and arbitrarily close to identity mappings $\tilde{f} : H^3 \longrightarrow H^3$ that transfer the tessellation of the space H^3 into locally topologically similar incomplete hyperbolic fillings (K_α, π_α), or more precisely, tessellations in a proper subdomain of H^3. Here the restriction of the mappings $\tilde{f}$ on the union of the polyhedra $\pi_W(P)$, $P \in K_W$, adjoining some tessellation polyhedron are homeomorphisms compatible with the tessellation groups. This, in particular, shows that, if the side-pairing for the model polyhedra $P_1, \ldots, P_m$ in the tessellation of H^3 specifies a three-manifold $M = H^3/G$ (necessarily non-compact), then the mappings $\tilde{f}$ become homeomorphisms $f : M \longrightarrow M'$. One obtains incomplete hyperbolic manifolds $M' = f(M)$ by side-pairing the model polyhedra $\tilde{f}(P_1), \ldots, \tilde{f}(P_m)$.

We show below that there is a superrigidity for the tessellations in n-dimensional hyperbolic space for $n \geq 4$, i.e., there are no deformations of a regular tessellation in H^n, $n \geq 4$, to locally topologically similar incomplete fillings.

THEOREM 6.10. *Let a polyhedra filling for the hyperbolic space* H^n, $n \geq 4$, *be specified by a regular complex* (K^n, π) *having finite volume model polyhedra* $P_1, \ldots, P_m \subset H^n$. *Then that filling is a regular tessellation in* H^n *if and only if it is locally topologically similar to a certain regular tessellation* (K_0, π_0) *of* H^n.

PROOF. Let O and O_0 be hyperbolic orbifolds corresponding to the regular complexes (K^n, π) and (K_0, π_0). From Corollary 5.48 and Bieberbach's Theorem 3.4 we see that a complete hyperbolic finite-volume orbifold O_0 is a geometrically finite orbifold, and therefore any end E_0 in it has a structure of some closed $(n-1)$-dimensional Euclidean orbifold E^{n-1}, while the neighbourhood $U(E_0)$ of such an end has the form (see Chapter 7 §1 for more details):

$$U(E_0) \approx E^{n-1} \times [0, \infty). \tag{6.5}$$

As we have the homeomorphism $f: \mathcal{O}_0 \longrightarrow \mathcal{O}$, we get from Corollary 6.8 and from the incompleteness of the orbifold $\mathcal{O} = f(\mathcal{O}_0)$ that there exists a limit vertex $p \in \bigcup P_i$ having the loxodromic (elementary) stabilizer-subgroup G_p of the holonomy group G for $\mathcal{O}$.

That action of the loxodromic subgroup G_p in the set of geodesic rays in H^n arising from the point p is equivalent to the action of an affine discrete group Γ in punctured Euclidean space $\mathbb{R}^{n-1}_*$ (without the origin), which is a finite extension of an Abelian group of homothetic transformations. This follows from the fact that the hyperbolic metric in H^n induces the Euclidean metric in the horosphere $S_p \subset H^n$ having its centre at $p \in \partial H^n$; the origin in $\mathbb{R}^{n-1}$ is here put into correspondence with the point of intersection between that horosphere and the axis of the loxodromic generator of the group G_p. At the same time, it is well known (see, for example, Goldman [6]) that the quotient space $\mathbb{R}^{n-1}_*/\Gamma$ has a finite-sheeted covering by the Hopf manifold $S^1 \times S^{n-2}$. Then it is clear that for $n \geq 4$ this manifold cannot be homeomorphic to the $(n-1)$-dimensional torus $T^{n-1} = S^1, \ldots, S^{n-1}$, which is a finite-sheeted covering of the complete end E_0 of $\mathcal{O}_0$, $E = f(E_0)$. This contradiction with (6.5) proves that $\mathcal{O}$ is a complete hyperbolic orbifold of finite volume. Then Theorem 6.9 and the condition of locally topologically similarity for (K, π) and (K_0, π^0) complete the proof of the tessellation properties in H^n for the filling (K, π).

The converse statement is trivial, which completes the proof of Theorem 6.10.

REMARK 6.11. The condition in Theorem 6.10 on the finiteness of the hyperbolic volumes for the model polyhedra $P_1, \ldots, P_m$ in (K^n, π) can be replaced by a weaker condition of finite-sidedness if one specifies additionally that the following condition holds:

the regular tessellation (K_0, π_0) in H^n, locally topologically similar to the filling (K^n, π), is such that each point p on the boundary at infinity of the polyhedra $\overline{\pi_0(P_i^0)}$, $i = 1, \ldots, m$, has rank of the stabilizer $G_p^0 \subset G^0$ either zero or at least three. To prove this, we need to transfer from $\mathcal{O}$ and $\mathcal{O}_0$ to the minimal convex, in the sense of Nielsen's hulls, retracts; the complete ends of these retracts are homeomorphic (up to a finite covering) to $T^{k_p} \times [0, 1]^{n-k_p-1}$, where $k_p = \operatorname{rank} G_p^0 \geq 3$. After this, the proof is completed with an observation analogous to the proof of Theorem 6.10, namely that an incomplete end in $\mathcal{O}$ of that type has a finite-sheeted covering which is homeomorphic to the product of a Hopf manifold $S^1 \times S^{k_p-1}$ on $[0, 1]^{n-k_p-1}$, so it cannot be homeomorphic to the end of the complete orbifold $\mathcal{O}_0$.

REMARK 6.12. Theorem 6.10 can be considered as a generalization of the fact (Garland and Raghunathan [1], Theorem 7.2) that for a discrete group $G \subset \operatorname{Isom} H^n$, $\operatorname{vol} H^n/G < \infty$, $n \geq 4$, there are no homomorphisms $\varphi: G \longrightarrow \operatorname{Isom} H^n$ arbitrarily close to the inclusion. Thurston's [1] non-rigidity theorem describes the directly opposite situation for $n = 3$ (see also Milnor [4]):

THEOREM 6.13. *Suppose that* $M = H^3/\Gamma$ *is an orientable hyperbolic 3-manifold which is non-compact but has finite volume. Then there exists an infinite sequence of hyperbolic manifolds*

$$M_j = H^3/h_j(\Gamma)$$

which have strictly smaller volume and which approximate the original manifold M *as* $j \longrightarrow \infty$ *in the sense that the homomorphisms*

$$h_j : \Gamma \longrightarrow PSL(2, \mathbb{C}) \cong \mathrm{Isom}\, H^3$$

tend to the inclusion $\Gamma \subset PSL(2, \mathbb{C})$ *in the topology of pointwise convergence. Furthermore, the volumes of the* M_j *tend to the volume of* M.

REMARK 6.14. Remark 6.7 shows that Theorem 6.9 (it gives conditions under which a hyperbolic filling is a tessellation) applies for fillings that are not regular (i.e., non-periodic). Whether Theorem 6.10 is correct for such fillings remains an open question, and it is also not known whether there are non-periodic fillings of H^n by non-compact finite-volume polyhedra and, if so, for what dimensions n.

§2. **Free amalgamated products of Kleinian groups**

In this section and the next one, we deal with definitions, properties and applications for products of discontinuous groups which give one of the major ways of constructing new discontinuous groups from simpler ones having familiar properties. Usually in this case one considers some family $\{G_i\}$, $i \in \mathcal{I}$, of discontinuous groups and the group $G = \langle G_i, i \in \mathcal{I}\rangle$, generated by that family. The task is to define conditions under which G is again discontinuous, and thus to elucidate the algebraic structure of G. The topic was first examined by Klein [1], who considered the case where G is a free product of the groups G_i, $i = 1, 2$. Maskit [3] - [5] gave an important development for the method of constructing discontinuous groups, which has been called the combination method. These constructions on the one hand are closely related to algebraic methods for group free products, the free amalgamated products of groups and HNN-extension of groups (see Magnus, Karrass, and Solitar [1] and Lyndon and Schupp [1]), while on the other hand they are of geometric type and have a deep relationship to topology and uniformization theory (see Zieschang, Vogt, and Coldewey [1], Chapter 7 §2, Chapter 8 §3 and Chapter 9).

THEOREM 6.15 (Klein combination). *Let* $\{G_i\}$, $G_i \subset \mathrm{M\ddot{o}b}(n)$ *be a discontinuous group family, where the groups have fundamental domains* F_i *and satisfy the conditions*

1) $\mathrm{ext}\, F_i \subset F_j$ for all $i \neq j$, $i, j \in \mathcal{I}$,

2) $F = \text{int}\left[\bigcap_{i\in\mathcal{J}} F_i\right] \neq \emptyset.$

Then the group $G = \langle G_i,\ i \in \mathcal{J}\rangle$, *generated by the union of the groups* G_i, *is discontinuous and is the free product of* G_i, *where* F *is a fundamental domain for* G.

The proof of this assertion is elementary and is independent of the space type and homeomorphism group (compare Ford [1]).

EXAMPLE 6.16. Consider the Fuchsian groups $G_1, \ldots, G_m \subset \mathcal{M}_n$ acting correspondingly in the balls $B_1, \ldots, B_m \subset \mathbb{R}^n$ sufficiently far apart (so that conditioin 1 holds in the theorem) and uniformizing the closed hyperbolic manifolds $M_i = B_i/G$, $i = 1, \ldots, m$. Then the Kleinian group $G = \langle G_1, \ldots, G_m\rangle \subset \mathcal{M}_n$ they generate is a function group with the invariant component $\Omega_0 \subset \bigcap_{i=1}^{m} \text{ext}\, B_i$ of the discontinuity set $\Omega(G)$ (Ω_0 is simply connected for $n \geq 3$). The group uniformizes the closed conformal manifolds

$$\Omega(G)/G = M_0 \cup M_1 \cup \ldots \cup M_m, \tag{6.6}$$

where $M_i = B_i/G_i$, $i = 1, \ldots, m$ and $M_0 = \Omega_0/G$, is the connected sum of the manifolds M_i, $M_0 \approx M_1 \# \ldots \# M_m$. If $n = 2$ and the surfaces M_i, $i = 1, \ldots, m$, are of genus g_i, the surface M_0 from (6.6) has genus $g_0 = g_1 + \ldots + g_m$.

EXAMPLE 6.17. Another major example is obtained by considering a family Σ of pairwise non-intersecting spheres in $\mathbb{R}^n$ whose common exterior has the non-empty interior F_Σ, compare (3.8). Then the group G_Σ, generated by inversions with respect to the spheres $\sigma \in \Sigma$ or involutions (i.e., elliptic transformations of order two), is discontinuous and has F_Σ as its fundamental domain. The group G_Σ is here the free product of cyclic second order groups. We have already used this approach in constructing the groups $G \subset \text{Möb}(n)$, $n \geq 3$, having given numbers of invariant components (see Theorem 3.6).

For any domain $D \subset \mathbb{R}^n$, we can consider a family Σ such that F_Σ and D have the same homotopy type. In that case, $\partial F_\Sigma \subset \Lambda(G_\Sigma)$, which shows, as Volynets [1] noted, that there is a way of constructing the Kleinian groups $G \subset \text{Möb}(n)$, whose connected components Ω_0 of the discontinuity set have a given homotopy type (for example, $\text{rank}\, \pi_1(\Omega_0) = k$, or $\text{rank}\, \pi_2(\Omega_0) = m$, where m and k are given integers).

EXAMPLE 6.18. As free cofactors $G_1, \ldots, G_m \subset \text{Möb}(n)$ in the Klein combination theorem, we consider free Abelian groups of rank n, $n \geq 3$, i.e. groups conjugate in $\text{Möb}(n)$ to the group G generated by the n translations $x \longmapsto x + a^i$, $i = 1, \ldots, n$, by linearly independent vectors $a^i \in \mathbb{R}^n \backslash \{0\}$. The manifolds $T^n = \mathbb{R}^n/G$, obtained by identifying the opposite sides of a parallelepiped having edges a^i, is an n-dimensional Euclidean torus (the Cartesian product of n circles).

We now select the elementary groups $G_1, \ldots, G_m \subset \text{Möb}(n)$ such that the conditions of Theorem 6.15 are satisfied, i.e., such that for all $i \neq j$ the spheres bounding the fundamental polyhedron P_j for the group $G_j \cong G$ lie outside the spheres bounding the fundamental polyhedron P_i for the group $G_i \cong G$. Then the function group $\Gamma = \mathop{*}_{i=1}^{m} G_i$ will have the fundamental polyhedron $P = \bigcap_{i=1}^{m} P_i$ and the representation

$$\Gamma = \langle a_{11}, a_{12}, \ldots, a_{1n}, \ldots, a_{mn} : [a_{11}, a_{12}] = \ldots = [a_{1n-1}, a_{1n}] = \text{id}, \ldots$$
$$[a_{m1}, a_{m2}] = \ldots = [a_{mn-1}, a_{mn}] = \text{id}\rangle. \qquad (6.7)$$

It is readily seen that the limit set $\Lambda(\Gamma)$ (a discontinuum) is a perfect, nowhere dense and disconnected set in $\mathbb{R}^n$. One can show (compare Beardon [1] for $n = 2$) that $\Lambda(\Gamma)$ has positive logarithmic capacity and for sufficiently large m has Hausdorff dimensions greater than $n/2$. The discontinuous set $\Omega(\Gamma) = \overline{\mathbb{R}}^n \setminus \Lambda(G)$ is the universal covering space for the compact manifold $M^n_m = \Omega(\Gamma)/\Gamma$, which is the connected sum of m copies of the n-dimensional tori T^n_i (see the definition in Chapter 7 §1).

As in the previous example, there is a difference from the case $n = 2$, where that connected sum is a Riemann surface of genus $m \geq 2$ and consequently a hyperbolic two-manifold, in that the manifold M^n_m, $n \geq 3$, is no longer aspheric, and it is even not hyperbolic. Figure 39 shows schematically the necessary gluing for the two-dimensional tori in $\mathbb{R}^3$ from which one obtains M^3_2.

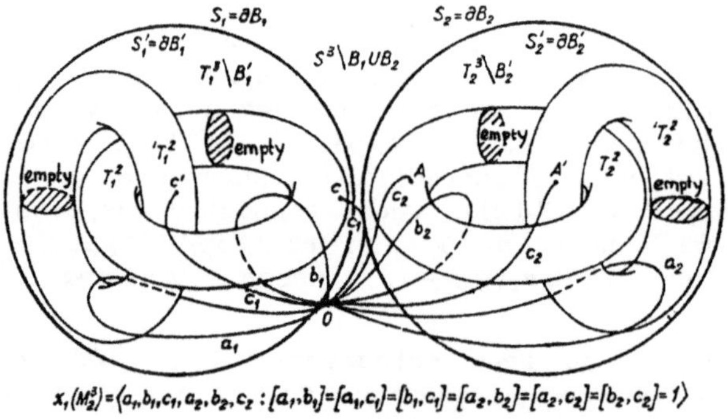

Figure 39

In what follows let us assume for convenience that the discontinuity set $\Omega(G)$ of any considered Kleinian group $G \subset \text{Möb}(n)$ contains the infinity. That condition does not restrict the generality because it is satisfied by the group hGh^{-1} for suitable $h \in \text{Möb}(n)$.

By $m_n(X)$ we denote the Lebesgue measure for a set X from the sphere $\overline{\mathbb{R}}^n$.

THEOREM 6.19 (Maskit first combination). *Let $G_1, G_2 \subset \text{Möb}(n)$ be discontinuous groups having a common subgroup H, with $S \subset \overline{\mathbb{R}}^n$ a Jordan hypersurface decomposing $\overline{\mathbb{R}}^n$ into two domains D_1 and D_2, whose closures are precisely invariant in G_1 and G_2 (correspondingly) with respect to the subgroup H, and let the following conditions hold:*

I) if Δ, F_1, and F_2 are fundamental domains for the groups H, G_1, and G_2 correspondingly, there exists a neighbourhood V of the surface S such that

$$\Delta \cap V \subset F_i, \qquad i = 1, 2; \tag{6.8}$$

II) $\Delta \cap \overline{D}_i = \overline{D}_i \cap F_i \neq F_i, \quad i = 1, 2.$

Then:

a) $G = \langle G_1, G_2 \rangle$ is a discontinuous group and is isomorphic to the free amalgamated product of the groups G_1 and G_2 with the amalgamated subgroup H;

b) $F = F_1 \cap F_2$ is a fundamental domain for the group G;

c) if $m_n(\Lambda(G_i)) = 0$, $i = 1, 2$, then $m_n(\Lambda(G)) = 0$;

d) if $g \in G$ is an elliptic or parabolic element, then g is conjugate in G with an element from $G_1 \cup G_2$.

The proof of the theorem follows Maskit's scheme [5], as he proved that theorem for Kleinian groups in the plane (compare Ivascu [1]). During the proof we need some preliminary assertions.

LEMMA 6.20. *Let $G \subset \text{Möb}(n)$ be a discontinuous group with $G_\infty = \{\text{id}\}$, $\infty \in \Omega(G)$, and let r_g be the radius of the isometric sphere of an element $g \in G \setminus \{\text{id}\}$. Then*

$$\sum_{g \in G} r_g^{2n} < \infty, \tag{6.9}$$

The proof of the lemma (see Apanasov [8], p.742) follows from the observation that the images $g(B)$, $g \in G$, of the exterior of a sufficiently large ball $B^n(x, r)$ lie within it and have radii kr_g^2/r.

LEMMA 6.21. *Let $S \subset \mathbb{R}^n$ be a Jordan hypersurface, invariant with respect to a subgroup H of the discontinuous group $G \subset \text{Möb}(n)$ and such that $\infty \in \Omega(G)$, $\infty \notin G(S)$, and let there exist a neighbourhood V of S such that $V \cap \Delta \subset F$, in which Δ and F are fundamental domains for H and G.*

Then for any coset decomposition

$$G = H \cup g_1 H \cup g_2 H \cup \dots \quad , \tag{6.10}$$

in which $g_i \in G$, $g_i H \cap g_j H = \varnothing$, the Euclidean diameters $d(g_i(S))$ are such that

$$\sum_{i=1}^{\infty} d^n(g_i(S)) < \infty. \tag{6.11}$$

PROOF. The basic concept in the proof goes back to Koebe, who used a similar lemma for the case $n=2$ to prove that $m_2(\Lambda(\Gamma))=0$ for a Schottky group Γ in the plane.

If for a given $g \in G$ we consider the set of centres $q(gh)$ for the isometric spheres of the transformations gh, in which $h \in H$, then it is evident that this set is invariant with respect to the group H:

$$q(gh) = (gh)^{-1}(\infty) = h^{-1}(q(g)). \tag{6.12}$$

As $\infty \in \Omega(G)$, for each g_i, $i=0, 1, \ldots,$ we get from (6.10) a unique element $h_i \in H$ such that $q(g_ih_i) \in \Delta$. One can assume that $q(g_i) \in \Delta$. We denote by δ_1 the distance between $q(g_i)$ and the surface S. Due to (6.12), the invariantness of S with respect to H, and the assertion in the lemma, we have that there is at least one i such that $q(g_i) \in \Delta \cap V \subset F$. Then we see that we can select $\delta_0 > 0$ such that for almost all i we have $\delta_i > \delta_0$.

Each element g_i has a representation $g_i = T_i\mathcal{J}_i$, where $\mathcal{J}_i$ is the inversion with respect to the isometric sphere $I(g_i)$ having radius r_i, while T_i is a Euclidean isometry. Then we can put

$$d(g_i(S)) = d(\mathcal{J}_i(S)) \le d[\mathcal{J}_i(S^{n-1}(q(g_i), \delta_i))] \le r_i^2/\delta_0.$$

These inequalities and Lemma 6.20 complete the proof of (6.11).

PROOF OF THEOREM 6.19. We first show that for any $g \in G$, written in normal form:

$$g = g_n \circ \ldots \circ g_1, \tag{6.13}$$

in which $n>0$ and either $g_{2i} \in G_2\backslash H$, $g_{2i+1} \in G_1\backslash H$, or $g_{2i} \in G_1\backslash H$, $g_{2i+1} \in G_2\backslash H$, and for any $x \in F = F_1 \cap F_2$ we have that $g(x) \notin F$.

We assume for definiteness that $g_1 \in G_1\backslash H$ in (6.13), i.e., the first case applies. As $\Delta \cap D_1 = F_1 \cap D_1$, for $x \in D_1$ and $g_i \in G_1\backslash H$ we must have $g_i(x) \in D_2$ and $g_i(x) \notin F$. Similarly, for $x \in D_2$ and for $g_i \in G_2\backslash H$ we have $g_i(x) \in D_1$ and $g_i(x) \notin F$.

We thus see that for any $x \in F$ we have: $g_1(x) \in D_2$, $g_1 \notin F$; $g_2g_1(x) \in D_1$, $g_2g_1(x) \notin F$, and so on, which shows that $g(x) \notin F$.

From condition II in the theorem it follows that

$$\emptyset \ne F_1 \cap D_2 \subset \Delta \cap D_2 = F_2 \cap D_2, \tag{6.14}$$

i.e., that $F \cap D_2 \ne \emptyset$. By analogy with (6.14) we get $F \cap D_1 \ne \emptyset$. As F is non-empty and as $g(F) \cap F = \emptyset$ for a non-trivial g from (6.13), it is clear that there are no non-trivial words having the (6.13) form and representing the identity in G, i.e., $G \cong G_1 \underset{H}{*} G_2$ is a free amalgamated product. Here, since $H(F) \cap F = \emptyset$, while the union $G_1(F_1 \cap D_2)$ is open in $\mathbb{R}^n$, we have simultaneously shown that the group G is discontinuous, i.e., the assertion a) in the theorem is true. In what follows, we can put that $\infty \in F$, which implies boundedness for S and its images $g(S)$ for $g \in G$.

We now prove the inclusions

$$\Delta \cap V \subset F \subset \Omega(G). \tag{6.15}$$

The first is obvious (see (6.8)). To prove the second, we note that condition II implies: $F \cap D_1 = F_2 \cap D_1$ and $F \cap D_2 = F_1 \cap D_2$. So we consider the normal form (6.13) for the elements $g \in G$ as above, to get that for any point $x \in F \cap (D_1 \cup D_2)$ there exists a neighbourhood U of it such that $g(U) \cap U \neq \varnothing$ only for a finite number of elements $g \in G$. It remains to prove an analogous property for the points from $F \cap S$. We have the obvious inclusions

$$F \cap S = \Delta \cap S \subset \Delta \cap V \subset F. \tag{6.16}$$

Let $x_0 \in F \cap S$ and U be its neighbourhood, $U \subset V$. For each point $x \in U$ there exists $h \in H$ and a point $y \in \overline{\Delta}$ such that $h(x) = y$. The neighbourhood U can be considered as so small that these points y can be taken from $\overline{\Delta} \cap V$. If this is not so, we can find a sequence of points x_m converging to x_0 and a sequence $\{h_m\} \subset H$ such that $\lim h_m(x_m) = y_0 \notin V$. This contradicts $H(S) = S$ and the fact that y_0 will then be a limit point for the group H. So we have shown that for $x_0 \in F \cap S$ there exists a neighbourhood U such that all its points are H-equivalent to points from $\overline{\Delta} \cap V \subset \overline{F}$. This and the property proved above ($g(F) \cap F = \varnothing$ for $g \in G \setminus \{\mathrm{Id}\}$) show that for the points $x_0 \in F \cap S$ there exist neighbourhoods U such that $g(U) \cap U \neq \varnothing$ only for a finite number of elements $g \in G$, which completes the proof of (6.15).

It is well known that, although (6.13) is not unique for $g \in G \setminus H$, the number of cofactors $n > 0$ is determined uniquely by the element g (we denote this by $|g|$, with $|g| = 0$ for $g \in H$). If here $g_1 \in G_1 \setminus H$ in the normal form (6.13) (or $g_1 \in G_2 \setminus H$)), we say that g is positive (correspondingly negative). We now consider a certain fixed decomposition of the group G:

$$G = H \cup (\bigcup_{n,m} a_{nm}H) \cup (\bigcup_{n,m} b_{nm}H), \tag{6.17}$$

where $|a_{nm}| = |b_{nm}| = n$, $a_{nm} > 0$, $b_{nm} < 0$, and for a given $n > 0$ we consider the sets:

$$T_n = \bigcup_m a_{nm}(D_1) \cup b_{nm}(D_2), \quad C_n = \overline{\mathbb{R}}^n \setminus T_n. \tag{6.18}$$

Now let $C = \bigcup_n C_n$ and $T = \overline{\mathbb{R}}^n \setminus C$. Our aim is to show that, if $x \in C$, then for a certain $g \in G$, $g(x) \in F \cup \Lambda(G_1) \cup \Lambda(G_2)$, and also that $T \subset \Lambda(G)$ and T has zero measure. These together will prove statements b) and c) in the theorem.

We describe C_1. Note that for $x \in T_1 \cap D_2$ there exists $g \in G_1 \backslash H$ such that $g(x) \in D_1$, so any point $x \in C_1 \cap D_2$ is either a limit point for G_1 or there exists $g \in G_1$ such that $g(x) \in F_1 \cap D_2 \subset F$. Similarly, if $x \in C_1 \cap D_1$, then either $x \in \Lambda(G_2)$ or there exists $g \in G_2$ such that $g(x) \in F$. If on the other hand $x \in C_1 \cap S$, then either $x \in \Lambda(H) \subset \Lambda(G_1) \cap \Lambda(G_2)$ or for some $h \in H$ we have $h(x) \in \Delta \cap S = F \cap S$. Therefore for $x \in C_1$ there exists $g \in G$ such that

$$g(x) \in F \cup \Lambda(G_1) \cup \Lambda(G_2).$$

At the same time, (6.18) gives us directly that $T_n \subset T_{n-1}$ for $n > 1$, i.e., $C_i \subset C_{i+1}$ for $i \geq 1$. But then, for any $x \in C$, there exists a number n such that $x \in C_n$, $x \notin C_{n-1}$. As the case of C_1 has been described above, we consider that $n > 1$ and $x \in T_{n-1}$. Then either for certain $y \in D_1$ and $a_{n-1,m}$ from (6.17) we have that $x = a_{n-1,m}(y)$, or for $y \in D_2$ and $b_{n-1,m}$ we have $x = b_{n-1,m}(y)$.

For definiteness, we consider the first case (the second is analogous). Then y is a point either in T_1 or C_1. If $y \in T_1$, the (6.18) definition shows that there are $z \in D_2$ and $g \in G_2 \backslash H$ such that $y = g(z)$. Therefore we have $x = a_{n-1,m} \circ g(z)$, where $|a_{n-1,m} \circ g| = n$ and $a_{n-1,m} \cdot g < 0$, which contradicts the fact that $x \notin T_n$. Then for $x \in C_n$, $n > 1$, there are $g_0 \in G$ and $y_0 \in C_1$ such that $g_0(x) = y_0$. From what we have proved above, we get that for $x \in C$ there is $g \in G$ having the property $g(x) \in F \cup \Lambda(G_1) \cup \Lambda(G_2)$.

We now characterize the set T. We note that $a_{nm}(D_1) \subset D_2$ and $b_{nm}(D_2) \subset D_1$ if and only if n is odd. Therefore, for a given n one has $a_{nm}(D_1) \cap b_{nm}(D_2) = \emptyset$. For two distinct representatives a_{nm} and a_{nk} of the decomposition (6.17) we write $a_{nm}^{-1} \cdot a_{nk} = g_r \cdot \ldots \cdot g_1$, where $r > 0$, $g_1 \in G_1 \backslash H$, $g_r \in G_1 \backslash H$, so $a_{nm}^{-1} \cdot a_{nk}(D_1) \subset D_2$ and consequently $a_{nm}(D_1) \cap a_{nk}(D_1) = \emptyset$. Similarly $b_{nm}(D_2) \cap b_{nk}(D_2) = \emptyset$ for $m \neq k$. At the same time, since the image $g(S)$ does not intersect S for $g \in (G_1 \cup G_2) \backslash H$, it is clear that this is also true for $g \in G \backslash H$: $g(S) \cap S = \emptyset$. We thus conclude that each connected component T_{nm} of the set T_n is a set having one of two forms: either $a_{nm}(D_1)$ or $b_{nm}(D_2)$. At the same time, T_n is bounded, as follows from the inclusion $\infty \in F \subset C_1$. Then we do not need to distinguish a_{nm} and b_{nm}. We can then write, instead of (6.17):

$$G = \bigcup_{i,j} c_{ij} H, \tag{6.19}$$

in which $|c_{ij}| = i$ and T_{ij} is the domain bounded by the surface $c_{ij}(S)$.

We apply Lemma 6.21 to (6.19) to get

$$\sum_{i,j} d^n(c_{ij}(S)) = \sum_{i,j} d^n(T_{ij}) = \sum_{i=1}^{\infty} \sum_j d^n(T_{ij}) < \infty. \tag{6.20}$$

The convergence of the (6.20) series implies that

$$\lim_{i\to\infty} \sum_j d^n(T_{ij}) = 0. \tag{6.21}$$

If now the point $x \in T$, then $x \in T_i$ for all i and thus one can state a number $j = j(i)$ such that $x \in T_{i,j(i)}$. Then (6.21) shows that

$$\lim_{i\to\infty} d(T_{i,j(i)}) = 0. \tag{6.22}$$

The difference in the elements $c_{i,j(i)}$ and (6.22) together show that the point $x \in T$ is a limit point for G, i.e. $T \subset \Lambda(G)$.

The description of the sets T and C, $T \cup C = \overline{\mathbb{R}}^n$, completes the proof of assertion b) in the theorem: F is a fundamental domain for G. Assertion c) is obtained from this, since from (6.21) for $i \to \infty$ we have

$$m_n(T) \le m_n(T_i) = \sum_j m_n(T_{ij}) \le \omega_n \sum_j d^n(T_{ij}) \to 0. \tag{6.23}$$

It follows directly from (6.23) that $m_n(\Lambda(G) \cap T) = 0$. The remaining part of the limit set $\Lambda(G) \cap C$ also has zero measure if $\Lambda(G_1)$ and $\Lambda(G_2)$ are the same, as follows from G being countable and the fact that

$$\Lambda(G) \cap C = \bigcup_{g\in G} g(\Lambda(G_1) \cup \Lambda(G_2)).$$

To complete the proof, it remains to demonstrate d). Let $g \in G$. We consider all elements in G conjugate to it and let g^* be an element of such type having minimum length $|g^*| = m$. It is clear that either $m = 1$ or m is even. If $m = 1$, the proof is complete, so we assume that m is even and assume for simplicity that $g^* > 0$. It is readily seen that $g^*(D_1) \subset D_1$, and this is possible only for loxodromic g^*, and thus for loxodromic g. Theorem 6.19 is proved.

REMARK 6.22. This combination theorem can be proved analogously for a countable number of combined groups, as can Theorem 6.24 (Maskit second combination).

REMARK 6.23. Although Theorem 6.19 has been formulated for Kleinian groups $G \subset \text{Möb}(n)$, it can obviously be reformulated for discrete isometry groups in the hyperbolic space H^n, $n \ge 2$: cf. Chapter 9 §3.3.

§3. HNN-extensions of Kleinian groups. Examples

In combinatorial group theory there is a well-known method of a group extension introduced in 1949 by G. Higman and by B. and H. Neumann, see for example Lyndon and Schupp [1]. The definition is as follows.

Given a group Γ, its subgroups A and B, and the isomorphism $\varphi : A \longrightarrow B$, we construct a group

$$G = \langle \Gamma,\ t : t^{-1}at = \varphi(a),\ a \in A\rangle$$

which we call the HNN-extension of Γ with respect to the subgroups A and B and the isomorphism φ.

For the case of Kleinian groups, geometric realization of this construction is known as Maskit's second combination. The conditions for it are given by the following theorem.

THEOREM 6.24 (Maskit's second combination). *Let* $\Gamma \subset \text{Möb}(n)$ *be a discontinuous group having subgroups* H_1 *and* H_2, *with respect to which there are precisely invariant closed domains* D_1 *and* D_2 *bounded by the Jordan surfaces* S_1 *and* S_2 *and such that there exist* $f \subset \text{Möb}(n)$, $f(\text{int}\, D_1) = \overline{\mathbb{R}}^n \setminus D_2$, *and let the following conditions hold*:

I) *If* Δ_1, Δ_2 *and* $F_0 \subset \Delta_1 \cap \Delta_2$ *are fundamental domains for* H_1, H_2, *and* Γ *correspondingly, there exist neighbourhoods* V_1 *and* V_2 *of* S_1 *and* S_2 *such that* $\Delta_i \cap V_i \subset F_0$, $i = 1, 2$;

II) $\Delta_i \cap D_i = F_0 \cap D_i$, $i = 1, 2$;

III) $F = \text{int}\{F_0 \cap \text{ext}(D_1 \cup D_2)\} \neq \varnothing$;

IV) $fH_1f^{-1} = H_2$;

V) $g(D_1) \cap D_2 = \varnothing$ *for all* $g \in \Gamma \setminus \{\text{id}\}$.

Then

a) *the group* $G = \langle \Gamma, f\rangle$ *is discontinuous and* F *is a fundamental domain for it*;

b) *each defining relation in* G *is a consequence of the defining relations in* Γ *and the relation in condition* IV;

c) *if* $m_n(\Lambda(\Gamma)) = 0$, *then* $m_n(\Lambda(G)) = 0$;

d) *if* $g \in G$ *is an elliptic or parabolic element, then* g *is conjugate in* G *with an element from* Γ.

The proof of the theorem is based on the concepts used in the proof of the previous theorem and is analogous to Maskit's proof [5] for Kleinian groups in the plane.

We give below some examples of the theorems applied. These indicate topological situations in which one gets free amalgamated products and HNN-extensions. In Chapter 7 §2, the topic is taken up again.

EXAMPLE 6.25. We consider in $\overline{\mathbb{R}}^3$ the discontinuous group obtained by the extension from $\mathbb{R}^2$ to $\mathbb{R}^3$ of the Picard group, i.e., the group $PSL_2(\mathbb{Z}(i))$. In the Picard group, we consider a finite-index subgroup G_1 generated by transformations

$$g_1 = \begin{pmatrix} 1 & 2 \\ 0 & 1 \end{pmatrix}, \quad g_2 = \begin{pmatrix} 1 & i \\ 0 & 1 \end{pmatrix}, \quad g_3 = \begin{pmatrix} 1 & 0 \\ -i-1 & 1 \end{pmatrix}.$$

It is readily seen that $\Lambda(G_1) = \overline{\mathbb{R}}^2$, and $\Omega(G_1) = \mathbb{R}^3 - \mathbb{R}^2$ splits up into two half-spaces. The fundamental polyhedron $F(G_1)$ can be represented as the intersection of the exterior of the isometric spheres $I(g_3)$ and $I(g_3^{-1})$ with the polyhedra

$\{x \in \mathbb{R}^3 : -1 < x_1 < 0,\ 0 < x_2 < 1\}$ and $\{x \in \mathbb{R}^3 : 0 < x_1 < 1,\ -1 < x_2 < 0\}$.

Figure 40 shows the orthogonal projection of $F(G_1)$ to the plane $\mathbb{R}^2$.

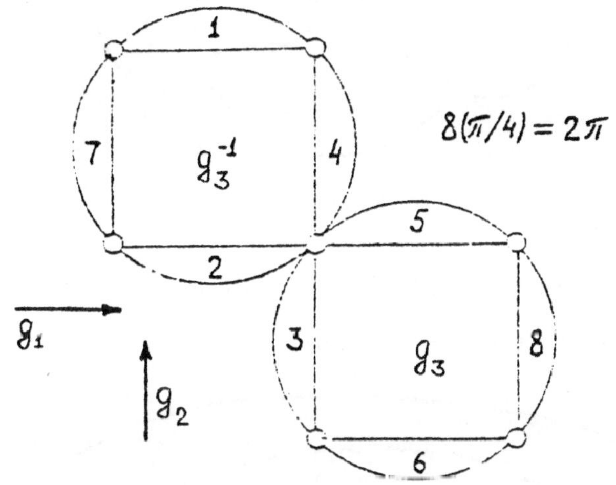

Figure 40

Translating $F(G_1)$ along x_3 by three we obtain a fundamental polyhedron $F(G_2)$ for the group $G_2 = \gamma G_1 \gamma^{-1}$, in which $\gamma(x) = x + e_3$.

It is clear that G_1 and G_2 satisfy all the conditions for Maskit first combination if the amalgamated subgroup H is taken as the group generated by the translations g_1 and g_2, while the surface S, precisely invariant with respect to H, is taken as the sphere $S = \{x \in \overline{\mathbb{R}}^3 : x_3 = 3/2\}$.

The combination theorem proof shows that the discontinuous group $G = G_1 \ast_H G_2$ has an invariant component Ω_0, $\Omega_0 \subset \{x \in \mathbb{R}^3 : 0 < x_3 < 3\}$, while all the non-invariant components Ω_i are equivalent either to the half-space $D_1 = \{x \in \mathbb{R}^3 : x_3 > 3\}$, or to the half-space $D_2 = \{x \in \mathbb{R}^3 : x_3 < 0\}$. Here Ω_0 is a contractible domain (homeomorphic to a ball - see Theorem 8.63). G_1 and G_2 have the presentation

$$\langle g_1, g_2, g_3 : g_3(g_1g_2^{-1})g_3g_2^{-1}g_3^{-1}g_2g_3^{-1}g_2^{-1} = [g_1, g_2] = 1\rangle. \qquad (6.24)$$

One can transform (6.24):

$$\langle g_2, g_3 : [g_2^{-1}g_3]\cdot[g_2g_3]\cdot[g_2, g_3^{-1}]\cdot[g_2^{-1}, g_3^{-1}] = 1\rangle. \qquad (6.25)$$

Then Riley's [1] theorem and the parabolicity of the generators g_2 and g_3 imply that the manifolds D_2/G_1 and D_1/G_2 are homeomorphic to the Whitehead link complement (Fig. 41), whose fundamental group is

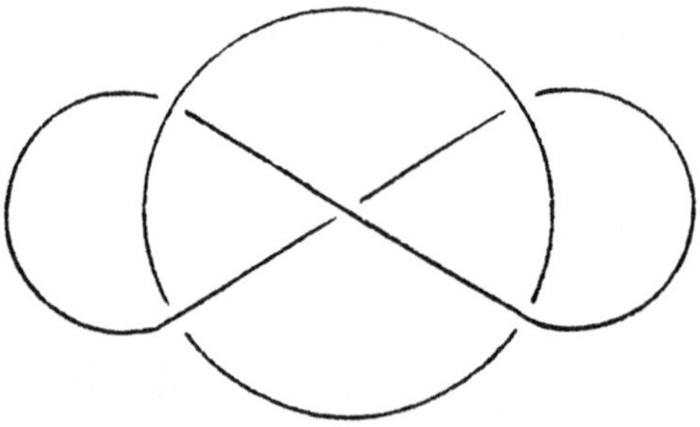

Figure 41

$$\langle a_1, a_2 : [a_2^{-1}, a_1]\cdot[a_2^{-1}, a_1^{-1}]\cdot[a_2, a_1^{-1}]\cdot[a_1, a_2] = 1\rangle \qquad (6.26)$$

and is isomorphic to the group (6.25).

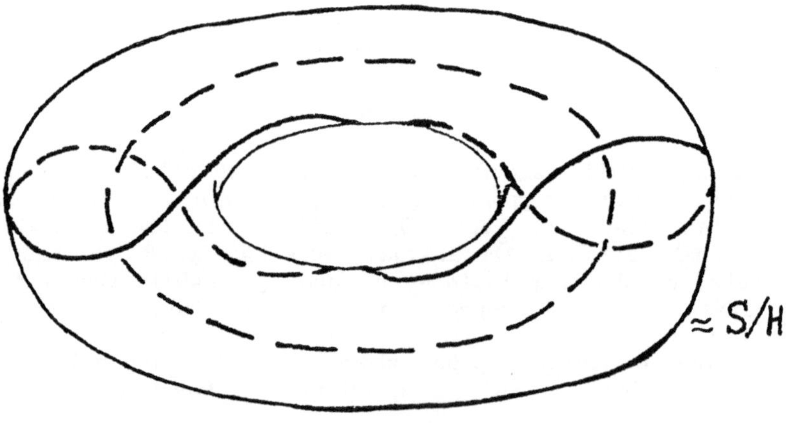

Figure 42

The natural projection $\pi : \Omega_0 \to \Omega_0/G$ specified a conformal structure in the manifold $M_0 = \Omega_0/G$, while the reflection in $\overline{\mathbb{R}}^3$ with respect to the plane S induces a conformal involution in M_0

(orientation reversing) $\varphi : M_0 \longrightarrow M_0$. The set of fixed points of the involution φ is the incompressible torus S/H in M_0, whose complement in M_0 splits up into two connected components homeomorphic to the Whitehead link complement in S^3 (Fig. 42), which shows that the manifold Ω_0/G is homeomorphic to the link complement shown in Fig. 43.

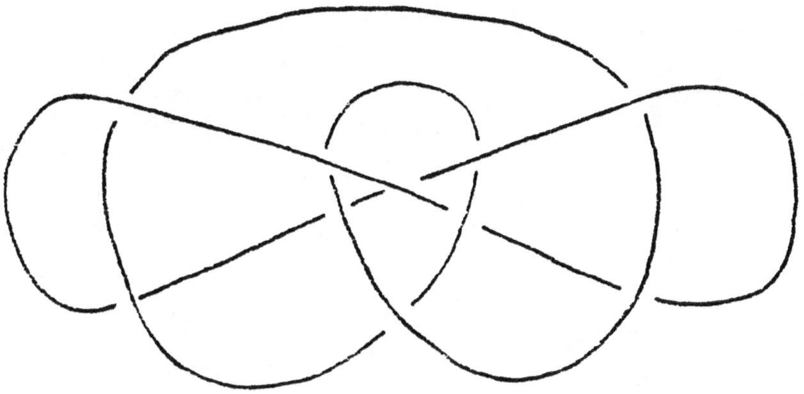

Figure 43

The existence of an incompressible torus S/H in Ω_0/G (which is not parallel to the boundary tori of ∂M_0) implies that $\pi_1(\Omega_0/G)$ contains a non-peripheral subgroup $\mathbb{Z} \oplus \mathbb{Z}$, so the manifold Ω_0/G is not hyperbolic (see Chapter 8 §3).

J.M. Montesinos pointed out that the manifold Ω_0/G is a fourfold branched covering of a torus link complement in S^3. This is evident from the scheme in Fig. 44 of two successive two-sheet coverings p_1 and p_2.

This shows that a torus sum of two copies of the Whitehead link complement (which is universal, see Chapter 7 §1) is a manifold that is not universal, namely as the complement to any link obtained by iteration from a torus link (Montesinos).

Then $\Omega(G)/G$ is the union of three $K(\pi, 1)$-manifolds, two of which are hyperbolic and universal (they are homeomorphic to the Whitehead link complement), while the third is homeomorphic to the link complement shown in Fig. 43 but neither hyperbolic nor universal.

In the group G, one can consider also an HNN-extension such that one identifies the parabolic subgroups that contain correspondingly the transformations g_3 and $\gamma g_3 \gamma^{-1}$. Topologically, this corresponds to the identification of two components of the boundary ∂M_0 of the manifold $\overline{M}_0 = \overline{\Omega_0/G}$. This identification further defines an embedding of an incompressible torus not homotopic to the torus S/H. The resulting closed manifold can be considered as the double $\hat{M}$ of the Whitehead link complement (see Chapter 8 §5 for more details).

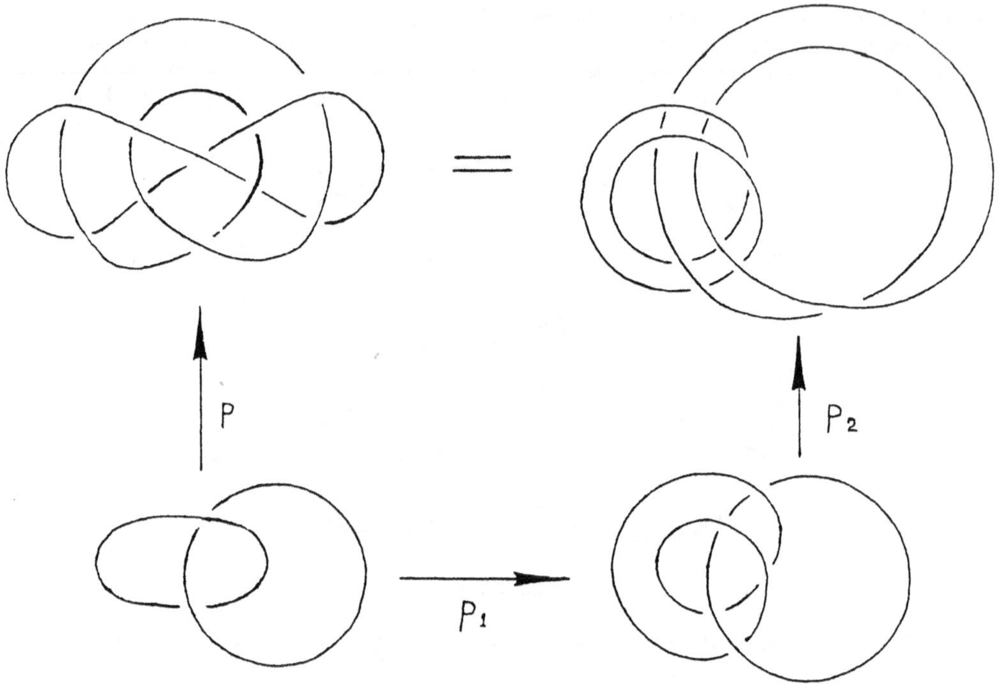

Figure 44

EXAMPLE 6.26. An example analogous to the previous is obtained if, as the subgroup G_1 in the Picard group, we take a group generated by the elements

$$g_1 = \begin{pmatrix} 1 & 4 \\ 0 & 1 \end{pmatrix}, \quad g_2 = \begin{pmatrix} 1 & 2i \\ 0 & 1 \end{pmatrix}, \quad g_3 = \begin{pmatrix} 1 & 0 \\ -1 & 1 \end{pmatrix}, \quad g_4 = \begin{pmatrix} 2+i & 2i \\ -1 & -i \end{pmatrix}$$

and having the presentation

$$G_1 = \langle g_1, g_2, g_3, g_4 : g_1g_2 = g_2g_1,\ g_4g_3g_4^{-1} = g_3g_1,\ g_4(g_2^{-1}g_3g_2)g_4^{-1} = g_2^{-1}g_3g_1g_2\rangle. \tag{6.27}$$

It can be shown (see Wielenberg [2]) that the intersection of the polyhedron

$$\{x \in \mathbb{R}^3 : -2 < x_1 < 2, \quad -1 < x_2 < 1\} \tag{6.28}$$

with the exterior of the isometric spheres $I(g_3)$, $I(g_3^{-1})$, $I(g_4)$, $I(g_4^{-1})$ and their images under the translations g_1 and g_2 gives the fundamental polyhedron $F(G_1)$ for the group G_1 (see Fig. 45), while

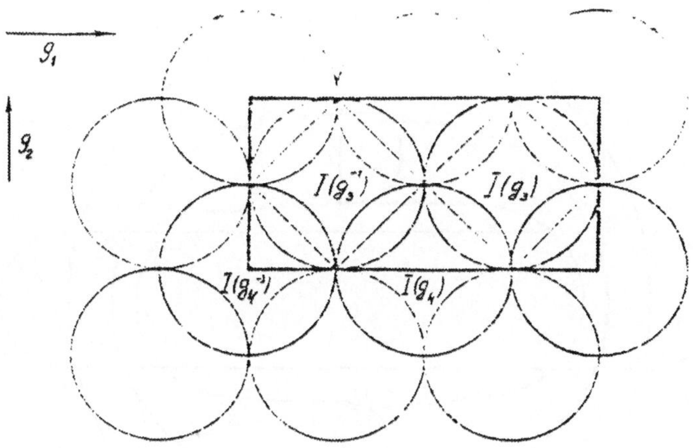

Figure 45

Figure 46

the manifold H^3/G_1 is homeomorphic to the Borromean rings complement (Fig. 46), whose fundamental group is isomorphic to (6.27) (with correspondence for the parabolic generators g_2, g_3, and g_4).

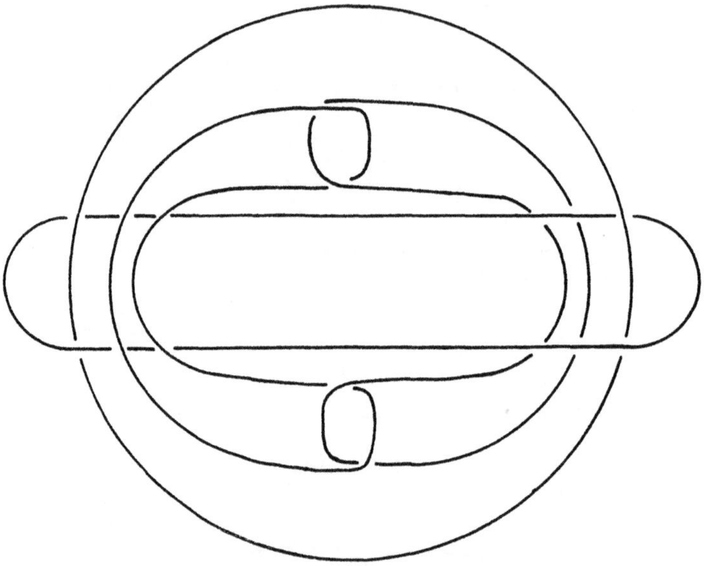

Figure 47

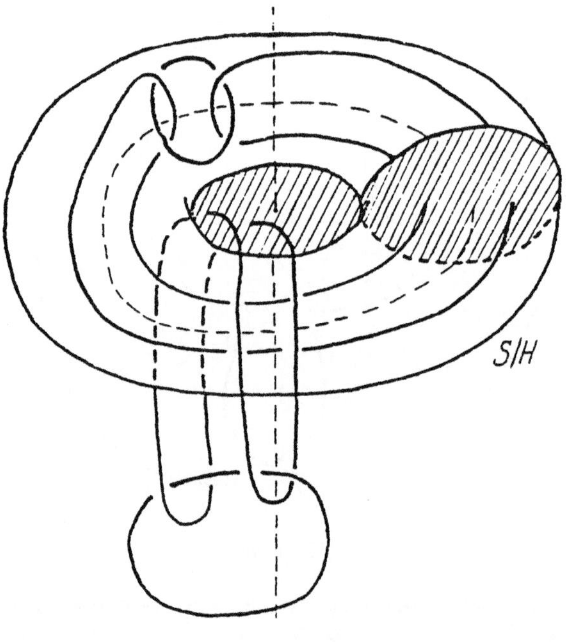

Figure 48

As in the previous example, the groups G_1 and $G_2 = \gamma G_1 \gamma^{-1}$ ($\gamma : x \longrightarrow x + 3e_3$) with the amalgamated subgroup $H = \langle g_1, g_2 : [g_1, g_2] = 1\rangle$ satisfy the Maskit first combination and the group $G = G_1 \underset{H}{*} G_2$ has an invariant contractible component Ω_0 and two classes of non-invariant components equivalent to the half-spaces D_1 and $D_2 = \mathbb{R}^3_-$. Here $\Omega(G)/G$ is the union of two hyperbolic manifolds (universal ones, see Hilden, Losano, and Montesinos [2]), which are homeomorphic to the Borromean rings complement and a conformal manifold $K(\pi, 1)$-type, that is homeomorphic to the link complement shown in Fig. 47. It is evident that the manifold Ω_0/G is not hyperbolic because reflection in the plane $S = \{x \in \mathbb{R}^3 : x_3 = 3/2\} \subset \Omega_0$ induces a conformal involution $\varphi : \Omega_0/G \longrightarrow \Omega_0/G$ (orientation reversing) with pointwise fixed incompressible embedded torus $\{x \in \Omega_0/G : \varphi(x) = x\} = S/H$, (Fig. 48). As the torus S/H is incompressible and is not parallel to the boundary tori $\partial(\Omega_0/G)$, $\pi_1(\Omega_0/G)$ contains a non-peripheral subgroup $\mathbb{Z} \oplus \mathbb{Z}$, so consequently the manifold Ω_0/G is not hyperbolic due to the hyperbolicity criterion in Chapter 8 §3.

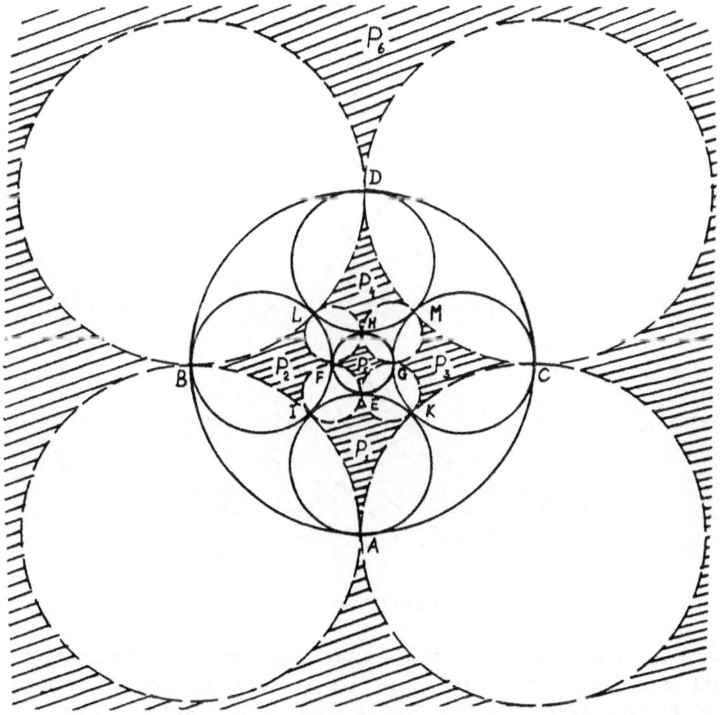

Figure 49

The group G can be extended to the group $\hat{G}$ by using the HNN-extension twice i.e., by identifying the parabolic subgroups containing correspondingly g_3 and g_4 with parabolic subgroups containing the elements $\gamma g_3\gamma^{-1}$ and $\gamma g_4\gamma^{-1}$. The resulting closed manifold $\hat{M} = \Omega_0(\hat{G})/\hat{G}$ will be the parabolic double for the Borromean rings complement (see Chapter 8 §5).

EXAMPLE 6.27. In the last example in this section we demonstrate the application of the second combination theorem.

We consider four circles in the boundary plane $\overline{\mathbb{C}} = \overline{\mathbb{R}}^2$ of the half-space $\mathbb{R}^3_+$:

$$C_1 = \{z \in \mathbb{C} : |z + i| = \sqrt{2}/2\}, \quad C_2 = \{z \in \mathbb{C} : |z + 1| = \sqrt{2}/2\},$$
$$C_3 = \{z \in \mathbb{C} : |z - 1| = \sqrt{2}/2\}, \quad C_4 = \{z \in \mathbb{C} : |z - i| = \sqrt{2}/2\}, \tag{6.29}$$

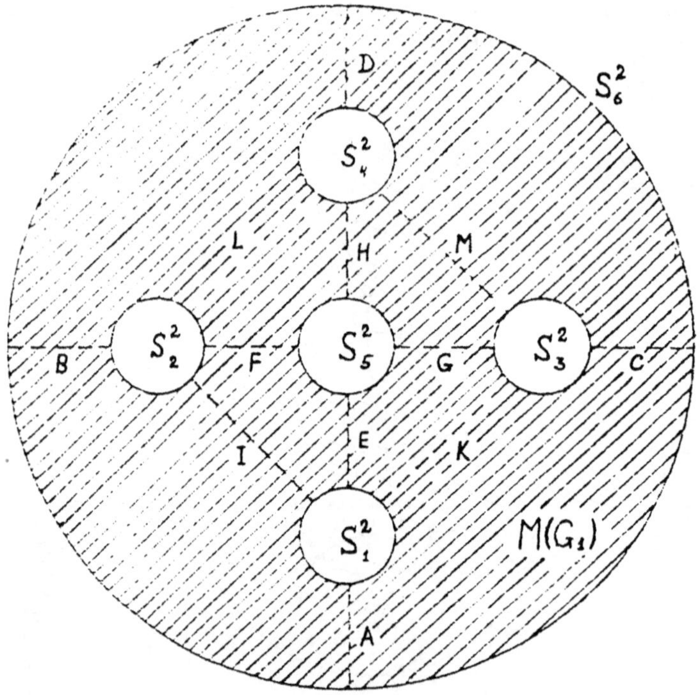

Figure 50

each of which is tangent to two of the others. Then in $\mathbb{R}^3$, we consider eight spheres σ_i, $i = 1, \ldots, 8$, having their centres in the plane $\mathbb{C}$ and intersecting it by circles orthogonal to the (6.29) circles, as shown in Fig. 49, where they are indicated by dashed lines, and with the C_i circle as solid lines. In each disc $\operatorname{int} C_i$, we consider a Fuchsian group H_i of the first kind consisting of compositions of an even number of inversions with respect to the spheres

σ_j intersecting C_1, and by G_1 we denote the group generated by those Fuchsian groups. From the first combination theorem, it follows that the group G_1 will be discontinuous in the plane. Here $\Omega(G_1)$ splits up into six classes of invariant components, which are spheres having four punctures each. If the fixed points of the parabolic elements are denoted by A, B, C, D, E, F, G, H, they correspond in the manifold $\mathbb{R}^3_+/G$ to the puntured solid cylinders. Figure 50 shows the scheme for $[\overline{\mathbb{R}}^3_+\backslash\Lambda(G_1)]/G_1$, in which $S_1^2, S_2^2, \ldots, S_6^2$ denote four-punctured spheres which are the components of the boundary $[\overline{\mathbb{C}}\backslash\Lambda(G_1)]G_1$ of that manifold.

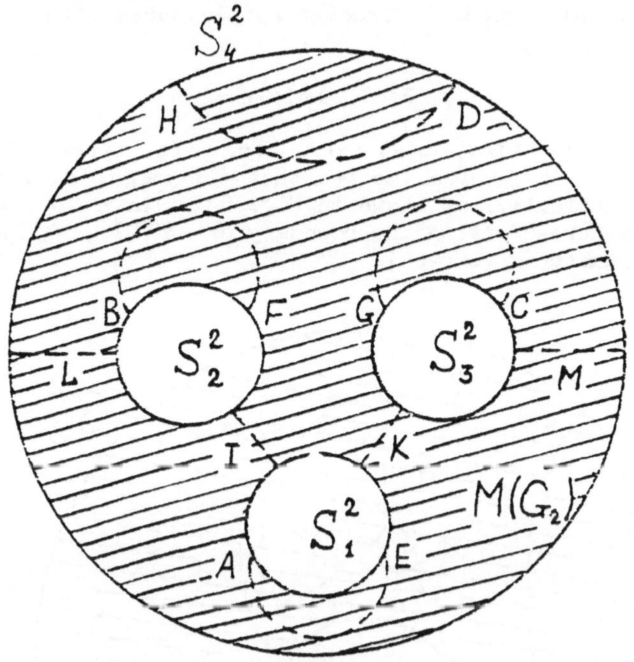

Figure 51

If we now consider these spheres S_5 and S_6 having their centres in $\mathbb{C}$ and passing correspondingly through the nearest (most remote) points to the origin on the circles C_1, while H_5 and H_6 denote the Fuchsian subgroups that leave them invariant, then they together with the mapping $f_{56}: x \longmapsto (2-\sqrt{2})^2 \cdot x/2$ satisfy the conditions of the second combination theorem, so the group $G_2 = \langle G_1, f_{56}\rangle$ is an HNN-extension of the group G by the element f_{56}, and is discontinuous, both in space and in the plane. Here $\Omega(G_2) \cap \overline{\mathbb{C}}$ is obtained from $\Omega(G_1) \cap \overline{\mathbb{C}}$ by rejecting the components corresponding to the punctured spheres S_5^2 and S_6^2. The manifold $[\overline{\mathbb{R}}^3_+\backslash\Lambda(G_2)]/G$ is obtained from the previous by identifying the boundary components S_5^2 and S_6^2 by the mapping $\tilde{f}_{56}$, and it therefore corresponds to the scheme given in Fig. 51. We continue this process and consider the spheres S_1 and S_4 orthogonally intersecting $\mathbb{C}$ on C_1 and C_4.

The Fuchsian groups H_1 and H_4 are conjugated by the hyperbolic transformation f_{14}, which transfers $\operatorname{ext} S_1$ into $\operatorname{int} S_4$. We thus have the conditions of the second combination theorem satisfied again, and therefore the HNN-extension of the group G_2 by the element f_{14}, i.e., the group $G_3 = \langle G_2, f_{14} \rangle$ is again discontinuous in $\overline{\mathbb{R}}^3$ and in the plane $\overline{\mathbb{C}}$. Its discontinuity set in $\overline{\mathbb{C}}$ consists of discs equivalent to $\operatorname{int} C_3$ and $\operatorname{int} C_2$ which universally cover the two boundary components $[\overline{\mathbb{R}}^3_+ \setminus \Lambda(G_3)]/G_3$. As there is symmetry in the action of the group G_3 with respect to the plane $H = \{x \in \mathbb{R}^3 : x_1 = 0\}$, we get hyperbolic symmetry for the manifold $M(G_3) = [\overline{\mathbb{R}}^3_+ \setminus \Lambda(G_3)]/G_3$ with respect to the totally geodesic incompressible torus with one puncture, $T_* = (H \cap \mathbb{R}^3_+)/\langle f_{56}, f_{14} \rangle$:

$$\varphi : \mathbb{R}^3_+/G_3 \longrightarrow \mathbb{R}^3_+/G_3.$$

The fixed set $\{x \in \mathbb{R}^3_+/G_3 : \varphi(x) = x\}$ for the hyperbolic isometry involution φ is T_*, which implies that the manifold $M(G_3)$ can be represented schematically, as shown in Fig. 52. Here S^2_2 and S^2_3 are four-punctured spheres forming the boundary of $M(G_3)$, while the dashed lined show the punctures (curves).

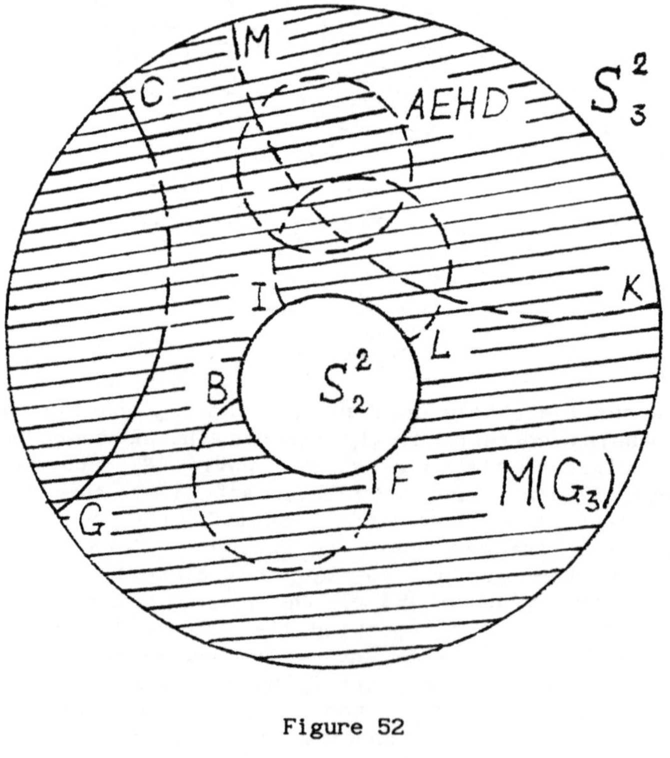

Figure 52

We continue the identification for the boundary components and consider the hyperbolic translation $f_{23} \in \text{Möb}(\mathbb{R}^3_+)$, whose restriction to $\partial\mathbb{R}^3_+ = \overline{\mathbb{C}}$ transforms $\text{ext}\, C_2$ into $\text{int}\, C_3$. Here the transformation f_{23} induces an isomorphism of the Fuchsian subgroups H_2 and H_3 in G_3, so one can apply Maskit's second combination theorem in the hyperbolic space $H^3 \cong \mathbb{R}^3_+$, which is an analogue of Theorem 6.24. Here, as the precisely H_i-invariant closed domains D_i, we must consider closed half-spaces in H^3 bounded by the planes S_i, $\overline{D}_i \cap \mathbb{C} = \overline{\text{int}_{\mathbb{C}}\, C_i}$, $i = 2, 3$. It is clear that $f_{23} : \text{int}\, D_2 \longrightarrow \text{ext}\, D_3$; similarly, the other conditions for Theorem 6.24 are satisfied.

Then the HNN-extension of G_3,

$$G_4 = \langle G_3, f_{23} \rangle$$

is the discrete isometry group in the hyperbolic space H^3, whose limit set is the whole sphere at infinity $\partial H^3 = \overline{\mathbb{C}}$ (the latter shows that this HNN-extension cannot be performed by applying Theorem 6.24 for groups in the plane $\overline{\mathbb{C}}$). The hyperbolic manifold H^3/G_4 has a finite volume and is homeomorphic to a three-component link complement in $\overline{\mathbb{R}}^3$.

REMARK 6.28. It is readily seen that the above HNN-extensions giving the group $G_4 \subset \text{Isom}\, H^3$:

$$G_4 = \langle G_1, f_{56}, f_{14}, f_{23} \rangle$$

are not unique, i.e., the mappings f_{56}, f_{14}, and f_{23} can be taken as loxodromic (i.e., as rotating the planes $S_i \subset H^3$ through the angles $\pi k/2$, $k = 1, 2, 3, 4$). The resulting complements in $\overline{\mathbb{R}}^3$ to the links $H^3/\tilde{G}_4$ will have the same hyperbolic volume as H^3/G_4, and they can be obtained from H^3/G_4 by corresponding twisting (see the analogous procedure in the proof of Theorem 8.23 for details).

REMARK 6.29. The groups G_2 and G_3 obtained in this example (just as $\tilde{G}_2$ and $\tilde{G}_3$) are web groups in the plane.

§4. Cayley graph and combinatorial structure of a discrete group

Let G be a finitely generated group and $\Sigma = \{g_1, \ldots, g_m\}$ be a set of generators. Then for the elements $g \in G$ we define the norm $|g|$ as the minimal length of a word in which g is written in terms of the elements Σ. This defines a left-invariant (word) metric $(*, *)$ in G, where $(a, b) = |a^{-1}b|$ for $a, b \in G$. This metric is dependent on the choice of the set Σ, but any two word metrics σ and σ' corresponding to the two generator sets Σ and Σ' are commensurable, i.e. the identity mapping $\text{id} : G \longrightarrow G$ is a quasi-isometry: $K \cdot \rho(a, b) \le \rho'(a, b) \le K' \cdot \rho(a, b)$, where $a, b \in G$ and the constants $K, K' > 0$ are dependent only on Σ and Σ'.

We define the Cayley graph $K(G, \Sigma)$ for the group G (see Cayley [1]) as an unoriented one-dimensional complex whose vertices a and b

correspond one-to-one to the elements a and b of the group G and are joined by an edge if $a = bg^{\pm 1}$ for some $g \in \Sigma$. To specify the metric in $K(G, \Sigma)$, we consider the decreasing summable function $f: \mathbb{N} \longrightarrow \mathbb{R}_+$ such that for any $k \in \mathbb{N}$ there exist $m, n \in \mathbb{N}$ for which $m \cdot f(r) \leq f(kr) \leq n \cdot f(r)$ (these conditions are satisfied by the function $f(r) = r^{-2}$, $f(0) = 1$). Then the length of an arc in that graph, having vertices a and b, is denoted by $\min\{f(a), f(b)\}$, while the distance $\langle a, b\rangle$ between any two vertices a and b in it is defined as the minimum of the length of the polygonal paths joining those vertices.

We shall see below that the Cayley graph is an internal tool for examining the combinatorial properties of the group G as well as for examining the action of the group G on the hyperbolic space H^n or on G-invariant subsets from $H^n \cup \partial H^n$.

1. *Visual boundary of the group G.* We denote by $\overline{K(G, \Sigma)}$ the Cauchy completion of the metric space $(K(G, \Sigma), \langle *, *\rangle)$; its elements are classes of convergent sequences of vertices in $K(G, \Sigma)$. Then the visual boundary $\partial G = \partial G(\Sigma, f)$ of the group G is the metric space $\overline{K(G, \Sigma)} \setminus K(G, \Sigma)$. Each point $W \in \partial G(\Sigma, f)$ can be represented as a Cauchy sequence $\{W_i\} \subset K(G, \Sigma)$, where $W_i \in G$. Such a sequence is called the shortest if $|W_i| = i$ and $(W_i, W_{i+1}) = 1$ for all $i \in \mathbb{N}$. Then the element W can be represented as an infinite word in the alphabet Σ, which at each finite step is written in the shortest form or, what is equivalent, as a shortest ("geodesic") path in $K(G, \Sigma)$. Writing each word W_i from the Cauchy sequence $\{W_i\}$ defining the element $W \in K(G, \Sigma)$ in the shortest form one can easily show by a Cantor diagonal construction that any element $W \in K(G, \Sigma)$ can be represented by a shortest Cauchy sequence.

EXAMPLE 6.30. Let $G = \mathbb{Z} * \mathbb{Z}$, $\Sigma = \{a, b\}$, and let $f(r) = r^{-2}$; then the Cayley graph $K(G, \Sigma)$ is a tree, and the boundary ∂G is homeomorphic to a Cantor discontinuum (just as for any free group of rank greater than one). Figure 53 shows a subset of $K(G, \Sigma)$ having vertices whose norm does not exceed four.

We get the following simple properties of the group boundary (Floyd [1]) directly from the properties of convergent sequences and the construction of the completion:

1. $\partial G(\Sigma, f)$ is a compact metric space.

2. If $H \xrightarrow{i} G \xrightarrow{j} J$ is an exact group sequence and H is a finite group, then ∂G and ∂J are Lipschitz-equivalent, i.e., there exists a homeomorphism $\varphi: \partial G \longrightarrow \partial J$ that is a quasi-isometry.

3. If $H \subset G$ is a finite-index subgroup, ∂h and ∂G are Lipschitz-equivalent.

The following theorem, due to Floyd [1], is less obvious:

LEMMA 6.31. *The boundary ∂G of a finitely generated Abelian group G of rank greater than one is a point.*

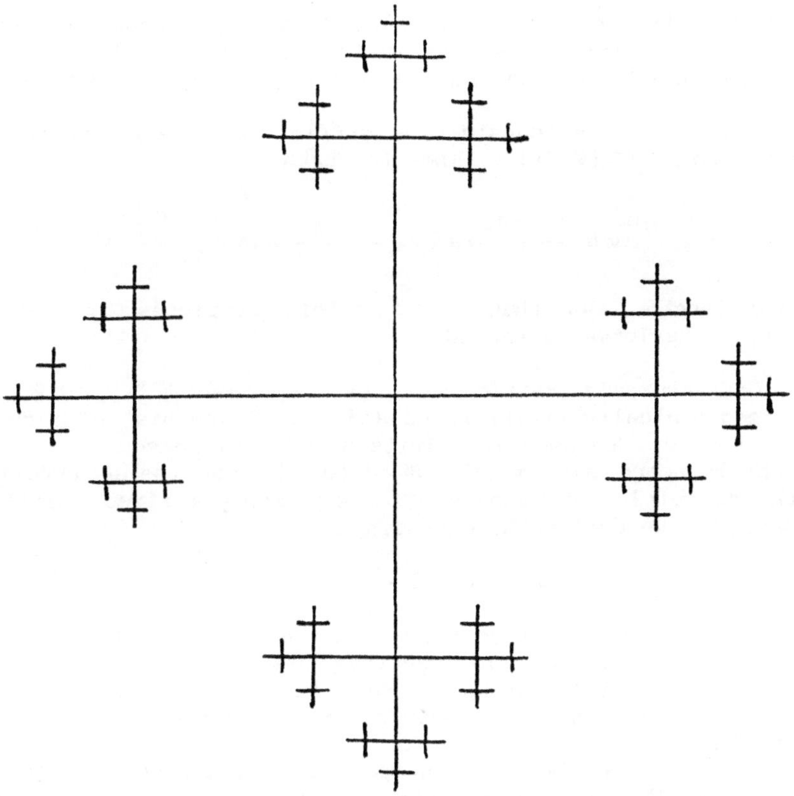

Figure 53

PROOF. It is sufficient to show that for any two infinite groups G and H, having finite generator sets Σ_1 and Σ_2 correspondingly, the boundary $\partial(G \oplus H)$ of their direct sum $G \oplus H$, having generator set $\Sigma_1 \cup \Sigma_2$, is a point. This will be demonstrated if, for any $\varepsilon > 0$, we can find a number $n \in \mathbb{N}$ and an element $W_n \in G \oplus H$ such that any element $g \in G \oplus H$, whose length $|g|$ is greater than n, lies in $K(G \oplus H, \Sigma_1 \cup \Sigma_2)$ at a distance less than ε from the element W_n.

We specify $W \in \partial G$ and $V \in \partial H$, which are defined by their shortest sequences $\{W_i\}$ and $\{V_i\}$. Let the number $n_1 \in \mathbb{N}$ be such that $\sum_{i=n_1}^{\infty} f(i) < \varepsilon/4$, and let $n = 2n_1$.

If the element $g = a \oplus b$ has length $|g| \geq n$, then for $|b| \geq n_1$ we consider a path in $K(G \oplus H, \Sigma_1 \cup \Sigma_2)$:

$$g = a \oplus b \xrightarrow{a^{-1}W_n} W_n \oplus b \xrightarrow{b^{-1}} W_n, \tag{6.30}$$

in which $\xrightarrow{a^{-1}W_n}$ denotes the path obtained by assigning the word

$a^{-1}W_n$ (written in the shortest form) and passing through all the edges corresponding to such an assignment in $K(G \oplus H, \Sigma_1 \cup \Sigma_2)$. It follows from the choice of n and n_1 that the path (6.30) has length less than ε.

If $|a| \geq n_1$, we consider a sufficiently large number $n_2 \in \mathbb{N}$ such that $f(n_2) < \varepsilon/4(W_n, a)$. Then the path

$$g = a \oplus b \xrightarrow{b^{-1}V_{n_2}} a \oplus V_{n_2} \xrightarrow{a^{-1}W_n} W_n \oplus V_{n_2} \xrightarrow{V_{n_2}^{-1}} W_n$$

again has length less than ε. This completes the proof that $<g, W_n> < \varepsilon$. The lemma is proved.

2. *Commensurable metrics.* Let $G \subset \operatorname{Isom} H^n$, $n \geq 2$, be a discrete geometrically finite group and let Σ consist of side-pairing elements for a fundamental finite-sided polyhedron $P = P(G) \subset H^n$. Taking the Poincaré ball model $B^n = B^n(0, 1)$ for the hyperbolic space H^n with the metric d and with 0 not being a fixed point for the group elements, we define the embedding

$$\varphi : K(G, \Sigma)) \hookrightarrow H^n. \tag{6.31}$$

That is, each vertex corresponding to an element $g \in G$ is mapped to $g(0)$; the edge joining the vertices $a, b \in G$ is mapped to the geodesic segment $[a(0), b(0)]_H$. Here we assume that the metric in $K(G, \Sigma)$ is either the word metric or is defined by means of the function $f(r) = r^{-2}$.

It is clear that φ is a proper G-equivariant embedding, i.e., the image $\varphi^{-1}(C)$ of a compactum $C \subseteq H^n$ is compact, $g \circ \varphi \circ g^{-1} = \varphi$ for any $g \in G$, and φ is injective at the set of vertices. Then the graphs $K(G, \Sigma)$ and $\varphi(K(G, \Sigma))$ can be identified with a graph dual to the tessellation of the hyperbolic space H^n by the images $g(P)$ of the fundamental polyhedron $P \subset H^n$.

THEOREM 6.32. *Let $G \subset \operatorname{Isom} H^n$, $n \geq 2$, be a geometrically finite discrete group without parabolic elements, while in the Cayley graph $K(G, \Sigma)$ we have defined the word metric $(*, *)$. Then the embedding φ is a pseudo-isometry, i.e., there exist constants $K > 0$ and $K' > 0$ such that*

$$K^{-1} \cdot (a, b) \leq d(\varphi(a), \varphi(b)) \leq K \cdot (a, b) \tag{6.32}$$

for all $a, b \in K(G, \Sigma)$ satisfying one of the conditions: $(a, b) \geq K'$ or $d(\varphi(a), \varphi(b)) \geq K'$.

PROOF. As the mapping φ is a proper one and G-equivarient, it is sufficient to show that there exist constants $K, K' > 0$ such that, for all $a \in G$, $|a| \geq K'$, the following inequalities apply (Floyd[1]):

$$K^{-1} \cdot |a| \leq d(0, \varphi(a)) \leq K \cdot |a|. \tag{6.33}$$

As $d(0, \varphi(a)) = d(0, a(0))$, the right-hand inequality in (6.33) follows from the triangle inequality and the finiteness of the set Σ. As K one can take the number

$$\max\{d(0, g(0)) : g \in \Sigma\}. \tag{6.34}$$

To prove the first inequality in (6.33), we consider the convex Nielsen hull $H_G \subset H^n$ for the limit set $\Lambda(G)$ (see Chapter 5 §2) on the assumption that $0 \in H_G$. Then from Theorem 5.39 and the condition that G lacks parabolic elements we have that the following polyhedron is compact:

$$P_h = P(G) \cap H_G.$$

Let $\alpha = \mathrm{diam}_H P_H$ and let $C = \max\{|a| : a \in G,\ d(0, a(0)) \le 7\alpha\}$; then for any $g \in G$ we decompose the geodesic segment $[0, g(0)]_H$ lying in H_G (by virtue of the convexity of H_G) into distinct intervals, whose lengths (apart from one short one) are 5α. As each end of any such interval lies at a distance not more than α from the orbit $G(0)$, the triangle inequality gives the bound

$$|g| \le C(1 + d(0, g(0))/\alpha). \tag{6.35}$$

If now $d(0, g(0)) \ge 1$, it follows from (6.35) that

$$|g| \le (C + C/\alpha)\cdot d(0, g(0)). \tag{6.36}$$

If on the other hand $|g| \ge C + C/\alpha$, it follows from (6.35) that $d(0, g(0)) \ge 1$, and therefore we have (6.36) applying again. This shows that the first inequality in (6.33) holds for the constants

$$K' = \max\{1, C + C/\alpha\} \qquad \text{and} \qquad K^{-1} = C + C/\alpha. \tag{6.37}$$

We now consider as our constant K the larger of the numbers (6.37) and (6.34), and we finally get the constants K and K' for which (6.33) and (6.32) apply. This completes the proof.

Theorem 6.32 extends a theorem due to Floyd [1] for geometrically finite subgroups in $PSL(2, \mathbb{C})$ and due to Cannon [1] for the co-compact groups $G \subset \mathrm{Isom}\, H^n$.[1] The inequalities (6.32) and (6.33) are incorrect if the group contains parabolic elements, but in that case one can use the scheme for the proof for the subgroups $PSL(2, \mathbb{C})$ due to Floyd [1] to show that the following bounds apply in that case:

---

[1] Actually we have in this way the proof of a more general fact: for a given space X with a path metric d and with isometric co-compact action of the group G, the embedding $\varphi : K(G, \Sigma) \hookrightarrow X$ is a pseudo-isometry, i.e. (6.32) holds.

THEOREM 6.33. *For a geometrically finite discrete group* $G \subset \operatorname{Isom} H^n$, $n \geq 2$, *there exist constants* K *and* K' *such that for all* $a \in G$. $|a| \geq K'$, *the following inequalities apply*:

$$2 \ln |a| - 2 \ln k \leq d(0, \varphi(a)) \leq K \cdot |a|. \tag{6.38}$$

PROOF. Let again $P \subset H^n$ be the fundamental polyhedron for the group G, which by virtue of the geometric finiteness of G has only a finite number of non-equivalent parabolic cusps $p_1, \ldots, p_m$ at its boundary $\partial P \cap \Lambda(G)$. We consider at these points p_i the horoballs $B_i = B_{p_i}$ having sufficiently small radii (see Remark 4.33) $\delta > 0$, which are precisely invariant with respect to the stabilizers $G_i = G_{p_i} \subset G$ and such that $G(B_i) \cap G(B_j) = \varnothing$ for $i \neq j$. This gives from Theorem 5.39 that the cut-off polyhedron

$$P_r = P_H \setminus \bigcup_{i=1}^{m} G(B_i) = P \cap (H_G \setminus \bigcup_{i=1}^{m} G(B_i))$$

is a compact set in H^n (but now not convex).

We define in the set $H_r = H_G \setminus \bigcup_{i=1}^{m} G(B_i)$ the metric $d_r(x, y)$ as the minimum of the hyperbolic lengths of the paths in H_r joining the points $x, y \in H_r$. It is clear that $d(x, y) \leq d_r(x, y)$ and the inequality becomes strict if the geodesic segment $[x, y]_H$ intersects some horoball $g(B_i)$, $g \in G$. We denote for $a \in G$ correspondingly $h(a) = d(0, a(0))$ and $h'(a) = d_r(0, a(0))$.

As the G-invariant set H_r decomposes into compact subdomains $g(P_r)$, $g \in G$, we obtain in the same way as in Theorem 6.32 that for certain constants $K, K' > 0$ and for all $a \in G$, $|a| \geq K'$, we have the inequalities (compare (6.33))

$$K^{-1} \cdot |a| \leq h'(a) \leq K \cdot |a|. \tag{6.39}$$

To compare $h(a)$ and $h'(a)$, i.e., to compare the metrics d and d_r, we conjugate the group G by hyperbolic isometry in such a way that H^n will be a half-space $\mathbb{R}^n_+$ having the length element $ds^2 = |dx|^2/x_n^2$. The points x and y lie at the intersection of the plane $\{x \in \mathbb{R}^n_+ : x_i = 0,\ i \neq 1, n\}$ with horosphere $\partial B_\infty = \{x \in \mathbb{R}^n_+ : x_n = 1\}$, with $x = (-\sqrt{t^2 - 1}, 0, \ldots, 0, 1)$, $y = (\sqrt{t^2 - 1}, 0, \ldots, 0, 1)$, in which t is the Euclidean radius of a geodesic arc in H^n joining those points. Then the path $[x, y]_e \subset \partial B_\infty$ in the horosphere ∂B_∞ has length $2\sqrt{t^2 - 1}$, while it is evident from Chapter 2 §3 or from Beardon [4] (7.2.6) in which $i = \sqrt{-1}$ and $x' = x_1 + ix_n$, $y' = y_1 + iy_n$,

$$d(x, y) = \ln\frac{(-t - y')(x' - t)}{(-t - x')(y' - t)} =$$

$$= \ln\frac{(-t - \sqrt{t^2 - 1} - i)(-\sqrt{t^2 - 1} + i - t)}{(-t + \sqrt{t^2 - 1} - i)(- \sqrt{t^2 - 1} + i - t)} =$$

$$= \ln[(t + \sqrt{t^2 - 1})/(t - \sqrt{t^2 - 1})] = 2\ln(\sqrt{t^2 - 1} + t).$$

Calculations imply that the length of the path $[x, y]_e \subset \partial B_\infty$ is less than $(\sqrt{t^2 - 1} + t) = e^{d(x,y)/2}$, and as $h'(a)$ is reduced by comparison with $h(a)$ only when the geodesic segment $[0, a(0)]_H$ intersects some horosphere $g(\partial B_i)$, $g \in G$, we see that

$$h'(a) \le e^{h(a)/2}.$$

Then it follows from (6.39) that with the above constants K and $K' > 0$ and for $a \in G$, $|a| \ge K'$, we have

$$K^{-1} \cdot |a| \le e^{h(a)/2},$$

which is equivalent to the first inequality in (6.38). The second inequality in (6.37) is given in (6.39) with the same constant $K > 0$. Theorem 6.33 is proved.

3. *Group boundary and the limit set.* We can now show, for the geometrically finite groups $G \subset \text{Möb}(n)$, $n \ge 1$, (repeating Floyd [1] the proof for $n = 2$) that the previously introduced boundary of the group G can be considered as an internal definition of the limit set $\Lambda(G)$. The following theorems apply:

THEOREM 6.34. *For a geometrically finite discrete group $G \subset \text{Möb}(n)$, $n \ge 1$, there exists a continuous G-equivariant surjective mapping $\Phi : \partial G \to \Lambda(G)$.*

THEOREM 6.35. *For a geometrically finite discrete group $G \subset \text{Möb}(n)$, $n \ge 1$, the mapping $\Phi : \partial G \to \Lambda(G)$ is a one-to-one mapping everywhere, apart from parabolic points of rank one, where it is a 2-1-valued one.*

We get two important corollaries directly from the theorem:

COROLLARY 6.36. *Let the geometrically finite group $G \subset \text{Möb}(n)$, $n \ge 1$, not contain maximal cyclic parbolic subgroups, and then $\Phi : \partial G \to \Lambda(G)$ is a G-equivariant homeomorphism.*

COROLLARY 6.37. *Let M be a closed hyperbolic n-dimensional manifold. Then there exists an $\pi_1(M)$-equivariant homeomorphism $\Phi : \partial(\pi_1(M)) \to S^{n-1}$ and therefore $\pi_1(M)$ is non-free.*

PROOF OF THEOREM 6.34. We assume again that H^n is the Poincaré ball model $B^n(0, 1)$ and that the constants K and K' are taken from Theorem 6.33.

Let $a \in G$ be some element written as a word of length n, with $|a| = n \geq K'$. If L is an edge of the graph $K(G, \Sigma)$ joining the vertices a and $b \in G$, in accordance with b, L has length either n^{-2} or $(n+1)^{-2}$.

If $r(a)$ is the Euclidean distance between the points 0 and $a(0)$, we have from (2.18) that

$$h(a) = \ln |(1 + r(a))/(1 - r(a))|, \quad r(a) = (e^{h(a)} - 1)/(e^{h(a)} + 1). \quad (6.39)$$

As for the embedding φ from (6.31), the hyperbolic length of a segment $\varphi(L) \subset H^n$ is bounded by the number $\max\{h(g) : g \in \Sigma\}$, there exists a constant $C > 0$ such that the Euclidean length $d_e(\varphi(L))$ of the segment $\varphi(L)$ is bounded by the number $C(1 - r(a)^2)$. Then from (6.39) and Theorem 6.33 we have that

$$d_e(\varphi(L)) \leq 4C(e^{h(a)} + 2 + e^{-h(a)})^{-1} \leq 4Ce^k n^{-2}.$$

This shows that $d_e(\varphi(L)) \leq 4C \cdot e^k \cdot$(the length of L), i.e., φ is a Lipschitz mapping. Therefore, φ induced a G-equivariant continuous mapping $\overline{\varphi} : \overline{K(G, \Sigma)} \longrightarrow \overline{\varphi(K(G, \Sigma)} \subset \overline{H}^n$. The restriction of the mapping $\overline{\varphi}$ to $\partial G = \overline{K(G, \Sigma)} \setminus K(G, \Sigma)$ gives the desired mapping $\Phi : \partial G \longrightarrow \Lambda(G)$. Surjectiveness for Φ follows from the fact that $\Lambda(G)$ is the set of accumulation points for the orbit $G(0)$; the G-equivariance of Φ is a consequence of the G-equivariance of φ. The proof is completed.

PROOF OF THEOREM 6.35. As neither ∂G nor $\Lambda(G)$ is altered by the transition to a finite-index subgroup, by virtue of Theorem 2.15 we can assume that G is torsion free. Also, Theorem 5.24 implies that $\Lambda(G)$ consists of approximation points and parabolic cusps, so we need to distinguish two cases: when $z = \Phi(W)$ for $W \in \partial G$ is an approximation point or $\Phi(W)$ is a parabolic cusp.

Let $z = \Phi(W)$ be an approximation point and let $\{W_i\} \subset G$ be the shortest sequence defining $W \in \partial G$. We show that there exists a subsequence in $\{W_i\}$ such that $W_i(0) = \varphi(W_i)$ converges to z (up to some finite distance) along a certain geodesic. For this purpose, in H^n we consider a path ν beginning at the point 0 and consisting of the union of the segments $[W_i(0), W_{i+1}(0)]_H$ and converging to the point z.

We see from (6.39) that, for any vertices $a, b \in \nu$, the distance between them along the path ν is commensurable with the distance $d_r(a, b)$, where d_r is the metric induced by the hyperbolic metric d in the set $H_r = H_G \setminus \bigcup_{i=1}^{m} G(B_i)$ (see the proof of Theorem 6.33). If now $\nu_i = [0, W_i(0)]_H$ are geodesic segments, as i increases, their ends converge to z, and therefore the number of vertices on the path ν between 0 and $W_i(0)$ and lying at bounded distances from ν_i increases infinitely with i. It follows from the fact that the length of a path in H^n increases exponentially by comparison with the distance by which that deviates from the geodesic, and also from the above commensurability of the length of the path ν and the distance

d_r. Thus we consider the subsequence of vertices $\{W_i(0)\}$ at bounded distances from the geodesic ray $[0, z)_H = \lim_{i\to\infty} \nu_i$ to get the required convergence of $W_i(0)$ to z along the ray $[0, z)_H$.

We now assume that $W' \in \partial G$ and $\Phi(W) = z = \Phi(W')$. We consider a subsequence in the shortest sequences $\{W_i\}$ and $\{W'_j\}$ for which $W_i(0)$ and $W'_j(0)$ converge to z (up to a finite distance) along the geodesic ray $[0, z)_H$. It is clear that then we can choose a sequence $\{a_i\} \subset G$ such that $\{a_i\}$ is contained in $\{W_i\} \cup \{W'_j\}$ and $a_i(0)$ converges to z along the ray $[0, z)_H$ and that for any arbitrarily large i_0

$$\{a_i : i \geq i_0\} \cap \{W_i\} \neq \emptyset, \quad \{a_i : i \geq i_0\} \cap \{W'_j\} \neq \emptyset. \tag{6.40}$$

Then the sequence $\{a_i\}$ is a Cauchy sequence in $K(G, \Sigma)$ and defines a certain point $\alpha \in \partial G$. It is evident from (6.40) that $W = \alpha = W'$, which shows that Φ is one-to-one on the inverse image of the approximation point set.

If $z = \Phi(W)$ is not an approximation point, i.e., a parbolic cusp for G, then, as above, let $\nu \subset H_r$ be a path constructed from the minimal sequence $\{W_i\}$ converging to z and (by the property of the parabolic point z) lying outside a certain horoball B_z having its centre at z.

Using the fact that the inclusion $i : G_z \hookrightarrow G$ is a quasi-isometry in the word metric $(*, *)$, we can show by analogy with the previous proof that $W \in \bar{i}(\partial G_z)$, where the embedding $\bar{i} : \partial G_z \longrightarrow \partial G$ is induced by the embedding i. As G_z is a free Abelian group having rank $k = k(z)$, $1 \leq k \leq n - 1$, Lemma 6.31 gives us that the boundary ∂G_z is a point for $\operatorname{rank} G_z > 1$. Therefore Φ is one-to-one on the inverse image of the set of parabolic cusps having rank greater than one.

If $\operatorname{rank} G_z = 1$, i.e., $G_z = \mathbb{Z}$, then $\partial G_z = \partial\mathbb{Z}$ consists of two points corresponding to the ends of $\mathbb{Z}$, and therefore in that case Φ is no more than two-to-one. To prove that here Φ is precisely two-to-one, it is sufficient to prove that the mapping $\bar{i} : \partial G_z \longrightarrow \partial G$ is injective. Let G_z be generated by an element g. Two points in ∂G_z are defined by two Cauchy sequences $\{g^n\}$ and $\{g^{-n}\}$. If we introduce into $\varphi(K(G, \Sigma))$ a metric specified by the Euclidean path length, the mapping φ becomes uniformly continuous. Then to prove that $\bar{i}$ is injective, it is sufficient to show that the distance between $g^m(0)$ and $g^{-m}(0)$ in that metric does not tend to zero. This follows clearly from the property of a parabolic cusp of rank one, since our situation is isometric to the one when $g^m(e_n)$ and $g^{-m}(e_n)$ tend to ∞, where $g : x \longrightarrow x + e_n$; $e_n = (0, \ldots, 0, 1) \in \mathbb{R}^n_+ \cong H$. The proof of Theorem 6.35 is completed.

4. *Applications to algorithmic problems.* By word problem (correspondingly, conjugation ones) in combinatorial group theory one means an algorithmic solution to whether, from any pair of words W and V in the group $G = \langle\Sigma; R\rangle$ one can say whether the elements they define in the group G are equal (or correspondingly conjugate) or not, where Σ and R are finite sets of generators and defining relations

in the group G. These problems were stated by Dehn [1], [2], [3], who solved them for the fundamental groups of closed hyperbolic surfaces. It is a well-known result of P.S. Novikov [1], [2] that there is a group $G = \langle\Sigma; R\rangle$ which does not have a solvable word problem, i.e., it is not possible to write computer programmes which carry out the kind of manipulations we need in general groups. However, we are studying groups which operate on some geometric space (they provide the central interest in mathematics). In practice, such groups have solvable problems (see Cannon-Holt-Epstein-Paterson-Thurston [1]). Often these solutions are based on the following Dehn algorithm which we describe for a surface group.

Let $p > 1$ and the group G take the form

$$G = \langle a_1, b_1, \dots, a_p, b_p : R \equiv [a_1, b_1] : . : [a_p, b_p] = 1\rangle. \tag{6.41}$$

Dehn proved that any non-empty word W in the generators a_i and $b_i \in \Sigma$ representing the identity element in G can be shortened by one of two processes:

1) delete a trivial word xx^{-1} from W, or

2) for some cyclic permutation $R' \equiv A \cdot B^{-1}$ of the relation R, with length $A >$ length B, replace the subword A in W by the word B.

Here the word W is called *Dehn-reduced* if it cannot be shortened by that process.

This algorithm also solves the conjugacy problem for the group (6.41). It reduces to considering the pair W and V of non-trivial Dehn-reduced words. That is, there exists an integer $k = 4p$ such that the cyclically reduced words W and V are conjugate in G if and only if there exist cyclic permutations W' and V' of W and V and a word u of length $|u| \le k$ such that $W' = uV'u^{-1}$ in G. In view of the solvability of the problem in the group (6.41) and the finiteness of the set of words u, $|u| \le k$, we have an answer on the conjugation of the words W and V.

In other words, Dehn's algorithm for the groups (6.41) can be described as the shortening of a non-empty word W representing the identity element in G by means of the relations xx^{-1}, having length two and the relations R' having length $K = 4p$.

A similar solution (Theorems 6.38 and 6.39) to the word and conjugacy problem applies for geometrically finite discrete groups G of isometries in the hyperbolic space H^n, $n \ge 3$ $(G \subset \text{Möb}(n-1))$ if those groups do not contain parabolic elements. Cannon [1] was first to note that such a solution can exist to co-compact discrete hyperbolic groups, whose proof can be rearranged to our case.

THEOREM 6.38. *Let $G \subset \text{Möb}(n)$, $n \ge 2$, be a geometrically finite discrete group free from parabolic elements; then there exists an integer $R > 0$ such that any non-empty word W in the generators $g_i \in \Sigma$ of the group G is shortened by means of the relations in G having length $\le R$.*

If the word W in the generators from the set Σ cannot be shortened as a result of the process from Theorem 6.38, we call it a R-reduced word.

THEOREM 6.39. *Let* $G \subset \text{Möb}(n)$, $n \geq 2$, *be a geometrically finite discrete group without parabolic elements and* $R > 0$ *an integer from Theorem 6.38. Then there exists an integer* $R' > 0$ *such that the cyclically R-reduced words* W *and* V *represent conjugate elements of the group* G *if and only if there exist cyclic permutations* W' *and* V' *of the words* W *and* V *and a word* u *of length* $|u| \leq R'$ *such that the words* $uV'u^{-1}$ *and* W' *represent the same element in* G.

The proof of Theorem 6.38 is based on the fact derived in Theorem 6.32, as is the proof of Theorem 6.39 (see Cannon [1], Theorems 5 and 6), namely that to the shortest words in a geometrically finite discrete group $G \subset \text{Möb}(n)$ the mapping $\varphi ; K(G, \Sigma) \longrightarrow H^{n+1}$ from (6.31) puts in correspondence the paths in H^{n+1} following along geodesics up to a finite distance. To detail this property, we give the following definition.

We specify the constants $K, K' > 0$. Then a path W in $K(G, \Sigma)$ is called a (K, K')-quasi-geodesic if the φ-image of any subpath $V \subset W$, $|V| \geq K'$, in the mapping $\varphi : K(G, \Sigma) \longrightarrow H^{n+1}$ has ends x_V and y_V having the distance

$$d(x_V, y_V) \geq K^{-1} \cdot |V|.$$

Then we get from the definition of quasi-geodesics in $K(G, \Sigma)$ and from the bound (6.32) in Theorem 6.32 not only that any shortest path in the Cayley graph $K(G, \Sigma)$ is a (K, K')-quasi-geodesic but also a more refined local property (see Cannon [1], Theorem 4):

> there exist positive constants K'', L, and L' (dependent on K and K') such that the (K, K')-quasi-geodesicity for any subpath V, $|V| \leq K''$, of the path W in $K(G, \Sigma)$ implies that W itself is (L, L')-quasi-geodesic, i.e., local quasi-geodesicity implies global.

We now assume that a certain word W, having length $|W| \geq R/2$, cannot be shortened by means of a relation having length $\leq R$, where we take R as larger than $2\max\{K'', L'\}$. This means precisely that each subword $V \subset W$, having length $|V| \leq R/2$ is a shortest one. Since then $K'' \leq R/2$, we get that W is K''-locally (K, K')-quasi-geodesic and consequently the word W with length greater than $R/2 > L'$ is (L, L')-quasi-geodesic. Therefore, the ends $\varphi(a)$ and $\varphi(b)$ of the path $\varphi(W) \subset H^{n+1}$, corresponding to the word W, are separated by the hyperbolic distance

$$d(\varphi(a), \varphi(b)) \geq L^{-1} \cdot (a, b) \geq L^{-1} \cdot L' > 0.$$

This shows that $a \neq b$ and therefore W is not the identity element of the group G.

We note finally in connection with a possible generalization of the small cancellations theory (see Lyndon and Schupp [1], Chapter 5), that the hypotheses there are obeyed globally for the class of geometrically finite discrete groups $G \subset \text{Möb}(n)$, $n \geq 1$, without parabolic elements, as Cannon [1] pointed out for co-compact groups. That is, the global combinatorial structure for such groups is obtained from the local structure by applying linear recursion, in the same way as the global structure for the integers $\mathbb{Z}$ is obtained from the local one by means of the simple linear recursion $n \longrightarrow n+1$ in $\mathbb{Z}$.

NOTES ADDED IN PROOF. The above ideas can also be used to give a solution to the word problem for all gemetrically finite groups of isometries of the hyperbolic space H^n, $n \geq 3$ (for the case $n \leq 3$ it is pointed out by Floyd-Hoare-Lyndon [1]).

Namely, we view this solution as an instance of the Todd-Coxeter coset enumeration algorithm (see Todd-Coxeter [1] and Johnson [1] for a good description), which we formulate as follows.

Let a group G have a presentation $\langle \Sigma; R \rangle$, with $G = F/N$, where F is a free group with basis Σ and N is the normal closure of the subset R in F. for any natural number n, let $R(n)$ be the set of all words in Σ of the form $w = uru^{-1}$ for $r \in R$ and $u \in F$ represented by a reduced word of length $|u| \leq n$. Let $F(n)$ be the set of all initial segments of words w in $R(n)$. Let $w \underset{n}{\sim} w'$ be the equivalence relation on $F(n)$ defined by setting $w_1w_2 \underset{n}{\sim} w_1w_0w_2$ whenever both are in $F(n)$ and w_0 is in $R(n)$.

Let f be any function $f : \mathbb{N} \longrightarrow \mathbb{N}$. The presentation $\langle \Sigma; R \rangle$ is said to be f-decidable if, given any element in F represented by a word w of length $|w| = m$, then $w \in N$ if and only if $w \underset{n}{\sim} 1$ for $n = f(m)$.

The Todd-Coxeter algorithm shows that every presentation is f-decidable for some function f, although, of course, for the general presentation it may not be possible to choose f to be recursive. Indeed, the word problem for G is decidable in the usual sense if and only if G has a presentation that is f-decidable for some recursive function f. If G is finite, it is possible to choose $f = \text{const}$. If the presentation satisfies the Dehn's algorithm (in the above extended sense - see Theorems 6.38 and 6.39), then f can be taken to coincide with the identity function $f(m) = m$ for all m greater than some M.

Now, in these terms, we have the following - see Floyd-Hoare-Lyndon [1]:

THEOREM 6.40. i) *Let G be a co-compact group of isometries of hyperbolic or Euclidean space of dimension $n \geq 2$, and let $\langle \Sigma; R \rangle$ be a presentation for G obtained from a Dirichlet polyhedron for G in the usual fashion. Then there is a natural number s such that $\langle \Sigma; R \rangle$ is f-decidable for the quadratic function $f(m) = sm^2$.*

(ii) *Let G be a discrete group of isometries of hyperbolic space H^n such that either $n \geq 2$ and H^n/G has finite volume or $n = 3$ and G is geometrically finite, and let $\langle \Sigma; R \rangle$ be a presentation for G obtained from a Dirichlet polyhedron in the usual fashion. Then there* *is a natural number* s *such that* $\langle\Sigma; R\rangle$ *is* f*-decidable for the function* $f(m) = [e^{sm}]$, *the greatest integer in* e^{sm}.

Note that this theorem is based on the finite presentation for the hyperbolic group G obtained from a Dirichlet polyhedron for G (which is finite-sided for the theorem's cases) and on estimates which are similar (6.32) and (6.38). We pointed out that application of the (finite-sided) pseudo-Dirichlet polyhedra (see Definition 3.14) instead of Dirichlet polyhedra in the proof of the above theorem, and Remark 5.34 after the proof of Theorem 5.24 give us the following solution to the word problem for all geometrically finite hyperbolic (or Möbius) groups (see Apanasov [34]):

Let G *be a geometrically finite group of isometries of the hyperbolic space* H^n, $n \geq 2$, *and let* $\langle\Sigma; R\rangle$ *be a presentation for* G *obtained from a pseudo-Dirichlet polyhedron for* G *in the usual fashion.*

Then there is a natural number s *such that* $\langle\Sigma; R\rangle$ *is* f*-decidable for the function* $f(m) = [e^{sm}]$, *the greatest integer in* e^{sm}.

Actually there are methods that should allow the word problem (and some other algorithmic problems) to be decided more quickly by means of automatic structures on groups defined by W. Thurston - see for good reference Epstein [1] and Cannon-Epstein-Holt-Paterson-Thurston [1]. In particular, negatively curved groups in the sense of M. Gromov [6] are automatic groups. In fact, the following is true.

THEOREM 6.41. *Any geometrically finite group* $G \subset \mathrm{Isom}\, H^n$, $n \geq 2$, *with a pseudo-Dirichlet presentation* $G = \langle G; R\rangle$ *has an automatic structure.*

This automatic structure is obtained by the following approach which is similar to that of Tatsuoka [1] for hyperbolic finite volume groups.

Namely, Theorems 5.39 and 7.9 show that a discrete group G of isometries in the hyperbolic space H^n (or a Möbius group) is geometrically finite if and only if there is a finite modulo the group G G-invariant family of distinct horoballs $B_p \subset H^n$ centred at cusp points $p \in \partial H^n$ such that the action of G in the space $H^n_r \subset H^n$,

$$H^n_r = H_G \setminus \bigcup_p \mathrm{int}\, B_p$$

is co-compact.

The convexity of the Nielsen hull H_G for $\Lambda(G)$ allows introducing in H^n_r the glancing geodesics, i.e., smooth broken geodesics in H^n_r. Here a broken geodesic between the points $x, y \in H^n_r$ can be described by a tuple of points $(x_1, x_2, \ldots, x_k)$ where

$$x_1 = x; \quad x_2, \ldots, x_{k-1} \in \partial H^n_r; \quad x_k = y$$

and $[x_1, x_2]$, $[x_2, x_3], \ldots, [x_{k-1}, x_k]$ are alternately hyperbolic geodesic arcs away from the horospheres and Euclidean geodesic arcs on the horospheres ∂B_p. The hyperbolic arcs of a glancing geodesic $(x_1, \ldots, x_k)$ are tangent to the horospheres at the endpoints, and the angle between adjoining hyperbolic arcs and Euclidean arcs is zero.

Given $x, y \in H^n_r$ there is a unique glancing geodesic $[x, y]$ between them together with a distance non-increasing retraction $r: H^n_r \longrightarrow [x, y]$ which contains two natural retractions in H^n. First, there is a "hyperbolic" (orthogonal) retraction onto hyperbolic geodesics and, second, a "horospherical" retraction onto horospheres.

Therefore (H^n_r, G) is a geodesic metric space with a distance function d_r and the isometric (co-compact) G-action. This gives us (see a remark for the proof of Theorem 6.32) that the Cayley graph $K(G, \Sigma)$ embedded in H^n_r by a pseudo-isometry. Then there exists $K > 0$ so that graph geodesics are K-geodesics in H^n_r, i.e., the paths $f: [a, b] \longrightarrow H^n_r$ such that for every subinterval $[a', b'] \subset [a, b]$ the length of $f([a', b'])$ is at most $K \cdot d_r(f(a'), f(b'))$. Moreover, using Tatsuoka's [1, Theorem 4.5] proof, we have for any K'-locally s-horospherical K-geodesic $f: [a, b] \longrightarrow H^n_r$, i.e., a path such that:

(i) for every subinterval $[a', b']$ with length $|f[a', b']| \leq K'$, the subpath $f|_{[a', b']}$ is a K-geodesic.

(ii) for every subinterval $[a', b']$ and every horosphere $H \subset \partial H^n_r$ with $f([a', b']) \subset \{x \in H^n_r : d(x, H) \leq s\} = N(H, s)$, the subpath $f|_{[a', b']}$ is a K-geodesic,

that there exists $L > 0$ such that f is an L-geodesic.

This observation gives us the following result allowing us to describe an algorithm for the automatic structure on G from Theorem 6.41 (cf. Tatsuoka [1, Theorem 5.1]:

THEOREM 6.42. *Let $G \subset \mathrm{Isom}\, H^n$ be a geometrically finite group. There exists a finite subgraph Γ of $K(G, \Sigma)$ and a finite collection $\{\Gamma_i\}$ of essentially $(n-1)$-planar subgraphs of $K(G, \Sigma)$ such that, if w is a word in G satisfying the conditions:*

(i) *w is a geodesic in Γ,*

(ii) *w is a geodesic in Γ_i,*

then $w \neq 1$.

§5. Geometric and algebraic convergence of groups

The limit of a discrete-group sequence is important in various aspects of the Kleinian-group theory, in particular in research on group deformation space. However, the limit may not be understood in the same sense. Troels Jørgensen was the first to analyse and understand this phenomenon (see Thurston [1], Chapter 9.1).

We say that a sequence $\{G_i\}$ of closed and, in particular, discrete subgroups in $\mathrm{Möb}(n)$ converges geometrically to a group $G \subset \mathrm{Möb}(n)$ if:

1) each element $g \in G$ is the limit of a sequence g_i, $g_i \in G_i$ and

2) the limit of any converging subsequence $\{g_{i_j}\}$, $g_{i_j} \in G_{i_j}$, is an element of the group G.

The geometric limit is automatically a closed subgroup in Möb(n) (compare Theorem 1.14). Geometric convergence of discrete hyperbolic-isometry (co-finite) groups corresponds to convergence in the Jørgensen sense (see Chapter 8 §3) of complete hyperbolic finite-volume manifolds. That is, such a manifold M is represented as $M_{(0,\varepsilon]} \cup M_{[\varepsilon,\infty)} = M$, in which $M_{(0,\varepsilon]}$ consists of those points through which passes a non-trivial loop having length $\leq \varepsilon$, while $M_{[\varepsilon,\infty)}$ contains points through which pass only non-trivial loops whose lengths are greater than ε (see Chapter 4 §4). If now M and N are two such manifolds, they are called close if, for a sufficiently small $\varepsilon > 0$, there exists a diffeomorphism approximated by the isometric mappings $M_{[\varepsilon,\infty)} \longrightarrow N_{[\varepsilon,\infty)}$.

Now let Γ be an abstract group and $\rho_i : \Gamma \longrightarrow$ Möb(n) be a representation sequence for Γ in Möb(n). One says that the sequence $\{\rho_i\}$ converges algebraically if, for any $\gamma \in \Gamma$, the sequence $\{\rho_i(\gamma)\}$ converges. The limit representation $\rho : \Gamma \longrightarrow$ Möb(n) is called the *algebraic limit* for the representation $\{\rho_i\}$.

This definition of convergence corresponds to the usual definition of group convergence by their generators.

Finally, let Γ be a countable group and $\{\rho_i\}$ the sequence of representations for it in the group Möb(n) with discrete images $\rho_i\Gamma$. The sequence $\{\rho_i\}$ converges strongly to a representation ρ if ρ is the algebraic limit of $\{\rho_i\}$ and the group $\rho\Gamma$ is the geometric limit of the sequence of groups $\{\rho_i\Gamma\}$.

It is clear that the geometric limit is greater than the algebraic one, i.e., if the representations ρ_i converge algebraically to ρ and the $\rho_i\Gamma$ converge geometrically to G, then $\rho\Gamma = G$. One can have $G \neq \rho\Gamma$, i.e., a difference between the algebraic and geometric convergences. This is evident from the following example concerned with the convergence of cyclic discrete groups $G_i \subset \mathrm{Isom}(\mathbb{R}^3)$.

EXAMPLE 6.42. We define the representations ρ_i of the cyclic group $\mathbb{Z}$ in the group of Euclidean isometries of $\mathbb{R}^3$. We put $x^i = (i, 0, 0)$ and by l_i we denote a straight line parallel to the third coordinate axis and passing through the point x^i. Then, as the generator of the group $\rho_i\mathbb{Z}$, we take the screw translation g_i along the straight line l_i that transfers the origin to a point having the second coordinate $(g_i(0), e_2) = 1$, and such that for some $n_i \in \mathbb{Z}$, $g^{n_i}(0) = e_3$.

The algebraic limit of the groups $\rho_i\mathbb{Z}$ is the cyclic group generated by translation by the vector e_2. At the same time, the groups $\rho_i\mathbb{Z}$ converge geometrically to the Abelian group of rank two generated by translations by the vectors e_2 and e_3.

It is clear that this example can be generalized in such a way that the algebraic limit $\rho\mathbb{Z} = \langle x \longmapsto x + e_2 \rangle$ is not changed but there is no geometric limit or the geometric limit is isomorphic to $\mathbb{Z} \times \mathbb{R}$, or else is an Abelian group of rank three, etc.

Also to obtain an analogous effect, the generators for the groups $\rho_i\mathbb{Z}$ can be taken as a loxodromic transformation (equivalent to a hyperbolic translation in H^3, see Thurston [1], §9.1.4). Then the algebraic limit to $\rho\mathbb{Z}$ will again be a cyclic group, while the geometric limit is a parabolic group of rank two. T. Jørgensen showed that it is possible to iterate this construction and produce examples as above where the algebraic limit is the fundamental group of the punctured torus, but the geometric limit is not even finitely generated (in fact, it is an infinite, free amalgamated product with each factor a free group of rank two).

THEOREM 6.43 (Thurston [1]). *For any Lie group $\mathcal{Y}$ (and in particular for $\mathcal{Y}=\text{Möb}(n)$), the space of closed subgroups in $\mathcal{Y}$ is compact (in geometric convergence topology).*

PROOF. We consider any sequence $\{G_i\}$ of closed subgroups in $\mathcal{Y}$, and first let there exist a lower bound for the distance of the elements $g\in G_i$ from the identity element e. Then, for any $r>0$, there exists an upper bound to a number of elements from G_i in the ball having radius $r>0$. We apply Tikhonov's theorem on the product of bicompacta to get that a sequence $\{G_{i_j}\}$ exists that converges geometrically to a discrete group.

In general, let $T_e(\mathcal{Y})$ be a tangent space for the group $\mathcal{Y}$ in the identity element e and let $S\subset T_e(\mathcal{Y})$ be a maximal subspace having the following property: for any $\varepsilon>0$, there exists a group G_i whose small elements correspond to the directions in S that cover the whole of S up to an angle ε.

The definition of S implies that S is closed under the Lie product. Further, if we take a sequence $\{G_{i_j}\}$ of groups whose small elements define tangents that cover the whole of S, then all the small elements of the groups G_{i_j} define tangents close to S. This shows, as in the previous case, that there exists a group sequence converging to a closed subgroup $G\subset\mathcal{Y}$ whose tangent space in the identity element e is the subspace S.

COROLLARY 6.44. *Let Γ be a countable group and $\rho_i:\Gamma\longrightarrow\text{Möb}(n)$ be discrete representations that converge algebraically to the representation ρ. If there is no finite-index Abelian subgroup in $\rho\Gamma$, then there exists a subsequence $\rho_{i_j}\Gamma$ that converges geometrically to a discrete group $G\subset\text{Möb}(n)$. Here $\rho\Gamma\subset G$ and, in particular, $\rho\Gamma$ is a discrete group.*

PROOF. This follows (Thurston [1]) from the fact that in a closed group $G=\lim_{j\to\infty}\rho_{i_j}\Gamma$ (which exists due to Theorem 6.43), the connected component containing the identity element should be Abelian. This follows directly from Margulis's lemma (Theorem 4.20). But as $\rho\Gamma\subset G$, and as $\rho\Gamma$ contains no finite-index Abelian subgroup, a connected component in the group G containing e is trivial, so consequently the groups G and $\rho\Gamma\subset G$ are discrete.

The converse of this assertion is incorrect. This is demonstrated by an example of any discrete group allowing an infinite order automorphism φ: the sequence of representations $\rho_i = \varphi^i$ has no algebraically convergent subsequence.

Thurston's [1] lamination technique on Riemann surfaces has given strong results on the limits of Kleinian groups in the plane and the related limits of hyperbolic three-manifolds. Thurston and Jørgensen observed in particular that the limit set of the algebraic limit $\rho\Gamma$ cannot increase sharply by comparison with the limit sets $\Lambda(\rho_i\Gamma)$ of the near groups $\rho_i\Gamma$, but at the same time it may be very decreased. This remark is based on the following observations.

If G is the geometric limit of the sequence G_i, then each point $x \in \Lambda(G)$ is a limit of the limit points $x_i \in \Lambda(G_i)$. In particular, for fixed points x of the elements $g \in G$ (such x are dense in $\Lambda(G)$, see Lemma 3.22), in G_i there are elements g_i having fixed points x_i close to x. Similar arguments apply for the algebraic limits ρG of the representation sequence $\rho_i G$. This shows that the limit set cannot be increased in the limit.

However, if $G \subset \text{Möb}(2)$ is any finitely generated group, then there exists a sequence $G_i \subset G$ of finite-index subgroups that converges geometrically to the trivial group $\{id\}$. Then $\Lambda(G_i) = \Lambda(G)$ is constant, but in the limit $\Lambda(\{id\}) = \varnothing$.

COROLLARY 6.45. *Let* $\Gamma \subset \text{Möb}(n)$, $n \geq 2$, *be a discrete non-elementary group without torsion and let the discrete representation sequence* $\rho_i : \Gamma \to \text{Möb}(n)$ *converge algebraically to the representation* ρ. *Then* $\rho\Gamma$ *is a discrete non-elementary group without torsion.*

The proof of the assertion follows directly from Corollary 6.44. Wielenberg [1] has given a direct proof.

We complete this section with a theorem (Apanasov [5]) on rank preservation for the algebraic limit of discrete groups, which has already been proved in the proof of Theorem 4.16 (about minimal sizes of fundamental polyhedra).

THEOREM 6.46. *Let a sequence of discrete groups* $G_m = \langle g_{1,m}, \ldots, g_{k,m} \rangle \subset \text{Möb}(n)$, $n \geq 2$, $2 \leq k \leq \infty$, *be such that all the* $g_{i,m}$ *have no common fixed points and are not elliptic. If* $G = \langle g_1, \ldots, g_k \rangle$ *is the algebraic limit of the groups* G_m, *then all the generators* $g_i = \lim_{m \to \infty} g_{i,m}$ *are different from the identity one.*

ADDITION. We briefly described major results on group convergence obtained in a recent work by Jørgensen and Marden [3]. The central technique there is the use of the geometry of "generic" fundamental Dirichlet polyhedra in H^3, which is stable under a change in the polyhedron centre, as Jørgensen and Marden [2] have shown. This involves the following new concept, which explains the properties of algebraic and geometric convergence.

The sequence of discrete groups $G_m \subset \text{Möb}(n)$ converges polyhedrally to the group H if H is discrete and, for some point $y \in H^{n+1}$, the Dirichlet polyhedra $D_y(G_m)$ converge to the polyhedron $D_y(H)$ uniformly on compacta from H^{n+1}. More precisely, we denote for any sufficiently large $r > 0$ by $D_{r,y}(G)$ the truncated polyhedron

$D_y(G) \cap B(y, r)$ where $B(y, r) \subset H^{n+1}$ is the ball of radius $r > 0$ centred at y. As its pairwise G-equivalent sides, we take the sides, which may be truncated, of the polyhedron $D_y(G)$. Then polyhedral convergence means that for $r > 0$ there exists a $N = N(r)$ such that:

a) for each side pairing element h of the polyhedron $D_{y,r}(H)$ and for all $m \geq N$ there exist side pairing elements g_m of the polyhedra $D_{y,r}(G_m)$ such that $\lim g_m = h$;

b) if g_m is a side pairing transformation for $D_{y,r}(G_m)$ then the limit h of any convergent subsequence of $\{g_m\}$ is a side, edge, or vertex pairing transformation of the polyhedron $D_{y,r}(H)$ (in particular, $h \neq \mathrm{id}$ - compare Theorem 6.46).

It can be shown that for any infinite sequence of discrete groups $G_m \subset \text{Möb}(n)$ there exists a sequence $\{A_m\} \subset \text{Möb}(n)$ such that the sequence $\{G_m^* = A_m G_m A_m^{-1}\}$ contains the polyhedrally convergent subsequence $\{G^*_{m_k}\}$, where the proof given by Jørgensen and Marden [3] can be applied on the basis of Theorem 4.16 on the universal size for fundamental polyhedra in any dimension. Here, if $\{G_m\}$ converges algebraically to the group G, the limit H of any such subsequence $G^*_{m_k}$ contains G.

Further, if $H \subset \text{Möb}(2)$ is finitely generated, then, for all large k, there exist homomorphisms ψ_k of the group H into the groups G_k such that for all $h \in H$ we have $\lim\limits_{k\to\infty} \psi_k(h) = h$. If, in addition, G is finitely generated, then $\varphi_k(H) = G_k$.

We note, however, that polyhedral convergence does not in general imply algebraic convergence, even for Fuchsian groups. The situation is different with geometric convergence:

I. *The sequence* $G_m \subset \text{Möb}(2)$ *of discrete groups converges geometrically to a discrete group if and only if* $\{G_m\}$ *converges polyhedrally to a discrete group, and then the geometric and polyhedral limits of* $\{G_m\}$ *are the same.*

II. *Let* $\varphi_m : \Gamma \longrightarrow G_m \subset \text{Möb}(2)$ *be a sequence of isomorphisms of a group* Γ *onto discrete groups* $G_m = \varphi_m(\Gamma)$ *converging algebraically to* φ, $\varphi : \Gamma \longrightarrow G = \varphi(\Gamma) \subset \text{Möb}(2)$. *Then* $\{G_m\}$ *also converges geometrically to* G *if and only if for every convergent subsequence the limit* $h = \lim \varphi_{m_k}(\gamma_{m_k})$ *belongs to* G *and* $\gamma_{m_k} = \varphi^{-1}(h)$ *for all large* k.

In particular, a sequence of cyclic loxodromic groups $\langle T_m \rangle$ converges polyhedrally to a discrete parabolic group Π if and only if it converges geometrically to Π and no subsequence of distinct elements converges to the identity.

This relation between algebraic and geometric convergence on the one hand and polyhedral convergence on the other can be employed with characteristic fundamental Dirichlet polyhedra, as by Jørgensen and Marden [3], who examined the relationship beteen group convergence and convergence in the sense of Caratheodory for the discontinuity sets of those groups, where that convergence is equivalent to convergence of the limit sets for groups in Hausdorff topology, see (5.31).

More precisely, an open-set sequence Ω_m converges in the sense of Caratheodory to the open set Ω if:

a) each compactum $K \subset \Omega$ lies in Ω_m for all sufficiently large m and

b) each open set U lying in Ω_m for an infinite subsequence of $\{m\}$ also lies in Ω.

In our situation, where $\Omega_m = \Omega(G_m)$, Ω are discontinuity sets for the discrete groups G_m and G^* such that each element $g \in G^*$ is the limit for certain elements $g_m \in G_m$: $g = \lim g_m$, condition b) is fulfilled automatically. Therefore, the discontinuity sets Ω_m converge to Ω in the sense of Caratheodory if and only if any compact $K \subset \Omega$ lies in all Ω_m for $m \geq m(K)$.

The following are two basic results:

CI. *Let $\varphi_m : \Gamma \longrightarrow G_m$ be isomorphisms of a finitely generated group Γ onto discrete groups $G_m \subset$ Möb(2) that converge algebraically to an isomorphism $\varphi : \Gamma \longrightarrow G$, in which G is a geometrically finite group having a non-empty discontinuity set $\Omega(G)$.*

Then $\{G_m\}$ converges geometrically to G if and only if $\Omega(G_m)$ converge to $\Omega(G)$ in the sense of Caratheodory.

CII. *Let Γ be a finitely generated torsion free group and $\theta_m : \Gamma \longrightarrow G_m$ be its isomorphisms onto discrete groups $G_m \subset$ Möb(2), which converge algebraically to $\theta : \Gamma \longrightarrow G$. Also, let $\{G_m\}$ converge geometrically to a geometrically finite group H. Then:*

1) *$\Omega(G_m)$ converge to $\Omega(H)$ in the sense of Caratheodory.*

2) *For all sufficiently large m, there exist homomorphisms $\psi_m : H \longrightarrow G_m$ such that $h = \lim \psi_m(h)$ for all $h \in H$ and $\psi_m(g) = \theta_m\theta^{-1}(g)$ for all $g \in G$.*

3) *Let $H_1, \dots, H_q$ be all rank two parabolic subgroups non-conjugate in H for which $\psi_m(H_k)$ are cyclic loxodromic groups. We denote by T^*_{km} a generator of the kernel* Ker ψ_m *for the homomorphism $\psi_m : H_k \longrightarrow \psi_m(H_k)$.*

Then Ker ψ_m *is the normal closure of the subgroup of H generated by the elements T^*_{km}, $1 \leq k \leq q$.*

4) *Assume, in addition, that each parabolic subgroup H_k contains an element of G. Then for some $T_m \in H_m \backslash G$ we have*

$$H = \langle G, T_1, \dots, T_m \rangle$$

REMARK. In 3) and 4) we allow the possibility that there are no such parabolic subgroups $H_k \subset H$. Then this result asserts that this is the case if and only if $H = G$.

§6. The limit set: Sullivan's ergodic theorems

In this section we continue the study of the action of a discrete group $G \subset$ Möb$(n-1)$ (acting isometrically in the hyperbolic space $H^n \cong \mathbb{R}^n_+$) on the limit set $\Lambda(G) \subset \partial H^n = \overline{\mathbb{R}}^{n-1} \equiv S^{n-1}$, which was begun in Chapter 3 §5. The results given below were obtained by Dennis Sullivan [1].

DEFINITION 6.47. The action of a discrete group $G \subset \text{Möb}(n-1)$ on the sphere $\partial H^n = S^{n-1}$ is split into two parts up to a set having measure zero: dissipative and conservative. The dissipative part is the union of pairwise non-intersecting measurable sets rearranged by elements of the group (the analogue of the action of G in H^n). The conservative part $K \subset S^{n-1}$ is characterized by the following: given any subset $X \subset K$, $\mu_{n-1}(X) > 0$, there exists a sequence of distinct elements $\{g_m\} \subset G$ such that the spherical $(n-1)$-dimensional Lebesgue measure $\mu_{n-1}(X \cap g_m(X)) > 0$ is non-zero for all $m = 1., 2, \ldots$.

DEFINITION 6.48. The group $G \subset \text{Möb}(n-1)$ acts on $S^{n-1} \times S^{n-1}$ (or what is the same, on the set of geodesics in H^n) ergodically if and only if for any measurable function φ in $S^{n-1} \times S^{n-1}$ its G-invariantness ($\varphi(x, y) = \varphi(g(x), g(y))$ for all $g \in G$) implies that φ is constant almost everywhere.

1. *Infinite solid angle of the orbit and ergodicity.* As has been shown in Chapter 1, any transformation $g \subset \text{Möb}(n-1)$ can be represented as a superposition of a hyperbolic translation in the ball H^n and a rotation of this (in the ball model for the hyperbolic space). We assume that in the sphere at infinity $\partial H^n = S^{n-1}$ we have the spherical metric $dx/(1 + |x|^2)$ and denote by $d_g(x)$ the linear derivative of a transformation $g|_{S^{n-1}}$ in that metric; it varies in an interval $[1/\lambda, \lambda]$, with $\lambda \geq 1$ (cf. Chapter 9 §2).

The discrete groups G form two classes: for the first of them, the series $\sum_G \lambda^{1-n}$ converges, but that is not so for the second. The condition is known as the finiteness or infiniteness condition for the solid angle. The meaning is evident from the following. In the ball H^n for a fixed point x and a point y, $y \neq x$, let $\theta(x, y)$ be the reciprocal of the area of the sphere $S^{n-1}(x, |x-y|)$ passing through y with centre x. It is clear that $\theta(x, y)$ is the density of the solid angle of one point, as viewed from the other.

LEMMA 6.49. *Given a discrete group* $G \subset \text{Möb}(n-1)$, $\sum_G \lambda^{1-n} = \infty$ *if and only if the total solid angle of our orbit* $\sum_{g \in G} \theta(x, g(y))$, *viewed from any point* $x \in H^n$ *not on the orbit* $G(y)$ *is infinite.*

PROOF. Consider a homothety $g : y \longmapsto y/\lambda$ in the upper half space model of H^n with a very small linear derivative $1/\lambda$. If x and y_0 are on the x_n-axis above zero, then clearly $\theta(x, y_o)$ and $\theta(x, y_o/\lambda)$ are in the approximate ratio $1/\lambda^{n-1}$. Thus, in general, the order of the term $\theta(x, g(y))$ for $\lambda(g)$ large and x fixed is in a bounded ratio to $1/\lambda^{n-1}$. This proves the lemma.

Another point of view on this condition is as follows. We denote by $g_x(y)$ Green's function of the hyperbolic space H^n with pole $x \in H^n$ (i.e., a function that is positive and harmonic in H^n and which tends to zero at infinity and is symmetric about x). Then on integrating along a radius (∞, y) we get

$$g_x(y) = \int_{\infty}^{y} \theta(x, z)dz. \tag{6.42}$$

As $\theta(x, y)$ tends to zero for $d(x, y) \longrightarrow \infty$, $g_x(y) \approx \theta(x, y)$ at large distances $d(x, y)$. Then the condition $\sum_G \lambda^{1-n} < \infty$ is equivalent to the existence of a bounded Green's function in the hyperbolic manifold $M = H^n/G$:

$$g_{\tilde{x}}(\tilde{y}) = \sum_{\gamma \in G} g_x(\gamma(y)). \tag{6.43}$$

This observation is the basis for the proof of the following theorem.

THEOREM 6.50. *If* $\sum_G \lambda^{1-n} = \infty$ *for a discrete group* $G \subset \text{Möb}(n-1)$, *the action of* G *on* $S^{n-1} \times S^{n-1}$ *(the set of geodesics in* H^n*) is ergodic.*

PROOF. For a point $p \in H^n$, we associate the Poisson measure μ_p on $\partial H^n = S^{n-1}$: if $X \subset S^{n-1}$, then $\mu_p(X)$ is the solid angle of X viewed from p. It is clear that μ_p is the spherical average of the measure μ_q, where the point q ranges over a sphere with centre p. Each of the two measures on the sphere at ∞ is invariant by the full rotation group about p and therefore they are equal. We can fill in the bounded measurable functions φ on $S^{n-1} = \partial H^n$ to bounded harmonic functions on H^n by means of the Poisson formula, namely we define

$$h(p) = \int \varphi d\mu_p \equiv \langle \varphi, \mu_p \rangle. \tag{6.44}$$

It is clear that the condition $\varphi \not\equiv \text{const}$ almost everywhere implies $h \not\equiv \text{const}$. In fact, let x and $y \in S^{n-1}$ be density points of sets where φ has values in disjoint intervals. For the points p in hyperbolic space near x, μ_p sees mostly the values in the interval associated, so $h(p)$ lies nearly in this interval. Similarly for the points q near y, $h(q)$ nearly lies in the disjoint interval.

We denote by P the averaging operator on functions and measures on hyperbolic space: $f \longmapsto Pf$, $\mu \longmapsto \mu P$, where the values are taken as the function $Pf(x)$, the average of the function f over a ball $D(x, \eta) \subset H^n$ having its centre at x and radius $\eta > 0$, and if δ_x is the Dirac mass at x, then $\delta_x P$ is the uniform measure on $D(x, \eta)$ of total mass one. Then in the (6.44) symbols: $\langle \mu P, f \rangle = \langle \mu, Pf \rangle$, while the functions h constructed by Poisson's formula (6.44) are P-harmonic ones, i.e. $Ph = h$, and therefore the operator P commutes with hyperbolic isometries.

LEMMA 6.51. *The density at the point* y *of the measure* $\sum_{i=1}^{N} \delta_x P^i$ *is, for sufficiently large* N, *at least a fixed constant times* $\theta(x, y)$.

PROOF. The sequence of measures $\delta_x P^i$ begins at x, spreads out symmetrically by steps of length at most η and converges to ∞ in the sense that, for sufficiently large i, almost all the mass is eventually outside any sphere centred at x. This occurs because balls having radius η and their centres in that sphere have more than half their volume outside it. Thus all the mass of the measure $\delta_x P^i$ for $i \longrightarrow \infty$ passes through any spherical shell of width η centred at x. As the density function of the measure $\sum_{i=1}^{\infty} \delta_x P^i$ have the following properties:

1) decreases for $d(x, y) \longrightarrow \infty$,
2) is dependent only on the radial distance $d(x, y)$, and
3) puts at least mass 1 in each shell of width η whose volume is proportional to $1/\theta(x, y)$ when y lies in the shell,

we get the required inequality.

We form the space T as the quotient $(H^n \times S^{n-1})/G$ by the diagonal action of G. We now provide the space T with an averaging operator $\tilde{P}$ on measures and functions, which is defined by using P in each H^n-level. We provide T with a natural smooth measure dm using the family of Poisson measures μ_p on the factors $\{p\} \times S^{n-1}$ and the natural measure on H^n to obtain a G-invariant volume element dm' on $H^n \times S^{n-1}$ and put $dm = dm'/G$. Let μ_p also denote the image measure resting on a sphere of T. We can now formulate the key lemma.

LEMMA 6.52. *The density of the measure* $\sum_{i=1}^{N} \mu_p P^i$ *relative to* dm *converges to infinity at almost all points of* T *as* $N \longrightarrow \infty$.

PROOF. We compute the density of the measure $\nu_N = \sum_{i=1}^{N} \mu_p (P \times \mathrm{id})^i$ in $H^n \times S^{n-1}$ relative to dm' and add up these densities up along a G-orbit:

$$(y, S), \quad (\gamma(y), \gamma(S)), \ldots, .$$

Let $g_x(y)$ be the density of Green's measure $\sum_{i=1}^{\infty} \delta_x P^i$ in terms of the natural hyperbolic volume dh. In terms of the product measure $dh \times \mu_p$, on the product $H^n \times S^{n-1}$, which is not G-invariant, the measure $\lim_{N\to\infty} \nu_N$ has density $g_p(y)$ at each point $(y, s) \in H^n \times S^{n-1}$. Rewriting in terms of dm' (which is G-invariant) introduces the Radon-Nicodim factor $(d\mu_p/d\mu_y)(S)$ at each point (y, s). Then the desired density takes the form

$$\sum_{\gamma\in G} g_p(\gamma(y)) \frac{d\mu_p}{d\mu_{\gamma(y)}}(\gamma(s)). \tag{6.45}$$

Using γ^{-1} to transform $(p, \gamma(s), \gamma(y))$ into $(\gamma^{-1}(p), s, y)$ and the symmetry $g_x(y) = g_y(x)$ changes the sum (6.45) to one of the form

$$\sum_{\gamma\in G} g_y(\gamma(p)) \frac{d\mu_{\gamma(p)}}{d\mu_y}(s). \tag{6.46}$$

It follows from Lemmas 6.49 and 6.51 that $\sum_{\gamma\in G} g_y(\gamma(p)) = \infty$, so the assertion in Lemma 6.52 is obtained from the following lemma.

LEMMA 6.53. *If* $\sum_{\gamma\in G} g_y(\gamma(p)) = \infty$, *then for almost all* $s \in S^{n-1}$ *the density* (6.46) *is infinite.*

PROOF. 1) Let B be a small ball centred at p and $G(B)$ be its orbit, namely the disjoint union of the balls $\gamma(B)$, $\gamma \in G$. We define a G-invariant function $\pi_B(y)$ to be the probability that a random walk, whose transition operator is P, starting at y, hits the orbit $G(B)$. It is clear that $\pi_B(y) \le 1$ and $P\pi_B \le \pi_B$. If $\pi_B \not\equiv 1$, then the G-invariant function $\pi_B - P\pi_B$ is greater than ε, $\varepsilon > 0$, on the G-orbit of some smaller ball. We consider the identity

$$<\delta_y + \delta_y P + \ldots + \delta_y P^N, \pi_B - P\pi_B> = <\delta_y, \pi_B> - <\delta_y P^{N+1}, \pi_B>. \tag{6.47}$$

The right-hand side in (6.47) is $\pi_B(y) - P^{N+1}\pi_B(y)$, so it is bounded with respect to N. On the other hand, the left-hand side for $N \to \infty$ is at least $\varepsilon \cdot \sum_{\gamma\in G} g_y(\gamma(p))$, so $\pi_B = P\pi_B$, or, what is the same, $\pi_B(y) = 1$. Thus almost all paths starting at y hit $G(B)$ if $\sum_{\gamma\in G} g_y(\gamma(p)) = \infty$.

2) From the averaging of μ_p over a sphere centred at p it follows that, for given s_0 and x_0, the function of x equal to $(d\mu_x/d\mu_{x_0})(s_0)$ is a P-harmonic function having limit values $+\infty$ at s_0 and zero at other points of the sphere S^{n-1}. One sees by a standard limiting procedure (see Kemeny, Knapp, and Snell [1]) that

$$g_x(y) \frac{d\mu_y}{d\mu_x}(s_0) \tag{6.48}$$

is the corresponding Green's density for the random walk conditioned so that the limit at infinity is s_0. From the first part and from Fubini's theorem it follows that, for almost all $s_0 \in S^{n-1}$, almost all paths starting at y and conditioned to end up at s_0 must also hit the orbit $G(B)$.

Then, if for one of these points s_0 we had

$$\sum_{\gamma \in G} g_y(\gamma(p)) \frac{d\mu_{\gamma(p)}}{d\mu_y}(s_0) < K < \infty, \tag{6.49}$$

then (6.49) would hold for all points p from a sufficiently small ball B. But then the conditioned Green's measure, having the density (6.48), would give a finite measure to $G(B)$. This contradicts the fact that almost all conditioned paths hit $G(B)$ starting from any point. Therefore, they hit $G(B)$ infinitely often by the Markov property, so the $G(B)$ measure is infinite. This proves Lemmas 6.53 and 6.52.

We now pass directly to the proof of Theorem 6.50. Let $\varphi(x, y)$ be a non-constant characteristic function on $S^{n-1} \times S^{n-1}$, invariant with respect to G and suppose $\sum_G \lambda^{1-n} = \infty$. We can suppose by Fubini's theorem (after interchanging x and y if necessary) that for a set $Y \subset S^{n-1}$, having positive measure $\mu_{n-1}(Y)$, $\varphi(x, y)$ is an a.e. non-constant function of x for all $y \in Y$.

Fill in each $\varphi(x, y)$ to a P-harmonic function on $S^{n-1} \times \{y\}$ by the Poisson formula (6.44). We obtain a P-harmonic function

$$h(p, y) = \int \varphi(x, y) d\mu_p(x)$$

on each level $H^n \times \{y\}$, so we get the function $h(x, y)$ on the space T that is harmonic on each level. This follows from the G-invariance of $\varphi(x, y)$ and the P-harmonicity, which immediately implies $\tilde{P}h = h$. As $\varphi(x, y)$ is not constant, the function h is non-constant on a set of levels of positive measure.

We break the function $h - 1/2$ into positive and negative parts h_+ and h_- and add the equality $\tilde{P}(h - 1/2) = h - 1/2$ to the inequality

$$\tilde{P}|h - 1/2| \geq |\tilde{P}(h - 1/2)| = |h - 1/2|. \tag{6.50}$$

Then (6.50) gives $\tilde{P}h_+ \geq h_+$. Using Lemma 6.52 and repeating the arguments in the first part of the proof of Lemma 6.53 we deduce that $\tilde{P}h_+ = h_+$ almost everywhere. The latter implies that subdomains in the set of levels, where h_+ is zero, do not communicate via posers of P with their complements. This contradiction proves Theorem 6.50.

2. *Conical approximation.*

DEFINITION 6.54. A point $s \in S^{n-1} = \partial H^n$ is called a *conical approximation point* for a discrete group $G \subset \text{Möb}(n-1)$ if infinitely many points of a certain orbit $G(y)$, $y \in H^n$, fall into a circular cone with vertex at s (see the equivalent definitions in Chapter 5 §3).

LEMMA 6.55. *If a density point of a G-invariant set $Y \subset S^{n-1}$ is a conical approximation point for a discrete group $G \subset \text{Möb}(n-1)$, then this set has full measure:* $\mu_{n-1}(S^{n-1} \setminus Y) = 0$.

PROOF. Let χ_Y be the characteristic function of Y, while the sequence $\{\gamma_i\} \subset G$ is such that for $p \in H^n$ the points $\gamma_i(p)$ have conical approach to a density point s_0 of the set Y. We again denote the Poisson measure by μ_p to get

$$\mu_{n-1}(Y) = \mu_p(Y) = \int \chi_Y d\mu_p = \int \gamma_i(\chi_Y) d\mu_p = \int \chi_Y d\mu_{\gamma_i(p)}. \tag{6.51}$$

It is clear that, when $\gamma_i(p)$ tends to s_0 in the cone, the right-hand side in (6.51) tends to the full measure of the sphere S^{n-1}, which proves the lemma.

COROLLARY 6.56. *If* $G \subset \text{Möb}(n-1)$ *has a conical approximation on a set* $Y \subset S^{n-1}$, $\mu_{n-1}(Y) > 0$, *there is a conical approximation on a set of full measure on the sphere.*

LEMMA 6.57. *If* $G \subset \text{Möb}(n-1)$ *has a conical approximation only on a set* $Y \subset S^{n-1}$, $\mu_{n-1}(Y) > 0$, *then the action of the group* G *on* $S^{n-1} \times S^{n-1}$ *(on the set of geodesics in* H^n*) is dissipative, (i.e., it has a fundamental domain* a.e).

PROOF. We consider the orbit $G(p)$ for a point $p \in H^n$ and for each point q of it we associate $D_q \subset S^{n-1} \times S^{n-1}$, those geodesics having distances to q less than those for all other points of $G(p)$; it is clear that $D_q \cap D_{q'} = \varnothing$ for $q \neq q'$, $q, q' \in G(p)$. For almost all pairs $(s, s') \in S^{n-1} \times S^{n-1}$ there is a $q \in G(p)$ such that $(s, s') \in D_q$ because, given a pair, we merely swell up a geodesic tubular neighbourhood of the connecting geodesic until it first meets the orbit. The condition $\mu_{n-1}(Y) = 0$ for the conical approximation set Y implies that this q is unique for almost all pairs (s, s') and also it follows from the fact that the set of pairs (s, s'), for which several points of $G(p)$ meet simultaneously in the swelling neighbourhood of the geodesic (s, s'), is a countable union of lower dimensional submanifolds in $S^{n-1} \times S^{n-1}$. Then D_q provide the desired fundamental set in $S^{n-1} \times S^{n-1}$ (up to a set of measure zero), i.e. the action of G in $S^{n-1} \times S^{n-1}$ is dissipative.

COROLLARY 6.58. *If a group* $G \subset \text{Möb}(n-1)$ *is such that* $\sum_G \lambda^{1-n} = \infty$, *then a conical* G*-approximation set has the full measure in* S^{n-1}.

The proof follows from Theorem 6.50, which gives that the action of G on $S^{n-1} \times S^{n-1}$ is ergodic, which could be contrary to the conclusion of Lemma 6.57 if G did not have full conical approach.

The above assertions imply directly the following fact, proved for S^2 by E. Hopf [1].

COROLLARY 6.59. *A group* $G \subset \text{Möb}(n-1)$ *has conical approximation either at a zero-measure set or at a full-measure one. In the first case, the* G*-action on* $S^{n-1} \times S^{n-1}$ *is dissipative, and in the second, ergodic.*

LEMMA 6.60. *If a group* $G \subset \text{Möb}(n-1)$ *has conical approximation at a set* $Y \subset S^{n-1}$, $\mu_{n-1}(Y) > 0$, *then*

$$\sum_G \lambda^{1-n} = \infty.$$

PROOF. This is obtained from the following observations. Lemma 6.55 gives $\mu_{n-1}(S^{n-1} \setminus Y) = 0$. We consider a positive measurable function on the sphere S^{n-1}, which is equal at a point s to the best angle for conical approximation at the point s, i.e., the least vertex angle for a circular cone having its vertex at s. That angle is greater than a certain $\alpha > 0$ for a positive-measure set Y_0. The angle α defines the size of a ball neighbourhood B around the point p whose orbit approximates points from Y_0 in cones with the angle α. If now the total solid angle of the orbit $G(B)$ were finite, we could cast out a finite number of balls from the orbit so the remaining solid angle would be arbitrarily small. Then the set $G(B)$ seen from a fixed point p at an arbitrarily small angle approximates the set Y_0 having positive measure $\mu_p(Y_0)$ in cones having the angle $\alpha > 0$. This contradiction proves the lemma.

The above description for the action of a discrete group on the sphere at infinity is now formulated as follows.

THEOREM 6.61. *The following conditions are equivalent for a discrete group* $G \subset \text{Möb}(n-1)$:

1) $\sum_G \lambda^{1-n} = \infty$;

2) *the set of points on the sphere at infinity* $\partial H^n = S^{n-1}$ *having a conical G-approximation has the full measure; and*

3) *the action of* G *on* $S^{n-1} \times S^{n-1}$ *is ergodic.*

PROOF. This is obtained by applying Theorem 6.50 ((1) $\Rightarrow$ 3)), Lemma 6.57 ((3) $\Rightarrow$ 2)) and Lemma 6.60 ((2) $\Rightarrow$ 1)).

REMARK. Sullivan's description for the action of a group $G \subset \text{Möb}(n-1) \cong \text{Isom}\, H^n$ on the sphere at infinity ∂H^n follows in the case of a Fuchsian group in the plane (i.e., $n=2$) from results obtained in the 1940s by E. Hopf [1], [2] and Tsuji [1].

3. *Horospherical limit points and conservative action.*

DEFINITION 6.63. A point $s \in \partial H^n = S^{n-1}$ is called a *horospherical* limit point for a discrete group $G \subset \text{Möb}(n-1)$ if the orbit $G(y)$ of a certain point $y \in H^n$ enters every horosphere in H^n centred at the point s. We denote the set of horospherical limit points by $\Lambda_h(G)$. It is geometrically clear that $\Lambda_h(G)$ is independent of the choice of reference orbit.

For the points $s \in S^{n-1} \setminus \Lambda_h(G)$ we can try to swell up the horospheres centred at s until one first hits a certain point q of a given orbit. If that point q is then unique, we call it the "closest orbit point" to s.

THEOREM 6.64. *For any discrete group* $G \subset \text{Möb}(n-1)$ *and any choice of reference orbit* $G(y) \subset H^n$, *the union* $\Lambda_h(G)$ *with the points in the sphere at infinity* $S^{n-1} = \partial H^n$ *which have closest orbit points form a set of full measure in the sphere* S^{n-1}. *Furthermore, this dichotomy is precisely the partition of the action of the group* G *on the sphere* S^{n-1} *into its conservative* ($= \Lambda_h(G)$) *and dissipative* ($= S^{n-1} \setminus \Lambda_h(G)$) *parts.*

PROOF. Assuming the first part for the moment, let us prove the second. To any point q of the orbit $G(y)$ we associate a set A_q consisting of those points $s \in S^{n-1}$ for which q is closest. It is clear that the sets A_q are freely permuted by G, so their union $\bigcup A_q$ is a dissipative set.

On the other hand, the set $\Lambda_h(G)$ is such that, for any point s in it, we have $\sup_{g \in G} |g'(s)| = \infty$, which shows that there is no dissipative part in $\Lambda_h(G)$, since along almost all dissipative orbits the sum of Jacobians is finite:

$$\sum_{g \in G} \det Dg(x) < \infty.$$

Convergence is demonstrated by analogy with Lemma 6.20 (it is based on the concept that the union of volumes is bounded).

Let us return to the first part of the theorem. We consider a set $\Lambda_G(G)$ in the sphere S^{n-1} obtained from the limit set $\Lambda(G)$ by removing the subset $\Lambda_h(G)$ of horospherical limit points and a countably many lower dimensional submanifolds of the sphere which points are bases of horospheres containing more than one point of the reference orbit, and the points $s \in S^{n-1}$ having closest orbit points. The set $\Lambda_G(G) \subset \Lambda(G)$ consists of the points $s \in S^{n-1}$ such that there is a critical horosphere $h(s)$, centred at s, that does not contain points of $G(y)$ but any larger horosphere contains an infinite subset from $G(y)$. Such points have been called *Garnett points*. We prove the theorem completely if we prove the following lemma.

LEMMA 6.65. *The set* $\Lambda_G(G)$ *of Garnett limit points has measure zero.*

PROOF. We assume the converse. Then from Luzin's theorem we find a subset $X \subset \Lambda_G(G)$, $\mu_{n-1}(X) > 0$ in which the critical-sphere radius is a continuous function. Let s be a point of density in the set X and $h(s)$ its critical horosphere. There are infinitely many points y^i from the orbit $G(y)$ lying outside $h(s)$ and converging to s but entering every larger horosphere. For each such y^i there exists an associate forbidden region on the sphere S^{n-1}, which does not contain Garnett points from X (since the radii of the critical horospheres at points in the forbidden region cannot be close to the radius of the horosphere $h(s)$). It is clear that the spherical measure for such an i-th region is not less, up to a certain constant, than the measure of the $(n-1)$-dimensional ball in S^{n-1}, centred at s, containing the radial (vertical) projection of the point y^i. This contradicts the fact that the point $s = \lim y^i$ is a Lebesgue density point of the set X.

COROLLARY 6.66. *Let* $P(G) \subset H^n$ *be a locally finite* (see Chapter 5 §1) *fundamental polyhedron for a discrete group* $G \subset \mathrm{Isom}\, H^n$. *Then the dissipative part of the action of* G *on the sphere at infinity* ∂H^n *is precisely the set*

$$\bigcup_{g \in G} g(\partial H^n \cap \overline{P(G)}).$$

COROLLARY 6.67. *If* $\mu_{n-1}(\overline{P(G)} \cap \partial H^n) = 0$, *then a discrete group* $G \subset \mathrm{Isom}\, H^n$ *acts conservatively on the sphere at infinity* ∂H^n.

By $M(r)$ with $r > 0$ we denote a subset in the hyperbolic orbifold $M = H^n/G$ consisting of points distant from a fixed point in M by not more than r, while by $D(r) \subset H^n$ we denote the ball of radius $r > 0$ in the hyperbolic space. Then the condition of the last corollary can be reformulated as follows.

THEOREM 6.68. *The following conditions are equivalent for a discrete group* $G \subset \mathrm{Isom}\, H^n$:

1) $\mu_{n-1}(\overline{P(G)} \cap \partial H^n) = 0$ *for a convex fundamental polyhedron* $P(G) \subset H^n$ *of* G;
2) $\lim_{r \to \infty} \mathrm{vol}\, M(r)/\mathrm{vol}\, D(r) = 0$;
3) G *acts conservatively on the sphere at infinity* ∂H^n; *and*
4) *the horospherical limit set* $\Lambda_h(G)$ *has full measure on the sphere at infinity* ∂H^n.

PROOF. We have already demonstrated equivalence for conditions 1), 3), and 4), so it remains to show that the second condition follows from the first (the converse is also simple, see Chapter 9 §2). We consider the solid angle $\omega(r)$ of the intersection between $P(G)$ and a sphere S_r^{n-1} having radius r and centre at a point q in the polyhedron $P(G)$. The convexity of $P(G)$ implies that the angle $\omega(r)$ decreases as r increases. Condition 1) gives $\lim_{r \to \infty} \omega(r) = 0$. If by $a(r)$ we denote the area of the sphere S_r^{n-1}, we have

$$\frac{\mathrm{vol}\ M(r)}{\mathrm{vol}\ D(r)} = \int_0^r a(r)\omega(r)dr \Big/ \int_0^r a(r)dr. \qquad (6.52)$$

As $\omega(r) < \varepsilon$ for $r > r_0$, (6.52) implies

$$\frac{\mathrm{vol}\ M(r)}{\mathrm{vol}\ D(r)} \le \int_0^{r_0} a(r)\omega(r)dr \Big/ \int_0^r a(r)dr + \varepsilon \cdot \int_{r_0}^r a(r)dr \Big/ \int_0^r a(r)dr. \qquad (6.53)$$

The first term on the right in (6.53) tends to zero as r tends to infinity, while the second is less than an arbitrary $\varepsilon > 0$, which implies that the second condition in the theorem holds.

CHAPTER 7

KLEINIAN MANIFOLDS

§1. Topological methods

Let G be a discrete group of isometries of the n-dimensional hyperbolic space H^n (Poincaré ball model $B^n(0, 1)$ or half-space model $\mathbb{R}^n_+$). Then on the sphere at infinity $\partial H^n = \overline{\mathbb{R}}^{n-1}$, it acts as a subgroup of the Möbius group $\text{Möb}(n-1)$. The action of G is discontinuous in $H^n \cup \Omega(G)$, where $\Omega(G) \subset \overline{\mathbb{R}}^{n-1}$ is the discontinuity set.

DEFINITION 7.1. A Kleinian manifold/orbifold is the quotient space by a discrete group $G \subset \text{Isom}\, H^n$:

$$M(G) = [H^n \cup \Omega(G)]/G = [\overline{H}^n \backslash \Lambda(G)]/G. \tag{7.1}$$

It is an orbifold if G is a group with torsion.

The boundary $\partial M(G)$ of $M(G)$ is not empty if and only if G is a Kleinian group in the sphere at infinity $\partial H^n = \overline{\mathbb{R}}^{n-1}$, i.e., if $\Omega(G) \neq \emptyset$; then $\partial M(G) = \Omega(G)/G$. If G does not contain finite-order elements, it is isomorphic to the fundamental group $\pi_1(M(G))$ of $M(G)$. If G is non-elementary, then (see Chapter 2) no component of the boundary $\partial M(G) = \Omega(G)/G$ is an elliptic, Euclidean or affine manifold (i.e., for the case of oriented 3-manifold $M(G)$, it is not a sphere, a sphere with one or two punctures, or a torus).

Standard cusp ends. In what follows, we assume (if the converse is not stated) that all these groups G are non-elementary torsion free geometrically finite groups. The latter condition is equivalent to finiteness for the hyperbolic volume of an ε-neighbourhood for the minimal convex retract of the manifold $M(G)$:

$$M_G = [H_G \backslash \Lambda(G)]/G \subset M(G), \tag{7.2}$$

and is also equivalent to compactness for the ε-thick submanifold $[M_G]_{[\varepsilon,\infty)} \subset M_G$ remaining when one has removed the ε-thin parts $[M_G]_{(0,\varepsilon]}$ from M_G. Each connected component of the ε-thin submanifold $[M_G]_{(0,\varepsilon]}$ for sufficiently small $\varepsilon>0$ is a cusp end, i.e., it is obtained by factorizing a parabolic ε-neighbourhood in H_G of some parabolic cusp p by its stabilizer $G_p \subset G$ (see Theorem 5.39).

More precisely (see Remark 4.22) the following is such a component of the submanifold $[M_G]_{(0,\varepsilon]}$ corresponding to the parabolic cusp $p=\infty$:

$$[H_G \cap V_\infty(G, \varepsilon)]/G_\infty = [H_G \cap \{('x, x_n) \in \mathbb{R}^{n-1}\times\mathbb{R}_+ : x_n \geq t_\varepsilon('x)\}]/G_\infty, \quad (7.3)$$

where the parabolic ε-neighbourhood $V_\infty(G, \varepsilon)$ is specified by the function $t_\varepsilon('x)$, defined in Remark 4.22 and satisfying (4.26)-(4.28). Also, if $\operatorname{rank} G_\infty = k$, $1 \leq k \leq n-2$, the Nielsen hull H_G lies in a layer of finite width:

$$\{x \in \mathbb{R}^n_+ : |x_i| \leq R, \quad k+1 \leq i \leq n-1\}.$$

Therefore, there are precisely G_∞-invariant half-spaces $\{x \in \mathbb{R}^n_+ : x_n > r'\} \subset \{x \in \mathbb{R}^n_+ : x_n > r\}$, i.e., the horoballs $B_{\delta'}(\infty) \subset B_\delta(\infty) \subset H^n$ having their centres at $p=\infty$ and radii δ, δ', $0<\delta'<\delta$ (see (4.61) and Remark 4.33) and such that

$$B_{\delta'}(\infty) \cap H_G \subset V_\infty(G, \varepsilon) \cap H_G \subset B_\delta(\infty) \cap H_G. \quad (7.4)$$

As a number of conjugacy classes of maximal parabolic subgroups in the group G is finite, this with (7.4) proves equivalence for the compactness of $[M_G]_{(0,\varepsilon]}$ and the compactness of the following submanifold:

$$M^*_G[\varepsilon, \infty) = [H_G \setminus \bigcup_p B_\delta(p)]/G. \quad (7.5)$$

Here the union is taken over all the parabolic points p of the group G, while by $B_\delta(p)$ we denote distinct precisely G_p-invariant horoballs with sufficiently small radius $\delta=\delta(\varepsilon)>0$ forming a G-invariant family. Here the homeomorphic manifolds $M^*_G(\varepsilon, \infty)$ and $(M_G)_{[\varepsilon,\infty)}$ differ one from the other by a compact submanifold $M_\varepsilon \subset M_G$. Also, the boundary

$$(H_G \cap S_\delta(p))/G_p$$

of the manifold

$$(\operatorname{int} H_G \cap \overline{B_\delta(p)})/G_p \subset M^*_G(0, \varepsilon] \quad (7.6)$$

has a complete Euclidean structure induced by the hyperbolic metric in H^n (this follows from the completeness of the manifold M_G and Corollary 6.8: $S_\delta(p) = \partial B_\delta(p) \cong \mathbb{R}^{n-1}$). Then Bieberbach's description (Theorem 3.4) of discrete Euclidean-isometry groups gives a description of components of an ε-neighbourhood (if $\dim M_G < n$) of the manifold $(M_G)_{(0,\varepsilon]}$, homeomorphic together with the components of (7.6) to manifolds which allow finite-sheeted covering by

$$\mathcal{P}^n = T^{n-1}\times[0,\infty), \quad \text{if} \quad \operatorname{rank} G_\infty = n-1;$$
$$C^{n,k} = T^k\times \overline{B}^{n-k-1}\times[0,\infty), \quad \text{if} \quad \operatorname{rank} G_p = k < n-1, \tag{7.7}$$

the first of which is called a punctured solid torus, and the second is called a punctured solid cylinder (of index k), where $T^k = S^1\times\ldots\times S^1$ is the k-dimensional torus.

We can now reformulate the assertion GF3 in Theorem 5.39 in the following form.

THEOREM 7.2. *A torsion free non-elementary group* $G\subset \operatorname{Isom} H^n$ *is geometrically finite if and only if the manifold* $U_\varepsilon(M_G)\subset M(G)$ *contains a finite number of distinct submanifolds which allow finite-sheeted coverings by punctured solid tori and cylinders and whose complement in* M_G *is compact.*

Instead of the manifolds (7.6), we consider the corresponding (homeomorphic to them) manifolds

$$\widetilde{U}_p(\delta)/G_p\subset M(G), \tag{7.8}$$

in which $\widetilde{U}_p(\delta)\subset H^n\cup\Omega(G)$ are the hyperbolic cusp domains for the parabolic cusps p (see Remark 5.18). Namely, if rank $G_p = k$ then, for a number $t = t(\delta) > 0$ and a transformation $h\in \text{Möb}(\mathbb{R}^n_+)$, $h(p) = \infty$, we have:

1) $\widetilde{U}_p$ is precisely G_p-invariant; and
2) $h(\widetilde{U}_p) = \{x\in\mathbb{R}^n : x_{k+1}^2+\ldots+x_n^2 > t, \quad x_n\geq 0\}$.

Here the sizes $t = t(\delta)$ of the cusp domains are such that the images of them $\widetilde{\rho}(\widetilde{U}_p(\delta))$, by the retraction $\widetilde{\rho}:\overline{H}^n\longrightarrow H_G$ are $H_G\cap B_\delta(p)$. Therefore the indicated submanifolds of the ε-neighbourhood of the minimal convex retract $M_G\subset M(G)$ in Theorem 7.2 correspond to neighbourhoods of cusp ends of the Kleinian manifold $M(G)$ that are conformally equivalent to the following manifolds:

$$\{x\in\mathbb{R}^n : x_{k+1}^2+\ldots+x_n^2 > \alpha^{-2}, \quad x_n\geq 0\}/\Gamma,$$

which give the base of neighbourhoods for a "standard cusp end", i.e., the unique end of the manifold $\{x\in\mathbb{R}^n : x_n\geq 0\}/\Gamma$ where $\Gamma\subset\operatorname{Isom}\mathbb{R}^{n-1}$ is a discrete torsion free group of rank k with the invariant subspace $\mathbb{R}^k\subseteq\mathbb{R}^{n-1}$. We note that the neighbourhoods of a standard cusp end have finite covering either by the punctured solid torus P^n or by the punctured solid cylinder $C^{n,k}$ ($k = n-1$ or $k < n-1$, correspondingly). Now we prove the statement GF5 of Theorem 4.39.

THEOREM 7.3. *A Kleinian group* $G\subset \text{Möb}(n-1)$ *is geometrically finite if and only if the Kleinian manifold/orbifold* $M(G)$ *contains only finitely many ends, and each such end is conformally equivalent to a standard cusp end, i.e., the end of* $(\overline{\mathbb{R}}^n_+\setminus\{\infty\})/\Gamma$ *for a discrete group* $\Gamma\subset\operatorname{Isom}\mathbb{R}^{n-1}$.

PROOF (*sufficiency*). Let $\tilde{U}_1, \ldots, \tilde{U}_k \subset H^n \cup \Omega(G)$ be distinct cusp domains which correspond to neighbourhoods of finitely many ends of $M(G)$ and $G_i \subset G$ be their stabilizers (isomorphic to discrete groups $\Gamma_i \subset \mathrm{Isom}\,\mathbb{R}^{n-1}$). Using (7.4), for given $\varepsilon > 0$, we take sufficiently small sizes of $\tilde{U}_i$ such that, for the retraction $\rho : M(G) \to M_G$, the inclusion holds:

$$\rho(M(G) \setminus \bigcup_i \tilde{U}_i/G_i) = M_G \setminus \bigcup_i \rho(\tilde{U}_i/G_i) \supset (M_G)_{[\varepsilon,\infty)}.$$

Then the compactness of the right ε-thick retract follows from the compactness of $M(G) \setminus \bigcup_i \tilde{U}_i/G_i$ and the continuity of ρ. Using a finite covering (for the orbifold case) and applying GF3 in Theorem 5.39, we complete the proof.

(*Necessity*). The finite-sidedness of a fundamental polyhedron $P \subset H^n$ for the group $G \subset \mathrm{Isom}\,H^n$ give us the finiteness of conjugacy classes for parabolic subgroups $G_i \subset G$ with fixed points that are parabolic cusps for G - see Theorem 5.24. Let $\tilde{U}_i \subset H^n \cup \Omega(G)$ be cusp domains for all the cusps $p_i \in \partial P$, $i = 1, \ldots, k$. Then $\tilde{U}_i/G_i$ are neighbourhoods of finitely many ends of $M(G)$ conformally equivalent to neighbourhoods of standard cusp ends (see above) and $(\bar{P} \setminus \bigcup_{i=1}^{k} \tilde{U}_i/G \subset M(G)$ is compact. This completes the proof.

For $n = 3$, the neighbourhood of the cusp ends in $M(G)$ (of rank 1) were defined in the following equivalent form (see Greenberg [2] and Marden [1]).

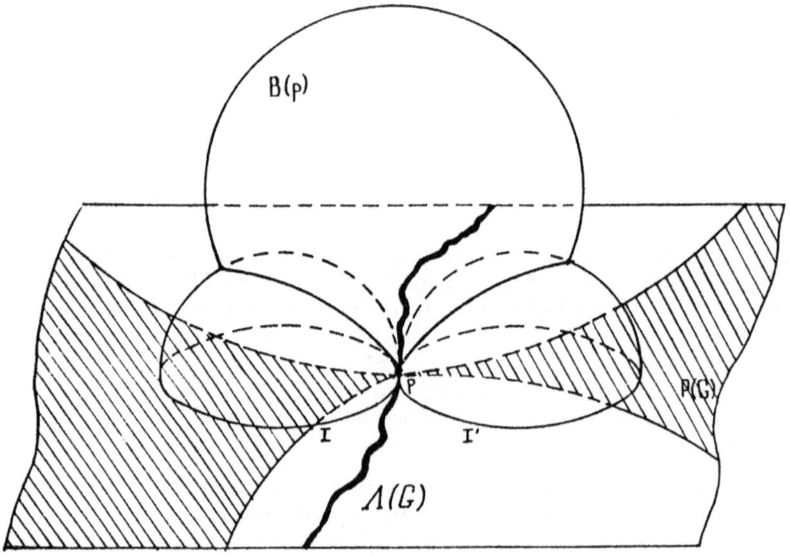

Figure 54

A parabolic cusp $p \in \partial H^3$ having a cyclic stabilizer G_p corresponds to a cusp domain in $\overline{\mathbb{R}}^2$ (Definition 5.15), which consists of two discs D_1 and D_2 tangent at the point p and precisely invariant with respect to G_p. Let $B(p)$ be a horoball having its centre at p and precisely invariant with respect to G_p, while $\overline{H}_i$ are closed sets bounded by surfaces $D_i \cup S_i$, where S_i are hyperbolic planes $\partial S_i = \partial D_i$, $H_i = \overline{H}_i \setminus S_i$, $i = 1, 2$ (Fig. 54). Then the punctured solid cylinder $C(p) \approx C^{3,1}$ is defined as follows:

$$\widetilde{U}_p/G_p \approx (B(p) \cup H_1 \cup H_2)/G_p = C(p).$$

COROLLARY 7.4 (Marden [1]). *A discrete torsion free group* $G \subset \mathcal{M}_2$ *is geometrically finite if and only if the manifold* $M(G)$ *contains a finite number of punctured solid tori* T_i *and punctured solid cylinders* C_j *such that the three-manifold* $M_0(G) = M(G) \setminus \bigcup C_j \cup T_i$, *which is the complement to it, is compact.*

It is clear from the results presented above that the analysis of the topological structure of the geometrically finite manifolds M_G and $M(G)$ leads to the study of properties of their compact submanifolds. We now state some well-known facts on 3-dimensional manifolds (denoted by M) that are important in the sequel; see also Chapter 2 §§1-2 and Chapter 5 §5.

THEOREM 7.5 (**the loop theorem and Dehn's lemma**, cf. Papakiriakopulos [1], [2]). *If* l *is a non-trivial loop on* ∂M *(a non-trivial element of the group* $\pi_1(\partial M)$*), which is trivial in* M *and* N, $\partial M \supset N \supset l$, *an arbitrarily small neighbourhood of* l, *then there is a simple loop* $l' \subset N$, *also non-trivial on* ∂M *and bounding a disc* $D \subset M : D \cap \partial M = l'$. *If* l *is a simple loop, it can be taken for* l'.

THEOREM 7.6 (**the cylinder theorem**, cf. Waldhausen [1]). *Let* l_1, l_2 *be disjoint loops on* ∂M *that are non-trivial and freely homotopic in* M, *and let* $N_1, N_2 \subset \partial M$ *be disjoint small neighbourhoods of them. Then there are simple loops* $l'_1 \subset N_1$ *and* $l'_2 \subset N_2$, *non-trivial in* M, *that are the components of the boundary of a topological cylinder* $C \subset M$, $C \cap \partial M = l'_1 \cup l'_2$. *If* l_1 *(or* l_2*) is a simple loop, then we can take it to be* l'_1 *(respectively,* l'_2*).*

THEOREM 7.7 (**the Seifert-van Kampen theorem**, cf. Massey [1]). *Let* $D \subset M$ *be a topological disc with non-trivial boundary loop in* ∂M. *If* $M \setminus D$ *has two components,* M_1 *and* M_2, *then* $\pi_1(M) = \pi_1(M_1) * \pi_1(M_2)$ *(the free product). If, on the other hand,* $M \setminus D$ *is connected, then* $\pi_1(M) = \pi_1(M \setminus D) * \mathbb{Z}$.

THEOREM 7.8 (**Kneser's theorem** - see Hempel [1]). *Let* M *be a compact 3-manifold and each component* $S \subset \partial M$ *is incompressible in* M, *i.e., the homomorphism* $\pi_1(S) \longrightarrow \pi_1(M)$, *induced by the inclusion* $S \subset M$, *is a monomorphism. If* $\pi_1(M)$ *is the non-trivial free product of the groups* G_1 *and* G_2, $\pi_1(M) \cong G_1 * G_2$, *then there exists a two-dimensional sphere embedded in* M *that makes* M *into a connected* *sum of manifolds* M_1 *and* M_2, $M = M_1 \# M_2$, *where* $\pi_1(M_i) \cong G_i$, $i = 1, 2$.

THEOREM 7.9 (**sphere theorem**, see Papakiriakopulos [1] and Whitehead [1]). *If* M *is an orientable 3-manifold such that* $\pi_2(\operatorname{int} M) \neq 0$, *then there exists a two-dimensional sphere embedded in* M *that represents a non-trivial element of* $\pi_2(M)$. *If* $\pi_2(\operatorname{int} M) = 0$, *and* $\pi_1(M)$ *is infinite, then* $\pi_3(\operatorname{int} M) = 0$.

If now $M = M(G)$ for some discrete torsion free group $G \subset \operatorname{Isom} H^3$, the following are added to the above topological properties of M.

1) Alexander's theorem on the unknottedness of a two-dimensional polyhedral sphere into the three-dimensional one (see Bing [1]) implies that each two-dimensional sphere embedded in $M(G)$ bounds a ball in $M(G)$, i.e., $M(G)$ is irreducible.

2) The sphere theorem and 1) imply that $M(G)$ is aspheric, i.e., all homotopy groups $\pi_i(M(G))$, $i > 1$, are trivial.

3) Similarly, 1) implies that $M(G)$ is a Poincaré manifold, i.e., each homotopy ball embedded in $M(G)$ is homeomorphic to the Euclidean ball. Therefore, although Poincaré's conjecture has not been proved for dimension three, this is not reflected in the study of Kleinian manifolds.

LEMMA 7.10. *Let* $G \subset \operatorname{Isom} H^n$ *be a discrete torsion free group and* S *a connected component of the boundary* $\partial M(G)$; *then the homomorphism* $\pi_1(S) \longrightarrow \pi_1(M(G)) \cong G$ *is a monomorphism if and only if all the components* $\pi^{-1}(S) \subset \Omega(G)$ *are simply connected.*

PROOF. This is obtained directly by considering the universal covering $\pi : H^n \cup \Omega(G) \longrightarrow M(G)$ and the simple connectivity of $H^n \cup \Omega(G)$.

It is also a simple exercise to obtain the proofs of the following two facts (cf. Gusevskii [1]).

LEMMA 7.11. *Let a discrete group* $G \subset \text{Möb}(n)$ *have an invariant component* $\Omega_0 \subset \Omega(G) \subset \overline{\mathbb{R}}^n$ and $S_0 = \Omega_0/G$; *then the homomorphism* $\pi_1(S_0) \longrightarrow \pi_1(M(G))$ *is an epimorphism.*

The converse assertion is also true:

LEMMA 7.12. *Let* S *be a connected component of the boundary* $\partial M(G)$ *such that the homomorphism* $\pi_1(S) \longrightarrow \pi_1(M(G))$ *is an epimorphism. Then* $\Omega_0 = \pi^{-1}(S) \subset \Omega(G) \subset \overline{\mathbb{R}}^n$ *is connected and is an invariant component of the group* G.

LEMMA 7.13 (Tucker [1]). *Let* $G \subset \operatorname{Isom} H^3$ *be a finitely generated discrete torsion free group, and* $\operatorname{int} M(G)$ *contains an embedded surface* S *having non-trivial fundamental group* $\pi_1(S)$, *and the homomorphism* $\pi_1(S) \longrightarrow \pi_1(M(G))$ *is a monomorphism. Then for any component* S^* *from* $\pi^{-1}(S)$, *the closures of* $H^3 \backslash S^*$ *are homeomorphic to* $\overline{H}^3$.

LEMMA 7.14. *Let* M_1 *and* M_2 *be 3-manifolds without boundaries with trivial homotopy groups* $\pi_2(M_i) = \pi_3(M_i) = 0$, $i = 1, 2$. *If* M_1

is compact and $\pi_1(M_1) \cong \pi_1(M_2)$, *then there are isomorphisms of the homology groups*

$$H_k(M_1, \mathbb{Z}) \cong H_k(M_2, \mathbb{Z}), \quad k = 0, 1, 2, 3, \tag{7.9}$$

and, in particular, M_2 *is also compact.*

PROOF. This follows from the existence of a continuous mapping $f : M_1 \longrightarrow M_2$ that induces the isomorphism $\pi_1(M_1) \longrightarrow \pi_1(M_2)$ (see Hu Sze-Tsen [1]), together with Whitehead's theorem (see Spanier [1], Chapter 7 §5), which asserts that this mapping will induce isomorphisms of the homology groups (7.9). Compactness for M_2 follows from $H_3(M_2, \mathbb{Z}) = 0$.

To conclude this section we recall some well-known methods for constructing manifolds, some of which allow a direct realization in terms of Kleinian groups and are closely related to the combination method, fundamental-polyhedron construction, etc.

1. *Connected sum.* Let M, M_1 and M_2 be n-manifolds (with or without boundary). Assume that there are n-dimensional topological balls $B_i \subset \operatorname{int} M_i$, $i = 1, 2$, and embeddings $h_i : N_i = M_i \setminus \operatorname{int} B_i \longrightarrow M$ such that $h_1(N_2) \cap h_2(N_2) = h_1(\partial B_1) = h_2(\partial B_2)$ and $M = h_1(N_1) \cup h_2(N_2)$. Then the manifold M is called a connected sum of manifolds M_1 and M_2 and denoted by $M = M_1 \# M_2$. If here M, M_1 and M_2 are orientable, it is required that the embeddings h_i preserve orientation.

The construction of manifolds by means of the connected sum (in the conformally uniformizable case) can be realized by the Klein combination (see Theorem 6.15, Example 6.18 and Theorem 8.45).

2. *Disc sum.* Let M, M_1 and M_2 be 3-manifolds with boundary, let D_i be a topological disc in the boundary of the manifold M_i, $i = 1, 2$, and let f be a homeomorphism mapping D_1 onto D_2. The manifold M obtained by gluing M_1 and M_2 according to the mapping f is called a disc sum of the manifolds M_1 and M_2. In the case of orientable manifolds, it is required that the homeomorphism f preserve orientation.

Let G_1 and $G_2 \subset M_2$ be Kleinian groups and let $G = G_1 * G_2$ be the group obtained by Klein combination. Then the manifold $[\mathbb{R}^3_+ \cup \Omega(G)]/G$ is the disc sum of the manifolds $M(G_i)$, $i = 1, 2$ - see §2.

3. *Excision of knots and links.* Let L be a knot (simple loop) or link (union of disjoint simple loops) in the sphere S^3 or in some 3-manifold M. Removing L from M, we obtain an open 3-manifold. Such manifolds play an important part in investigations on hyperbolic manifolds (see Chapter 8 §5). Here one also gets an interesting class of universal knots and links introduced by Thurston [5] cf. Example 6.25.

Namely the link $L \subset S^3$ is called *universal* if every closed oriented 3-manifold M is a covering S^3 branched over L. The first example of a universal link was the Whitehead link (Thurston [5]). Also the Borromean rings and many knots in S^3 are universal (Hilden-Lozano-Montesinos [1]-[3]. Moreover, Hilden, Lozano, Montesinos and Whitten [1] show that every closed oriented 3-manifold is a covering of S^3 branched over the Borromean rings and such that the branching indices of the components are (1, 2, 4), (1, 2, 4) and (2, 4). As a consequence of this result, they obtained a universal co-compact group $U \subset \operatorname{Isom} H^3$ generated by 90°-rotations in the three edges of the regular hyperbolic dodecahedron $D \subset H^3$. Namely, those edges corresponding to three symmetry lines in three faces of a cube from Example 2.30 (see Fig. 15) having a common vertex (here we deform D into this cube). More exactly, the following theorem holds:

THEOREM 7.15. *Every closed, oriented 3-manifold M is orientation-preserving homeomorphic to the quotient H^3/G, where $G = G(M)$ is a subgroup of the universal group U of finite index. Here the fundamental group $\pi_1(M)$ is isomorphic to G/F, where F is the normal subgroup of G generated by the elements of finite order.*

Note that, in particular, this result offers a new approach to the Poincaré Conjecture since the simply connected 3-manifolds correspond precisely to the subgroups $G \subseteq U$ of finite index that are generated by elements of finite order.

4. *Dehn surgery.* Let L be a knot or an m-component link in a 3-manifold N. We excise a tubular neighbourhood of some component of L (whose boundary is clearly a torus) and glue it back (solid) by means of a certain homeomorphism of the torus into itself. That procedure is known as Dehn surgery.

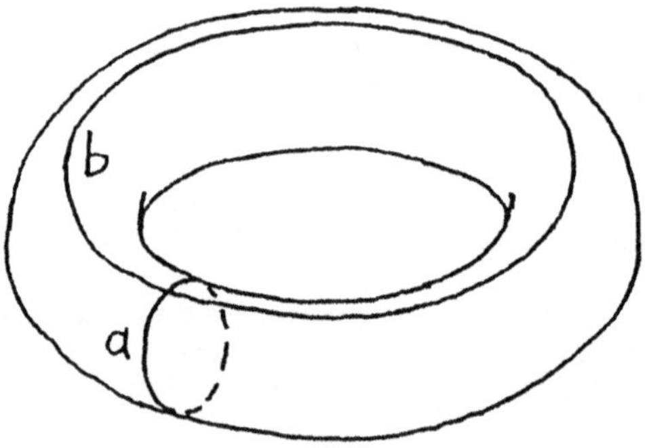

Figure 55

More precisely, let P_i be a tubular neighbourhood in N for the i-th component of the link L, $T_i^2 = \partial P_i$, with a_i the meridian ($a_i \simeq 0$ in $\bar{P}_i$), and b_i the parallel in T_i^2 (see Fig. 55), which we consider as generators $H_1(T_i, \mathbb{R})$. Then the generalized Dehn surgery in $M = N \setminus L$, with parameters $(d_1, \ldots, d_m)$, $(d_i = (\alpha_i, \beta_i) \in \mathbb{R}^2)$, is a procedure that gives the manifold $M_{d_1,\ldots,d_m}$, obtained by gluing solid tori to $M \setminus \bigcup P_i$ in such a way that the meridian in the glued

i-th torus goes over to $\alpha_i a_i + \beta_i b_i$. If, on the other hand, the i-th link component is not involved, the pair $d_i = (\alpha_i, \beta_i)$ is denoted by infinity (Thurston [1], see also Culler, Gordon, Luecke, and Shalen [1]).

5. *Heegard splitting*. Let M be a closed 3-manifold. The Heegard splitting of the manifold M is defined to be a pair (N_1, N_2), where N_i, $i = 1, 2$, is a handlebody, while the manifold M is the union $N_1 \cup N_2$, in which $N_1 \cap N_2 = \partial N_1 = \partial N_2$ (here and subsequently, a handlebody is understood to be an orientable manifold with boundary that is homeomorphic to a disc sum of solid tori).

It is known (Hempel [1]) that each closed 3-manifold has a Heegard splitting, i.e., can be glued together from two homeomoprhic handlebodies. As a set of gluing homeomorphisms in the general case has not lent itself to description, it is impossible, as a rule, to say anything about the resulting glued manifold, apart from certain simple cases such as solid tori splitting. That is, let a closed 3-manifold M have a Heegard splitting of genus one. Then in the orientable case M is a lens space $L(p, q)$, in which p and q are some coprime integers, while, in the non-orientable case, M is homeomorphic to the non-orientable fibre on two-dimensional spheres over the circle (Hempel [1]).

6. *Gluing of polyhedron faces*. Let there be a polyhedron $P \subset S^3$, homeomorphic to a ball, and let its boundary be a Jordan surface split up into polygons satisfying the usual triangulation conditions. Therefore, we have vertices, edges, and faces of P. By baricentric subdivision, from P one can obtain a simplicial complex (see Rourke and Sanderson [1]). After this, we glue in pairs the faces of P by means of orientation-preserving homeomorphisms in such a way that the vertices and edges of any face become the vertices and edges of another face. The resulting complex K is a 3-manifold if and only if its characteristic is zero (see Seifert and Threlfall [1]).

It has been mentioned in Chapter 2 §5 that, if one takes the gluing homeomorphisms from the isometry group of some space (hyperbolic, Euclidean, elliptic, affine, and so on), the corresponding geometric structure is introduced into K. If P is a fundamental polyhedron of a Kleinian group G, then in $K = \Omega(G)/G$ one introduces at least a conformal structure (see also Chapter 6 §1).

§2. Topological aspects of combination theorems

Chapter 6 §2 and §3 deal with certain constructions (the Klein-Maskit combination), which enable one to construct more complicated Kleinian groups from simple ones. Our purpose here is to demonstrate what occurs in such a construction for a 3-manifold $M(G)$ for the constructed group G, following Marden [1]. The multi-dimensional case for one class of manifolds will be considered in Chapter 8 §6, too.

1. *Free products.* We are given the manifolds $M(G_1)$ and $M(G_2)$ having closed Euclidean discs $D_i \subset \partial M(G_i)$, $i = 1, 2$; if $M(G_1) = M(G_2)$, then we specify $D_1 \cap D_2 = \emptyset$. By the Euclidean disc $D \subset \partial M(G)$ we mean a domain such that each component $\pi^{-1}(D)$ (by natural projection $\pi : H^3 \cup \Omega(G) \longrightarrow M(G)$) can be mapped into the unit disc $U \subset \mathbb{R}^2$ by a Möbius transformation.

We demonstrate that there exists a group G_3 such that $M(G_3)$ is homeomorphic to the manifold obtained from $M(G_i)$, $i = 1, 2$, by identifying the discs D_1 and D_2 (the disc sum). To construct G_3, we assume initially that $M(G_1) \neq M(G_2)$ (the possibility $G_1 = G_2$ remains). We denote by D_1^* a fixed component of $\pi^{-1}(D_1)$ and find the group G_2' conjugate to G_2 $(M(G_2) = M(G_2'))$ such that one of the components $\pi^{-1}(D_2)$ coincides with the complement to D_1^*. Then the group $G_3 = \langle G_1, G_2' \rangle \cong G_1 * G_2$ (Fig. 56), obtained by the Klein combination from G_1 and G_2', has the necessary property.

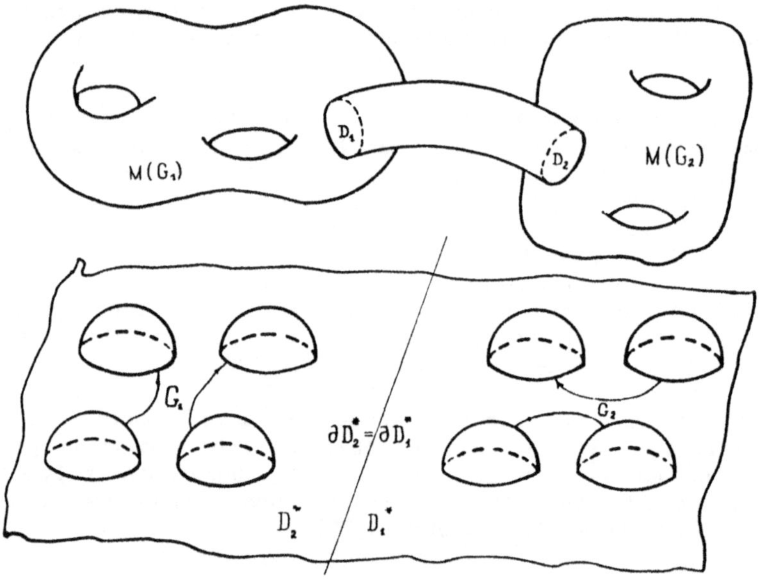

Figure 56

If we have $M(G_1) = M(G_2)$, we specify along with D_1^* a certain component $D_2^* \subset \pi^{-1}(D_2)$; it is clear that $D_1^* \cap D_2^* = \emptyset$. Then, if g is a Möbius transformation that transfers the exterior of D_1^* into the interior of D_2^*, while the groups $G_1 \cong G_2 \cong G$ correspond to the manifolds $M(G_1) = M(G_2) = M(G)$, then as G_3 one can take $G_3 = \langle G, g \rangle \cong G * \langle g \rangle$. We note that, if D_1 and D_2 lie on the same component of the boundary $\partial M(G)$, the process gives a new handle on that component; if the discs D_i lie on different components of the boundary $\partial M(G)$, the gluing combines these components into one component of the boundary of the resulting manifold $M(G_3)$ (Fig. 57).

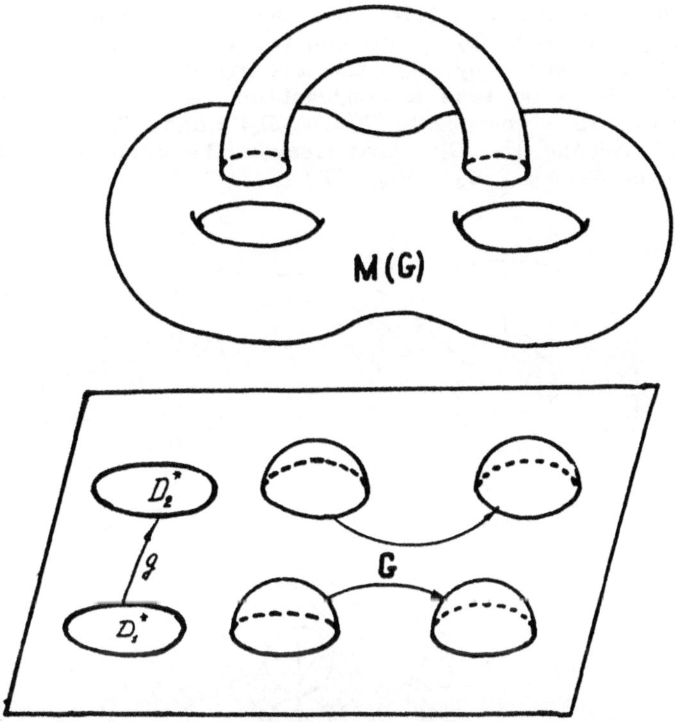

Figure 57

If we assume that D_1 and D_2 are tangent, $D_1 \cap D_2 = \{p\}$, then as the mapping g one can take a parabolic transformation with the fixed point $p^* = D_1^* \cap D_2^*$. The effect of gluing D_1 and D_2 then gives a pair of punctures in the corresponding component of the boundary $\partial M(G_3)$.

A Schottky group of rank g corresponds to gluing together $2g$ discs in the sphere at infinity $S^2 = \partial H^3$ ($M(G)$ is a handlebody of genus g). If G_1 and G_2 are Fuchsian groups that uniformize surfaces of genus g_i, having n_i punctures, then $G_3 = G_1 * G_2$ is a functional group having an invariant component $\Omega_0(G_3)$, and the surface $\Omega_0(G_3)/G_3$ has genus $g_1 + g_2$ and has $n_1 + n_2$ punctures.

2. *Free products with amalgamated cyclic subgroups.* By a horocyclic disc D for a puncture p in $\partial M(G)$ we understand the following: D is conformally equivalent to a disc with a puncture and each component $\pi^{-1}(D)$ is a Euclidean disc or half-plane, invariant

with respect to the maximal parabolic subgroup, defined by the puncture p.

Now let there be two manifolds $M(G_1)$ and $M(G_2)$ having punctures p_1 in $\partial M(G_1)$ and p_2 in $\partial M(G_2)$, and horocyclic discs D_1 and D_2. Then there exists a group $G_3 \subset \mathrm{Isom}\, H^3$ such that $M(G_3)$ is homeomorphic to the manifold obtained from $M(G_1)$ and $M(G_2)$ by identifying D_1 and D_2. To construct G_3, we specify a component $D_1^* \subset \pi^{-1}(D)$ and denote by G_p the maximal cyclic parabolic subgroup in G_1, $G_p(D_1^*) = D_1^*$; we take a conjugation G_2' of the group G_2 such that $\mathrm{ext}\, D_1^*$ is a component $D_2^* \subset \pi^{-1}(D_2)$ and G_p is a maximal cyclic parabolic subgroup in G_2' that leaves the component D_2^* invariant. Then $G_3 = \langle g_1, G_2' \rangle \cong G_1 \underset{G_p}{*} G_2'$ (Fig. 58).

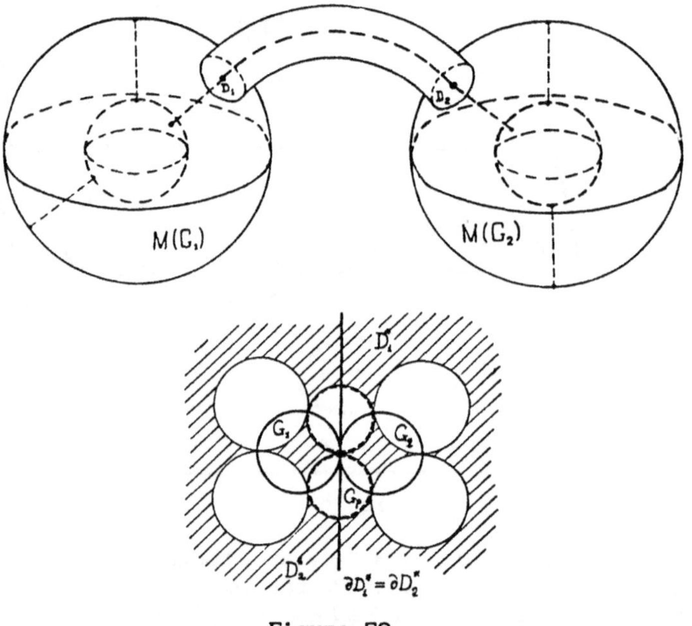

Figure 58

This construction can also be performed for two distinct parabolic punctures p_1 and p_2 at the boundary of a manifold $M(G)$; for that purpose, for $i = 1, 2$, we specify the components $D_i^* \subset \pi^{-1}(D_i)$, where D_i is a horocyclic disc in $\partial M(G)$ in the puncture p_i, and let $G_i \subset G$ be a maximal parabolic subgroup corresponding to p_i, $i = 1, 2$. We consider a Möbius transformation g that transforms $\mathrm{ext}\, D_1^*$ into $\mathrm{int}\, D_2^*$ and which conjugates the subgroups $G_i : G_2 = gG_1g^{-1}$. Then the group $G_3 = \langle G, g \rangle$ corresponds to the manifold $M(G_3)$, homeomorphic to a manifold obtained from $M(G)$ identifying the horocyclic discs D_1 and D_2.

In particular, for the Fuchsian groups G_1 and G_2 that uniformize surfaces of genus g_i having n_i punctures, the G_3 obtained by gluing one puncture in $\partial M(G_1)$ to one puncture in $\partial M(G_2)$ is a B-group. Here $\Omega_0(G_3)/G_3$ is a surface of genus (g_1+g_2) having (n_1+n_2-2) punctures. The other two components $\partial M(G_3)$ are surfaces of genus g_i having n_i punctures (Fig. 58). When one glues two punctures in one component $\partial M(G_1)$, one gets a B-group G_3, whose surface $\Omega_0(G_3)/G_3$ has genus g_1+1 and has n_1-2 punctures. The other component $\partial M(G_3)$ has genus g_1 and has n_1 punctures. In that way, we can eliminate all the punctures in $\partial M(G)$ and obtain a manifold $M(G_3)$ having only punctured solid tori - see Example 6.27.

3. *Double of the manifold* $M(G)$ *with respect to the boundary* $\partial M(G)$. Let the boundary $M(G)$ consist of the surfaces S_i, $i=1,\ldots,m$, for which all the connected components in $\pi^{-1}(S_i)$ are Euclidean discs (up to Möbius transformations). Then when we exclude the case of a Fuchsian group G, we can obtain a group $\Gamma\subset \operatorname{Isom} H^3$ such that $M(\Gamma)$ is homeomorphic to the double of $M(G)$ with respect to the surfaces $S_1,\ldots,S_m$.

To construct that Γ, we specify the component $D_1^*\subset\pi^{-1}(S_1)$ and let $\mathcal{T}_1$ be a reflection with respect to the circle ∂D_1^*. As a first step, we consider the group $G_1=\langle G, \mathcal{T}_1G\mathcal{T}_1\rangle$. The manifold $M(G_1)$ is homeomorphic to the double of $M(G)$ with respect to S_1. The other components S_i of the boundary $\partial M(G)$ at the boundary of the manifold $M(G_1)$ will correspond each to two surfaces S_{ij}, $j=1,2$, having components $\pi^{-1}(S_{ij})$ that are Euclidean discs.

The next step is to identify the surfaces S_{21} and S_{22}. For this purpose, we specify the component $D_2^*\subset\pi^{-1}(S_{21})$. Then $\mathcal{T}_1(D_2^*)\subset\pi^{-1}(S_{22})$. We denote reflection with respect to ∂D_2^* by $\mathcal{T}_2$. Then $\mathcal{T}_1\mathcal{T}_2: D_2^*\longrightarrow \operatorname{ext}\mathcal{T}_1(D_2^*)$ conjugate the stabilizers in G_1 of the components D_2^* and $\mathcal{T}_1(D_2^*)$. Therefore, the manifold $M(G_2)$, in which $G_2=\langle G_1, \mathcal{T}_1\mathcal{T}_2G_1\mathcal{T}_2\mathcal{T}_1\rangle$, is homeomorphic to the manifold glued from $M(G_1)$ by the identification of S_{21} and S_{22}.

We continue this gluing of the components of S_{i1} and S_{i2}, $i=3,\ldots,m$, which gives the desired group Γ. Here $\mathcal{T}_1\Gamma\mathcal{T}_1=\Gamma$, and the reflection $\mathcal{T}_1$ specifies an anticonformal involution in $M(\Gamma)$ that permutes the two sheets in the double.

That construction does not work only in the case of a Fuchsian group G. In that case $\mathcal{T}_1\mathcal{T}_2=\mathrm{id}$. It is evident from other considerations that, for that case, the double $\hat{M}(G)$ is not homeomorphic to $M(\Gamma)$ for any $\Gamma\subset\operatorname{Isom} H^3$ because $\hat{M}(G)$ is homeomorphic to that product $S\times S^1$, which has a non-trivial centre in the fundamental group $\pi_1(S\times S^1)$. However, the double $\hat{M}(G)$ is homeomorphic to the quotient $[\overline{\mathbb{R}}^3\backslash\Lambda(G)]/G$, where we assume that $G\subset \text{Möb}(\mathbb{R}^3_+)$.

§3 Kleinian manifold of a function group

In this section we show how the topological technique presented in the previous sections and in Chapter 5 §5 applies to a description of the structure of a manifold $M(G) = H^{n+1} \cup \Omega(G)$, in which $G \subset \text{Möb}(n)$, $n \geq 2$, is a geometrically finite Kleinian group having an invariant component $\Omega_0 \subset \Omega(G)$ (i.e., G is a function group). We shall see that the properties of such manifolds in dimensions $n+1 \geq 4$ and $n+1 = 3$ differ substantially.

1. *Three-dimensional manifolds.* We begin with the structure of a three-dimensional manifold $M(G)$, where we follow Marden [1]. We put $S_0 = \Omega_0/G$ and let $S_1, \dots, S_m$ denote the remaining components of the boundary $\partial M(G)$. Function groups are characterized by the property that the homomorphism $\pi_1(S_0) \longrightarrow \pi_1(M(G))$ is an epimorphism (see Lemmas 7.11 and 7.12).

Case 1. The group $G \subset \mathcal{M}_2$ has two invariant components $\Omega_0, \Omega_1 \subset \Omega(G)$. Then the manifold $M(G)$ is homeomorphic to a surface layer $S_0 \times [0, 1]$, while the group G is quasi-Fuchsian. This is true for finitely generated G (geometric finiteness is a consequence).

To prove this, we note first that Ω_0 and Ω_1 should be simply connected (cf. Accola [1]). In fact, if $\pi_1(\Omega_0) \neq 0$, Theorem 7.5 implies that there exists a loop, non-trivial in S_0, bounding a disc D, embedded in $M(G)$. D divides the manifold $M(G)$, since otherwise one could find a loop $\gamma \subset \text{int}\, M(G)$ having a non-zero intersection index with D and consequently not homologic to a loop on the surface $S_1 = \Omega_1/G$, $S_1 \cap D = \emptyset$. This contradicts the epimorphism $\pi_1(S_1) \longrightarrow \pi_1(M(G))$. Now let N be a component of $M(G) \setminus D$, not adjoining the surface S_1. Then any loop in N is freely homotopic to a loop in S_1, and therefore is freely homotopic to a loop in D, i.e., $\pi_1(N) = 0$. Then we see that each component $\pi^{-1}(N)$ is homeomorphic to N and therefore $N \cap S_0$ is a disc, while $\partial D = \partial(N \cap S_0)$ is a trivial loop in S_0. This contradicts the choice of the disc D.

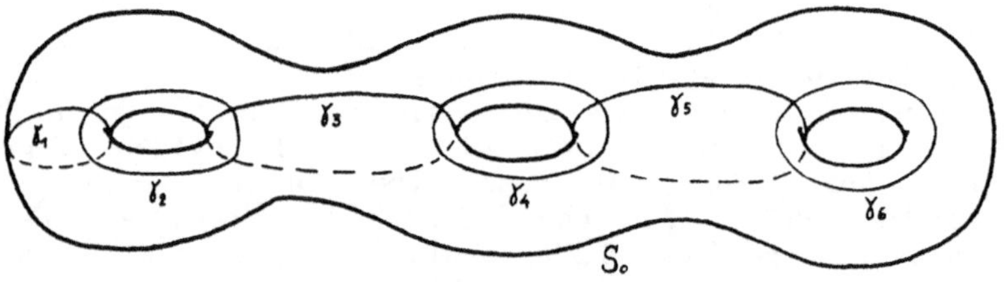

Figure 59

We now show for the compact case (S_0 is a closed surface of genus p) how to prove that $M(G)$ and the surface layer $S_0 \times [0, 1]$ are homeomorphic (for details and the non-compact case, see Marden [1]). Namely, we specify a system of simple loops $\gamma_1, \dots, \gamma_{2p}$ in S_0 such that γ_i intersects only γ_{i-1} and γ_{i+1} (at unique points and transversally) while not intersecting the other loops γ_j (Fig. 59). Then the surface $S_0 \setminus \bigcup \gamma_i$ is a disc. From Theorem 7.6, each loop γ_i

corresponds to a simple loop $\gamma'_i \subset S_1$, together with which it bounds a cylinder C_i in $M(G)$. One can assume that C_i intersects only the cylinders C_{i-1} and C_{i+1} (transversally, on one arc in each cylinder). We denote by $C'_i = C_i \times (-\varepsilon, \varepsilon)$ the ε-neighbourhood of C_i. Then the surface $\partial(\bigcup C'_i)$ is a cylinder, and $S_0 \setminus [S_0 \cap (\bigcup C'_i)]$ is a disc. Similarly, as $\bigcup \gamma'_i$ bounds a disc in S_1, it follows that $S_1 \setminus [S_1 \cap (\bigcup C'_i)]$ is also a disc. Then $\operatorname{int} M(G) \setminus \bigcup C'_i$ is a ball, and therefore $M(G)$ is homeomorphic to the layer $S_0 \times [0, 1]$.

If now G_0 is a Fuchsian group that uniformizes a closed surface of genus g, we can consider a piecewise linear (and therefore quasi-conformal) homeomorphism $M(G_0) \longrightarrow M(G)$.

Its lifting in $H^3 \cup \Omega(G_0)$ is again a quasi-conformal homeomorphism, which is extended to a quasi-conformal automorphism $f : \overline{H}^3 \longrightarrow \overline{H}^3$ (see Gehring [1]). The restriction f in the plane $\overline{\mathbb{C}}$ gives a quasi-conformal conjugation of the group G_0, whose result is the group $G = fG_0f^{-1}$.

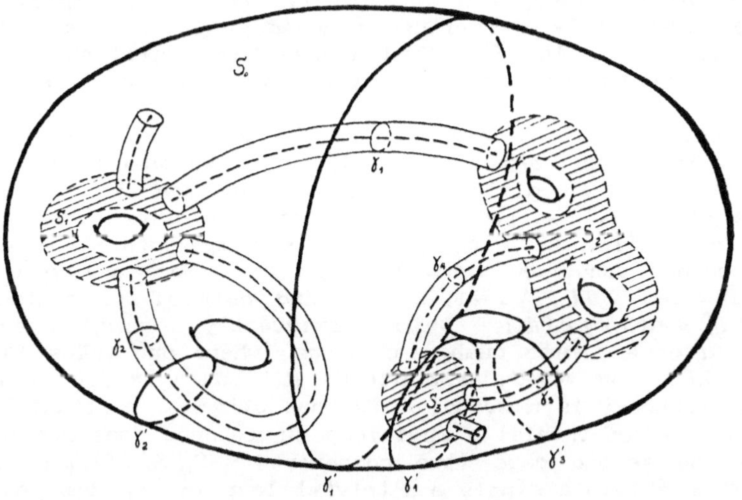

Figure 60

Case 2. The group G is a B-group, i.e. the invariant component $\Omega_0 \subset \Omega(G)$ is simply connected. Let the manifold M_0 be obtained from $M(G)$ by cutting out all punctured solid cylinders C_i, $1 \le i \le n$, that do not intersect the surface $S_0 = \Omega_0/G$. When we examine the inclusion of the homology groups $H_1(\partial M_0) \longrightarrow H_1(M_0)$, we can show that $\partial M_0 \ne S_0$. As in case 1, it is shown that M_0 is homeomorphic to the surface layer $S_0 \times [0, 1]$. We now establish how the cylinders C_i, $1 \le i \le n$, correspond to elements of G. From Theorem 7.6, in S_0 there exist simple non-intersecting loops γ'_i, $1 \le i \le n$, that, together with the loops γ_i around C_i, bound cylinders C'_i in M_0. Here the cylinder C'_i divides M_0 if and only if the loop γ_i divides S_0. Let N be a component of $M_0 \setminus \bigcup C'_i$, and N^* be a component of $\pi^{-1}(N)$. Then the stabilizer $G_0 \subset G$ of the component N^* is a quasi-Fuchsian group, and N is the retract of $M(G_o)$. The element in the group G, determined by the loop γ_i, is an accidental parabolic one (it is parabolic, but in Ω_0 it acts as a hyperbolic element in the unit disc, i.e., it does not define a puncture in S_0, see Fig. 60).

Case 3. The free groups $G \subset \mathcal{M}_2$ (Schottky and Schottky-type groups). Let M_0 be a compact manifold obtained from $M(G)$ by cutting out all punctured solid cylinders (a finite number); we assume that ∂M_0 consists of one surface, genus g. As G is free, the homomorphism $\pi_* : \pi_1(\partial M_0) \longrightarrow \pi_1(M_0) \cong G$ cannot be a monomorphism. Theorem 7.5 shows that there is a simple loop γ_1, non-trivial in δM_0, bounding the disc D_1 in M_0. We assume first that γ_1 does not divide up ∂M_0, so D_1 does not divide up M_0. We remove from M_0 the neighbourhood $D_1^* = D_1 \times (-\varepsilon, \varepsilon)$ of D_1 to get a compact manifold M_1, whose boundary is a surface of genus $g-1$. We obtain the subsequent loops γ_i in ∂M_1, which have the properties of the loop γ_1, and continue the process until at the step g we obtain a closed ball M_g.

The proof that it is possible to choose these loops $\gamma_1, \gamma_2, \ldots$ that do not divide up ∂M_0 is as follows. We assume, for example, that all the loops γ_1 divide up ∂M_0. There are two possibilities: either the disc D_1 splits up M_0 into two parts, M_1 and M_2, different from balls, or $\partial(M_0 \setminus D_1)$ has two components. In the first case, the common genus of $\partial M_1 \cup \partial M_2$ is g. Here a simple loop in $\partial M_i \cap \partial M_0$ divides ∂M_i, $i = 1, 2$, if and only if it divides ∂M_0. Also $\pi_1(M_0) \cong \pi_1(M_1) * \pi_1(M_2)$, and consequently $\pi_1(M_i)$ are free. Therefore, after a finite number of steps (where one takes loops that divide up ∂M_0), we obtain a manifold M_n having a free fundamental group whose boundary is a torus. We thus obtain a contradiction with the fact that a non-trivial simple loop in a torus does not divide it up. In the second case (two components $S_1, S_2 \subset \partial(M_0 \setminus D_1)$), from Theorem 7.5 we obtain a simple non-trivial loop in S_1 that divides up S_1 and bounds a disc in $M_0 \setminus D_1$ and so on. We continue this process to get a submanifold $M \subset M_0$ that has a component ∂M as a sphere. That sphere is a component of ∂M_0, which is impossible.

Thus, we have shown that, if G is a free group, $M_0(G)$ is a handlebody of genus g (a solid sphere with g handles). Such groups, in particular, are represented by all G having connected $\Omega(G)$. If G is also a purely loxodromic group, then $M_0(G) = M(G)$, and we see that G is a Schottky group (see Maskit [1]). In the general case, one can show that G is a Schottky-type group.

Case 4. A group $G \subset \mathcal{M}_2$ is isomorphic to the fundamental group of a closed surface S. If $M_0(G)$ is again the manifold obtained from $M(G)$ by cutting out all punctured solid cylinders, then it is homeomorphic to the layer $S\times[0, 1]$. The proof is analogous to that for case 2 on the basis of the following two facts:

1) if a subgroup $H\subset G$ is isomorphic to the fundamental group of a closed surface whose genus does not exceed the genus of S, then $H=G$; and

2) the group G is not a non-trivial free product of its subgroups.

The general case is obtained from the above. Namely, let $S_0=\Omega_0/G$. We apply Theorem 7.5 to get the pairwise non-intersecting simple loops γ_i, non-trivial in S_0, that are not boundaries of rings in S_0 and which bound discs $D_i\subset M(G)$. We take the $\{\gamma_i\}$ loop system as maximal: for each component $S'\subset S_0\setminus\bigcup\gamma_i$, the homomorphism $\pi_*:\pi_1(S^1)\longrightarrow\pi_1(M(G))$ is a monomorphism.

Now let N be a component of $M(G)\setminus\bigcup D_i$. We specify a component $N^*\subset\pi^{-1}(N)$. Then its stabilizer G_0 is either a B-group or a free Abelian group of rank two, or a cyclic parabolic or loxodromic group. Here the manifold N is the retract of $M(G_0)$. Then in general G can be represented as a free product of a certain finite number of B-groups, free Abelian groups of rank two, and cyclic parabolic and cyclic loxodromic groups. The structure of $M(G)$ is clearly understood by means of the glueings described in §2 corresponding to such a combination of groups.

2. *Multi-dimensional Kleinian manifolds.* The manifolds $M(G)=[H^{n+1}\cup\Omega(G)]/G$ for the geometrically finite groups $G\subset \text{Möb}(n)$, $n\geq 3$, have to a certain extent ananalogous structure when they have the invariant contractible components $\Omega_0\subset\Omega(G)$.

Namely, the three-dimensional manifold $M(G)$ for a geometrically finite $G\subset M_2$ having a simply connected invariant component $\Omega_0\subset\Omega(G)$ is converted into a surface layer $S_g\times[0, 1]$, as we have seen above, after eliminating some punctures and removing neighbourhoods of others. For $n\geq 3$, analogous procedures, starting with an $(n+1)$-dimensional manifold $M(G)$, give, instead of $S_g\times[0, 1]$, a certain analogue of the surface layer.

We can consider analogies of the $(n+1)$-dimensional surface layer $(n\geq 3)$ with various degrees of generality:

a) the product of an n-dimensional manifold $N_0=\Omega_0/G$ and the segment $[0, 1]$;

b) a manifold M, whose boundary ∂M consists of two components N_0 and N_1 and such that the triple $(M; N_0, N_1)$ is an h-cobordism; and

c) a manifold M with the boundary components N_0 and N_1, which is a homology cobordism.

Here we use the following cobordism definition. Let M be a compact manifold whose boundary ∂M consists of two disjoint connected closed manifolds N_0 and N_1, $\partial M=N_0\cup N_1$.

Then the triple $(M; N_0, N_1)$ is called a homology (homologically trivial) cobordism if all the relative homology groups are trivial:

$$H_*(M, N_0) = H_*(M, N_1) = 0. \qquad (7.10)$$

The triple $(M; N_0, N_1)$ of compact manifolds with boundary is called a homology cobordism with boundary if $N_0, N_1 \subset \partial M$, $N_0 \cap N_1 = \emptyset$, and for the boundary $\delta M = \partial M \setminus (N_0 \cup N_1)$ we have

$$H_*(\delta M, \partial N_0) = H_*(\delta M, \partial N_1) = H_*(M, N_0) = H_*(M, N_1) = 0. \qquad (7.11)$$

If above we replace the requirement for triviality of the relative homology groups by that for triviality of the corresponding relative homotopy groups, we get definitions of an h-cobordism and an h-cobordism with boundary.

We shall see below that an $(n+1)$-dimensional manifold

$$M(G) = [H^{n+1} \cup \Omega(G)]/G$$

for $n \geq 3$ inherits the surface-layer structure from the case $n = 2$ in the following form (for simplicity we assume here that M is compact):

1. In the sense of (a), if the group G is a quasi-conformal conjugation of a certain Fuchsian group in $\overline{\mathbb{R}}^n = \partial H^{n+1}$;
2. In the sense of (b), if G has two invariant contractible components $\Omega_0, \Omega_1 \subset \Omega(G)$, see Theorem 7.18.
3. In the sense of (c), if G has an invariant contractible component $\Omega_0 \subset \Omega(G)$, see Theorem 7.15.

Moreover, it is shown below that there are four-dimensional compact manifolds $M(G)$ that are homology cobordisms, i.e., they realize (c), but which are not h-cobordisms, i.e., do not satisfy (b).

We now give precise formulations.

THEOREM 7.15 (Tetenov [3], [4]). *Let G be a geometrically finite torsion free Kleinian group in $\overline{\mathbb{R}}^n$, $n \geq 2$, having an invariant contractible component $\Omega_0 \subset \Omega(G)$ and let $N_0 = \Omega_0/G$. Then in the Kleinian manifold $M(G) = [H^{n+1} \cup \Omega(G)]/G$ there is a compact $(n+1)$-dimensional submanifold M_C with the following properties*:

a) *M is obtained from M_C by attaching an open collar $\delta M_C \times [0, 1)$ on the boundary $\delta M_C = \partial M_C \setminus \partial M$ of the submanifold M_C in M;*

b) *connected components of the collar $\delta M_C \times [0, 1)$ are finite-sheeted covered by punctured solid cylinders*

$$T^{n-k} \times B^k \times [0, 1), \qquad 1 \leq k \leq n-1,$$

(here B^k is a closed k-dimensional ball and T^{n-k} is an $(n-k)$-dimensional torus).

c) *the boundary ∂M_C contains N_{0C} and N_{1C}, which are connected disjoint n-dimensional manifolds with boundary, such that*

$$\pi_*(M_C, N_{0C}) = 0 \quad \text{and} \quad H_*(M_C, N_{1C}) = 0$$

and the cobordism with boundary $(M_c;\ N_{0c},\ N_{1c})$ *is a homology cobordism.*

In this case,

$$N_{0c} = N_0 \cap M_c, \qquad N_{1c} \supset M_c \cap (\partial M \setminus N_0),$$

$$\partial N_{0c} \approx \partial N_{1c}, \qquad \partial M_c = N_{0c} \cup N_{1c} \cup (\partial N_{0c} \times [0,\ 1]).$$

From this theorem and Theorm 5.39 we get directly:

COROLLARY 7.16. *Suppose the Kleinian group* G *from Theorem 7.15 does not have parabolic elements. Then the compact Kleinian manifold* $M(G)$ *has two boundary components* $N_0 = \Omega_0/G$ *and* $N_1 = (\Omega(G)\setminus\Omega_0)G$, *and the triple* $(M(G);\ N_0,\ N_1)$ *is a homology cobordism in the sense of* (7.10).

One can also give a direct proof of the latter without using Theorem 7.15, which on the one hand is simple and on the other casts light on the basic ideas in proving Theorem 7.15.

PROOF OF COROLLARY 7.16. The condition that the group G contains no parabolic elements and Theorem 5.39 imply that $M(G)$ is a compact manifold. That manifold and its boundary component $N_0 = \Omega_0/G \subset \partial M(G)$ are spaces of $K(G,\ 1)$ type. At the same time, the inclusion $N_0 \subset M(G)$ induces an isomorphism of the fundamental groups $\pi_1(N_0) \longrightarrow \pi_1(M(G))$ and thus it is a homotopy equivalence: $\pi_*(M(G),\ N_0) = 0$. Then $H_*(M(G),\ N_0) = 0$. Therefore, using Poincaré duality and the theorem on universal coefficients (it expresses cohomology via homology, see Dold [1], VI.7.10 and VIII.7.2) we obtain that $H_*(M(G),\ \partial M(G)\setminus N_0) = 0$, and, in particular, that $H_0(\partial M(G)\setminus N_0) = \mathbb{Z}$. Then $\partial M(G)\setminus N_0$ consists of only one component N_1, $N_1 = (\Omega(G)\setminus\Omega_0)/G$ and here $H_*(M(G),\ N_1) = 0$. This completes the proof of (7.10), which demonstrates the homology triviality of the cobordism $(M(G);\ N_0,\ N_1)$.

PROOF OF THEOREM 7.15. Here we use the symbols, constructions, and a lot of major facts obtained in Chapter 5 in the proof of Theorem 5.57, which is closely related to this theorem. Assume that the maximal set of non-equivalent parabolic cusps p_i, $0 \le i \le s$, has been chosen such that for a certain integer $r \le s$ we have

$$\operatorname{rank} G_{p_i} = \begin{cases} k_i < n-1 & \text{for } \ i \le r, \\ k_i = n-1 & \text{for } \ r < i \le s, \end{cases}$$

(here always $k_i < n$, due to G-invariance and contractibility of Ω_0). Here by W_i we denote the neighbourhoods of parabolic ends in the manifold $M(G)$, having (modulo a finite-sheeted covering) the following form (see (5.44)):

$$W_i = \tilde{U}_i/G_{p_i} \approx B^{n-k_i} \times (0,\ 1) \times T^{k_i},$$

in which $\tilde{U}_i \subset H^{n+1} \cup \Omega(G)$ are cusp domains for p_i. When these neighbourhoods are deleted from $M(G)$, there remains a compact manifold M_C, for which we put $N_{0C} = M_C \cap N_0$, where $N_0 = \Omega_0/G$ is the boundary component of $M(G)$.

We have the following possibilities (see the proof of Theorem 5.57):

(1) $\operatorname{rank} G_{p_i} < n-1$ for $i \le r$. In this case $\bar{W}_i \cap \partial M$ are connected and $\bar{W}_i \cap N_o = \varnothing$.

(2) For a certain q, $r < q \le s$, and for all i, $r < i \le q$, we have $\operatorname{rank} G_{p_i} = n-1$ and $\bar{W}_i \cap N_o \neq \varnothing$. Then $\bar{W}_i \cap (\partial M(G) \setminus N_o) \neq \varnothing$.

(3) For any i, $q < i \le s$, we have $\operatorname{rank} G_{p_i} = n-1$ and $\bar{W}_i \cap N_o = \varnothing$.

We change nothing in the cases (1) and (3), while in case (2) we consider a finite-sheeted cover of $M(G)$ by a manifold $M(G')$, where the subgroup $G' \subset G$ of finite index is such that for any i, $r < i \le q$, the stabilizers $G'_{p_i} \subset G'$ are free Abelian groups of rank $n-1$. Keeping the notation unchanged, we act as follows. We glue up to the compact manifold M_C a manifold $\hat{T}^n \times [0, 1]$ (where $\hat{T}^n$ is a solid torus $T^{n-2} \times [0, 1]^2$) instead of each parabolic neighbourhood W_i, $r < i \le q$.[1] The resulting manifold we denote by $\hat{M}$, and its boundary component containing N_{0C} we denote by $\hat{N}_0$. Then, as has already been proved (see (5.50)), the boundary $\partial\hat{M}$ consists of two components - the second we denote by $\hat{N}_1$. Besides the inclusions $\hat{N}_0 \hookrightarrow \hat{M}$ and $\hat{N}_1 \hookrightarrow \hat{M}$ are homology equivalences (see (5.48) and (5.51)). The last fact, combined with the Mayer-Vietoris exact sequence (2.7) for pairs of triads

$$(\hat{M};\ M_c,\ \bigcup_{j=1}^{q-r}(\hat{T}^n \times [0,\ 1])_j) \supset (\hat{N}_1;\ \hat{N}_1 \cap M_c,\ \bigcup_{j=1}^{q-r}(\hat{T}^n \times \{1\})_j)$$

imply (compare the proof of (5.48)) that the pair

$$(M_C,\ N_{1C}) = (M_C,\ \hat{N}_1 \cap M_C)$$

is homology trivial. The complement in ∂M_C to the union of subsets N_{0C} and N_{1C} is a set of the form

---

[1] We can avoid using the finite-sheeted cover if we use the Seifert fibre structure on closed Euclidean manifolds E^{n-1} (for $n \le 5$ compare Charlap [1]; for $n \le 4$ also see Scott [4], Theorem 4.3). Namely, instead of the described glueing up of $\hat{T}^n \times [0, 1]$ to the finite-sheeted covering $M(G')$ one has to glue up to the original manifold M_C a manifold $\hat{W}_i$ instead of W_i. Here $\hat{W}_i$ is obtained from $W_i \approx E^{n-1} \times [0, \infty)$ by changing the punctured discs $S^1 \times [0, \infty)$ by the closed discs $[0, 1]^2$ in the fibration of W_i.

$$\left[\bigcup_{j=1}^{q-r}(T^{n-1})_j\right]\times(0,\ 1)$$

and, therefore, the manifold M_C is a homology cobordism with boundary

$$\bigcup_{j=1}^{q-r}(T^{n-1}\times[0,\ 1])_j = \bigcup_{j=1}^{q-r}(\overline{W}_j\cap M_C).$$

Now we return on to the original manifold $M(G)$ (i.e., before taking its finite cover), with the theorem already for its finite cover. One easily checks the properties a) and b) of the theorem for $M(G)$. As far as the property c) is concerned, we obtain $\pi_*(M(G)\,N_0)=0$ from the fact that $M(G)$ and N_0 are of the type $K(G,\ 1)$. We also obtain $H_*(M(G),\ N_1)=0$ by applying the proof scheme for the fact $H_*(\hat{M},\ \hat{N}_1)$ (see (5.49)) to a manifold obtained by glueing two copies of M_C with their $\partial N_{0C}\times[0,\ 1]$ subsets identified. Further using the Mayer-Vietoris sequence for a couple of pairs shows the homology triviality of the pair $(M_C,\ N_{1C})$.

Thus we obtain that the manifold $M_C\subset M(G)$ is a homology cobordism with a boundary

$$\bigcup_{j=1}^{q-r}(\overline{W}_j\cap M_C)$$

each of whose components is finitely covered by a cylinder homeomorphic to $T^{n-1}\times[0,\ 1]$. This completes the proof.

REMARK 7.17. The proof of Theorem 7.15 (or Corollary 7.16) implies (compare Rourke and Sanderson [1]) that contractibility for all components $\Omega_i\subset\Omega(G)$ means that the above homology cobordism with boundary is an h-cobordism with boundary. In particular, the following theorem applies (compare Theorem 5.56):

THEOREM 7.18. *Let a torsion free Kleinian group G in $\overline{\mathbb{R}}^n$, $n\geq 2$, have two invariant contractible components $\Omega_1,\ \Omega_1\subset\Omega(G)$ having compact quotient manifolds $N_0=\Omega_0/G$ and $N_1=\Omega_1/G$. Then $M(G)$ is also compact, and the triple $(M(G);\ N_0,\ N_1)$ is an h-cobordism:*

$$\pi_*(M(G),\ N_0) = \pi_*(M(G)\mathrm{m}\,N_1) = 0.$$

However the following important question remains open:
is the h-cobordism from Theorem 7.18 trivial (i.e., the product), or more generally, is the $(n+1)$-dimensional h-cobordism $(M(G);\ \Omega_0/G,\ \Omega_1/G)\,n\geq 3$, which corresponds to a Kleinian geometrically finite group G with two invariant contractible components $\Omega_0,\ \Omega_1\subset\Omega(G)$ homeomorphic to the product.

This is a particular case of the S.P. Novikov conjecture - see Farrell-Jones [1], [2] and Farrell-Hsiang [1], [2].

3. *Non-trivial four-dimensional cobordisms having geometrically finite hyperbolic structures.*

In this section, we construct the four-manifolds $M(G)$ having a geometrically finite hyperbolic structure in $\text{int}\, M(G) = H^4/G$ which are homology cobordisms but are not h-cobordisms (see Apanasov and Tetenov [2]).[2]

For this purpose we construct below the geometrically finite groups $G \subset \text{Möb}(3)$ without parabolic elements whose discontinuity sets consist each of two G-invariant components Ω_0 and Ω_1 (where one of them, Ω_0, is a quasi-conformal ball, and $\pi_1(\Omega_1) \neq 0$), which adjoin one another on a wild sphere $S^2 \approx \Lambda(G) \subset S^3$.

For such a G, we have, from Corollary 7.16, that the triple $(M(G); \Omega_0/G, \Omega_1/G)$ is a homology cobordism. However, as Ω_1 is not simply connected, while $\pi_1(H^4 \cup \Omega(G)) = 0$, the kernel of the homomorphism $\pi_1(\Omega_1/G) \longrightarrow \pi_1(M(G))$ induced by embedding, is non-trivial. This means that the second relative homotopy group $\pi_2(M(G), \Omega_1/G) \neq 0$ is non-trivial. This shows that the cobordism $(M(G); \Omega_0/G, \Omega_1/G)$ is not an h-cobordism.

Moreover, we shall see below in constructing these groups that the group G is isomorphic to a certain cocompact Fuchsian group in $\mathbb{R}^3$, or in other words, it is isomorphic to the fundamental group of a certain closed hyperbolic 3-manifold M^3. It is known for such a group (Farrell and Jones [1]) that the Whitehead group $\text{Wh}\, G$ is trivial (see Cohen [1]), and also that as $Wh_2 G = 0$, $\tilde{K}_0(\mathbb{Z}G) = 0$, $K_{-m}(\mathbb{Z}G) = 0$ for $m > 0$ and $Wh_m G \otimes \mathbb{Q} = 0$ for all m, see Farrell and Jones [1].

These facts show that there must be non-algebraic additional conditions on G to provide homotopy triviality of the cobordism $M(G)$ or coincidence of it with a product.

Figure 61

EXAMPLE 7.19 (*Wild spheres as the limit sets for geometrically finite groups*).

We can construct geometrically finite Kleinian groups G in $\mathbb{R}^3$ having wildly knotted limit spheres $\Lambda(G)$ on an idea of periodicity in knotting, used for the first time by the author for the construction of the wildly knotted curve $\Lambda(G) \subset S^3$ (see Apanasov [11], compare Example 9.54).

Let us consider some periodically knotted Fox-Artin arc $d \subset \mathbb{R}^3$ having ends x and y (see Bing [1], IV.9 and Fig. 61). By periodicity we mean that the arc d is invariant for the action of some cyclic group generated by a hyperbolic transformation $h \in \text{Möb}(3)$ such

[2] For further development see Apanasov [34], [35].

that $h(x) = x$ and $h(y) = y$. Moreover, if $I(h) = \{x \in \mathbb{R}^3 : |h'(x)| = 1\}$ and $I(h^{-1})$ are the isometric spheres of h and h^{-1}, $h(\text{ext}\, I(h)) = \text{int}\, I(h^{-1})$, then

$$I(h) \cap d = \{x_1, x_2, x_3\} \quad \text{and} \quad I(h^{-1} \cap d = \{x'_1, x'_2, x'_3\},$$

where $h(x_i) = x'_i$ and the points x_i and x'_i are placed on d in the following order:

$$x_1, \quad x_2, \quad x_3, \quad x'_1, \quad x'_2, \quad x'_3.$$

The intersection d_h of the arc d and

$$\text{ext}\, I(h) \cap \text{ext}\, I(h^{-1})$$

consists of three arcs:

$$d_h = (x_1, x_2) \cup (x_3, x'_1) \cup (x'_2, x'_3)$$

and forms the knotting period of the arc d, shown in Figs. 62 and 64.

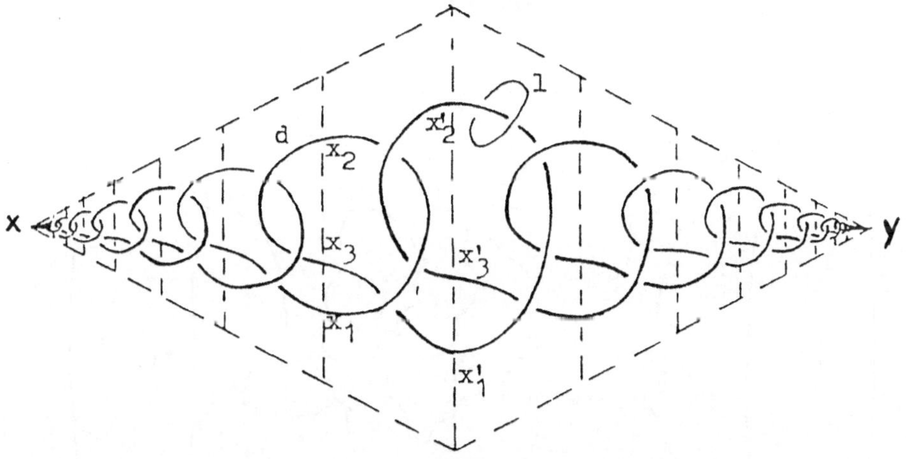

Figure 62

We now take a neighbourhood U_h of the three arcs of d_h in the domain $\text{ext}\, I(h) \cap \text{ext}\, I(h^{-1})$ consisting of three disjoint tubes formed by the union of finitely many sequentially intersecting balls B_i. Then, in accordance with the established periodicity of d, we manifest here that $h(B_i \cap I(h)) = B_j \cap I(h^{-1})$ if

$$x_k \in B_i \cap I(h), \quad x'_k \in B_j \cap I(h^{-1}), \quad h(x_k) = x'_k; \quad k = 1, 2, 3.$$

It is clear that the closure of the union of spherical annuli

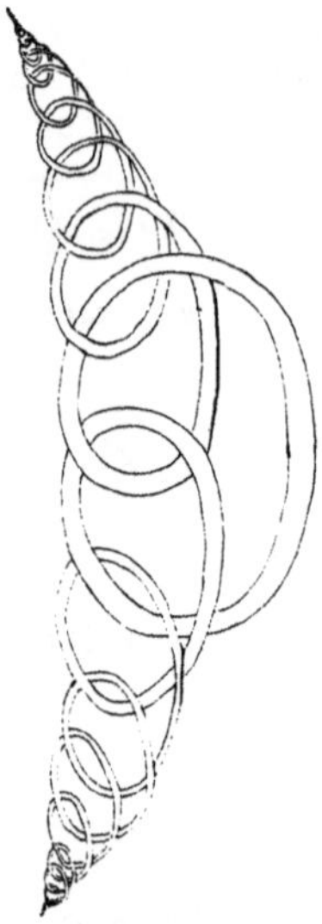

Figure 63

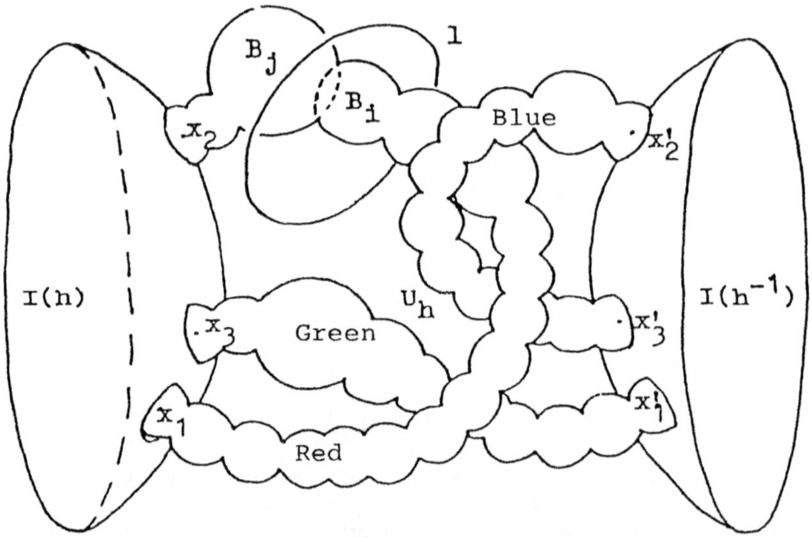

Figure 64

$$X_i = \partial B_i \setminus (\bigcup_{j \neq i} B_j \cup \text{int } I(h) \cup \text{int } I(h^{-1}))$$

and their h^m-images, $m \in \mathbb{Z}$, is the boundary of the fattening

$$U(d) = \bigcup\{h^m(\overline{U}_h) : m \in \mathbb{Z}\} \cup \{x, y\}$$

of the arc d, and is a wild sphere S^* in $\mathbb{R}^3$ (Fig. 63).

We can now form a finite family $\mathscr{C}$ of spheres S_j, $j \in J$, (contained in some regular neighbourhood of the tubes boundary), possessing the following properties:

1. The union of the annuli X_i is covered by interiors of the spheres S_j.
2. For any i, j either $S_j \cap B_i = \varnothing$ or S_j is orthogonal to the sphere ∂B_i. This is also so for $I(h)$ and $I(h^{-1})$ taken instead of ∂B_i.
3. If $S_j \cap S_k \neq \varnothing$, then the dihedral (external) angle between S_j and S_k is equal to π/m, $m \in \mathbb{Z}$.
4. If $S_j \cap S_k \neq \varnothing$, then there exists a common annulus X_i such that $S_j \cap X_i \neq \varnothing$ and $S_k \cap X_i \neq \varnothing$.
5. There is a one-to-one correspondence between spheres $S_j \in C$ crossing $I(h)$ and spheres $S'_j \in C$ crossing $I(h^{-1})$, which is such that

$$h(S_j) = S'_j.$$

In other words we form a finite "bubble cover" of ∂U_h with good angles between the bubbles and right angles between the bubbles and the spheres ∂B_i, and respecting the periodicity condition. The freedom in choosing the balls B_i (as well as the arcs d and mappings h) permit us to vary moduli of the spherical annuli X_i and thus enables us to obtain the required family $\mathscr{C}$.

However, taking into account the rigidity of circular coverings of a two-dimensional sphere (this follows from the rigidity of compact hyperbolic polyhedra (see Andreev [1] and Thurston [1], Ch. 13)), we shall, besides the above-mentioned arguments of existence, give a detailed construction of such a covering $\mathscr{C}$ for the chosen type of Fox-Artin arc knotting (it is evident that our method is applicable to wild arcs of other types).

4. *Block-building construction.* To construct the desired family C, let us consider a right prism P in $\mathbb{R}^3$ with height 17, whose base is a polygon equal to the union of 55 equal regular hexagons with unit sides and non-intersecting interiors (a tesselation). These hexagons cover the regular triangle with sides $9\sqrt{3}$ (Fig. 65), whose vertices are the centres of the extreme hexagons. Let us enumerate all the hexagons as shown in Fig. 65: that the extreme ones have numbers one, 10, and 55 and the central one 31.

We consider P as split up into 55×17 small (unit) prisms $P(k, n)$ having unit height, which are enumerated by the pairs (k, n), in which k, $1 \leq k \leq 17$, denotes the "floor" of the prism P containing $P(k, n)$, and n, $1 \leq n \leq 55$, is the number of the hexagon at the base of P which is the orthogonal projection of $P(k, n)$.

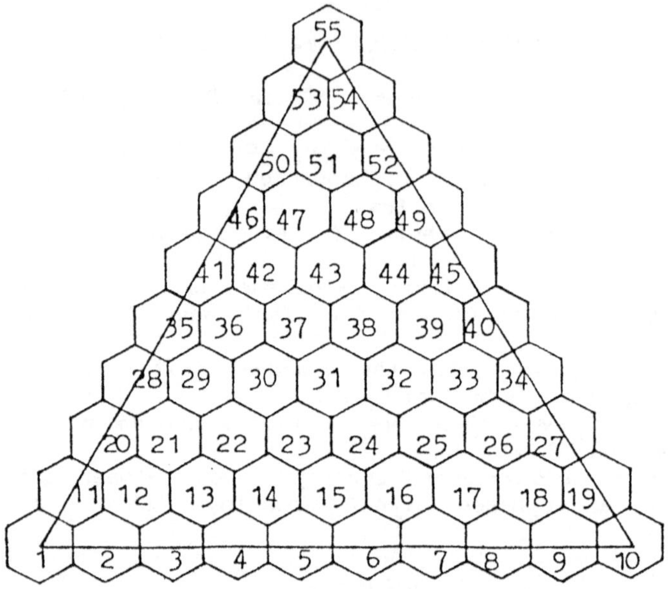

Figure 65

Now we shall put in corresponce to the three tubes forming the neighbourhood U_h of the link period $d_h = (x_1, x_2) \cup (x'_2, x'_3) \cup (x_3, x'_1)$ three disjoint domains (fractured tubes) $D(x_1, x_2)$, $D(x'_2, x'_3)$, $D(x_3, x'_1)$ obtained as the union of small prisms $P(k, n)$ having numbers (k, n) from the following pair sets:

(17,33) (17,34) (17,27) (17,19) (17,10)
(16,10)
(15,10)
(14,10)
$D(x_3, x'_1)$: (13,10)
(12,10)
(11,10)
(10,10)
(9,46) (9,42) (9,37) (9,31) (9,24) (9,16) (9,7) (9,8) (9,9) (9,10)
(8,46)
(7,46)
(6,46)
(5,46)
(4,46)
(3,46)
(2,46)
(1,46) (1,42)

(12,28) (12,29) (12,30) (12,31) (12,32) (12,33) (12,34)
(11,28) (11.34)
(10,28) (10,34)
(9,28) $D(x_1, x_2)$ (9,34)
(8,28) (8,34)
(7.28) (7,34)
(6,28) (6,20) (6,11) (6,1) (6,34)
(5,1) (5,34)
(4,1) (4,34)
(3,1) (3,34)
(2,1) (2,34)
(1,14) (1,4) (1,3) (1,2) (1,1) (1,34) (1,33)

(17,14) (17,4) (17,55) (17,53) (17,50) (17,46) (17,42)
(16,4) (16,55)
(15,4) (15,55)
(14,4) (14,55)
(13,4) (13,55)
(12,4) (12,55) (12,54) (12,52) (12,49)
(11,4) (11,49)
(10,4) (10,49)
(9,4) $D(x'_2, x'_3)$ (9,49)
(8,4) (8,49)
(7,4) (7,49)
(6,4) (6,14) (6,23) (6,31) (6,38) (6,44) (6,49)

Now let S_i be the spheres of radii $\sqrt{3}/3$ with the centres in the vertices of the prisms $P(k, n)$ forming our domains $D(*, *)$. If such spheres intersect, $S_i \cap S_j \neq \varnothing$, then their centres are the adjacent vertices of some prism $P(k, n)$ and their angle of intersection is $\pi/3$. Denote by $B(k, n)$ the ball centred at the centre of the prism $P(k, n)$ with radius $\sqrt{11/12}$. Its boundary sphere $S(k, n)$ is orthogonal to each of the spheres S_i, centred at the vertices of that prism $P(k, n)$.

After this, as the balls B_i, whose union should form the three cylindrical components of U_h, we consider the set of balls $B(k, n)$ corresponding to the prisms $P(k, n)$ from the domains $D(*, *)$. As the isometric spheres $I(h)$ and $I(h^{-1})$, we consider the spheres $I(h) = S^2(z_1, \sqrt{83/12})$ and $I(h^{-1}) = S^2(z_{17}, \sqrt{83/12})$, centred at the centres of floors one and 17 (i.e., at the centres of the prisms $P(1, 31)$ and $P(17, 31)$) with radii $\sqrt{83/12}$). Here the radii of these spheres are such that they intersect (orthogonally) only the spheres S_i centred at the vertices of the prisms $P(1, j)$ and correspondingly $P(17, j)$, $j = 14, 33, 42$, which are the nearest to z_1 and z_{17} respectively.

We note that the interiors of the spheres S_i centred at the vertices of the prisms cover not the whole boundary ∂U_h, i.e., not all the spherical annuli $X \subset S(k, n)$; there remain hexagonal and quadrangular domains (regular hexagons and quadrangles with angles $\pi/3$ in the hyperbolic metric of some spherical discs) on these annuli that

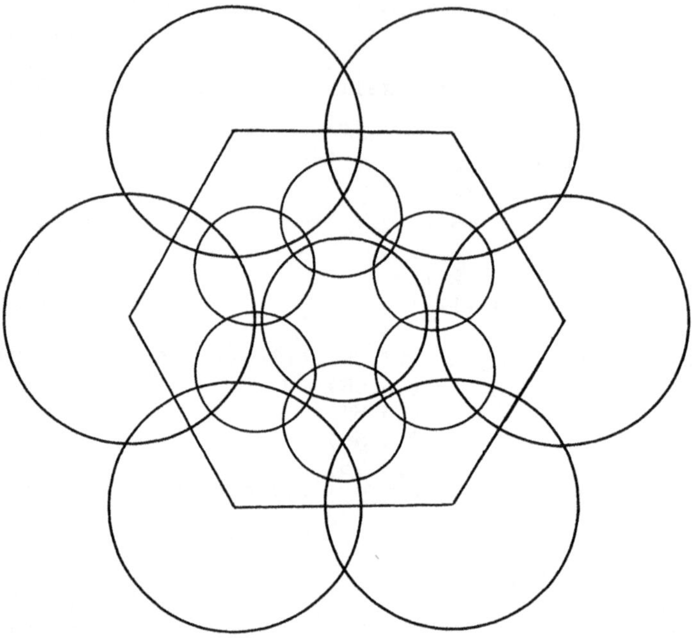

(a)

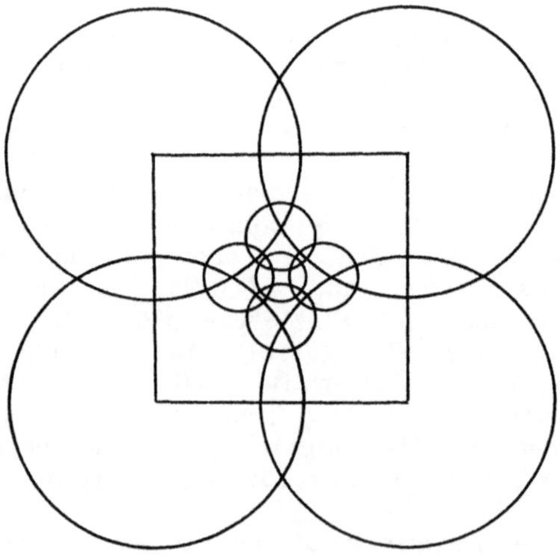

(b)

Figure 66

are uncovered, which correspond to faces of the prisms $P(k, n)$. We cover each hexagonal domain in $S(k, n)$ by the interiors of seven spheres orthogonal to the sphere $S(k, n)$. Six of them have equal radii and are orthogonal to the spheres S_i and consequently intersect at angles of $\pi/3$; the seventh sphere intersects the other six at right angles (see the case a) in Fig. 66) and does not intersect the spheres S_i.

We cover each quadrangular domain on $S(k, n)$ by the interiors of five spheres orthogonal to $S(k, n)$; four of them have equal radii and are orthogonal to the S_i and consequently intersect at angles of $\pi/3$; the fifth sphere intersects the other four at right angles and does not intersect the S_i (see the case b) in Fig. 66).

Direct examination shows that this covering by spheres for ∂U_h has all the properties (1)-(5) of the family $\mathscr{C}$, and this completes the construction of it.

We note finally that characteristics (2) and (4) of the covering $\mathscr{C}$ (where (2) for the case $S_k \cap B_i \neq \varnothing$ and $S_k \cap B_j \neq \varnothing$ means that S_k is orthogonal to the circle $\delta_{ij} = \partial B_i \cap \partial B_j$) allow us to "bend" the cylindrical surfaces forming ∂U_h and thus to bend the whole surface S^* along the circles bounding the spherical annuli $h^m(X_i)$, $i \in i$, $m \in \mathbb{Z}$, without changing the dihedrall angles between the spheres in our covering $h^m(\mathscr{C})$, $m \in \mathbb{Z}$.

Denote by G_1, G_2, and G_3 Möbius groups generated by reflections with respect to the spheres S_i, whose interiors intersect correspondingly the domains $D(x_1, x_2)$, $D(x'_2, x'_3)$ and $D(x_3, x'_1)$. Maskit's first combination theorem (Theorem 6.19) applied in $\overline{\mathbb{R}}^3$ shows that the groups G_k are amalgamated free products:

$$G_k = G_{k1} \underset{H_{12}}{*} G_{k2} \underset{H_{23}}{*} G_{k3} \underset{H_{34}}{*} \cdots\cdots \underset{H_{n_k-1,n_k}}{*} G_{kn_k}; \qquad k = 1, 2, 3,$$

where the groups G_{ki} are generated by reflections with respect to spheres from the family $\mathscr{C}$ that intersect the balls B_i, $B_i \cap D(*, *) \neq \varnothing$, while the amalgamated subgroups H_{ij} are generated by reflections with respect to spheres from $\mathscr{C}$ whose interiors cover the circles $\delta_{ij} = \partial B_i \cap \partial B_j$.

Using the fact that the spheres S_i covering the distinct domains $D(*, *)$ do not intersect, i.e., that the groups G_1, G_2, and G_3 do not intersect, we see that the discrete Möbius group G_0, generated by reflection in the spheres from the family $\mathscr{C}$, (discrete due to property (3) of the family $\mathscr{C}$ and Theorem 6.9) is the free product of the groups G_k, $G_0 = G_1 * G_2 * G_3$ (the Klein combination, Theorem 6.15).

We denote by H_i (correspondingly, H'_i) the subgroups of G_0, generated by reflections in the spheres from $\mathscr{C}$ covering circles $I(h) \cap \partial B_j$ (correspondingly, $I(h^{-1}) \cap \partial B_j$), which contain inside them the points x_i (correspondingly, the points x'_i), $i = 1, 2, 3$; also let

$$H = H_1 * H_2 * H_3, \qquad H' = H'_1 * H'_2 * H'_3.$$

Then the group G generated by G_0 and the hyperbolic element h, $h(\text{ext } I(h)) = \text{int } I(h^{-1})$ is a discrete group obtained by HNN-extension of the group G_0 by means of the element h that conjugates the subgroups H and H' of the group G_0. This is obtained by direct application of Maskit's second combination theorem (Theorem 6.24), since the closed balls D and D', bounded by the spheres $I(h)$ and $I(h^{-1})$, are precisely invariant in G_0 with respect to the subgroups H and H' correspondingly. We thus have the following representation of the group G:

$$G = \langle G_1 * G_2 * G_3,\ h : hH_ih^{-1} = H'_i, \quad i = 1, 2, 3\rangle. \tag{7.12}$$

The group G has, as a fundamental polyhedron, the following finite-sided sphereical polyhedron F, with four connected components: an unbounded one F_∞ and three bounded ones F_1, F_2, and F_3:

$$F = \text{ext } I(h) \cap \text{ext } I(h^{-1}) \cap \{\text{ext } S_i\ ;\ S_i \in \mathscr{C}\}. \tag{7.13}$$

We can extend the bounding spheres S_i, $I(h)$, and $I(h^{-1})$ up to hyperbolic 3-dimensional planes in $H^4 = \mathbb{R}^4_+$ (having these spheres as the spheres at infinity) to obtain a finite-sided convex polyhedron $F^4 \subset H^4$ similar to (7.13) and fundamental for the action of G in H^4. This shows that G is a geometrically finite group.

If we take in G, instead of the generators contained in the group G_1, their conjugation by the element h (i.e., elements from the group hG_1h^{-1}), we can use the Tietze transformation (see Tietze [1] or Lyndon and Schupp [1], Chapter 2 §2) to get that the representation (7.12) is rewritten in the new generators as

$$G = \langle G_2 \underset{H_2}{*} hG_1h^{-1} \underset{H_1}{*} G_3,\ h : hH_3h^{-1} = H'_3\rangle. \tag{7.14}$$

In other words, (7.14) means that G is an HNN-extension of the group $\hat{G}_0$ by means of the element h, where the group $\hat{G}_0$ is generated by reflections in spheres from the following family $\hat{\mathscr{C}}$: $\hat{\mathscr{C}} = h(\mathscr{C}_1) \cup \mathscr{C}_2 \cup \mathscr{C}_3$. Here the subfamilies $\mathscr{C}_i$ consist of the spheres $S_j \in \mathscr{C}$ related to the groups G_i, i.e., intersecting the i-th domain in $D(*, *)$, $i = 1, 2, 3$.

It is readily seen that, instead of the polyhedron F from (7.13), we can take the following polyhedron as the fundamental polyhedron for the group G:

$$\hat{F} = \text{int}(\overline{F}_\infty \cup \overline{F}_2 \cup \overline{F}_3 \cup h(\overline{F}_1)), \tag{7.15}$$

which consists of the unbounded component F_∞ and the bounded (connected) one $F_0 = \text{int}(\overline{F}_2 \cup \overline{F}_3 \cup h(\overline{F}_1))$containing the subarc (x_3, x'_3) of the Fox-Artin arc d.

From the construction of the group G, we see (see Lemma 3.7) that the domains

$$\Omega_0 = \bigcup\{g(\overline{F}_0) : g \in G\}, \ \Omega_1 = \bigcup\{g(\overline{F}_\infty) : g \in G\} \tag{7.16}$$

are G-invariant components of the discontinuity set $\Omega(G)$ and their union gives the whole discontinuity set $\Omega(G)=\Omega_0\cup\Omega_1$ (due to $\hat{F}$ being fundamental, as (7.13), for the group G).

Then it remains to establish that the following lemma is valid.

LEMMA 7.20. *The limit set* $\Lambda(G)$ *for the above geometrically finite group* $G\subset \text{Möb}(3)$ *is a two-dimensional sphere, wildly embedded in* $\mathbb{R}^3$, *that splits up* $\Omega(G)$ *into two* G*-invariant components, one of which is a quasi-conformal image of the ball.*

PROOF. This consists of constructing a homeomorphism $\hat{f}:\overline{\Omega}_0=\Omega_0\cup\Lambda(G)\longrightarrow\overline{B}$ of the closure of the component Ω_0 (from (7.16)) onto the closed ball $\overline{B}\subset\mathbb{R}^3$ that is quasi-conformal in the domain Ω_0 (quasi-symmetric in $\overline{\Omega}_0$ - see Apanasov [33]) and is compatible with the group G.

Let B_i and B_j be any two adjacent intersecting balls from our construction (or h-images of such balls) and let $\delta_{ij}=\partial B_i\cap\partial B_j$. Then there exists a quasi-conformal homeomorphism f_{ij} that transfers $B_i\cup B_j$ onto the ball B_i and which is conformal in a neighbourhood of the spherical discs $\partial B_i\backslash\overline{B}_j$ and $\partial B_j\backslash\overline{B}_i$. Namely, let B_i and B_j be half-spaces, where the planes ∂B_i and ∂B_j contain the origin and intersect at an angle α, $0<\alpha<\pi$, on the third coordinate axis $\delta_{ij}=\{x\in\mathbb{R}^3,\ x_1=x_2=0\}$. Also let there be the natural complex structure in the plane $\mathbb{R}^2=\{x\in\mathbb{R}^3: x_3=0\}$ and let ζ be a fixed number such that

$$0<\zeta<\pi/2,\quad 0<\alpha<\pi-2\zeta.$$

Then the quasi-conformal homeomorphism f_{ij} is described by its behaviour projection in the plane $\mathbb{C}=\mathbb{R}^2$ (compare Fig. 78), where

$$f_{ij}^{-1}(z)=\begin{cases} z, & |\arg z|\geq\pi-\zeta\\ z\cdot\exp(i\alpha), & |\arg z|\leq\zeta\\ z\cdot\exp(i\alpha\cdot(1-(\arg z-\zeta)/(\pi-2\zeta))), & \zeta<\arg z<\pi-\zeta\\ z\cdot\exp(i\alpha\cdot(1+(\arg z-\zeta)/(\pi-2\zeta))), & \zeta-\pi<\arg z<-\zeta \end{cases}\tag{7.17}$$

Here, for the case when ∂B_i is equal to either $I(h)$ or $I(h^{-1})$, the homeomorphism f_{ij} should map the sphere ∂B_i to the sphere $f_{ij}(\partial B_i)$ orthogonal to ∂B_j. Its description is similar to (7.17).

We consider the superposition of a finite number of quasi-conformal homeomorphisms, as in (7.17), to get a quasi-conformal homeomorphism f_0 of the polyhedron F_0 into a ball B, which transfers the sides of F_0 into the sides of a certain polyhedron $F_0'=f_0(F_0)\subset B$ which lie on spheres orthogonal to the sphere ∂B. As the homeomorphisms f_{ij} are conformal near the discs $\partial B_i\backslash\overline{B}_j$ and $\partial B_j\backslash\overline{B}_i$ the properties (2) and (4) of the family $\mathscr{C}$ give that all the dihedral angles W and $f_0(W)$ of the polyhedra F_0 and F_0' are equal. This proves the following important fact.

Let G' be a Möbius group generated by reflections in the sides of the polyhedron F_0' (except for the sides on $f_0(I(h))$ and $f_0(I(h^{-1}))$)

and by the hyperbolic transformation h_0 that transfers the side of the polyhedron F_0' lying in $f_0(I(h))$ (regular hyperbolic quadrangular with angles $\pi/3$) into the side of F_0' lying in $f_0(I(h^{-1}))$ (regular hyperbolic quadrangular with angles $\pi/3$). Then we use Maskit's combination Theorems 6.19 and 6.24 to get that G' is the HNN-extension of the subgroup $G_0' \subset G'$ generated by reflections in the sides of F_0' by the element h_0 (as for the group $G_0 \subset G$). This subgroup G_0' is the free product of its subgroup G_i' and G_j' with amalgamated subgroups $G_{ij}' = G_i' \cap G_j'$. Then the information on the dihedral angles $f_0(W)$ of the polyhedron F_0' (property (3) of the family $\mathscr{C}$) and the representation (7.14) for G show that $G' \subset \text{Möb}(3)$ is a discrete group in the ball B having compact quotient B/G' and isomorphic to the group G.

Moreover, extending the mapping $f_0 : \overline{F}_0 \longrightarrow \overline{F}_0'$ compatible with the actions of the groups G and G' upon the images $G(\overline{F}_0)$, i.e., upon the whole component Ω_0, we obtain a quasi-conformal mapping $f : \Omega_0 \longrightarrow B$ conjugating the actions of the groups G and G': $fGf^{-1} = G'$.

It now remains to extend the mapping f up to a homeomorphism $\hat{f} : \overline{\Omega}_0 \longrightarrow \overline{B}$ of closed domains. This mapping $\hat{f}$ is obtained by the usual extension by continuity where Tukia's theorem on homeomorphisms of limit sets, induced by the isomorphism $G \longrightarrow G'$ (see Theorem 7.25), shows that the resulting $\hat{f}$ is a homeomorphism (even a quasi-symmetric one).

To complete the proof of the lemma, we need only to show that the topological sphere $\Lambda(G) = \partial\Omega_0 = \partial\Omega_1 = \hat{f}^{-1}(S^2)$ is wildly embedded in $\mathbb{R}^3$. This follows from Alexander's theorem on unknottedness of any polyhedral two-dimensional sphere in $\mathbb{R}^3$ (see Bing [1], VIII.4) and the fact that $\pi_1(\Omega_1) \neq 0$.

In fact, we consider a non-trivial loop l in the unbounded component $F_\infty \subset \Omega_1$ of the polyhedron F (see Fig. 64) and assume that it is contractible in Ω_1, see (7.16); then Dehn's lemma (Theorem 7.5) shows that there exists a disc $D \subset \Omega_1$ such that $\partial D = l$. As D is compact, it is covered by a finite number of polyhedra $g_i(\overline{F}_\infty)$, $g_i \in G$. On the other hand, as the loop l is not trivial in $\mathbb{R}^3 \backslash d$ we have $D \cap d \neq \emptyset$. Therefore, there exists $g_0 \in G$ such that $g_0(\overline{F}_\infty) \cap d \neq \emptyset$. The contradiction completes the proof of the lemma and thus completes our construction of the group G with the wild properties explained above (up to a torsion free subgroup of finite index).

REMARK 7.21. The set of points $z \in \Lambda(G)$ near which the topological sphere $\Lambda(G)$ is wildly knotted is a dense subset in $\Lambda(G)$, which follows from the density of the G-orbit $G(\{x, y\})$ of the points x, y (the ends of the arc d fixed for the hyperbolic element $h \in G$) in the set $\Lambda(G)$ (see Lemma 3.22).

So one cannot extend topologically a quasi-symmetric embedding $\overline{B}^3 \hookrightarrow \mathbb{R}^3$ of a closed ball $\overline{B}^3$ to any neighbourhood of any boundary point $x \in S^2 = \partial B^3$. The analogous inextensible embedding effect (only for two boundary points $x \in S^2$) was discovered by F.W. Gehring [3] (for the detailed proof of this fact see Martin [1]).

§4. Isomorphism theorems

The Fenchel-Nielsen theorem asserts that an isomorphism of the fundamental group of Riemann surfaces is induced by a homeomorphism and can be formulated in Fuchsian-group language as follows.

THEOREM 7.22. *Let* G *and* $H \subset \mathcal{M}_2$ *be non-cyclic finitely generated Fuchsian groups in the disc* U *and* $\varphi : G \longrightarrow H$ *an isomorphism having the following properties*:

1) φ *is type-preserving*;

2) $\varphi(g)$ *leaves invariant an arc from* $\Omega(H) \cap \partial U$ *if and only if* g *leaves invariant an arc from* $\Omega(G) \cap \partial U$.

Then there exists a topological automorphism $f : \overline{U} \longrightarrow \overline{U}$ *such that* $fg(x) = \varphi(g) \cdot f(x)$ *for all* $g \in G$ *and* $x \in \overline{U}$.

Here the isomorphism φ is called *type-preserving* if it transfers bijectively parabolic elements of G to parabolic elements of H. The proof of this theorem was given by Fenchel and Nielsen in their unpublished book "Discontinuous groups of non-Euclidean motions". A rigorous proof can be found in Marden [4] or in the book by Zieschang, Vogt, and Coldewey [1].

Marden [1] gave the first generalization of this result for Kleinian groups. To formulate it, we give the following definition.

DEFINITION 7.23. The isomorphism of two discrete groups G, $H \subset \text{Möb}(n)$ is called a *geometrical isomorphism* if there exists a homeomorphism $f : \overline{H}^{n+1} \longrightarrow \overline{H}^{n+1}$ such that $f \cdot g(x) = \varphi(g) \cdot f(x)$ for all $g \in G$ and $x \in \overline{H}^{n+1}$ (f induces φ).

THEOREM 7.24. *Let* $G \subset \mathcal{M}_2$ *be a geometrically finite torsion free group and* $\varphi : G \longrightarrow H$ *an isomorphism in the discrete group* $H \subset \mathcal{M}_2$ *that preserves parabolic elements. We assume that* φ *is induced by an orientation-preserving homeomorphism* $f : \Omega(G) \longrightarrow \Omega(H)$, i.e., $f \cdot g = \varphi(g) \cdot f(x)$ *for all* $g \in G$ *and* $x \in \Omega(G)$. *Then* φ *is a geometric isomorphism that is induces by a quasi-conformal automorphism in* H^3 *in the case where* f *is quasi-conformal and by an isometry in* H^3 *if* f *is conformal.*

We give the basic points of the proof of this (see Marden [1]). Applying Theorem 7.5 and the technique in §2 we can assume that all components of $\Omega(G)$ are simply connected. We show that the group H is geometrically finite. The punctured tori in the manifold $M(H)$ correspond to free Abelian subgroups in H and there correspond in a one-to-one fashion (φ is an isomorphism) to punctured tori in $M(G)$. It is also clear that the mapping f defines the correspondence between the punctured solid cylinders in $M(G)$ and $M(H)$. We remove these punctured solid tori and cylinders from the manifolds to get the manifolds $M_0(G)$ and $M_0(H)$, the first of which is compact. As f induces the homeomorphism $f_0 : \partial M_0(G) \longrightarrow \partial M_0(H)$, an isomorphism is induced for the fundamental group $\pi_1(\hat{M}_0(G))$ of the double $\hat{M}_0(G)$ with respect to the boundary $\partial\hat{M}_0(G)$ in the fundamental group of the double $\hat{M}_0(H)$ with respect to $\partial\hat{M}_0(H)$. As the doubles $\hat{M}_0(G)$ and $\hat{M}_0(H)$ are manifolds without boundary and have trivial homotopy groups π_2 and

π_3, while $\hat{M}_0(G)$ is compact, Lemma 7.14 gives that $\hat{M}_0(H)$ is compact, so the group H is geometrically finite (Corollary 7.4).

The main part of the proof is to construct a homeomorphism $f^* : M(G) \longrightarrow M(H)$ that extends f_0. That extension is obtained from a profound result due to Waldhausen [3], which in our situation asserts that the isomorphism $\varphi_* : \pi_1(M_0(G)) \longrightarrow \pi_1(M_0(H))$ (which acts with allowance for the monomorphism of the embeddings $\partial M_0(G)$ and $\partial M_0(H)$) is induced by the piecewise linear homeomorphism $M_0(G) \longrightarrow M_0(H)$. Waldhausen's proof method consists of the following. He shows that the following finite sequence of surfaces $\{S_i\}$ exists.

The surface $S_1 \subset M_0(G)$ has the properties:

1) $S_1 \cap \partial M(G) = \partial S_1 \neq \varnothing$ (if $\partial M_0 \neq \varnothing$);

2) S_1 does not divide up $M_0(G)$;

3) $\pi_1(S_1) \longrightarrow \pi_1(M_0)$ is a monomorphism.

We take the neighbourhood $S_1^* = S_1 \times (-\varepsilon, \varepsilon)$ of the surface S_1 and the set $M_1 = M_0(G) \setminus S_1^*$ to select a surface $S_2 \subset M_1$ having the same properties as $S_1 \subset M_0$ (but now always $\partial M_1 \neq \varnothing$), and so on. At the last step, we get a ball as M_n. We use the isomorphism φ^* to perform the corresponding decomposition for $M_0(H)$. Then f^* is obtained as a piecewise linear homeomorphism by recovering from f_0 on the corresponding parts of the decomposition. It is clear that, if we take f as quasi-conformal, we get the quasi-conformal extension $f^* : M(G) \longrightarrow M(H)$. We lift the homeomorphism f^* to the universal covering $H^3 \cup \Omega(G)$ to get from Gehring's results [1] its unique extension in $\overline{H}^3$ (and thus we define a homeomorphism in $\Lambda(G)$), which proves that φ is geometric.

In the case $\Omega(G) = \varnothing$, the theorem follows from Mostow's rigidity theorem (Theorem 9.8). Here φ is realized in $\mathrm{Isom}_+ H^3 \cong \mathcal{M}_2$.

If $\Omega(G) \neq \varnothing$, then $\mu_2(\Lambda(G)) = 0$ (Corollary 5.32), so, if f is conformal, we again get that φ is realized in $\mathrm{Isom}\, H^3$. Here in fact one can specify that f is conformal not in all $\Omega(G)$ (if G is not Fuchsian) but only in the components different from Euclidean discs and that are transferred into Euclidean discs. A detailed proof for this is obtained by considering the double $\hat{M}(G)$, see §2.3.

Another approach to an isomorphism for geometrically finite discrete groups is obtained from the Tukia [5] construction of a limit-set homeomorphism, which plays a key role in his theorem below. We formulate this result for $n \geq 2$ (compare Tukia [5], Theorems 3.3 and 3.8) in the following partial form:

THEOREM 7.25. *Let G and $H \subset \mathrm{M\ddot{o}b}(n)$, $n \geq 2$, be non-elementary geometrically finite discrete groups and let $\varphi : G \longrightarrow H$ be an isomorphism. Then we have:*

a) *if φ is type-preserving, there exists the unique homeomorphism*

$$f_\varphi : \Lambda(G) \longrightarrow \Lambda(H) \tag{7.18}$$

that induces the isomorphism φ, where $f_\varphi \in \mathrm{M\ddot{o}b}(n)$ if $\Omega(G) = \varnothing$;

b) *if $f : \Omega(G) \longrightarrow \Omega(H)$ is a homeomorphism that induces the isomorphism φ, then φ is always type-preserving, and then the* *mapping* $\hat{f}$ *such that*

$$\hat{f}(x) = \begin{cases} f(x), & \text{if } x \in \Omega(G) \\ f_\varphi(x), & \text{if } x \in \Lambda(G) \end{cases} \tag{7.19}$$

is a homeomorphism of the sphere $\overline{\mathbb{R}}^n$;

c) *if the homeomorphism* $f : \Omega(G) \longrightarrow \Omega(H)$ *that induces the isomorphism* φ *is* K*-quasi-conformal for some* $K \geq 1$, *then* $\hat{f}$ *is* K*-quasi-conformal, too.*

In dimension $n = 2$, this approach enables one to remove certain restrictions in Theorem 7.24 because of the following profound result due to Reimann [1] and Thurston [1] on the extension of quasi-conformal mappings (see also Tukia [4] in the case $n = 1$, which also allows a multi-dimensional approach).

THEOREM 7.26. *Let* G *and* $H \subset \text{Möb}(n)$, $n \leq 2$, *and let* $f : \overline{\mathbb{R}}^n \longrightarrow \overline{\mathbb{R}}^n$ *be a quasi-conformal homeomorphism that induces an isomorphism* $\varphi : G \longrightarrow H$. *Then there exists a quasi-conformal homeomorphism* $F : \overline{H}^3 \longrightarrow \overline{H}^3$ *that is an extension of* f *and that induces* φ.

COROLLARY 7.27. *Let* G *be a geometrically finite discrete Möbius group in the plane and let* $\varphi : G \longrightarrow H$ *be its isomorphism onto a discrete group* H *induced by a homeomorphism* $f : \Omega(G) \longrightarrow \Omega(H)$. *Then* φ *is a geometric isomorphism induced by a homeomorphism* $F : \overline{H}^3 \longrightarrow \overline{H}^3$, $F|_{\Omega(G)} = f$. *In addition,* F *is quasi-conformal if* f *is quasi-conformal and* $F \in \text{Isom}\, H^3$ *if* f *is conformal or if* $\Omega(G) = \varnothing$. *Here quasi-conformality means that* f *is an orientation preserving or reversing quasi-conformal embedding (independently in each component of* $\Omega(G)$) *and in the conformal case it is allowed that* f *is anti-conformal in some components of* $\Omega(G)$.

The proof of this result of Tukia [5] for conformal or quasi-conformal f is a direct consequence of Theorems 7.25 and 7.26 for a geometrically finite group H, but the geometric finiteness condition is superfluous, since the transition to torsion free finite index subgroups and Theorem 7.24 imply geometric finiteness for H and allows us to use Theorem 7.25 in that case.

For an arbitrary homeomorphism f, the corollary amounts to applying Theorem 7.26 by means of certain PL-reconstructions for homeomorphisms of the surfaces $\Omega(G)/G$ and $\Omega(H)/H$ of finite type.

To describe the conditions on the groups G and H and on φ for which the homeomorphism $f : \Omega(G) \longrightarrow \Omega(G)$ exists, we give some definitions.

DEFINITION 7.28. The stabilizer subgroup $G_\Delta \subset G$ for a component $\Delta \subset \Omega(G)$ is called *non-maximal* if $G_\Delta \subset G_{\Delta'} \subset G$ for a certain component $\Delta' \subset \Omega(G) \setminus \Delta$.

DEFINITION 7.29. An element g in the Kleinian group G is called a *boundary element* if it belongs to the stabilizer $G_\Delta \subset G$ of a certain component $\Delta \subset \Omega(G)$. Here g is called *inexact* if g is contained in a certain non-maximal stabilizer G_Δ.

We note that a boundary element is defined by a loop in $\partial M(G) = \Omega(G)/G$, while the non-maximal stabilizer G_Δ corresponds to a component $\partial M(G)$ retractable in $M(G)$ to a proper part of another component of $\partial M(G)$. In those terms, Marden and Maskit [1] obtain the following generalization of the Fenchel-Nielsen theorem.

THEOREM 7.30. *Let $G \subset M_2$ be a geometrically finite torsion free Kleinian group that has more than two components, all of which are simply connected. Let $\varphi : G \longrightarrow H$ be an isomorphism onto a Kleinian group having the properties that, for all $g \in G$, the elements $\varphi(g)$ and g simultaneously are or are not parabolic, boundary, or inexact boundary elements. Then φ is a geometric isomorphism.*

The proof is based on the following lemma obtained by induction on the number of generators in H_0 (Marden and Maskit [1]):

LEMMA 7.31. *Let $H \subset M_2$ be a finitely generated torsion free Kleinian group and let $H_0 \subset H$ be a finitely generated subgroup. If each element in H_0 leaves invariant at least $k \geq 1$ components from $\Omega(H)$, then H_0 leaves invariant at least k components from $\Omega(H)$.*

We now consider a certain component $\Delta \subset \Omega(G)$ and assume initially that the stabilizer G_Δ is a maximal subgroup of G. Applying Lemma 7.31, we get that there exists at least one component $\Delta' \subset \Omega(H)$ invariant for $H_0 = \varphi(G_\Delta)$. Let $H_{\Delta'} \supset H_0$ be its stabilizer in H. We apply Lemma 7.31 again but this time in the group G to get that there exists a component $\Delta^* \subset \Omega(G)$, invariant for $\varphi^{-1}(H_{\Delta'})$, where $G_\Delta \subset \varphi^{-1}(H_{\Delta'})$. Then $H_0 = H_{\Delta'}$.

The component $\Delta' \subset \Omega(H)$ should be simply connected. If this is not so, $H_{\Delta'}$ and consequently G_Δ, should be represented as the free product $G_\Delta = G_1 * G_2$, where all parabolic elements of G_Δ are conjugate to parabolic elements either in G_1 or in G_2. It is impossible, since G_Δ is either quasi-Fuchsian or a B-group. Also, the theorem condition implies that H has more than two components, and therefore $\Omega(H)$ has only one component (= Δ') invariant for $H_0 = H_{\Delta'}$. Then we can apply Theorem 7.22 for the components Δ and Δ' and the isomorphism $\varphi : G_\Delta \longrightarrow H_{\Delta'}$ and construct a homeomorphism $\Delta \longrightarrow \Delta'$ that realizes φ.

We now assume the second possibility: $G_0 = G_\Delta$ is a non-maximal stabilizer subgroup in G. In that case, G_0 leaves invariant exactly two components $\Delta, \Delta_1 \subset \Omega(G)$ (see Chapter 3 §2 or Accola [1]), and the group $G_1 = G_{\Delta_1}$ is maximal. Lemma 7.31 implies that $H_0 = \varphi(G_0)$ also has at least two invariant components Δ' and Δ'_1 from $\Omega(H)$, one of which (= Δ'_1) is invariant for $\varphi(G_1) = H_1$. As above, we apply the theorem conditions to get that $H_{\Delta'}$ cannot be a maximal subgroup. Therefore, its maximal extension is the group H_1. Moreover, Lemma 7.31 implies (as above) that $H_{\Delta'} = H_0$.

Then independently of the maximality of G_Δ, each component $\Delta \subset \Omega(G)$ is put into correspondence (uniquely) with a component

$\Delta' \subset \Omega(H)$ such that $\varphi(G_\Delta) = H_{\Delta'}$ and the restriction $\varphi|_{G_\Delta}$ is realized by a homeomorphism $\Delta \longrightarrow \Delta'$, or in other words, we have a homeomorphism $f : \Omega(G) \longrightarrow \Omega(H)$ compatible with the isomorphism $\varphi : G \longrightarrow H$.

In Theorem 7.24, it was specified that f had one type of orientation in all components (compare Corollary 7.27). However, Waldhausen's proof, mentioned above, does not use that hypothesis. The only fact needed is that there is one type of orientation in the components Δ_1 and Δ_2 connected by a parabolic cusp, i.e., in Δ_1 and Δ_2 there exist discs precisely invariant with respect to the common parabolic subgroup $G_{\Delta_1} \cap G_{\Delta_2}$. However, we have that property for the homeomorphism f, so the proof of the theorem is completed by applying Theorem 7.24 or Corollary 7.27.

REMARK 7.32. It is evident that the type-preserving condition for the isomorphism φ is necessary for φ to be geometric. The condition that the group G should be torsion free (compare Corollary 7.27) is unessential for the existence of the homeomorphism $f : \Omega(G) \longrightarrow \Omega(H)$ that induces the isomorphism $\varphi : G \longrightarrow H$ in Theorem 7.30, and the same applies to the simply connected condition for components of $\Omega(G)$ and the existence of three distinct ones amongst them. We introduce some definitions to formulate that generalization of this theorem.

DEFINITION 7.33. An isomorphism $\varphi : G \longrightarrow G'$ of two discrete Möbius groups in $\overline{\mathbb{R}}^2$ is called *strongly type-preserving* if it is type-preserving and if, whenever $g \in G$ is an elliptic orientation-preserving element, $\varphi(g)$ also is and they correspond to rotation through angles of the same absolute magnitude.

DEFINITION 7.34. Following Tukia [5], we say that a loxodromic transformation $g \in \text{Möb}(2)$ and a "quasi-Fuchsian" group Γ intersect if and only if the fixed points of g lie in different components of $\overline{\mathbb{R}}^2\backslash\Lambda(\Gamma)$ (i.e., the axis l_g of the element g intersects the Nielsen hull H_Γ transversely). A type-preserving isomorphism $\varphi : G \longrightarrow G'$ of two geometrically finite Möbius groups in $\overline{\mathbb{R}}^2$ is called *intersection-preserving* if, whenever $g \in G$ is loxodromic and $\Gamma \subset G$ is a quasi-Fuchsian subgroup, the element g and Γ intersect if and only if $\varphi(g)$ and $\varphi(\Gamma)$ intersect.

This definition is correct since point a) of Theorem 7.25 shows that $\Lambda(\varphi(\Gamma)) = f_\varphi(\Lambda(\Gamma))$ is a topological circle (i.e., $\varphi(\Gamma)$ is a "quasi-Fuchsian" group). This definition extends the fact for the hyperbolic plane H^2 (which is given by the Fenchel-Nielsen theorem on axis intersection, cf. Theorem 7.22), that an isomorphism $\varphi : G \longrightarrow G'$ of two groups from $\text{Isom}\, H^2$ is induced by some homeomorphism $f : H^2 \longrightarrow H^2$, i.e. $\varphi(g) = fgf^{-1}$, if and only if φ preserves intersection between axes of hyperbolic elements (i.e., the hyperbolic elements $g, h \in G$ have intersecting axes if and only if the axes of the elements $\varphi(g)$ and $\varphi(h)$ intersect).

The importance of the intersection-preserving condition for the isomorphism $\varphi: G \longrightarrow G'$ occurs because such isomorphisms in the non-quasi-Fuchsian case define bijective correspondence between components of $\Omega(G)$ and components of $\Omega(G')$. The following lemma, due to Tukia [5], demonstrates this fact:

LEMMA 7.35. *Let* $\varphi: G \longrightarrow G'$ *be an intersection-preserving isomorphism of two geometrically finite, non-quasi-Fuchsian Möbius groups in* $\bar{\mathbb{R}}^2$. *Then for any component* $D \subset \Omega(G)$ *there exists a unique component* $D' \subset \Omega(G')$ *such that* $f_\varphi(\partial D) = \partial D'$. *The stabilizers of these components satisfy*

$$\varphi(G_D) = G'_D. \tag{7.20}$$

Further, if D' is a component of $\Omega(G')$, then there exists a unique component $D \subset \Omega(G)$ such that $f_\varphi^{-1}(\partial D') = \partial D$ and (7.20) is true.

In addition to the above isomorphism properties, we need the following concept (see Tukia [5], 4D) if $\Lambda(G)$ is non-connected.

DEFINITION 7.36. A type-preserving isomorphism $\varphi: G \longrightarrow G'$ of geometrically finite groups in $\bar{\mathbb{R}}^2$ is called *orientation-preserving* if the following conditions are met:

a) for any loxodromic $g \in G$ and for any "quasi-Fuchsian" subgroup $\Gamma \subset G$ there exists an orientation-preserving homeomorphism $f: \bar{\mathbb{R}}^2 \longrightarrow \bar{\mathbb{R}}^2$ extending $f_\varphi|_{\Lambda(\Gamma)}$ such that $f(p(g) = p(\varphi(g))$ (here $p(g)$ is the attracting fixed point for g); and

b) if p is some parabolic fixed point of the group G for which $\operatorname{rank} G_p = 2$, then there exists an orientation-preserving homeomorphism $f: \bar{\mathbb{R}}^2 \longrightarrow \bar{\mathbb{R}}^2$ inducing $\varphi|_{G_p}$.

The isomorphism φ is called *orientation-reversing* if a) and b) are true with the words "orientation-preserving" in the above definition replaced by the words "orientation-reversing". φ is called *orientation-consistent* if it is either orientation-preserving or an orientation-reversing isomorphism. (An orientation-consistent isomorphism φ is intersection- preserving.)

The following theorem, due to Tukia [5] (and due to B. Maskit in the case when $\Lambda(G)$ is non-connected) extends the above Fenchel-Nielsen, Marden, and Maskit theorems and gives the fullest description of geometric isomorphisms for geometrically finite Möbius groups in the plane.

THEOREM 7.37. *Let* $\varphi: G \longrightarrow G'$ *be a type-preserving isomorphism of two non-elementary geometrically finite Möbius groups in* $\bar{\mathbb{R}}^2$. *In addition,*

1) *if* $\Lambda(G)$ *is a circle, assume that* $g \in G$ *is orientation-reversing if and only if* $\varphi(g)$ *is;*

2) *if* $\Lambda(G)$ *is connected but is not a circle, assume that the isomorphism* φ *preserves intersection; and*

3) *if* $\Lambda(G)$ *is non-connected, assume that*

a) *elements of* G *and* G' *are orientation-preserving;*

b) *each finite subgroup in* G *is cyclic, and*

c) *the isomorphism* φ *is strongly type-preserving and orientation-consistent.*

Then φ *is induced by a quasi-conformal homeomorphism*

$$F : \overline{H}^3 \longrightarrow \overline{H}^3.$$

We note that, when the group G does not contain parabolic elements, we have for each component $D \subset \Omega(G)$ that its stabilizer $G_D \subset G$ is a quasi-Fuchsian group (see Maskit [2], Theorem 4, and Maskit [9], Theorems 9 and 1.11). Therefore, it would suffice in this case to assume in Theorem 7.37 that φ preserves intersection for all loxodromic $g \in G$ and all quasi-Fuchsian $\Gamma \subset G$ of the form G_D or $\varphi^{-1}(G'_{D'})$ for some components $D \subset \Omega(G)$ or $D' \subset \Omega(G')$. That condition comes near to the Marden and Maskit [1] condition that φ preserves the boundary elements (see Definition 7.29) in their theorem:

THEOREM 7.38. *Let* G *be a geometrically finite purely loxodromic Kleinian group having more than two components, all of which are simply connected. Let* $\varphi : G \longrightarrow G'$ *be an isomorphism onto a purely loxodromic Kleinian group that preserves the property of elements to be boundary elements. Then* φ *is a geometric isomorphism.*

To prove this, we note that the group G cannot be a non-trivial free product of its subgroups, while each stabilizer subgroup $G_0 \subset G$ not only cannot be non-maximal but even cannot be contained in any subgroup $G_1 \subset G$, isomorphic to $\pi_1(S)$, for a closed surface S (G_1 is not necessarily G_{Δ_1} for $\Delta_1 \subset \Omega(G)$). Namely, if that is not so, the index of G_0 in G_1 is finite, $\Lambda(G_0) = \Lambda(G_1)$, and consequently G_1 is quasi-Fuchsian (as is G_0) and leaves invariant the same component as G_0, which is impossible.

This gives that G_0 leaves invariant only one component Δ, and the component $\Delta' \subset \Omega(G')$, corresponding to Δ, is simply connected. Since $\varphi(G_0)$ cannot be contained in a large subgroup isomorphic to $\pi_1(S')$ for a closed surface S', $\varphi(G_0) = G_{\Delta_1}$. We apply Theorems 7.22 and 7.24 to complete the proof.

It follows from the Theorem 7.38 conditions that $M(G)$ is a compact 3-manifold with boundary; if further we put $G = G'$, we obtain information (see Marden and Maskit [1]) on the automorphisms of $\pi_1(M(G))$ that are induced by the homeomorphisms of $M(G)$, where one has in mind that any boundary element $g \in G \cong \pi_1(M(G))$ corresponds to a loop in $M(G)$ freely homotopic to a loop in $\partial(M(G)$.

COROLLARY 7.39. *Let* M *be a compact 3-manifold with non-empty boundary that admits a geometrically finite hyperbolic structure and let for each component* $S \subset \partial M$ *the inclusion* $\pi_1(S) \longrightarrow \pi_1(M)$ *be a monomorphism. Then any automorphism* φ *of the fundamental group* $\pi_1(M)$ *that preserves boundary elements is induced by some homeomorphism* $f : M \longrightarrow M$.

§5. **Geometrically tame ends, Kleinian manifold compactification, and the Ahlfors problem on limit-set measure**

We say that a non-compact manifold M can be compactified if there exist a compact manifold $\hat{M}$ with boundary and an embedding $h: M \to \hat{M}$ induces a homotopy equivalence such that $h(\mathrm{int}\, M) = \mathrm{int}\, \hat{M}$.

We have seen in §1 (Theorem 7.2 and Remark 7.3) that any Kleinian manifold $M(G) = [H^n \cup \Omega(G)]/G$ for a geometrically finite group $G \subset \mathrm{M\ddot{o}b}(n-1)$, $n \geq 2$, admits natural compactification; here to neighbourhoods of non-compact pieces (ends) in the manifold $M(G)$ (of which there are a finite number and they are finitely covered by manifolds of the (7.7) form, i.e., punctured solid tori $\mathcal{P}^n$ and punctured solid cylinders $\mathcal{C}^{n,k}$) correspond compact submanifolds, which are finitely covered by manifolds of the following form:

$$T^{n-1} \times [0, 1] \quad \text{and} \quad T^k \times \bar{B}^{n-k-1} \times [0, 1], \qquad 1 \leq k \leq n-2.$$

The problem of compactification of a Kleinian manifold $M(G)$, i.e., of describing the end structure, was first considered in 1965 by Ahlfors [4], where he conjectured that such a structure is simple in some sense. The first major advance in resolving this, which confirmed Ahlfors's hypothesis in a certain class of manifolds, was made in 1977 by Thurston [1]. The main tool in those researches, which plays a major part in proving the uniformization theorem for 3-manifolds (see Chapter 8), is based on a study of geodesic laminations on a surface S admitting realization in the homotopy class $f: S \to M(G)$. An important part is played here by Thurston's geometrically tame manifolds (in particular, geometrically finite manifolds). Thurston's research has been completed in a certain sense by a recent result obtained by Bonahon [1], who proved Thurston's conjecture on geometrically tame manifolds in the basic case. We need some definitions for exact formulations.

DEFINITION 7.40. By an end of a non-compact manifold M we mean a family $E = \{U_i\}_{i \in I}$ of the subsets $U_i \subset M$ such that:

1) each $U_i \subset E$ is a connected non-empty open set having compact boundary $\partial U_i = \mathrm{cl}_M U_i \setminus U_i$;
2) if $U_i, U_j \in E$, then there exists $U_k \in E$, $U_k \subset U_i \cap U_j$;
3) $\bigcap_{i \in I} U_i = \varnothing$;
4) if the open connected set $U \subset M$ has a compact boundary ∂U and $U_i \subset U$ for a certain $U_i \in E$, then $U \in E$.

Any subset of M that contains a certain element from E is called a *neighbourhood of the end* E. An end E is called *isolated* if it is not contained in any other end.

LEMMA 7.41 (Husch and Price [1]). *Let M be a 3-manifold having an isolated end E. Then for any neighbourhood N of the end E there exists also a neighbourhood $\hat{N}$ of the end E such that*:

E1) $\hat{N} \subset N$;
E2) $\hat{N}$ *is closed in* M;
E3) $\hat{N}$ *is a connected piecewise linear 3-manifold whose boundary*

is a connected surface; and

E4) $\hat{N}$ *has precisely one end.*

Let $M = M_{(0,\varepsilon]} \cup M_{[\varepsilon,\infty)}$ be the Margulis decomposition of a complete hyperbolic manifold M into an ε-thin part and its complement - see the Margulis Theorem 4.20 in Chapter 4 §4. There is the Margulis constant $\varepsilon = \varepsilon(n)$ that is dependent only on the dimension $n = \dim M$ and is such that, for any $x \in M$, a subgroup in $\pi_1(M, x)$, generated by loops of length $\le\varepsilon$, is parabolic or loxodromic (without excluding the trivial case). For this constant $\varepsilon = \varepsilon(n)$, we denote the ε-thin part $M_{(0,\varepsilon]}$ by M_t, and the ε-thick part $M_{[\varepsilon,\infty)}$ by M_h. The components of the submanifold M_t are either Margulis tubes, i.e., quotients of hyperbolic cylinders $D(l, r) = \{x \in H^n : d(l, x) < r\}$ by the action of its stabilizer $G_l \subset \pi_1(M)$ (a cyclic loxodromic group) or are parabolic cusps, i.e., quotients of the parabolic neighbourhood $V_p(\varepsilon) \subset H^n$ (see Chapter 4 §4) by the action of its stabilizer $G_p \subset \pi_1(M)$ (a virtually free Abelian group of rank k, $1 \le k \le n-1$). The cusps in the hyperbolic manifold M are homeomorphic (up to finite-sheeted coverings) to the punctured solid tori $\mathcal{P}^n$ or punctured solid cylinders $\mathcal{C}^{n,k}$ (see (7.7) and (4.26)-(4.28)), while their boundaries in M are homeomorphic to the corresponding Euclidean $(n-1)$-dimensional manifolds $\mathbb{R}^{n-1}/\Gamma$, where torsion free groups $\Gamma \subset \mathrm{Isom}\,\mathbb{R}^{n-1}$ are discrete (see Theorem 3.4). In dimension $n = 3$, the boundary of the ε-thin part of M is formed from tori, Klein bottles, rings, and Möbius bands.

For a hyperbolic manifold M, we denote by M_o the submanifold that is the complement in M to the cusp components of the ε-thin part M_t.

LEMMA 7.42. *Let M be an acyclic three-dimensional manifold whose boundary ∂M consists of a finite number of surfaces of finite type and having zero Euler characteristic. Also let $\pi_1(M)$ be a finitely generated group that satisfies the following condition:*

for any non-trivial decomposition $\pi_1(M)$ to a free product $\pi_1(M) = A * B$, there exists a closed curve in ∂M whose class is not conjugate in $\pi_1(M)$ to any element from A or B.

Then M contains a compact submanifold M_C of codimension 0 such that the embedding $(M_C, M_C \cap \partial M) \longrightarrow (M, \partial M)$ is a homotopy equivalence.

The proof of this lemma (see Bohahon [1], Lemma 1.2) employs the usual 3-manifold topological technique (see §1) and is based on two facts proved by Scott [1], [2]: firstly, $\pi_1(M)$ is finitely presented and secondly, for a 3-manifold M having a finitely presented fundamental group $\pi_1(M)$, there exists a compact submanifold M_C such that the embedding $M_C \longrightarrow M$ is a homotopy equivalence. We note that here the pair (N, P), in which N is a 3-manifold and $P \subset N$, $\dim P = m$, is locally modelled on $(\mathbb{R}^3_+, \mathbb{R}^m_+)$, where $\mathbb{R}^m_+ = \{x \in \mathbb{R}^m : x_1 > 0\}$ in the case $\operatorname{codim} P \ge 1$, and is locally modelled on $(\mathbb{R}^3, \mathbb{R}_+ \times \mathbb{R}^2)$ in the case $\operatorname{codim} P = 0$, in such a way that the topological boundary δP in N is a submanifold of codimension one in N ($\delta P = \overline{\partial P \setminus \partial N}$ if P is topologically closed).

THEOREM 7.43. *Let* M *be a complete hyperbolic 3-manifold having with the finitely generated fundamental group* $\pi_1(M)$ *that satisfies the following condition (*):*

For any non-trivial decomposition of $\pi_1(M)$ *to a free product* $A * B$, *there exists a parabolic element that is not conjugate in* $\pi_1(M)$ *to any element from* A *or* B. (*)

Also, let M_0 *be the complement in* M *to cusp components from* $M_t \subset M$; *then* M_0 *contains a compact submanifold* M_C *having codimension zero for which the embedding*

$$(M_C, M_C \cap \partial M_0) \longrightarrow (M_0, \partial M_0) \tag{7.21}$$

is a homotopy equivalence, while the frontier δM_C *is an incompressible surface (i.e.,* $\pi_1(\delta M_C) \longrightarrow \pi_1(M_0)$ *is a monomorphism).*

The proof of the first part (see Bonahon [1], Theorem 1.4 and Lemma 1.5) follows directly from Lemma 7.42 and Sullivan's theorem on the finiteness of the number of cusps (Theorem 5.38). The second assertion that δM_C is incompressible follows from the fact that, if this is not so, Dehn's lemma (Theorem 7.5) implies that there is a disc that splits up M_C and gives a non-trivial decomposition of the group $\pi_1(M)$ to a free product $A * B$ (Theorem 7.7), in which each parabolic element is conjugate either to an element from A or from B. As then $\pi_1(M_C) \longrightarrow \pi_1(M)$ is an isomorphism, the contradiction between that fact and condition (*) shows that δM_C is incompressible.

Waldhausen's results [3] on sufficiently large manifolds enable one to show that the submanifold $M_C \subset M_0 \subset M$ is unique up to isotopy in the manifold M_0. We call that submanifold M_C the compact core of M_0.

THEOREM 7.44. *Under the conditions of Theorem 7.43 each component of the complement* $M_0 \backslash M_C$ *contains precisely one end* E *and one component* S_E *of the boundary* δM_C. *There is thus a one-to-one correspondence between the ends of the manifold* M_0 *and the components of the boundary* δM_C *of its compact core.*

This theorem (see Bonahon [1], Theorem 1.6) enables one to give a classification for the ends of a hyperbolic 3-manifold M: that is, we say that an end E is geometrically finite if the corresponding component M_E, included in $M_0 \backslash M_C$ has a geometrically finite structure (see Definition 5.44). The end E is called *geometrically infinite* if it is not geometrically finite.

Amongst geometrically infinite ends, Thurston [1] distinguished an important class of geometrically tame ends, where he used his concept of a pleated surface in the hyperbolic manifold M.

DEFINITION 7.45. A pleated surface in a complete hyperbolic 3-manifold M is a complete hyperbolic surface S having finite area together with an isometry $f : S \longrightarrow M$ that transfers interior points of any geodesic segment in S to a geodesic segment in M, and each cusp of S into a cusp of the manifold M.

DEFINITION 7.46. As above, let M_0 be a complement in the hyperbolic manifold M to its cusps, while N is the minimal convex retract of the manifold M (see Theorem 5.12).

Let E be the end of the manifold M_0 defined by a component U_E of the complement $M_0 \backslash M_C$ (or by the component S_E of the boundary δM_C) of the compact core $M_C \subset M_0$. Then the end E is called *geometrically tame* if one of the following conditions is fulfilled:

1) $N \cap U_E$ is compact;

2) the set of pleated surfaces S', homotopic to S_E and such that $S'_h = S'_{[\varepsilon,\infty)} \subset U_E$, is non-compact.

The end E is called *almost geometrically tame* if a certain finite-sheeted covering for it (more precisely, a covering for U_E) up to a compact set is a geometrically tame end.

DEFINITION 7.47. Let a hyperbolic manifold M have a compact submanifold M_C lying in the complement M_0 to the cusps in M and homotopically equivalent to the manifold M. If $M_0 \backslash M_C$ is the union of disjoint geometrically tame ends, then the manifold M and its fundamental group $\pi_1(M)$ are called *geometrically tame*.

Geometrically tame ends play a key part in research on hyperbolic manifolds. Thurston [1], Chapter 9, has proved the following major theorems on them.

THEOREM 7.48. *Let M be a hyperbolic 3-manifold and $\hat{M}$ a covering of it such that $\hat{M}_0$ has a geometrically tame end E bounded by the surface $(S_E)_t$ and which is geometrically infinite. Then either* $\operatorname{vol} M < \infty$ *and a certain finite-sheeted covering of the manifold M fibre over the circle with fibres S_E or the image of the end E in the manifold M_0 is an almost geometrically tame end of the manifold M up to a compact set.*

The manifolds from Theorem 7.48 are given by the examples of geometrically infinite discrete groups $G \subset \mathcal{M}_2$, mentioned in Chapter 5 and constructed by Jørgensen [3] and Jørgensen and Marden [1]; they are, on the one hand, geometrically tame and, on the other, are infinite cyclic coverings of finite-volume manifolds that fibre over the circle (see also Krushkal', Apanasov, and Gusevskii [3], Examples 43, 53, and 63). Other examples of geometrically infinite manifolds are given by the hyperbolic 3-manifolds M whose fundamental groups act in the hyperbolic space H^3 as degenerate groups, i.e., $M = H^3/G$, in which $G \subset \mathcal{M}_2 \cong \operatorname{Isom}_+ H^3$ is a non-elementary Kleinian group having the connected and simply connected discontinuity set $\Omega(G)$, see Chapter 3 §3.5. The boundary of Teichmüller space (see Chgapter 8 §1) for a geometrically finite Kleinian group consists, in the main, of discrete groups that contain degenerate subgroups. However, the following fundamental result (Thurston [1], 9.6.1) implies that the geometrically tame manifolds H^3/G correspond to all these boundary groups G.

Let $G \subset \mathcal{M}_2$ be a torsion free discrete group that gives a geometrically tame manifold $M(G) = [H^3 \cup \Omega(G)]/G$. We consider some sequence $\{\rho_i\}$ of discrete faithful representations $\rho_i : G \longrightarrow \mathcal{M}_2$ that preserve parabolic and non-parabolic elements and are such that the manifolds $M(\rho_i G)$ corresponding to the discrete groups $\rho_i G$ are geometrically tame and admit homeomorphisms to the submanifold $M_0 \subset M(G)$ (in the correct homotopy class that preserves cusps and ends).

THEOREM 7.49. *If the representations* ρ_i *converge algebraically to a representation* ρ *(see Chapter 6 §5), then the limit manifold* $M(\rho G)$ *is geometrically tame and admits a homeomorphism (in the correct homotopy class) to the submanifold* $M_0 \subset M(G)$ *which is the complement to cusps in* $M(G)$.

This notable result of Thurston [1] nevertheless does not allow one to obtain a description of any Kleinian manifold $M(G)$ because we do not know whether an arbitrary degenerate group G lies at the boundary of Teichmüller space $T(\Gamma)$ for a certain Fuchsian group $\Gamma \subset \mathcal{M}_2$, or even for a geometrically tame one (see Theorem 7.59).

For the particular case of a torsion free degenerate group $G \subset \mathcal{M}_2$ without parabolics that uniformizes a closed oriented surface $S = \Omega(G)/G$ ($\Omega(G)$ is connected and simply connected), the description of $M(G)$, without the assumption that the group G is a boundary group, gives the following result due to Gusevskii [1] (compare Corollary 7.58).

THEOREM 7.50. *Let* $G \subset \mathcal{M}_2$ *be a torsion free degenerate group and* $S = \Omega(G)/G$ *be a closed oriented surface; then* $M(G)$ *is homeomorphic to* $S \times [0, \infty)$.

PROOF. As the group G is not geometrically finite (Greenberg [2]), the manifold $M(G)$ cannot be compact (see §1). We show first that $M = M(G)$ has precisely one end (geometrically infinite). For this it is sufficient to show that all relative homotopy groups $\pi_i(M, S)$, $i = 1, 2, \ldots$, are trivial. For this we write the homotopy sequence of the pair $(M(G), S)$:

$$\longrightarrow \pi_2(M(G), S) \longrightarrow \pi_1(S) \longrightarrow \pi_1(M(G)) \longrightarrow \pi_1(M(G), S) \longrightarrow 0. \tag{7.22}$$

As $\pi_i(S) = \pi_i(M(G)) = 0$ for $i \geq 2$, we get from the exactness of this sequence at once that $\pi_i(M(G), S) = 0$ for $i \geq 3$; the remaining equalities $\pi_2(M(G), S) = \pi_1(M(G), S) = 0$ also follow from the exactness of the (7.22) sequence and the isomorphism $\pi_1(M(G) \cong \pi_1(S)$.

To prove that $M(G) \approx S \times [0, \infty)$, we first construct a decreasing sequence $\{M_i\}$ of submanifolds $M_i \subset M = M(G)$ having the following properties:

1) $M_0 = M$; M_i, $i \geq 1$, is a piecewise linear 3-manifold having one end;

2) M_i is closed in M, $M \setminus \operatorname{int} M_i$ is compact;

3) $M_0 \supset M_1 \supset \ldots$, $\bigcap_i M_i = \varnothing$;

4) the boundary ∂M_i is a closed orientable surface, $\partial M_i \cap \partial M_j = \varnothing$ for $i \neq j$; and

5) the homomorphism $\pi_1(\partial M_i) \longrightarrow \pi_1(M)$ is a monomorphism.

Firstly we put $M_0 = M$ and construct $M_1 \subset M_0$ possessing properties 1) - 5); we apply Lemma 7.41 to find the neighbourhood

$N \subset \text{int}\, M_0$ of the end in M_0 having properties E1 - E4 and demonstrate how to change N to meet the condition 5).

This is done as follows. Suppose that the homomorphism $\pi_1(\partial N) \longrightarrow \pi_1(M_0)$ has a non-trivial kernel. Then, as in Lemma 7.10, there exists a component $S^* \subset \pi^{-1}(\partial N)$ that is not simply connected. We apply the generalized van Kampen theorem (see Hempel [1], compare Theorem 7.7) to S^* and to two components $H^3\backslash S^*$ to get that there exists a non-trivial loop $\alpha^* \subset S^*$ that is contractible to a point in one of the components $H^3\backslash S^*$. Then Dehn's lemma (Theorem 7.5) implies that there exists a simple loop β^*, non-trivial in S^*, that bounds the disc D^*, $\text{int}\, D^* \subset H^3\backslash S^*$. D^* intersects only a finite number of the components of $\pi^{-1}(\partial N)$. One can assume that all the intersections $D \cap \pi^{-1}(\partial N)$ consist of simple loops in D^*, from which we select the loop γ^* not containing other loops within it. Then γ^* bounds a disc $D_1^* \subset D^*$ and is a simple non-trivial loop in some component $S_1^* \subset \pi^{-1}(\partial N)$. We project γ^* and D_1^* into $\text{int}\, M_0$ to obtain a simple loop γ, non-trivial in ∂N, that bounds D whose interior lies in $M_0\backslash\partial N$.

Now let $K = M_0\backslash\text{int}\, N$. There are four possibilities:

(1) $D \subset K$, γ does not divide ∂N;
(2) $D \subset N$, γ does not divide ∂N;
(3) $D \subset N$, γ divides ∂N; and
(4) $D \subset K$, γ divides ∂N.

In the first case, we put $N' = N \cup V(D)$, in which $V(D)$ is a regular neighbourhood of the disc D in K; in the second case, we put $N' = \overline{N\backslash\text{int}\, V(D)}$, in which $V(D)$ is a regular neighbourhood of D in N. In the third and fourth cases, one can show that D divides N (correspondingly K). As M has one end, the closure of one of the components $N\backslash D$ in (3) is compact. The closure of the other component we take as N'. In the fourth case, as N' we take the union of N with the closure of the component $K\backslash D$ that does not intersect S.

In each of these cases, it is readily verified that $\chi(\partial N') > \chi(\partial N)$, i.e., the genus of the surface ∂N is reduced as a result of the reconstruction, so that the reconstriction associated with the non-trivial kernel of the homomorphism $\pi_1(\partial N) \longrightarrow \pi_1(M)$ finishes after a finite number of steps, and we obtain a manifold $M_1 = N'$ for which conditions 1) - 5) hold.

We now assume that the manifolds M_i with the necessary characteristics have been constructed for all $i \le n$. We apply Lemma 7.41 to find the manifold $N \subset \text{int}\, M_n$ having properties 1) - 4). If the homomorphism $\pi_1(\partial N) \longrightarrow \pi_1(M)$ has a non-trivial kernel, then we take a loop γ, non-trivial in ∂N, that is homotopoic to a point in M. We denote by ∂N^* any given component $\pi^{-1}(\partial N)$ and by M_n^* the component $\pi^{-1}(\text{int}\, M_n)$ containing ∂N^*. Let γ^* be the lifting of γ, beginning at the point $0^* \in \partial N^*$. As γ is trivial in M, γ^* is a loop in ∂N^*. Lemma 7.13 gives that M_n^* is a topology ball. Therefore, γ^* is homotopic to a point in M_n^*, so consequently γ is homotopic to a point in $\text{int}\, M_n$. Then we see that all the above reconstrictions of the manifold N can be performed within M_n. As a result, we get the manifold $N' \subset \text{int}\, M_n$ having the property that the homomorphism $\pi_1(\partial N') \longrightarrow \pi_1(M)$ is a monomorphism. We put $M_{n+1} = N'$ and thus complete the proof that there exists a sequence $\{M_i\}$ having properties 1) - 5).

We now put $K_i = M(G)\setminus \mathrm{int}\, M_i$, $i = 1, 2, \ldots$. It is evident that the homomorphism $\pi_1(K_i) \longrightarrow \pi_1(M(G))$ is an epimorphism by virtue of the isomorphism $\pi_1(S) \cong \pi_1(M(G))$, so $\pi^{-1}(K_i) = K_i^*$ is connected. Lemma 7.13 implies that K_i^* is contractible, so $\pi_j(K_i) = 0$ for all $j \geq 2$. At the same time, it is readily checked that one can use the simple connectedness of $\Omega(G) = \pi^{-1}(S)$ and K_i^* to show that the homomorphism $\pi_1(S) \longrightarrow \pi_1(K_i)$ is an isomorphism. Therefore S and K_i have the same homotopy type and so their Euler characteristics are equal: $\chi(K_i) = \chi(S)$.

As $M(G)$ is an irreducible orientable manifold, Waldhausen's [3] theorem shows that K_i is homeomorphic to $S \times [0, 1]$. Therefore, the Euler characteristics of the surfaces ∂M_i coincide with $\chi(S)$. As $\pi_1(\partial M_i) \longrightarrow \pi_1(M)$ are monomorphisms, it follows that the homomorphisms $\pi_1(K_i) \longrightarrow \pi_1(K_{i+1})$ are also monomorphisms. Moreover, if $\pi_1(\partial M_i) \longrightarrow \pi_1(M_i\setminus \mathrm{int}\, M_{i+1})$ is not an epimorphism, then $\pi_1(K_i) \longrightarrow \pi_1(K_{i+1})$ is not, and so $\pi_1(K_i)$ is a proper subgroup in $\pi_1(K_{i+1})$. If this occurs for an infinite subset of numbers i, the group $\pi_1(M(G))$ is clearly isomorphic to the direct limit of the groups $\pi_i(K_i)$ and will be the union of an increasing sequence of its proper subgroups. This immediately implies infinite generation for the group $\pi_1(M(G) \cong G$. Therefore for any sufficiently large $i \geq i_0$, the homomorphism $\pi_1(\partial M_i) \longrightarrow \pi_1(M_i\setminus \mathrm{int}\, M_{i+1})$ is an epimorphism.

We consider the double $\hat{M}_i$ of the compact manifold $M_i\setminus \mathrm{int}\, M_{i+1}$, which by virtue of Poincaré duality has the Euler characteristic $\chi(\hat{M}_i) = 0$. We have

$$0 = \chi(\hat{M}_i) = 2\chi(M_i\setminus \mathrm{int}\, M_{i+1}) - \chi(\partial(M_i\setminus \mathrm{int}\, M_{i+1})) = 2\chi(M_i\setminus \mathrm{int}\, M_{i+1}) - 2\chi(\partial M_i).$$

This means that $\chi(\partial M_i) = \chi(M_i\setminus \mathrm{int}\, M_{i+1})$, so we apply the product theorem and get a homeomorphism of the manifold $M_i\setminus \mathrm{int}\, M_{i+1}$ to the surface layer $\partial M_i \times [0, 1]$ for all $i \geq i_0$, so the manifold M_{i_0} is homeomorphic to the product $\partial M_{i_0} \times [0, \infty)$. However, the manifold $M(G) = K_{i_0} \cup M_{i_0}$, so $M(G)$ is homeomorphic to $S \times [0, \infty)$. The proof is complete.

The property of the manifold $M(G)$ for a degenerate group G that uniformizes a closed surface S in fact applies for all geometrically tame manifolds. Thurston ([1], Theorem 9.4.1) observed the following fact:

THEOREM 7.51. *A geometrically tame end E of the submanifold M_0 for a complete hyperbolic manifold M (see Theorem 7.43) is topologically tame. In other words, its neighbourhood $U_E \subset M_0\setminus M_C$ is homeomorphic to the product $S_E \times [0, \infty)$, where the surface S_E is the component of the boundary δM_C corresponding to E.*

The proof of this is based on a geometric construction (Thurston [1], 9.5.13) for the following proper mapping:

THEOREM 7.52. *In any complete hyperbolic 3-manifold M having a geometrically tame end E which is bounded by a hyperbolic surface $S_E(\subset \partial M_C)$ there is a proper homotopy:*

$$F : S_E \times [0, \infty) \longrightarrow M, \tag{7.23}$$

that moves the surface S_E to infinity at the end E.

These theorems show how the neighbourhoods for geometrically tame ends of hyperbolic manifolds are constructed. Almost geometrically tame ends here in no way differ from geometrically tame ones, as Theorem 7.51 implies.

Theorem 7.51 shows that geometrically tame manifolds admit a compactifiaction and Theorem 7.49 implies that any hyperbolic manifold corresponding to boundary points of Teichmüller space of a geometrically tame Kleinian group can be compactified. These results enabled Thurston ([4] p.379) to formulate the hypothesis that all Kleinian groups $G \subset \mathcal{M}_2$ are geometrically tame. We shall describe Bonahon's [1], [2] solution to this problem in the basic case of finitely generated Kleinian groups, not decomposable into free products. Before this, we give a simple example of a non-tame (geometrically infinite) manifold. That manifold is naturally defined by an infinitely generated Kleinian group $G \subset \mathcal{M}_2$.

EXAMPLE 7.53 (*the non-tame 3-manifold $M(G)$*). Let the coordinate axis $\mathbb{R}\cdot e_1$ be the union of segments $[16l, 16(l+1)]$, $l \in \mathbb{Z}$, with the centres z_i and integer points x_{ij}, $1 \le j \le 4$, that are not multiples of four. They are the centres of the circles

$$\sigma_i = \{x \in \mathbb{R}^2 : |z_i - x| = 15/2\} \quad \text{and} \quad \sigma_{ij} = \{x \in \mathbb{R}^2 : |x - x_{ij}| = 1\}.$$

Now let $g_{i1}, g_{i2} \in \mathcal{M}_2$ $i \in \mathbb{Z}$, be hyperbolic transformations such that

$$g_{i1} : \text{ext}\,\sigma_{i1} \longrightarrow \text{int}\,\sigma_{i3}, \quad g_{i2} : \text{ext}\,\sigma_{i2} \longrightarrow \text{int}\,\sigma_{i4}.$$

They generate Kleinian groups $G_i = \langle g_{i1}\rangle * \langle g_{i2}\rangle$ (classical Schottky groups). Then the group

$$G = \mathop{\ast}_{i\in\mathbb{Z}} G_i = \mathop{\ast}_{i,j} \langle g_{ij}\rangle \tag{7.24}$$

is the required Kleinian group, since it is not geometrically tame.

The group G leaves the circle $\overline{\mathbb{R}}^1 \supset \Lambda(G)$ invariant and is a Fuchsian group in the plane and leaves the hyperbolic plane $H^2 = \{x \in \mathbb{R}^3_+ : x_2 = 0\} \subset H^3$ invariant. Therefore, the manifold $M(G) = [H^3 \cup \Omega(G)]/G$ is homeomorphic to the product of the surface $S(G) = [\overline{H}^2 \backslash \Lambda(G)]/G$ and the segment $I = [0, 1]$, in which $H^2/G = \text{int}\, S(G)$ is a totally geodesic surface.

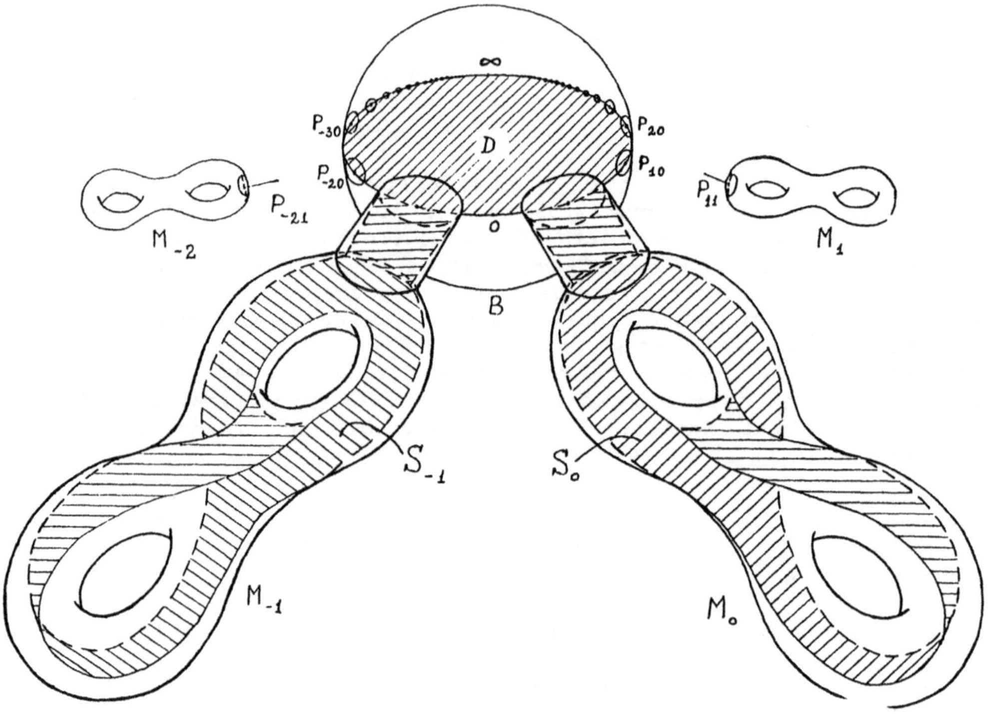

Figure 67

Let $C_i \subset H^3$, $i \in \mathbb{Z}$, be hyperbolic planes spanned on the circles S_i (i.e., $\partial C_i = S_i$) and splitting up $H^3 \cup \Omega(G)$ to an unbounded component B and components B_i, $i \in \mathbb{Z}$, (bounded in the Euclidean sense). Taking their quotients by the action of the group G we obtain the submanifolds $M = B/G = B$ and $B_i/G = B_i/G_i = M_i$ of the manifold $M(G)$; here $M(G)$ itself is obtained by identifying the closed discs $p_{i0} \subset \partial M = \partial B$ and $p_{i1} \subset \partial M_i$, corresponding to the planes $C_i \subset H^3$, $i \in \mathbb{Z}$, (the infinite disc sum). Each of the manifolds M_i, $i \in \mathbb{Z}$, is homeomorphic to a body with two handles. At the same time, this decomposition of $M(G)$ is performed along with an analogous decomposition of the totally geodesic surface $S(G) \subset M(G)$ as a sum of an infinite number of surfaces $S_i = (\overline{H}^2 \cap B_i)/G_i \subset M_i$, $i \in \mathbb{Z}$, and the disc $D = B \cap H^2$, where each of the S_i is homeomorphic to a torus with a deleted open disc. Figure 67 shows the scheme for this decomposition of a three-dimensional manifold $M(G)$ and surface $S(G)$ to a disc sum of manifolds M_i, $i \in \mathbb{Z}$, and correspondingly a sum of surfaces S_i, $i \in \mathbb{Z}$. These manifolds M_i are homeomorphic to the products $S_i \times I$ of a torus without disc and the closed segment I, while the embeddings $S_i \hookrightarrow M_i$ (totally geodesic ones in the hyperbolic metrics $\operatorname{int} S_i$ and $\operatorname{int} M_i$) induce isomorphisms of the fundamental groups $\pi_1(S_i) \longrightarrow \pi_1(M_i)$, $i \in \mathbb{Z}$.

If the (unique) end E of the manifold $M(G)$ were topologically tame, then the complement $M(G)\backslash K$ of a certain compact $K \subset M(G)$ would be homeomorphic to $M' \times [0, \infty]$, in which $M' \subset M(G)$ is a certain compact three-dimensional submanifold with boundary and consequently the neighbourhood $U_E = \operatorname{int} M(G)\backslash K$ would have the homotopy type of the compact manifold $M' \subset M(G)$. At the same time, up to a certain compact, the complement $M(G)\backslash K$ is the infinite disc sum of the manifolds $M_i \approx S_i \times I$, $|i| \geq i_0$, and therefore has the homotopy type of the infinite bouquet of circles. This contradiction shows that the end E of $M(G)$ cannot be topologically tame. Theorem 7.51 shows that $M(G)$ and the Kleinian group G from (7.24) are not geometrically tame.

We now formulate the basic result of Bonahon [1], [2], which gives a solution to the above Thurston conjecture:

THEOREM 7.54. *Let M be a complete hyperbolic 3-manifold having a finitely generated fundamental group $\pi_1(M)$ and let condition* (*) *be satisfied:*

for any non-trivial decomposition $\pi_1(M) = A * B$ as a free product there exists a parabolic element in $\pi_1(M)$ that is not conjugate to any element from A and B. (*)

Then M is a geometrically tame manifold.

Much of the technique Bonahon used to prove this theorem extends ideas due to Thurston [1] - [3]. A new idea is the geodesic currents concept in a hyperbolic surface S, which extends the concept of closed geodesics in S. The space of geodesic currents consistutes the completion of the space of homotopy classes of closed curves in S; for the subspace formed by simple curves, that completion gives the space $\mathcal{GL}(S)$ of geodesic laminations introduced by Thurston [1]:

DEFINITION 7.55. A lamination L in the manifold M^n is a foliation of a closed subset in M^n, or more precisely, if (U_i, φ_i) is a set of charts covering the closed subset $A \subset M^n$ (the support of L), $\varphi_i : U_i \longrightarrow \mathbb{R}^{n-k} \times \mathbb{R}^k$, then $\varphi_i(A \cap U_i)$ has the form $\mathbb{R}^{n-k} \times B$, in which $B \subset \mathbb{R}^k$. Then the changes of charts φ_{ij} should be represented as

$$\varphi_{ij}(x, y) = (f_{ij}(x, y), g_{ij}(y)),$$

in which $y \in B$. The leaves of the lamination L are defined the same way as the leaves of a foliation. A lamination whose leaves are geodesics is called *geodesic*.

If the surface S is a boundary component of the manifold $M(G)$, any lamination γ in S cannot have the whole surface S as a support, i.e., it is not a foliation in S (this follows from considering the Euler characteristic). The complement $S\backslash\gamma$ consists of domains bounded by closed geodesics or infinite ones.

We note that the subset $\gamma \subset S$, which is a union of bending lines of a pleated surface S is a geodesic lamination in the surface S. (Here the bending lines are the geodesics which consist of such points for which, in Definition 7.45, there is a unique direction of geodesic intervals containing them.) Moreover the isometry $f : S \longrightarrow M$ in the complement $S\backslash\gamma$ transfers all geodesic segments to geodesics ones in M.

However, Bonahon's concept of geodesic currents is best adapted to researching the ends of a manifold $M(G)$ than is Thurston's realization of geodesic laminations by means of pleated surfaces. From this viewpoint, the notion of a geometrically tame manifold (Definition 7.47) can be formulated as follows (Bonahon [1], [2]).

DEFINITION 7.56. Let a geometrically infinite end E of a hyperbolic manifold M be determined by the component S_E of the boundary δM_C of the compact core in the manifold $M_0 \subset M$ (see Theorem 7.44). Then the end E is called simply degenerate if, for each of its neighbourhoods U_E, there exists a simple closed curve $\gamma \subset S_E$, homotopic in M to a closed geodesic $\gamma^* \subset U_E$.

The manifold M, having finitely generated $\pi_1(M)$, is geometrically tame if and only if each of its ends is either geometrically finite or simply degenerate.

We now describe the main points in the proof of Bonahon's theorem, which consists in proving the following:

> If E is a geometrically infinite end of the manifold $M_0 \subset M$ and determined by the component S_E of the boundary δM_C, then there exists a family (α_j) of closed simple loops in S_E such that each neighbourhood U_E of the end E contains closed geodesics α_j^* for sufficiently large j that are homotopic to α_j in M (we say that α_j^* tend to E for $j \longrightarrow \infty$).

As the first step in proving this, we construct a set of closed curves α_j having that property on S_E, but which are not in general simple curves; that part of the proof is not dependent on the dimensions and is substantially simplified if M has no parabolic ends.

As these curves α_j are not simple, it is necessary to demonstrate that one can select them as j increases to be increasingly simple, i.e., such that the ratio between the number of double points and the square of the length tends to zero as j increases. This is obtained from the following extension of Thurston's Lemma 9.35 in [1] concerning the intersectionn index (see Bonahon, Theorem 3.4 in [2], also [1]):

THEOREM 7.57. *Under the conditions of Theorem 7.54, there is a constant* C *having the following property:*

Given $d>0$, *let* α_1^* *and* α_2^* *be closed geodesics in* M *at distances* $\geq d$ *from the surface* S_E *and homotopic to two curves* $\alpha_1, \alpha_2 \subset S_E$, *where the both homotopies do not intersect* S_E *and approach* S_E *from the same side. Also let each of these geodesics either not intersect* M_t *or be contained within a certain component of the thin submanifold* $M_t \subset M$. *Then*

$$i(\alpha_1, \alpha_2) \leq C \cdot e^{-d} \cdot l(\alpha_1) \cdot l(\alpha_2) + 4,$$

where i *is the intersection index in* S_E *and* $l(\alpha)$ *is the length of the curve* α *in* M.

A refinement of the meaning of the fact that the α_j on S_E become more simple for $j \longrightarrow \infty$ is obtained from the concept of geodesic currents in S_E; the space $\mathscr{C}(S_E)$ of geodesic currents in S_E is the completion (in a certain uniform structure) of the set of homotopy classes for closed curves in S_E. Introducing a geometrical number of intersections of these geodesic currents, which extends the analogous concept for closed curves, we conclude that the space of geodesic currents having zero self-intersection coincides with the space of measurable geodesic laminations (in addition to Definition 7.55 for such laminations, we introduce a transverse measure invariant with respect to translations along the leaves). That space is also the closure in $\mathscr{C}(S_E)$ of the set of simple-loop homotopy classes.

In the next step, we consider those α_j which, on a certain normalization, converge in $\mathscr{C}(S_E)$ to a geodesic current α_∞. Then it follows from considerations of continuity of the intersection number and the estimate in Theorem 7.57 that α_∞ is a measurable geodesic lamination in S_E. Here α_∞ is also the limit of a set of simple loops β_k in S_E (with appropriate normalization). In practice, only one component β_∞ of α_∞ is the limit for β_k. Actually β_∞ and α_∞ coincide.

In the last step, it is shown from a detailed argument that these simple loops β_K are homotopic in M to closed geodesics β_k^* that tend to E, i.e., each neighbourhood U_E of the end E contains all the β_k^* up to a finite number of them. This proves that E is a simply degenerate end and therefore M is a geometrically tame manifold.

COROLLARY 7.58. *Let* M *be a complete hyperbolic 3-manifold whose finitely generated fundamental group* $\pi_1(M)$ *acts on the sphere at infinity* $\partial H^3 = \overline{\mathbb{R}}^2$ *of the hyperbolic space* H^3 *as a function group (i.e., it has an invariant component* Ω_0 *in the discontinuity set). Then* M *is a geometrically tame manifold.*

The proof of this (pointed out by G.A. Swarum, see pp. 3-4 of Bonahon [2]) is based on the following observation. Let $M = H^3/G$, in which $G \cong \pi_1(M)$, $G \subset \mathscr{M}_2$, and let $\Omega_0 \subset \Omega(G)$ be an invariant component for the Kleinian group G. Then, from Ahlfors's finiteness theorem (Theorem 5.36), the submanifold $M_0 \subset M$ (the complement to the cusps in M) has a geometrically finite end corresponding to the surface Ω_0/G (of finite type). Then it is readily verified that all the other ends of the manifold M_0 are such that their fundamental groups satisfy condition (*) in Bonahon's theorem (Theorem 7.54). Bonahon's theorem then applied completes the proof.

Bonahon's theorem has many other consequences, of which we describe only two, which are connected with manifold compactification and Ahlfors's problem on Kleinian-group limit-set measure.

THEOREM 7.59. *Let M be a complete hyperbolic 3-manifold with finitely generated fundamental group $\pi_1(M)$ and let one of the following conditions be satisfied*:

1) *for any non-trivial decomposition $\pi_1(M) = A * B$ there exists a parabolic element in $\pi_1(M)$ not conjugate to elements from A and B or*

2) *$\pi_1(M)$ acts on the sphere at infinity of H^3, $\partial H^3 = \overline{\mathbb{R}}^2$, as a function group.*

Then M is diffeomorphic to the interior of a compact manifold.

PROOF. The proof follows directly from Bonahon's theorem (Theorem 7.54 and Corollary 7.58) and Theorem 7.51, which was proved by Thurston [1] (Corollary 9.4.2) by means of pleated surfaces. However, Bonahon's technique enables one to give a short proof of Theorem 7.51 based on a result from the classical 3-manifold topology obtained by Freedman, Hass, and Scott [1], which does not require the refined approximations by pleated surfaces made by Thurston.

From §1 we get the following theorem on geometrically finite ends of a Kleinian manifold $M(G) = [H^{n+1} \cup \Omega(G)]/G = M \cup \partial M$, where $G \subset \text{Möb}(n)$, $G \cong \pi_1(M)$:

THEOREM 7.60. *If E is a geometrically finite end of a complete hyperbolic n-manifold M, then there exists a neighbourhood of it diffeomorphic to $(S, \partial S) \times [0, \infty)$, in which S is a certain compact submanifold.*

An analogous result is needed also for simply degenerate ends. It is provided by the following theorem (see Theorem 1.8 in Bonahon [1]).

THEOREM 7.61. *Let E be a simply degenerate end of a complete hyperbolic 3-manifold M satisfying condition* 1) *in Theorem* 7.59; *then E is contained in the minimal convex retract $C(M) \subset M$ of the manifold M. Also, there exists a neighbourhood U of the end E for which the pair $(U, U \cap \partial M_0)$ is diffeomorphic to the pair $(S_E, \partial S_E) \times [0, \infty)$.*

PROOF. Without loss of generality, one can assume that $\pi_1(M) = \pi_1(S_E)$, in which S_E is a component of the frontier δM_C (see Theorem 7.43). In fact, let $\widetilde{M}$ be a covering of the manifold M such that $\pi_1(\widetilde{M}) = \pi_1(S_E)$. Then the component of the complement $M_0 \backslash M_C$, containing the end E, is lifted diffeomorphically to $\widetilde{M}_0$, and the end E naturally corresponds to the end $\widetilde{E}$ of $\widetilde{M}_0$. That $\widetilde{E}$ is simply degenerate (as is E) because the component of the frontier $\delta\widetilde{M}_C$

adjoining it is diffeomorphic to the surface $S_E \subset \delta M_C$. The latter follows from $(S_E, \partial S_E)$ and $(S_{\tilde{E}}, \partial S_{\tilde{E}})$ having the homotopy type of the pair $(V_E, \partial V_E)$, in which V_E is the component of the complement $M_0 \setminus M_C$ containing E. As the minimal convex retract $C(M)$ contains a projection of the corresponding retract $C(\tilde{M})$ of the manifold $\tilde{M}$, the end E in M_Q satisfies the conclusions in the theorem, as does the end $\tilde{E}$ in $\tilde{M}_0$. Therefore, we can restrict ourselves to the case $M = \tilde{M}$, i.e., when $\pi_1(M) = \pi_1(S_E)$.

Also, changing the Margulis constant $\varepsilon = \varepsilon(n)$, $n = 3$, does not change the topological type of the manifold $M_0 = M_0(\varepsilon) \subset M$, so we make it sufficiently small and we can then assume that the distances between the Margulis tubes in $M_t = M_t(\varepsilon)$ are not less than one.

We denote by $\hat{S}_E$ the surface obtained from S_E by gluing the collars $\cong S^1 \times [0, \infty]$ along each component of the boundary ∂S_E; as E is simply degenerate, one can construct a family $\varphi_i : \hat{S}_E \longrightarrow M$ of hyperbolically simplicial embeddings such that each neighbourhood U of the end E intersects at least one of the images $\varphi_i(\hat{S}_E)$. Also, the diameter $\varphi_i(\hat{S}_E)$ modulo the thin submanifold M_t is bounded by a constant, independent of i, while the images $\varphi_i(\hat{S}_E)$ are contained in the minimal convex retract $C(M)$ and are pairwise homotopic there. The construction of such embeddings is based on the following bounded diameter lemma; see Bonahon [1] (Theorem 1.8) and compare Thurston [1], 8.8.5.

LEMMA 7.62. *Let $\varphi : S \longrightarrow M$ be an embedding of a compact connected surface S in a hyperbolic 3-manifold M for which $\varphi_* : \pi_1(S) \longrightarrow \pi_1(M)$ is a monomorphism, while each component of ∂S corresponds to a parabolic element from $\pi_1(M)$. Let $\gamma \subset S$ be a simple closed curve, while $\varphi(\gamma)$ is homotopic in M to a closed geodesic γ^*. Then φ is homotopic to a mapping $\varphi' : S \longrightarrow M$ such that $\varphi'(\gamma) = \gamma^*$, $\varphi'(\partial S) \subset M_t$ and $\mathrm{diam}_\varepsilon(\varphi'(S))$ is bounded above by a constant C dependent only on ε and on the topological type of S.*

Here $\mathrm{diam}_\varepsilon X$ of a set $X \subset M$ is defined as $\sup_{x, y \in X} d_\varepsilon(x, y)$, where the semi-metric $d_\varepsilon(x, y)$ is defined by the hyperbolic metric in M modulo the thin part $M_t = M_t(\varepsilon)$, i.e., as the minimum in the lengths of curves $(\gamma \cap M_0(\varepsilon))$, where the curve $\gamma = \gamma(x, y) \subset M$ joins the points x and y.

We note that the assumption $\pi_1(M) = \pi_1(S_E)$ implies that each curve in $\hat{S}_E$ that is homotopic in M to a curve in the interior of a component M_E (= the union of cusp components of M_t intersecting S_E) will be homotopic in $\hat{S}_E$ to the end of that surface.

Let U be a neighbourhood of E. We need to show that $U \cup M_E$ contains at least one $\varphi_i(\hat{S}_E)$. Without loss of generality, one can assume that the frontier δU in M_0 is compact and that U does not intersect parabolic components from $M_t \setminus M_E$.

As the action of the group $\pi_1(M)$ on the universal covering H^3 of the manifold M is properly discontinuous, there is only a finite number of Margulis tubes at distances $\leq c_1$ from the compactum δU. Using induction and the fact that distances between the Margulis tubes are ≥ 1, we conclude that there exists no more than a finite number of Margulis tubes that can be joined with δU by a path in which the part going beyond the Margulis tubes has length $\leq c_1$. Consequently, U contains the neighbourhood U' of the end E in M_0 such that U' cannot be joined with δU by a path whose length outside the Margulis tubes is $\leq c_1$. Then for any i for which $\varphi_i(\hat{S}_E)$ intersects U', we have in fact that $\varphi_i(\hat{S}_E)\setminus M_E$ is contained in U.

We also note that the construction of the simplicial embeddings φ_i in Lemma 7.61 is such that the images $\varphi_i(\hat{S}_E)$ are contained in the minimal convex retract $C(M)$ and are pairwise homotopic there (by "radiation" of their homotopies from M).

Then, if U_0 is a connected neighbourhood of the end E that does not intersect $\varphi_0(\hat{S}_E)$, then this neighbourhood is wholly contained in $C(M)$. Then, if $x \in U_0$, there exists a neighbourhood $U \subset \operatorname{int} U_0$ of the end E for which $U_0\setminus U$ is connected and contains x. As $\varphi_i(\hat{S}_E)$ is contained in $U \cup M_E$, each homotopy from φ_0 into φ_i passes at least once through x and therefore lies in $C(M)$. We have thus shown that U_0, and therefore the end E, will be contained in $C(M)$, which was what we set out to prove.

Before we prove the last part of Theorem 7.60, namely that E has a neighbourhood in M_0 in the form of a product, we note the following. As E is simply degenerate, each component of δS_E preserves the orientation of M, i.e., as $\pi_1(M) = \pi_1(S_E)$, the components of ∂M_E are rings, but not Möbius bands. Consequently, we conclude that each neighbourhood U of E contains $\varphi_i(\hat{S}_E)\setminus M_E$. In particular, we consider the inverse image $\varphi_i^{-1}(\partial M_E)$ to get that, for any neighbourhood U of E, ∂S_E can be made homotopic in $U \cap \partial M_E$, which would be impossible if even one of the components of ∂M_E were a Möbius band.

To prove that there is a product neighbourhood of the end E in M_0, we consider a neighbourhood U of E. We require the existence of $\varphi_i : \hat{S}_E \longrightarrow M$ whose image is contained in $U \cup M_E$. We deform φ_i in such a way that the intersections $\varphi_i(\hat{S}_E)$ and ∂M_E become transverse. We then take the composition of φ_i with an isotopy of the surface $\hat{S}_E$ to produce the homotopy of φ_i in φ'_i such that $\varphi'_i(\partial S_E) \subset \partial M_E$ and $\varphi'_i(\hat{S}_E) \subset U \cup M_E$. Also, as each component of M_E is diffeomorphic to $S^1 \times \mathbb{R} \times [0, \infty)$, one can move $\varphi'_i(S_E)$ outside the interior of M_E and make it such that $\varphi'_i(S_E)$ is contained in U.

From this construction of φ'_i we have that $\varphi'_i|_{S_E} : S_E \longrightarrow M_0$ is homotopic to the embedding $S_E \longrightarrow M_0$, namely to the natural embedding. We apply the following corollary of a Freedman-Hass-Scott result [1] to that embedding (this has been proved inexplicitly in §2 of this work in the case $\partial S = \varnothing$, while the general case is given by passing to doubles):

LEMMA 7.63. *Let $\psi : (S, \partial S) \longrightarrow (N, \partial N)$ be a homotopy equivalence, where S is a compact surface and N is a 3-manifold. If ψ is homotopic to an embedding, then ψ is homotopic to an embedding into a small neighbourhood of the image $\psi(S)$.*

Using this lemma, we can assume additionally that $\varphi'_i(S_E)$ is a submanifold embedded in M_0 and contained in U.

Then any neighbourhood U of the end E contains an embedding $\varphi'_i : S_E \longrightarrow M_0$ homotopic to the embedding $\varphi'_0 : S_E \longrightarrow M_0$. Here that homotopy preserves the image of the boundary ∂S_E in ∂M_0. We assume, without loss of generality, that $\varphi'_i(\hat{S}_E)$ do not intersect and tend monotonically to E in such a way that $\varphi'_i(S_E)$ lie homotopically between $\varphi'_{i-1}(S_E)$ and $\varphi'_{i+1}(S_E)$. As φ'_i and φ'_{i+1} are homotopic, Waldhausen's result [3] shows that $\varphi'_{i-1}(S_E)$ and $\varphi'_{i+1}(S_E)$ are separated in M_0 by a collar diffeomorphic to $S_E \times [i, i+1]$. The union of these collars gives a neighbourhood W of the end E in M_0 diffeomorphic to the product $S_E \times [0, \infty)$. This completes the proof of Theorem 7.61 and thus that of Theorem 7.59.

Another important application of Bonahon's theorem is the following one, which makes a major advance in solving the Alhfors problem.

THEOREM 7.64. *Let G be a discrete finitely generated Möbius group in the plane that is either a function group or that satisfies the following condition:*

> *for any non-trivial HNN-extension $G = A *_C$ or decomposition $G = A\,{}_C{*}\,B$ to a free product with an amalgamated subgroup C, where C is a finite group (cyclic of dihedral), there exists a parabolic element in G not conjugate to any element from A or B.*

Then either the limit set $\Lambda(G)$ has zero spherical measure or $\Lambda(G) = \overline{\mathbb{R}}^2$.

The proof of this theorem for a torsion free group G follows directly from Bonahon's theorem (Theorem 7.54 and Corollary 7.58) and the following theorem of Thurston (Thurston [1], Theorem 8.12.4) on geometrically tame groups, which extends Ahlfors's [3] theorem on geometrically finite groups, see Corollary 5.32 and also Theorem 6.61:

THEOREM 7.65. *If the fundamental group $\pi_1(M)$ of a geometrically tame hyperbolic 3-manifold M acts in $\overline{\mathbb{R}}^2$ as a discrete Möbius group G, then either $\Lambda(G)$ has zero spherical measure or $\Lambda(G) = \overline{\mathbb{R}}^2$. In the latter case, G acts ergodically in $\overline{\mathbb{R}}^2$.*

To prove Theorem 7.65, given a G-invariant measurable set $A \subset \Lambda(G)$, we consider a harmonic function h in H^3 which is an extension of the characteristic function for the set A. That G-invariant extension is provided by means of the visual Poisson measure (the volume of the solid angle from which a set in ∂H^3 is seen), as is evident from the detailed construction in (6.44), and therefore it is a harmonic function in the hyperbolic space H^3 (the gradient flow $\operatorname{grad} h$ preserves the hyperbolic volume):

$$\Delta h = \operatorname{div} \operatorname{grad} h = 0. \tag{7.25}$$

Then we have in the minimal convex retract $C(M) = M_G \subset M$, $M_G = [H_G \backslash \Lambda(G)]/G$ the harmonic function h that (due to (7.25) and also because M is a geometrically tame manifold (see Thurston [1], Theorem 8.12.3)) has the following property:

$$\inf_{C(M)} h = \inf_{\partial C(M)} h. \tag{7.26}$$

If $\Lambda(G) \neq \bar{\mathbb{R}}^2$, then for all $x \in H^3 \setminus H_G$ we have $h(x) < 1/2$ because x lies in a certain half-space whose closure does not contain points from $\Lambda(G)$. Then we get from (7.26) that

$$\inf_{C(M)} (1 - h) \geq 1/2. \tag{7.27}$$

If A contains a density point p, $h(x)$ tends to one for $x \longrightarrow p$, which contradicts (7.27) and thus shows that A has zero spherical Lebesgue measure. This shows that $\Lambda(G)$ has zero measure in $\bar{\mathbb{R}}^2$.

If G contains elements of finite order, one can prove Theorem 7.64 by reducing it to the case of torsion free groups by considering a torsion free subgroup of finite index $G' \subset G$ (see Theorem 2.15). That reduction, in which $\Lambda(G') = \Lambda(G)$ is possible because of the following observation due to Bonahon [1], Lemma 1.24:

Let $M = H^3/G$ and let $M_0 \subset M$ be the complement to the parabolic components of the ε-thin part of $M_t \subset M$, as in Theorem 7.43. Then the manifold M contains a compact submanifold M_C such that the embedding $(M_C, M_c \cap \partial M_0) \longrightarrow (M_0, \partial M_0)$ is a homotopy equivalence, while the frontier δM_C is an incompressible surface.

This observation shows that $G' = \pi_1(M_C)$ satisfies the condition (*) in Theorem 7.54 (compare Hempel [1], Chapter 7, where in Theorem 7.1, instead of ∂M we need to take $M_c \cap \partial M_0$), which thus completes the proof of Theorem 7.64.

CHAPTER 8

UNIFORMIZATION OF MANIFOLDS

§1. The classical approach

The concept of uniformization of Riemann surfaces and more general manifolds arose during the last century when a detailed study of multi-valued analytic functions was beginning. Poincaré considered uniformization of algebraic and general analytic manifolds, and Hilbert included it in the major problems he formulated in his outstanding address at the Second International Congress of Mathematicians at Paris in 1900.

Some simple examples demonstrate the essence of uniformization: one has to construct a universal covering for a given manifold, where (since it is simply connected) the corresponding functions are single-valued. Construction of the covering leads to consideration of domains in Euclidean space and the corresponding discrete groups of automorphisms of these domains. When the domains are suitably chosen, these automorphisms are Möbius transformations, while the groups are discontinuous and even have an invariant simply connected component of the discontinuity set. If one denotes such a component by Ω_0 and the automorphism group by G, then the original (uniformizable) manifold M is conformally equivalent to the manifold Ω_0/G obtained by factorizing Ω_0 under the action of the group G. A multi-valued analytic function f on M in terms of the universal conformal covering $\pi : \Omega_0 \longrightarrow \Omega_0/G$, i.e., the function $f \circ \pi$, becomes single-valued in $\tilde{M} \cong \Omega_0$. In application to functions, one says that the set $A \subset \mathbb{C}^m$ (or $\mathbb{CP}^m$) is uniformized by a system $f = (f_1, \ldots, f_m)$ of functions meromorphic in a domain $D \subset \mathbb{C}^n$ if f is a holomorphic covering $D_0 \longrightarrow A$ of a dense subset A_0 $(D_0 \subset D)$ having discrete fibre on which a (discrete) group G of automorphisms of D acts transitively, i.e. $f(z') = f(z'')$ only for the G-equivalent points $z', z'' \in D_0$.

No complete solution has yet been obtained, with the exception of the classical one-dimensional case. Here the Klein-Poincaré solution is that each conformal structure of an orientable surface is induced by a constant-curvature Riemannian metric. In the basic case, where the Euler characteristic is negative and, in particular, if the genus of the surface is greater than one, that metric defines a hyperbolic structure: small neighbourhoods on the surface are isometric to neighbourhoods in the hyperbolic plane, while the surface S itself is a quotient of the hyperbolic plane $H^2 \cong \tilde{S}$ under the action of the torsion free discrete group $\Gamma \cong \pi_1(S)$. In the exceptional case of the sphere and torus, that metric specifies spherical or Euclidean structures, and the surface is presented in the form $\tilde{S}/\Gamma$, where $\tilde{S}$ is the sphere $\overline{\mathbb{C}} = \overline{\mathbb{R}}^2$ or the plane $\mathbb{C} = \mathbb{R}^2$, with the discrete group $\Gamma \cong \pi_1(S)$ of $\tilde{S}$ motions either trivial or a translation group with generator $z \longmapsto z + w$ ($w \in \mathbb{C}\setminus\{0\}$), or else it is a free Abelian group of rank 2 with generators $z \longmapsto z + w_1$, $z \longmapsto z + w_2$, in which $w_1, w_2 \in \mathbb{C}\setminus\{0\}$ are such that $\operatorname{Im}(w_2/w_1) > 0$.

The natural projection $\pi : \tilde{S} \longrightarrow \tilde{S}/\Gamma$ is an unbranched covering and uniformizes all the functions f on the Riemann surface S because the functions $f \circ \pi$ are single-valued in $\tilde{S}$.

That result can be extended to branched coverings of Riemann surfaces. Let us associate with the points $p \in S$ a function $\nu(p)$ equal to zero outside a certain discrete set $V \subset S$ and taking positive integer values or infinity on V. We assume further that, if $S = \overline{\mathbb{C}}$, then V must not reduce to a single point, and if V consists of two points p_1 and p_2, then $\nu(p_1) = \nu(p_2)$.

The Klein-Poincaré uniformization theorem can be formulated in general as follows:

THEOREM 8.1. *Any Riemann surface S with a discrete set V distinguished on it and a given function $\nu(p)$ can be represented up to conformal equivalence in the form $S = \hat{S}/\Gamma$, where $\hat{S}$ is either $\overline{\mathbb{C}}$ or $\mathbb{C}$ or else $B^2(0, 1)$; Γ is a discrete group of Möbius automorphisms $\hat{S}$; and the covering $\hat{S} \longrightarrow \hat{S}/\Gamma = S$ is branched over S at the points $p \in V$ and has there branching orders $\nu(p)$.*

The branch points $p \in V$, $\nu(p) < \infty$, correspond to fixed points of stationary elliptic subgroups of Γ of orders $\nu(p)$.

There are various proofs of this fundamental theorem. Their scheme is as follows. Using the topological method of Chapter 2 §1, one constructs a universal covering $\hat{S}_0$ of the surface $S_0 = S\setminus\{p \in V : \nu(p) = \infty\}$ such that the covering $\pi : \hat{S}_0 \longrightarrow S$ has the required branching orders over the points $p \in V \cap S_0$ (when such points are present). Then the conformal structure is lifted from S_0 to $\hat{S}_0$ as a covering (see Chapter 2 §1), so that the covering π becomes holomorphic and reduces (to within an equivalence) to taking the quotient of $\hat{S}_0$ by a group Γ of conformal automorphims of it. As the surface $\hat{S}_0$ is simply connected, one can use the Riemann theorem on conformal mapping of simply connected Riemann surfaces onto canonical domains.

Another approach to uniformization was proposed roughly at the same time by Koebe. The general uniformization principle, going back to Koebe, asserts that, if a Riemann surface $\tilde{S}$ is topologically equivalent to a planar domain $D \subset \overline{\mathbb{C}}$, then there also exists a conformal homeomorphism $f : \tilde{S} \longrightarrow D$ such that the group of deck transformations $G(\tilde{S}, S)$ goes over to a Kleinian group $G \subset \mathcal{M}_2$,

$G = fG(\tilde{S}, S)f^{-1}$, having an invariant component D of the discontinuity set $\Omega(G)$. The same problem of (analytic) uniformization reduces to the topological problem of finding all the (branched, in general) planar coverings $\tilde{S} \longrightarrow S$ of a given Riemann surface S. The following results, due to Maskit [1], solve this problem.

In fact, each planar regular covering can be characterized by a certain finite set of simple disjoint loops on the surface S.

Let $v_1, \ldots, v_n, \ldots$ be a set of loops on S at points $O_1, \ldots, O_n, \ldots$ and let u_i besome path joining O_i to a fixed point O, and let f_i be the element of $\pi_1(S, O)$ containing the loop $u_i v_i u_i^{-1}$. Denote by N the smallest normal subgroup of $\pi_1(S, O)$ containing all the elements f_i, and consider the regular covering $\pi : \tilde{S} \longrightarrow S$ of the surface S, whose defining subgroup coincides with N. It is readily checked that the covering $(\tilde{S}, \pi)$ determined in this way depends only on the free homotopy classes of the loops $v_1, \ldots, v_n, \ldots$, (i.e., the classes of homotopy loops without initial-point specification); the normal subgroup N is correctly defined for any point $O \in S$ and also depends only on the free homotopy classes of the loops $v_1, \ldots, v_n, \ldots$. In this case, we say that the normal subgroup N is spanned by the loops $v_1, \ldots, v_n, \ldots$ and we write $N = \langle v_1, \ldots, v_n, \ldots \rangle$.

THEOREM 8.2. *Let S be an orientable surface and $v_1, \ldots, v_n, \ldots$ a set of simple disjoint loops on S. If $(\tilde{S}, \pi)$ is the regular covering with defining subgroup $N = \langle v_1^{\alpha_1}, \ldots, v_n^{\alpha_n}, \ldots \rangle$, where $\alpha_1, \ldots, \alpha_n, \ldots$ are positive integers, then $\tilde{S}$ is a planar surface.*

THEOREM 8.3. *Let $(\tilde{S}, \pi)$ be a regular planar covering of an orientable surface S with defining subgroup N. If S is a surface of finite type, i.e., $\pi_1(S)$ is finitely generated, then there exists a finite set of simple disjoint loops $v_1, \ldots, v_n$ and positive integers $\alpha_1, \ldots, \alpha_n$ such that $N = \langle v_1^{\alpha_1}, \ldots, v_n^{\alpha_n} \rangle$.*

In particular, if S is a closed orientable surface of genus $p \geq 1$, its fundamental group is generated by $2p$ elements:

$$\pi_1(S) = \langle a_1, b_1, \ldots, a_p, b_p : \prod_{i=1}^{p} [a_i, b_i] = 1 \rangle, \tag{8.1}$$

in which $[*, *]$ is the commutator, and the normal subgroup N defining S can be taken as the smallest normal subgroup spanned by $a_1, \ldots, a_p$. Then S is uniformized by a Schottky group G having p generators: $S = \Omega(G)/G$.

Koebe proved this result, which is known as the classical theorem on cuts.

In conclusion, we mention that there are also more general uniformizations, for example, by Kleinian groups with torsion and with accidental parabolic elements. A description of them can be found in surveys by Bers [5] - [8] and Krushkal', Apanasov, and Gusevskii [3].

§2. Gromov's and Thurston's geometric uniformization concepts

We have seen above that any two-dimensional surface admits a geometric structure to be introduced into it: spherical, Euclidean, or (in the main case) hyperbolic. Even three-dimensional manifolds include ones on which one cannot introduce any geometry locally modelled on some homogeneous space.

1. *Length spaces and convexity.* To formulate this problem in the class of convex spaces, we follow Gromov [2], [3]. Let a length function on curves be defined in the space X, which thus defines a distance $\rho(x, y)$, $x, y \in X$, as the length of the shortest curve in X joining the points $x, y \in X$. Then a length space X is called convex if the distance function $\rho(x, y)$ is convex. In other words, for any two geodesic segments

$$\alpha : [a, b] \longrightarrow X, \qquad \beta : [c, d] \longrightarrow X$$

the function $\rho(\alpha(t_1), \beta(t_2))$ is convex on $[a, b] \times [c, d]$.

X is called locally convex if for a certain covering of X by open sets U_i, $\rho(x, y)$ is convex when $x, y \in U_i$. Note that a Riemannian manifold M is locally convex if and only if its sectional curvature $K \leq 0$. To provide convexity, it is necessary and sufficient to add the condition for triviality of the fundamental group $\pi_1(M)$. In the context of length spaces one has, by adapting the classical arguments the Cartan-Hadamard theorem:

A simply connected locally convex space is convex. Every convex space is straight (any two points can be joined by the unique geodesic segment) and hence it is contractible.

In the context of convex length spaces, geometric uniformization can be formulated as the following two questions (Gromov [2]):

(I) Let M be an n-dimensional manifold of $K(\pi, 1)$ type. Does there exist a locally convex space (non-compact but preferably finite-dimentional), homotopically equivalent to M?

(II) Let M be a certain n-dimensional manifold. Does there exist a convex space X and a discrete group (with torsion) G of its isometries such that M is homeomorphic to X/G?

Answers are available only for dimension two. A positive answer to the first for a wide class of 3-manifolds containing all the known examples is given by Thurston's theory [1] - [4] (see also Bass and Morgan [1]), whose results enabled him to formulate the following hopeful hypothesis with considerable confidence.

2. *Thurston's structural conjecture*:

The interior of any compact 3-manifold M can be cut up canonically into pieces that allow one to introduce a geometric

(X, G)-structure of one of eight types.

That canonical cutting is performed either on two-dimensional spheres embedded in M and decomposing it into pieces different from three-dimensional balls (this process involves, as Kneser has shown (see Hempel [1], Chapter 7) only a finite number of steps) or on two-dimensional tori incompressibly embedded in M. The latter process (as we shall see later from results by Jaco and Shalen [1]) also is finite.

This canonical cutting of a 3-manifold M gives submanifolds whose geometric structures should be determined by Thurston's conjecture by geometries of the canonical homogeneous spaces X of one of the eight types, i.e., the (X, G)-structures (see Chapter 2), which have the following properties:

1. The space X is simply connected (in the opposite case, one can consider, in the (X, G)-manifold M, a structure defined by the universal covering space $\tilde{X}$).

2. The group G is unimodular, i.e. there exists a measure on G invariant by multiplication on the right or the left. Otherwise, X would possess a G-invariant vector field which expands volume, so there could be no (X, G)-manifolds which are compact or which even have finite volume.

3. G is a maximal group of homeomorphisms of the space X with compact stabilizers. In the opposite case, if G were contained in a larger group G', then any (X, G)-manifold would be a (X, G')-manifold, so (X, G) would be redundant.

Before we describe these eight geometries, we note that the most interesting, most complicated, and most frequently encountered amongst them is the hyperbolic one. The other seven are exceptional geometries, as Euclidean and spherical geometries in the two-dimensional case. In particular, for six of them (with the exception of the Sol-geometry), all closed 3-manifolds are Seifert fibre spaces, see Definition 8.5 (and any Seifert manifold admits a geometric structure).

Scott [4] has given a notable discussion of these exceptional seven geometries together with a classification of all 3-manifolds having geometry in one of these seven types, and there one can find a sketch of Thurston's proof of the existence of only eight types of three-dimensional geometries (described below) and of the fact that, if a closed 3-manifold admits a geometric structure, the geometry defining that structure is unique (although the geometric structure itself is not necessarily unique).

When we describe these eight canonical three-dimensional geometries (homogeneous spaces), we shall indicate a simply connected space X and a group G of orientation-preserving diffeomorphisms of X having compact stabilizers G_x, $x \in X$, that are isomorphic either to $SO(3)$ or $SO(2)$ or the trivial group $SO(1)$.

For the first three geometries, $G_x \cong SO(3)$.

T1. *Hyperbolic geometry* (see Chapter 2 §3).

T2. *Spherical* or *elliptic geometry* (see Chapter 2 §3), which is obtained for $X = S^3$ and $G \cong S^3 \times S^1/\mathbb{Z}$. See Seifert [1] or Wolf [1] for a classification of all spherical 3-manifolds.

The most beautiful is the Poincaré dodecohedral space, obtained from a dodecahedron by identifying opposite faces with a 1/10 right-handed rotation. This identification pattern forces the edges to be identified in triples. Since dihedral angles of a regular dodecahedron in Euclidean space are roughly equal to 117°, it is clear that a spherical convex dodecahedron with dihedral angles $2\pi/3$ may be taken and may identify its faces in the given pattern. Note that the universal cover of the Poincaré dodecahedral space is S^3, with a tessellation by 120 dodecahedra and with the deck transformations group of order 120 with the presentation

$$\langle a, b, c : a^5 = b^2 = c^3 = abc \rangle$$

(i.e., the binary icosahedron group).

T3. *Euclidean geometry* corresponds to $X = \mathbb{R}^3$ and $G \cong \mathbb{R}^3 \times SO(3)$. There are only 10 non-homeomorphic compact three-dimensional Euclidean manifolds. Each of them has a finite-sheeted covering by the three-dimensional torus $T^3 = S^1 \times S^1 \times S^1$ (see Wolf [1], Chapter 3).

For the next four classes, the identity component of $G_x \subset G$ is $SO(2)$.

T4. *The space* X *is* $S^2 \times \mathbb{R}$. The group G is the product of the isometry group of the sphere S^2 and the isometry group of the one-dimensional Euclidean space $\mathbb{R}$. There are only two non-homeomorphic examples of compact orientable manifolds having that geometry: the Hopf manifold $S^2 \times S^1$ and $\mathbb{RP}^3 \# \mathbb{RP}^3$ which is the unique closed 3-manifold that can be represented as a non-trivial connected sum and which has a geometrical structure.

T5. The space X is equal to *the product* $H^2 \times \mathbb{R}$ of the hyperbolic plane H^2 and the Euclidean 1-space $\mathbb{R}$. The group G of its diffeomorphisms is the product of the H^2-isometry group ($\cong SL(2, \mathbb{R})/\{+1\}$) and the $\mathbb{R}$-isometry group. Any closed orientable manifold admits such a $(H^2 \times \mathbb{R}, \operatorname{Isom} H^2 \times \operatorname{Isom} \mathbb{R})$-structure if and only if it has a finite-sheeted covering by the product $S_g \times S^1$, in which S^1 is the circle and S_g is a closed surface of genus $g > 1$ (see Scott [4]). Therefore, in what follows, such manifolds will be called almost trivial Seifert manifolds (see §5 for more details).

T6. *The space* $X = PSL(2, \mathbb{R})$ is the universal covering space for the set T_1H^2 of unit-length vectors in the hyperbolic plane. Here $PSL(2, \mathbb{R})$ is homeomorphic to an open solid torus $B^2 \times S^1$ in which there exists a constant-curvature Lorentz metric, see Goldman [2], §2, for details. The group G is the product of $\mathbb{R}$ and the universal covering of the group of isometries of the hyperbolic plane H^2. Here the H^2-isometries act by their derivatives (rotations) and $\mathbb{R}$ acts as a simultaneous rotation of all the vectors that fixes all points in H^2. An example of a 3-manifold with that (X, G)-structure is the space of unit tangent vectors to any hyperbolic surface. Another example is a manifold obtained by factorizing $T_1(H^2)$ by the action of a group of orientation-preserving automorphisms of a certain regular tessellation in H^2, e.g., a (2, 3, 7)-tessellation, see Fig. 68.

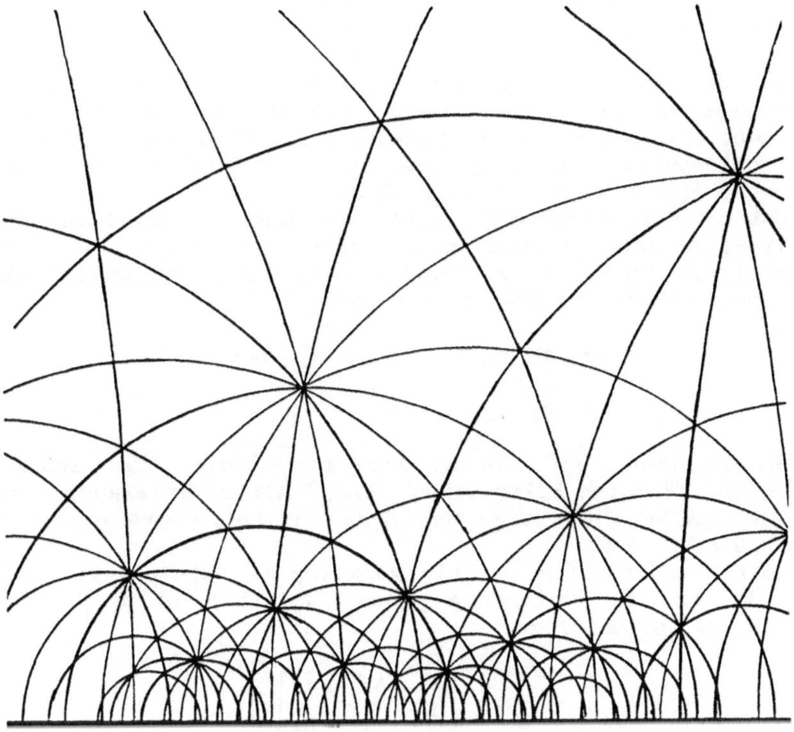

Figure 68

T7. *Nil-geometry (Heisenberg group).* The space X is a twisted product of $\mathbb{R}$ and $\mathbb{R}^2$, while the group G is a $\mathbb{R}$-extension of the $\mathbb{R}^2$ isometry group such that we have the exact sequence:

$$\mathbb{R} \longrightarrow G \longrightarrow \operatorname{Isom} \mathbb{R}^2. \tag{8.2}$$

To represent X more clearly, we consider $\mathbb{R}^3$ with a certain 2-plane field τ constructed as follows. At points of the axis $\mathbb{R}\cdot e_3 = \{x \in \mathbb{R}^3 : x_1 = x_2 = 0\}$ the planes in the field τ are orthogonal to the axis $\mathbb{R}\cdot e_3$. At points of any ray emanating from the axis $\mathbb{R}\cdot e_3$ and orthogonal to it, the field τ consists of planes tangent to that ray and having slopes in the direction orthogonal to that ray and increasing linearly, so this slope always equals half the distance to the axis $\mathbb{R}\cdot e_3$. In other words, at a point x with coordinates (x_1, x_2, x_3), the plane from the field τ is spanned by the vectors

$$e_1 - (x_2/2)\cdot e_3 \quad \text{and} \quad e_2 + (x_1/2)\cdot e_3. \tag{8.3}$$

Elements of the group G from (8.2) are defined as affine transformations in $\mathbb{R}^3$ obtained by lifting the Euclidean isometries of the plane $\{x \in \mathbb{R}^3 : x_3 = 0\}$ to affine transformations of $\mathbb{R}^3$ that preserve the field τ (as specified by the vectors (8.3)) and also preserve the distances along lines parallel to the axis $\mathbb{R}\cdot e_3$. In fact, there exists a whole one-parameter family of such lifts. We note that an orientation-reversing isometry of $\mathbb{R}^2$ lifts only to maps which reverse the direction of the axis $\mathbb{R}\cdot e_3$.

That (X, G)-structure occurs in any 3-manifold that is a non-trivial oriented circle bundle over a 2-torus T^2 (i.e., one distinct from T^3). If we consider a discrete nilpotent subgroup Γ in G whose generators are the mappings

$$\begin{aligned} (x_1, x_2, x_3) &\longmapsto (x_1 + 1,\ x_2,\ x_3 + x_2/2), \\ (x_1, x_2, x_3) &\longmapsto (x_1,\ x_2 + 1,\ x_3 - x_1/2), \end{aligned} \tag{8.4}$$

then the quotient X/Γ is at once a manifold (the simplest circle bundle over T^2 with Euler class 1). All other examples of compact manifolds having that structure are finitely covered by manifolds homeomorphic to X/Γ.

Another description of this geometry is obtained by defining the space X as the Heisenberg group H, that is a nilpotent group consisting of the matrices

$$H = \left\{ \begin{bmatrix} 1 & x_1 & x_3 \\ 0 & 1 & x_2 \\ 0 & 0 & 1 \end{bmatrix} \right\}.$$

Here G is the semidirect product of the group H by S^1 acting as a group of automorphisms of H which rotate the plane $\{(x_1, x_2, 0)\}$. With that description of the (X, G)-geometry, the discrete subgroup Γ with the generators (8.4) corresponds to a subgroup in H, in which x_1, x_2 and x_3 are integers (see also §4).

T8. *Sol-geometry*. The last (X, G)-geometry has the trivial identity components of the stabilizers G_x, $x \in X$. Here, as previously, X is a Lie group, namely the solvable Lie group corresponding to the exact sequence

$$\mathbb{R}^2 \longrightarrow X \longrightarrow \mathbb{R} \tag{8.5}$$

where $\mathbb{R}$ acts on $\mathbb{R}^2$ (by conjugation) with the formula

$$(x_1, x_2) \longmapsto (e^t\cdot x_1,\ e^{-t}\cdot x_2). \tag{8.6}$$

The group G is an extension of X by $\mathbb{Z}_2 \times \mathbb{Z}_2$ acting as a group of automorphisms, whose three non-trivial elements are the 180° rotations:

$$(x_1, x_2, t) \longmapsto (-x_1, -x_2, t),$$

$$(x_1, x_2, t) \longmapsto (x_2, x_1, -t), \tag{8.7}$$

$$(x_1, x_2, t) \longmapsto (-x_2, -x_1, -t).$$

A manifold with that (X, G)-structure is provided by any torus bundle over S^1 whose monodromy is a linear mapping with distinct real eigenvalues.

We now turn to the canonical cutting of compact 3-manifolds mentioned in Thurston's structural conjecture. We give some definitions in order to describe it exactly.

DEFINITION 8.4. A three-dimensional manifold M is called *irreducible* if each two-dimensional sphere embedded in M bounds a ball in M; M is called sufficiently large in the sense of Waldhausen [3] if M contains an incompressible two-sided surface (i.e., the normal bundle to it is oriented) which is not a 2-sphere and whose boundary (if not empty) lies at the boundary ∂M of the manifold. A 3-manifold M is called a *Haken manifold* if it is prime and sufficiently large.

We note that, with the exception of the trivial ball case, a Haken manifold is provided by any prime 3-manifold M having non-empty boundary ∂M. Another condition for an irreducible 3-manifold M to be a Haken manifold is that the rank of the first homology group $H_1(M)$ is positive.

DEFINITION 8.5. Let (μ, ν) be a pair of relatively prime integers, $I = [0, 1]$, and let $B^2 \subset \mathbb{R}^2$ be a unit disc in polar coordinates: $B^2 = \{(r, \theta) : 0 \le r \le 1\}$. A fibred solid torus of type (μ, ν) is the manifold obtained by the following identification for bases of the cylinder $B^2 \times I$:

$$(r, \theta, 1) = (r, \theta + 2\pi\nu/\mu, 0). \tag{8.8}$$

The fibres in this resulting solid torus P are images of the arcs $\{x\} \times I$ by gluing (8.8); here each fibre (with the exception of the central one) consists of μ arcs. We assume (up to a homeomorphism that preserves the fibres) that $\mu > 0$ and $0 < \nu < \mu/2$. The integer μ is called the *index* of the fibred solid torus. If $\mu > 1$, the central fibre (= core) is called an *exceptional fibre*; but otherwise we call the core a *regular fibre*.

A Seifert fibred 3-manifold Σ is a 3-manifold which is a union of pairwise disjoint simple closed curves (fibres) and such that each fibre τ has a closed neighbourhood T consisting of a union of fibres, which is homeomorphic to a fibred solid torus P via a fibre-preserving homeomorphism. Such a neighbourhood of a fibre is called a fibred-neighbourhood. A fibre $\tau \subset \Sigma$ is called *exceptional* if it goes over to an exceptional fibre of the fibred solid torus P for that fibre-preserving homeomorphism $T \longrightarrow P$, or otherwise a fibre τ

is called a *regular fibre* of the Seifert fibred manifold in Σ.

By identifying each fibre τ in Σ to a point, we get a 2-manifold, which is called the *orbit-manifold of* Σ. A natural mapping from Σ to the orbit-manifold is called the *projection*.

After Jaco and Shalen's [1] description of Seifert fibrations in 3-manifolds and Johannson's [1] classification results, it became clear that it is necessary to split up 3-manifolds along non-trivially embedded tori, on which one obtains 3-manifolds whose bounderies consist of tori.

A fundamental role in this description is played by the mapping theorem (Jaco and Shalen [1], Chapter 3), which we formulate as follows.

THEOREM 8.6. *In the Haken 3-manifold M, there exist pairwise distinct submanifolds $\Sigma_1, \ldots, \Sigma_m \subset \operatorname{int} M$, that are Seifert fibre spaces having boundary components incompressible in M and having the following property: any mapping $f: S \longrightarrow M$ for any Seifert fibre space S with the non-cyclic fundamental group $\pi_1(S)$ that induces the fundamental group monomorphism $f_*: \pi_1(S) \longrightarrow \pi_1(M)$ is homotopic in M to a mapping whose image of S is contained in a certain Σ_i. If here Σ_i in M are such that for $i \neq j$ the inclusion mappings $\Sigma_i \longrightarrow M$ are not homotopic to the mappings $\Sigma_i \longrightarrow \Sigma_j$, then the set $\{\Sigma_1, \ldots, \Sigma_m\}$ is unique up to ambient isotopy.*

Subsequently, in particular in Thurston's hyperbolization theorem, an important part has been played by the following atoroidality concept: a 3-manifold M is called geometrically atoriodal if each torus T^2, incompressibly embedded in M, is isotopic to a component of the boundary ∂M. The manifold M is called homotopically atoroidal if each mapping $f: T^2 \longrightarrow M$ that induces a monomorphism $\pi_1(T^2) \longrightarrow \pi_1(M)$ is homotopic to a mapping $T^2 \longrightarrow \partial M$.

For a manifold M having an incompressible boundary ∂M, homotopic atoriodality is equivalent to the fundamental group $\pi_1(M)$ not being decomposable into the free product of its non-trivial subgroup (compare the condition (*) in Chapter 7 §5) and any subgroup of it that is a free Abelian group of rank two is conjugate to a subgroup of the fundamental group of a certain boundary torus from ∂M. Examples of homotopically atoroidal manifolds are those from Examples 2.22 and 2.23.

The above mapping theorem on Seifert fibre spaces in a 3-manifold implies an important splitting theorem (Jaco and Shalen [1], Corollary V.5.1).

THEOREM 8.7. *In a closed Haken 3-manifold M, there exists a family $\{T_1, \ldots, T_m\}$ of pairwise distinct incompressible tori (unique up to ambient isotopy) such that:*

1) *each component of the manifold obtained from M by cutting it along the tori of the family $\{T_1, \ldots, T_m\}$ is either a Seifert fibre space or a homotopically atoroidal manifold; and*

2) *no proper subfamily in the family $\{T_1, \ldots, T_m\}$ has the first property.*

We can now discuss (see also Scott [4]) main positions on

Thurston's geometrization conjecture, which asserts that there exists a reasonable canonical way for decomposing any compact 3-manifold M into a union of geometric pieces. The point of Thurston's structural conjecture is that the pieces should all be geometric, since the needed decomposition does exist.

Firstly, we have the decomposition of M into prime manifolds M_i that cannot be represented as non-trivial connected sums: $M = \#_i M_i$ (Theorem 7.8 gives such finite decomposition). Here the M_i are not submanifolds in M but instead are the result of surgery in M along non-trivial embedded (separating) 2-spheres and then adding three-dimensional balls to each of the manifolds obtained. We note that, although the factors M_i in the orientable case are unique (see Milnor [5]), the family of 2-spheres defining that decomposition is not unique up to isotopy. If M is non-orientable, one can describe precisely the ambiguity which arises. It is caused by the fact that, if a manifold N is non-orientable, then $N \# (S^2 \times S^1)$ and $N \# (S^2 \tilde{\times} S^1)$ are homeomorphic.

The obtained compact prime 3-manifolds M are of three basic types in the orientable case:

1) having finite fundamental groups $\pi_1(M)$;

2) having infinite $\pi_1(m)$ and reducible, i.e., admitting a non-separating embedded 2-sphere; and

3) having infinite $\pi_1(m)$ and irreducible.

Each manifold of type 1 is either a 3-ball or is closed and then is covered by a homotopy sphere (Hempel [1]). In that case, Thurston's structural conjecture can be stated as follows:

CONJECTURE 8.8. If a closed orientable 3-manifold M has the finite fundamental group $\pi_1(M)$, then it possesses spherical geometry (modelled on S^3).

This conjecture asserts, first, that Poincaré's conjecture is true and, second, that all free actions of finite groups on the sphere S^3 are conjugate to orthogonal actions. For a discussion on the progress made on the second part of the conjecture, see Bass and Morgan [1], Meyers [1], and Scott [4].

A unique orientable manifold of the second type is provided by the Hopf manifold $S^2 \times S^1$ (see Hempel [1]) having T4 geometry.

We now consider M of the third type, where it is known (Theorem 7.9) that they are aspheric: $\pi_2(M) = \pi_3(M) = 0$. M may or may not have boundary.

If M has a boundary $\partial M \neq \emptyset$, then it is either compressible or it is not so; in the first case, one can decompose M into pieces with the boundary of each piece incompressible. Such a decomposition (cutting along discs) is obtained due to the loop Theorem 7.5. These pieces, which are distinct from 3-balls, are here defined uniquely up to isotopy in M (but the splitting 2-discs are not unique up to isotopy).

Then the discussion of the third type of manifold reduces to that of manifolds having incompressible or empty boundary. The further decomposition of such manifolds is obtained by cutting along embedded tori (unique up to isotopy) in accordance with Theorems 8.4 and 8.5.

Here there are two cases to distinguish: Haken and non-Haken manifolds. Thurston's conjecture has been proved for Haken manifolds (and hence, of course, for all connected sums of Haken manifolds), which is discussed in the next section.

A non-Haken orientable irreducible 3-manifold M having infinite $\pi_1(M)$ is always closed. For such manifolds, Thurston's structural conjecture can be stated as follows:

CONJECTURE 8.9. A closed, orientable and irreducible non-Haken 3-manifold with infinite fundamental group is either a Seifert fibre space or admits a hyperbolic structure.

Many examples of such manifolds are now known, and they all satisfy that conjecture, but there has been almost no progress in proving the conjecture: see Hempel [2] and Scott [5].

In the next sections we describe some important results, of which the main one is Thurston's uniformization theorem, which partially confirms the notable hypothetical picture for structures on 3-manifolds arising from Thurston's structural conjecture in the situation where no classification for such manifolds has been found.

§3. Thurston's hyperbolization theorem and the volumes of hyperbolic manifolds

What kind of properties have hyperbolic manifolds, i.e., complete connected Riemannian manifolds having constant negative sectional curvature -1? We recall that these manifolds are precisely quotients H^n/G of the hyperbolic space H^n by free actions of discrete groups $G \subset \operatorname{Isom} H^n$ (see Chapters 4 and 7).

The topological characteristics of the hyperbolic 3-manifold M are at least the following:

1) M is irreducible, i.e., each 2-sphere embedded in M bounds a ball in M;

2) M is aspheric, i.e., $\pi_2(M) = \pi_3(M) = 0$;

3) M is a Poincaré manifold, i.e., each homotopic ball embedded in M is homeomorphic to a Euclidean ball;

4) if $G \cong \pi_1(M)$ is a non-elementary group, $\pi_1(M)$ has the trivial centre, and for each component S of the boundary ∂M, its fundamental group $\pi_1(S)$ is non-Abelian; and

5) if M is compact, $\pi_1(M)$ does not contain a subgroup isomorphic to the free Abelian group of rank 2, i.e. $\mathbb{Z} \oplus \mathbb{Z}$.

These necessary conditions are not independent, and in addition to them we note that most hyperbolic manifolds are sufficiently large and are thus Haken manifolds, but there are also "small" hyperbolic manifolds, including the hyperbolic dodecahedron space, see Seifert and Weber [1] or Krushkal', Apanasov, and Gusevskii [3], Example 58.

Thurston observed a notable fact that is his notable theorem on hyperbolization of manifolds, in that these necessary topological conditions for M are in a certain sense sufficient for M to have a hyperbolic metric. We formulate this more precisely as two theorems (see Thurston [4]):

THEOREM 8.10. *The interior of a compact 3-manifold M with a non-empty boundary has a hyperbolic structure (i.e., admits introducing a metric having constant sectional curvature -1) if and only if M is an irreducible homotopically atoroidal manifold non-homeomorphic to the manifold from Example 2.23.*

THEOREM 8.11. *A closed Haken manifold has a hyperbolic structure if and only if it is homotopically atoroidal.*

See Morgan [1] for a fairly detailed exposition of the main points in Thurston's proof of the hyperbolization theorem (or, as it is also called, Thurston's uniformization theorem), which nevertheless is extremely complicated.

At present, the first three (main) papers by Thurston [2], [3], [7] have appeared out of a series in which Thurston has planned to publish the complete proof of his results on the uniformization of the three-dimensional manifold. In particular, the second of these papers contains a detailed survey of the case where the manifold is fibred over a circle (see also Sullivan [8]):

THEOREM 8.12 (hyperbolic mapping torus). *Let M^3 be a compact 3-manifold, possibly with boundary, which fibres over S^1, and whose fibre is a compact surface of negative Euler characteristic.*

Then the interior of M^3 either

1) *has a complete $H^2\times\mathbb{R}$ structure of finite volume, and can be described as a Seifert fibration over some hyperbolic 2-orbifold; or*

2) *contains an embedded incompressible torus not isotopic to a boundary component, and splits along this torus into two simpler 3-manifolds; or*

3) *(the generic case) has a complete hyperbolic structure of finite volume.*

Cases 1 and 2 are not mutually exclusive, but case 3 includes the other two.

We note that, if M fibres over S^1 and $\partial M\neq\emptyset$, then ∂M is a union of tori and (in the non-orientable case) Klein bottles.

The following theorem, due to Thurston [4], follows directly from the hyperbolization Theorems 8.10 and 8.11 and from the splitting Theorem 8.7, which gives a strong confirmation for his structural conjecture from the previous section:

THEOREM 8.13. *Thurston's structural conjecture is true for all Haken manifolds and also for compact irreducible 3-manifolds with non-empty boundaries.*

Another large class of 3-manifolds, in which Thurston's structural conjecture is true, is given by knot complements in S^3 (always irreducible and homeomorphic to interiors of manifolds whose boundaries in each case consist of a torus). For knot complements in S^3, we have the following criterion (Thurston [4]), which was first conjectured by Riley [2]:

THEOREM 8.14. *A knot complement* $S^3\backslash K$ *has a geometric structure if and only if the knot* K *is not a satellite knot. The resulting geometric structure is hyperbolic if and only if, in addition,* K *is not a torus knot.*

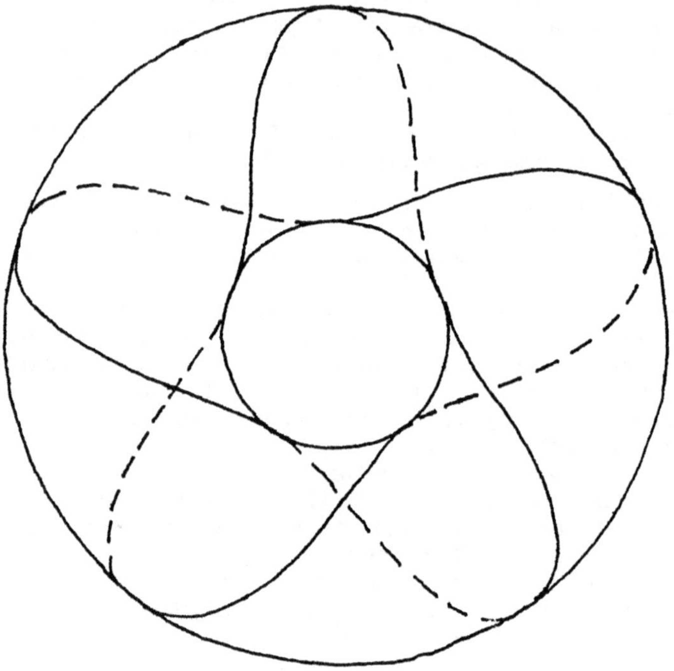

Figure 69

Here by torus knot we mean a knot that can be placed on an ordinary torus T^2 in $\mathbb{R}^3$. Figure 69 shows, for example, a torus knot with parameters (2, 5): it winds 2 times around the short way while going 5 times around the long way. Each non-trivial knot $K \subset S^3$ corresponds to a family of other knots called *satellites of* K. These are knots which are obtained by a non-trivial embedding of a circle S^1 in a small solid torus neighbourhood of the knot K. Here by "non-trivial" is meant an embedding non-isotopic to K itself and such that the image of the circle is not contained in a ball inside the solid torus. In general, a knot is a satellite knot if it is a satellite of a certain non-trivial knot $K \subset S^3$.

We note also that the toral decomposition of $S^3\backslash K$ in Theorem 8.7 is non-trivial if and only if K is a satellite knot.

If $S^3\backslash K$ has a complete hyperbolic structure, then automatically $\mathrm{vol}(S^3\backslash K) < \infty$. Adams [1] used a computer to derive such hyperbolic structures for a large set of knots and links in S^3 together with computing the corresponding volumes.

Hyperbolic finite-volume manifolds. These manifolds M are completely determined by their fundamental groups, as follows from Mostow's rigidity theorem, see Chapter 9 §2. Here for all even $n = \dim M$, the volume $V(M)$ takes only a discrete set of values $V(M) = c_n\chi(M)$, as follows from the Gauss-Bonnet theorem, which in the case $n = 2$ gives that the area of a complete hyperbolic surface M can take only the values $V(M) = 2\pi k$, $k = -1\ -2, \ldots$. Moreover, for $n \geq 4$, there is even greater rigidity (Wang [1]):

THEOREM 8.15. *For $n \geq 4$ and any $C > 0$, there exists only a finite set of pairwise non-diffeomorphic (or equivalently, pairwise non-isometric) n-dimensional complete hyperbolic manifolds having volumes not exceeding C.*

We note that all such manifolds M, $\operatorname{vol} M < \infty$, are necessarily geometrically finite (Corollary 5.48), so Theorem 7.2 gives a description of them:

THEOREM 8.16. *A complete oriented hyperbolic manifold M, $n = \dim M \geq 3$, having finite volume is the union of a compact submanifold (whose boundary, if non-empty, is a union of Euclidean space forms E_i which have finite-sheeted coverings by $(n-1)$-dimensional tori) and possibly a finite set of punctured solid Euclidean space forms $E_i \times [0, \infty)$.*

Wang's theorem [1] does not apply for $n = 3$ (see Thurston's non-rigidity theorem), and in that case one can give a more exact description of all hyperbolic manifolds whose volumes are bounded by a constant. For that purpose, we consider the set $\mathcal{H}$ of all complete finite-volume hyperbolic 3-manifolds and on it define the geometric convergence topology as follows (see Thurston [1], Chapter 5).

Two manifolds $M, N \in \mathcal{H}$ are called *close* if for a sufficiently small $\varepsilon > 0$ there exists a diffeomorphism approximated by isometric mappings $M_{[\varepsilon,\infty)} \longrightarrow N_{[\varepsilon,\infty)}$ of complements in the manifolds M and N to their ε-thin parts. This is equivalent to the following more formal definition (Burago [1]):

DEFINITION 8.17. The manifolds $M_i \in \mathcal{H}$ converge to $M \in \mathcal{H}$ if for certain points $x_i \in M_i$, $x \in M$, and for any numbers $\delta > 0$, $R > 0$, there exists $i_0 = i_0(\delta, R)$ such that for all $i \geq i_0$ there exist the mappings

$$f_i : \{y \in M_i : d_i(x_i, y) \leq R\} = M_i(R) \longrightarrow M$$

having the following properties:

1) $f(x_i) = x$,
2) $M(R - \delta) = \{y \in M : d(x, y) \leq R - \delta\} \subset f_i(M_i(R))$,
3) for all $y, z \in M_i(R)$

$$(1 - \delta)d_i(y, z) \leq d(f_i(y), f_i(z)) \leq (1 + \delta)d_i(y, z).$$

In terms of the space $\mathcal{H}$ of hyperbolic finite-volume manifolds, we have (see Thurston [1], Chapter 5) the following Jørgensen description for such manifolds:

THEOREM 8.18. *The continuous mapping* $\mathrm{vol}: \mathcal{H} \longrightarrow \mathbb{R}_+$ *is a proper mapping, i.e., each sequence from* $\mathcal{H}$ *having uniformly bounded volumes has a convergent subsequence. Moreover for a constant* $C > 0$, *there exists a finite set* $\{M_1, \ldots, M_m\} \subset \mathcal{H}$, $\mathrm{vol}\, M_i \le C$, *such that any manifold* $M \in \mathcal{H}$, $\mathrm{vol}\, M \le C$, *is obtained from* M_{i}, $1 \le i_0 \le m$, *by hyperbolic Dehn surgery.*

COROLLARY 8.19. *For a constant* $C > 0$, *there exists a link* L_c *in* S^3 *such that each complete 3-manifold* M, $\mathrm{vol}\, M \le C$, *is obtained from the link complement* $S^3 \setminus L_c$ *by Dehn surgery (the limit case can occur, i.e. non-compactness).*

These two facts are connected with the effect that, if M_1 and M_2 are hyperbolic manifolds, and M_2 is obtained from M_1 by Dehn surgery, i.e., by replacing a punctured solid torus $T^2 \times [0, \infty)$ by the solid torus $S^1 \times B^2(0, 1)$ (see §1.4), then $\mathrm{vol}\, M_1 > \mathrm{vol}\, M_2$.

How many hyperbolic manifolds are there amongst the manifolds obtained by Dehn surgery from an m-component link complement (in a 3-manifold) having a hyperbolic structure?

If the Dehn-surgery parameters are finite, $(d_1, \ldots, d_m) \in \mathbb{R}^2 \times \ldots \times \mathbb{R}^2$, the surgery gives a closed manifold $M_{d_1,\ldots,d_m}$. When $(d_1, \ldots, d_m)$ tend to $(\infty, \ldots, \infty)$ in the Cartesian product $S^2 \times \ldots \times S^2$ topology, the manifolds $M_{d_1,\ldots,d_m}$ become close to the manifold $M = M_{\infty,\ldots,\infty}$. This closeness is shown by the following result due Thurston [1], which demonstrates how large is the class $\mathcal{H}$:

THEOREM 8.20. *Let an* m*-component link* $L \subset N$ *in a 3-manifold* N *be such that the complement* $M = N \setminus L$ *has a complete hyperbolic structure. Then there exists a neighbourhood* $U \subset S^2 \times \ldots \times S^2$ *of the point* $(\infty, \ldots, \infty)$ *such that any manifold* $M_{d_1,\ldots,d_m}$, $(d_1, \ldots, d_m) \in V$, *has a complete hyperbolic structure.*

As an example, we consider the figure-eight knot complement, see Example 2.17 and Fig. 4. In that case, one can give a complete description for all the manifolds M_{d_1}. It is obtained by means of the lamination technique, see Thurston [1]:

THEOREM 8.21. *Generalized Dehn surgery along the figure-eight knot complement gives the closed irreducible manifolds* M_d *which, except the six manifolds, are hyperbolic. The six non-hyperbolic ones correspond to the following surgery parameters* $d = (\alpha, \beta)$:

1) $(1, 0)$: *trivial surgery,* $M_{(1,0)} = S^3$;

2) $(1, 1)$, $(2, 1)$, $(3, 1)$: *surgery giving Seifert fibre spaces and two sufficiently large manifolds (each containing an incompressible two-sided surface)*;

3) $(0, 1)$: *a manifold fibres over the circle with fibre a torus*;

4) (4, 1): *a graph manifold.*

EXAMPLE 8.22. As another example, we consider the knot complements (Fig. 70) obtained by means of Dehn surgery along one component of the Whitehead link L (which admits a complete hyperbolic structure, see Example 2.18 and Fig. 8). The first amongst these knots is the figure-eight knot with the second surgery parameter d_2 tending to infinity $(d_1 = \infty)$, the knot complements which are the manifolds M_{∞, d_2}, tend to the manifold $M_{\infty,\infty} = S^3 \setminus L$, which is the Whitehead link complement. In fact, Thurston's [1] theorem applies:

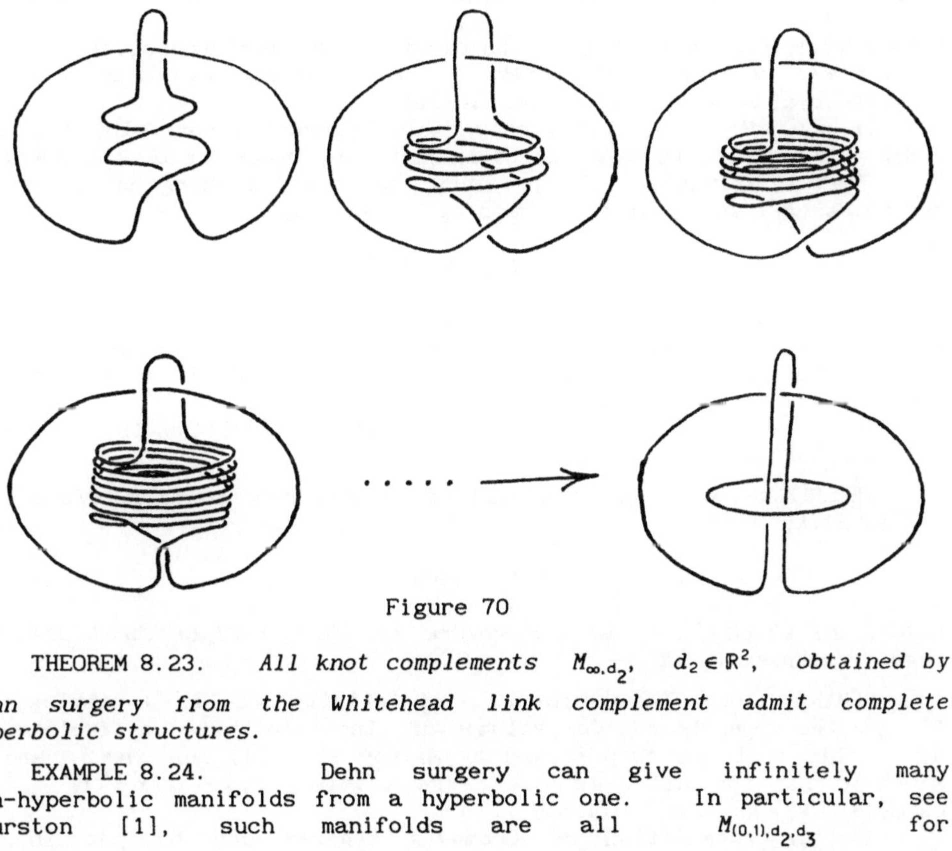

Figure 70

THEOREM 8.23. *All knot complements* M_{∞, d_2}, $d_2 \in \mathbb{R}^2$, *obtained by Dehn surgery from the Whitehead link complement admit complete hyperbolic structures.*

EXAMPLE 8.24. Dehn surgery can give infinitely many non-hyperbolic manifolds from a hyperbolic one. In particular, see Thurston [1], such manifolds are all $M_{(0,1), d_2, d_3}$ for $(d_2, d_3) \in S^2 \times S^2$, in which M is the Borromean rings complement (see Example 2.19 and Fig. 12). This follows from the fact that $M_{(1,0)\infty,\infty} = S^3 \setminus (S_1^1 \cup S_2^1)$, where the circles S_1^1, S_2^1 are not linked. Therefore, for any $(d_2, d_3) \in \mathbb{R}^2 \times \mathbb{R}^2$, the manifold $M_{(1,0), d_2, d_3}$ is the connected sum of lens spaces. It is universally covered by the discontinuity set $\Omega(G) \subset S^3$ obtained by deleting from S^3 the perfect nowhere dense and disconnected limit set $\Lambda(G)$:

$$M_{(1,0),d_2,d_3} = \Omega(G)/G.$$

The volume of a complete hyperbolic manifold M is completely determined by the topology of M in any dimensions $n = \dim M \geq 3$, not merely even, which can be seen also from the following invariant, introduced by Gromov.

Let X be a topological space and $C_k(X)$ vector spaces of singular k-chains having as bases all continuous mappings of the standard simplexes Δ^k into X. We define a norm of a chain $C = \sum_t a_i\sigma_i$, in which $\sigma_i : \Delta^k \longrightarrow X$, as $\|C\| = \sum_i |a_i|$. Then the Gromov norm for real singular homology is defined as follows. For $a \in H_k(X, \mathbb{R})$ we put $\|a\| = \inf\|z\|$, where infimum is taken over all singular cycles representing the element a.

If now M is a closed orientable hyperbolic manifold, the norm $\|[M]\| = \|M\|$ of its fundamental class gives an invariant of the manifold M. That invariant has the property that the following holds for the manifolds M_1 and M_2 and the mapping $f : M_1 \longrightarrow M_2$:

$$\|M_1\| \geq |\text{the degree } f| \cdot \|M_2\|. \tag{8.9}$$

As an example, we note that for an n-dimensional sphere S^n, $n \geq 1$, allowing the mappings $f : S^n \longrightarrow S^n$ with degree two or more, the Gromov invariant $\|S^n\|$ is zero. On the other hand, it is two for the zero-dimensional sphere S^0.

Gromov's theorem can now be formulated as follows (see Thurston [1], Chapter 5).

THEOREM 8.25. *For a closed orientable hyperbolic manifold M^n, the equality*

$$\text{vol } M^n = V_n \cdot \|M^n\| \tag{8.10}$$

holds, in which V_n is the volume of the n-dimensional infinite regular simplex in H^n.

REMARK 8.26. The volumes V_n of infinite regular n-simplexes in H^n are the upper bounds for volumes of linear infinite n-simplexes in H^n. This fact has been proved by Milnor [3], [4] for $n = 3$ and by Munkholm [1] for all $n \geq 3$. Here always $V_k < V_{k-1}/(k-1)$. For example, $V_2 = \pi$, $V_3 = 1{,}0149416\ldots$

Another formulation of Gromov's theorem can be put in the following form (compare (8.9) and (8.10)):

THEOREM 8.27. *Let $f : M_1 \longrightarrow M_2$ be a mapping of non-zero degree between closed orientable hyperbolic manifolds. Then*

1) $\text{vol } M_1 \geq |\textit{the degree } f| \cdot \text{vol } M$.

2) *if the equality holds in* 1), *the mapping f is homotopic to a covering map that is a local isometry.*

3) *if* $\operatorname{vol} M_1 < \infty$, *while the* 3-*manifold* M_2 *contains a link* L *such that* M_1 *is homeomorphic to the complement* $M_2 \setminus L$, *then* $\operatorname{vol} M_1 > \operatorname{vol} M_2$.

These theorems, due to Jørgensen and Gromov, and hyperbolic Dehn surgery imply some major corollaries:

COROLLARY 8.28. *The set of volumes of complete hyperbolic 3-manifolds is a well-ordered subset of* $\mathbb{R}$ *and has the order type* ω^ω.

COROLLARY 8.29. *Volume is a finite-valued function on the set* $\mathcal{H}$ *of complete hyperbolic* 3-*manifolds*.

The following observation, due to Wielenberg [3], is of interest in connection with the last corollary:

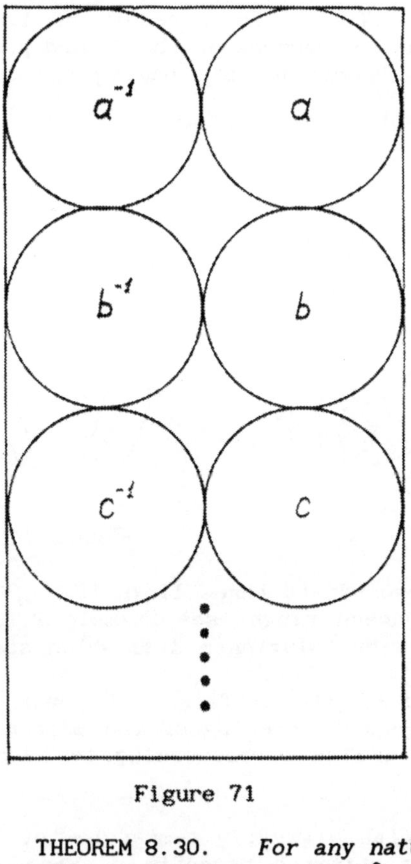

Figure 71

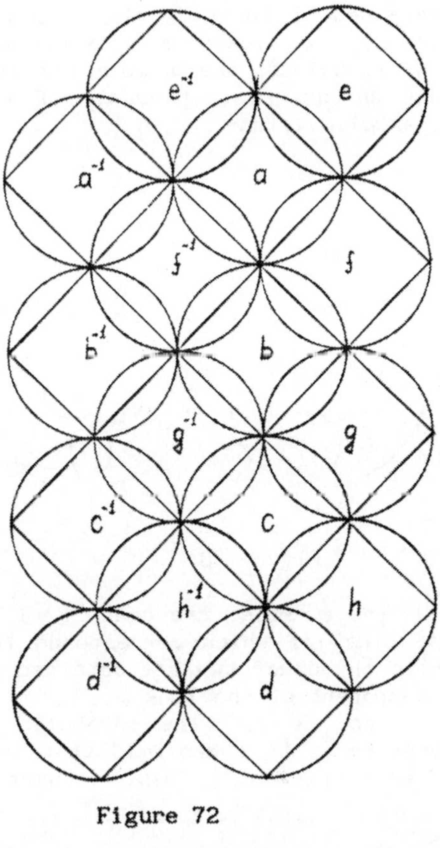

Figure 72

THEOREM 8.30. *For any natural number* N, *there exists a convex, unbounded polyhedron* $P_N \subset H^3$, $\operatorname{vol} P_N < \infty$, *that is a fundamental polyhedron for at least* N *distinct non-compact manifolds of the same finite volume.*[1]

The proof of this involves the following scheme. We first take a sufficiently "large" polyhedron $P \subset H^3$, the projection of which on $\partial H^3 = \overline{\mathbb{R}}^2$ is shown in Fig. 71. This corresponds to the subgroup G of the Picard group $SL(2, \mathbb{Z}(i))$ having $k+2$ parabolic generators. The boundary of the manifold $M(G) = [\overline{H}^3 \setminus \Lambda(G)]/G$ consists of an even number $(2k)$ of thrice-punctured spheres. We consider the corresponding Fuchsian subgroups $G_i, G_i' \subset G$ and hyperbolic planes $H_i, H_i' \subset H^3$ spanned by the circles $\Lambda(G_i)$, $\Lambda(G_i')$. So we can take the distinct totally geodesic surfaces $S_i = H_i/G_i$, $S_i' = H_i'/G_i'$ in the manifold $M(G)$ parallel to the surfaces (thrice-punctured spheres) of the boundary $\partial M(G)$. Pairwise isometric gluing of these surfaces in various ways provides certain HNN-extensions of the group G by means of the corresponding isometries H^3 (this gluing corresponds to gluing the sides at infinity of the polyhedron P (see Fig. 72) by means of parabolic or loxodromic isometries of H^3). As a result of these HNN-extensions, one obtains non-isomorphic subgroups of the Picard group that uniformize complements of various links in S^3 having the same hyperbolic volume.

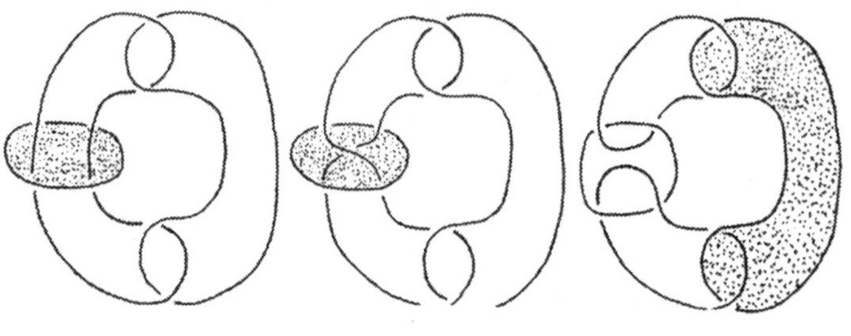

Figure 73

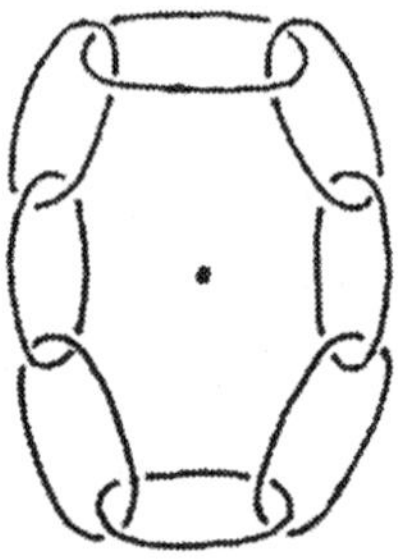

Figure 74

For example, the case $k=1$ corresponds to three links (Fig. 73), the first of which corresponds to Borromean rings (see Example 2.19), while the other two are obtained from it by twisting a disc which spans a component of the link.

For $k>1$, the parabolic gluing of the surfaces S_i and S_i' corresponds to a manifold that is a k-sheet covering of the Borromean rings complement; Fig. 74 shows such a link corresponding to $k=4$,

[1] For closed hyperbolic 3-manifolds with a greatly symmetric totally geodesic surface, a similar result was obtained recently by means of HNN-extensions $\langle G_N, h_i \rangle$, $i = 1, \ldots, N$, of certain web-groups $G_N \subset$ Möb(2) (the Klein manifolds $M(G_N)$ have two boundary components) in the recent work by Apanasov-Gutsul [1].

where the point at the centre denotes the ninth component, which passes through infinity. Replacing one parabolic gluing by a loxodromic one corresponds to twisting an open disc spanning the distinguished link component. The resulting links are shown in Fig. 75. In the general case, replacing the parabolic gluings by loxodromic ones for $k=4$ will give complements to various links having six, seven, eight, or nine components but having the same hyperbolic volume.

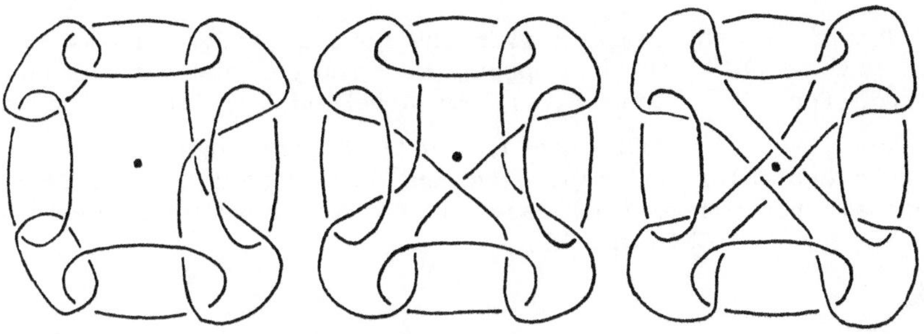

Figure 75

REMARK 8.31. In Corollary 8.28, the notation ω^{ω} is that of ordinals, not cardinals: there is some lowest volume $v_1 = \mathrm{Vol}\, M$; some second lowest volume v_2; and so on until the first accumulation point of volumes v_{ω}, which is the smallest volume of a hyperbolic manifold with one cusp. There is a next highest volume $v_{\omega+1}$, and so on until the second accumulation point $v_{2\omega}$. The first accumulation point of accumulations points is called v_{ω^2}, and it is the smallest volume of a hyperbolic manifold with two cusps. And so on.

The structure of the set $V \subset \mathbb{R}$ of volumes is significant, and it would be interesting to know a few simple cases like $v_1, v_{\omega}, \ldots$. So, which are they? Thurston [4] conjectured that v_{ω} would be the volume of the figure-eight knot complement, $v_{\omega} = 2.029883\ldots$, and v_1 could be the volume $(0.98139\ldots)$ of a manifold obtained by Dehn surgery along the figure-eight knot, but there is a closed manifold with smaller volume - see Fomenko-Matveev's approach in the next section and Theorems 8.38 and 8.40.

Meyerhoff [1] has solved a similar minimization problem for orientable non-compact hyperbolic 3-orbifolds, for which results similar to those above for manifolds apply. He showed that the minimum-volume orbifold, among such orbifolds is an orbifold H^3/G, where G_1 is the discrete group $G_1 = PGL(2, \mathcal{O}_3)$ and $\mathcal{O}_3$ is the ring of integers in $\mathbb{Q}(-\sqrt{3})$. The group G_1 is the orientatioin-preserving subgroup of index two in the group generated by reflection with respect to the sides of the 1/24 part (symmetrical decomposition) of the regular infinite simplex. Here $\mathrm{vol}\, H^3/G_2 = V_3/12 \approx 0.0846\ldots$. Without the condition for orientability, the minimal volume for a non-compact 3-orbifold is $V_3/24 \approx 0.0423\ldots$.

Also hyperbolic-volume minimization has been solved completely for arithmetic 3-orbifolds - see Chinburg and Friedman [1]. In fact, a minimum-volume orientable arithmetic hyperbolic 3-orbifold is the quotient of the hyperbolic space H^3 by the action of the group $\Gamma_a \subset \mathrm{Isom}_+$, which is the index two subgroup of the group generated by reflections in the sides of the simplex having the Coxeter diagram (see Chapter 4 §6 and Fig. 19): . Here

$$\mathrm{vol}\, H^3/\Gamma_a = \mathrm{vol}(P_a) = (275)^{3/2}\cdot\zeta_k(2)\cdot 2^{-7}\cdot\pi^{-6} = 0.03905\ldots,$$

where $P_a \subset H^3$ is the simplex with the Coxeter diagram  (it is the double of the previous simplex and is a fundamental polyhedron for Γ_a) and $\zeta_k(s)$ is Dedekind's ζ-function of the base field for Γ_a, $k = \mathbb{Q}(\sqrt{3 + 2\sqrt{5}})$, with discriminant $= -275$. However, arithmetic orbifolds and manifolds are fairly rare: a difference from the general (non-arithmetic) case is that their volumes form a discrete subset in $\mathbb{R}$, see Chinburg [1].

§4. Complexity of 3-manifolds, integrable Hamiltonian systems and hyperbolic volumes

A new approach to hyperbolic geometry was obtained by A.T. Fomenko and S.V. Matveev [1] on the basis of an interesting interaction between Fomenko's theory of constant energy surfaces of integrable Hamiltonian systems (Fomenko [3], [4]) and Matveev's theory of complexity of 3-manifolds (Matveev [3], [4]).

Let $(\mathbb{H})$ denote the class of all three-dimensional isoenergy surfaces of integrable Hamiltonian systems. The computer enumeration of all closed orientable irreducible 3-manifolds with complexity ≤ 5 (see Matveev-Savvateev [1]) shows that all such manifolds belong to the class $(\mathbb{H})$ and hence are not hyperbolic. Further complete enumeration is very difficult because of the enormous amount of work. Thus it is reasonable to continue the modulo $(\mathbb{H})$ enumeration only. The realization of the above idea led Fomenko and Matveev to the following results:

THEOREM 8.32. 1. *All closed orientable 3-manifolds with complexity* ≤ 8 *belong to the class* $(\mathbb{H})$;

2. *all compact irreducible atoroidal 3-manifolds having the complexity* ≤ 3 *and only the toral boundary components belong to the class* $(\mathbb{H})$, *except exactly two manifolds* Q_1, Q_2 *with complexity* 2 *and nine manifolds* Q_i, $3 \leq i \leq 11$, *with complexity* 3;

3. *the result of computation is the set of about* 1000 *volumes of closed hyperbolic 3-manifolds having the type* $(Q_i)_{p,q}$, $1 \leq i \leq 11$, (see Theorem 8.40);

4. *Among the manifolds* $(Q_i)_{p,q}$ *the remarkable manifold* $(Q_1)_{5,-2}$ *has the smallest volume, the next one - the manifold* $(Q_2)_{5,1}$

with the greater volume. [2]

It must be noted that the manifold $(Q_2)_{5,1}$ (= (5, 1)-Dehn surgery on the figure-eight knot complement) was known to W. Thurston [1], who proposed it as a canditate for the smallest volume hyperbolic manifold. As the volume of $(Q_1)_{5,-2}$ (≈ 0.94) is less than that of $(Q_2)_{5,1}$ (≈ 0.98), this conjecture is not true - see Remark 8.31.

Let V denote the set of all hyperbolic volume values.

CONJECTURE 8.32 (Fomenko and Matveev). The distribution of volumes of manifolds $(Q_i)_{p,q}$, $1 \leq i \leq 11$, shows the initial part of the set V. In particular, the manifold $(Q_1)_{5,-2}$ has the smallest hyperbolic volume $\approx 0.94272\ldots$.

The complexities of $(Q_1)_{5,-2}$ and $(Q_2)_{5,1}$ are equal to 9. Thus $(Q_1)_{5,-2}$ and $(Q_2)_{5,1}$ are the simplest closed hyperbolic 3-manifolds (in the sense of complexity). These manifolds are the simplest 3-manifolds with the following property: the arbitrary Hamiltonian system is non-integrable on $(Q_1)_{5,-2}$ and $(Q_2)_{5,1}$ (and in their neighbourhoods). These facts serve as an indirect confirmation of the conjecture.

Let us describe these results in more detail.

1. *The Hamiltonian systems and the class* ($\mathbb{H}$). Let M^4 be a symplectic manifold and $v = s\,\mathrm{grad}\,H$ be a Hamiltonian field (system) on M^4, with the Hamiltonian H (see, for example, Fomenko-Trofimov [1]). It appears that topological obstructions to the complete integrability of this system exist and can be effectively described. Let the system v be Liouville integrable on some level surface Q which is given by the equation $H = \mathrm{const}$. This means that in a neighbourhood of Q there exists a second integral f commuting with H on Q and independent of H. It is known that in the case of the connected non-singular level surfaces of f, restricted to the 3-manifold Q, there are two-dimensional tori (so-called Liouville tori).

DEFINITION 8.33 (see Fomenko [3], [4]). The integral f is called a Bott integral if the set of critical points of f on Q is the union of non-degenerate critical submanifolds.

In almost all cases which are known to be integrable, f is the Bott integral (Fomenko [3]).

THEOREM 8.34 (Fomenko [3], [4]). *Let Q be a three-dimensional constant energy surface of a Hamiltonian system v which is integrable with the help of a Bott integral f. Then Q can be obtained by gluing some number of manifolds $D^2 \times S^1$ and $N^2 \times S^1$ along their*

---

2 Recently T. Chinburg and T. Jørgensen proved that $(Q_1)_{5,-2}$ is arithmetic, as also $(Q_2)_{5,1}$, Chenburg [2], where he noted that $(Q_1)_{5,-2}$ was discovered by J. Weeks [1] too.

boundaries, where N^2 *denotes the disc with two holes and* D^2 *denotes the disc.*

DEFINITION 8.35. The compact orientable 3-manifold Q belongs to the class $(\mathbb{H})$ if and only if it can be obtained by gluing some number of manifolds $D^2 \times S^1$ and $N^2 \times S^1$ along some components of their boundaries.

Fomenko's theorem implies that any three-dimensional non-singular constant energy surface Q of a Hamiltonian system v which is integrable with the help of a Bott integral is contained in the class $(\mathbb{H})$. The converse is also true (A.V. Brailov and A.T. Fomenko): any 3-manifold $Q \in (\mathbb{H})$ can be realized as a constant energy surface of a Hamiltonian system with the Bott integral (in the case $\partial Q \neq \varnothing$ additional conditions on the motion of the system must be satisfied). As Fomenko and Zieschang [1], [2] proved, the class $(\mathbb{H})$ coincides with the class of so-called graph-manifolds in the sense of Waldhausen [4]. It is necessary to remark that the class $(\mathbb{H})$ is much smaller than the class of all compact 3-manifolds, see Fomenko [3]. Exactly this important fact implies the existence of the above-mentioned topological obstructions to integrability - if a constant energy surface of a Hamiltonian system does not belong to the class $(\mathbb{H})$, then the system is not integrable with the help of a Bott integral. Thus the problem of constructing manifolds outside the class $(\mathbb{H})$ (including the hyperbolic ones) appears to be very urgent.

2. *The complexity of three-dimensional manifolds.* A compact CW-subcomplex P of a 3-manifold M is called a *spine* of M if M (or M without a 3-ball if $\partial M = \varnothing$) collapses to P. A finite two-dimensional CW-complex is called a *special polyhedron* if each point of its 1-skeleton has a regular neighbourhood homeomorphic to the cone over the circle with 2 or 3 radii. A compact CW-complex P is called *almost special* if it embeds in some special polyhedron. The point $v \in P$ is called a vertex if the link of v is homeomorphic to the circle with 3 radii  .

DEFINITION 8.36 (Matveev [3], [4]). The complexity $d(M)$ of a compact 3-manifold M is equal to k if M has an almost special spine of M with k vertices and does not have almost special spines with less than k vertices.

THEOREM 8.37 (see Matveev-Savvateev [1]). *For any integer* d *there are only a finite number of distinct closed orientable irreducible 3-manifolds with the complexity* d.

This theorem (at least, its constructive proof) makes it possible to enumerate (with the help of a computer) all closed orientable irreducible 3-manifolds whose complexities are not greater than a given number d. It turns out, that all closed manifolds with complexity $d \leq 8$ belong to the class $(\mathbb{H})$ (for $d \leq 5$ see Matveev-Savvateev [1]). Further complete enumeratioin is very difficult because of the enormous amount of calculations required. On the other hand the class (H) is well studied, thus it is reasonable to try the mod($\mathbb{H}$)-enumeration by recognizing and casting the element of ($\mathbb{H}$) in the early stage of its construction. Although all closed 3-manifolds with small complexity belong to the class ($\mathbb{H}$), this is not the case for manifolds with toral boundaries. The mod($\mathbb{H}$)-enumeration of all 3-manifolds whose boundaries consist of tori and whose complexities are not greater than 3 enables one to prove the following theorem.

THEOREM 8.38 (see Fomenko-Matveev [1]). *All compact irreducible atoroidal orientable 3-manifolds with the complexity* ≤ 3 *and with the toral boundary components only belong to the class* ($\mathbb{H}$) *except two manifolds* Q_1 *and* Q_2 *with complexity* 2 *and nine manifolds* Q_i, $3 \leq i \leq 11$, *with complexity* 3.

The 1-skeletons of the special simplest spines of Q_1 and Q_2 are homeomorphic to the graph S which is the union of two circles with two common points. The 1-skeletons of the simplest special spines of the remaining manifolds Q_i are homeomorphic to a circle with inscribed triangle or to the graph S with an additional loop on one of its edges. From a topological viewpoint the manifolds Q_i, $1 \leq i \leq 11$, are homeomorphic to the complements to certain knots in:

$$L_{5,1},\quad S^3,\quad RP^3,\quad L_{6,1},\quad L_{7,2},\quad S^3,\quad L_{9,2},\quad L_{6,1},\quad L_{5,1},\quad L_{3,1},\quad S^3,$$

respectively (here the spaces $L_{p,q}$ are the lens spaces). The manifold Q_2 is the figure of eight knot complement, the Q_{11} is the complement of the knot 5_2, the Q_6 is the complement of a mutant of the knot 5_2 having 16 crossings. Plugging in a solid torus to the Q_i (Dehn surgery) leads to the closed manifold M whose special spine can be easily constructed from the minimal special spine P_i of Q_i: we must add to P_i new 2-cells - the meridional disc of the solid torus. It turns out that all manifolds with complexity $d \leq 8$ which can be obtained in such a manner belong to the class ($\mathbb{H}$). For the complexity $d = 9$ this is not the case because the manifolds $(Q_1)_{5,-2}$ and $(Q_2)_{5,1}$ have complexity 9 and do not belong to the class ($\mathbb{H}$). Thus we have the following criterion:

> If the special spine P of the closed 3-manifold M has ≤ 8 vertices and the closure of one of its 2-cells contains all but ≤ 3 vertices, then M belongs to the class ($\mathbb{H}$).

The use of the criterion made it possible to enumerate all closed orientable irreducible 3-manifolds with complexity ≤ 8 (modulo ($\mathbb{H}$)). It turns out that the following theorem is true.

THEOREM 8.39 (Matveev [4]). *All compact orientable 3-manifolds with complexity* ≤ 8 *belong to the class* ($\mathbb{H}$).

There is now a new proof of this theorem without a computer.

So the above-mentioned closed manifolds $(Q_1)_{5,-2}$ and $(Q_2)_{5,1}$ are the simplest ones outside the class ($\mathbb{H}$).

3. *Computation of volumes of hyperbolic 3-manifolds.* The above constructed manifolds Q_i, $1 \leq i \leq 11$, do not belong to the class ($\mathbb{H}$)

because their interiors admit complete hyperbolic structures. In the case of the manifolds Q_1 and Q_2 these structures can be obtained by gluing two regular ideal tetrahedra. It is not hard to see that the consistency conditions are satisfied and that the holonomies $H_a(\sqrt[3]{-1})$ and $H_b(\sqrt[3]{-1})$ of meridian a and parallel b are Euclidean - see Thurston [1]. The hyperbolic structures for the manifolds Q_i, $3 \le i \le 11$, can be obtained by gluing three ideal tetrahedra which are not necessarily regular. Using a computer, Fomenko and Matveev calculated the volumes of many closed hyperbolic 3-manifolds of the type $(Q_i)_{p,q}$, $1 \le i \le 11$. This calculation was based on the Thurston [1] method and on Fomenko's and Matveev's theories described above. It is convenient to bring their results together in the following theorem.

THEOREM 8.40. (i) *For each* i, $1 \le i \le 11$, *the number of coprime integer points* (p, q) *for which the manifold* $(Q_i)_{p,q}$ *is not hyperbolic is less than* 24.

(ii) *The minimal volumes of the hyperbolic manifold* $(Q_i)_{p,q}$ *and its limiting values (i.e., the volumes of non-compact hyperbolic 3-manifolds* $\operatorname{int} Q_i$*) are given in the table*:

i	$\min_{(p,q)} \operatorname{vol} (Q_i)_{p,q}$	$\lim_{(p,q)\to\infty} \operatorname{vol} (Q_i)_{p,q} = \operatorname{vol} \operatorname{int} Q_i$
1	0.94	2.02
2	0.98	2.02
3	1.41	2.66
4	1.29	2.66
5	1.54	2.82
6	1.54	2.82
7	1.42	2.78
8	1.58	2.94
9	1.26	2.56
0	1.44	2.56
1	1.39	2.82

§5. Conformal uniformization of manifolds

The classical Klein-Poincaré uniformization theorem (Theorem 8.1) shows that studying conformal structures on two-dimensional manifolds is equivalent to studying the geometric structures: hyperbolic, Euclidean, or spherical.

The geometric structures of n-dimensional manifolds for $n \geq 3$ have such properties that this approach becomes impossible; they have firstly a known strong rigidity, in contrast to the conformal structures (see Chapter 9 §3), and secondly, it is often impossible to introduce a geometric structure in a conformal manifold, as it is necessary to cut (along spheres and (or) tori) the manifold into pieces, in which one then introduces one of the eight geometries appearing in Thurston's structural conjecture (see §2). Therefore, the conformal structures on an n-manifold are for $n \geq 3$ more flexible and more useful tools of uniformization than geometric ones, which leads us to hope for unified uniformization. We give some definitions to give an exact sense to these words.

By a conformal n-manifold (similarly, orbifold), as defined in Chapter 2 §5, we mean an n-dimensional manifold modelled on the n-dimensional Möbius sphere $S^n = \overline{\mathbb{R}}^n$, in which the group $\text{Möb}(n)$ of all Möbius automorphisms acts. For $n \geq 3$, this is equivalent to the notion of conformally flat manifolds - see Kulkarni [2]. As for simplicity, we shall often deal with the orientable case, the conformal orientable manifold M is precisely a $(S^n, \mathcal{M}_n)$-manifold.

By a conformal structure on such a (smooth) manifold M we mean the following.

DEFINITION 8.41. Let two conformal atlases C and C' (with conformal changes of charts) be specified in M (manifold or orbifold). These are said to be equivalent if their union is again a conformal atlas in M. An equivalence class for conformal atlases is called a conformal structure on M.

Note that for $n = 2$, this definition coincides with the projective structure definition on a 2-dimensional surface (see Gunning [1]) since $(S^2, \mathcal{M}_2) \cong (\mathbb{CP}^1, PSL(2, \mathbb{C}))$.

A homeomorphism for orbifolds M_1 and M_2 (in the orbifold category) $f : M_1 \longrightarrow M_2$ with conformal structures C_1 and C_2 is called a *conformal homeomorphism*, while those orbifolds M_1 and M_2 are called *conformally equivalent* if f is a $(S^n, \text{Möb}(n)$-mapping, i.e., f is a conformal mapping in certain atlases in M_1 and M_2 that define the structures C_1 and C_2. In other words, the conformal structure in M_1, obtained by lifting by means of the mapping f from the conformal structure C_2, coincides with the structure C_1.

The developing map (or development) d, $d : \widetilde{M} \longrightarrow S^n$ and the holonomy homomorphism (or holonomy) $d_* : \pi_1(M) \longrightarrow \text{Möb}(n)$ are defined for a conformal structure C on M, as for any (X, G)-structure (see Theorem 2.35). Here $\widetilde{M}$ is the universal covering space with projection $\pi : \widetilde{M} \longrightarrow M$, while $\pi_1(M)$ is the deck transformation group for that covering (this coincides with the fundamental group of M in the manifold case), while the mapping d is a local $(S^n, \text{Möb}(n))$-diffeomorphism. The group $G = d_*(\pi_1(M)) \subset \text{Möb}(n)$ is called the *holonomy group* for the structure C on M. The development d for the structure C in M is defined up to composition with elements of the Möbius group $\text{Möb}(n)$, while the holonomy group G is defined up to conjugation by elements of $\text{Möb}(n)$.

There is the following evidence for the place of conformal manifolds in the class of n-dimensional ones:

1. All Riemannian conformally flat manifolds are conformal. Here by conformally flat we mean a Riemannian manifold in which each point has a neighbourhood homeomorphic and conformally equivalent to a Euclidean-space domain.

2. The subclasses of conformally flat manifolds form Riemannian manifolds having constant sectional curvature: hyperbolic, Euclidean, and spherical.

3. Hopf manifolds are conformal (conformally flat), i.e., manifolds covered by the punctured Euclidean space $\mathbb{R}^n \setminus \{0\} = \mathbb{R}^n_*$. Such manifolds are obtained as quotients $\mathbb{R}^n_* / \langle A^n : n \in \mathbb{Z} \rangle$, in which $A \in GL(n, \mathbb{R})$, $\det A > 1$, and are diffeomorphic to $S^1 \times S^{n-1}$.

4. The connected sum $M = M_1 \# M_2$ of conformal manifolds M_1 and M_2 is a conformal manifold (Kulkarni [2]).

5. An important class of conformal manifolds is provided by n-dimensional ones that are components of the boundary of $(n+1)$-dimensional Kleinian manifolds (see Chapter 7), i.e., n-manifolds obtained as quotients of the discontinuity set $\Omega(G) \subset \overline{\mathbb{R}}^n$ by the actions of Kleinian groups $G \subset \text{Möb}(n)$. The natural conformal structure in such a manifold is defined by the natural projection $\pi : \Omega(G) \to \Omega(G)/G$, since the group G is a group of conformal automorphisms of $\Omega(G)$.

DEFINITION 8.42. The conformal structure C in an n-dimensional manifold M is called *uniformized* if there exists a Kleinian group $G \subset \text{Möb}(n)$ such that the conformal manifold (M, C) is conformally equivalent to a connected component of $\Omega(G)/G$.

It is extremely difficult to describe the whole class of conformal manifolds and its subclass of uniformized conformal ones even for $n = 3$; the complexities arising here are evident from the early results, which will be discussed further, and in particular Goldman's [6] examples for closed three-dimensional manifolds do not admit any conformal structures (Theorem 8.52). However, those same results have enabled us to make the following conjectural picture of conformal uniformization of n-manifolds:

CONJECTURE 8.43. Let a compact manifold M satisfy Thurston's structural conjecture and by canonical cutting of $\operatorname{int} M$, it splits up into pieces, non-homeomorphic to a product of a torus and segment, and which admit uniformized conformal structures. Then the interior of the manifold M (or its finite-sheeted covering) admits a conformal structure that is also uniformizable.

Partial arguments for this conjecture are the following:

1. Kuiper-Fried-Goldman theorems (see Theorems 8.46 - 8.49).

2. Kulkarni-Pinkal uniformization theorems for geometric structures:

THEOREM 8.44 (Kulkarni-Pinkal [1]). *Let* M *be a compact space with an* (X, G)*-structure with a development map* $d : \tilde{M} \to X$ *and holonomy representation* d_*, $\operatorname{im} d_* = H \subset G$. *Let* N_0 *be the union of those components of the domain of normality of* H *which intersect* $\operatorname{im} d$.

Then $d|_{d^{-1}(N_0)} : d^{-1}(N_0) \to N_0$ *is a covering map.*

3. Non-rigidity of conformal structures (see Chapter 9) and therefore the existence of variations of some geometric invariants of manifolds.

4. For uniformized conformal structures there holds an analogue of the famous Klein-Maskit combination theorems (see Chapter 6) which is a sharpening of the above-mentioned Kulkarni [2] result. Namely, it is contained in the following theorem (Kulkarni-Pinkal [1]):

THEOREM 8.45. *A connected sum* $M = M_1 \# M_2$ *of manifolds* M_1 *and* M_2 *with uniformized conformal structures admits a uniformized conformal structure.*

5. The existence of a uniformized conformal structure on a manifold $M = M_1 \cup_f M_2$ obtained by gluing together the torus boundaries of some uniformized conformal (hyperbolic) manifolds M_1 and M_2 by means of some conformal map f (Apanasov [25], see §6).

6. Recently M. Kapovich announced the validity of this conjecture in the following case: if the 3-manifold M is obtained by gluing together torus boundaries of uniformized conformal 3-manifolds M_i, $1 \le i \le k$, by means of homeomorphisms and any adjacent M_i and M_j are simnultaneously neither hyperbolic manifolds nor Seifert manifolds.[1]

The first studies on conformal manifolds (namely, conformally flat Riemannian manifolds) were made in 1949-1950 by Kuiper [1] - [3], in which he defined the developing map and the holonomy homomorphism for a conformal manifold, and in particular proved the following two theorems.

THEOREM 8.46. *Let* M *be an* n*-dimensional* $(n \ge 3)$ *closed conformal manifold with finite fundamental group* $\pi_1(M)$; *then its development* $d : \tilde{M} \to S^n$ *is a homeomorphism onto the sphere* S^n, *while the manifold* M *is conformally equivalent to* S^n/G, *where the holonomy group* $G = d^*(\pi_1(M))$ *acts freely on* S^n.

---

[1] Added in proof: While I was preparing this book a brief proof of this result (for some finite cover $\hat{M}$ over M) appeared by M. Kapovich in preprint "Flat conformal structures on 3-manifolds", Novosibirsk, 1987. This preprint contains also an example of a 3-manifold M admitting no uniformized conformal structure, while some finite covering $\hat{M}$ of M does admit such a structure.

THEOREM 8.47. *Let M be an n-dimensional $(n \geq 3)$ closed conformal manifold with an infinite Abelian fundamental group. Then its development $d : \tilde{M} \longrightarrow S^n$ is a homeomorphism either onto $\mathbb{R}^n$ or onto $\mathbb{R}^n \backslash \{0\}$. In the first case, the holonomy group $G = d^*(\pi_1(M))$ consists of Euclidean motions, while the manifold M is conformally equivalent to the n-dimensional Euclidean torus $T^n = \mathbb{R}^n/G$. In the second case, G consists of similarities, while M is conformally equivalent to a Hopf manifold $(\mathbb{R}^n \backslash \{0\})/G \approx S^1 \times S^{n-1}$.*

A generalization of the last result to a manifold with similarity structure is the following theorem, which was first formulated by Kuiper [3]. Other proofs were given by Fried [1] and Goldman [6].

THEOREM 8.48. *Let a closed n-dimensional conformal manifold M be a $(\mathbb{R}^n, \mathrm{Sim}\,\mathbb{R}^n)$-manifold, in which $\mathrm{Sim}\,\mathbb{R}^n$ is the similarity group of $\mathbb{R}^n$. Then the development $d : \tilde{M} \longrightarrow S^n$ is a homeomorphism either onto $\mathbb{R}^n$ or onto $\mathbb{R}^n \backslash \{0\}$. In the first case, the holonomy group $G = d^*(\pi_1(M))$ is a finite extension of a free Abelian rank n group of Euclidean translations, while M is a Bieberbach manifold, i.e., it has a finite-sheeted covering by the Euclidean torus T^n. In the second case, G is a finite extension of an infinite cyclic similarity group (that does not preserve the Euclidean metric), while M is an almost Hopf manifold, i.e., it has a finite-sheeted covering by the Hopf manifold $S^1 \times S^{n-1}$.*

In 1983, Goldman [6] obtained the following generalization of the above Kuiper and Fried results:

THEOREM 8.49. *Let M be a closed conformal n-dimensional manifold having a virtually nilpotent or virtually polycyclic fundamental group $\pi_1(M)$ of rank ≤ 3. Then M is finitely covered either by the sphere S^n or by the Euclidean torus T^n or by the Hopf manifold $S^1 \times S^{n-1}$.*

Here a group G is called a virtually nilpotent (correspondingly, Abelian, polycyclic, and so on) if it contains a nilpotent subgroup $G' \subset G$ of a finite index $\mu(G', G) < \infty$. Theorem 8.48 gives a corresponding result for the manifold M with a virtually solvable $\pi_1(M)$ if we use the following result due to Goldman [6], Lemma 1.4:

LEMMA 8.50. *Let the group $G \subset \mathrm{M\ddot{o}b}(n)$ be virtually solvable. Then there exists a finite-index subgroup $G' \subset G$ conjugate in $\mathrm{M\ddot{o}b}(n)$ to a subgroup from $\mathrm{Sim}\,\mathbb{R}^n$.*

REMARK 8.51. In relation to Theorem 8.49, there is the following important observation, due to Goldman [6]. Let the interior of a three-dimensional manifold split up, by Thurston's canonical decomposition (§2), into pieces such that those pieces M are not covered by the Hopf manifold $S^1 \times S^2$ and which have infinite almost solvable fundamental groups $\pi_1(M)$. Then they belong to one of three classes:

(E) $\pi_1(M)$ is virtually Abelian.

(N) $\pi_1(M)$ is virtually nilpotent but is not virtually Abelian, and

(S) $\pi_1(M)$ is virtually solvable but it not virtually nilpotent.

Here class (E) consists of conformal manifolds finitely covered by the Euclidean torus T^3, while (N) consists of 3-manifolds finitely covered by T^2-bundles over the circle S^1 whose attaching map is a Dehn twist; and class (S) consists of 3-manifolds finitely covered by hyperbolic torus bundles (see also Evans and Moser [1]). All these manifolds have geometric structures in the Thurston sense (for the class (N), that structure is defined by the metric of the Heisenberg group, as follows from the paper by Evans and Moser [1], which must be combined in the case of non-Haken manifolds with the paper by Hass and Scott [1]).

This observation and Theorem 8.49 give directly a description of a certain class of closed 3-manifolds into which one cannot introduce any conformal structures but which have Heisenberg-group structure (nil-geometry), namely such manifolds are obtained by gluing two boundary components of the torus layer $T^2 \times [0, 1]$, in which the identifying diffeomorphism

$$f : T^2 \times \{0\} \longrightarrow T^2 \times \{1\}$$

(the attaching map of the torus bundle) is not periodic, which is demonstrated by the following theorem (Goldman [6], Corollary C):

THEOREM 8.52. *Let the 3-manifold M be a 2-torus bundle over the circle S^1. Then M admits a conformal structure if and only if the attaching map of this bundle is periodic.*

Uniformized structures. The further study of conformal structures is related to the uniformization problem. The first result here is for two-dimensional structures, i.e., projective structures on surfaces, see Gunning [1] and Kra [1].

THEOREM 8.53. *For a projective structure C on a compact Riemann surface S of genus greater than 1, the following conditions are equivalent*:

a) $D = d(\tilde{S}) \neq \overline{\mathbb{C}}$, *i.e., the development d is not surjective,*

b) *the development $d : \tilde{S} = H^2 \longrightarrow D$ is a covering, and*

c) *the action of the holonomy group $G = d^*(\pi_1(S)) \subset \mathcal{M}_2$ is discontinuous in the image D, and D is an invariant component of the discontinuity set $\Omega(G)$.*

At the same time, Maskit [8] (see also Hejhal [1]) constructed example of non-uniformized projective structures on surfaces of any genus greater than 1, i.e., structures for which $d(\tilde{S}) = \overline{\mathbb{C}} = S^2$ and the development d is not a covering mapping.

These non-uniformized projective structures on surfaces have enabled Sullivan and Thurston [1] (see also Smilli [1]) to construct affine structures in certain compact 3-manifolds (= S^1-bundles over surfaces) for which the developing maps are not coverings. Goldman [7] gave a further extension of that method to almost trivial Seifert fibre spaces.

One can obtain conformal structures in n-dimensional manifolds ($n \geq 3$) with surjective developing maps (or, equivalently, they are not covering maps on the images $d(\tilde{M}) = S^n$) by means of connected sums of conformal structures, where Kulkarni [2] has shown that one again gets a conformal structure. However, there are such structures on manifolds (even hyperbolic manifolds) that do not decompose into connected sums, which have been constructed by Goldman [7]. Namely, such structures having the property $d(\tilde{M}) = S^n$ are introduced on any hyperbolic manifold M^n, $n \geq 3$, having a closed totally geodesic hypersurface. For $n = 3$, even less may be required of the manifold M^3 because of Thurston's Theorem 8.11: it is sufficient for the manifold M^3 to be a closed atoroidal manifold having an incompressible surface which is not a fibre if M^3 fibres over S^1.

One can say that these conformal structures with the property $d(\tilde{M}) = S^n$ are obtained from the natural conformal structure of a hyperbolic manifold M^n as a result of bending along that hypersurface through an angle 2π, compare Chapter 9 §3.

The above Kulkarni-Pinkal uniformization theorem (Theorem 8.44) combined with Kuiper-Fried's theorem (Theorem 8.48) implies a direct extension of Gunning's Theorem 8.53:

THEOREM 8.54 (Kulkarni-Pinkal [1]). *Let M be a compact conformal n-manifold with infinite fundamental group, let $d : \tilde{M} \longrightarrow S^n$ be its development map and $d_* : \pi_1(M) \longrightarrow G = d_*(\pi_1(M)) \subset$ Möb(n) be the corresponding holonomy. Then the following properties are equivalent:*

a) *the development d is not surjective: $D = d(\tilde{M}) \neq S^n$;*

b) *the development d is a covering onto $D \subset S^n$.*

PROOF. The implication b) $\Rightarrow$ a) is evident since $\tilde{M}$ must be non-compact in the case b) but the sphere S^n is compact and simply connected.

a) $\Rightarrow$ b). If $D = d(\tilde{M})$, then the boundary ∂D is G-invariant. If ∂D is equal either to a point x or two points $\{x, y\}$, then regarding the point x as ∞, G may be considered as a group of similarity transformations in $R^n = S^n \setminus \{\infty\}$. In this case the result follows from Theorem 8.48 which, in fact, asserts that M has a finite covering which is conformal, equivalent to a flat space-form or else to a Hopf manifold and the development d is a homeomorphism.

Now suppose that ∂D contains more than two points. Then, as is well known, $\partial D \supset L(G)$ and the family of elements $g \in G$ is normal in D (see Beardon [4], Theorem 3.6.4). Therefore it follows from the uniformization Theorem 8.44 that d is a covering onto D. The proof is complete.

It is obvious that we are in a category where the covering space theory is valid then the (X, G)-structure on the manifold M is uniformized (see Definition 8.42 and Kulkarni-Pinkal [1]) if and only if the development $d : \tilde{M} \longrightarrow D = d(\tilde{M}) \subset X$ is a covering map and the holonomy group $d_*(\pi_1(M))$ acts freely and is properly discontinuous on D. We shall say that this (X, G)-structure on M is almost uniformized if only the development d is a covering map.

So Theorem 8.54 says that any conformal structure (except for the manifolds conformally equivalent to the spherical space-forms) is almost uniformized if and only if its development is not surjective.

Here is another quite different criterion (see Kulkarni-Pinkal [1]) for a conformal structure to be uniformized or almost uniformized:

THEOREM 8.55. *Let M be a compact conformal n-manifold and let the domain $O(G)$ of properness of the holonomy group $G = d_*(\pi_1(M))$ be connected and its fundamental group $\pi_1(O(G))$ be finitely generated. Then this conformal manifold is almost uniformized.*

Here the properness domain $O(G)$ coincides with the set of all wandering points $x \in S^n$ (i.e., such that there is a neighbourhood U of x with compact set $Z(U) = \{g \in G : g(U) \cap U \neq \emptyset\}$). It is easy to see that $S^n \setminus O(G)$ can be identified with the limit set, as defined in the classical situation (see Beardon [4]), i.e., with the closure of loxodromic fixed points of the group G - cf. Lemma 3.22.

REMARK 8.56. Perhaps it should be pointed out that not every conformal structure on the manifold M is uniformized or even almost uniformized. Moreover the hypothesis in Theorems 8.55 and 8.54 cannot be entirely dropped. There are some very interesting examples illustrating various phenomena (see Kulkarni-Pinkal [1] and Kamishima [1]).

Namely, let M be a Seifert manifold $S_g \times S^1$ with the following $(H^2 \times \mathbb{R})$-structure (S_g is a surface of genus $g > 1$).

Let the group $\pi_1(S_g) \times Z$ act isometrically on $H^2 \times \mathbb{R}$ such that $\pi_1(S_g)$ acts on the hyperbolic plane H^2 as a discrete isometry group and Z acts on $\mathbb{R}$ as a cyclic group with generator h, $h(t) = t + a$.

Then the manifold

$$(H^2 \times \mathbb{R})/(\pi_1(S_g) \times Z)$$

is homeomorphic to $S_g \times S^1$ and the geometry of $H^2 \times \mathbb{R}$ defines a conformal structure on the product $S_g \times S^1$.

It is clear (cf. Lemma 8.58 and (8.14)) that for $a \in \mathbb{R}$ such that $a/\pi \notin \mathbb{Q}$ the above conformal structure on $S_g \times S^1$ defines the holonomy group $G = d_*(\pi_1(S_g) \times Z)$ which is non-discrete but virtually discrete (i.e., the group $G|_{\mathbb{R}} \subset \text{Möb}(1)$ is discrete but the rotations subgroup in G is an infinite cyclic group). Therefore this conformal structure on the manifold $M = S_g \times S^1$ is a non-uniformized but almost uniformized one.

Moreover, if $M = M_1 \# M_2$ be the connected sum of two compact manifolds with conformal structures, one of which is almost uniformized but non-uniformized (as above) and the other is non-homeomorphic to the sphere S^n, then M admits a conformal structure with surjective development d, so this structure is non-almost-uniformized (Theorem 8.54). Note that here $\pi_1(O(G))$ is non-finitely generated.

However, the above almost uniformized but non-uniformized conformal structure on $S_g \times S^1$ plays some exclusive role. Namely, there holds the following criterion (this was proved by Kamishima [1] for the general case $n \geq 3$).

THEOREM 8.57. *Let M be a closed conformal 3-manifold with infinite fundamental group and let it be not conformal to any manifold with a $(H^2 \times \mathbb{R})$-structure (i.e., with T5-geometry). Then the properties* a) *and* b) *of Theorem* 8.54 *are equivalent to the following property*:

c) *the holonomy group* $G = d_*(\pi_1(M))$ *acts discontinuously on* $D = d(\tilde{M})$.

PROOF. c) ⇒ a). If $d(\tilde{M}) = S^3$, the discontinuity property of the holonomy group G on the sphere S^3 implies that G is finite, which shows that G is a subgroup of the orthogonal group $O(4)$, while M is a $(S^3, O(4))$-manifold. This fact and the compactness of M contradicts the fundamental group $\pi_1(M)$ being infinite. It shows that $d(\tilde{M}) \neq S^3$.

a) ⇒ c). As above in the proof of Theorem 8.54, due to Theorem 8.48, one can assume that $S^3 \backslash D$ contains more than two points. Then, if the holonomy group $G = d^*(\pi_1(M))$ is discrete, it acts discontinuously in D (see Chapter 3 §1). Then either condition c) applies or the group G is not discrete. In the latter case, we apply Greenberg's [1] classification for the closed Lie subgroups in Möb(3) (see Theorems 1.16 and 1.20) to get that only the following cases are possible if G is not discrete (one can assume that they are mutually exclusive):

1) G is dense in Möb(3);
2) there exists a point $x \in S^3$ fixed for the elements $g \in G$;
3) a certain two-point set $\{x, y\} \subset S^3$ is G-invariant;
4) a certain circle $l \subset S^3$ is G-invariant;
5) a certain sphere $S^2 \subset S^3$ is G-invariant; and
6) G is a subgroup of the orthogonal group $O(4)$.

We show consecutively that all these cases are impossible, which thus proves condition c) and completes the proof of the theorem (see Gusevskii-Kapovich [1]).

If $G \subset O(4)$, then M is an $(S^3, O(4))$-manifold. Compactness of M means that $\pi_1(M)$ is finite. The contradiction of this and the theorem conditions shows that case 6) is impossible.

Let case 3) apply. Then we change the coordinates and transfer (if necessary) to an index two subgroup and assume that $G(0) = 0$ and $G(\infty) = \infty$. As case 6) is impossible, G contains non-elliptic elements, and therefore G has a loxodromic element g such that $g(0) = 0$, $g(\infty) = \infty$. This shows that $\{0, \infty\} \not\subset D$, since otherwise $\lim g^n(x) \in \{0, \infty\}$ for $x \in S^3 \backslash D$, which contradicts D being open and G-invariant. Then one can assume that $\infty \notin D$. Then the arguments used in the proof of Theorem 8.54, a) ⇒ b), show that the development $d : \tilde{M} \longrightarrow D$ is a homeomorphism compatible with G, which contradicts G being non-discrete and shows that the third case is impossible.

Case 2) is excluded in the same way as case 3).

Let case 1) apply. Then the transitive action of Möb(3) in S^3 implies that for any points $x, y \in S^3$ there exists a sequence $\{g_m\} \subset G$ such that $\lim g_m(x) = y$. Taking a point x in $S^3 \backslash D$ and a point y in D, we again get a contradiction to D being open and G-invariant, which shows that case 1) is impossible.

If case 5) applies, one can assume (taking an index two subgroup if necessary) that:

$$S^2 = \{x \in \overline{\mathbb{R}}^3 : x_3 = 0\}$$

and the upper and lower half-spaces $\mathbb{R}^3_+$ and $\mathbb{R}^3_-$ are G-invariant. Therefore, the restriction of the action of G to the sphere $S^2 = \partial\mathbb{R}^3_+$ gives a subgroup of the group Möb(2), which we denote by Γ. The group Γ is discrete or not at the same time as G, and, as we can assume that Γ does not have an invariant circle and does not leave one or two points invariant (it is the content of the other cases), we get that Γ is dense in Möb(2). As the action of Möb(2) in S^2 is transitive, so for any points $x, y \in \mathbb{R}^3_+$, there exists a sequence $\{g_m\} \subset G$ such that $\lim g_m(x) = y$. This effect and the arguments used in case 1) show that only the following cases are possible: $D = \mathbb{R}^3_+$, $D = \mathbb{R}^3_-$ or $D = S^3$. The last is excluded due to condition a) of the theorem. If on the other hand D is a half-space, as D is simply connected and d is a covering, with a) $\Rightarrow$ b), we get that d is a homeomorphism. This contradiction to G being non-discrete shows that case 5) is impossible.

The last case 4) is considered on the basis of the following:

LEMMA 8.58. *Let a conformal 3-manifold M be such that the image $D = d(\widetilde{M})$ in the development d is the domain $X = \{x \in \mathbb{R}^3 : x_2^2 + x_3^2 \neq 0\}$. Then M is an $(H^2 \times \mathbb{R}, \operatorname{Isom} H^2 \times \operatorname{Isom} \mathbb{R})$-manifold, i.e., it has a geometry of type* T5.

PROOF. Consider the following Riemannian metric in the half-space $Y = \{(x, r, y) \in \mathbb{R}^3 : r > 0\}$:

$$ds_Y^2 = (dx^2 + dr^2)/r^2 + d\varphi^2, \tag{8.11}$$

which transforms Y into the space $H^2 \times \mathbb{R}$ (see the geometry T5 from §2). In the domain X, we introduce the Riemannian metric

$$ds_X^2 = (dx_1^2 + dx_2^2 + dx_3^2)/(x_2^2 + x_3^2), \tag{8.12}$$

which is conformally Euclidean. The isometry group of the space (X, ds_X^2) coincides with the group of Möbius automorphisms of the domain X (a subgroup in Möb(3)).

Let a map $q : Y \longrightarrow X$ be defined by

$$q(x, r, \varphi) = (x, r\cos\varphi, r\sin\varphi). \tag{8.13}$$

It is clear that the map q is a covering with a cyclic deck-transformation group generated by translation by 2π:

$$(x, r, \varphi) \longrightarrow (x, r, \varphi + 2\pi). \tag{8.14}$$

Also, q is a local isometry, since the lifting by means of q of the metric ds_X^2 coincides with the metric ds_Y^2, as (8.11) and (8.13) show. This implies that, if $d(\widetilde{M}) = X$, the manifold $\widetilde{M}$ with a Riemannian metric lifted from the space (X, ds_X^2) by means of the development d is isometric to the space (Y, ds_Y^2), so M is conformally equivalent to a certain $(H^2 \times \mathbb{R}, \operatorname{Isom} H^2 \times \operatorname{Isom} \mathbb{R})$-manifold. The lemma is proved.

We now assume that case 4) applies. We denote by P the half-plane $\{x \in \mathbb{R}^3 : x_3 = 0,\ x_2 > 0\}$. Also let $\mathcal{M}(l) = \{g \in \mathcal{M}_3 : g(l) = l\}$ and $\mathcal{M}_+(P) = \{g \in \mathcal{M}_3 : g(P) = P$ and g is orientation-preserving on $P\}$; here without loss of generality we assume that $l = \{x \in \overline{\mathbb{R}}^3 : x_2 = x_3 = 0\}$. Transferring, if necessary, to an index two subgroup, we can also assume that the restriction $G|_l = \Gamma$ is a subgroup of $\mathcal{M}_1$. As cases 2) and 3) have already been considered, there are only two possibilities for Γ: either Γ is dense in $\mathcal{M}_1 \cong \mathcal{M}_+(P)$ or it is a discrete group.

Assume that the first applies. Then because $D \neq S^3$, we have $D \subset X = \mathbb{R}^3 \setminus l$. The theorem condition and Lemma 8.58 imply that $D \neq X$, i.e., there exists a point $x \in X \setminus D$. Then the density of Γ in $\mathcal{M}_1 \cong \mathcal{M}_+(P)$ and the transitiveness of the action of the group $\mathcal{M}_+(P)$ on P imply (by analogy with the previous case 1)) that in any circle

$$C(a, r) = \{x \in X : |x - a|^2 = r\}, \qquad a \in l \setminus \{\infty\},$$

there is a point $p = p(a, r) \in X \setminus D$.

At the same time, it is known (Greenberg [1]) that the density of Γ in $\mathcal{M}_1$ implies that there are two infinite-order elliptic elements $e_1, e_2 \in \Gamma$ such that $e_1 e_2 \neq e_2 e_1$. By $h_1 = \tau_1 e_1$ and $h_2 = \tau_2 e_2$ we denote the corresponding elements of the group G. If $C(a_i, r_i) \subset X$, $i = 1, 2$, are pointwise fixed circles for the elliptic elements e_i, considered as elements in $\mathcal{M}_3$, the action of h_i on the circle $C(a_i, r_i)$ coincides with the action on it of the elliptic element τ_i. If τ_i has infinite order, then for a point $p_i = p(a_i, r_i) \in C(a_i, r_i) \cap (X \setminus D)$, the closure of the set

$$\bigcup_{n \in \mathbb{Z}} \tau_i^n(p_i)$$

coincides with the circle $C(a_i, r_i)$ and therefore $C(a_i, r_i) \subset S^3 \setminus D$. The transitiveness of the action of $\mathcal{M}_+(P)$ on P, the density of Γ in $\mathcal{M}_1 \cong \mathcal{M}_+$, and the G-invariantness of D then show that $S^3 = D$, which is impossible.

Therefore, there are integers n_1 and n_2 such that $\tau_i^{n_i} = \mathrm{id}$. As $\tau_i e_i = e_i \tau_i$, we see that $u_i = h_i^{n_i} = e_i^{n_i} \in \mathcal{M}_1$, $i = 1, 2$, are infinite-order elliptic elements and $u_1 u_2 \neq u_2 u_1$. This gives that the group $\langle u_1, u_2 \rangle$ is dense in $\mathcal{M}_1 \cong \mathcal{M}_+(P)$.

If now $\psi(x)$, $x \in X$, is the angle of the vector $(0, x_2, x_3)$ with the half-plane P, we get for the point $x \in X \setminus D$, found above, that the whole half-plane

$$\{y \in X : \psi(y) = \psi(x)\}$$

lies in $X\setminus D$. This shows that D is either non-connected or simply connected or else empty. Condition b) following from condition a) gives that $d:\widetilde{M}\longrightarrow D$ is a covering, and therefore, the first and last cases are impossible. If D is simply connected, d is a homeomorphism, which is impossible because G is not discrete and d is compatible with the action of G.

We now assume that the last possibility applies: $\Gamma = G|_l \subset \mathcal{M}_1$ is a discrete group having limit set $\Lambda(\Gamma)\subset l$. As $D\neq S^3$, then $D\cap\Lambda(\Gamma)=\emptyset$. If here, in addition, $D\neq S^3\setminus\Lambda(\Gamma)$, then the quotient of the open in $S^3\setminus\Lambda(\Gamma)$ set D by the action of the group G, we get a non-compact quotient D/G, which contradicts the compactness of the manifold M. This shows that $D=S^3\setminus\Lambda(\Gamma)$, so we have either that $D=X$ (i.e., $\Lambda(\Gamma)=l$) or that D is the complement in S^3 to the discontinuum $\Lambda(\Gamma)\subset l$ and is thus simply connected. Lemma 8.58 implies that the first case is impossible. Simple connectedness of D in the second case implies that the covering d is a homeomorphism. This contradicts G being non-discrete, and shows that the last case 5) is impossible and thus completes the proof of Theorem 8.57.

REMARK 8.59. We have seen in §2 that 3-manifolds M of type T5 modelled on the homogeneous space $H^2\times\mathbb{R}$ are Seifert fibred manifolds. Therefore the conformal structures for which conditions a) and b) in Theorem 8.54 hold, but condition c) in Theorem 8.57 does not hold, can exist only on Seifert fibred manifolds with geometries modelled on $H^2\times\mathbb{R}$. Moreover, it follows directly from Scott [4] that, if M is a closed 3-manifold with conformal structure C equivalent to a certain $(H^2\times\mathbb{R})$-structure, then the holonomy group $G=d^*(\pi_1(M))$ of structure C has an invariant circle $l\subset S^3$, $d(\widetilde{M})=S^3\setminus l$; here G is either a discrete or virtually discrete group (i.e., $G|_l\subset\mathcal{M}_1$ is discrete, while the subgroup in G of rotations around the circle l is an infinite cyclic group). Both these cases may be realized, and the manifold M has a finite-sheeted covering homeomorphic to $S_g\times S^1$, in which S_g is a closed surface of genus $g\geq 2$, while S^1 is the circle.

This realization of the virtually discrete holonomy group G corresponds to the almost uniformized conformal structure on $S_g\times S^1$ and is contained in Remark 8.56.

Unsolved problems. To complete this section, we consider certain open questions which are important in the light of the above conjecture on conformal uniformization of manifolds.

1. Let M be a closed n-manifold with a uniformized conformal structure. Does each n-manifold M' homotopically equivalent to M have a uniformized conformal structure?

Here, for geometric finite structures on M and M', is there a simultaneous uniformization of these structures, i.e., such a Kleinian (or maybe quasi-Fuchsian) group $G\subset \text{Möb}(n)$ that M and M' are components of $\Omega(G)/G$ - compare Bers [5]?

2. What are the conditions for the existence of a conformal (or a

uniformized conformal) structure on a manifold M which has a finite-sheeted covering $\hat{M}$ with such a structure?

3. Let $M(g, f)$ be a manifold obtained from the surface layer $S_g \times [0, 1]$ by identifying the closed orientable surfaces $S_g \times \{0\}$ and $S_g \times \{1\}$ of genus $g \geq 2$ by means of a diffeomorphism $f: S_g \rightarrow S_g$. What are the classes of diffeomorphisms f for which the manifold $M(g, f)$ has a uniformized conformal structure (compare Theorems 8.12 and 8.52)?

4. Let a manifold $M(f)$ be a torus sum of the manifolds M_1 and M_2 with uniformized conformal structures, i.e., obtained by gluing together the boundaries of solid tori removed from M_1 and M_2 by means of some diffeomorphism f. What f will result in $M(f)$ having a conformal structure (or a uniformized one)?

A partial positive answer to the last question (for conformal f) will be given in the next section. In the general case (for a non-conformal diffeomorphism f), to obtain a similar positive answer it is sufficient, apparently, that a (quasi-conformal) deformation, transforming a given Euclidean lattice of rank $n-1$ (as a subgroup of a Kleinian group) to the standard lattice $\mathbb{Z}^{n-1}$ exists - cf. Maskit [11].

§6. Conformal uniformization of doubles of atoroidal manifolds

Let M be an irreducible homotopically atoroidal compact n-dimensional manifold (orbifold), $n \geq 3$, whose boundary is a union of a finite number of Bieberbach $(n-1)$-manifolds (they are finitely covered by $(n-1)$-dimensional tori T^{n-1}), and let $\hat{M}$ be a closed manifold (orbifold) that is the double of M (with a conformal identification of the boundary components). The following task arises in relation to conformal uniformization: how by a uniformized conformal structure in $\operatorname{int} M$ to determine a uniformized conformal structure in the double $\hat{M}$?

We note that, for three-dimensional manifolds M having those properties, the uniformized conformal structure in $\operatorname{int} M$ can be specified by means of a hyperbolic metric that transforms $\operatorname{int} M$ into a complete hyperbolic manifold of finite volume. This follows from Thurston's uniformization theorem (Theorem 8.10). However, the double $\hat{M}$ here is not a hyperbolic manifold for all $n \geq 3$ because $\pi_1(\hat{M})$ contains free Abelian subgroups of rank $n-1$, while $\hat{M}$ is closed (see Theorem 8.11).

If the uniformized conformal structure in $\operatorname{int} M$ is "strongly" quasi-conformally equivalent to the unique conformal structure defined by the complete hyperbolic metric (i.e., this quasi-conformal equivalence is induced by a quasi-conformal automorphism of the n-sphere), there exists a quasi-Fuchsian geometrically finite group $G \subset \text{Möb}(n)$ having contractible components Ω_0 and Ω_1 of the discontinuity set $\Omega(G_0) = \Omega_0 \cup \Omega_1$ and such that $\operatorname{int} M$ is conformally equivalent to the manifold (orbifold) Ω_0/G_0 with natural conformal structure. A generalization of this situation is provided by conformal structures in $\operatorname{int} M$ that are uniformized by non-degenerate B-groups $G_0 \subset \text{Möb}(n)$, i.e., structures conformally equivalent to the natural structure of Ω_0/G_0 in which $\Omega_0 \subset \Omega(G_0)$ is an invariant contractible component of a geometrically finite Kleinian group $G_0 \subset \text{Möb}(n)$ (by analogy with the planar case, see Chapter 3 §3, we call such groups non-degenerate B-groups in space).

The following theorem solves the problem of conformal uniformization of doubles $\hat{M}$ for such conformal structures in M.

THEOREM 8.60. *Let M be an irreducible homotopically atoroidal compact n-dimensional $(n \geq 3)$ manifold/orbifold. Let its boundary ∂M be a union of m, $m \geq 1$, Euclidean space forms finitely covered by $(n-1)$-dimensional tori. If on the interior $\operatorname{int} M$ there is specified a conformal structure C_0, uniformized by a non-degenerate B-group $G_0 \subset \text{Möb}(n)$, then on a double $\hat{M}$ of M specified by conformal identifications h_i, $i = 1, \ldots, m$, of boundary components ∂M, there exist a conformal structure C that is uniformized by a certain non-degenerate B-group $G \subset \text{Möb}(n)$. Moreover, in $(\hat{M}, C)$ there exists a family of conformally embedded pairwise non-intersecting $(n-1)$-dimensional Euclidean space forms $E_1, \ldots, E_m$ such that $(\hat{M} \setminus \bigcup_i E_i, C)$ is the union of two conformal manifolds/orbifolds conformally equivalent to (M, C_0).*

To prove the theorem, we need a certain construction which we call a parabolic combination for B-groups, and which is shown below to give a Kleinian group that is also a B-group.

Parabolic combination. Let $G_0 \subset \text{Möb}(n)$ be a non-degenerate B-group, i.e., a geometrically finite Kleinian group having an invariant contractible component $\Omega_0 \subset \Omega(G_0)$ and having parabolic cusps p whose cusp domains U_p have a non-empty intersection with Ω_0. The last condition is equivalent (see Theorem 5.42) to the orbifold Ω_0/G_0 having cusp ends. Then the contractibility condition and the G_0-invariantness of Ω_0 imply (see Lemmas 5.59 and 5.61) that the rank of all these parabolic cusps p is $n-1$, and moreover that the cusp domain U_p consists of two non-intersecting balls (tangent at a point p), of which only one lies in Ω_0. We denote that ball by $B_p(r)$ in which $r \geq 0$ is its "radius", which is defined in Chapter 4 §5 - see (4.61), while $S_p = S_p(r)$ denotes an n-dimensional plane in the hyperbolic space H^{n+1} such that

$$\partial S_p(r) = \partial B_p(r). \tag{8.15}$$

In $H^{n+1} \cup \Omega(G_0)$ we consider a certain fundamental polyhedron $P^{n+1} = P^{n+1}(G_0)$ not having equivalent parabolic cusps at the boundary (as such a P^{n+1}, one can take a certain Dirichlet polyhedron or pseudo-Dirichlet one, see Definition 3.15). Let $p_1, \ldots, p_m$ be all parabolic cusps on ∂P^{n+1}, where Theorem 5.39 implies that the number of them is finite. Then the same property is possessed by the polyhedron $P^n = \operatorname{int}(\overline{P}^{n+1} \cap \overline{\mathbb{R}}^n)$ for the action of the group G_0 in the sphere at infinity $\overline{\mathbb{R}}^n = \partial H^{n+1}$. For the cusps p_i, $1 \le i \le m$, we denote by B_i and S_i correspondingly the balls $B_{p_i}(r_i) \subset \overline{\mathbb{R}}^n$ and the planes $S_{p_i}(r_i) \subset H^{n+1}$ that satisfy (8.15) and have "radii" $0 < r_i < K$, where K is such that the G_0-orbits of these balls and planes do not intersect.

We denote by $\mathcal{T}_i$ reflections in the hyperbolic space H^{n+1} with respect to the n-dimensional planes S_i, $i = 1, \ldots, m$, and consider the Kleinian group $\mathcal{T}_1 G_0 \mathcal{T}_1 = \text{Möb}(n)$. Then that group and the group G_0 have the amalgamated subgroup G_{p_1}, and taken together with the sphere ∂B_1 and the fundamental domains P^n and $\mathcal{T}_1(P^n)$, they satisfy the conditions of the combination Theorem 6.19, whose application gives a Kleinian group

$$G_1 = G_1(r_1) = G_0 \underset{G_{p_1}}{*} \mathcal{T}_1 G_0 \mathcal{T}_1, \tag{8.16}$$

which has an invariant component $\Omega_0(G_1) \subset \Omega_0 \cap \mathcal{T}_1(\Omega_0)$. The fundamental polyhedron

$$P^{n+1}(G_1) = P^{n+1}(G_0) \cap \mathcal{T}_1(P^{n+1}(G_0)) \tag{8.17}$$

has at its boundary (non-equivalent) parabolic cusps $p_{i1} = p_i$ and $p_{i2} = \mathcal{T}_1(p_i)$, $i = 2, \ldots, m$, which are vertices also of the polyhedron $P^n(G_1) \cap \Omega_0(G_1)$. The balls $B_{i1}(r_i) = B_i$ and $B_{i2}(r_i) = \mathcal{T}_1(B_i)$ are connected components of the cusp domains at those cusp points lying in $\Omega_0(G_1)$.

Observe that the balls B_{i1} are transformed by the transformation $\mathcal{T}_1\mathcal{T}_i \in \text{Möb}(n)$ to the exteriors of the closed balls $\overline{B}_{i2}$ for all $i = 2, \ldots, m$; the stabilizer of a point p_{i1} in the group G_1 goes over on conjugation by the transformation $\mathcal{T}_1\mathcal{T}_i$ to the stabilizer of the point p_{i2}. It is also readily verified that for all $i = 2, \ldots, m$, the conglomerates

$$G_{p_{i1}},\ G_{p_{i2}} \subset G_1, \quad f_i = \mathcal{T}_1\mathcal{T}_i, \quad S_{i1} = \partial B_{i1}, \quad S_{i2} = \partial B_{i2},$$

which are dependent on the parameter $r = (r_1, r_2, \ldots, r_m) \in (0, K)^m$, satisfy the conditions in Theorem 6.24. When that theorem is applied $m - 1$ times, we get a Kleinian group $G \subset \text{Möb}(n)$,

$$G = G(r) = \langle G_0, \mathcal{T}_1 G_0 \mathcal{T}_1, \mathcal{T}_1\mathcal{T}_2, \ldots, \mathcal{T}_1\mathcal{T}_m \rangle, \tag{8.18}$$

which has the invariant component $\Omega_0(G) \subset \Omega_0(G_1) \subset \Omega_0 \cap \mathcal{T}_1(\Omega_0)$.

REMARK. 8.61. Parabolic combination is applicable to any function group $G_0 \subset \text{Möb}(n)$ having fixed parabolic points p at which strictly G_p-invariant balls are tangent externally. Such points are called parabolic Ω-peaks for the group (see Chapter 4 §5).

REMARK. 8.62. It is clear that there exists the following

modification of our construction of parabolic combination. Namely, we can take, instead of "identical" gluings (8.16) and (8.18) (which we defined by reflections J_i with respect to the n-planes $S_i \subset H^{n+1}$, $p_i \in \overline{S}_i$, $i = 1, \ldots, m$), some "conformal gluings" which are defined by "slide reflections" with respect to the n-planes S_i, i.e., by superpositions of the reflections J_i and non-loxodromic transformations $\hat{h}_i \in \text{Möb}(n)$ which are commuting with the parabolic subgroups G_{p_i}:

$$\hat{h}_i G_{p_i} \hat{h}_i^{-1} = G_{p_i}$$

and are preserving the half-spaces $H^{n+1} \setminus S_i$. These transformations $\hat{h}_i$ define conformal automorphisms of the boundary components ∂M, i.e., of the i-th $(n-1)$-dimensional Euclidean space form E_i:

$$h_i : (\partial S_i \setminus \{p_i\})/G_{p_i} \longrightarrow (\partial S_i \setminus \{p_i\})/G_{p_i}.$$

The following two theorems show that the class of non-degenerate B-groups is closed with respect to the above parabolic combination.

THEOREM 8.63. *Let non-degenerate B-groups $G_1, G_2 \subset \text{Möb}(n)$ containing invariant contractible components Ω_1 and Ω_2 correspondingly have an amalgamated maximal parabolic subgroup $H = H_p = G_1 \cap G_2$, and these together with an $(n-1)$-dimensional sphere $S \subset \overline{\mathbb{R}}^n$, $S \setminus \{p\} \subset \Omega_1 \cap \Omega_2$, which splits up $\overline{\mathbb{R}}^n$ into balls D_1 and D_2, satisfy the conditions of the combination Theorem* 6.19.

Then the group $G = G_1 \ast_H G_2$ is a non-degenerate B-group.

THEOREM 8.64. *Let a non-degenerate B-group $G \subset \text{Möb}(n)$ have maximal parabolic subgroups H_1 and H_2 with respect to which the balls B_1 and B_2 from the invariant contractible component $\Omega_0 \subset \Omega(G)$ are precisely invariant. Let also there exist $A \in \text{Möb}(n)$ such that $A(B_1) = \text{ext}\,\overline{B}_2$ and the group $G = \langle G_1, A\rangle$ is a* HNN*-extension of the group G_1, which is obtained by means of Maskit's second combination (Theorem* 6.24). *Then G is a non-degenerate B-group.*

PROOF OF THEOREM 8.63. From the conditions in Theorem 6.19, the intersection between a certain neighbourhood of a sphere S and a fundamental domain of the subgroup H lies in a certain component $F_0 \subset \Omega_1 \cap \Omega_2$ of a fundamental domain $G(G)$ of the group G. Then because Ω_1 and Ω_2 are invariant with respect to the groups G_1 and G_2 correspondingly, it follows directly that for any given point $x_0 \in S \cap F_0$ and for any generator g of the group G, the points x_0 and $g(x_0)$ can be joined by a path lying in $\Omega(G)$. Then Lemma 3.7 shows that the domain $\Omega_0 = G(\overline{F}_0)$ is a G-invariant component of the discontinuity set $\Omega(G)$. We show that Ω_0 is a contractible domain.

We write all the $g \in G \setminus H$ in the normal form:

$$g = g_n \cdot \ldots \cdot g_1, \tag{8.19}$$

in which $n>0$ and either $g_{2i+1}\in G_1\backslash H$, $g_{2i}\in G_2\backslash H$, or $g_{2i+1}\in G_2\backslash H$ and $g_{2i}\in G_1\backslash H$. It is well known that the length $n>0$ is uniquely defined by an element g; we denote it by $|g|$ and for $g\in H$ we put $|g|=0$. We say g is positive $(g>0)$ or negative $(g<0)$ if in (8.19) $g_1\in G\backslash H$ or correspondingly $g_1\in G_2\backslash H$.

For a fixed decomposition of the group G

$$G=H\cup(\bigcup_{n,m}a_{nm}H)\cup(\bigcup_{n,m}b_{nm}H), \tag{8.20}$$

in which $|a_{nm}|=|b_{nm}|=n$, $a_{nm}>0$, $b_{nm}<0$, and for a fixed $n>0$, we consider a set

$$T_n=\bigcup_m(a_{nm}(D_1)\cup b_{nm}(D_2)) \tag{8.21}$$

and will assume without loss of generality that $\infty\in\Omega(G)$ and $\infty\notin G(S)$. Then the proof of Theorem 6.19 gives us the following properties:

1. $T_n\subset T_{n-1}$ for $n>1$;
2. The connected components T_{nm} of the set T_n are disjoint balls of the form $a_{nm}(D_1)$ or $b_{nm}(D_2)$;
3. $\sum_{i,j}d^n(T_{ij})=\sum_{i=j}^{\infty}\sum_j d^n(T_{ij})<\infty$ and therefore the Euclidean diameters $d(T_{ij})$ satisfy

$$\lim_{i\to\infty}\sum_j d^n(T_{ij})=0. \tag{8.22}$$

We put $\Omega_i'=\Omega(G_i)\backslash\Omega_i$, $i=1,2$. The theorem conditions imply that $\Omega_1'\cap\Omega_2'=\varnothing$, $\Omega_1'\subset D_2$, $\Omega_2'\subset D_1$, $\Omega_0\cap(\Omega_1'\cup\Omega_2')=\varnothing$. Also, one can say that Ω_1' is tangent to Ω_2' at a fixed point p of the parabolic subgroup H in the sense that there are balls $B_1\subset\Omega_1'$, $B_2\subset\Omega_2'$ precisely H-invariant in G_1 and G_2 and which are tangent one to the other at p.

We now consider the following sets:

$$O_n=\bigcup_m a_{nm}(\Omega_2')\cup b_{nm}(\Omega_1'). \tag{8.23}$$

The properties of Ω_1' and T_n imply directly the following ones for the sets O_n: $O_n\subset T_n\subset T_{n-1}\backslash O_{n-1}$; the connected components of the set O_n form disjoint families O_{nm} of the form $a_{nm}(\Omega_2')\subset a_{nm}(D_1)$ or $b_{nm}(\Omega_1')\subset b_{nm}(D_2)$. Here a unique component $a_{nm}(\Omega_2')$ (which contains the ball $a_{nm}(B_2)$) is tangent to a unique component $b_{n-1,m}(\Omega_1')$ (which contains the ball $b_{n-1,m}(B_2)$) at a single point, the image of the point p.

It follows from Theorem 6.19 that the fundamental domain $F(G)$ of G splits up into three parts: $F=F_0\cup F_1\cup F_2$, $F_0\subset\Omega_0$, $F_1\subset\Omega_1'$, $F_2\subset\Omega_2'$. As here $\Omega_i'=\operatorname{int}G_i(\overline{F}_i)$, $i=1,2$, $\Omega_0=G(\overline{F}_0)$, the above properties of the sets O_i, $i\geq 1$, and $O_0=\Omega_1'\cup\Omega_2'$ imply that

$$\Omega_0 = \operatorname{int}(\overline{\mathbb{R}}^n \setminus \bigcup_{i=0}^{\infty} O_i), \tag{8.24}$$

From the condition in the theorem, $S\setminus\{p\} \subset \Omega_1 \cap \Omega_2 = \operatorname{int}(\overline{\mathbb{R}}^n \setminus O_0) \equiv \Omega_*$. This and the contractibility of the invariant components Ω_1 and Ω_2 would imply contractibility of the domain Ω_*. From this and the conditions of Theorem 6.19 we obtain that the balls D_1 and D_2 contain precisely H-invariant balls $D'_i \subset D_i$ tangent one to another at the point p and such that $\partial D'_i \setminus \{p\} \subset \Omega_*$. Assuming $p = \infty$, we can choose balls D'_1 and D'_2 (where we conjugate the group G if necessary), in such a way that their boundary spheres are at Euclidean distance 1 from S, while their complement is a neighbourhood V of the sphere S mentioned in condition I in Theorem 6.19.

Denote by T'_n subsets in T_n obtained by replacing the balls D_i by the balls D'_i in Definition (8.21). We have the following obvious inclusion analogous to the above properties of T_n:

$$T_{n+1} \subset T'_n \subset T_n \subset T'_{n-1} \quad \text{and} \quad O_n \subset T'_n \subset T_n. \tag{8.25}$$

Consider closed balls in $\overline{\mathbb{R}}^n$ tangent one to another (they are called standard):

$$\begin{gathered} \Delta_1 = \{x \in \overline{\mathbb{R}}^n : x_n \le 0\} \subset \{x \in \overline{\mathbb{R}}^n : x_n \le 1\} = \Delta, \\ \Delta_2 = \{x \in \overline{\mathbb{R}}^n : x_n \ge 2\}. \end{gathered} \tag{8.26}$$

For these balls, we define a "standard" diffeomorphism φ that commutes with the Euclidean isometry group in $\mathbb{R}^n$ that leaves the half-space $\mathbb{R}^n_+$ invariant,

$$\varphi : \operatorname{ext}(\Delta_1 \cup \Delta_2) \longrightarrow \operatorname{ext} \Delta_1 \tag{8.27}$$

and which is such that $\varphi|_{\Delta} = \mathrm{id}$ and for all $y = \varphi(x)$, $y' = \varphi(x')$ we have

$$y_i = y'_i = x_i \quad \text{for} \quad i \ne n \quad \text{and} \quad y_n = y'_n = \varphi_0(x_n),$$

where φ_0 is a diffeomorphism of the interval $(0, 2)$ onto $\mathbb{R}_+$, and $\varphi_0(x) = x$ for $0 < x \le 1$.

We can now define a sequence of diffeomorphisms φ_k, $k \ge 1$,

$$\varphi_k : \operatorname{int}(\overline{\mathbb{R}}^n \setminus \bigcup_{i=0}^{k} O_i) \longrightarrow \operatorname{int}(\overline{\mathbb{R}}^n \setminus \bigcup_{i=0}^{k-1} O_i), \tag{8.28}$$

Each of the diffeomorphisms φ_k is an identity in $\overline{\mathbb{R}}^n \setminus T_k$, and in each of the disjoint balls T_{km} (= a component of T_k tangent to a certain component of $O_{k-1,l}$ and containing a family $O_{k,n}$) is obtained (see Fig. 76) from the standard diffeomorphism (8.26) by conjugation by means of a certain Möbius transformation h_{km} such that

$$h_{km}(T_{km}) = \overline{\mathbb{R}}^n \setminus \Delta_1, \quad h_{km}(T'_{km}) = \text{int}\,\Delta_2, \quad h_{km}(\overline{T}_{km} \cap \overline{T}'_{km}) = \infty. \tag{8.29}$$

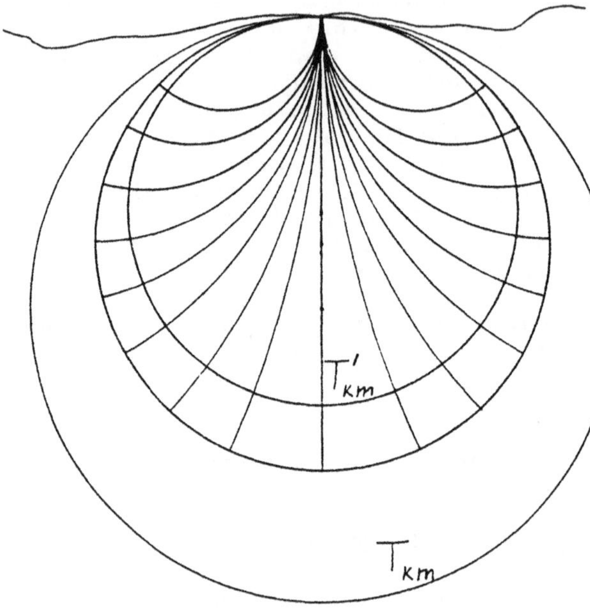

Figure 76

Diffeomorphisms φ_k define a limit diffeomorphism

$$f = \lim_{k\to\infty} \varphi_1 \circ \varphi_2 \circ \ldots \circ \varphi_k. \tag{8.30}$$

It is readily seen from (8.24) that this diffeomorphism f maps the invariant component Ω_0 of the group G onto the contractible domain $\Omega_* = \Omega_1 \cap \Omega_2$, which shows that G is a non-degenerate B-group, whose geometric finiteness is evident from $P^{n+1}(G) = P^{n+1}(G_1) \cap P^{n+1}(G_2)$.

Then to prove the theorem, it remains to show that the diffeomorphisms $f_k = \varphi_1 \circ \varphi_2 \circ \ldots \circ \varphi_k$ have a limit mapping and that the limit mapping of (8.30) is smooth (diffeomorphic).

For this purpose we consider for any $\varepsilon > 0$ a number k such that for all $i \geq k$, for all j, and for the Lebesgue measure μ_n

$$\mu_n(T_i) < \varepsilon, \quad d(T_{ij}) \leq \varepsilon \tag{8.31}$$

holds.

The existence of such k follows from (8.22); (8.31) implies that for any $m > l \geq k$ the mapping $f_l^{-1} \circ f_m = \varphi_{l+1} \circ \ldots \circ \varphi_m$ is different from the identity only in the set T_{l+1} whose measure is less than ε. Also, as that mapping transfers each component $T_{l+1,j}$ into itself, (8.31) gives

$$|f_l^{-1} \circ f_m(x) - x| < \varepsilon \quad \text{for} \quad x \in T_{l+1} \tag{8.32}$$

(8.31) and (8.31) imply that for any $\varepsilon > 0$ there exists a number k such that for all $m > l \geq k$ the homeomorphisms f_m and f_l are different only in the set T_{l+1}, $\mu_n(T_{l+1}) < \varepsilon$, where they differ from each other by not more than ε, while $T_{i+1} \subset T_i$ for all i. This demonstrates the uniform convergence of the diffeomorphisms f_i to the homeomorphism (8.30). Smoothness of the mapping follows from the fact that f coincides with a certain diffeomorphism f_i having sufficiently large i in a sufficiently small neighbourhood of any point x. This completes the proof of Theorem 8.63.

PROOF OF THEOREM 8.64. We shall use the concepts in the proof of the previous theorem and Maskit's [5] algebraic technique. Namely each element $g \in G \setminus \{id\}$ is written (not uniquely) in normal form:

$$g = f^{\alpha_n} \circ g_n \circ \ldots \circ f^{\alpha_1} \circ g_1, \tag{8.33}$$

in which $g_i \in G_1$; $g_2, \ldots, g_n \neq id$; $\alpha_1, \ldots, \alpha_{n-1} \neq 0$; if $\alpha_i < 0$ and $g_{i+1} \in H_1$, then $\alpha_{i+1} \leq 0$; if $\alpha_i > 0$ and $g_{i+1} \in H_2$, then $\alpha_{i+1} \geq 0$. For any such word, the following number is unaltered:

$$|g| = \sum_{i=1}^{n} |\alpha_i|,$$

which we call the length of the element g. It is clear that the length of $g \in G$ does not alter on superposition of g with any element $g_1 \in G_1$. We call (8.33) a positive normal form $(g > 0)$ if in it $g_1 \in H_1$ and $\alpha_1 > 0$. If this is not so, we call the form (8.33) non-positive $(g \leq 0)$. Similarly, $g < 0$ if $g_1 \in H_2$ and $\alpha_1 < 0$ in (8.33); if that is not so, g is non-negative $(g \geq 0)$. These properties of an element $g \in G \setminus \{id\}$ as also its length do not depend on the choice of (8.33), see Maskit [5].

We now consider a decomposition of G modulo the subgroup H_1, compare (8.20); then all elements of a fixed class gH_1 have the same length and are simultaneously positive or non-positive. From each non-positive class of length n, we select an element a_{nm}. Similarly, for a decomposition of G modulo subgroup H_2, from each non-negative class gH_2 of length n we select an element b_{nm}, and for each given $n = 0, 1, 2, \ldots$ we consider the following sets (compare (8.21)):

$$T_n = \bigcup_m a_{nm}(B_1) \cup b_{nm}(B_2). \tag{8.34}$$

Maskit [5] showed that $T_n \subset T_{n-1}$ for $n > 0$ and also that all connected components T_{nm} of the set T_n, $n \geq 0$, are disjoint balls of the form $a_{nm}(B_1)$ or $b_{nm}(B_2)$. Assuming, without loss of generality, that $\infty \in \Omega(G)$ and $\infty \notin G(\partial B_1 \cup \partial B_2)$ we have (see Lemma

6.21) additionally the following property of the Euclidean diameters $d(T_{ij})$ of the components $T_{ij} \subset T_i$, which is analogous to (8.22):

$$\lim_{i\to\infty} \sum_j d^n(T_{ij}) = 0. \tag{8.35}$$

As the sets (8.34), we consider the set $\Omega' = \Omega(G_1)\setminus\Omega_0 \equiv O_0$ and for $n \geq 1$ the sets

$$O_n = \bigcup_m a_{nm}(\Omega') \cup b_{nm}(\Omega'). \tag{8.36}$$

Also, we consider in the balls B_1 and B_2 smaller balls $B'_i \subset B_i$ precisely invariant with respect to the subgroups $H_i \subset G_1$ (see the conditions in Theorem 6.24) and such that their images $A(B'_1)$ and $A^{-1}(B'_2)$ are precisely invariant with respect to the subgroups H_2 and H_1 correspondingly. Then we define subsets $T'_n \subset T_n$, $n \geq 0$, by replacing the balls B_i by the smaller balls B'_i in the definition (8.34).

The definitions of T_n, T'_n and O_n give the following inclusions for $n \geq 0$:

$$O_{n+1} \subset T'_n \subset T_n, \quad T_{n+1} \subset T'_n \setminus O_{n+1}. \tag{8.37}$$

We note also that the domain family O_n has subfamilies O_{nm} of the form $a_{nm}(\Omega')$ or $b_{nm}(\Omega')$. Let us denote by Ω_1 and Ω_2 components from Ω' tangent to the balls B_1 and B_2 correspondingly (in the sense that there are balls $B^*_i \subset \Omega'$ at the parabolic cusps p_i fixed for the subgroups H_i, $i = 1, 2$). The subfamilies O_{nm} have the following property.

For each $n \geq 0$, in $a_{n+1,m}(\Omega')$ there is a unique component ($= a_{n+1,m}(\Omega_2)$) tangent to the family $a_{nm}(\Omega)$, namely its unique component $= a_{nm}(\Omega_1)$. Similarly $b_{n+1,m}(\Omega')$ "tangent" to the family $b_{nm}(\Omega')$ at a unique point, and that tangency is provided by components $b_{n+1,m}(\Omega_1)$ and $b_{nm}(\Omega_2)$. The points of tangency are parabolic fixed points of subgroups conjugate in G to the groups H_1 and H_2.

Now let F_* be a component of the fundamental set $F(G) = \text{int}\{F_0 \cap \text{ext}(\overline{B}_1 \cup \overline{B}_2)\}$ that adjoins the balls B_1 and B_2. Lemma 3.7 shows that the domain $\Omega_* = G(\overline{F}_*)$ is a G-invariant component of $\Omega(G)$. We shall show that Ω_* is diffeomorphic to the contractible domain Ω_0 and therefore G is a non-degenerate B-group.

As in the proof of the previous Theorem 8.63 (compare (8.24)), (8.36) and (8.37) for O_n give

$$\Omega_* = \text{int}(\overline{\mathbb{R}}^n \setminus \bigcup_{i=0}^{\infty} O_i). \tag{8.38}$$

By analogy with (8.28) and (8.29), for $k \geq 1$ we specify a sequence of diffeomorphisms

$$\varphi_k : \text{int}(\overline{\mathbb{R}}^n \setminus \bigcup_{i=0}^{k} O_i) \longrightarrow \text{int}(\overline{\mathbb{R}}^n \setminus \bigcup_{i=0}^{k-1} O_i),$$

each of which is an identity in the set $\overline{\mathbb{R}}^n \setminus T_{k-1}$ and is obtained (Fig. 76) on each of the disjoint balls $T_{k-1,m}$ (components of the set T_{k-1}) from the standard diffeomorphism (8.27) by conjugation by means of a certain Möbius transformation $h_{km} \subset \text{Möb}(n)$ such that

$$h_{km}(T_{k-1,m}) = \overline{\mathbb{R}}^n \setminus \Delta_1, \quad h_{km}(T'_{k-1,m}) = \text{int}\,\Delta_2, \quad h_{km}(\overline{T}_{k-1,m} \cap \overline{T}'_{k-1,m}) = \infty.$$

Then these diffeomorphisms φ_k define a limit mapping

$$f = \lim_{i\to\infty} \varphi_1 \circ \varphi_2 \circ \dots \circ \varphi_i : \Omega_* \longrightarrow \Omega_0. \tag{8.39}$$

Here the mapping f that is the limit of the sequence of diffeomorphisms $f_i = \varphi_1 \circ \varphi_2 \circ \dots \circ \varphi_i$ is also a diffeomorphism. This is proved similarly to the proof of the diffeomorphism for the mapping (8.30), with property (8.22) replaced by the corresponding property (8.35) of the sets T_n from (8.34) in this case. This completes the proof of Theorem 8.64.

REMARK 8.65. Theorems 8.63 and 8.64 remain correct even without the assumption of geometric finiteness for the B-groups G_1 and G_2, see Apanasov [25]. This demonstrates closure of the whole class of B-groups in $\overline{\mathbb{R}}^n$, i.e., for the Kleinian groups G that $\Omega(G)$ have invariant contractible components, with respect to the process of parabolic combination.

PROOF OF THEOREM 8.60. From the theorem condition, we have that the conformal manifold/orbifold $(\text{int}\,M, C_0)$ is conformally equivalent to the quotient Ω_0/G_0, in which $G_0 \subset \text{Möb}(n)$ is some non-degenerate B-group containing an invariant (contractible) component $\Omega_0 \subset \Omega(G_0)$. We apply the above parabolic combination to the group G_0. (Here we are applying Remark 8.62 where the transformations $\hat{h}_i \in \text{Möb}(n)$ should be determined by our conformal automorphisms h_i of i-th boundary components of ∂M.) This construction gives a Kleinian group $G \subset \text{Möb}(n)$ specified in (8.18) and that, according to Theorems 8.63 and 8.64, is a non-degenerate B-group with an invariant contractible component $\Omega_* \subset \Omega(G)$. The construction of G shows that the closed manifold/orbifold $\hat{M} = \Omega_*/G$ is the double of M, $\text{int}\,M = \Omega_0/G_0$. Let C be a conformal structure on $\hat{M}$ defined by the natural projection π:

$$\pi : \Omega_* \longrightarrow \Omega_*/G.$$

We note that disjoint G-orbits of the punctured precisely G_{p_i}-invariant spheres $\partial B_i(r_i)\setminus\{p\}$, $1 \le i \le m$, (see (8.15)) are projected by π to disjoint non-parallel $(n-1)$-dimensional Euclidean space forms $E_1, \dots, E_m$ conformally embedded in $\hat{M}$ (and incompressible). Discarding these punctured-sphere G-orbits from Ω_*

we get a decomposition of $\Omega_*\setminus \bigcup_{i=1}^{m}(\partial B_i)$ into connected components R_j that form two non-intersecting G-orbits. These two G-orbits correspond to two components into which the double $\hat{M}$ decomposes when one has discarded the conformally Euclidean manifolds/orbifolds $E_1, \ldots, E_m$. As two non-equivalent components from $G(\bigcup_j R_j)$, one can take the components R_1 and R_2 adjoining one another on the punctured sphere $\partial B_1\setminus\{p_1\}$ and such that $\partial R_1 \cap \Lambda(G_0) \neq \emptyset$, $\partial R_2 \cap \Lambda(\tilde{h}_1\mathcal{I}_1 G_0 \mathcal{I}_1 \tilde{h}_1^{-1}) \neq \emptyset$. Denote the stabilizers for R_1 and R_2 in the group G by Γ_1 and Γ_2 correspondingly; (8.18) shows that $\Gamma_1 = G_0$ and $\Gamma_2 = \tilde{h}_1\mathcal{I}_1 G_0 \mathcal{I}_1 \tilde{h}_1^{-1}$.

As the standard diffeomorphism (8.27) commutes with the action of Euclidean isometries in the half-space $\mathbb{R}^n_+$, by analogy with the construction of the diffeomorphisms φ_k and f in the proof of Theorems 8.63 and 8.64 (see (8.28)-(8.30) and (8.39)) one constructs diffeomorphisms

$$f_1 : R_1 \longrightarrow \Omega_0, \quad f_2 : R_2 \longrightarrow \tilde{h}_1\mathcal{I}_1(\Omega_0) \tag{8.40}$$

compatible with the action of the groups Γ_1 and Γ_2.

This shows that the groups Γ_i, $i = 1, 2$, act in the conformal structure in R_i lifted by the diffeomorphism (8.40) from Ω_0 or from $\tilde{h}_1\mathcal{I}_1(\Omega_0)$ as conformal automorphism groups. Therefore, the conformal structure of the orbifold $(R_1/\Gamma_1, C)$ is conformally equivalent to the structure C_0 of the orbifold $\operatorname{int} M = \Omega_0/G_0$, while the orbifold $(R_2/\Gamma_2, C)$ is anticonformally equivalent (with orientation reversing) to the orbifold $(\operatorname{int} M, C_0)$, since $(R_2/\Gamma_2, C)$ is conformally equivalent to $\tilde{h}_1\mathcal{I}_1(\Omega_0)/\tilde{h}_1\mathcal{I}_1 G_0 \mathcal{I}_1 \tilde{h}_1^{-1}$. This completes the proof of uniformization Theorem 8.60.

COROLLARY 8.66. *Let under the conditions of Theorem* 8.60 *the conformal structure* C_0 *in the interior of the manifold/orbifold* M *be specified by a complete hyperbolic structure* τ_0. *Then restrictions* C_1 *and* C_2 *of a conformal structure* C *in the double* $\hat{M}$ *for* M *on the components* $\hat{M}\setminus \bigcup_{i=1}^{\infty} E_i$ *are specified by complete hyperbolic structures* τ_1 *and* τ_2 *isometric to* τ_0.

To prove the corollary, it is sufficient to note additionally that the diffeomorphisms (8.40) define complete hyperbolic metrics in R_1 and R_2 (lifted from $\mathbb{R}^n_+ = \Omega_0$ and $\mathcal{I}(\mathbb{R}^n_+)$) in which the groups Γ_1 and Γ_2 act isometrically. The isometric action follows from compatibility of f_1 and f_2 with Γ_1 and Γ_2 and from the commutation of the standard homeomorphism φ (8.27) with $\operatorname{Isom}\mathbb{R}^{n-1} \subset \operatorname{Isom} H^n$, in which $H^n \cong \mathbb{R}^n_+$ is the hyperbolic space.

CHAPTER 9

DEFORMATION SPACES OF GROUPS AND STRUCTURES

§1. Deformations of locally homogeneous structures

1. *Marked locally homogeneous structures.* Let M be a smooth n-dimensional manifold/orbifold with a locally homogeneous structure modelled on some homogeneous space $X = G/H$. In other words, M has a (X, G)-structure s defined by a maximal (X, G)-atlas - see Chapter 2. An equivalence class of (X, G)-structures on M (such that two structures s_1 and s_2 are called *equivalent* if there exists a (X, G)-diffeomorphism $f : (M, s_1) \longrightarrow (M, s_2)$ free-homotopic to the identity) is called a *marked* (X, G)-structure on M. Denote the space of marked (X, G)-structures on M by $\mathcal{S}(M) = \mathcal{S}(M; X)$. This space is covered by the space $\mathcal{D}(M)$ of developments $d : \widetilde{M} \longrightarrow X$ with a metrizable topology defined by uniform convergence of developing mappings on compacta (see Lok [1], Thurston [1, Ch.5].

In particular, if X is the hyperbolic space H^n and $G = \operatorname{Isom} H^n \cong SO(n, 1)$ is its isometry group, we get the definition of the space $\mathcal{T}(M)$ of marked hyperbolic structures in M; if X is the conformal sphere $S^n = \overline{\mathbb{R}}^n$, while G is the Möbius transformation group, $G \cong SO(n+1, 1)$, we have the space of marked conformal structures $\mathcal{C}(M)$.

In this chapter, we are mainly interested in how uniquely one can determine a marked (X, G)-structure, and in particular a hyperbolic or conformal structure on a manifold/orbifold M. To handle this more clearly, we briefly consider deformations of conformal structures on a closed Riemann surface of genus $p > 1$ (for details, see Bers [4] - [7] or Krushkal', Apanasov, and Gusevskii [3]).

2. *Teichmüller space of a Riemann surface.* As we have already mentioned in Chapter 8, each conformal structure on a surface S of genus $p > 1$ is defined by a complete hyperbolic metric, and as Riemann observed in his work "Abelian function theory" (1857), the classes of conformal equivalence of surfaces depend on $3p - 3$ complex parameters which he called moduli. The problem of studying the manifold of these parameters and its structure, known as the classical problem of moduli,

was solved for $n=2$. It had been noted already in the classical studies that the solution of the problem becomes simpler if conformal equivalence is replaced by a weaker equivalence by imposing certain topological constraints, for example, by fixing generators of the fundamental group. This leads to a branched covering of a space of Riemann surfaces, which can be described by means of quasi-conformal mappings. A programme of investigations in this direction was sketched in the papers of Teichmüller.

Thus we consider a marked closed Riemann surface S of genus $p>1$, i.e., the pair (S, Σ), in which

$$\Sigma = \{a_1, b_1, \ldots, a_p, b_p\} \subset \pi_1(S), \quad \prod_{i=1}^{p} [a_i, b_i] = 1 \tag{9.1}$$

is a singled out system of generators of $\pi_1(S)$ to within an inner automorphism of $\pi_1(S)$. Since two homeomorphisms $S \longrightarrow S'$ are homotopic if and only if the corresponding isomorphisms $\pi_1(S) \longrightarrow \pi_1(S')$ coincide (up to inner automorphisms of the fundamental groups), it follows that considering homeomorphisms of marked Riemann surfaces $(S, \Sigma) \longrightarrow (S', \Sigma')$ is equivalent to specifying a homotopy class of homeomorphisms $S \longrightarrow S'$.

Further, any two marked surfaces of the same genus are homeomorphic and even quasi-conformally equivalent. This follows from the topological equivalence of closed surfaces of the same genus and Nielsen's theorem, which states that every automorphism of the fundamental group $\pi_1(S)$ of a closed surface S is realized geometrically, i.e., is induced by some topological automorphism of S which is determined up to homotopy; with a homeomorphism $f: S \longrightarrow S'$ at hand it is possible to construct a quasi-conformal homeomorphism homotopic to it (for details see, for example, Krushkal' [1], Chapter 1 §6). That construction may be extended to any finite-type surface, i.e., one obtained from a closed surface by deleting finitely many points.

The homeomorphisms $f_1: S \longrightarrow S_1$ and $f_2: S \longrightarrow S_2$ are regarded as equivalent if $f_2 f_1^{-1}$ is homotopic to a conformal mapping $h: S_1 \longrightarrow S_2$. Passing to universal coverings of those surfaces (take the coverings to be the unit disc U), we get that f_1 and f_2 can be lifted to automorphisms $\hat{f}_1$ and $\hat{f}_2$ of the disc U such that the following diagram is commutative:

$$\begin{array}{ccc} U & \xrightarrow{\hat{f}_1} & U' \\ {\scriptstyle \pi}\downarrow & & \downarrow{\scriptstyle \pi_1} \\ S & \xrightarrow{f_1} & S_1 \end{array}$$

(as does the analogous diagram for f_2), where $\hat{f}_1$ and $\hat{f}_2$ coincide on the boundary (the circle) ∂U.

The Teichmüller space $T(S)$ of the surface S is formed by the set of equivalence classes $[S]$ of marked surfaces of genus p with n punctures or, what is the same, by the set of corresponding equivalence classes $[f]$ of quasi-conformal homeomorphisms. In it we introduce the metric

$$d([f],\ [f_1]) = \frac{1}{2} \cdot \inf_{f\in[f],\ f_1\in[f_1]} \ln K(f_1 \circ f^{-1}), \tag{9.2}$$

where $K(f_1 \circ f^{-1})$ is the quasi-conformal dilatation for the map $f_1 \circ f^{-1}$. If, instead of S, we take another marked surface S_1, we obtain an isometry of the spaces $T(S_1) \longrightarrow T(S)$ by the formula $[f \circ h] \longrightarrow [f]$, where h is a homeomorphism $S_1 \longrightarrow S$. Therefore, one can speak of a simple space $T(p, n)$ (of equivalence classes) of surfaces of type (p, n), in which the isometries $[f \circ h] \longrightarrow [f]$ generate the so-called modular group $\operatorname{Mod} T(p, n)$ of $T(p, n)$, which acts discontinuously there. Here the space $R(p, n)$ of Riemann surfaces themselves is obtained as the quotient

$$T(p,\ n)/\operatorname{Mod} T(p,\ n).$$

A basic result here is that provided by Ahlfors [1] and Bers [1], which solves the moduli problem: into the space $T(p, n)$, which is a cell of real dimension $6p - 6 + 2n$, one can introduce a complex analytic structure, and the structure compatible with the metric (9.2) is unique.

3. *Stable representations and the holonomy theorem.* In dimension $n \geq 3$, the main tool for researching (X, G)-structures on M, $\dim M = n$, becomes the developing map (development) $d : \widetilde{M} \longrightarrow X$ (this is a continuous extension of coordinate charts in M around elements of $\pi_1(M)$, see Chapter 2), which induces representation (= holonomy homomorphism = holonomy)

$$d_* : G(\widetilde{M},\ M) \cong \pi_1(M) \longrightarrow G \tag{9.3}$$

and thus defines a map

$$\operatorname{hol} : \mathcal{S}(M) \longrightarrow \operatorname{Hom}(\pi_1(M),\ G)/G. \tag{9.4}$$

The map hol puts each structure $\Delta \in \mathcal{S}(M)$ into correspondence with the representation class $G \cdot d_*$ (where the group G acts in $\operatorname{Hom}(\pi_1(M), G)$ by conjugation: $g \cdot d_* = g d_* g^{-1}$ for all $g \in G$).

However, the holonomy d_* in general does not determine the (X, G)-structure on M; for example, the two immersions of an annulus in $\mathbb{R}^2$, one of which has self-intersection, define Euclidean structures on the annulus as both have trivial holonomy but are not equivalent in any reasonable sense. This shows that it is necessary to examine local properties of the space $\mathcal{S}(M)$.

For this purpose, we define a usual algebraic convergence topology (see Chapter 6 §5) in the representation space $\operatorname{Hom}(\pi_1(M), G)$ and define in the space $\mathcal{S}(M)$ a topology of development convergence on compacta from $\widetilde{M}$, i.e., the topology that lifted to the covering space $\mathcal{D}(M)$ (of the developments $d : \widetilde{M} \to X$) to the topology of development convergence on compacta.

An important property of the map hol from (9.4) is given by the following holonomy theorem (Lok [1], Theorem 1.11, see also Goldman [1] and Thurston [1], section 5.2):

THEOREM 9.1. *The map* hol *in* (9.4) *is an open map which lifts to a local homeomorphism from the space* $\mathcal{D}(M)$ *of* (G, X)*-developments to the representation space* $\mathrm{Hom}(\pi_1(M), G)$.

Unfortunately, hol is not necessarily a local homeomorphism, which is the case in particular for the space $\mathcal{C}(M)$ of conformal structures in a neighbourhood of a structure c on a closed n-manifold M, $\dim M = n = 2m+1 \geq 3$, for which $d(\widetilde{M}) = \overline{\mathbb{R}}^n \backslash S^{n-2}$. In that case, the holonomy representation $d_* : \pi_1(M) \to \mathrm{M\ddot{o}b}(n)$ is not stable in the following sense.

DEFINITION 9.2. Let Γ be a finitely generated group and $\mathbf{G}$ a simple linear algebraic group (G consists of its real points), which is defined over $\mathbb{R}$ and acts algebraically on $\mathrm{Hom}(\Gamma, \mathbf{G})$ by means of conjugation. Here the set $\mathrm{Hom}(\Gamma, \mathbf{G})$ is the set of complex points of an affine variety defined over $\mathbb{R}$ with real points $\mathrm{Hom}(\Gamma, G)$. A representation $\rho \in \mathrm{Hom}(\Gamma, \mathbf{G})$ is called *stable* if the orbit $\mathbf{G}\cdot\rho$ is closed in $\mathrm{Hom}(\Gamma, \mathbf{G})$ and if the stabilizer subgroup $Z(\rho)$ of the representation ρ in the group $\mathbf{G}$ is finite. A stable representation ρ is called good if $Z(\rho)$ is the centre $Z_{\mathbf{G}}$ of the group $\mathbf{G}$.

Since $\mathbf{G}$ is reductive, there is a quotient variety $X(\Gamma, \mathbf{G})$ that is a quotient of the space $\mathrm{Hom}(\Gamma, \mathbf{G})$ by the action on it by conjugation of the group $\mathbf{G}$ (see Newstead [1]). However, that variety $X(\Gamma, \mathbf{G})$ is not isomorphic to the orbit space $\mathrm{Hom}(\Gamma, \mathbf{G})/\mathbf{G}$. In particular, we have the following situation (see Johnson and Millson [1]) for quasi-Fuchsian representations of the fundamental group $\Gamma = \pi_1(M)$ of an n-dimensional closed hyperbolic manifold M in the group $SO(n+1, 1)$:

For even n, the space $\mathrm{Hom}(\Gamma, SO(n+1, 1))/SO(n+1, 1)$ is an open subset of a real algebraic set $X(\Gamma, SO(n+1, 1)) \subset X(\Gamma, \mathbf{SO}(n+1, 1))$. In the odd case, this result is not necessarily correct: there is another component of the real points in $X(\Gamma, \mathbf{SO}(n+1, 1))$ passing through the uniformizable representions ρ of the group Γ in $SO(n, 1)$ corresponding to deformations in the group $SO(n, 2)$ of the representation ρ (composed with the inclusion $SO(n, 1) \subset SO(n, 2)$, in which $\mathbf{SO}(n, 2)$ is identified with $\mathbf{SO}(n+1, 1)$ by conjugating by the diagonal matrix with diagonal entries $(1, 1, \ldots, 1, i)$).

The quotient variety $X(\Gamma, \mathbf{G})$ is defined as follows. Let $\{\gamma_1, \ldots, \gamma_N\}$ be a set of generators for Γ and let $\{f_1, \ldots, f_m\}$ be a set of generators for the algebra of invariant polynomials on $\mathbf{G}^N$, which are chosen such that f_i take real values on G^N. Then a map $F : \mathrm{Hom}(\Gamma, \mathbf{G})/\mathbf{G} \to \mathbb{C}^m$ (not necessarily injective) is defined such that

$$F(\mathbf{G}\cdot\rho) = (f_1(\rho(\gamma_1), \ldots, \rho(\gamma_N)), \ldots, f_m(\rho(\gamma_1), \ldots, \rho(\gamma_N))). \tag{9.5}$$

Then $X(\Gamma, \mathbf{G})$ is the affine variety corresponding to the ring of $\mathbf{G}$-invariant polynomials on $\mathrm{Hom}(\Gamma, \mathbf{G})$, i.e., this is the image under F contained in an affine variety determined by relations among the generators $\{\gamma_1, \ldots, \gamma_N\}$ and the relations among the invariants $\{f_1, \ldots, f_m\}$. $X(\Gamma, \mathbf{G})$ is defined over $\mathbb{R}$. The set $X(\Gamma, G)$ of real points is the image under F of the classes of representations on which $\{f_1, \ldots, f_m\}$ take real values.

For stable representations $\rho \in \mathrm{Hom}(\Gamma, G)$, the following criterion applies (Johnson and Millson [1], Theorem 1.1):

THEOREM 9.3. *A representation* $\rho \in \mathrm{Hom}(\Gamma, G)$ *is stable if and only if the image* $\rho(\Gamma)$ *is not contained in any proper (complex) parabolic subgroup of the group* $\mathbf{G}$.

Here by the (complex) parabolic subgroup P of a semisimple Lie group G we mean the full normalizer of a parabolic subalgebra, i.e., an algebra whose complexification contains a maximal solvable subalgebra, see Varadarayan [1], 279-288.

We denote by $S = S(\Gamma)$ a set of all stable representations ρ from $\mathrm{Hom}(\Gamma, \mathbf{G})$; that set is Zariski open in $\mathrm{Hom}(\Gamma, \mathbf{G})$ (see Newstead [1], Proposition 3.8) and has the important property that the map F from (9.5) induces a homeomorphism from $S/\mathbf{G} \subset \mathrm{Hom}(\Gamma, \mathbf{G})/\mathbf{G}$ onto an open subset of the affine variety $X(\Gamma, \mathbf{G})$. Here the actions of the group $\mathbf{G}$ on S and G on $S \cap \mathrm{Hom}(\Gamma, \mathbf{G})$ are proper and admit analytic slices through any representation ρ (Johnson and Millson [1], Theorem 1.2); see Borel and Wallach [1], p.277, for the definition of a slice to a group action through a point.

In what follows, we are interested in so-called quasi-Fuchsian representations, which are obtained as follows.

Let Γ be a torsion-free group and $\mathbf{H}$ a classical simple algebraic group defined over $\mathbb{R}$ with real points H and let $\rho_0 : \Gamma \longrightarrow H$ be an embedding of Γ into H in which $\rho_0\Gamma$ is a uniform discrete subgroup or lattice, whose rigidity for H locally non-isomorphic to $PSL_2(\mathbb{R})$ follows from Mostow's theorem [2], compare Theorem 9.12 in §2. In addition, let us suppose that $\mathbf{H}$ is represented in another algebraic group $\mathbf{G}$ over $\mathbb{R}$ so that the representation $\mathbf{H} \longrightarrow \mathbf{G}$ is defined over $\mathbb{R}$ and the image of $\mathbf{H}$ is not contained in a proper parabolic subgroup of $\mathbf{G}$ (see Theorem 9.3). Then the composition of the representations $\mathbf{H} \longrightarrow \mathbf{G}$ with ρ_0 gives discrete subgroups in G and forms a deformation space $\mathrm{Hom}(\Gamma, \mathbf{G})$. Here it is found (see also §3) that this space is degenerate, apart from the case where the group H is locally isomorphic to $SO(n, 1)$ or $SU(n, 1)$. Namely, the G-orbit of the standard representation $\rho \in \mathrm{Hom}(\Gamma, G)$ is open in $\mathrm{Hom}(\Gamma, G)$ (see Johnson and Millson [1], Theorem 1.3).

We now put $\mathbf{H} = \mathbf{SO}(n, 1)$ and $\mathbf{G} = \mathbf{SO}(n+1, 1)$, in which the group $\mathbf{SO}(n, 1)$ consists of the complex points of the algebraic group of orientation-preserving isometries of the quadratic form f in $\mathbb{C}^{n+1}$ given by

$$f(x_1, \ldots, x_{n+1}) = -x_1^2 + x_2^2 + \ldots + x_{n+1}^2.$$

Then $SO(n, 1)$ consists of the real points of $\mathbf{SO}(n, 1)$. The embedding $\mathbf{SO}(n, 1) \subset \mathbf{SO}(n+1, 1)$ is defined as the stabilizer subgroup of the last standard basis vector.

DEFINITION 9.4. A representatiion $\rho \in \mathrm{Hom}(\Gamma, SO(n+1, 1))$ defines a group $\rho\Gamma \subset \mathrm{Isom}\, H^{n+1}$ acting on the sphere at infinity S^n of the hyperbolic space H^{n+1}. We call ρ a Fuchsian representatioin if it is conjugate to the representation

$$\rho_0 : \Gamma \longrightarrow SO(n, 1) \subset SO(n+1, 1), \tag{9.6}$$

where Γ is embedded by ρ_0 as a co-compact discrete subgroup of $SO(n, 1)$.

A representation $\rho \in \mathrm{Hom}(\Gamma, SO(n+1, 1))$ is said to be quasi-Fuchsian if the action of the group $\rho\Gamma$ on the sphere $S^n = \partial H^{n+1}$ is quasi-conformally conjugate to the action of the Fuchsian group $\rho_0\Gamma$. We denote by $R_n(\Gamma)$ the space of all quasi-Fuchsian representations in $\mathrm{Hom}(\Gamma, SO(n+1, 1))$ and by $T_n(\Gamma)$ we denote the space of conjugacy classes of quasi-Fuchsian representations:

$$T_n(\Gamma) = R_n(\Gamma)/SO(n+1, 1).$$

In §3, we shall give the proof for the non-triviality of the space $T_n(\Gamma)$.

Johnson and Millson [1] used the stability criterion (Theorem 9.3) and some results due to Lok [1] to prove the following major properties of the spaces $R_n(\Gamma)$ and $T_n(\Gamma)$, which we formulate as follows (see also §3):

THEOREM 9.5. (1) *The space* $R_n(\Gamma)$ *is open in* $\mathrm{Hom}(\Gamma, SO(n+1, 1))$ *and consists of stable representations that are good ones, apart from the Fuchsian ones (conjugate to* ρ_o*).*

(2) *The space* $T_n(\Gamma)$ *is an open subset of the real quasi-algebraic set defined by the image of* $\mathrm{Hom}(\Gamma, SO(n+1, 1))$ *in* $X(\Gamma, SO(n+1, 1))$ *by the projection* π *induced by the mapping* (9.5).

(3) $T_n(\Gamma)\backslash\pi(\rho_0)$ *is an open subset of the real algebraic set* $X(\Gamma, SO(n+1, 1))$.

(4) *If* n *is even, then* $T_n(\Gamma)$ *is an open subset of the real algebraic set* $X(\Gamma, SO(n+1, 1))$.

We note that here we mean by a quasi-algebraic subset in $\mathbb{R}^m$ a subset defined by polynomial equatioins and inequalities; for n odd, statement (4) in the theorem is in general incorrect, as noted above.

4. *Infinitesimal deformations and obstructions.* Let $\Delta_0 \in \mathcal{S}(M)$ be a certain (X, G)-structure on the manifold/orbifold M. By local deformation of Δ_0 we understand the one-parameter family of (X, G)-structures $\Delta_t \in \mathcal{S}(M)$, which can be described by a continuous one-parameter family of transition functions $\varphi_{ij} : (-\varepsilon, \varepsilon) \longrightarrow G$,

satisfying the cocycle conditioin relative to some open cover $\{U_i\}$ of M up to the group G (see (2.40)):

$$\varphi_{ij}\varphi_{jk} = \varphi_{ik}.$$

The local deformation $\{\Delta_t\}$ is trivial if and only if Δ_t define a single (X, G)-structure up to (X, G)-diffeomorphisms, see the definition of $\mathcal{S}(M)$ in §1.

To examine deformations of locally homogeneous marked structures in M, in the case of their stable holonomy representation $d_* \in \mathrm{Hom}(\pi_1(M), G)$, one naturally (due to the holonomy Theorem 9.1) transfers to deformations of representations from

$$\mathrm{Hom}(\Gamma, G), \quad \Gamma = \pi_1(M).$$

DEFINITION 9.6. As in the above subsection, let Γ be some group and G be a Lie group, while $\rho_0 \in \mathrm{Hom}(\Gamma, G)$ is a certain injective homomorphism having a discrete image $\rho_0\Gamma$. A mapping $R: W \longrightarrow \mathrm{Hom}(\Gamma, G)$ of a neighbourhood W of the origin in $\mathbb{R}^m$ is called a deformation of the representation ρ_o if $R(0) = \rho_0$, and for every $t \in W$, $R(t) = \rho_t$ is an injective homomorphism $\rho_t \in \mathrm{Hom}(\Gamma, G)$ with discrete image, while the mapping $t \longmapsto \rho_t\gamma$ is differentiable in W for all $\gamma \in \Gamma$ (or, if especially mentioned, continuous in W).

Two deformations $R: W \longrightarrow \mathrm{Hom}(\Gamma, G)$ and $R' : W' \longrightarrow \mathrm{Hom}(\Gamma, G)$ are called equivalent if there exists a diffeomorphism $\psi: W \longrightarrow W'$, $\psi(0) = 0$, and a differentiable mapping $\Psi: W \longrightarrow G$ such that $R'(\psi(t)) = \Psi(t)_*R(t)$; here for $h \in G$ by h_* we denote the induced inner automorphism of G, namely for all $g \in G$

$$h_* : g \longmapsto hgh^{-1}. \tag{9.7}$$

The deformation R is called *trivial* if it is equivalent to a constant deformation.

If Γ is a discrete subgroup of the Möbius group $\mathrm{Möb}(n-1)$ $(\mathrm{Isom}\, H^n)$, we shall consider deformations that are type-preserving. For that purpose, we can consider the quasi-conformal deformations R.

Namely, a family $\{f_t\}$ of quasi-conformal self-homeomorphisms of the sphere S^n (or the ball H^{n+1}) compatible with the group Γ:

$$f_t\gamma f_t^{-1} \in \mathrm{Möb}(n) \quad \text{for all} \quad \gamma \in \Gamma, \tag{9.8}$$

defines a quasi-conformal deformation R of the inclusion $\rho_0 : \Gamma \hookrightarrow \mathrm{Möb}(n-1) \subset \mathrm{Möb}(n)$, in which $R(t) = \rho_t = (f_t)_*$.

The following two results on quasi-conformal mappings show that this definition is correct.

THEOREM 9.7 (Gehring [1], Väisälä [1], and Reshetnyak [3]). *If a sequence of K-quasi-conformal homeomorphisms $f_i : X \longrightarrow Y_i$ of metric spaces converges pointwise to a homeomorphism $f: X \longrightarrow Y$, then f is a K-quasi-conformal mapping.*

THEOREM 9.8 (Gehring [1], Reshetnyak [1]). *If* $f: X \longrightarrow Y$ *is a 1-quasi-conformal homeomorphism of domains in* $\mathbb{R}^n$, $n \geq 3$, *then* f *is the restriction of a Möbius transformation.*

To introduce infinitesimal deformations, we begin with a defintion of Eilenberg-MacLane cohomology. Namely, let V be a vector space and $A: \Gamma \longrightarrow \mathrm{Aut}\, V$ a representation. Then a 1-cocycle on the group Γ with coefficients in A is a map $c: \Gamma \longrightarrow V$ such that for all $\gamma, \delta \in \Gamma$ we have

$$c(\gamma\delta) = c(\delta) + A(\delta)\cdot c(\gamma). \tag{9.9}$$

We let $Z^1(\Gamma, V)$ denote the space of 1-cocycles on Γ with values in V. The elements of $Z^1(\Gamma, V)$ are also called crossed-homomorphisms (with values in V). If here for the 1-cocycle c there exists $v \in V$ such that for all $\gamma \in \Gamma$

$$c(\gamma) = A(\gamma)\cdot v - v, \tag{9.10}$$

then c is called a 1-*coboundary*. The set of 1-coboundaries is denoted by $B^1(\Gamma, V)$.

The first Eilenberg-MacLane cohomology group of Γ with values in V is defined by

$$H^1(\Gamma, V) = Z^1(\Gamma, V)/B^1(\Gamma, V). \tag{9.11}$$

In the same way, one defines the cohomology groups $H^p(\Gamma, V)$ for $p \geq 2$, see Eilenberg and MacLane [1].

Now let there be a local deformation $R: I \longrightarrow \mathrm{Hom}(\Gamma, G)$ of a representation $\rho \in \mathrm{Hom}(\Gamma, G)$, in which I is an open interval containing the origin (see Definitiion 9.6), and $R(t) = \rho_t$. Differentiating the equality $\rho_t(\gamma\delta) = \rho_t(\gamma)\cdot\rho_t(\delta)$ we get

$$\left.\frac{d}{dt}\right|_{t=0} \rho_t(\gamma\delta) = \rho(\gamma)\cdot\left[\left.\frac{d}{dt}\right|_{t=0} \rho_t(\delta)\right] + \left[\left.\frac{d}{dt}\right|_{t=0} \rho_t(\gamma)\right]\cdot\rho(\delta). \tag{9.12}$$

This equation applies in the tangent space to G at the point $\rho(\gamma\delta)$, with the points on the right denoting correspondingly left and right translations. As each vector on G can be extended uniquely to a left-invariant vector field on G, we can follow Weil [1] and define an element $c(\gamma)$ of the Lie algebra $\mathcal{G}$ of the group G that is an extension of the tangent vector $\left.\frac{d}{dt}\right|_{t=0} \rho_t(\gamma)$. This and (9.12) give

$$c(\gamma\delta) = c(\delta) + \mathrm{Ad}\,\rho(\gamma)\cdot c(\gamma), \tag{9.13}$$

which defines a crossed homomorphism $c = c_R: R \longrightarrow \mathcal{G}$, twisted by the adjoint endomorphism $\mathrm{Ad}\,\rho$ of the representation ρ. If here R and

R' are conjugate deformations, then $c_R - c_{R'}$ is a principal crossed homomorphism, i.e., a 1-coboundary or, equivalently, tangent to a trivial deformation. This enables us to give the following definition (compare (9.9) - (9.11)).

DEFINITION 9.9. An infinitesimal deformation of the representation $\rho = R(o) : \Gamma \longrightarrow G$ is a cohomology class in the first cohomology group $H^1(\Gamma, \mathcal{Y})$.

When $G = \text{Möb}(n)$ (or equivalently $G = \text{Isom}\, H^{n+1}$), while $R(o)$ is the inclusion of a finitely generated discrete subgroup $\Gamma \subset \text{Möb}(n-1) \subset \text{Möb}(n)$, in that way we can define infinitesimal deformations corresponding to a quasi-conformal deformation of the group Γ, i.e., a deformation that is specified by a family $\{f_t\}$ of quasi-conformal automorphisms of the sphere $\overline{\mathbb{R}}^n$ (or the hyperbolic space H^{n+1}) compatible with the group Γ. For this purpose we introduce some definitions.

Let $F : X \times I \longrightarrow \mathbb{R}^n$, $F(x, o) \equiv x$, be a 1-parameter family of almost everywhere differentiable homeomorphisms of $X \subseteq \mathbb{R}^n$ onto a certain domain in $\mathbb{R}^n$. We assume additionally that the mapping $t \longmapsto F(x, t) = f_t(x)$ is differentiable in the origin for almost all $x \in X$ and define

$$\xi(x) = \left.\frac{d}{dt}\right|_{t=0} f_t(x). \tag{9.14}$$

Then $F(x, t) = x + t\xi(x) + o(t)$ and, if $\xi(x)$ is differentiable almost everywhere, then $D_1F(x, t) = E + tD\xi(x) + o(t)$ and the derivative of

$$|\det D_1F(x, t)|^{-2/n} \cdot (D_1F(x, t))^{\mathrm{T}} \cdot D_1F(x, t)$$

with respect to t at $t = 0$ is a double value of the operator

$$S_\xi = \frac{1}{2}[D\xi + (D\xi)^{\mathrm{T}}] - \frac{1}{n}\operatorname{tr} D\xi \cdot E,$$

in which $\det A$ denotes the determinant of a $n \times n$ matrix A, while A^{T} is the transposed matrix, tr is the trace, and E is the unit matrix, while a norm $\|A\|$ is defined as $(\operatorname{tr} A^{\mathrm{T}} \cdot A)^{1/2}$. The operator S_ξ was introduced by Ahlfors [4], [6], [7], [10] as a generalization of the derivative operator with respect to $\bar{z}$. In mechanics, this operator S_ξ is known as the deviator of the tensor of deformations, and the equation $S_\xi = 0$ is called the Liouville condition for conformality of $F(x, 0)$ - see J. Serrin [1]. The properties of S_ξ enable us to give the following definition (Ahlfors [10]):

A K-quasi-conformal vector field on a domain $X \subset \mathbb{R}^n$ is a continuous mapping $\xi : X \longrightarrow \mathbb{R}^n$ from the class $W^1_{1,loc}(X)$ (i.e., it has locally integrable first order distributional derivatives $D_j\xi_i$) such that the function $x \longmapsto \|S_\xi(x)\|$ is in $L_\infty(X)$ and $\|S_\xi\| \le K$ almost everywhere.

THEOREM 9.10 (Ahlfors [7], [10], Reiman [1], [2], and Semenov [1]). *Let* ξ *be a* K*-quasi-conformal vector field in* $\mathbb{R}^n$ *with compact support. Then there exists a unique 1-parameter group of homeomorphisms* F_t *generated by the field* ξ, *which satisfies*

$$F_t(x) = x + \int_0^t \xi(F_s(x))ds,$$

and the homeomorphisms F_t *are* $e^{K\cdot |t|}$*-quasi-conformal ones.*

We now assume that a family $F(x, t) = f_t(x)$ of quasi-conformal self-homeomorphisms of the $(n+1)$-dimensional ball H^{n+1} is compatible with a discrete group $\Gamma \subset \text{Möb}(n)$ and defines a deformation $R = (f_t)$, $R : t \longmapsto (f_t)_*$. The group $\text{Möb}(n)$ acts in H^{n+1} as the isometry group and one can therefore define its action on the set of continuous vector fields η in H^{n+1} by a rule:

$$\gamma\cdot\eta(x) = D\gamma(\gamma^{-1}(x))\cdot\eta(\gamma^{-1}(x)). \tag{9.15}$$

Now let the vector field ξ in H^{n+1} be given in (9.14) with respect to the family $\{f_t\}$; by $\mathfrak{M}_n$ and $\mathcal{H}_{n+1}$ we denote isomorphic Lie algebras correspondingly for the group $\text{Möb}(n)$ of the Möbius automorphisms of the sphere $\overline{\mathbb{R}}^n$ and the isometry group $\text{Isom}\, H^{n+1}$ of the hyperbolic space. Then the compatibility of the family $\{f_t\}$ with the group Γ implies that a vector field $c(\gamma)$ such that for all $x \in H^{n+1}$ we have

$$c(\gamma)(x) = \xi(x) - \gamma\cdot\xi(x) \tag{9.16}$$

is compactible with Γ and defines a class $c : \Gamma \longrightarrow \mathcal{H}_{n+1}$ of the cohomology group $H^1(\Gamma, \mathcal{H}^{n+1})$, and thus a class of $H^1(\Gamma, \mathfrak{M}_n)$. In fact we have for all $\gamma \in \Gamma$, $x \in h$, $x \in H^{n+1}$:

$$c(\gamma)(x) = \xi(x) - \gamma\cdot\xi(x) = \xi(x) - D\gamma(\gamma^{-1}(x))\cdot\xi(\gamma^{-1}(x)) =$$

$$= D_2F(x, o)\cdot E + D_1F(x, o)\cdot D\gamma(\gamma^{-1}(x))\cdot(-D_2F(\gamma^{-1}(x), o)\cdot E =$$

$$= \frac{d}{dt}\bigg|_{t=0} (f_t \circ \gamma \circ f_t^{-1} \circ \gamma^{-1})(x)$$

and therefore $c(\gamma)$ is an infinitesimal isometry of the space H^{n+1}.

We have seen in subsection 2 above that the map $\text{hol} : \mathcal{S}(M) \longrightarrow \text{Hom}(\pi_1(M), G)/G$ is not in general a local homeomorphism, but for a (X, G)-structure $\Delta \in \mathcal{S}(M)$ whose image $\rho \in \text{hol}(\Delta)$ is a stable representation, there exist a neighbourhood $U \subset \mathcal{S}(M)$ of the structure Δ, a neighbourhood $V \subset \text{Hom}(\pi_1(M), G)/G$ of the representation ρ, and finite stabilizer subgroups H_1 and H_2 of Δ and ρ correspondingly ($H_1 \subset H_2$ in the natural sense) that specify finite coverings $\widetilde{U}$ and $\widetilde{V}$ ($U = \widetilde{U}/H_1$, $V = \widetilde{V}/H_2$) of the neighbourhoods U and V such that hol is lifted to a local homeomorphism

$\widetilde{\text{hol}}: \widetilde{U} \longrightarrow \widetilde{V}$ (Johnson and Millson [1]). In particular, when the representation ρ is good (Definition 9.2) the map hol itself is a local homeomorphism.

In that situation, the infinitesimal deformations $H^1(\Gamma, \mathcal{Y})$ are identified with those of the (X, G)-structure on M. In §3, we shall prove the non-triviality of the space $H^1(\Gamma, \mathfrak{M}_n)$, $n \geq 3$.

However, one naturally expects that not every infinitesimal deformation will correspond to a certain deformation of a representation $\rho \in \text{Hom}(\Gamma, G)/G$ (or a structure $\Delta \in \mathcal{S}(M)$), i.e., not every infinitesimal deformation is integrable. The first obstruction to that integration is the second Eilenberg-MacLane cohomology group $H^2(\Gamma, \mathcal{Y})$, in which $\mathcal{Y}$ is the Lie algebra of the group G. To formulate this exactly, we use the Kolmogorov-Alexander product (it is equal to the cup-product which is a well-defined map from $H^1(\Gamma, \mathcal{Y}) \otimes H^1(\Gamma, \mathcal{Y})$ to $H^2(\Gamma, \mathcal{Y})$ - see Dold [1], Chapter 7 §8) to define, for any element $c \in Z^1(\Gamma, \mathcal{Y})$ that specifies an infinitesimal deformation for a curve (ρ_t) in $\text{Hom}(\Gamma, G)$, $\rho_0 = \rho$, the following 2-cocycle $c \cup c \in Z^2(\Gamma, \mathcal{Y})$ (the cup-square of c) that is defined by values in any $\gamma, \delta \in \Gamma$:

$$c \cup c(\gamma, \delta) = [c(\gamma), \text{Ad}\,\rho(\gamma) \cdot c(\delta)], \tag{9.17}$$

where $[*, *]$ on the right denotes the Lie bracket in $\mathcal{Y}$.

Then the necessary condition for integrability can be formulated as follows (see Goldman and Millson [1], Lemma 2.4, and Johnson and Millson [1], Proposition 2.1).

THEOREM 9.11. *If $c \in Z^1(\Gamma, \mathcal{Y})$ defines the infinitesimal deformation for a curve (ρ_t) in $\text{Hom}(\Gamma, G)/G$ and ρ_0 is good, then $c \cup c$ is the zero element in the second cohomology group $H^2(\Gamma, \mathcal{Y})$.*

That obstruction to integration in the form of $H^2(\Gamma, \mathcal{Y})$ is the first of an infinite sequence of such obstructions to the existence of a deformation (ρ_t) tangent to c. The construction of them follows the general scheme of the Kodaira-Spencer deformation theory. We shall return to these topics in §3.5 in studies on deformation space of conformal and hyperbolic structures that are specified by deformations of the corresponding discrete Möbius groups.

§2. Rigidity theorems for complete hyperbolic structures and isometry groups

1. *Finite hyperbolic-volume case.* Mostow's rigidity theorem gives a completely accurate description of the space $\mathcal{T}(M)$ of hyperbolic structures on an n-dimensional $(n \geq 3)$ closed manifold/orbifold M (or $\text{vol}\, M < \infty$), according to which $\mathcal{T}(M)$ consists of one point:

THEOREM 9.12 (algebraic form). *Let G_1 and $G_2 \subset \text{Isom}\, H^n$ be* *discrete isometry groups in the hyperbolic space* H^n, $n \geq 3$, *having* $\operatorname{vol} H^n/G_i < \infty$, $i = 1, 2$, *and let* $\varphi : G_1 \to G_2$ *be an isomorphism. Then* G_1 *and* G_2 *are conjugate in* $\operatorname{Isom} H_n$.

As a complete hyperbolic manifold M has the space H^n as its universal covering, in which the fundamental group $\pi_1(M)$ acts via holonomy as a discrete group, this theorem may be reformulated as:

THEOREM 9.13 (geometric form). *If* M_1 *and* M_2 *are complete hyperbolic* n*-dimensional* $(n \geq 3)$ *manifolds/orbifolds of finite volume,* $\operatorname{vol} M_1 < \infty$, $i = 1, 2$, *then any isomorphism of the fundamental groups* $\varphi : \pi_1(M_1) \to \pi_1(M_2)$ *is induced by a unique isometry* $M_1 \to M_2$ *(up to an inner automorphism).*

As hyperbolic manifolds are all of the type $K(\pi, 1)$, since they are universally covered by H^n, and for such manifolds homotopic equivalence is equivalent to the isomorphism of the fundamental groups, we have:

COROLLARY 9.14. *If* M_1 *and* M_2 *are complete hyperbolic manifolds of finite volume, then they are homeomorphic if and only if they are homotopically equivalent.*

COROLLARY 9.15. *If* M *is a complete hyperbolic* n*-dimensional manifold* $(n \geq 3)$ *of finite volume* $\operatorname{vol} M < \infty$, *the outer automorphism group* $\operatorname{Out}(\pi_1(M)) = \operatorname{Aut}(\pi_1(M))/\operatorname{Inn}(\pi_1(M))$ *is finite and is isomorphic to the isometry group of* M.

Note that this result is not true in dimension $n = 2$: here $\operatorname{Out} \pi_1(M)$ is always an infinite group.

The proof of the rigidity theorem was first obtained by Mostow [1] for the isomorphisms $\pi_1(M_1) \to \pi_1(M_2)$ defined by quasi-conformal homeomorphisms. Prasad [1] gave a complete proof for non-compact manifolds. Mostow [2] proved the above formulations of Theorems 9.12 and 9.13 (as extensions to locally symmetric spaces), although a proof had been given by Margulis [1] for closed manifolds. In addition to those proofs, there are geometric proofs due to Gromov and Thurston [1] and proofs based on examining the action of the fundamental group (via holonomy) on the absolute of the hyperbolic space coinciding with the limit set of the action: Sullivan [1], and Tukia [7], [8]. Margulis's [7] super-rigidity theorem is of different type (see the exposition by Zimmer [1]) for lattices in a semisimple Lie group (cf. also Goldman and Millson [1]).

However, we show below that the class of hyperbolic manifolds for which the rigidity theorem applies does not consist solely of manifolds/orbifolds of finite volume, i.e., manifolds with geometrically finite structures.

To explain the theorem more clearly, we examine the main stages of proofs.

DEFINITION 9.16. A continuous mapping $f : X \to X'$ of the metric spaces (X, d) and (X', d') is called a pseudo-isometry if there exist positive constants C_1 and C_2 such that

$$d'(f(x),\ f(y)) \le C_1 d(x,\ y) \quad \text{for all} \quad x,\ y \in X;$$
$$d'(f(x),\ f(y)) \ge d(x,\ y)/C_1 \quad \text{for all} \quad x,\ y \in X, \quad d(x,\ y) \ge C_2. \tag{9.18}$$

A quasi-conformal homeomorphism is a particular case of pseudo-isometry for the hyperbolic space H^n, $n \ge 3$.

Let some fundamental-group isomorphism $\pi_1(M) \longrightarrow \pi_1(M')$ be induced by a psuedo-isometry; without mentioning what that pseudo-isometry is, we give the following definition of deformation rigidity.

DEFINITION 9.17. A complete hyperbolic n-dimensional orbifold $M = H^n/G$, $G \subset \text{Isom}\, H^n$, is called *rigid* in the sense of Mostow if any pseudo-isometry between M and another complete hyperbolic orbifold M' is homotopic to an isometry.

DEFINITION 9.18. A discrete group G of isometries of H^n is called *rigid* in the sense of Mostow if any isomorphism of it onto a discrete group G' of isometries of H^n induced by a pseudo-isometry of H^n is inner in the group $\text{Isom}\, H^n$, i.e., is a conjugation by isometry.

Examining the proof, we note first that, since complete hyperbolic manifolds are $K(\pi,\ 1)$-manifolds $(\pi_i(M) = 0,\ \ i \ge 2)$, then the isomorphism $\pi_1(M) \longrightarrow \pi_1(M')$ is induced by a homotopy equivalence, i.e., there exist continuous mappings $f_1 : M \longrightarrow M'$ and $f_2 : M' \longrightarrow M$ such that $f_1 f_2$ and $f_2 f_1$ are homotopic to identity mappings. One can pass from the mappings f_i to their liftings $\tilde{f}_i$ in the universal covering H^n having the same property.

The first step in proving Theorems 9.12 and 9.13 is that the mappings $\tilde{f}_1$ and $\tilde{f}_2$ that induce a group isomorphism can be taken as pseudo-isometries.

The second step is that one can extend the pseudo-isometries $\tilde{f}_i : H^n \longrightarrow H^n$ to the sphere at infinity $S^{n-1} = \partial H^n$. This profound result was obtained (in a more general form) by Efremovich and Tikhomirova [1]. For $n \ge 3$, the extended mappings in the sphere ∂H^n are quasi-conformal homeomorphisms (Gehring [1], [2]).

The third step consists of proving that there is only one measurable conformal structure in the sphere ∂H^n (up to a set of measure zero) invariant with respect to the discrete group Γ of isometries of H^n, $n \ge 3$, if the group Γ has a finite-volume fundamental polyhedron (i.e., Γ is a co-finite group).

The fourth (last) step is given by the fact that a quasi-conformal automorphism of the sphere S^{n-1} having dilatation almost everywhere equal to one is a Möbius transformation. The proof of that fact is given in final form by Reshetnyak [2], [4], see Theorem 9.8.

2. *Infinite-volume hyperbolic structures.* Examining the above proof scheme, Mostow [2], and independently Krushkal' [3] - [5] using other arguments (see Chapter 3 §5), conjectured that the rigidity could be extended to the class of complete hyperbolic manifolds M without boundary (no matter what their volume) or, what is equivalent, to the

class of discrete isometry groups in H^n whose action on ∂H^n is not discontinuous, i.e., $\partial H^n = \Lambda(G)$. Naturally, the basic point here is a possibility of extending the property formulated above in the third step: uniqueness of the conformal structure in the sphere ∂H^n compatible with the group.

The following result (Apanasov [7], [11]) decides that question negatively in the general case.

THEOREM 9.19. *The property of deformation rigidity is not possessed by the class of discrete isometry groups in the n-dimensional hyperbolic space* H^n, $n \geq 3$, *having* ∂H^n *as the limit set.*

PROOF. It is sufficient to construct a group G from that class together with its non-trivial quasi-conformal deformation (f_t), $t \in I$, i.e., a family of quasi-conformal automorphisms f_t of the space H^n such that the groups $G_t = f_t G f_t^{-1} \subset \mathrm{Isom}\, H^n$ and G are not conjugate in the isometry group of H^n. We proceed to such a construction. Let

$$Q = \{x \in \mathbb{R}^{n-1} : |x_i| < 1/2\} \tag{9.19}$$

be a unit cube in $\mathbb{R}^{n-1}$ and $\{k_m\}$ an increasing sequence of odd integers. Split up Q into k_1^{n-1} equal cubes Q_{1i} and cast out the interior of the central cube. Similarly, in the second step, we split up each of the remaining cubes Q_{1i} into k_2^{n-1} equal cubes Q_{2j} and cast out the interiors of the central ones. Continuing this process infinitely and choosing the numbers k_m suitably (e.g., $k_m = 3^m$), we make the $(n-1)$-dimensional Lebesgue measure $m_{n-1}(K)$ of a continuum K remaining at the limit in the cube (7.42) any number between zero and one (see, for example, Hausdorff [1]).

Consider the set Δ_1, the projection of the set $\Delta = Q \setminus K$ to the first coordinate axis, and specify the mapping $f_t : Q \times \mathbb{R} \longrightarrow \mathbb{R}^n$, $t > 0$, by the formulae

$$\begin{gathered} f_t(x_1, \ldots, x_n) = (\psi_t(x_1), x_2, \ldots, x_n), \\ \psi_t(x_1) = \int_0^{x_1} \varphi_t(y)\,dy, \end{gathered} \tag{9.20}$$

where the measurable function $\varphi_t(y)$ is

$$\varphi_t(y) = \begin{cases} 1, & \text{if } y \in \Delta_1; \\ t, & \text{if } y \in [-\frac{1}{2}, \frac{1}{2}] \setminus \Delta_1. \end{cases} \tag{9.21}$$

It is clear that f_t is a quasi-conformal mapping, and its quasi-conformality coefficient (dilatation) is t for $t \geq 1$ and $1/t$ for $0 < t \leq 1$, and transforms $Q \times \mathbb{R}$ to the product

$$Q_{t_1} \times \mathbb{R} = f_t(Q \times \mathbb{R}) = \{x \in \mathbb{R}^n : |x_1| \leq t_1/2, \quad |x_i| \leq 1/2, \quad i \neq 1, n\}, \tag{9.22}$$

obtained from $Q \times \mathbb{R}$ by stretching along the $\mathbb{R}e_1$ axis by a factor t_1 (t_1 is dependent only on t, with $t_1 = 1$ only for $t = 1$). The faces of $Q \times \mathbb{R}$ go over to the faces of (9.22) and the mapping f_t from (9.20) is conformal on the set $\Delta \times \mathbb{R}$ (translation by a vector whose length is dependent on a component $Q_{ij} \subset \Delta$, see Fig. 77).

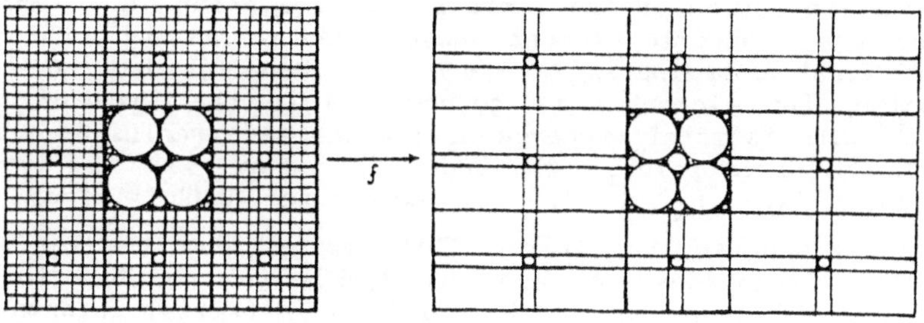

Figure 77

To construct the discrete groups G and G_t from our class, we consider a system $\{B_i\}$ of closed disjoint n-dimensional balls in $\mathbb{R}^n$ having centres in $\mathbb{R}^{n-1}$, $B_i \cap \mathbb{R}^{n-1} \subset \Delta$, and forming a dense subset of Δ. That family on the mapping f_t goes over to a family of balls $\{B_i^t\}$ having analogous properties. We then define the groups Γ and Γ_t as the groups generated by reflections with respect to the spheres $S_i = \partial B_i$ and correspondingly $S_i^t = \partial B_i^t$. We also define free Abelian groups H and H_t of rank $n-1$ whose generators are translations by the coordinate vectors $e_1, e_2, \ldots, e_{n-1}$ and correspondingly $t_1 e_1, e_2, \ldots, e_{n-1}$. Then the groups G and $G_t \subset \operatorname{Isom} H^n$ from our class are defined as the free products

$$G = \Gamma * H, \quad G_t = \Gamma_t * H_t. \tag{9.23}$$

The construction shows that G and G_t are discrete and act in $\mathbb{R}^n_+ \cong H^n$ as hyperbolic isometry groups, and $\overline{\mathbb{R}}^{n-1} = \partial H^n = \Lambda(G) = \Lambda(G_t)$; (9.23) also implies that, as fundamental polyhedra $P(G)$ and $P(G_t) \subset \overline{\mathbb{R}}^n$, we can take the polyhedra:

$$P(G) = (Q \times \mathbb{R}) \setminus \overline{\bigcup_i B_i}, \quad P(G_t) = (Q_{t_1} \times \mathbb{R}) \setminus \overline{\bigcup_i B_i^t}. \tag{9.24}$$

Here f_t transfers $P(G)$ to $P(G_t)$, with all the sides corresponding. We extend f_t to the mapping $\Phi_t : \overline{\mathbb{R}}^n \to \overline{\mathbb{R}}^n$ (on account of $\overline{\mathbb{R}}^{n-1}$ being removable for it) by the rule

$$\begin{aligned} &\Phi_t(x) = f_t(x) \quad \text{for} \quad x \in \overline{P(G)}, \\ &\Phi_t \circ g(x) = g_t \circ \Phi_t(x), \quad g \in G, \quad g_t \in G_t, \end{aligned} \tag{9.25}$$

in which g_t corresponds to g in the $G \to G_t$ isomorphism. Φ_t is clearly a t-quasi-conformal mapping for $t \ge 1$ (or a $1/t$-quasi-conformal one for $0 < t \le 1$), and, due to (9.25), is compatible with the action of the group $G = G_t : \Phi_t G \Phi_t^{-1} \subset \text{Möb}(n)$. We thus have a family of quasi-conformal homeomorphisms that specifies a deformation Φ of the group G.

It follows that Φ is a non-trivial deformation when one consideres its action on the limit set of the group G. Namely, let $g \in G$ be any loxodromic element whose fixed points x and y have distinct first coordinates: $x_1 \ne y_1$. Then in the $G \longrightarrow G_t = \Phi_t G \Phi_t^{-1}$ isomorphism, the element g corresponds to the loxodromic element $g_t = \Phi_t g \Phi_t^{-1}$ with isometric spheres having the same radius as that of g and with the fixed points x^t and y^t, where $x_i^t = x_i$, $y_i^t = y_i$ for $i \ne 1$ and $|x_1 - y_1| \gtrless |x_1^t - y_1^t|$ in accordance with whether t is less than or greater than one. This implies that the length of the hyperbolic translations g and $g_t = \Phi_t g \Phi_t^{-1}$ (in the hyperbolic space $H^n = \mathbb{R}_+^n$) are different for $t \ne 1$. But for a trivial deformation, the lengths of hyperbolic translations are invariant. This completes the proof that the quasi-conformal deformation $\{\Phi_t\}$ of the group G is non-trivial and thus proves the theorem.

REMARK 9.20. In our arguments, the main part has been played by the property that the $(n-1)$-dimensional measure of the set of limit vertices $m_{n-1}(\overline{P(G)} \cap \mathbb{R}^{n-1})$ of the polyhedron $P(G)$ is positive. We shall see below that this is the defining property of non-rigid groups in H^n, $n \ge 3$.

REMARK 9.21. The construction in the proof of Theorem 9.19 implies a simple solution (see Apanasov [11]) to another problem stated by Bers [6] and Krushkal' [3] - [5] (compare subsection 3 in this section):

> whether there can be Beltrami G-differentials $\mu(z)dz/d\overline{z}$, with supports in $\Lambda(G) \subset \mathbb{C}$, $m_2(\Lambda(G)) > 0$, or whether any quasi-conformal deformation of a Kleinian group $G \subset \text{Möb}(2)$ should be conformal in $\Lambda(G)$.

To obtain a positive solution to this problem for Kleinian groups $G \subset \text{Möb}(n-1)$, $n \ge 3$, and in particular for groups in the plane, it is sufficient to take a Kleinian group $G_0 \subset \text{Möb}(n-1)$ and quasi-conformal mappings $\Phi_t^0 : \overline{\mathbb{R}}^{n-1} \longrightarrow \overline{\mathbb{R}}^{n-1}$, $t \in \mathbb{R}_+$, obtained from the group G and the mapping Φ_t in the proof of Theorem 9.19 as follows.

In the definition of G and of Φ_t, we exclude a certain finite subset $\{B_{i_j}\}$ from the system of balls $\{B_i\}$. We then denote the group then constructed and its quasi-conformal deformation by $\tilde{G}$ and $\tilde{\Phi}_t$. Then the family of quasi-conformal mappings

$$\Phi_t^0 = \tilde{\Phi}_t\Big|_{\overline{\mathbb{R}}^{n-1}} : \overline{\mathbb{R}}^{n-1} \longrightarrow \overline{\mathbb{R}}^{n-1} \tag{9.26}$$

is compatible with the action of the orientation-preserving group $G_0 \subset \mathcal{M}_{n-1}$ which is the restriction on $\overline{\mathbb{R}}^{n-1}$ of the action of the two-index subgroup of the group $\widetilde{G} \subset \text{Möb}(n)$. It is readily seen from this construction that G_0 is a Kleinian group in the $(n-1)$-dimensional sphere, while Φ_t^0, $t \in \mathbb{R}_+$, specifies a non-trivial quasi-conformal deformation of G_0 conformal on the whole discontinuity set $\Omega(G_0) \subset \overline{\mathbb{R}}^{n-1}$ of G_0.

In particular, if $n=3$, we get the Kleinian group $G_0 \subset \mathcal{M}_2$ in the complex plane $\overline{\mathbb{C}}$ together with the non-trivial deformation that specifies the family of quasi-conformal mappings Φ_t^0, $t \in \mathbb{R}_+$. The complex characteristics $\mu^t(z)$ of Φ_t^0 correspond to Beltrami differentials $\mu^t(z)dz/d\bar{z}$, invariant with respect to G_o, whose supports are contained in $\Lambda(G_0)$, $m_2(\Lambda(G_0)) > 0$.

As in Theorem 9.19, the decisive part in this effect is played by the property $m_{n-1}(\Lambda(G_0) \cap P(\widetilde{G})) > 0$ for $G_0 \subset \text{Möb}(n-1)$, where $P(\widetilde{G}) \subset H^n$ is the fundamental polyhedron of it.

We now transfer from the general class of complete hyperbolic manifolds without boundaries, whose fundamental groups (via holonomy d_*) act non-discontinuously in ∂H^n to its subclass of manifolds M whose fundamental groups via holonomy d_* have fundamental polyhedra in H_n with limit-vertex set in ∂H^n having zero $(n-1)$-dimensional spherical Lebesgue measure.

If we put $M(r) = \{x \in M : d(x, 0) \le r\}$ and let $D(r) \subset H^n$ denote a ball of radius $r > 0$, Theorem 6.68 shows that this subclass of such manifolds is characterized by any of the following conditions:

1) $\lim_{r\to\infty} \text{vol}\, M(r)/\text{vol}\, D(r) = 0$;
2) the action of $G = d_*(\pi_1(M))$ on the sphere ∂H^n is conservative; and
3) the horospheric limit set of the holonomy group $G = d_*(\pi_1(M))$ has the complete measure in ∂H^n.

Sullivan [1] gave a remarkable solution to the rigidity problem in this manifold class from his description of the action of a discrete hyperbolic isometry group in the sphere at infinity ∂H^n (conservativeness and dissipativeness, see Chapter 6 §6). The combination of his result with Theorem 9.19 gives a complete picture of the general case $(\Lambda(G) = \partial H^n)$.

THEOREM 9.22. *Let M be a complete hyperbolic n-manifold, $n \ge 3$, which satisfies one of the following equivalent conditions:*

1) *a fundamental polyhedron of the discrete holonomy group $G = d_*(\pi_1(M)) \subset \text{Isom}\, H^n$ has zero measure in ∂H^n;*

2) $\lim_{r\to\infty} \text{vol}\, M(r)/\text{vol}\, D(r) = 0$;

3) *the action of the holonomy group G on ∂H^n is conservative; and*

4) *the horospheric limit set of the holonomy group G has the complete measure in ∂H^n.*

Then the manifold M and the holonomy group G are rigid in the sense of Mostow.

Before the proof we make some remark.

REMARK 9.23. We have set aside the case $n=2$, which has some special properties caused by a quasi-conformal automorphism of the 2-disc in general not being extensible to an absolutely continuous homeomorphism of the boundary circle (in contrast to the spatial situation). In that case, one can give examples of discrete groups $G \subset \mathrm{Isom}\, H^2$ whose fundamental polygon has zero limit-vertex measure but which does not have deformation rigidity because for such groups one can construct automorphic (and even bounded) functions in ∂H^2. In 1918-1921 V. Golubev ([11], pp 322-351 and 401-403) constructed such bounded automorphic functions for discrete groups of the following types:

1. for a group $G \subset \mathrm{Isom}\, H^2$ having fundamental polygon $P = P(G) \subset H^2$ whose set of limit vertices $P^* = \overline{P} \cap \Lambda(G)$ has a positive 1-measure;

2. for a group $G \subset \mathrm{Isom}\, H^2$ for which a fundamental polygon $P(G)$ has the limit-vertex set with continuum cardinality and zero 1-measure; and

3. for a group $G \subset \mathrm{Isom}\, H^2$ for which the limit-vertex set P^* of a fundamental polygon $P(G)$ is countable.

PROOF OF THEOREM 9.22. This is obtained by replacing step 3 in the proof of Mostow's rigidity theorem by the following one.

THEOREM 9.24. *Let $G \subset \mathrm{Möb}(n)$, $n \geq 2$, be a discrete group and let ν be a measurable conformal structure in S^n (on the tangent space) invariant almost everywhere with respect to G. Then ν agrees almost everywhere with the standard conformal structure on the conservative part of the action of the group G on the sphere S^n.*

Sullivan [1] gave a detailed proof of this theorem for $n = 2$ on the basis that the conservative part of the action of G on S^2 does not contain a measurable almost everywhere G-invariant tangent vector field. When one compares a conformal structure ν on S^n, $n \geq 3$, with the standard conformal structure, one gets almost everywhere a field of ellipsoids, each defined up to similarity. We can work on an invariant set of positive measure where the smallest axes form a proper subspace of constant dimension. We find then an invariant tangent k-plane field invariant by G, $k < n$. This shows that Theorem 9.24 for $n \geq 3$ is reduced to the following.

THEOREM 9.25. *Let $G \subset \mathrm{Mob}(n)$, $n \geq 2$, be a discrete group. Then the conservative part of the action of G on S^n does not have a measurable tangent field of k-dimensional planes $(1 \leq k \leq n-1)$, that is G-invariant almost everywhere.*

PROOF. The proof of this theorem splits up into two cases: the infinite solid angle $\sum_G \lambda^{-n}$ of the G-orbit of a ball in H^{n+1} and the finite solid angle $\sum_G \lambda^{-n} < \infty$.

In the first case, $\sum_G \lambda^{-n} = \infty$, the proof follows from Sullivan's [1] theorem on the ergodicity of the action of G on geodesics in H^{n+1} (Theorem 6.50): if $\varphi(x, y) = \varphi(g(x), g(y))$ for a measurable function $\varphi(x, y)$ in $S^n \times S^n$ and for all $g \in G$, then φ is a constant almost everywhere. In fact, if the conservative part in S^n had a tangent k-plane field almost everywhere G-invariant for almost all pairs of points on S^n, we would have a measurable angle difference function (angles between k-planes measured along the geodesic connecting those points in H^{n+1}). The ergodicity implies that this function is almost everywhere constant, so consequently we have a continuous tangent k-plane field that is G-invariant. This is obviously impossible.

In the second case, $\sum_G \lambda^{-n} < \infty$, Sullivan's [1] proof for $n=2$ does not apply completely for $n \geq 3$, which relates in particular to the inapplicability for $n \geq 3$ of his Lemma 6 on a variation of $\log g'(z)$ for $g \in G$, $z \in \mathbb{C}$ (its real and imaginary parts). However, we shall extend the essentially geometric ideas on which Sullivan's proof is based to give a detailed proof of the theorem in this case too.

Namely, let $\sum_G \lambda^{-n} < \infty$. Here, as in Chapter 6 §6, we assume that we decompose each $g \in \text{Möb}(n)$ into a superposition of a hyperbolic translation with antipodal (poles of S^n) fixed points and an orthogonal transformation in S^n. Then the linear derivative d_g of a mapping g in the spherical metric varies in an interval $[1/\lambda, \lambda]$ with $\lambda \geq 1$, and is constant on concentric spheres $S_\theta^{n-1} \subset S^n$ (interpolating between these fixed points of the hyperbolic translation), which are parametrized by the angle θ, $0 \leq \theta \leq \pi$, in which $\theta = 0$ corresponds to the point at which $d_g = \lambda$.

We now consider a sufficiently large $k > 1$ and for all $g \in G$ having $\lambda \geq k^2$ we define a spherical layer A_g (concentric annuli) consisting of points from S^n lying on the spheres S_θ^{n-1} with parameter θ from the interval $\pi k/\lambda \leq \theta \leq \pi k$. The complement $S^n \backslash A_g$ is the union of two components: the smaller one E_g, in which $0 \leq \theta \leq \pi k/\lambda$, and the larger one C_g, in which $\pi K < \theta \leq \pi k$. We define a continuous piecewise approximation D_g for the linear derivative d_g by putting

$$D_g(x) = \begin{cases} \lambda/k^2, & \text{if } x \in E_g; \\ \pi^2/\lambda\theta^2, & \text{if } x \in A_g \cap S_\theta^{n-1}; \\ k^2/\lambda, & \text{if } x \in C_g. \end{cases} \tag{9.27}$$

Subsequently, some measurements will be made in the Euclidean space $\mathbb{R}^n$. We demonstrate what happens for the mapping $\gamma : x \longmapsto \lambda x$. For this we transfer from $\overline{\mathbb{R}}^n = \mathbb{R}^n \cup \{\infty\}$ to the sphere S^n of diameter one (by stereographic projection). Then, in $\overline{\mathbb{R}}^n$, the spheric metric is specified in the form $ds = |dx|/(1 + |x|^2)$. Therefore, the linear derivative is $d_\gamma(x) = \lambda(1 + |x|^2)/(1 + |\lambda x|^2)$. The domain E_γ corresponds to the ball $\{x \in \mathbb{R}^n : |x| \leq k/\lambda\}$, in which $d_\gamma(x)$ varies in the interval

$$[\lambda(1+k^2/\lambda)/(1+k^2),\ \lambda] \subset \lambda[1/(1+k^2),\ 1]. \tag{9.28}$$

The domain C_γ is $C_\gamma = \{x \in \overline{\mathbb{R}}^n : 1/k \le |x|\}$. Here $d_\gamma(x)$ varies in the interval

$$[1/\lambda,\ \lambda(1+1/k^2)/(1+\lambda^2k^2)] \subset 1/\lambda\cdot[1,\ 1+k^2]. \tag{9.29}$$

In the annulus $A_\gamma = \{x \in \mathbb{R}^n : k/\lambda \le |x| \le 1/k\}$, the ratio $d_\gamma(x)/D_\gamma(x) = d_\gamma(x)\cdot\lambda|x|^2$ varies in the interval

$$[|\lambda x|^2/(1+|\lambda x|^2), \lambda(1+1/k^2)\cdot\lambda|x|^2/|\lambda x|^2] \subset$$
$$\subset [1/(1+1/k^2),\ 1+1/k^2]. \tag{9.30}$$

We use (9.28)-(9.30) with transfer to the parameter θ in S^n to see that for all $g \in G$ having $\lambda \ge k^2$, the ratio $d_g(x)/D_g(x)$ is uniformly bounded in the domains E_g and C_g, while in A_g it is arbitrarily close to one for sufficiently large k. Then we get that there are parameters b_k and c_k that are dependent only on k, such that $\lim_{k\to\infty} c_k = 1$, and

$$\begin{aligned} &\operatorname{vol} E_g \le b_k/\lambda^n, \quad \operatorname{vol} g(C_g) \le b_k/\lambda^n, \\ &\operatorname{diam} A_g \le 2c_k/k, \quad \operatorname{diam} g(A_g) \le 2c_k/k. \end{aligned} \tag{9.31}$$

The last bound in (9.31) follows from the fact that the spheres $S_\theta^{n-1} \subset A_g$ pass on the mapping g to spheres of the same size (up to a factor near 1 for k large) as the spheres corresponding to the angular parameter $\theta_1 = \pi^2/\lambda\theta$. As $\pi k/\lambda \le \theta \le \pi/k$ for the annulus A_g, θ_1 lies in the interval $[\pi k/\lambda,\ \pi/k]$ whose size determines the number c_k.

Now let K denote a set in S^n, where the discrete group G acts conservatively: for any $X \subset K$, $\mu_n(X) > 0$, there exist infinitely many elements $g \in G$ such that $\mu_n(X \cap g(X)) > 0$. The logarithm of the ratio of radii for the boundary spheres is taken by definition as the modulus of a concentric annulus. Then:

LEMMA 9.26. *Let a set* $X \subset K$, $\mu_n(X) > 0$ *and two positive constants* δ *and* Δ *be given. Then for a discrete group* $G \subset \text{Möb}(n)$, $\sum_G \lambda^{-n} < \infty$, *there exist a point* x, *an element* $g \in G$, *and a concentric annulus* A *such that*

1) $x \in X \cap A$, $g(x) \in X$;
2) A *and* $g(A)$ *have diameters less than* δ;
3) A *and* $g(A)$ *have moduli as close as we like to* Δ; *and*
4) *the linear derivative* $d_g(x)$ *is constant on the concentric spheres of* A *and varies in* A *by an amount as close as we like to* $\exp(2\Delta)$.

PROOF. Fix k so large that $2c_k/k < \delta$ in (9.31). Using the discreteness of G we remove finitely many elements from it to make

$\lambda/k^2 > e^\Delta$ for the remaining elements. From $\sum_G \lambda^{-n} < \infty$ and (9.31) we see that on removing finitely many elements of G one can make that $\sum \mathrm{vol}(E_g \cup g(C_g))$ for the remaining elements $g \in G$ will be less than $m_n(X)$. Then removing this infinite union $E_g \cup g(C_g)$ from X we get a set X', $m_n(X') > 0$. We use the conservativeness of the action of G in X' to find an element $g \in G$ from the remaining ones (outside the finite sets above), which is such that $m_n(g^{-1}(X') \cap X') > 0$. The previous construction implies that, if a point $x \in g^{-1}(X') \cap X'$, then $x \in A_g$. From (9.30) we see that on the annulus A_g the linear distortion $d_g(x)$ varies in the interval $[k^2/\lambda, \lambda/k^2]$, which due to the property $\lambda/k^2 > e^\Delta$ (after removing a finite subset of G) contains an interval with length $\exp(2\Delta)$ which contains $d_g(x)$. Let such an interval define the annulus $A \subset S^n$ in question, which then satisfies the conditions 1) and 4). From (9.31) and the inclusion $A \subset A_g$, we have that 2) applies for A. To prove 3), we note that (9.27) and (9.30) imply that $d_g(x)$ behaves in A_g and thus in A as $\pi^2/\lambda\theta^2$. At the same time, $\ln d_g(x)$ varies in A by 2Δ, up to small quantities of higher order. This implies that the modulus of the annulus A is close to $\Delta = \ln \theta_2/\theta_1 = (\ln \theta_2^2/\theta_1^2)/2$. As the annulus $g(A)$ is bounded by spheres with sizes defined by parameters $r\theta_1 e^{2\Delta}$ and $r\theta_2$ (for some r), the modulus of $g(A)$ is

$$\ln r\theta_1 e^{2\Delta}/r\theta_2 = 2\Delta - \ln \theta_2/\theta_1 \approx \Delta. \tag{9.32}$$

The proof of the lemma is complete.

Now we pass from the sphere S^n to the space $\mathbb{R}^n$. Here let there be given a set B and its subset A with a proportion η of volume $m_n(A)/m_n(B) = \eta$, and let $B' = g(B)$, $g \in \mathrm{Mob}(n)$. Then the set $A' = g(A)$ is contained in B' in proportion at least η' of volume:

$$m_n(g(A))/m_n(g(B)) \geq \eta' = 1 - d\cdot(1 - \eta), \tag{9.33}$$

where d is the maximum ratio of volume distortion at various points of B (in fact, that property applies for any absolutely Lesbegue continuous mapping $g : \mathbb{R}^n \longrightarrow \mathbb{R}^n$). This is evident from the following observation. By an affine scaling we can assume that $m_n(B) = 1$, while the low value of volume distortion is 1, and the high value is d. Then the worst case occurs when 1 occurs on all of A and d occurs on all of the complement of A. Then

$$m_n(g(A))/m_n(g(B)) = \eta/(\eta + d(1 - \eta)) \geq 1 - d(1 - \eta). \tag{9.34}$$

It is clear that (9.33) follows from the bound (9.34).

We now define a sector box of shape Δ, which is a set V_Δ obtained by the intersection of an annulus bounded by spheres having common centre $p_0 \in \mathbb{R}^n$ and radii r_0 and $r_0 \cdot e^\Delta$ with a circular cone whose vertex is the point p_0, whose axis is parallel to the vector $v_0 \in \mathbb{R}^n$ and whose ruling forms an angle $\Delta/2$ with v_0.

LEMMA 9.27. *Let* $X \subset \mathbb{R}^n$,, $m_n(X) > 0$, *and let there be given positive numbers* η *and* Δ. *Then there exist* $\delta > 0$ *and* $X' \subset X$, $m_n(X') > 0$, *such that each sector box* V_Δ *of shape* Δ *and diameter* $< \delta$ *containing a point of* X' *also contains at least the proportion* η *of points of* X:

$$m_n(V_\Delta \cap X)/m_n(V_\Delta) \geq \eta. \tag{9.35}$$

The proof of the lemma follows from Lebegue's theorem, which states that almost all points in a positive-measure set are density points (see E. Stein [1], Chapter 1 §2). For this theorem, we can get sector boxes V_Δ instead of round balls, since they are generated by similarity transformation from one of them. Namely, for almost all $x \in X$, there exists a positive maximum number δ_x such that the proportion of X in the sector boxes V_Δ containing x of diameter not exceeding δ_x is at least η. Then the function $x \longmapsto \delta_x$ is a measurable positive function in X, which has to be greater than some $\delta > 0$ on a set $X' \subset X$, $m_n(X') > 0$. This proves (9.35) for points $x \in X'$.

We continue the proof of Theorem 9.25 on the basis of these results. We assume that the conservative part of $K \subset \mathbb{R}^n$ has a tangent k-plane field Σ, $1 \leq k \leq n-1$, invariant with respect to G almost everywhere. We consider $X \subset K$, $m_n(X) > 0$, on which this field is almost parallel, i.e., it varies only in an interval of inclinations of length $\Delta/2$. We specify a number η, $0 < \eta < 1$. Then, according to Lemma 9.27, there exist a $\delta_1 > 0$ and a positive-measure subset $X' \subset X$ such that each sector box V_Δ of shape Δ (diameter of $V_\Delta < \delta_1$), containing some point from X', consists of not less than the proportion η of the X points, i.e., (9.35) applies. We consider any density point in X' and consider the stereographic projection of $\mathbb{R}^n$ to the sphere S^n with respect to that point. We denote by Y the intersection between X' and the ball B' centred at that point whose radius δ_2 is sufficiently small so that the linear derivative of the stereographic projection on the ball $2B'$ is as close to 1 as we need for the following. By δ we denote the smaller of δ_1 and δ_2.

Then, with the given Δ and δ as well as $Y \subset K \subset S^n$, $\mu_n(Y) > 0$, we find the point x, the element $g \in G$, and the concentric annulus $A \subset S^n$ centred at the centre of the isometric sphere $I(g)$ that satisfy Lemma 9.26 (and put A back on $\mathbb{R}^n$). In A, we consider the sector boxes V_Δ of shape Δ containing the point x and centred at the centre of A. As the diameters of V_Δ and $g(V_\Delta)$ are less than $\delta \leq \delta_1$, which follows from (9.31), our construction gives us that V_Δ and $g(V_\Delta)$ contain an η-portion of the set X ($x \in V_\Delta \subset A$, $g(x) \in g(V_\Delta) \subset g(A)$, x and $g(x)$ belong to X'). From property 4) in Lemma 9.26, we see that the ratio of volume distortion for $g: V_\Delta \longrightarrow g(V_\Delta)$ is at most $\exp(2\Delta)$, so (9.34) gives us that the set of points $y \in V_\Delta$ such that $g(y) \in X$, is contained in V_Δ in the following proportion:

$$\mu_n(V_\Delta \cap g(X))/\mu_n(V_\Delta) \geq \eta' = 1 - e^{2\Delta}(1-\eta). \tag{9.36}$$

From (9.34) and (9.36) we get that

$$\frac{\mu_n(V_\Delta \cap X \cap g^{-1}(X))}{\mu_n(V_\Delta)} \geq 1 - [(1-\eta) + (1-\eta')] \geq \eta - e^{2\Delta}(1-\eta), \qquad (9.37)$$

i.e., the set $X \cap g^{-1}(X)$ has at least the portion $\eta_0 = \eta - e^{2\Delta}(1-\eta)$ of the sector box V_Δ. If now we consider the vectors joining the centre of V_Δ to points from $X \cap g^{-1}(X)$, the variation in the angles between these vectors in any direction will be not less than $\Delta \cdot \eta_0$. Namely, for the worst case (when the set $V_\Delta \cap X \cap g^{-1}(X)$ coincides with the intersection $V_\Delta \cap V$, in which V is a round cone with the same centre as V_Δ and some rolling angle t), we have

$$\int_0^{t/2} \mathrm{Sin}^{n/2}\varphi d\varphi \geq \eta_0 \int_0^{\Delta/2} \mathrm{Sin}^{n-2}\varphi d\varphi.$$

This gives that $t \geq \Delta \cdot \eta_0^{1/(n-1)} > \Delta \cdot \eta_0$.

Then the initial almost everywhere G-invariant field Σ of tangent k-planes has angles between those k-planes not greater than Δ. A direct calculation shows that inversion with respect to the isometric sphere $I(g)$, or, what is the same, the mapping g, also maps that field Σ into the field $g(\Sigma)$ in such a way that it will contain k-planes between which the angles vary in the interval $(2\Delta\eta_0 - a,\ 2\Delta\eta_0 - a + \Delta)$, where a, $0 \leq a \leq \Delta$, is a normalization. However, as the group G acts conservatively in X, that interval points should be separated from zero by a distance not exceeding Δ. However, that is impossible, since for η sufficiently close to one, η_0 is also close to one. That contradiction completes the proof of Theorem 9.25 and with it the proofs of Theorems 9.24 and 9.22.

3. *Rigidity of manifolds with boundary.* The above hyperbolic rigidity theorems lead to the following natural question:

> Let M be an n-dimensional $(n \geq 3)$ conformal manifold/orbifold with boundary ∂M whose conformal structure in $\operatorname{int} M = M \setminus \partial M$ is defined by a complete hyperbolic metric $(\mathrm{vol}(\operatorname{int} M) = \infty)$. Also, let there be a quasi-conformal mapping $f : M \longrightarrow M'$ of the orbifold M onto a similar orbifold M' whose restriction at the boundary $\partial M \longrightarrow \partial M'$ is a conformal map. Is there a hyperbolic isometry
>
> $$F : \operatorname{int} M \longrightarrow \operatorname{int} M',$$
>
> that induces the same fundamental-group isomorphism $F_* = f_* : \pi_1(M) \longrightarrow \pi_1(M')$?

This question is a reformulation of a problem stated by Bers [6] and Krushkal' [3] - [5] within the framework of Kleinian-group deformation theory, see Remark 9.21.

In Remark 9.21 we have shown by modifying the proof of Theorem 9.19

that, in general, the complete hyperbolic orbifolds $\operatorname{int} M$ and $\operatorname{int} M'$ are not isometric. However, the following positive solution to this problem has been obtained (Apanasov [16], compare Sullivan [1], [4]).

THEOREM 9.28. *Let M and M' be n-dimensional ($n \geq 3$) conformal orbifolds with boundary in which conformal structures in the interiors are defined by complete hyperbolic metrics d_M and $d_{M'}$. Let $f: M \longrightarrow M'$ be a quasi-conformal mapping (in the orbifold category) conformal at the boundary ∂M. If for some (any) $\varepsilon > 0$ and for the ε-neighbourhood $C_\varepsilon \subset M$ of the minimal convex retract $C(M) \subset M$ (see Theorem 5.12) we have the asymptote*

$$\lim_{r \to \infty} \frac{\operatorname{vol}\{x \in C_\varepsilon \,:\, d_M(0, x) \leq r\}}{\operatorname{vol}\{x \in H^n \,:\, d(0, x) \leq r\}} = 0, \tag{9.38}$$

then there exists a hyperbolic isometry $F: \operatorname{int} M \longrightarrow \operatorname{int} M'$ which induces the same isomorphism $F_* = f_*: \pi_1(M) \longrightarrow \pi_1(M')$.

PROOF. We consider the developments $d: \widetilde{M} \longrightarrow \overline{\mathbb{R}}^n$ and $d': \widetilde{M}' \longrightarrow \overline{\mathbb{R}}^n$ and the discrete holonomy groups $G = d_*(\pi_1(M))$, $G' = d'_*(\pi_1(M')) \subset \text{Möb}(n)$ to get from the theorem conditions that the groups G and G' may be considered as groups of isometries of the hyperbolic space model $H^n \cong \mathbb{R}^n_+$ that act discontinuously in $\mathbb{R}^n_+ \cup \Omega(G) = d(\widetilde{M})$ and in $d(\widetilde{M}') = \mathbb{R}^n_+ \cup \Omega(G')$ correspondingly. The quasi-conformal mapping $f: M \longrightarrow M'$ is lifted to a quasi-conformal mapping $\tilde{f}: \overline{\mathbb{R}}^n_+ \longrightarrow \overline{\mathbb{R}}^n_+$, conformal in $\Omega(G) \subset \overline{\mathbb{R}}^{n-1}$ and compatible with the group $G: \tilde{f}G\tilde{f}^{-1} = G'$. Let $P \subset \mathbb{R}^n_+ \cup \Omega(G)$ be a fundamental polyhedron for the group G (convex in H^n). We shall prove that condition (9.38) is equivalent to the condition of equality to zero for the $(n-1)$-dimensional spherical Lebesgue measure of the set $P^* = (\overline{P} \cap \Lambda(G))$ of limit vertices of the polyhedron $P: \mu_{n-1}(\overline{P} \cap \Lambda(G)) = 0$.

For that purpose, we first recall (see Chapter 5) that the minimal convex retract $C(M) \subset M$ is obtained by G-identification of the sides of the cut off polyhedron $P_H = P \cap H_G$, where $H_G \subset H^n \cong \mathbb{R}^n_+$ is the convex Nielsen hull for the limit set $\Lambda(G)$.

If now $\mu_{n-1}(P^*) > 0$, then almost all points $x \in P^*$ are density points for P^*, i.e., if $B^{n-1}(x, r)$ is a ball in $\mathbb{R}^{n-1}$ centred at such a point x with radius $r > 0$, we have

$$\lim_{r \to \infty} \frac{\mu_{n-1}(P^* \cap B^{n-1}(x, r))}{\mu_{n-1}(B^{n-1}(x, r))} = 1. \tag{9.39}$$

We now fix the point $x_0 \in P_H$, corresponding to the distinguished point $0 \in C(M) \subset M$, appearing in (9.38). Then the spherical measure of the set P^* is equal to the solid angle under which that set is seen from the point x_0. Then (9.39) implies that the limit in (9.38) will tend to the measure $\mu_{n-1}(P^*)$, which on the theorem conditions is positive. That contradiction shows that $\mu_{n-1}(P^*) = 0$.

Conversely, if $\mu_{n-1}(P^*) = 0$, we consider the solid angle $\omega(r)$ for that part of a sphere with radius $r > 0$ and centred at the point

$x_0 \in P_H$, that lies on the intersection $U_\varepsilon(P_H)$ of the ε-neighbourhood of the polyhedron P_H with polyhedron P. Then the convexity of P_H and $\mu_{n-1}(\overline{P}_H \cap \Lambda(G)) = 0$ imply that, when the radius r increases, $\omega(r)$ decreases and tends to zero. If by $a_{n-1}(r)$ we denote the hyperbolic $(n-1)$-volume of the $(n-1)$-dimensional sphere with radius $r > 0$, we have

$$\frac{\operatorname{vol}\{x \in C_\varepsilon : d_H(0, x) \le r\}}{\operatorname{vol}\{x \in H^n : d(x_0, x) \le r\}} = \frac{\operatorname{vol}(x \in U_\varepsilon(P_H) \cap P : d(x_0, x) \le r\}}{\operatorname{vol}\{x \in H^n : d(x_0, x) \le r\}} = \tag{9.40}$$

$$= \int_0^r a_{n-1}(r)\omega(r)dr \Big/ \int_0^r a_{n-1}(r)dr.$$

Thus, since for any $\delta > 0$ one can find an $r_\delta > 0$ beginning with which (for $r \ge r_\delta$) the solid angle $\omega(r)$ is less than δ, we have that the left-hand side of (9.40) does not exceed

$$\frac{\int_0^{r_\delta} a_{n-1}(r)\omega(r)dr}{\int_0^r a_{n-1}(r)dr} + \delta \cdot \frac{\int_{r_\delta}^r a_{n-1}(r)dr}{\int_0^r a_{n-1}(r)dr}. \tag{9.41}$$

The first term in (9.41) tends to zero as r increases, while the second term does not exceed an arbitrarily selected $\delta > 0$, which shows that (9.38) applies.

Then to complete the proof, it remains to note a point following from Corollary 6.66, namely that the dissipative part of the action of the group G on $\partial H^n = \overline{\mathbb{R}}^{n-1}$ is precisely the set $G(\partial H^n \cap \overline{P})$, which shows that G acts conservatively in the limit set $\Lambda(G)$ of positive measure only if $\mu_{n-1}(\overline{P} \cap \Lambda(G)) = 0$, and therefore Theorem 9.24 shows that the restriction of the mapping $\tilde{f}$ to $\overline{\mathbb{R}}^{n-1} = \partial H^n$ is a 1-quasi-conformal homeomorphism of the sphere at infinity $\overline{\mathbb{R}}^{n-1}$, i.e., it is a Möbius transformation (Theorem 9.8). Then we see that G and G' are conjugated by an isometry of the hyperbolic space H^n, which is projected to an isometry $F : \operatorname{int} M \longrightarrow \operatorname{int} M'$ that induces the isomorphism $f_* : \pi_1(M) \longrightarrow \pi_1(M')$. The theorem is proved.

Theorem 9.28 for $n = 3$ gives (see Sullivan [1]) the following:

COROLLARY 9.29. *For a finitely generated Kleinian group $G \subset \text{Möb}(2)$, the quasi-conformal deformation classes there form the Teichmüller space $T(G)$ for the finite union $\Omega(G)/G = S_1 \cup \ldots \cup S_m$ of finite-type Riemann surfaces S_i simultaneously uniformized by the group G, i.e.,*

$$T(G) \cong T(\Gamma_1) \times \ldots \times T(\Gamma_m), \tag{9.42}$$

in which Γ_i are the Fuchsian equivalents for G-orbits of components

$\Omega_i \subset \Omega(G)$ *which correspond to the surfaces* S_i, $i = 1, \ldots, m$.

This theorem is known as the decomposition theorem and was first obtained by Bers [3] subject to an additional condition on the stability of G. Then Kra [4] eliminated that condition, and Bers [6] proved (9.42) without making any finiteness assumption, where one assumes that on the right in (9.42) in general one has an infinite number of cofactors and suitable topology.

REMARK 9.30. The examples of Kleinian groups $G \subset \text{Möb}(n-1)$, $n \geq 3$, constructed in Remark 9.21, show that the condition (9.38) in Theorem 9.28 is exact. However, that condition is not necessary for $n \geq 4$, which is obtained from the following example of rigid groups (Apanasov [16] and [19, Example 9.11]).

EXAMPLE 9.31 (*Rigid group and manifold without condition* (9.39)). For $n \geq 4$, we consider a rigid Kleinian group $G \subset \text{Möb}(n-1)$ acting in $H^n \cong \mathbb{R}^n_+$ as a subgroup of $\text{Isom}\, H^n$, whose discontinuity set $\Omega(G) \subset \overline{\mathbb{R}}^{n-1}$ consists of two invariant components Ω_0 and Ω_1, while the limit set $\Lambda(G) = \partial\Omega_0 = \partial\Omega_1$ has a positive $(n-1)$-dimensional Lebesgue measure $\mu_{n-1}(\Lambda(G)) > 0$.

As that G one can consider a group, infinitely generated by reflections with respect to a family of spheres in $\overline{\mathbb{R}}^n$, whose disjoint interiors cover a Peano hypersurface S, $\mu_{n-1}S > 0$, that splits up $\overline{\mathbb{R}}^{n-1}$ into two domains. That construction is performed by exact analogy with Abikoff's [1] construction, which uses the Peano curve in the plane.

In the G thus constructed, the set P^* of limit vertices of the convex fundamental polyhedron $P(G) \subset H^n \cup \Omega(G)$, $P^* = \overline{P} \cap \Lambda(G) \subset S$ has a positive measure $\mu_{n-1}(P^*) > 0$, which we have shown above to be equivalent to the difference from zero of the limit in (9.38).

Now let $G' \subset \text{Möb}(n-1)$ be a Kleinian group such that $M' = [H^n \cup \Omega(G')]/G'$ is a quasi-conformal image $f(M)$, where the homeomorphism $f: M \longrightarrow M'$ is conformal at the boundary ∂M. Then the lifting of the homeomorphism f to the universal covering is extended to a quasi-conformal homeomorphism $\tilde{f}: \overline{H}^n \longrightarrow \overline{H}^n$, conformal in $\Omega(G)$ and compatible with the group $G: \tilde{f}G\tilde{f}^{-1} = G'$. As $n \geq 4$, the conformal mappings $f_0 = \tilde{f}|_{\Omega_0}$ and $f_1 = \tilde{f}|_{\Omega_1}$ are restrictions on Ω_i for Mobius transformations that are obliged to coincide on the set of fixed points of elements of the group G, dense in $\Lambda(G)$. This implies that $f_0(x) = f_1(x)$ for all $x \in \Lambda(G)$ and therefore the restriction $\tilde{f}$ in $\partial H^n = \overline{\mathbb{R}}^{n-1}$ is a unit Möbius transformation.

Then G and G' are always conjugate in $\text{Möb}(n-1)$, while H^n/G and H^n/G' are isometric, which shows the rigidity, although (9.38) is not obeyed. Similar rigidity occurs for any function group, as is evident from the proof (compare Krushkal' [4]).

§3. Deformation space for a Fuchsian group

Let $G \subset \text{Möb}(n) \cong SO(n+1, 1)$ be a Fuchsian group in the n-dimensional conformal sphere $S^n = \overline{\mathbb{R}}^n$, i.e., a group that leaves invariant some n-dimensional ball B^n. In §1, we defined the spaces $R_n(G)$ and $T_n(G) = R_n(G)/\text{Möb}(n)$ for such groups (or, what is equivalent, for Fuchsian representations) in which $R_n(G)$ consists of all representations $\rho \in \text{Hom}(G, \text{Möb}(n))$ for which the action of the group ρG in the sphere S^n is quasi-conformally conjugate to the action of G, i.e., for which there exist quasi-conformal self-homeomorphisms $f: S^n \to S^n$ compatible with G and such that $\rho G = fGf^{-1}$.

The space $T_n(G)$ is naturally related to the space of uniformized (quasi-Fuchsian) conformal structures on the orbifold B^n/G and for $n=2$ is called the Teichmüller space of the Fuchsian group G. We explain this relation in more detail, initially for the planar case.

1. *Teichmüller space $T(G)$ of a Fuchsian group in the plane.*

Let $G \subset \mathcal{M}_2$ be a Fuchsian group of the first kind that leaves invariant the half-plane $H^2 \subset \mathbb{C}$ and which for simplicity does not have elliptic elements. Then the necessary and sufficient condition for compatibility of a quasi-conformal automorphism f of $\overline{\mathbb{C}}$ and the group G is the following condition for its complex quasi-conformality characteristic $\mu(z)$:

$$\mu(\gamma(z))\overline{\gamma'(z)}/\gamma'(z) = \mu(z) \quad \text{for all} \quad \gamma \in G, \quad z \in \mathbb{C}. \tag{9.43}$$

In other words, $\mu(z)d\overline{z}/dz$ should be G-invariant (a Beltrami differential). There are two approaches to examining the relation between the deformation space of the group G and the Teichmüller space for the Riemann surface $S = H^2/G$. The first is as follows.

Consider the complex Banach space $M^S(\mathbb{C}, G)$ of functions $\mu \in L_\infty(\mathbb{C})$ such that $\mu(z)d\overline{z}/dz$ is a Beltrami G-differential (condition (9.43)) and also

$$\mu(\overline{z}) = \mu(z), \tag{9.44}$$

with the norm $\|\mu\|_{M^S} = \|\mu\|_{L_\infty}$ and let $M_1^S(\mathbb{C}, G) = \{\mu \in M^S(\mathbb{C}, G) : \|\mu\|_{M^S} < 1\}$ be its unit ball. Let $Q(G)$ be the group (with respect to superposition) of quasi-conformal automorphisms of H^2 with fixed points 0, 1, ∞ compatible with G, i.e., automorphisms f^μ with complex characteristic $\mu \in M_1^S(\mathbb{C}, G)$, and let $Q_0(G)$ be the normal subgroup of $Q(G)$ consisting of all the $f^\mu \in Q(G)$ that leave fixed all the points on the real axis $\overline{\mathbb{R}} = \partial H^2$. The quotient

$$T(G) = Q(G)/Q_0(G) \tag{9.45}$$

is called the Teichmüller space of the group G, whose elements (equivalence classes) are denoted by $[f^\mu]$. The Teichmüller space $T(G)$ of G can be identified with the Teichmüller space $T(S)$ for the Riemann surface $S = H^2/G$ because condition (9.43) of compatibility of f^μ with G is equivalent to specifying the corresponding quasi-conformal mapping on the surface S, while the equivalence relation with respect to the subgroup $Q_0(G)$ corresponds to the homotopy equivalence relation (see §1.1) for the mappings of the surface S.

Similarly to (9.2), we introduce the Teichmüller metric into $Q(G)$, which in accordance with (9.45) takes the following form in $T(G)$:

$$d([f_1], [f_2]) = \frac{1}{2} \inf_{f \in [f_1 f_2^{-1}]} \ln K(f) \tag{9.46}$$

and turns $T(G)$ into a complete connected metric space.

The second approach to $T(G)$ is as follows. Instead of the space $M^s(\mathbb{C}, G)$, we consider the space $M(H, G)$ of functions $\mu \in L_\infty(\mathbb{C})$ such that $\mu(z)d\bar{z}/dz$ is a Beltrami G-differential and $\mu(z) = 0$ in the lower half-plane $\mathbb{R}^2_-$ (in place of property (9.44)), with the norm $\|\mu\|_M = \|\mu\|_{L_\infty}$. Let again $M_1(H, G)$ be the unit ball in $M(H, G)$ and f_μ be quasi-conformal automorphisms of $\overline{\mathbb{C}}$ having complex characteristics $\mu \in M_1(H, G)$ and fixed points 0, 1, and ∞; all such f_μ are conformal in $\mathbb{R}^2_-$.

It is readily seen that the equality $f_{\mu_1} = f_{\mu_2}$ holds on $\mathbb{R}$ if and only if $f^{\tilde{\mu}_1} = f^{\tilde{\mu}_2}$ on $\mathbb{R}$ for the corresponding $\tilde{\mu}_i \in M^s_1(\mathbb{C}, G)$, obtained from the μ_i by extension from $\mathbb{R}^2_+$ to $\mathbb{R}^2_-$ by symmetry - see (9.44). Therefore there is a one-to-one correspondence between the restrictions $f_\mu|_{\mathbb{R}}$ and the classes $[f^{\tilde{\mu}}]$, i.e., the points of $T(G)$ - see (9.45).

In fact, $f^{\tilde{\mu}_i}(z) = h_i \circ f_{\mu_i}(z)$ for $z \in \mathbb{R}^2_+$, where h_i is a conformal mapping of the domain $f_{\mu_i}(\mathbb{R}^2_+)$ onto $\mathbb{R}^2_+$, so if $[f^{\tilde{\mu}_1}] = [f^{\tilde{\mu}_2}]$, then the function

$$h(w) = \begin{cases} h_2 \circ h_1^{-1}(w), & w \in f_{\mu_1}(\mathbb{R}^2_+); \\ f_{\mu_2} \circ f_{\mu_1}^{-1}(w), & w \in f_{\mu_1}(\mathbb{R}^2_-), \end{cases}$$

extends to a quasi-conformal automorphism of $\overline{\mathbb{C}}$ that is conformal in $\overline{\mathbb{C}} \setminus f_{\mu_1}(\overline{\mathbb{R}})$. Consequently, $h \in M_2$ and, by the accepted normalization, reduces to the identity.

Thus if, instead of the classes $[f^\mu]$ from (9.45) we start from the equivalence classes $[f_\mu]$, then we get (up to an isometric isomorphism) the same Teichmüller space $T(G)$.

In that second approach, the points in $T(G)$ correspond to the quasi-Fuchsian groups $G_f = f_\mu G f_\mu^{-1}$ that leave the components of $\overline{\mathbb{C}} \setminus l$

invariant, in which $l = f_\mu(\overline{\mathbb{R}})$ is a quasi-circle in $\overline{\mathbb{C}}$.

2. *The space* $T_n(G)$ *of conjugacy classes of quasi-Fuchsian representations.*

Continuing the above two approaches to Teichmüller space $T(G)$ for a Fuchsian group $G \subset \text{Möb}(n)$, $n \geq 3$, i.e., for a discrete subgroup $G \subset \text{Möb}(n-1) \subset \text{Möb}(n)$, we go to the problems of describing the following two representation spaces (with discrete images):

$$\text{Hom}(G, \text{Möb}(n-1))/\text{Möb}(n-1) \quad \text{and} \quad \text{Hom}(G, \text{Möb}(n))/\text{Möb}(n).$$

Mostow's rigidity Theorem 9.12 shows that the first space (as well as the space $\mathcal{T}(H^n/G)$ of hyperbolic structures) is trivial. But the second problem (a question of Borel and Wallach [1, p.221-224] about the triviality of infinitesimal deformations $H^1(G, \mathfrak{M}_n)$ (see §1.3) is a refinement of the problem) has a non-trivial solution in dimension $n \geq 3$, too.

The first results showing the non-triviality of the second space for $n \geq 3$ were obtained by Apanasov [11], [12] (for finitely generated G) and Tetenov [2] (for infinitely generated G); see also Apanasov and Tetenov [1]. The matter was greatly clarified by Thurston's [1] idea of bending a Fuchsian group - see Mickey Mouse Example 5.11 and Sullivan [4]. It demonstrated that the deformations obtained (curves in the representation space) for the 2-dimensional case correspond to "bending" the Riemann surface H^2/G along closed geodesics. For the n-dimensional case, this corresponds to bending the hyperbolic n-manifold H^n/G along its totally geodesic hypersurface. We shall see below how that idea implies a lower bound for a dimension of $T^n(G)$ by three independent methods: by the geometry of the hyperbolic space H^{n+1} and a fundamental polyhedron $P(G) \subset H^{n+1}$ (Apanasov [24], [25], [27]), by the geometry of the manifold H^{n+1}/G, and using some limit process in the construction of the bending (Kourouniotis [1], [2]) and by the algebraic properties of group representations (Johnson and Millson [1]).

Before we examine the general properties of $T_n(G)$, we give an example of a co-compact Fuchsian group G in $\overline{\mathbb{R}}^3$ that is a modification of the first example of a group in space (Apanasov [12]) having non-trivial bending deformations (in the sense of Definition 9.6).

EXAMPLE 9.32 (*a smooth curve in* $T_3(G)$). Let $Q = \{x \in \mathbb{R}^3 : |x_i| < 1/2, \quad i = 1, 2, 3\}$ be a unit cube. From its vertices we draw spheres S_i, $1 \leq i \leq 8$, having radii $\sqrt{3}/3$, which intersect at angles of $\pi/3$, while they are orthogonal to the sphere $S_0 = S^2(0, \sqrt{15}/6)$. We add six spheres S_i, $9 \leq i \leq 14$:

$$\begin{aligned} &S_9 = S^2(re_1, \rho) \qquad S_{10} = S^2(-re_1, \rho), \qquad S_{11} = S^2(re_2, \rho), \\ &S_{12} = S^2(-re_2, \rho), \qquad S_{13} = S^2(re_3, \rho), \qquad S_{14} = S^2(-re_3, \rho), \end{aligned} \tag{9.48}$$

in which $\{e_i\}$ is the orthogonal basis of $\mathbb{R}^3$, and we put $r = 5/6$, $\rho = \sqrt{10}/6$. Those parameters mean that the spheres in (9.48) are orthogonal to S_0 and to S_i, $1 \le i \le 8$.

We consider the group $G \subset \mathcal{M}_3$, which is an index 2 subgroup in the group generated by the inversions $\mathcal{I}_i$ with respect to the spheres S_i, $1 \le i \le 14$. As the generators for the group G one can take the transformations $g_i = \mathcal{I}_1\mathcal{I}_i$, $2 \le i \le 14$. It is clear that G leaves invariant the ball $B = B^3(0, \sqrt{15}/6)$. It follows from Corollary 3.20 that G is a Fuchsian group of the first kind having a compact fundamental polyhedron in B.

We consider also the sphere system $\{\sigma_i(t)\}$, which is obtained from the $\{S_i\}$ by replacing the sphere S_{14} (having centre $-e_3 \cdot 5/6$) by the sphere $\sigma_{14}(t)$ having centre $-e_3 \cdot t$ and which is orthogonal to the corresponding spheres $\sigma_i(t) \equiv S_i$, $i \ne 14$. It is clear that $\sigma_{14}(t)$ has radius $\sqrt{t^2 - t + 5/12}$. As above, we consider a group $G(t) \subset \mathcal{M}_3$ generated by the transformations $g_i^t = \mathcal{I}_1^t\mathcal{I}_i^t$, $2 \le i \le 14$. Then for the parameters t satisfying

$$t_0 = \frac{5(1 + \sqrt{5/6})}{11 + 12\sqrt{5/6}} < t < 5(1 + \sqrt{5/6}) = t_1, \tag{9.49}$$

$G(t)$ will be Kleinian, which follows from Corollary 3.20 applied to the extension of $G(t)$ in the half-space H^{n+1}. We also get that the fundamental domain of $G(t)$ is

$$F^t = \operatorname{int}(\overline{P^t \cup \mathcal{I}_1(P^t)}), \qquad P^t = \bigcap_{i=1}^{14} \operatorname{ext} \sigma_i(t) \tag{9.50}$$

and splits up into two components: the bounded one F_0^t and the unbounded one F_1^t.

The discontinuity set $\Omega^t = \Omega(G(t))$ splits up into two invariant components Ω_0^t and Ω_1^t. In fact, if we assume that Ω^t is connected and thus has linear connectivity, since Ω^t is open, we can take the curve $\delta \subset \Omega^t$ joining points $0 \in F_0^t$ and $\infty \in F_1^t$ such that it intersects with the images of the polyhedra F_0^t and F_1^t while bypassing their edges and intersecting them not more than once each. As δ is compact, the number of intersected images of the two polyhedra is finite. However, amongst them there are polyhedra $g, g' \in G(t)$ with a common face, which is impossible because $G(t)$ is isomorphic to the Fuchsian group G. We apply Lemma 3.7 and use the fact that $F_0^t \subset F_1^t$ is a fundamental domain for $G(t)$ to get that Ω^t splits up into two invariant components:

$$\Omega_0^t = G(t)(\overline{F_0^t}) \qquad \text{and} \qquad \Omega_1^t = G(t)(\overline{F_1^t}),$$

that have the common boundary $\partial\Omega_0^t = \partial\Omega_1^t = \Lambda(G(t))$ and which are homeomorphic to a ball.

Subsequently the objects corresponding to the group $G(5/6)$, i.e., to G, will be distinguished by discarding the parameter t. Our construction and (9.50) show that $F_0^t \subset F_0$ for $t \in (t_0, 5/6)$, and $F_0 \subset F_0^t$ for $t \in (5/6, t_1)$. For definiteness we assume the last and consider the ε-layer $c_\varepsilon = F_0 \backslash F_0^{(5/6-\varepsilon)}$ for a specified ε, $0 < \varepsilon < 5/12$. We now define for $t \in (5/6, t_1)$ the mapping f_0^t:

$$f_0^t(x) = \begin{cases} x, & \text{for } x \in F_0^{(5/6-\varepsilon)}, \\ \varphi_\varepsilon(x), & \text{for } x \in c_\varepsilon, \end{cases}$$

in which φ_ε is the uniform stretching along a family of circles orthogonal simultaneously to the spheres S_{14}, $\sigma_{14}(5/6-\varepsilon)$ and $\sigma_{14}(t)$. The mapping φ_ε maps quasi-conformally the layer c_ε onto the layer $C_\varepsilon^t = F_0^t \backslash F_0^{(5/6-\varepsilon)}$, where the quasi-conformality coefficient $q(t)$ is equal to the ratio of the thickness of the layer C_ε^t to the thickness of the ε-layer C_ε. It is clear that $q(t)$ is continuously dependent on t and $\lim_{t\to 5/6} q(t) = 1$.

We define an analogous mapping f_1^t for the unbounded polyhedra F_1 and F_1^t. The thus constructed $q(t)$-quasi-conformal mappings

$$f_0^t : F_0 \longrightarrow F_0^t, \qquad f_1^t : F_1 \longrightarrow F_1^t$$

translate the faces, edges, and vertices of the polyhedra F_i into the corresponding ones for F_i^t, $i = 1, 2$; here the corresponding dihedral angles are equal. We can therefore extend these mappings by symmetry to the images of F_0 and F_1, where we get the $q(t)$-quasi-conformal mappings

$$\Phi_0^t : \text{int } S_0 \longrightarrow \Omega_0^t \qquad \text{and} \qquad \Phi_1^t : \text{ext } S_0 \longrightarrow \Omega_1^t.$$

The compatibility of the mappings Φ_i^t with G implies that their boundary values in the sphere S_0 coincide for $i = 0, 1$. To prove this, we consider the set C of sequences $\zeta = \{g_m\}$ for the elements of the group G such that there exist limits $\lim g_m(F) = x_\zeta \in \Lambda(G)$ and in it we define the equivalence relation θ:

$$\zeta_1 \overset{\theta}{\approx} \zeta_2 \iff x_{\zeta_1} = x_{\zeta_2}.$$

It is clear that each limit point in G is defined uniquely by its equivalence class from C/θ. We do the same for the group $G(t)$, where we have that a converging sequence $\zeta \in C$ in the isomorphism $G \longrightarrow G(t)$ corresponds to a converging sequence $\xi^t \in C(t)$. This means that there are equal radial limits:

$$\lim_{\substack{r\to 1\\ r<1}} \Phi_0^t(rx) = \lim_{\substack{r\to 1\\ r>1}} \Phi_1^t(rx) = x(t), \qquad x \in S_0. \tag{9.51}$$

We thus have that the mapping $\Phi^t : \overline{\mathbb{R}}^n \longrightarrow \overline{\mathbb{R}}^n$,

$$\Phi^t(x) = \begin{cases} \Phi_0^t(x), & x \in \operatorname{int} S_0; \\ x(t), & x \in S_0; \\ \Phi_1^t(x), & x \in \operatorname{ext} S_0, \end{cases}$$

is absolutely continuous on the lines (the ACL^n property), since it is quasi-conformal in $\overline{\mathbb{R}}^n \setminus S_0$ and the limits (9.51) exist. It is well known (Deny and Lions [1]) that then the mapping Φ^t belongs to the class W_n^1, i.e., its coordinate functions have generalized first derivatives (in the sense of Sobolev) locally integrable to the power n. This implies that Φ^t is a quasi-regular mapping. Reshetnyak's results [3] at once then give that Φ^t is a $q(t)$-quasi-conformal homeomorphism of the space $\overline{\mathbb{R}}^n$ for all $t \in (t_0, t_1)$, see (9.49). As the Φ^t are by construction compatible with the group G, $G(t) = \Phi_*^t(G) \subset \mathcal{M}_3$ and $\Phi_*^{5/6} = \mathrm{id}$, we get that the family (Φ^t) specifies a deformation of G in the sense of Definition 9.6.

The following observation shows that this deformation is non-trivial. In the isomorphism $\Phi_*^t : G \to G(t)$ an element $\mathcal{T}_{13}\mathcal{T}_{14} \in G$ (hyperbolic translation in H^4) goes over to a hyperbolic translation in H^4 equal to $\mathcal{T}_{13}\mathcal{T}_{14}^t \in G_t$; it is readily seen that a geodesic $l \subset H^4$ joining fixed points x and y of the mapping $\mathcal{T}_{13}\mathcal{T}_{14}$ is orthogonal to the 3-planes in H^4 spanned on the spheres S_{13} and S_{14}. The length $d(13, 14)$ of the geodesic segment between those planes is the length of the hyperbolic translation $\mathcal{T}_{13}\mathcal{T}_{14}$. In the mapping Φ^t, the geodesic l corresponds to the geodesic l_t joining fixed points of the translation $\mathcal{T}_{13}\mathcal{T}_{14}^t$. Here the length $d_t(13, 14)$ of the geodesic segment orthogonal to the 3-planes in H^4 spanned on S_{13} and $\sigma_{14}(t)$, i.e., the length of the translation $\mathcal{T}_{13}\mathcal{T}_{14}^t$ in H^4 for $t > 5/6$ is more than $d(13, 14)$, while for $t < 5/6$ it is less than $d(13, 14)$.

However, a trivial deformation, i.e. conjugation in Möb(3), should preserve the hyperbolic translation lengths, so the continuous family (Φ^t), $t_0 < t < t_1$, specifies a non-trivial deformation of the group G.

We shall see later (Theorem 9.34) that the family (Φ^t) is dependent smoothly on the parameter t and defines a curve in the space $T_3(G)$, $\dim_{\mathbb{R}} T_3(G) \geq 6$.

3. *Properties of the space $T_n(G)$: polyhedral point of view*

We assume that a Fuchsian group $G \subset \text{Möb}(n)$ acts in the invariant ball $B^n \subset \overline{\mathbb{R}}^n$ as a group of hyperbolic space isometries simultaneously with its isometric action in the hyperbolic space $H^{n+1} \cong \mathbb{R}_+^{n+1}$ and in the hyperbolic n-dimensional plane $H^n \subset H^{n+1}$, $\partial H^n = \partial B^n$. Let $P^{n+1} = P^{n+1}(G) \subset H^{n+1} \cup \Omega(G)$ be a hyperbolically convex fundamental polyhedron for the group G and let $P^n = P^{n+1} \cap H^n$ and $P = P^{n+1} \cap \overline{\mathbb{R}}^n \subset \Omega(G)$ be fundamental polyhedra for G correspondingly in the n-plane H^n and in $\overline{\mathbb{R}}^n$.

For a side s of codimension 1 in one (any) of the polyhedra P, P^n, or P^{n+1}, we denote by S an n-dimensional hyperbolic plane in

H^{n+1}, whose closure $\bar{S} = S \cup \partial S$ contains the side s.

DEFINITION 9.33. The sides s_1 and s_2 of codimension 1 in one of the fundamental polyhedra P, P^n, or P^{n+1} of the group G are called $\theta(G)$-equivalent if the corresponding n-planes $S_1, S_2 \subset H^{n+1}$ $(s_i \subset \bar{S}_i)$ are G-equivalent: $G(S_1) = G(S_2)$.

Amongst the $(n-1)$-dimensional sides of the polyhedron P or P^n we distinguish the set $\Pi = \Pi(G)$ of those sides s that are orthogonal to all the other $(n-1)$-dimensional sides s' of the polyhedron that intersect s in the sense that $\bar{s} \cap \bar{s}' \neq \emptyset$ and are not $\theta(G)$-equivalent to them.

We are now concerned by the groups G having fundamental polyhedra P or P^n for which $\Pi(G) \neq \emptyset$ and the set $\Pi(G)/\theta(G)$ does not contain adjacent sides of the polyhedron in different classes. Then the following theorem applies (Apanasov [24]).

THEOREM 9.34. *Let* $G \subset \text{Möb}(n-1)$, $n \geq 2$, *be a geometrically finite discrete group which has a fundamental polyhedron* P *without obtuse angles. Let* $\Pi(G)/\theta(G) \neq \emptyset$ *and the distinct elements correspond to non-adjacent sides of* P. *Then there exists a smooth embedding (class* C^∞*) of a* $(\text{card}\,\Pi(G)/\theta(G))$*-dimensional ball in the space* $T_n(G) \subset \text{Hom}(G, \text{Möb}(n))/\text{Möb}(n)$.

PROOF. We put $m = \text{card}\,\Pi(G)/\theta(G)$ and consider an m-dimensional cube $I^m = \prod_{i=1}^{m} (-a_i, b_i)$, where the numbers $a_i, b_i > 0$ are defined later. The basis of the proof extends an idea of Example 9.32 and lies in constructing for each $\alpha = (\alpha_1, \ldots, \alpha_m) \in I^m$ a quasi-conformal mapping $f_\alpha : \bar{\mathbb{R}}^n \longrightarrow \bar{\mathbb{R}}^n$ compatible with the group G and obtained as a result of "displacements" depending on α_i for all the $(n-1)$-sides $s \in \Pi(G)$ lying in the i-th class of the set $\Pi(G)/\theta(G)$. With these "displacements", which preserve the orthogonality to the adjacent sides, the polyhedron P becomes the polyhedron $P(\alpha)$, fundamental for a group $G(\alpha) = f_\alpha G f_\alpha^{-1}$. However, these α_i-displacements of sides will be carried out, in contrast to Example 9.32, by using the geometry for the polyhedron P^{n+1} and the geometry of the space H^{n+1}.

Then let S_i be the n-plane in H^{n+1} orthogonal to H^n and whose closure contains a side s from the i-th class of $\Pi(G)/\theta(G)$. Then the α_i-"displacement" of the sphere ∂S_i (i.e., the α_i-displacement of the side $s \longmapsto s(\alpha_i)$), which preserves the angles of intersection with the adjacent sides of the polyhedron P, corresponds to rotation in H^{n+1} of the plane S_i around the $(n-1)$-dimensional plane $S_i \cap H^n$ through an angle α_i, $S_i \longmapsto S_i(\alpha_i)$ conjugate in $\text{Isom}\,H^{n+1}$ to the orthogonal transformation specified by the matrix

$$A(\alpha_i) = \begin{pmatrix} E_{\cos\alpha_i} & 0_{\sin\alpha_i} \\ 0_{-\sin\alpha_i} & \cos\alpha_i \end{pmatrix} \in O(n+1). \tag{9.52}$$

The upper and lower bounds a_i and b_i for the parameters α_i should be chosen so that the closures of all the planes $\overline{S}_i(\alpha_i)$ do not intersect the closure of the planes $\overline{S}_j(\alpha_j)$ for any distinct $i, j = 1, \ldots, m$. The possibility of such a choice $a_i > 0$, $b_i > 0$, follows from the geometric finiteness of the group G and from the non-intersection for $i \neq j$ of the closed planes $\overline{S}_i$ and $\overline{S}_j$. The latter is a direct consequence of the condition that the polyhedron P is acute angled, and thus P_n is also, and of Andreev's [3] theorem on the non-intersection of planes of non-adjacent sides for acute angled hyperbolic polyhedra.

We now consider any $\alpha = (\alpha_1, \ldots, \alpha_m) \in I^m \setminus \{0\}$ and perform for all $i = 1, \ldots, m$ the α_i rotations in H^{n+1} of all n-dimensional planes $S \subset H^{n+1}$ (a finite number of them) corresponding to the i-th classes of $\Pi(G)/\theta(G)$. The planes $S(\alpha)$ obtained as a result of these rotations along with the planes of fixed sides of P^{n+1} bound a convex polyhedron $P^{n+1}(\alpha)$ of the same combinatorial type as P^{n+1}. We consider the Möbius group $G(\alpha)$, whose generators are transformations that identify sides of the polyhedron $P^{n+1}(\alpha)$. These generators are elements of G if they identify the sides of $P^{n+1}(\alpha)$ invariant for the α_i-rotations; if this is not so (if the sides are moved by them), the generators of $G(\alpha)$ are superpositions of the generators of the group G with the corresponding α_i-rotations.

For the polyhedron $P^{n+1}(\alpha)$, as for P^{n+1}, the conditions for the H^{n+1}-tessellation criterion (see Theorem 6.9) are satisfied. In fact, the condition of one-to-one filling of a neighbourhood of each edge of codimension 2 in the polyhedron $P^{n+1}(\alpha)$ follows from the analogous property of P^{n+1} and from the preservation of the polyhedron angles under our deformations. The condition for local finiteness of the filling of H^{n+1} by images of $P^{n+1}(\alpha)$ also follows from an analogous property of P^{n+1} and from the fact that by the choice of the size of the parameter cube I no additional fixed points of the group $G(\alpha)$ (in comparison with the group G) can arise at the boundary of the polyhedron $\overline{P}^{n+1} \cap \overline{\mathbb{R}}^n$, and particularly no fixed points for loxodromic elements.

Then we get that $G(\alpha)$ is discrete and there exists a natural isomorphism $\chi_\alpha : G \longrightarrow G(\alpha)$ specified on the generators of the group G (which identify the sides of P^{n+1}).

We consider the $q(\alpha)$-quasi-conformal mapping $\varphi_\alpha : P^{n+1} \longrightarrow P^{n+1}$, $\lim_{\alpha\to 0} q(\alpha) = 1$, which in neighbourhoods of the $(n-1)$-dimensional sides $S \in \Pi(G)$ of P^{n+1} lying in the n-dimensional planes $S \subset H^{n+1}$ is a uniform stretching along the family of circles orthogonal simultaneously to the planes S and $S(\alpha)$ and which are the trajectories for the α-rotations around the corresponding $(n-1)$-dimensional planes $S \cap H^n$. Outside all such neighbourhoods $\varphi_\alpha = \mathrm{id}$. It is evident that φ_α is compatible with G at the boundary of P^{n+1} and therefore it can be extended to a $q(\alpha)$-quasi-conformal mapping $\hat{f}_\alpha : H^{n+1} \longrightarrow H^{n+1}$ compatible with G if we put

$$\hat{f}_\alpha(x) = \varphi_\alpha(x) \quad \text{for} \quad x \in \bar{P}^{n+1}$$
$$\hat{f}_\alpha(x) = \chi_\alpha(g) \circ \varphi_\alpha \circ g^{-1}(x) \quad \text{for} \quad x \notin P^{n+1}, \tag{9.53}$$

where for the element g we take some element $g \in G$ for which $x \in g(\bar{P}^{n+1})$. The mapping $\hat{f}_\alpha$ can be extended to the sphere at infinity $\partial H^{n+1} = \bar{\mathbb{R}}^n$ to a $q(\alpha)$-quasi-conformal homeomorphism $f_\alpha : \bar{\mathbb{R}}^n \longrightarrow \bar{\mathbb{R}}^n$ compatible with G and defining the quasi-Fuchsian representation $f_\alpha^* = f(\alpha)_* \in R_n(G)$,

$$f_\alpha^* : G \longrightarrow G(\alpha) : g \longmapsto f_\alpha g f_\alpha^{-1}. \tag{9.54}$$

We now demonstrate the smooth dependence of this deformation of G on the parameter $\alpha \in I^m$ and that it is non-trivial; the first property is formulated as the following lemma.

LEMMA 9.35. *Under the conditions of Theorem 9.34, the mapping $\beta : I^m \longrightarrow R_n(G)$ with values $\beta(\alpha) = f_\alpha^*$ is differentiable (of class C^∞).*

PROOF. We first show that G can be decomposed into a non-trivial product of its subgroups.

For this purpose, we consider for any given $i = 1, \ldots, m$ an n-dimensional plane $S \subset H^{n+1}$ containing an n-dimensional side s of the polyhedron P^{n+1} from the i-th class of $\Pi(G)/\Theta(G)$. The orbit $G(S)$ splits up the space H^{n+1} to connected components D_k, amongst which we distinguish the component $D_1 \supset P^{n+1}$ and the component adjoining D_1 along the plane S, namely D_2. Let the stabilizers of the domains D_1 and D_2 and the plane S in G be denoted by G_1, G_2, and G_S correspondingly.

Two cases are possible: the quotient $[H^{n+1} \setminus G(S)]/G$ is either connected or consists of two components. In the first case, there exists an element $\gamma_i \in G$ such that $\gamma_i(D_1) = D_2$. We show that in this case G is the HNN-extension of the group G_1 by the element γ_i. We leave aside the obvious case where $\gamma_i^2 = \text{id}$ (γ_i is a rotation through π around a subspace in S) and assume that $\gamma_i \notin G_S$. We put $S' = \gamma_i^{-1}(S)$ and denote by $G_{S'}$ the stabilizer of the plane S' in the group G. Let D_3 be a component of the set $H^{n+1} \setminus G(S)$ adjoining the domain D_1 by S'. We also denote by Δ and Δ' fundamental polyhedra for the subgroups $G_S \cap G_1$ and $G_{S'} \cap G_1$, which contain P^{n+1}, and by $P_1 \subset \Delta \cap \Delta'$ we denote the fundamental polyhedron for the group G_1 (= the union of the polyhedra $\Delta \cap \bar{D}_2$, $\Delta' \cap \bar{D}_3$ and P^{n+1}). We have:

1. $\bar{D}_2$ and $\bar{D}_3$ are precisely invariant in G_1 with respect to the subgroups $G_S \cap G_1$ and $G_{S'} \cap G_1$ correspondingly, and $g(D_2) \cap D_3 = \varnothing$ for all $g \in G_1$;
2. $\gamma_i(D_3) \subset H^{n+1} \setminus \bar{D}_2$ is the domain D_1 and $G_1 \cap G_{S'} = \gamma_i^{-1}(G_1 \cap G_S)\gamma_i$;
3. there are r-neighbourhoods V_S and $V_{S'}$ of the planes S and S' such that $V_S \cap \Delta$, $V_{S'} \cap \Delta' \subset P_1$, where $0 < r \le d(S, G(S) \setminus S)/4$;

4. $\Delta \cap D_2 = P_1 \cap D_2$, $\Delta' \cap D_3 = P_1 \cap D_3$;
5. $\operatorname{int}\{P_1 \cap (H^{n+1} \setminus (\bar{D}_2 \cup \bar{D}_3))\} = P^{n+1} \neq \varnothing$.

The properties 1. - 5. allow one to apply Maskit's second combination theorem in the hyperbolic space H^{n+1} (see Theorem 6.24) to G_1 and γ_i, which implies on the basis of property 5. that the group obtained by the combination is G.

The resulting HNN-decomposition of the group $G = \langle G_1, \gamma_i \rangle$ shows that the representation $f^*_{\alpha_i} : G \longrightarrow G(\alpha_i)$ in that case can be written as

$$f^*_{\alpha_i}(g) = g, \qquad f^*_{\alpha_i}(\gamma_i) = A_S(\alpha_i) \circ \gamma_i \circ A^{-1}_{S'}(\alpha_i), \tag{9.55}$$

in which $g \in G_1$, $S' = \gamma_i^{-1}(S)$, while the elliptic transformations (rotations in H^{n+1}) $A_S(\alpha_i)$ and $A_{S'}(\alpha_i)$ have fixed $(n-1)$-dimensional planes in the n-planes S and S' correspondingly and are conjugate in $\text{Möb}(n) \cong \operatorname{Isom} H^{n+1}$ to the rotation $A(\alpha_i)$ from (9.52) through the angle $\alpha_i \in (-a_i, b_i)$.

In the second case, when $[H^{n+1} \setminus G(S)]/G$ has two connected components, no element in G transforms D_1 into D_2. Then $G_S = G_1 \cap G_2$ and we have:

1. S splits up H^{n+1} into two domains B_1 and B_2 ($D_1 \subset B_2$ and $D_2 \subset B_1$), whose closures are precisely invariant with respect to the stabilizer - the subgroup G_S in the groups G_1 and G_2 correspondingly.

2. If Δ, P_1, and P_2 are fundamental polyhedra for the groups G_S, G_1, and G_2 correspondingly that contain P^{n+1}, then for $i = 1, 2$, the polyhedra $\Delta \cap \bar{B}_i = \bar{B}_i \cap P_i$ do not contain P^{n+1} and therefore are not equal to the polyhedra P_i. Also $P_1 \cap P_2 = P^{n+1}$.

3. There is an r-neighbourhood V_S of the plane S such that $\Delta \cap V_S \subset P_i$, $i = 1, 2$, in which $r = d(S, G(S) \setminus S)/4 > 0$.

Maskit's first combination theorem (see Theorem 6.19) and these properties give us that G is a free product of G_1 and G_2 with the amalgamated subgroup G_S. Then the representation $f^*_{\alpha_i} : G \longmapsto G(\alpha_i)$ in that case can be written as

$$f^*_{\alpha_i}(g_1) = g_1, \qquad f^*_{\alpha_i}(g_2) = A_S(\alpha_i) \circ g_2 \circ A_S^{-1}(\alpha_i), \tag{9.56}$$

in which $g_1 \in G_1$, $g_2 \in G_2$, while the elliptic transformation $A_S(\alpha_i) \in \mathcal{M}_n$ is a rotation in H^{n+1} around an $(n-1)$-dimensional plane in S conjugate in $\mathcal{M}_n$ to the rotation $A(\alpha_i)$ from (9.52) through the angle $\alpha_i \in (-a_i, b_i)$.

From (9.55) and (9.56) we see that the smooth (class C^∞) dependence of the representations f^*_α on the parameter $\alpha \in I^m$ follows from the real analytic dependence on α_i for $A(\alpha_i)$ in (9.52). This completes the proof of the lemma.

To complete the proof of the theorem, it remains to demonstrate non-triviality and independence for the representations $f^*_{\alpha_i}$,

$i=1, \dots, m$. It follows from the definition of the set $\Pi(G)/\theta(G)$ and the proof of the lemma that there exists a non-elementary subgroup $G_0 \subset G$ (which is generated by the identifications of the sides of the polyhedron P^{n+1} which are not contained in $\Pi(G)$) from which the group G is obtained by a finite combination process, namely free amalgamated products and HNN-extensions. The fact that G_0 is non-elementary implies that for any $i=1, \dots, m$ an n-plane $S_i \subset H^{n+1}$ $(S_i \cap P^{n+1} \in \Pi(G))$ corresponding to the i-th class is spanned on the sphere ∂S_i, which splits up the limit set $\Lambda(G)$ (i.e., outside these spheres there are fixed points of certain loxodromic elements in the group G_0). Then by Theorem 3.23 on the density of the set of loxodromic fixed points in $\Lambda(G) \times \Lambda(G)$ we get that for each i there exists a loxodromic element $g^i \in G$ whose axis $l_i \subset H^{n+1}$ intersects the plane S_i transversely. For this hyperbolic translation g^i, let $d_i = d(g^i)$ be the translation length (in the hyperbolic metric d), namely for any point $z \in l_i$

$$d_i = d(z, g^i(z)) = \inf_{x \in H^{n+1}} d(x, g^i(x)).$$

In the deformation $f^*_{\alpha_i} : G \longrightarrow G(\alpha_i)$, the element g^i goes over to the element $f_{\alpha_i} g^i f^{-1}_{\alpha_i}$, which is a hyperbolic translation in H^{n+1} through a distance $d(\alpha_i) \gtrless d_i$ (in accordance with the sign of α_i) because for $z \in l_i$ due to (9.55) or (9.56) we have

$$d(z, g^i(z)) \gtrless d(z, f^*_{\alpha_i}(g^i)(z)).$$

Since the lengths of hyperbolic translations remain unchanged by conjugation in the group $\operatorname{Isom} H^{n+1}$, it follows that $f^*_{\alpha_i}$ is non-trivial for all $i=1, \dots, m$.

Independence for the representations $f^*_{\alpha_i}$, $i=1, \dots, m$, follows again from (9.55) and (9.56) and from the fact that n-planes $S_i, S_j \subset H^{n+1}$ corresponding to distinct i-th and j-th classes from the quotient $\Pi(G)/\theta(G)$ are such that their closures do not intersect. Theorem 9.34 is proved.

REMARK 9.36. It is readily seen that for the group $G \subset \text{Möb}(3)$ from Example 9.32 we have $\operatorname{card} \Pi(G)/\theta(G) = 6$; the representatives of the six distinct classes of that set are polyhedron sides lying in the non-intersecting spheres S_i, $9 \le i \le 14$.

REMARK 9.37. The above deformations $f^*_{\alpha_i}$ can be usefully explained from the viewpoint of the geometry in the hyperbolic manifold/orbiford H^n/G (hyperbolically rigid, due to Mostow's theorem since $\operatorname{vol} H^n/G < \infty$) and the manifold/orbifold H^{n+1}/G. Namely, if an n-plane $S \subset H^{n+1}$ corresponds (as previously) to the i-th class of $\Pi(G)/\theta(G)$, then the $q(\alpha_i)$-quasi-conformal, compatible with G, homeomorphism $\hat{f}_{\alpha_i} : H^{n+1} \longrightarrow H^{n+1}$, constructed in (9.53), projects to a

$q(\alpha_i)$-quasi-conformal homeomorphism of the orbifolds

$$F_{\alpha_i} : H^{n+1}/G \longrightarrow H^{n+1}/G(\alpha_i)$$

that turns the n-dimensional totally geodesic suborbifold $S/G_S \subset H^{n+1}/G$ around the $(n-1)$-dimensional totally geodesic suborbifold $(H \cap S)/G_S \subset H^n/G \subset H^{n+1}/G$ to the angle α_i. Here the image $F_{\alpha_i}(H^n/G)$ of the hyperbolic rigid orbifold (totally geodesic in H^{n+1}/G) is an orbifold bent along the $(n-1)$-dimensional totally geodesic suborbifold $(H^n \cap S)/G_S$. By bending we mean here (see Example 5.11 and Figs. 23 - 25) that the orbifold

$$F_{\alpha_i}[(H^n \backslash G(S \cap H^n))/G] = F_{\alpha_i}(H^n/G) \backslash (H^n \cap S)/G_S \tag{9.57}$$

is a totally geodesic orbifold, but the geodesics δ in H^n/G, orthogonal to the totally geodesic suborbifold $(H^n \cap S)/G_S$, go over F_{α_i} to geodesics in the orbifold (9.57), whose ends meet at points from $(H^n \cap S)/G_S$ at an angle $\pi - \alpha_i$.

Therefore the constructed deformations are naturally called bending ones following Thurston's term for surfaces.

REMARK 9.38. One can also establish the converse relation, namely from $(n-1)$-dimensional totally geodesic suborbifolds in H^n/G to quasi-Fuchsian deformations. For example, for $n=2$, i.e., for a Fuchsian group $G \subset \mathcal{M}_2$ that uniformizes a closed surface of genus $p>1$, one can construct a fundamental set $P(G)$ consisting of hyperbolic polygons $P_1, \ldots, P_{2p-2}$ containing right angles for which $\operatorname{card} \Pi(G)/\theta(G) = 3p-3$. This number is equal to the number of non-intersecting closed geodesics on the surface H^2/G.

4. *Properties of the space* $T_n(G)$ *defined by the geometry of the orbifold* H^nG.

Here we give another but equivalent definition of bending deformations for the Fuchsian group $G \subset \text{Möb}(n)$, which is due to Kourouniotis [1], [2]. The approach enables one to relate directly each totally geodesic $(n-1)$-orbifold in H^n/G to a bending deformation from $T_n(G)$ (see Remark 9.38), which coincides with some deformation $f^*_{\alpha_i}$ from the previous subsection. Although this definition is less obvious, it has some technical advantages, and in particular it enables one to find the bounds of the parameters of deformations and the corresponding infinitesimal deformations.

Let, as previously, the Fuchsian group G in the sphere S^n (the boundary of the unit ball $B^{n+1}(0, 1)$ be considered as a discrete isometry group in the hyperbolic space H^{n+1} $(\partial H^{n+1} = S^n)$ and its n-dimensional plane $H^n = B^n(0, 1) \subset B^{n+1}(0, 1)$. Let the hyperplane P_0 in H^{n+1} intersect the H^n orthogonally on a hyperplane $P \subset H^n$ with stabilizer $G_P \subset G$, $G_P \neq G$, and let there hold

$$\inf\{d(g(P), P) : g \in G\backslash G_P\} = a > 0. \tag{9.58}$$

We also assume without loss of generality that the origin lies in one of the components of $H^n\backslash G(P)$ adjoining the plane P. When all these conditions are satisfied, we call P a bending hyperplane for G and denote by $\mathcal{P}$ the orbit $G(P)$ consisting of an infinite number of distinct planes (due to $G_P \neq G$).

For each plane $R \in \mathcal{P}$, we consider (as for P) the hyperplane $R_0 \subset H^{n+1}$, $R_0 \cap H^n = R$, and denote by $R_0^+ \subset H^{n+1}$ the half-space bounded by R_0 and disjoint from 0. We define an order relation in $\mathcal{P}$ by

$$R \gg Q \iff R_0^+ \subset Q_0^+.$$

We note that for any $R, Q, \in \mathcal{P}$ the half-spaces R_0^+ and Q_0^+ either do not intersect, or one contains the other (i.e., we have a partial ordering in $\mathcal{P}$).

We consider the sequence of balls $B_i = B^{n+1}(0, 1 - 1/i) \subset B^{n+1}(0, 1)$ centred at the origin. Then we can use the finite subsets $\{R \in \mathcal{P} : R_0 \cap B_i \neq \emptyset\}$, $i \geq 2$, (due to the discreteness of G) to define a bijection $q : \mathbb{N} \longrightarrow \mathcal{P}$ such that q^{-1} preserves order.

Now let there be given real numbers α and ζ such that

$$0 < \zeta < \pi/2, \qquad 0 \leq \alpha < \pi - 2\zeta. \tag{9.59}$$

For these, we define the bending homeomorphism $\varphi = \varphi_{\alpha,\zeta} : \mathbb{C} \to \mathbb{C}$ (see Fig. 78):

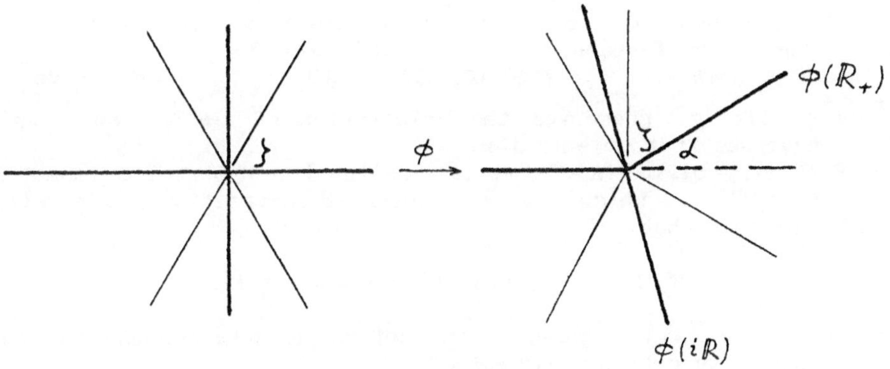

Figure 78

$$\varphi_{\alpha,\zeta}(z) = \begin{cases} z, \text{ if } \pi - \zeta \le |\arg z| \\ z\cdot\exp(i\alpha), \text{ if } |\arg z| \le \zeta \\ z\cdot\exp(i\alpha(1-(\arg z - \zeta)/(\pi - 2\zeta))), \\ \qquad \text{if } \zeta < \arg z < \pi - \zeta \\ z\cdot\exp(i\alpha(1+(\arg z + \zeta)/(\pi - 2\zeta))), \\ \qquad \text{if } \zeta - \pi < \arg z < -\zeta, \end{cases} \tag{9.60}$$

in which $\arg z$ is the principal value of the argument of a complex number $z \in \mathbb{C}$, $-\pi < \arg z \le \pi$. Then for the linear distortion $K(\varphi, z)$ we have:

$$K(\varphi, z) = \begin{cases} 1, \text{ if } \pi - \zeta \le |\arg z| \\ 1, \text{ if } |\arg z| \le \zeta \\ (\pi - 2\zeta)/(\pi - 2\zeta - \alpha), \text{ if } \zeta < \arg z < \pi - \zeta \\ (\pi - 2\zeta + \alpha)/(\pi - 2\zeta), \text{ if } \zeta - \pi < \arg z < -\zeta. \end{cases}$$

Then we see that for the α and ζ from (9.59), the map $\varphi = \varphi_{\alpha,\zeta}$ is quasi-conformal (and is conformal near the real axis)

For negative α, we put for $2\zeta - \pi < \alpha < 0$:

$$\varphi_{\alpha,\zeta}(z) = \overline{\varphi_{-\alpha,\zeta}(\bar{z})}.$$

To define the bending $\psi^{\pm}_{\alpha,\zeta} : \mathbb{R}^{n+1} \longrightarrow \mathbb{R}^{n+1}$, we identify the plane $(\mathbb{R}e_n + \mathbb{R}e_{n+1})$ with the complex plane $\mathbb{C} = \mathbb{R}e_n + i\mathbb{R}e_{n+1}$ and put

$$\psi^{+} = \psi^{+}_{\alpha,\zeta} : (x_1, \ldots, x_{n+1}) \longmapsto (x_1, \ldots, x_{n-1}, \varphi_{\alpha,\zeta}(x_n + ix_{n+1}))$$
$$\psi^{-} = \psi^{-}_{\alpha,\zeta} : (x_1, \ldots, x_{n+1}) \longmapsto (x_1, \ldots, x_{n-1}, j \circ \varphi_{\alpha,\zeta} \circ j(x_n + ix_{n+1})), \tag{9.61}$$

where j denotes the involution $j : \mathbb{C} \longrightarrow \mathbb{C} : z \longmapsto -\bar{z}$.

The mappings ψ^{+} and ψ^{-} leave invariant the ball $B^{n+1}(0, 1)$ and the half-space $\{x \in \mathbb{R}^{n+1} : x_1 > 0\}$; also, for $g \in \operatorname{Isom} H^n \subset \operatorname{Isom} H^{n+1}$, $g(B^{n-1}(0, 1)) = B^{n-1}(0, 1)$ we have that $g\psi^{\pm} = \psi^{\pm}g$ if g preserves the orientation of B^{n-1}, and $g\psi^{\pm} = \psi^{\mp}g$ if g reverses that orientation.

For every plane $R \in \mathcal{P}$ we now define the hyperbolic isometry γ_R of $H^{n+1} = B^{n+1}$, which is an orientation-preserving isometry of $H^n = B^n$ and such that

$$\gamma_R(B^{n-1}) = R, \quad \gamma_R(\{x \in \bar{B}^{n+1} : x_n \ge 0\}) = R^{+}_{0}. \tag{9.62}$$

If here $R = q(j)$, $j \in \mathbb{N}$, we define a quasi-conformal bending $\psi_j : H^{n+1} \longrightarrow H^{n+1}$ by putting $\psi_j = \gamma_R \psi^{+} \gamma_R^{-1}$.

The properties of ψ^{+} and ψ^{-} show that this definition does not depend on the choice of γ_R. The mapping ψ_j bends H^n within H^{n+1} along the hyperplane R and fixed the half-space $H^n \backslash R$ containing 0.

Kourouniotis [1] and [2, Theorem 4.7] showed that iterations of these mappings ψ_j converge to a quasi-conformal homeomorphism $H^{n+1} \longrightarrow H^{n+1}$ compatible with G, which he obtained by studying the supports of the mappings ψ_j and of the sets where the linear distortions $K(\psi_j, x)$ are different from 1. More precisely:

THEOREM 9.39. *Let* $G \subset \mathrm{Isom}\, H^n$ *be a discrete group and* $P \subset H^n$ *its bending hyperplane with the condition* (9.58). *If* α *and* ζ *satisfy* (9.59) *and*

$$\cosh(a/2) > 1/\sin \zeta, \tag{9.63}$$

then the mapping $f_{\alpha,\zeta} = \lim_{j \to \infty} \psi_1 \circ \psi_2 \circ \ldots \circ \psi_j : H^{n+1} \longrightarrow H^{n+1}$ *is a quasi-conformal homeomorphism of* H^{n+1} *compatible with the group* G.

We note that the homeomorphism $f_{\alpha,\zeta}$ does not depend on the choice of the bijection $q : \mathbb{N} \longrightarrow \mathcal{P}$ in which q^{-1} preserves order, while the homomorphism

$$f^*_\alpha : G \longrightarrow G(\alpha) = f_\alpha G f_\alpha^{-1} \subset \mathrm{Isom}\, H^{n+1}$$

does not depend on the choice of ζ satisfying (9.59) and (9.63).

The bound to the parameter α is defined by a number η, $0 < \eta < \pi$, which is such that $\mathrm{Cos}(\eta/2) = 1/\cosh(a/2)$. In fact, for any θ, $0 < \theta < \eta$, there is the above constructed family of quasi-conformal homeomorphisms f_α compatible with G that defines the quasi-conformal bending deformation

$$\beta : (-\theta, \theta) \longrightarrow R_n(G) : \alpha \longmapsto f^*_\alpha.$$

REMARK 9.40. A bending hyperplane $P \subset H^n$ defines (as in the proof of Lemma 9.35) a decomposition of the group $G : G = \langle G_1, g_P \rangle$ or $G = G_1 *_{G_P} G_2$ in accordance with whether the quotient $[H^n \setminus G(P)]/G$ is connected or not. Here G_1, G_2, and G_P are stabilizers in the group G of the components D_1, D_2, and the plane P, in which $D_1, D_2 \subset H^n \setminus G(P)$ adjoin one another on the plane P and $D_2 \subset P_0^+$, while the element $g_P \in G$ transfers D_1 to D_2. Kourouniotis [1] (Theorems 4.1 and 4.2) gave an algebraic proof of this; see Lemma 9.35 for the geometric proof.

This decomposition of G shows at once that the representations (9.55) and (9.56) apply also for the homomorphism f^*_α that is defined by the homeomorphism $f_{\alpha,\zeta}$. In fact, for a HNN-extension $G = \langle G_1, g_P \rangle$, we have:

$$f^*_\alpha(g) = g \quad \text{for} \quad g \in G_1, \quad f^*_\alpha(g_P) = A_P(\alpha) \circ g_P \circ A_P^{-1}(\alpha),$$

and for the case of an amalgamated product $G = G_1 *_{G_P} G_2$, we have:

$$f^*_\alpha(g_1) = g_1 \quad \text{for} \quad g_1 \in G_1, \quad f^*_\alpha(g_2) = A_P(\alpha) \circ g_2 \circ A_P^{-1}(\alpha),$$

in which $P' = g_P^{-1}(P)$, while A_P is the hyperbolic rotation in H^{n+1} around P through an angle α, $A_P|_P = \mathrm{id}$, conjugate in $\mathrm{Isom}\, H^{n+1}$ to the rotation $A(\alpha)$ from (9.52).

Then from (9.55) and (9.56) it is clear that the bending deformations constructed in Theorem 9.34 coincide with the bending deformations in this subsection.

REMARK 9.41. It is important that a bending hyperplane $P \subset H^n$ for the group $G \subset \mathrm{Isom}\, H^n$ corresponds to a totally geodesic suborbifold P/G_P of the orbifold H^n/G, whose r-neighbourhood for some $r > 0$ is universally covered by the r-neighbourhood of the hyperplane P. The converse also applies: to each totally geodesic suborbifold $M_0 \subset H^n/G$ having an r-neighbourhood of the form $M_0 \times [-r/2, r/2]$ in H^n/G there corresponds a hyperplane $P \subset H^n$ (which is a universal covering of M_0), and this is a bending hyperplane for the group G $(G_P = d_*(\pi_1(M_0)))$. Then these bending deformations of the group G correspond to bendings of the orbifold H^n/G along totally geodesic suborbifolds of codimension 1, compare Remark 9.37.

We now describe conditions for non-triviality and independence of bending deformations, where we say that a set $X \subset S^n$, $\mathrm{card}\, X \geq 2$, spans an m-dimensional sphere $S \subset S^n$ if S is the sphere of lowest dimension which contains the set X.

THEOREM 9.42 (Kourouniotis [2], Theorem 5.3). *Let $P \subset H^n$ be a bending hyperplane for a non-elementary discrete group $G \subset \mathrm{Isom}\, H^n$. Then the bending deformation β of G along the hyperplane P is trivial if and only if P does not intersect the interior of the convex Nielsen hull H_G.*

The proof is based on the existence of a loxodromic element $g \in G$ whose axis $l_g \subset H^n$ transversely intersects the plane P (a consequence of Theorem 3.23) and on the fact that the hyperbolic translation lengths are invariants by conjugations in the group $\mathrm{Isom}\, H^{n+1}$ (compare the proof of non-triviality for deformations in Theorem 9.34).

Let us assume in general that we have not one but several disjoint bending hyperplanes $P_1, \dots, P_m \subset H^n$ for the group $G \subset \mathrm{Isom}\, H^n$, $G(P_i) \cap G(P_j) = \emptyset$, $i \neq j$, such that for all P_i and for $\mathcal{P} = G(\bigcup_{i=1}^m P_i)$

$$a_i = \inf\{d(P_i, Q) : Q \in \mathcal{P} \setminus P_i\} > 0.$$

As above, taking a bijection $q : \mathbb{N} \longrightarrow \mathcal{P}$ and the order relation $\gg$ on $\mathcal{P}$ defined by it (for which q^{-1} preserves the order), we can define (as in Theorem 9.39) a family $\{f_w : H^{n+1} \longrightarrow H^{n+1}\}$ of quasi-conformal homeomorphisms compatible with G, $w = (w_1, \dots, w_m) \in W^m \subset \mathbb{R}^m$ and specifying the deformation

$$\beta : W^m \longrightarrow R_n(G) \subset \mathrm{Hom}(G, \mathrm{Isom}\, H^{n+1}) : w \longmapsto f_w^*.$$

Here W^m is any neighbourhood of the origin in $\mathbb{R}^m$ whose closure is contained in the set

$$U = \{u \in \mathbb{R}^m : \cos(u_i/2) > 1/\cosh(a_i/2), \quad -\pi < u_i < \pi\},$$

while the non-triviality of β is established by Theorem 9.42.

THEOREM 9.43. *Let $G \subset \mathrm{Isom}\, H^n$, $n \geq 3$, be a discrete group with a set $\mathcal{P}$ of disjoint bending hyperplanes P, finite $(= m)$ modulo G, and let*

$$p : R_n(G) \longrightarrow T_n(G) = R_n(G)/\mathrm{Isom}\, H^{n+1}$$

be the natural projection (the quotient map).

If $\Lambda(G) \cap \partial P$ spans ∂P for each $P \in \mathcal{P}$ and the deformation defined by bending along each $G(P)$ is not trivial, then the map

$$p\beta : W^m \longrightarrow T_n(G)$$

is an embedding.

PROOF. Let P_1 and $P_2 \subset H^n$ be bending hyperplanes for the discrete group $G \subset \mathrm{Isom}\, H^n$ that define non-trivial deformations $\beta_i : (-\theta, \theta_i) \longrightarrow R_n(G)$, $0 < \theta_i < a_i$, and are such that, for each $i = 1, 2$, $\Lambda(G) \cap \partial P_i$ spans the sphere ∂P_i. It is sufficient to show that, if for $\alpha_1 \in (-\theta_1, \theta_1)$ and $\alpha_2 \in (-\theta_2, \theta_2)$, $\alpha_1 \neq 0$, $\alpha_2 \neq 0$, there exists $\gamma \in \mathrm{Isom}\, H^{n+1}$ such that for all $g \in G$

$$\beta_1(\alpha_1)(g) = \gamma \cdot \beta_2(\alpha_2)(g) \cdot \gamma^{-1}, \tag{9.64}$$

then $\alpha_1 = \alpha_2$ and $P_2 \subset G(P_1)$.

We denote by $f_{\alpha_i} : H^{n+1} \longrightarrow H^{n+1}$ quasi-conformal mappings that define the representations $\beta_i(\alpha_i)$. Then (9.64) shows that the quasi-conformal mapping $f_{\alpha_i}^{-1} \circ \gamma \circ f_{\alpha_2}$ commutes with elements of the group G and therefore its extension in $\partial H^{n+1} = S^n$ fixes the points of $\Lambda(G)$:

$$\gamma f_{\alpha_2}(x) = f_{\alpha_1}(x), \quad x \in \Lambda(G). \tag{9.65}$$

As $\Lambda(G) \cap \partial P_1$ spans the $(n-2)$-dimensional sphere $\partial P_1 \subset \partial H^n$ and $\Lambda(G) \not\subset \partial P_1$ $(G_{P_1} \neq G)$, the limit set $\Lambda(G)$ spans the sphere $S^{n-1} = \partial H^n$. We now show that this with (9.65) implies that (9.65) holds for all $x \in H^n$.

In fact, the conformality for all $g \in G$ of the restriction $f_{\alpha_1}|_{\partial(gP_1)}$ implies that the mapping $f_{\alpha_1}|_{G(\partial P_1)}$ is determined by the mapping $f_{\alpha_1}|_{\Lambda(G)}$. If also U is any component of the set $H^n \backslash G(P_1)$ and R and Q are images of the plane P_1 adjoining U, then U is conformally mapped by f_{α_1} into the sphere spanned by the set $f_{\alpha_1}(Q \cup R)$. This shows that the mapping $f_{\alpha_1}|_{H^n}$ is determined by its values in $G(\partial P_1)$ and thus by its values in $\Lambda(G)$. Similarly, $f_{\alpha_2}|_{H^n}$ is determined by its values in $\Lambda(G)$. This implies directly that (9.65) applies for all $x \in H^n$.

From (9.63) in particular we see that γ maps totally geodesic pieces of the hypersurface $f_{\alpha_2}(H^n)$ to totally geodesic pieces of $f_{\alpha_1}(H^n)$, which shows that $\alpha_1 = \alpha_2$ and implies that the mapping $f_{\alpha_1}^{-1}\gamma f_{\alpha_2}$ maps the set $G(P_2)$ to the set $G(P_1)$. The identity (9.65) completes the proof that $G(P_1) = G(P_2)$ and also the proof of the theorem.

COROLLARY 9.44 (Kourouniotis [2], Johnson and Millson [1]). *If $G \subset \operatorname{Isom} H^{n+1}$ is a discrete torsion free group for which the hyperbolic manifold H^n/G is compact and contains m disjoint totally geodesic $(n-1)$-dimensional connected submanifolds, then there is a smooth embedding of an m-dimensional ball in the space $T_n(G)$.*

5. *Infinitesimal deformations for bendings and singularities of the spaces $R_n(G)$ and $T_n(G)$.*

To determine infinitesimal deformations for bendings defined by totally geodesic hypersurfaces in H_n/G (see previous subsections), we begin by calculating vector fields corresponding to the family of quasi-conformal homeomorphisms $f_\alpha : H^{n+1} \longrightarrow H^{n+1}$.

We have for the mapping φ_α from (9.60) and the involution $j : \mathbb{C} \longrightarrow \mathbb{C}$, $j(z) = -\bar{z}$:

$$\frac{d}{d\alpha}\bigg|_{\alpha=0} (\varphi_\alpha(z)) - \frac{d}{d\alpha}\bigg|_{\alpha=0} (j \circ \varphi_\alpha \circ j(z)) = iz.$$

Then for the mappings $\psi_\alpha^{\pm}$ from (9.61) we get the vector field

$$\chi(x) = \frac{d}{d\alpha}\bigg|_{\alpha=0} (\psi_\alpha^+(x)) - \frac{d}{d\alpha}\bigg|_{\alpha=0} (\psi_\alpha^-(x)) = (0, \ldots, 0, -x_{n+1}, x_n).$$

We are also give the vector field $\xi : x \longmapsto \frac{d}{d\alpha}\Big|_{\alpha=0} (\psi_\alpha^+(x))$.

If now $R \in \mathcal{P} = G(P)$ and $\psi_{R,\alpha} = \gamma_R \cdot \psi_\alpha^+ \cdot \gamma_R^{-1}$, in which $P \subset H^n$ is a bending hyperplane for G and γ_R is defined in (9.62), then using the action of $\operatorname{Isom} H^{n+1}$ given in (9.15) on vector fields, we write

$$\frac{d}{d\alpha}\bigg|_{\alpha=0} (\psi_{R,\alpha}(x)) = \gamma_R \cdot \xi(x).$$

This property enables one to determine a vector field from (9.14) for the mappings f_α:

$$\theta(x) = \frac{d}{d\alpha}\bigg|_{\alpha=0} (f_\alpha(x)) = \sum_{R\in\mathcal{P}} \gamma_R\cdot\xi(x).$$

As we have seen in §1.3, this vector field defines the infinitesimal deformation for the deformation $(f^*_\alpha)_{\alpha\in I}\subset R_n(G)$, i.e., an element $c\in H^1(G, \mathcal{H}_{n+1})$ ($\mathcal{H}_{n+1}$ is the Lie algebra of $\operatorname{Isom} H^{n+1}$), which is defined by the vector field (compare (9.16)):

$$c(\delta)(x) = \frac{d}{d\alpha}\bigg|_{\alpha=0} (\beta_\alpha(\delta)(x) = \theta(x) - \delta\cdot\theta(x) =$$

$$= \sum_{R\in\mathcal{P}} [\gamma_r\cdot\xi(x) - (\delta\gamma_R)\cdot\xi(x)] = \sum_{R\in\mathcal{P}} [\gamma_r\cdot\xi(x) - (\delta\cdot\gamma_{\delta^{-1}(R)})\cdot\xi(x)],$$

in which $\delta\in G$.

The latter sum can be simplified. Namely, if $\delta(\delta^{-1}(R)^+_0) = R^+_0$. Then $\gamma_R^{-1}\delta\gamma_{\delta^{-1}(R)}$ preserves the orientation of the ball B^{n-1} and we have

$$\delta\gamma_{\delta^{-1}(R)}\cdot\xi(x) = \gamma_R\cdot\xi(x).$$

If on the other hand $\delta(\delta^{-1}(R)^+_0) \neq R^+_0$, then $g = \gamma_R^{-1}\delta\gamma_{\delta^{-1}(R)}$ reverses the orientation of B^{n-1} and $g\cdot\psi^{\pm} = \psi^{\mp}\cdot g$, as noted above, so

$$(\delta\gamma_{\delta^{-1}(R)})\cdot\xi(x) = \gamma_R\cdot\left[\frac{d}{d\alpha}\bigg|_{\alpha=0}(\psi^-_\alpha)\right](x)$$

and

$$\gamma_R\cdot\xi(x) - (\delta\gamma_{\delta^{-1}(R)})\cdot\xi(x) = \gamma_R\cdot\chi(x).$$

This completes the proof of the following theorem (Kourouniotis [2]):

THEOREM 9.45. *Let $G\subset\operatorname{Isom} H^n$ be a finitely generated discrete group and let $P\subset H^n$ be a bending hyperplane for it. Then the infinitesimal deformation for bending along P is presented by the crossed homomorphism $c_p\in H^1(G, \mathcal{H}_{n+1})$,*

$$c_p(\delta)(x) = \Sigma\{\gamma_R\cdot\chi(x) : R\in G(P) \quad\text{and}\quad R^+_0 \neq \delta(\delta^{-1}(R)^+_0)\}.$$

Now we describe briefly the relation between the properties of infinitesimal deformations and the existence of singularities of the spaces $R_n(G)$ and $T_n(G)$.

If $R_n(G)$ lacks singularities, to each two deformations β_1 and β_2 of the group G with infinitesimal deformations $c_1, c_2\in H^1(G, \mathcal{H}_{n+1})$ there should correspond a deformation β that defines the infinitesimal deformation $c_1 + c_2\in H^1(G, \mathcal{H}_{n+1})$. However, examining co-compact discrete torsion free groups $G\subset\operatorname{Isom} H^n$ given by

Venkov-Borel's arithmetic construction (see Chapter 4 §2), Johnson and Millson [1] have shown that, if β_P and β_R are bending deformations with respect to intersecting totally geodesic hyperplanes P, $R \subset H^n$, $P \cap R \neq \emptyset$ (which correspond to HNN-extensions of G - see (9.55)), i.e., with respect to intersecting totally geodesic hyperplanes in the manifold H^n/G (not dividing it), then for $n \geq 4$, the infinitesimal deformation $c_P + c_R \in H^1(G, \mathcal{H}_{n+1})$ is not in general integrable. Then the space $R_n(G)$ of quasi-Fuchsian representations has the inclusion representation $\rho_0 : G \subset \mathrm{Isom}\, H^{n+1}$ as a singular point of it. Further, if a hyperplane $Q \subset H^n$, $G(Q) \cap (G(P) \cap G(R)) = \emptyset$, is also a bending plane for the group G, then we have seen above that the bending β_Q along Q does not alter the restrictions of the representation ρ_0 on the stabilizers G_p and G_R. Therefore, a curve in $R_n(G)$ determined by the deformation β_Q consists entirely of singular points.

These facts are proved by using the necessary integrability conditions for infinitesimal deformations given by Johnson-Millson's Theorem 9.11. Namely, one checks that for the cocycle $c_P + c_R \in H^1(G, \mathcal{H}_{n+1})$ the cup-product $(c_p + c_R) \cup (c_p + c_R)$ given by (9.17) and cohomologous to $2(c_p + c_R)$ is a non-zero element of the second cohomology group $H^2(G, \mathcal{H}_{n+1})$ for $n \geq 4$. For $n = 3$, that approach becomes impossible because then $c_p \cup c_R = 0$ always, since $H^2(G, \mathcal{H}_4) = 0$.

We note also that in this approach the following questions remain open: are these singularities of the spaces $R_n(G)$ and $T_n(G)$ topological and are there such singularities in the spaces $R_3(G)$ and $T_3(G)$?

Another approach to these questions will be described in the next subsection.

6. *Stamping deformations*

It is well known that in the case of a closed surface H^2/G of genus $g > 1$ the Teichmüller space $T(H^2/G)$ is a $(6g-6)$-dimensional cell, and bending deformations yield it a $(3g-3)$-dimensional submanifold: here $(3g-3)$ is the number of closed non-intersecting geodesics on the surface H^2/G. The remainder of the deformations are assigned by Dehn twists of these geodesics. However, in contrast to the closed geodesic on H^2/G, a number of conformal homeomorphisms of a totally geodesic hypersurface in a manifold H^n/G, $G \subset \mathrm{Isom}\, H^n$, is finite. So the number of conformal uniformized structures on H^n/G corresponding to different conformal gluings along a bending surface in H^3/G is at most finite. This and the impossibility of simultaneous bendings along intersecting hypersurfaces $R, P \subset H^n/G$ (i.e., non-integrability of $c_P + c_R \in H^1(G, \mathcal{H}_{n+1})$ - see subsection 5) implied a conjecture (C. Kourouniotis, Oberwolfach, September 1985) that spatial deformations H^n/G, $n \geq 3$, are exactly bendings.

Below we shall describe two new kinds of spatial deformations and discuss their connections with singularity questions.

We shall begin with the stamping deformation of a special hyperbolic 3-manifold $M = H^3/\Gamma$, $\mathrm{vol}\, M < \infty$, with the following properties. The manifold M contains at least three totally geodesic surfaces $S_i \subset M$ intersecting along a common geodesic $l = \bigcap_i S_i$. Here S_i are separated surfaces. Namely, if $\Gamma_i \subset \Gamma$ are their fundamental groups, i.e., $S_i = H_i/\Gamma_i$ where H_i are hyperbolic 2-planes universally covering the surfaces S_i, $\bigcap_i H_i = \tilde{l}$, and $\Gamma_l = \Gamma_i \cap \Gamma_j$ is a loxodromic group with the axis $\tilde{l}$, then the group Γ is the free amalgamated product:

$$\Gamma = \Gamma_1 \ast_{\Gamma_l} \Gamma_2 = \Gamma_2 \ast_{\Gamma_l} \Gamma_3 = \Gamma_1 \ast_{\Gamma_l} \Gamma_3.$$

We shall explain such a kind of deformations by the following example.

EXAMPLE 9.46 (*stamping deformation*). Let $G_0 \subset \mathrm{Isom}\, H^3$ be the discrete group of finite volume $\mathrm{vol}\, H^3/G < \infty$ from Example 5.83. Its fundamental polyhedron $P_0 = P_0(G) \subset H^3$ has all dihedral angles $\pi/3$ and is bounded by eight 2-planes spanned on the circles shown in Fig. 35. Let $H_1, H_2, H_3 \subset H^3$ be planes whose circles at infinity $\partial P_i \subset \partial H^3 = \mathbb{R}^3$ are the following:

$$\{x \in \overline{\mathbb{R}}^2 : x_1 = 0\}, \quad \{x \in \overline{\mathbb{R}}^2 : x_1 = \sqrt{3}\cdot x_2\}, \quad \{x \in \overline{\mathbb{R}}^2 : x_1 = \sqrt{3}\cdot x_2\}.$$

It is readily seen that (9.58) is obeyed for the planes H_i and therefore Theorem 9.39 gives the bending deformations $\beta_i : (-\theta_i, \theta_i) \longrightarrow R_3(G_0) \subset \mathrm{Hom}(G_o, \mathrm{Isom}\, H^4)$ having the infinitesimal deformations $c_i = c_{p_i}$ (see Theorem 9.45).

We now consider for any α from the interval $(\alpha_0, 2\pi/3]$ with some $\alpha_0 < \pi/2$ a quasi-Fuchsian "pea-pod" group $G(\alpha) \subset \mathrm{Isom}\, H^4$ defined by the regular trihedral angle $D_{0\infty}(\alpha) \subset \overline{\mathbb{R}}^3$ with the side angles α at the vertex at the origin and unit vectors $v_1(\alpha) = (1, 0, 0)$, $v_2(\alpha)$ and $v_3(\alpha)$ such that the scalar product $(v_i(\alpha), v_j(\alpha) = \cos\alpha$ for all $i \neq j$.

That is, by analogy with Example 5.83, we specify for $G(\alpha)$ eight spheres in $\overline{\mathbb{R}}^3$ bounding a fundamental polyhedron $P_\alpha \subset \overline{\mathbb{R}}^3$:

$$\begin{aligned} S_k(\alpha) &= S^2(\rho(\alpha)\cdot(v_i(\alpha) + v_j(\alpha))/\sqrt{2 + 2\cos\alpha},\ r(\alpha)), \\ S_{ij}(\alpha) &= S^2(v_k(\alpha)\cdot\rho(\alpha),\ r(\alpha)), \qquad\qquad (9.66) \\ I(\alpha) &= S^2(0,\ R(\alpha)), \quad I'(\alpha) = S^2(0,\ R'(\alpha)), \end{aligned}$$

in which the numbers (i, j, k) are obtained by all permutations of the numbers $(1, 2, 3)$ and the numbers $R(\alpha)$, $R'(\alpha)$, $\rho(\alpha)$, and $r(\alpha)$ are determined (modulo a positive factor $\lambda > 0$) by the equalities:

$$\begin{aligned} 2\rho(\alpha)\cdot \mathrm{Sin}(\alpha/4) &= \sqrt{3}\cdot r(\alpha), \quad R'(\alpha) = (\rho^2(\alpha) - r^2(\alpha))/R(\alpha), \qquad (9.67) \\ R(\alpha) &= (-\rho(\alpha) + \sqrt{4\rho^2(\alpha) - 3r^2(\alpha)})/2. \end{aligned}$$

Taking a normalization $\rho(\alpha)=\sqrt{3}$ we obtain:

$$R(\alpha) = \sqrt{3}\cos(\alpha/4) - \mathrm{Sin}(\alpha/4)$$

$$R'(\alpha) = \sqrt{3}\cos(\alpha/4) + \mathrm{Sin}(\alpha/4)$$

$$r(\alpha) = 2\,\mathrm{Sin}(\alpha/4).$$

Direct calculations give that all dihedral angles of the polyhedron P_α,

$$P_\alpha = \bigcap_{i,j}(\mathrm{ext}\, S_{ij}(\alpha) \cap \mathrm{ext}\, S_i(\alpha)) \cap \mathrm{int}\, I'(\alpha) \cap \mathrm{ext}\, I(\alpha)$$

do not depend on α, $\alpha_0<\alpha<2\pi/3$, and are equal to $\pi/3$ where the value $\alpha_0<\pi/2$ corresponds to the tangency of the spheres $S_{ij}(\alpha_0)$ and $S_{ik}(\alpha_0)$, $j\neq k$. Note that for $\alpha=2\pi/3$ we have the fundamental polyhedron $P=P_{2\pi/3}$ of the group $G_0=G(2\pi/3)$ acting isometrically in the Poincaré half-space model $\mathbb{R}^3_+$ of the hyperbolic space H^3.

Let us now consider the polyhedra P^4 and P^4_α in the hyperbolic space $H^4\cong\mathbb{R}^4_+$ which are bounded by 3-planes spanned on 2-spheres (9.66) and are fundamental polyhedra in H^4 for G_0 and $G(\alpha)$ respectively.

Then, as in the proof of Theorem 9.39, we construct the family of $q(\alpha)$-quasi-conformal mappings $f_\alpha : H^4 \longrightarrow H^4$ compatible with the group G_0, which we do by extending (compatible with G_0) the quasi-conformal mappings $\varphi_\alpha : P^4_0 \longrightarrow P^4_\alpha$ of the fundamental polyhedra of the isomorphic groups G_0 and $G(\alpha)$. The family of those mappings (f_α) is dependent smoothly (real analytically) on α, as (9.66) and (9.67) show, and specifies a quasi-conformal deformation for G_0 (a smooth curve in the space of quasi-Fuchsian representations of the group G_0):

$$\sigma : (\alpha_0, 2\pi/3] \longrightarrow R_3(G_0) \subset \mathrm{Hom}(G_0, SO(4, 1) : \alpha \longmapsto f^*_\alpha.$$

With this deformation, there are both simultaneous bendings of the group G_0 along the hyperbolic planes H_1, H_2, and H_3 through equal angles $\alpha_1=\alpha_2=\alpha_3=h(\alpha)$ and a stretching along the geodesic $H_1\cap H_2\cap H_3$. It shows that the infinitesimal deformation c_σ of σ is such that the crossed homomorphism

$$c_\sigma - c_1 - c_2 - c_3 \in H^1(G_0, \mathfrak{M}_3)$$

determines a non-zero class.

The following observations point to a topological singularity in $R_3(G_0)$ at the point $\rho_0 : G_0 \hookrightarrow \mathrm{Isom}\, H^4$.

LEMMA 9.47. *For any $\alpha\in(\alpha_0, 2\pi/3)$, the representation $\sigma(\alpha)=f^*_\alpha$ has an open neighbourhood in $R_3(G_0)\subset Hom(G_0, Isom\, H^4)$ that does not intersect a subset in $R_3(G_0)$ corresponding to bending deformations.*

PROOF. We note that for any group $G = \rho G_0$ in which a representation $\rho \in R_n(G_0)$ is obtained by bending along a certain family of disjoint hyperplanes in H^n, the boundary of its convex Nielsen hull $H_G \subset H^{n+1}$ is isometrically developed into the hyperbolic n-plane $H^n \subset H^{n+1}$, see Chapter 5 §6. At the same time, we have observed in Theorem 5.84 and Remark 5.85 that for each $\alpha \in (\alpha_0, 2\pi/3)$, the boundary of the Nielsen hull $H_{G(\alpha)}$ for the group $G(\alpha)$ can be developed into the hyperbolic plane H^3 only by means of a quasi-isometry of H^4 expanding its dihedral 3-dimensional angles of the values α up to the values $2\pi/3$. This shows that the quasi-isometry coefficient of any such mapping $f : H^4 \longrightarrow H^4$ that conjugates the groups G_0 and $G(\alpha)$ cannot be less than $2\pi/3\alpha > 1$, which completes the proof.

The quasi-conformal deformation σ of the co-finite Fuchsian group G_0 in the ball, constructed in Example 9.46, i.e., deformation of the natural conformal structure of the hyperbolic orbifold H^3/G_0 is called a stamping deformation of H^3/G_0 along its closed geodesic that is the intersection of the totally geodesic suborbifolds $S_i = H_i/G_{H_i} \subset H^3/G_0$.

We use here the term stamping because such a deformation in a Riemann surface (bending with respect to two intersecting geodesics) corresponds to stamping that surface at the point of intersection of the geodesics (Fig. 79).

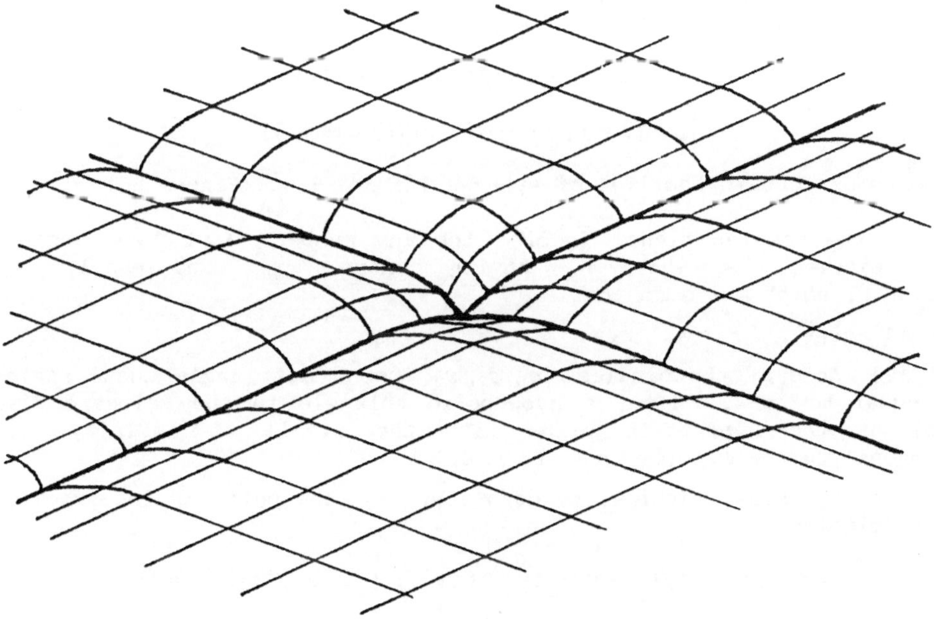

Figure 79

REMARK 9.48. The constructed stamping deformation on H^3/G of finite volume along a closed geodesic can also be realized in the case of a closed hyperbolic 3-manifold as well as in the case of 4-dimensional hyperbolic manifolds. The question of the existence of stamping deformations along submanifolds of codimension 2 in H^n/G, $n \geq 4$, remains open.

7. *Stamping-with-torsion along a closed geodesic*

Here we prove the existence of a third kind of deformations of a conformal structure on a closed hyperbolic 3-manifold H^3/G which are distinct from bendings and stampings. To determine this class of deformations we consider hyperbolic manifolds H^3/G of a special kind, associated with some quasi-Fuchsian groups on S^3 similar, in a certain sense, to the Jørgensen [3] group G_J. Namely, both the Jørgensen group and our groups have a form of semidirect product of geometrically infinite normal subgroups and subgroups of similarities with co-presentation

$$\langle a, b : a^k b^m = 1 \rangle$$

where k and m are certain integers. The construction of such groups was given by A.V. Tetenov and the author - see Apanasov [32], [35].

As in Example 9.46, we shall define a co-compact group $G \subset \mathrm{Isom}\, H^3 \cong \text{Möb}(2)$ and its deformation

$$\sigma_{\mathrm{tors}} : (-\varepsilon, \varepsilon) \longrightarrow \mathrm{Hom}(G, \text{Möb}(3))/\text{Möb}(3),$$

$$\sigma_{\mathrm{tors}}(t) = f_t^* : G \longrightarrow G_t = f_t G f_t^{-1} \subset \text{Möb}(3)$$

by means of certain families of spheres in $\mathbb{R}^3$ - see Figure 80.

In the complex plane $\overline{\mathbb{C}} = \partial H^3$ for any number $\lambda > 1$ we consider three circles $s_i = S^1(z_i, r_i)$ having centre z_i and radii r_i, $i = 1, 2, 3$, which are such that:

(1) $r_1 = 1$, $r_2 = \lambda$, $r_3 = \lambda^3$;

(2) Arcs of the circles s_1, s_2, s_3 bound a triangle whose angles are equal to $\pi/4$, i.e., a hyperbolic triangle in the Poincaré disc model of the hyperbolic plane with the circle at infinity s_4 orthogonal to the circles s_i, $i = 1, 2, 3$.

For the chosen circles, we denote by z_0 the point of intersection of two circles

$$\{z \in \mathbb{C} : |z_2 - z|/|z_1 - z| = 1\}, \quad \{z \in \mathbb{C} : |z_3 - z|/|z_1 - z| = \lambda^3\},$$

farther from z_1.

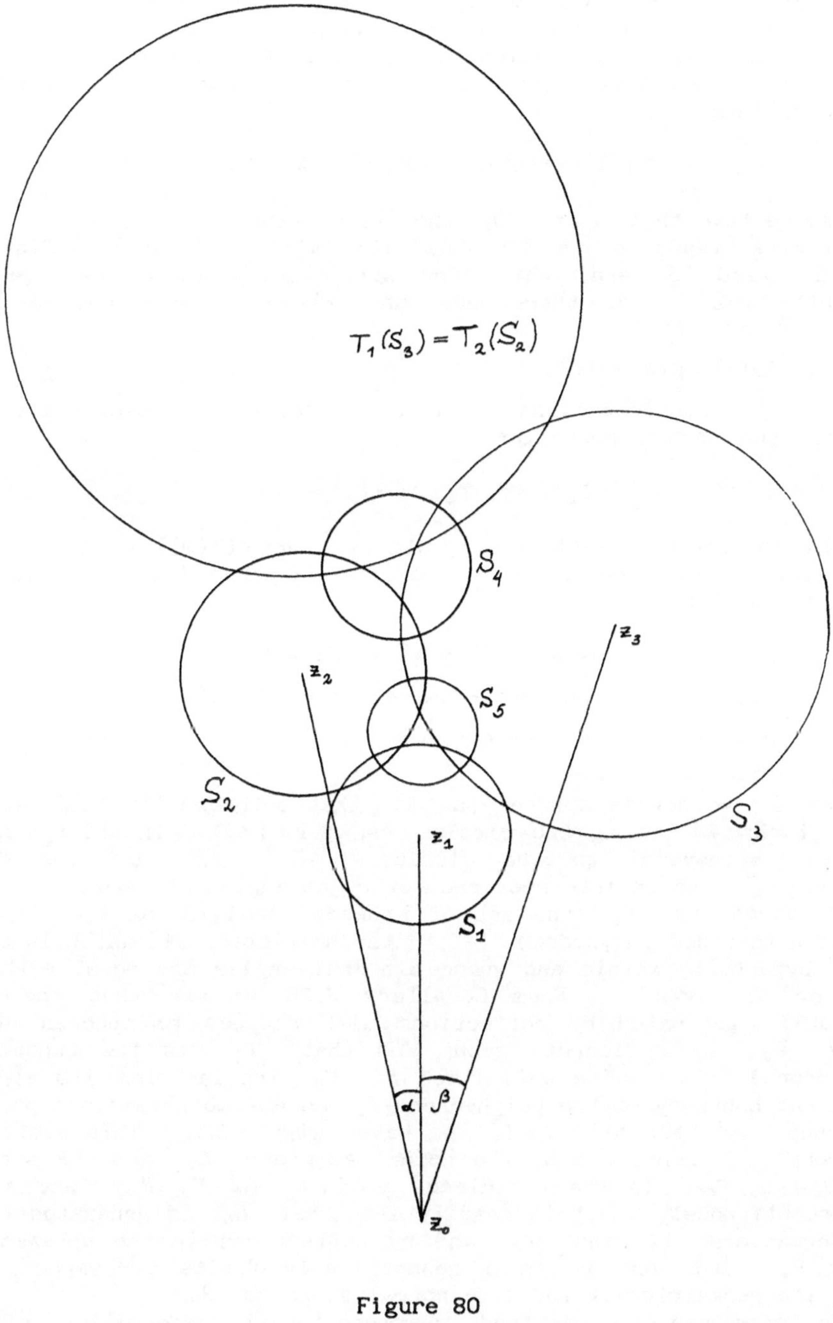

Figure 80

Denote by $\alpha > 0$ and by $\beta > 0$ respectively the angles between the vectors $(z_2 - z_0)$ and $(z_3 - z_0)$ with a vector $(z_1 - z_0)$. Applying a Euclidean translation in the plane $\mathbb{C}$, we can assume that the point z_0 coincides with the origin. Now define loxodromic transformations

$$T_1(z) = \lambda \cdot e^{i\alpha} \cdot z, \qquad T_2(z) = \lambda^3 \cdot e^{-i\beta} \cdot z.$$

For them we have that $T_1(s_1) = s_2$ and $T_2(s_1) = s_3$.

When λ tends to one, the functions $\alpha(\lambda)$ and $\beta(\lambda)$ defining α and β tend to zero, while for sufficiently large λ we have $\alpha(\lambda) + \beta(\lambda) > \pi/2$, so there are an integer m and a value $\lambda = \lambda(m) > 1$ such that:

(3) $3\alpha(\lambda) + \beta(\lambda) = 2\pi/m$.

(4) The transformations T_1 and T_2 generate an elementary group Γ_0 having the co-representation

$$\Gamma_0 = \langle T_1, T_2 : T_1^{3m} \cdot T_2^{-m} = 1 \rangle.$$

The numbers $m \in \mathbb{N}$, $\lambda = \lambda(m)$, $\alpha = \alpha(\lambda(m))$ and $|z_1|$ $(= |z_1 - z_0|)$ obtained by our geometric constructions satisfy the system of equations

$$\begin{cases} |z_1|^2(1 - 2\lambda \cos \alpha + \lambda^2) = \lambda^2 + \lambda\sqrt{2} + 1 \\ |z_1|^2(1 - 2\lambda^3 \cos(2\pi/m - 3\alpha) + \lambda^6) = \lambda^6 + \lambda^4\sqrt{2} + 1 \\ |z_1|^2(\lambda^2 - 2\lambda^4 \cos(2\pi/m - 2\alpha) + \lambda^6) = \lambda^6 + \lambda^4\sqrt{2} + \lambda^2 \end{cases} \qquad (*)$$

due to the law of cosines.

By S_i we denote spheres in $\mathbb{R}^3$ that orthogonally intersect the plane $\mathbb{C} = \mathbb{R}^2 \subset \mathbb{R}^3$ along the circles s_i, $1 \le i \le 5$, in which s_5 is a circle orthogonal to the circles s_2, s_3 and the circle $T_1(s_3) = T_2(s_2)$, which intersect each other at angles of $\pi/4$.

We denote by $\mathscr{C}_0$ the set of spheres $S \in \Gamma_0(S_1, S_4, S_5)$ bounding the infinite-sided polyhedron P_0 in the half-space $\mathbb{R}^3_+$ which is convex in the hyperbolic metric and whose dihedral angles are equal either to $\pi/4$ or to $\pi/2$. From Corollary 3.20 we see that the group $H_0 \subset \text{Möb}(3)$ generated by reflections with respect to spheres of the family $\mathscr{C}_0$ is a discrete group and that P_0 is its fundamental polyhedron in $\mathbb{R}^3_+$. The definition of Γ_0 implies that its elements act at the boundary of the polyhedron P_0 as automorphisms and preserve the group H_0: for all $\gamma \in \Gamma_0$ we have $\gamma H_0 \gamma^{-1} = H_0$. This shows that the group G_0 generated by the normal subgroup H_0 and the subgroup Γ_0, $G_0 = \langle H_0, \Gamma_0 \rangle$ is the semidirect product $G_0 = \Gamma_0\ H$ and acts in $\mathbb{R}^3_+$ discontinuously. It is readily seen that G_0 is generated by the transformations T_1 and T_2 and by reflections in the spheres S_i, $i = 1, 4, 5$, and that it is a geometrically finite co-compact group having the geometrically infinite normal subgroup H_0.

To construct a required quasi-conformal deformation of the co-compact group $G_0 \subset \text{Isom}\, H^3$, we construct a family of quasi-Fuchsian groups $G_t \subset \text{Möb}(3)$ depending smoothly on the parameter t in a certain ε-neighbourhood of zero (ε to be defined below). Namely, as for G_0, we consider for any t, $-\varepsilon < t < \varepsilon$, and any $\lambda > 1$ the spheres $S_i(t) = S^2(x_t^i, r_i(t))$, $i = 1, 2, 3$, with the centres $x_t^i \in \mathbb{C} \times \mathbb{R} = \mathbb{R}^3$ and radii $r_i(t)$ such that:

(1_t) $r_1(t) = 1, \quad r_2(t) = \lambda, \quad r_3(t) = \lambda^3;$

$$x_t^1 = (z_t^1, t), \quad x_t^2 = (z_t^2, \lambda t), \quad x_t^3 = (z_t^3, \lambda^3 t);$$

(2_t) Each sphere $S_i(t)$ intersect two other spheres at the angles $\pi/4$.

In the exterior of the spheres $S_i(t)$ consider an analogue of the point z_0, i.e., the point $x_t^0 = (z_t^0, 0)$ farther from the intersection point x_t^1 of the plane $\mathbb{C} \times \{0\}$ with the circle of intersection of two spheres:

$$\{x \in \mathbb{R}^3 : |x_t^3 - x|/|x_t^1 - x| = \lambda^3\}, \quad \{x \in \mathbb{R}^3 : |x_t^2 - x|/|x_t^1 - x| = \lambda\}.$$

The point x_t^0 depends smoothly on the parameter t and $\lim_{t\to 0} x_t^0 = z_0$.

Let $\alpha_t = \alpha(t, \lambda) > 0$ and $\beta_t = \beta(t, \lambda) > 0$ be the angles between the orthogonal projection of the vectors $(x_t^2 - x_t^0)$, $(x_t^1 - x_t^0)$ and $(x_t^3 - x_t^0)$, $(x_t^1 - x_t^0)$, respectively, onto the plane $\mathbb{C} \times \{0\}$. Applying a translation $x \longmapsto (x - x_t^0)$ we can assume that $x_t^0 = 0$. Let us consider the orthogonal matrices

$$A_t = \begin{pmatrix} \cos\alpha_t & \sin\alpha_t & 0 \\ -\sin\alpha_t & \cos\alpha_t & 0 \\ 0 & 0 & 1 \end{pmatrix}, \quad B_t = \begin{pmatrix} \cos\beta_t & -\sin\beta_t & 0 \\ \sin\beta_t & \cos\beta_t & 0 \\ 0 & 0 & 1 \end{pmatrix}$$

that specify the loxodromic transformations

$$T_{1t}(x) = \lambda\cdot A_t\cdot x, \quad T_{2t}(x) = \lambda^3\cdot B_t\cdot x$$

and are such that

$$T_{1t}(S_1(t)) = S_2(t), \quad T_{2t}(S_1(t)) = S_3(t).$$

Moreover, for the integer m, determined while defining the group Γ_0 (condition (3)) one can find a number $\lambda_t = \lambda(m, t) > 1$ such that:

(3_t) $3\alpha(t, \lambda(m, t)) + \beta(t, \lambda(m, t)) = 2\pi/m;$

(4_t) the transformations T_{1t} and T_{2t} generate an elementary group

$$\Gamma_t = \langle T_{1t}, T_{2t} : T_{1t}^{3m}\cdot T_{2t}^{-m} = 1\rangle.$$

The numbers $\lambda_t = \lambda(m, t)$, $\alpha_t = \alpha(t, \lambda(m, t))$ and $|z_t^1|$ $(= |x_t^1 - x_t^0|)$ are solutions of the system of equations obtained (as in (*)) from our geometric construction by applying the law of cosines:

$$\begin{cases} \lambda_t^2 + \lambda_t\sqrt{2} + 1 = |z_t^1|^2(1 - 2\lambda_t \cos \alpha_t + \lambda_t^2) + t^2(\lambda_t - 1)^2 \\ \lambda_t^6 + \lambda_t^3\sqrt{2} + 1 = |z_t^1|^2(1 - 2\lambda_t^3 \cos(2\pi/m - 3\alpha_t) + \lambda_t^6) + t^2(\lambda_t^3 - 1)^2 \\ \lambda_t^6 + \lambda_t^4\sqrt{2} + \lambda_t^2 = |z_t^1|^2(\lambda_t^2 - 2\lambda_t^4 \cos(2\pi/m - 2\alpha_t) + \lambda_t^6) + t^2(\lambda_t^3 - \lambda_t)^2 \end{cases}$$

It should be noted that, due to the smoothness of this system with respect to t, the above parameters of the family $\{S_i(t)\}$ and of the group Γ_t depend smoothly on t and they tend to the parameters from (*) determining the group Γ_0 when t approaches zero.

Now let $S_4(t)$ and $S_5(t)$ be disjoint spheres orthogonal to the sphere triples

$$S_2(t),\quad S_3(t),\quad T_{1t}(S_3(t)) = T_{2t}(S_2(t)) \quad \text{and} \quad S_2(t),\quad S_3(t),\quad S_1(t)$$

correspondingly and not intersecting other spheres from the Γ_t-orbit of spheres $S_i(H)$, $i = 1, 2, 3$, (lower and upper bounds for the parameter t are defined by that property). We consider a family $\mathscr{C}_t$ of spheres in $\mathbb{R}^3$ that is the Γ_t-orbit of the above five spheres $S_i(t)$. The spheres in that family bound a infinite-sided polyhedron P_t with dihedral angles of $\pi/2$ and $\pi/4$ having the same combinatorial type as P_0. Let H_t be the discrete group generated by reflections with respect to the spheres of the family $\mathscr{C}_t$, which is preserved by elements of the group Γ_t, and, as above, let the discrete group G_t be the semidirect product of its subgroups:

$$G_t = \Gamma_t \quad H_t.$$

Clearly G_t is isomorphic to the group G_0 and that isomorphism may be induced in a standard fashion by a quasi-conformal homeomorphism f_t of $\overline{\mathbb{R}}^3$. Namely, by extending spheres of the families $\mathscr{C}_0$ and $\mathscr{C}_t$ to 3-planes of the hyperbolic space $H^4 \cong \mathbb{R}^4_+$ which bound polyhedra P^4 and $P^4(t)$ in H^4, we can construct in a standard way a quasi-conformal homeomorphism $f_t : \overline{H}^4 \longrightarrow \overline{H}^4$ compatible with G_0 and inducing the isomorphism $f_t^* : G_0 \longrightarrow G_t = f_t G_0 f_t^{-1}$. We have thus a family of quasi-Fuchsian representations $f_t^* : G_0 \longrightarrow G_t$ defined by quasi-conformal mappings $f_t : \overline{\mathbb{R}}^3 \longrightarrow \overline{\mathbb{R}}^3$ compatible with G_0, which defines a quasi-conformal deformation of G_0:

$$\sigma_{tors} : (-\varepsilon, \varepsilon) \longrightarrow T_3(G_0) : t \longmapsto f_t^*.$$

This deformation σ_{tors} specifies the deformation of the conformal structure of the orbifold H^3/G_0, which we call a stamping-with-torsion along the closed geodesic $l \subset H^3/G_0$ not lying in any totally geodesic surface from H^3/G_0. Here the geodesic $l \subset H^3/G_0$ is covered by the axis $\tilde{l} = \{x \in \mathbb{R}^3_+ : x_1 = x_2 = 0\}$ of the loxodromic group $\Gamma_0 \subset G_0$.

REMARK 9.49. We use the name "stamping-with-torsion" for two reasons. First, the construction of σ_{tors} is close to that of the stamping σ from Example 9.46 (vector fields defining infinitesimal deformations for both stampings leave those closed geodesics invariant), and second, the discontinuity set $\Omega(G_t)$ of the quasi-Fuchsian group $G_t \subset \text{Möb}(3)$ contains a maximal cone-domain $C(0, \infty)$ (a circular cone with axis $\tilde{l}$, compare Chapter 5 §6 and Example 5.73) i.e., the boundary of the Nielsen hull $H_{G_t} \subset H^4$ has an isometrically non-developable singularity of the circular cone type, compare Lemma 9.47.

The following conjecture is excited by the properties of the above quasi-Fuchsian deformations of Fuchsian groups and local properties of the limit set of a geometrically finite group (compare Theorem 5.75).

CONJECTURE 9.50. Let the quasi-Fuchsian group $G \subset \text{Möb}(n)$, $n \geq 3$, be the result of quasi-conformal deformation of a Fuchsian co-finite group $\Gamma \subset \text{Isom}\, H^n \cong \text{Möb}(n-1)$ (Γ is isomorphic to the fundamental group of the hyperbolic n-orbifold H^n/Γ of finite volume). Then G is obtained from Γ by the superposition of three types of deformations (at least for $n = 3$):

a) bending along totally geodesic hypersurfaces $M_i \subset H^n/\Gamma$;

b) stamping along totally geodesic suborbifolds of dimension $\leq n-2$, which are intersections of the hypersurfaces $M_i \subset H^n/\Gamma$; and

c) stamping-with-torsion along totally geodesic suborbifolds of dimension $\leq n-2$.

It is possible that here the deformation space $T_n(\Gamma)$ for the group Γ, or even the space of marked uniformized conformal structures on the orbifold H^n/Γ, are such that the complement in it to the set of Fuchsian representations of the group Γ splits up into connected components corresponding to different types of deformations, as Lemma 9.47 would indicate.

§4. The boundary of deformation space for a Fuchsian group

Let $G \subset \text{Möb}(n)$ be a Fuchsian group of the first kind (finitely generated) acting in the half-space R^n_+ as a hyperbolic isometry group; §1 and §3 show that for it there are defined deformation spaces $R_n(T)$ and $T_n(\Gamma) = R_n(\Gamma)/\text{Möb}(n)$ consisting of quasi-Fuchsian representations $\rho : \Gamma \longrightarrow \text{Möb}(n)$, which are specified by quasi-conformal conjugations of Γ. Considering the set of sequences $\{\rho_i\} \subset T_n(\Gamma)$, $\rho_i : \Gamma \longrightarrow G_i = f_i \Gamma f_i^{-1}$ converging in the topology $T_n(\Gamma)$ (algebraic convergence topology), which is equivalent to the Teichmüller topology defined by the metric (9.46), we obtain the closure $\overline{T_n(\Gamma)}$. The boundary points of $\overline{T_n(\Gamma)} \backslash T_n(\Gamma)$ correspond to injective representations

$\rho \in \mathrm{Hom}(\Gamma, \mathrm{M\ddot{o}b}(n))$ [1] that are not quasi-conformal conjugations but are their limits.

For $n = 2$, properties of boundary points (in the Teichmüller sense) of the space of quasi-Fuchsian representations for Γ are described by results due to Bers [4], Maskit [2], and Abikoff [3]: the boundary $\partial T(\Gamma)$ consists of the so-called B-groups, i.e., Kleinian groups each having one invariant (simply connected) component, which are different from quasi-Fuchsian ones. Here there are B-groups of three types: non-degenerate, partially degenerate, and degenerate (see Chapter 3 §3), where the last consistute in a certain sense the majority of the boundary groups. It has been proved that:

1) for a Riemann surface $S = B^2/\Gamma$ of finite type there exists a continuum set of degenerate groups $G \subset \partial T(\Gamma)$, non-conjugate in $\mathcal{M}_2$, such that $\Omega(G)/G \approx S$;

2) a non-degenerate group $G \in \partial T(\Gamma)$ uniformizes the main surface $S = S_0 = \Omega_0/\Gamma$, in which Ω_0 is an invariant simply connected component in $\Omega(G)$, together with a finite number of surfaces $S_1, \ldots, S_m$, corresponding to non-invariant components in $\Omega(G)$ and derived from S by contraction to a point for certain non-homotopic loops in S; and

3) the partially degenerate groups occupy a position intermediate between the non-degenerate and degenerate ones; they are obtained by the Maskit combination from the quasi-Fuchsian and degenerate groups (for Thurston's compactification $\overline{T(\Gamma)}$, see Fathi et al. [1]).

For $n \geq 3$, the boundary groups $G \in \partial T_n(\Gamma)$ acquire properties of a purely spatial character (see Apanasov [11], [12]).

THEOREM 9.51. *For $n \geq 3$, there exist boundary groups $G = \rho\Gamma$ of the quasi-Fuchsian representation space $T_n(\Gamma)$ having the following properties:*

1) *the limit set $\Lambda(G)$ splits up the discontinuity set $\Omega(G)$ into two invariant components Ω_0 and Ω_1;*

2) *the component Ω_0 is quasi-conformally equivalent to a ball; and*

3) *the homotopy groups $\pi_i(\Omega_i) \neq 0$ for all $i = 1, 2, \ldots, n-2$.*

THEOREM 9.52. *At the boundary of the quasi-Fuchsian representation space $T_n(\Gamma)$, $n \geq 3$, there exist points ρ such that Kleinian groups $G = \rho\Gamma \subset \mathrm{M\ddot{o}b}(n)$ have the following properties:*

1) *the limit set $\Lambda(G)$ splits up the discontinuity set $\Omega(G)$ into two invariant components Ω_0 and Ω_1; and*

2) $\pi_i(\Omega_j) \neq 0$ *for all* $i = 1, \ldots, n-2$; $j = 0, 1$.

The proof of Theorems 9.51 and 9.52 will be shown by examining the properties of the groups $G(t_1, 5/6)$ and $G(t_1, \tau_0)$ which are the boundary groups of the space $T_3(\Gamma)$, where Γ is the Fuchsian co-compact group from Example 9.32.

---

[1] These boundary representations $\rho \in \mathrm{Hom}(\Gamma, \mathrm{M\ddot{o}b}(n))$ are not only injective but also discrete representations. This fact is a corollary of a compactness theorem by Thurston-Morgan-Shalen, - see Apanasov [35].

Then let Γ be the Fuchsian group generated by inversions with respect to the spheres S_i, $1 \le i \le 14$, constructed in Example 9.32 (see (9.48)). We consider the groups $G(t, \tau)$, isomorphic to Γ and corresponding to the system of spheres $\{\sigma_i(t, \tau)\}$, obtained from the system $\{S_i\}$ by replacing the spheres $S_{14} = S^2(-re_3, \rho)$ and $S_{13} = S^2(re_3, \rho)$, $\rho = \sqrt{10}/6$, $r = 5/6$, by the spheres $\sigma_{14}(t, \tau)$ and $\sigma_{13}(t, \tau)$ having their centres correspondingly at $-te_3$ and τe_3 and having radii $\sqrt{t^2 - t + 5/12}$ and $\sqrt{\tau^2 - \tau + 5/12}$ correspondingly (they are orthogonal to the corresponding spheres $\sigma_i(t, \tau) = S_i$, $i = 1, \dots, 12$). We have shown in Example 9.32 that the values of t and τ satisfying $t_0 < t < t_1$, $\tau_0 < \tau < \tau_1$, in which $t_0 = \tau_0$ and $t_1 = \tau$ are defined in (9.49), correspond to the quasi-Fuchsian groups $G(t, \tau) = \Phi^*_{t,\tau}(\Gamma)$ defined by the quasi-conformal representations $\Phi^*_{t,\tau} : \gamma \longmapsto \Phi_{t,\tau}\gamma\Phi^{-1}_{t,\tau}$. Here the quasi-conformal mappings $\Phi_{t,\tau} : \overline{\mathbb{R}}^3 \to \overline{\mathbb{R}}^3$ are constructed by analogy with the mappings Φ_t from Example 9.32 or as in the proof of Theorem 9.34.

We now consider the limit groups $G(t_1, \tau)$ and $G(t_1, \tau_0)$, obtained for $(t, \tau) \to (t_1, \tau)$ and $(t, \tau) \to (t_1, \tau_0)$. We note firstly that they are Kleinian, which is demonstrated by analogy with the case of non-limit parameters. The fundamental polyhedra $F_{t_1,\tau}$, F_{t_1,τ_0} have two components each: bounded ones $F^0_{t_1,\tau}$ and $F^0_{t_1,\tau_0}$ and unbounded ones $F^1_{t_1,\tau}$ and $F^1_{t_1,\tau_0}$. $F^0_{t_1,\tau}$ is simply connected, as are all the F^0_t from (9.50) for non-limit values t, while $G(t_1, \tau)$ is an invariant component, $\Omega_0(t_1, \tau)$ is an image of the ball $B(0, \sqrt{15}/6)$ by the quasi-conformal mapping $\Phi_{t_1,\tau}$. At the same time, the fundamental groups of the remaining polyhedra are such that

$$\pi_1(F^1_{t_1,\tau}) \cong \pi_1(F^0_{t_1,\tau_0}) \cong \pi_1(F^1_{t_1,\tau_0}) \cong \mathbb{Z}^4. \tag{9.68}$$

This will be clear if we note that the spheres $\sigma_{14}(t_1, \tau)$, $\sigma_{13}(t_1, \tau_0)$ and $\sigma_{14}(t_1, \tau_0)$ are tangent to the four spheres $\sigma_i = S_i$, $9 \le i \le 12$, and the tangent points are found by direct examination to lie outside the other spheres $\sigma_j = S_j$, $1 \le j \le 8$. Therefore, these tangent points are fixed points for some parabolic elements of the groups. Lemma 3.7 implies that the discontinuity sets $\Omega(t_1, \tau)$ and $\Omega(t_1, \tau_0)$ of the groups $G(t_1, \tau)$ and $G(t_1, \tau_0)$ split up each into two pairs of invariant components:

$$\Omega(t_1, \tau) = \Omega_0(t_1, \tau) \cup \Omega_1(t_1, \tau), \qquad \Omega(t_1, \tau_0) = \Omega_0(t_1, \tau_0) \cup \Omega_1(t_1, \tau_0). \qquad (9$$

At the same time, due to Lemma 3.22, the orbits of these parabolic (tangent) points are dense in the corresponding limit sets $\Lambda(t_1, \tau)$ and $\Lambda(t_1, \tau_0)$. Then (9.68) implies that, for all $\tau_0 < \tau < \tau_1$, the group $G(t_1, \tau)$ has the connected components (9.69) which satisfy Theorem 9.51, while $G(t_1, \tau_0)$ has the components (9.69) that satisfy Theorem 9.52. These groups belong to the boundary of $T_3(\Gamma)$ because they are not quasi-Fuchsian but at the same time are the limits of groups corresponding to points of $T_3(\Gamma)$ (to the parameters (t, τ), which tend to (t_1, τ) and (t_1, τ_0)).

REMARK 9.53. For $n \geq 3$, a natural generalization for the non-degenerate B-groups in the plane is provided by geometrically finite Kleinian groups $G \subset \text{Möb}(n)$ each having contractible invariant components $\Omega_0 \subset \Omega(G)$, as Chapter 9 §5 shows from their properties. However, Theorem 9.52 shows that, for $n \geq 3$, not all the boundary points on $\partial T_n(\Gamma)$ correspond to the groups G having contractible invariant components $\Omega_0 \subset \Omega(G)$. Moreover, Example 7.19 (the wild limit sphere in $\overline{\mathbb{R}}^3$) gives an example of a spatial non-degenerate B-group $G \subset \text{Möb}(3)$ not lying[2] in the quasi-Fuchsian representation space $T_n(\Gamma)$ for any Fuchsian group $\Gamma \subset \text{Möb}(3)$.

The parabolic elements of the boundary groups $G(t_1, \tau)$ and $G(t_1, \tau_0)$ in Theorems 9.51 and 9.52 are analogues to accidental parabolic elements of groups on the plane: for limit isomorphisms $\Gamma \to G(t_1, \tau)$ and $\Gamma \to G(t_1, \tau_0)$ they correspond to hyperbolic elements in Γ.

The following important questions remain open.

What are the properties of the boundary points $\rho \in \partial T_n(\Gamma)$ for $n \geq 3$ for which the discrete groups $\rho\Gamma$ do not have accidental parabolic elements?

Do such groups $\rho\Gamma$ include Kleinian groups whose discontinuity sets each consist of one connected contractible component (as degenerate groups for $n = 2$)?

It appears likely that the answer to the second question is negative.

The following Example 9.55 of a deformation for a Fuchsian group $\Gamma \subset \text{Möb}(3)$ is related to the first question, which shows that the bounds to the bending parameters in general are not determined by the values for which the corresponding hyperplanes are tangent. This example also gives a proof of the following theorem.

THEOREM 9.54. *The homotopy groups of the isometric fundamental domain for a quasi-Fuchsian group $G \subset \mathcal{M}_n$ are not invariant on quasi-conformal deformation.*

This theorem for $n = 2$ can be obtained in two ways. First, it follows from considering the dimension of the boundary of the Teichmüller space $T(G)$ and the lower dimension of a subset of it corresponding to the cusp points, see Jørgensen [1]. Second, one can quote a detailed example constructed by the author in Krushkal', Apanasov, and Gusevskii [1] (Example 10) and [3] (Example 19).

---

[2] This is related to non-connectedness of the space of uniformized conformal structures on a closed hyperbolic 3-manifold H^3/Γ - see Apanasov [35], [36]. Recently the author proved (Apanasov [36]) that this group does not lie in the closure of bendings and stampings subsets in $T_3(\Gamma)$.

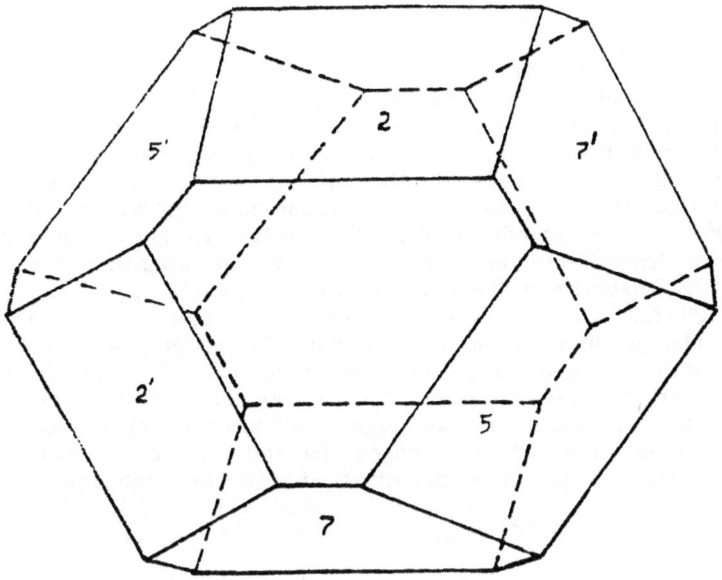

Figure 81

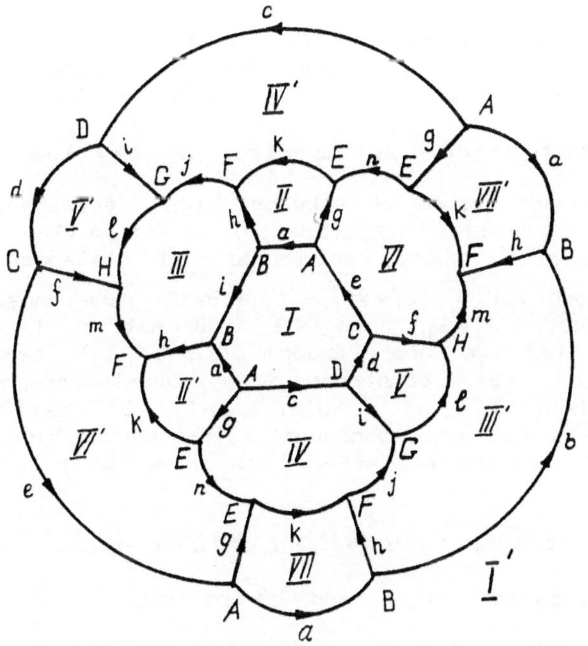

Figure 82

EXAMPLE 9.55. Let $n=3$ and $Q=\{x\in\mathbb{R}^3 : |x_i|\le 1/2\}$ be a unit cube, while $\{S_i\}_{1\le i\le 14}$ is a system of spheres with radii $\rho_i=\sqrt{3}/3$ centred at the vertices Q for $i=1,\dots,8$, while for $9\le i\le 14$, the spheres S_i are defined in (9.48). All these spheres are orthogonal to the sphere $S_0=S(0,\sqrt{15}/6)$. If the spheres S_i and S_j intersect for $i,j\le 8$, then they do so at angles of $\pi/3$; if the spheres S_j intersect the spheres S_i, $1\le i\le 8$, $9\le j\le 14$, then they do so at angles of $\pi/2$. These spheres bound the polyhedron $P\subset B^3$, $\partial B^3=S_0$, as shown in Fig. 81, which is convex and bounded in the Poincaré hyperbolic metric. If now we consider the group G generated by loxodromic transformations $g_1,\dots,g_7$ that glue the sides of P as shown in the spherical development of the polyhedron (Fig. 82), the group will be a co-compact Fuchsian group due to Theorem 3.19. Namely, in such a gluing the edges of P split up into 14 classes: four classes each of four edges having dihedral angles of $\pi/2$, four classes each of two edges having dihedral angles of $\pi/3$, and six classes each of two edges having dihedral angles of $\pi/3$ (Fig. 82). Then we get a representation for the Fuchsian group G:

$$G = \langle g_1,\dots,g_7 : (g_6^{-1}g_5)^2 = (g_4^{-1}g_5)^2 = (g_1^{-1}g_5)^2 = (g_3g_5)^2 =$$
$$= (g_4g_3)^3 = (g_6g_3)^3 = (g_6g_4)^3 = (g_3^{-1}g_1)^3 = (g_4^{-1}g_1)^3 = (g_6^{-1}g_1)^3 = \qquad (9.70)$$
$$= g_1^{-1}g_7^{-1}g_1g_2 = g_3^{-1}g_7^{-1}g_3g_2 = g_6^{-1}g_7^{-1}g_4g_2 = g_4g_7^{-1}g_6^{-1}g_2 = 1\rangle.$$

We now consider a deformation $\Phi=\Phi(t_1,t_2,\dots,t_6)$ of the group G, which is dependent on six parameters t_i. Namely, if the generators g_2, g_5, and g_7 glue the spheres S_i, $i=9,\dots,14$, as follows:

$$g_2 : S_{13}\longmapsto S_{10},\quad g_5 : S_{12}\longmapsto S_{11},\quad g_7 : S_{14}\longmapsto S_9, \qquad (9.71)$$

we consider a new system of spheres $\{\sigma_i\}$ obtained from $\{S_i\}$ by replacing the spheres $S_i=S_2(\pm re_k,\rho)$, $9\le i\le 14$, by the spheres $\sigma_i(t_i)=S^2(\pm t_i\cdot e_k,\rho_i)$, whose centres are at distances t_i from the origin and whose radii $\rho_i=\sqrt{t_i^2 - t_i + 5/12}$ are chosen such that the $\sigma_i(t_i)$ intersect $\sigma_j=S_j$, $1\le j\le 8$, at angles of $\pi/2$. By the sphere system $\{\sigma_i\}$ we form a group $G(t_1,\dots,t_6)$ isomorphic to (9.70) and as in §3 construct a quasi-conformal deformation $\beta : D^6\longrightarrow \mathrm{Hom}(G,\mathrm{M\ddot{o}b}(3))$, $\beta(t_1,\dots,t_6)=\Phi^* : G\longrightarrow \Phi G\Phi^{-1}=G(t_1,\dots,t_6)$, where $\Phi:\overline{\mathbb{R}}^3\longrightarrow\overline{\mathbb{R}}^3$ are quasi-conformal self-homeomorphisms.

To produce the desired effect, we allow the parameters to tend as follows

$$t_9=t_{10}=t_{11}=t\longrightarrow t^1,\quad t_{13}=t_{14}=\tau\longrightarrow\tau_0, \qquad (9.72)$$

where the limit parameters t^1 and τ^0 satisfy

$$\tau^0 = \frac{5}{4(3+\sqrt{3})},\qquad t^1<t_1=5(1+\sqrt{5/6}). \qquad (9.73)$$

A direct calculation shows that (9.73) will give the spheres σ_i, $9 \le i \le 14$, that for the parameters (9.72) do not intersect pairwise, while the spheres $\sigma_{13}(\tau^0)$ and $\sigma_{14}(\tau^0)$ are tangent at the origin.

To complete the proof, we note that P is an isometric fundamental polyhedron for the Fuchsian group G, since all the spheres S_i are isometric for the generators g_k and for the inverse transformations g_k^{-1}, $k = 1, \ldots, 7$. On the other hand, for the quasi-Fuchsian group Φ^*G, the isometric spheres $I(g_2)$ and $I(g_7)$ are such that

$$I(g_2^{-1}) \subset \operatorname{int} \sigma_{10}(t^1), \quad \sigma_{13}(\tau^0) \subset \operatorname{int} I(g_2),$$

$$\sigma_{14}(\tau^0) \subset \operatorname{int} I(g_7), \quad I(g_7^{-1}) \subset \operatorname{int} \sigma_9(t^1).$$

These inclusions follow from the radii of the spheres $\sigma_{13}(\tau^0)$ and $\sigma_{14}(\tau^0)$ being $\sqrt{(\tau^0)^2 - \tau^0 + 5/12}$ and less than the radius t^1 of the spheres $\sigma_{10}(t^1)$ and $\sigma_9(t^1)$ for t^1 sufficiently close to t_1 from (9.73) (see Fig. 82). It is thus clear that there is the non-trivial fundamental group of the isometric fundamental polyhedron for the limit group Φ^*G (not a boundary group for $T(G)$) obtained by quasi-conformal conjugation (by $\Phi = \Phi(t^1, t^1, t^1, t^1, \tau^0, \tau^0)$) from G.

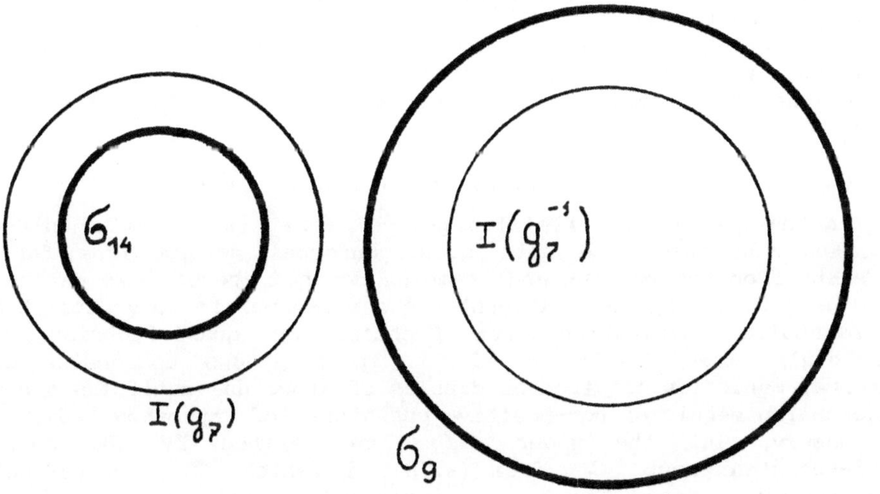

Figure 83

§5. Deformations of Kleinian groups and conformal structures

Let $G \subset \text{Möb}(n) \cong SO(n+1, 1)/\{\pm 1\}$ be a certain Kleinian group whose inclusion homomorphism is a stable good representation in the sense of Definition 9.2. Theorems 9.5 and 9.3 show that for the group G to meet this condition, it is sufficient for G to have a subgroup quasi-conformally conjugate to a Fuchsian group in the n-dimensional sphere $\bar{\mathbb{R}}^n$. For G one can define by analogy with §1 varieties $\mathrm{Hom}(G, \mathrm{Möb}(n))$ and $\mathrm{Hom}(G, \mathrm{Möb}(n))/\mathrm{Mob}(n)$ of injective type-preserving representations $\rho : G \longrightarrow \mathrm{Möb}(n)$ with the topology given by pointwise convergence of homomorphisms (for finitely generated G, it is equal to the compact-open topology). Considering connected components of the set of its inner points, we can apply to them the Sullivan [7] stability theorem:

THEOREM 9.56. *Geometrically finite, torsion free Kleinian groups* $G \subset \mathrm{Möb}(n)$ *without parabolic subgroups of ranks* $\leq n-2$ *are quasi-conformally stable. In other words, if* $\rho \in \mathrm{Hom}(G, \mathrm{Möb}(n))$ *is sufficiently near to the inclusion* $G \subset \mathrm{Möb}(n)$, *then* ρ *is a* K*-quasi-conformal conjugation of the group* G *with a small* $(K-1)$.

This theorem shows that for considered groups G, the space $T_n(G) = R_n(G)/\mathrm{Möb}(n)$ of quasi-conformal conjugations $f_* : G \longrightarrow fGf^{-1}$, where $f : S^n \longrightarrow S^n$ are quasi-conformal self-homeomorphisms of S^n, is a connected component of the space $\mathrm{int}[\mathrm{Hom}(G, \mathrm{Möb}(n))/\mathrm{Möb}(n)]$ of injective representations with discrete images acting discontinuously on the sphere S^n and, the topology of $T_n(G)$ equals the Teichmüller topology - see (9.46).

Considering in the variety of representations $\rho : G \longrightarrow \mathrm{Möb}(n)$ continuous or smooth curves, we obtain (as in Definitions 9.6 and 9.9) the notions of deformation and infinitesimal deformation of the Kleinian group G.

Quasi-conformal deformations of a Kleinian group.

In the case of the Kleinian groups $G \subset \mathrm{Möb}(n)$, $n \geq 3$, different from Fuchsian ones, one gets quasi-conformal deformations for them different from the bending and stamping defined above. We define this class of "stretching deformations" for non-degenerate B-groups obtained by parabolic combination from Fuchsian or quasi-Fuchsian groups $G_0 \subset \mathrm{Möb}(n)$, see Chapter 8 §5. These groups uniformize closed conformal manifolds obtained as doubles of atoroidal manifolds and allow a Riemannian metric of non-positive curvature (cf. Apanasov [25]).

Namely, let the group G be defined by the co-finite quasi-Fuchsian group G_0 from (8.18), in which $\mathcal{T}_i$ are reflections with respect to the hyperbolic n-dimensional planes $S_i = S_i(r_i^0) \subset H^{n+1}$ precisely invariant with respect to the parabolic subgroups $G_i \subset G$ (maximal set: $i = 1, \ldots, m$). These planes $S_i(r_i^0)$ are spanned on $(n-1)$-dimensional punctured spheres $\partial S_i \setminus \{z_i\} \subset \Omega_0$ that are horospheres with "radii" r_i^0, $0 < r_i^0 < K(G)$, at the parabolic points z_i, see (8.15).

If, instead of the parameter $r^0 = (r_1^0, \ldots, r_m^0) \in I^m = (0, K)^m$, we consider any other parameter $r = (r_1, \ldots, r_m) \in I^m \setminus \{r^0\}$, the group obtained for it in (8.18) is denoted by $G(r)$:

$$G(r) = \langle G_0, \mathcal{T}_1(r)G_0\mathcal{T}_1(r), \mathcal{T}_2(r)\mathcal{T}_2(r), \ldots, \mathcal{T}_1(r)\mathcal{T}_m(r)\rangle. \tag{9.74}$$

For the groups $G = G(r^0)$ and $G(r)$ one can use a fixed fundamental polyhedron $P^{n+1}(G_0) \subset H^{n+1}$ having only non-equivalent parabolic vertices $z_1, \ldots, z_m$ at its boundary, see (8.17), to construct convex fundamental polyhedra $P^{n+1} = P^{n+1}(G)$, $P_r^{n+1} = P^{n+1}(G(r)) \subset H^{n+1}$ having the following properties:

1) P^{n+1} and P_r^{n+1} are symmetric with respect to the n-planes $S_1 = S_1(r^0)$ and $S_1(r)$ correspondingly;

2) each of the planes S_i, $\mathcal{T}_1\mathcal{T}_i(S_i)$, $2 \le i \le m$, contains an n-side of the polyhedron P^{n+1}; and

3) each of the planes $S_i(r)$, $\mathcal{T}_1(r)\mathcal{T}_i(r)(S_i(r))$, $2 \le i \le m$, contains an n-side of the polyhedron P_r^{n+1}.

Now let $\varphi_r : P^{n+1} \longrightarrow P_r^{n+1}$ be a K-quasi-conformal mapping compatible in ∂P^{n+1} with the action of the group G. This mapping φ_r is defined as follows.

In neighbourhoods of the planes stated in 1) - 3), φ_r is a uniform stretching along the oricycles of the space H^{n+1} that are orthogonal simultaneously to the planes S_i and $S_i(r)$, $\mathcal{T}_1\mathcal{T}_i(S_i)$ and $\mathcal{T}_1(r)\mathcal{T}_i(r)(S_i(r))$, correspondingly, $1 \le i \le m$. Outside these neighbourhoods $\varphi_r = \mathrm{id}$ in one of the complement components, while in the other component there is a parabolic transformation $h(r_1) = \mathcal{T}_1 \circ \mathcal{T}_1(r)$ having a fixed point z_1; if $z_1 = \infty$ and $e_n \perp S_1$, we have $h(r_1) : x \longmapsto x + 2e_n \ln(r_1^0/r_1))$. Here $\varphi_r(S_i \cap P^{n+1}) \subset S_i(r)$, $\varphi_r(\mathcal{T}_1\mathcal{T}_i(S_i)) \subset \mathcal{T}_1(r)\mathcal{T}_i(r)(S_i(r))$, while the quasi-conformality coefficient $K(r)$ is continuously dependent on r and $\lim_{r \to r_0} K(r) = 1$.

The mapping φ_r is extended compatible with the group G in the whole space $\overline{H}^{n+1} = H^{n+1} \cup \partial H^{n+1}$ to a $K(r)$-quasi-conformal mapping:

$$f_r : \overline{H}^{n+1} \longrightarrow \overline{H}^{n+1},$$

that specifies a quasi-conformal deformation $\lambda : I^m \longmapsto R_n(G)$, namely a stretching deformation:

$$\lambda(r) = f_r^* : G \longmapsto G(r) = f_r G f_r^{-1}. \tag{9.75}$$

LEMMA 9.57. *The deformation* $\lambda : I^m \longrightarrow R_n(G)$ *is smooth (class* C^∞*).*

PROOF. (9.74) shows that the deformation $\lambda : r \longrightarrow f_r^*$ is generated by m independent deformations $r_i \longmapsto f_{r_i}^*$, $1 \le i \le m$. Here for $r = (r_1, r_2, \ldots, r_m) \in I^m$:

$$f_{r_1}^* : G_1 = \langle G_0, \mathcal{T}_1 G_0 \mathcal{T}_1\rangle \longmapsto \langle G_0, \mathcal{T}_1(r) G_0 \mathcal{T}_1(r)\rangle,$$

$$f_{r_i}^* : \langle G_0, \mathcal{T}_1 G_0 \mathcal{T}_1, \mathcal{T}_1\mathcal{T}_i\rangle \longmapsto \langle G_0, \mathcal{T}_1 G_0 \mathcal{T}_1, \mathcal{T}_1\mathcal{T}_i(r_i)\rangle.$$

We use the transformation $h(r_1)$ to write the first of these deformations as

$$f^*_{r_1} = (g_0) = g_0, \quad f^*_{r_1}(g) = h(r_1) \circ g \circ h(r_1)^{-1},$$

in which $g_0 \in G_0$, $g \in \mathcal{J}_1 G_0 \mathcal{J}_1$, while $h(r_1) = \mathcal{J}_1 \circ \mathcal{J}_1(r_1)$ is conjugated by means of the transformation $x \longmapsto U_1((x - z_1)/|x - z_1|^2)$, $U_1 \in O(n)$ to the translation

$$x \longmapsto x = 2e_n \cdot \ln(r_1^0/r_1). \tag{9.76}$$

Similarly, if the parabolic transformations $h(r_i)$ have fixed points z_i and translate S_i to $S_i(r)$, i.e., are conjugated by means of the transformations $x \longmapsto U_i((x - z_i)/|x - z_i|^2)$, $U_i \in O(n)$, to the translations

$$x \longmapsto x + e_n \cdot \ln(r_i^0/r_i), \tag{9.77}$$

then the representations $f^*_{r_i}$, $2 \le i \le m$, are written as

$$f^*_{r_i}(g_1) = g_1, \quad f^*_{r_i}(\mathcal{J}_1\mathcal{J}_i) = \mathcal{J}_1 \cdot h(r_i) \circ \mathcal{J}_i \circ h(r_i)^{-1},$$

in which $g_1 \in G_1$.

After this, the proof follows from the real analytic dependence on the parameter $r \in I^m$ for the translations (9.76) - (9.77).

THEOREM 9.58. *Let a Fuchsian group* $G_0 \subset \text{Möb}(n))$ *of the first kind have* k *orbits of bending hyperplanes in the Poincaré ball model of* H^n *and* m *equivalence classes of parabolic cusps. Then there exists a neighbourhood of the origin* $W^{m+k} \subset \mathbb{R}^{m+k}$ such that the *B-group* $G \subset \text{Möb}(n)$ *specified in* (9.74) *from the group* G_0 *has a non-trivial deformation* $D : W^{m+k} \longrightarrow R_n(G)$. *If here* $p : R_n(G) \longrightarrow T_n(G) = R_n(G)/\text{Möb}(n)$ *is the natural projection, the mapping* $pD : W^{m+k} \longrightarrow T_n(G)$ *is an embedding.*

PROOF. The condition of precise invariance with respect to the stabilizers of the bending planes P_j, $1 \le j \le k$, (see 9.58)) and of the planes S_i, $1 \le i \le m$, shows that the bending deformation $\beta : W^k \longrightarrow R_n(G)$ specified by the bending deformation of the Fuchsian group G_0 from Theorem 9.43 (or Theorem 9.34) and the stretching deformation $\lambda : I^m \longrightarrow R_n(G)$ from Lemma 9.58 are independent. Therefore, the superposition of them gives the desired deformation

$$D = (\beta, \lambda') : W^{m+k} = W^k \times \{r - r^0 : r \in (0, K)^m\} \longrightarrow R_n(G),$$

in which $\lambda'(r) = \lambda(r + r^0)$.

To complete the proof, it is now sufficient to use Theorem 9.43 and the fact that $p\lambda : I^m \longrightarrow T_n(G)$ is an embedding. The last follows by analogy with the proof of Theorem 9.43 from the deformation representation in the proof of Lemma 9.58 and from the fact that the limit sets of the groups G and $G(r) = f^*_r(G)$ are symmetric with respect to the spheres ∂S_i and $\partial S_i(r)$, $i = 1, \ldots, m$, correspondingly.

Groups with limit curve; the wild-knot effect.

Now we consider deformations of the Kleinian groups $\Gamma \subset \text{Möb}(n)$ which are not equal quasi-conformal conjugations. Moreover, we consider the case of non-good groups $\Gamma : \Lambda(G) = S^{n-2}$. For simplicity, we consider the case of dimension $n = 3$ where Γ acts in a disc as a subgroup of $\text{Isom}\, H^2$. The first result on wild-knottedness of the limit set of $\rho\Gamma$ for some $\rho \in \text{Hom}(\Gamma, \text{Möb}(3))$ was Apanasov [10], [11].

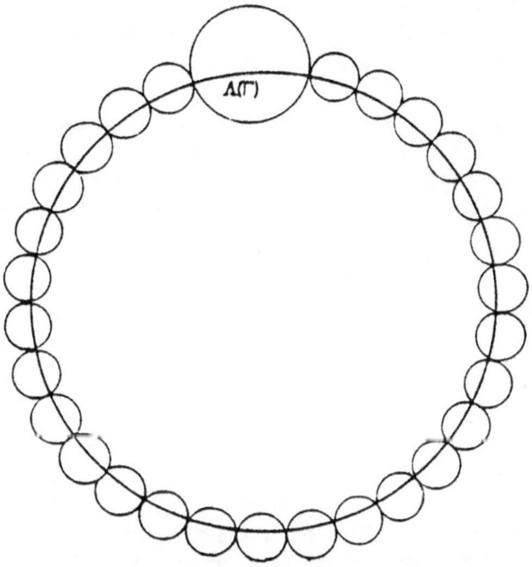

Figure 84

EXAMPLE 9.59 (*a wild-knot group*). Let the circle $S^1 \subset \mathbb{R}^2$ be covered by finitely many discs b_i, $1 \le i \le m$, tangent one to another at the points $p_i \in S^1$ and whose boundaries are orthogonal to S^1 (see Fig. 84), and let the spheres $S_i \subset \mathbb{R}^3$ orthogonally intersect the plane $\mathbb{R}^2$ along circles ∂b_i. We consider the discrete group $\Gamma \subset \text{Möb}(3)$ generated by the reflections γ_i with respect to the spheres S_i, and by Γ' we denote the index two subgroup of orientation-preserving elements ($\Gamma' \subset \mathcal{M}_3$, while the restriction $\Gamma'|_{\mathbb{R}^2}$ is a Fuchsian group in the plane). As fundamental polyhedra for Γ and Γ' we can take

$$P(\Gamma) = \bigcap_i \text{ext}\, S_i, \quad P(\Gamma') = \text{int}\, \overline{P(\Gamma) \cup \gamma_1(P(\Gamma))}. \tag{9.78}$$

We now take some polygonal knot $l \subset \mathbb{R}^3$ whose vertices are points $q_i \in l$ of sequential contact between the spheres σ_i, $1 \le i \le m$, $\text{int}\,\sigma_i \cap \text{int}\,\sigma_j = \emptyset$ for $i \ne j$. Let G be the discrete Möbius group generated by reflections $\mathcal{J}_i$ with respect to the spheres σ_i and let $G' = G \cap \mathcal{M}_3$ be the index 2 subgroup of it.

We show that G and G' are obtained from Γ and Γ' as a result of a deformation $F : [0, 1] \to \mathrm{Hom}(\Gamma, \mathrm{M\ddot{o}b}(4))$, in which $F(t) = f_t^*$ is defined by a quasi-conformal homeomorphism $f_t : \overline{\mathbb{R}}^4 \to \overline{\mathbb{R}}^4$ compatible with the group Γ and $f_0^* = \mathrm{id}$, $f_1^*(\Gamma) = G$. Without changing the symbols, we shall assume that all the spheres defined above are obtained by orthogonal intersection of $\overline{\mathbb{R}}^3$ with the three-dimensional spheres S_i, $\sigma_i \subset \mathbb{R}^4$, where the reflections γ_i and $\mathcal{J}_i$ with respect to them generate the discrete groups Γ and $G \subset \mathrm{M\ddot{o}b}(4)$.

As l is a polygonal circle unknotted in $\mathbb{R}^4$, there exists an isotopy $\varphi : \mathbb{R}^4 \times [0, 1] \to \mathbb{R}^4$ that transforms the circle $S^1 \subset \mathbb{R}^4$ to the topological circle $l \subset \mathbb{R}^4$ in such a way that for any t, $0 \le t \le 1$, the homeomorphism $\varphi_t : \mathbb{R}^4 \to \mathbb{R}^4$ is K-quasi-conformal and transfers the family of tangent balls B_i bounded by the spheres S_i into a family of disjoint balls $B_i^t = \varphi_t(B_i)$ tangent one to another only at points $p_i^t = \varphi_t(p_i)$, while for $t = 1$ we get $B_i^1 = \mathrm{int}\, \sigma_i$, $p_i^1 = q_i$. We extend for any $t \in [0, 1]$ the homeomorphism $\varphi_t|_{P(\Gamma)}$ of the polyhedron (9.78) compatible with Γ to the quasi-conformal mapping f_t, $f_t : \Omega(\Gamma) \to \Omega(\Gamma_t)$ of the whole discontinuity set $\Omega(\Gamma) = \mathbb{R}^4 \backslash S^1$, where a discrete group $\Gamma_t \subset \mathrm{M\ddot{o}b}(4)$ is generated by reflections in the spheres $S_i^t = \partial B_i^t$, $1 \le i \le m$. Here we note that the property of $\bigcap_i B_i^t$ being a fundamental polyhedron for the discrete reflection group Γ_t follows from Corollary 6.8 and the fact that tangent points are parabolic fixed points. The arguments used in Example 9.32 (or independently from Theorem 7.25) give us that S^1 is a removable set for the homeomorphism f_t, which therefore is a quasi-conformal automorphism of the sphere $\overline{\mathbb{R}}^4$ compatible with Γ. The family $\{f_t\}$, $t \in [0, 1]$ defines the desired deformation F of the group Γ, $F(t) = f_t^* : \Gamma \longmapsto f_t \Gamma f_t^{-1}$. Moreover, the representation $f_1^* \in \mathrm{Hom}(\Gamma, \mathrm{M\ddot{o}b}(4))$ defines the desired representation $\rho : \Gamma \to \mathrm{M\ddot{o}b}(3)$, $\rho\Gamma = \Gamma$.

We show that $\Lambda(G) = \Lambda(G')$ is a wildly knotted circle in $\overline{\mathbb{R}}^3$. It follows that $\Lambda(G)$ is a topological circle because $\Lambda(G) = f_1(S^1) \subset \overline{\mathbb{R}}^3$ is a quasi-circle in $\mathbb{R}^4$ (as a matter of fact, it belongs to $\mathbb{R}^3$).

We denote the genus of the knot $l \subset \mathbb{R}^3$ by $h(l)$, i.e., the minimal integer $h(l)$ such that on a knot l one can span an oriented surface of genus $h(l)$ ($h(l) = 0 \iff l$ is the trivial knot). Now, for each i, let $\mathcal{J}_i$ be the inversion with respect to the sphere σ_i. Then for the knot

$$l_1 = l \# \mathcal{J}_i(l) = (l \cap \overline{\mathrm{ext}\ \sigma_i}) \cup \mathcal{J}_i(l \cap \overline{\mathrm{ext}\ \sigma_i}),$$

which is the composition of knots of the same genus $h(l)$, the genus $h(l_1)$ is $2h(l)$, which follows from the fact (Crowell and Fox [1], appendix 1, or Fox [1], §7) that the genus of a composition of knots is equal to the sum of their genera.

Figure 85

Now let there be given the natural partial order (by the inclusions) on the set of spheres $\Sigma = \bigcup_i G(\sigma_i)$ and a bijection $b : \Sigma \longrightarrow \mathbb{N}$ for which b^{-1} preserves that order. Then the resulting enumeration for the spheres in the set Σ specifies the knot composition process (Fig. 85):

$$l_k = l_{k-1} \# g_k(l_{k-1}),$$

in which g_k is a reflection with respect to the k-th sphere from Σ, whose limit is the knot $\Lambda(G)$. From what has been said, we see that $h(l_k) = 2^k h(l)$, which shows that the genus of the knot $\Lambda(G)$ cannot be finite, i.e., $\Lambda(G)$ is a wild knot, even locally wild at that, due to Lemma 3.22.

REMARK 9.60. The above construction of the group G can be generalized.[1] Namely, let us assume that for S^1 the covering balls

---

1 A recent work by Gromov-Lawson-Thurston [1] contains an independent and different approach to this construction related to an S^1-bundle over compact surfaces and hyperbolic structures on a 2-plane bundle over a compact surface. For the case of tangent balls see also a recent book by Maskit [10]. Ohshika [1] contains a detailed description of corresponding Teichmüller spaces (they are non-connected for $u \neq 1$ in (9.81)).

B_i, $1 \le i \le m$, consequently intersect on angles α_i ($= \alpha_{i,i+1}$) and the Fuchsian group $G \subset \mathcal{M}_2$ is generated by involutions γ_i which leave invariant the i-th sides of $P(\Gamma) = \bigcap_i \operatorname{ext} B_i$. Also let σ_i, $1 \le i \le m$, be 2-spheres such that $\sigma_i \cap \sigma_j \ne \emptyset \iff \operatorname{int} \sigma_i \cap \operatorname{int} \sigma_j \ne \emptyset \iff i - j = \pm 1 \pmod m$ where the intersection angle of σ_i and σ_{i+1} is the same α_i. Denote $\operatorname{int} \sigma_i = B'_i$.

Then the boundary of the solid torus $\bigcup_i B'_i$ is the union of spherical annuli $\sigma'_i \subset \sigma_i$. Denote by $\mathcal{T}_i$ an involution that changes the position of the boundary circles of σ'_i, $\mathcal{T}_i(\sigma'_i) = \sigma'_i$, and denote by G, $G \subset \text{Möb}(3)$, $G \cong \Gamma$, the group generated by the involutions $\mathcal{T}_i$, $1 \le i \le m$.

Let us examine when $\Gamma(P(\Gamma))$ and $G(P(G))$, $P(G) = \bigcap_i B'_i$, are tessellations of topological circle complements in S^3. For this purpose we shall use Theorem 6.9 for the action of Γ and G in the hyperbolic space H^4. For the present case, the theorem conditions are:

First, $\Gamma(P(\Gamma) \cap \mathbb{R}^2)$ defines a regular tessellation of the disc H^2, $\partial H^2 = S^1$, each vertex of which belongs to the same number (denoted by n) of polygons. Let l be the greatest common divisor of the number m and n. Then for some $u, v \in \mathbb{N}$, $n = lu$, $m = lv$, we have:

$$(\gamma_m \cdot \ldots \cdot \gamma_1)^u = (\gamma_n \cdot \ldots \cdot \gamma_1)^v = 1 \tag{9.79}$$

where $\gamma_i = \gamma_j \iff i = j \pmod m$. This gives a condition on α_i: $(\alpha_1 + \ldots + \alpha_m) \cdot u = 2\pi \cdot v$ or, equivalent:

$$(\alpha_1 + \ldots + \alpha_m) \cdot n = 2\pi \cdot m. \tag{9.80}$$

Second, the theorem conditions for the tessellation $G(P(G))$ include (in addition to the above angle condition (9.81)) the condition of local finiteness for the filling by polyhedra, i.e., for our case, we need for a finite order of the transformation $(\mathcal{T}_m \cdot \ldots \cdot \mathcal{T}_1)$ or exactly, that $(\mathcal{T}_m \cdot \ldots \cdot \mathcal{T}_1)^u$ is the identity on the 2-plane in H^4 spanned on the circle $\sigma_1 \cap \sigma_m$, i.e.,

$$(\mathcal{T}_m \cdot \ldots \cdot \mathcal{T}_1)^u(x) = x \quad \text{for all} \quad x \in \sigma_1 \cap \sigma_m. \tag{9.81}$$

With the conditions (9.79) - (9.81) we obtain an isomorphic tessellation of H^4 and therefore, isomorphic tessellations of the discontinuity sets $\Omega(\Gamma) = S^3 \setminus S^1$ and $\Omega(G) \approx S^3 \setminus S^1$ in $S^3 = \partial H^4$.

For such a generalization of the construction in Example 9.59, we have both wild knotted limit circles $\Lambda(G) \subset S^3$ and unknotted ones in correspondence with knottedness or unknottedness of a solid torus $\bigcup B'_i \subset S^3$. For the unknotted case, it is interesting that the order of the elliptic element (9.81) is the topology invariant of the construction.[2]

REMARK 9.61. Wild knotting of the limit circle $\Lambda(G)$ (for the knotted case) arises by necessity. Kulkarni [1] has shown that the limit set $\Lambda(G)$ of any Kleinian group $G \subset \text{Möb}(3)$ cannot be a tame knot. An analogous generalization of the construction in Example 9.59 is given by M. Kapovich. Using the Möbius band for the construction, he has a group G with topology invariant corresponding to order 2 of the elliptic element (9.81).

Deformations of conformal structures.

In conclusion we briefly consider the relation between the space of deformations of a Kleinian group $G \subset \text{Möb}(n)$ and the space of deformations of uniformized conformal structures on the orbifold $\Omega(G)/G$. For $n = 2$, the relationship is obvious from Corollary 9.29 and §1.1. For $n \geq 3$, a basic fact that establishes the relationship is provided by the holonomy Theorem 9.1 which gives conditions under which the space of uniformized marked conformal structures $\mathcal{C}(\Omega/G)$ on the orbifold $\Omega(G)/G$ can be identified locally with the space $T_n(G)$ of classes of quasi-conformal representations of the Kleinian group G. A basic condition here is that of stability and goodness for the inclusion representation $G \subset \text{Möb}(n)$. As the above Kleinian, and in particular Fuchsian, groups $G \subset \text{Möb}(n)$ have those properties, we can apply the results obtained for them to the corresponding conformal structures. In particular, one can readily reformulate the results on the following properties of conformal structures.

1. There are defined constructions for bending, stamping (stamping-with-torsion), and stretching deformations for a conformal structure along the corresponding conformal suborbifolds $M_i \subset \Omega(G)/G$, or what is equivalent, deformations of the hyperbolic structure H^{n+1}/G along the totally geodesic suborbifolds $N_i \subset H^{n+1}/G$, $\partial N_i = M_i$, compare §3 Remark 9.49 and Theorem 9.60.

2. There is an embedding in $\mathcal{C}(\Omega/G)$ of the ball B^N whose dimension is determined by the number N of disjoint totally geodesic n-dimensional suborbifolds $N_i \subset H^{n+1}/G$ for which $\partial N_i \subset \Omega(G)/G$ are conformal suborbifolds in $\Omega(G)/G$ and in particular $(n-1)$-dimensional Euclidean space forms and cylinders, compare Theorems 9.34, 9.43, and 9.58.

[2] Note added in proof. The recent papers by N. Kuiper [4], [5] contain a similar approach to the study of Kleinian groups in a 3-sphere with the limit circle. This is related to the investigation of tessellation parameters and the condition $|\chi(E)| \leq |\chi(\Sigma)|$ for the 2-plane bundle $E \longrightarrow \Sigma$ over a compact surface Σ where χ is the Euler number (cf. footnote on page 421 and the parameter conditions (9.79-9.81).

3. There are conformal n-manifolds (orbifolds) M (hyperbolic and non-hyperbolic) such that $\mathcal{C}(M)$ has singularities that may be non-isolated, compare §3.5 and §5.[3]

---

[3] Note added in proof. In recent works Apanasov [35], [36] the author discovered a set of non-standard uniformized conformal structures even on closed hyperbolic 3-manifolds M (with a sufficiently large set of disjoint totally geodesic surfaces). These structures $s \in \mathcal{C}(M)$ are such that s is not contained in the inverse image by the map hol (see Theorem 9.1) of the space $T_3(\pi_1(M))$ of quasi-Fuchsian representations and even likely in its closure (at least this is true for the closure of bending and stamping representations). Also there is an extremely interesting correspondence between the set of such non-standard structures and the set of non-trivial hyperbolic 4-dimensional homology cobordisms W, of which one boundary component is homologically equivalent to W and homeomorphic to M (compare Chapter 7 §3).

REFERENCES

Abels H.

[1] Geometrische Erzeugung von diskontinuerlichen Gruppen, *Schr. Math. Inst. Univ. Münster* No.33, 1966.

Abikoff W.

[1] Some remarks on Kleinian groups, Advances in the Theory of Riemann Surfaces, Edited by L.V. Ahlfors et al., Proc. Conf. Stony Brook, 1969, *Ann. of Math. No.*66, *Princeton Univ. Press*, 1971, pp. 1 - 5.

[2] The residual limit set of Kleinian groups, *Acta Math.*, **130** (1973), 127-144.

[3] On boundaries of Teichmüller spaces and Kleinian groups, Part 3, *Acta Math.*, **134** (1975), 211-237.

[4] The real analytic theory of Teichmüller spaces, Lecture Notes in Math., No.820, *Springer-Verlag, Berlin*, 1980.

Abikoff W. and Maskit B.

[1] Geometric decompositions of Kleinian groups, *Amer. J. Math.* **99** No.4 (1977), 687-697.

Accola R.D.M.

[1] Invariant domains for Kleinian groups, *Amer. J. Math.* **88** (1966), 329-336.

Adams C.C.

[1] Hyperbolic structures on link complements, Ph.D. Thesis, Univ. of Wisconsin-Madison (1983).

[2] Thrice-punctured spheres in hyperbolic 3-manifolds, *Trans. of Amer. Math. Soc.* **287** (1985), 654-656.

[3] The non-compact hyperbolic 3-manifold of minimal volume, *Proc. Amer. Math. Soc.* **100** (1987), 601-606.

Adian S.I.

[1] Defining relations and algorithmic problems for groups and semigroups, *Trudy Inst. Math. Steklov,* **85** (1966) (= *Amer. Math. Soc. Translation, Providence,* 1967).

Agard S.
[1] A geometric proof of Mostow's rigidity theorem for groups of divergence type, *Acta Math.* **151** (1983), 231-252.

Ahlfors L.V.
[1] On quasi-conformal mappings, *J. Analyse Math.* **3** (1953-4), 1-58.
[2] Finitely generated Kleinian groups, *Amer. J. Math.* **86** (1964), 413-429; **87** (1965), 759.
[3] Fundamental polyhedra and limit point sets of Kleinian groups, *Proc. Mat. Acad. Sci. USA* **55** (1966), 251-254.
[4] Kleinische Gruppen in der Ebene und im Raum, Festband zum 70. Geburtstag von R. Nevanlinna, *Springer-Verlag, Berlin*, 1966, 7-15.
[5] Hyperbolic motions, *Nagoya Math. J.* **29** (1967), 163-166.
[6] Invariant operators and integral representations in hyperbolic space, *Math. Scand.* **36** (1975), 27-43.
[7] Quasi-conformal deformations and mappings in $\mathbb{R}^n$, *J. Analyse Math.* **30** (1976), 74-97.
[8] Quasi-conformal mappings, Teichmüller spaces and Kleinian groups, Proc. Intern. Congress Math., Helsinki (1978), *Acad. Sci. Fenn., Helsinki*, 1980, 71-84.
[9] Ergodic properties of groups of Möbius transformations, Analytic Functions, Edited by J. Lawriniwicz, Kozubnik (1979); Lecture Notes in Math. No.798 *Springer Verlag, Berlin*, 1980, 1-9.
[10] Möbius transformations in several dimensions, Ordway Professorship Lectures in Math., Univ. of Minnesota (1981).
[11] Old and new in Möbius groups, *Ann Acad. Sci. Fenn. Ser.* A1 **9** (1984), 93-105.
[12] On the fixed points of Möbius transformations in $\mathbb{R}^n$, *Ann. Acad. Sci. Fenn. Ser.* A1 **10** (1985), 15-27.

Aleksandrov A.D.
[1] On the filling space by polyhedra, *Vestnik Leningradskogo Universiteta* No.2 (1954), 33-43 (Russian).

Al-Jubori N.K.
[1] On non-orientable hyperbolic 3-manifolds, *Quart. J. Math., Oxford Ser* (2) **31**:121 (1980), 9-18.

Andreev E.M.
[1] Convex polyhedra in Lobachevskii spaces, *Mat. Sb.* **81** (1970), 445-478.
[2] Convex polyhedra of finite volume in a Lobachevskii space, *Mat. Sb.* **83** (1970), 256-260. (= Math. USSR-Sb. **12** (1970), 255-259.)
[3] The intersections of the planes of faces of polyhedra with sharp angles, *Mat. Zametki* **8** (1970), 521-527. (= Math. Notes **8** (1970), 761-764.)

Anosov D.V.

[1] Geodesic flows on closed Riemannian manifolds of negative curvature, *Trudy Mat. Inst. Steklov* **90** (1967). (= Amer. Math. Soc. Trans., Providence, 1968.)

Apanasov B.N.

[1] Certain characteristics of conformal mappings in $\mathbb{R}^n$, Metric questions of the theory of functions and mappings, Issue 5 (Proceedings of the 3rd Colloquium on the theory of quasi-conformal mappings and extensions, Donetsk, 1972), *Naukova Dumka, Kiev*, 1974, 3-13 (Russian).

[2] A certain class of Kleinian groups in $\mathbb{R}^n$, *Dokl. Akad. Nauk SSSR* **215** (1974), 509-510. (= Soviet Math. Doklad. **15** (1974), 515-517.)

[3] A certain analytic method in the theory of Kleinian groups for multi-dimensional Euclidean space, *Dokl. Akad. Nauk SSSR* **222** (1975), 11-14. (= Soviet Math. Dokl. **16** (1975), 553-557.)

[4] Kleinian groups in space, *Sibirsk. Matem Zh.* **16** (1975), 891-898. (= Siberian Math. J. **16** (1975), 679-684.)

[5] A certain universal property of Kleinian groups in the hyperbolic metric, *Dokl. Akad. Nauk SSSR* **225** (1975), 15-18. (= Soviet Math. Dokl. **16** (1975), 1418-1421.)

[6] The relation between the group $SL(n, \mathbb{Z})$ with the n-dimensional modular group, *Dokl. Akad. Nauk SSSR* **231** (1976), 1033-1036. (= Soviet Math. Dokl. **17** (1976), 1670-1674.)

[7] On Mostow's rigidity theorem, *Dokl. Akad. Nauk SSSR* **243** (1978), 829-832. (= Soviet Math. Dokl. **19** (1978), 1408-1412.)

[8] Entire automorphic forms in $\mathbb{R}^n$, *Sibirsk. Mat. Zh.* **19** (1978), 735-748. (= Siberian Math. J. **19** (1978), 518-528.)

[9] On the problem of the existence of automorphic mappings in n-dimensional space. Questions in the metric theory of mappings and its applications (Proceedings of the 5th Colloquium Quasi-conformal mappings and Generalizations, Donetsk, 1976), *Naukova Dumka, Kiev*, 1978, 3-9 (Russian).

[10] On invariants of quasi-conformal deformations for discontinuous groups, Komplexe Analysis und ihre Anwendungen auf partielle Differentialgleichungen, Part 3, Martin-Luther-Universität, Halle (1980), 207-209.

[11] Kleinian groups, Teichmüller spaces, and Mostow's rigidity theorem, *Sibirsk. Mat. Zh.* **21** (1980), 3-15. (= Siberian Math. J. **21** (1980), 483-491.)

[12] Non-triviality of Teichmüller space for Kleinian groups in space, Riemann surfaces and related topics: Proceedings of the 1978 Stony Brook Conference, Edited by I. Kra and B. Maskit (Ann. of Math. Studies, No.97), *Princeton University Press*, 1981, 21-31.

[13] Geometrically finite groups of transformations of space, *Sibirsk. Mat. Zh.* **23**:6 (1982), 16-27. (= Siberian Math. J. **23** (1982), 771-780.)

[14] Geometrically finite hyperbolic structures on manifolds, *Ann. of Glob. Analysis and Geom.* **1**:3 (1983), 1-22.

[15] Discrete transformations groups and structures on manifolds, *Nauka, Novosibirsk*, 1983 (Russian).
[16] Condition of conformal rigidity of hyperbolic manifolds with boundaries, Analytic functions, Blažejewko, 1982, Edited by J. Lawrynowicz (Lecture Notes in Math. No.1039), *Springer-Verlag Berlin-New York*, 1983, 1-9.
[17] Incomplete hyperbolic manifolds and local finiteness of a space tessellation by polyhedra, *Dokl. Akad. Nauk SSSR* **273**:4 (1983), 777-781. (= Soviet Math. Dokl. **28** (1983), 686-690.)
[18] Limit set and geometric finiteness of Kleinian groups, Complex analysis and applications (Varna 1981), Edited by L. Iliev and V. Andreev, *Bulg. Acad. Sci., Sofia*, 1984, 14-28.
[19] Parabolic vertices and finiteness properties for Kleinian groups in *n*-dimensional space, *Sibirsk. Matem. Zh.* **25**:4 (1984), 9-27. (= Siberian Math. J. **25** (1984), 516-530.)
[20] Cusp neighbourhoods for Kleinian groups in space, Complex analysis and its applications, Edited by W. Tutschke, Kongress und Tagungsberichte der Martin-Luther-Universität, 1984/85 (M 35), Halle, 1984, 37-39.
[21] Cusp ends of hyperbolic manifolds, *Ann. of Glob. Analysis and Geom.* **3**:1 (1985), 1-11.
[22] Deformations of hyperbolic structures without the completeness condition, *Zeits. Anal. und Anwendungen* **5**:2 (1986), 99-110.
[23] The effect of dimension 4 in Aleksandrov's problem of filling a spaceby polyhedra, *Ann. of Glob. Analysis and Geom.* **4**:2 (1986), 1-22.
[24] Thurston's bends and geometric deformations of conformal structures, Complex Analysis and Applications 1985, Edited by L. Iliev and I. Ramadanov, *Publ. House of Bulgarian Acad. Sci., Sofia* 1986, 14-28.
[25] Doubles of atoroidal manifolds, their conformal uniformization and deformations, Deformations of mathematical structures, Edited by J. Lawriniowicz (Proc. Varna, 1985, Lublin, 1986), *Kluwer Academic Publishers, Dordrecht* 1989, 93-114.
[26] Tessellations of space by polyhedra and deformations of incomplete hyperbolic structures, *Sibirsk. Matem. Zh.* **27**:4 (1986), 3-19. (= Siberian Math. J. **27** (1986))
[27] Deformation of conformal structures on manifolds related to totally geodesic submanifolds, *Dokl. Akad. Nauk SSSR* **293**:1 (1987), 14-17. (= Soviet Math. Dokl. **35** (1987), 242-245.)
[28] The geometry of Nielsen's hull for Kleinian groups in space and quasi-conformal mappings, *Ann. of Glob. Analysis and Geom.* **6**:3 (1988), 207-230.
[29] The method of maximal balls in the theory of Kleinian groups and its applications, *Novosibirsk Institute of Mathematics, Siberian Division, Academy of Sciences of the USSR, Preprint* No.19 (1986) (Russian).
[30] Hyperbolic manifolds with non-tame ends, Dep. VINITI, No.1646-B 86 Dep. (*Sibirsk. Matem. Zh.*) (1986) (Russian).

[31] New kind of fundamental polyhedra for discrete groups, Geometry of discrete groups, *Kishinev, Stiinza Acad. Publ.*, 1990, 3-12 (Russian).
[32] Two kinds of deformations of conformal structures distinct from bendings, in Complex Analysis and Applications, 1987, *Publ. House Bulgarian Acad. Sci., Sofia,* 1989, 5-10.
[33] Quasi-symmetric embeddings of a closed ball inextensible in neighbourhoods of any boundary points, *Ann. Acad. Sci. Fenn.*, Ser.A1 **14** (1989), 243-255.
[34] Pseudo-Dirichlet polyhedra and the word problem for geometrically finite hyperbolic groups, *Math. Sci. Res. Inst., Berkeley*, Preprint 02523-89, 1989.
[35] Deformations of conformal structures and Lobachevski geometry, *Math. Sci. Res. Inst., Berkeley*, Preprint 02723-89, 1989 (*J. of Diff. Geom.*, 1991, to appear).
[36] Hyperbolic cobordisms and conformal structures, Mittag-Leffler Inst. Preprint, Djursholm, 1989 (*Topology* 90, W. de Gruyter, 1991, to appear).
[37] On non-connectedness of the space of uniformized conformal structures on hyperbolic manifolds, Mittag-Leffler Inst. Preprint, Djursholm, 1989 (*Invent. Math.*, 1991, to appear).
[38] Bending and stamping deformations of hyperbolic manifolds, *Ann. Glob. Anal. Geom.* **8** (1990), N1, 3-12.
[39] Kobayashi conformal metric on manifolds, Chern-Simons and η-invariants, Univ. Autonoma de Barcelona, Preprint 25/1990 (*Intern. J. of Math.*, **2** (1991), no.4, to appear).

Apanasov B.N. and Gutsul I.S.
[1] Closed hyperbolic 3-manifolds with the same volume. Geometry of discrete groups, *Kishinev, Stiinza Acad. Publ.*, 1990, 13-28 (Russian).
[2] Greatly symmetric totally geodesic surfaces and closed hyperbolic 3-manifolds which share a fundamental polyhedron, *Topology* 90, Proc. Res. Semester in Low Dimensional Topology at Ohio St. Univ., W. de Gruyter, 1991, to appear.

Apanasov B.N., Neumann W.D., Reid A.W. and Siebenmann L.C.
[1] (Editors), Topology 90: Proc. Res. Semester in Low Dimensional Topology at Ohio State Univ., *W. de Gruyter, Berlin-New York*, to appear.

Apanasov B.N. and Tetenov A.V.
[1] The existence of non-trivial quasi-conformal deformations of Kleinian groups in space, *Doklad. Akad. Nauk SSSR* **239** (1978), 14-17 (= Soviet Math. Dokl. **19** (1978), 242-245.)
[2] Non-trivial cobordisma with geometrically finite hyperbolic structures, *J. of Diff. Geom.* **28** (1988), 407-422.

[3] The topology of hyperbolic geometrical finite 4-manifolds with boundary, Topology and its applications: Proceedings International Topology Conference (Baku, October 3rd-8th 1987), (Trudy Steklov Inst. of Math., **193**), *Nauka, Moscow*, 1990, to appear.

Armstrong M.A.
[1] The fundamental group of the orbit space of a discontinuous group, *Proc. Cambridge Phil. Soc.*, **64** (1968), 299-301.

Ballmann W.
[1] Manifolds of non-positive sectional curvature and manifolds without conjugate points, *Proc. Intern. Congress of Math., Berkeley, California,* 1986, 484-490.

Bass H. and Morgan J.
[1] (Editors), The Smith Conjecture, *Academic Press* (1984).

Beardon A.F.
[1] The Hausdorff dimension of singular sets of properly discontinuous groups, *Amer. J. Math.*, **88** (1966), 722-736.
[2] Fundamental domains for Kleinian groups, In: Discontinuous groups and Riemann surfaces, edited by L. Greenberg (*Ann. of Math. Studies* **79**), *Princeton Univ. Press, Princeton,* 1974, 31-41.
[3] Lie products, closed geodesics and Fuchsian groups, *Proc. Amer. Math. Soc.*, **85** (1982), 87-90.
[4] The geometry of discrete groups, Graduate texts in Math., Vol.91, *Springer-Verlag, New York-Berlin,* 1983.
[5] Inequalities for certain Fuchsian groups, *Acta Math.*, **127** (1971), 221-258.

Beardon A.F. and Jørgensen T.
[1] Fundamental domains for finitely generated Kleinian groups, *Math. Scand.*, **36** (1975), 21-26.

Beardon A.F. and Maskit B.
[1] Limit points of Kleinian groups and finite sided fundamental polyhedra, *Acta Math.*, **132** (1974), 1-12.

Beardon A.F. and Nicholls P.J.
[1] On classical series associated with Kleinian groups, *J. London Math. Soc.*, **5**:2 (1972), 645-655.
[2] Ford and Dirichlet regions for Fuchsian groups, *Canad. J. Math.*, **34** (1982), 806-815.

Beem J. and Erlich P.
[1] Global Lorentzian geometry, Monographs and Textbooks in Pure and Applied Math. **67**, *Dekker, New York*, 1981.

Belinskii P.P.

[1] Normality of families of quasi-conformal mappings, *Doklad. Akad. Nauk SSSR* **128**:4 (1959), 16-17 (Russian).

[2] General properties of quasi-conformal mappings, *Nauka, Novosibirsk*, 1974 (Russian).

[3] (Editor: Some aspects of modern function theory, *Institut. Matem. Sibirsk. Otd. AN SSSR, Novosibirsk* (1976) (Russian).

Belinskii P.P and Lavrent'ev M.A.

[1] On locally quasi-conformal mappings in $(n \geq 3)$-space, Contribution to Analysis (a collection of papers dedicated to Lipman Bers), edited by L.V. Ahlfors et al., *Academic Press*, 1974, 27-30.

Berger M.

[1] Géométrie, *Cedic, Paris, Nathan Information, Paris*, 1977.

Bers L.

[1] Quasi-conformal mappings and Teichmüller's theorem. Analytic functions, *Princeton Univ. Press, Princeton*, 1960, 89-120.

[2] Inequalities for finitely generated Kleinian groups, *J. Anal. Math.* **18** (1967), 23-41.

[3] Spaces of Kleinian groups. Several complex variables, I, *Springer-Verlag*, 1970, 9-34.

[4] On boundaries of Teichmüller spaces and on Kleinian groups, Part 1, *Ann. of Math.*, **91** (1970), 570-600.

[5] Unimorphization, moduli and Kleinian groups, *Uspekhi Mat. Nauk*, **28**:4 (1973), 153-198. (= Bull. London Math. Soc. **4** (1974), 257-300.)

[6] The moduli of Kleinian groups, *Uspekhi Mat. Nauk* **29**:2 (1974), 86-102. (= Russian Math. Surveys **29**:2 (1974), 88-102.)

[7] On Hilbert's 22nd problem. Mathematical developments arising from Hilbert problems, edited by F. Browder (Proc. Symp. Pure Math., **281**), *Amer. Math. Soc., Providence*, 1976, 559-609.

[8] Finite-dimensional Teichmüller spaces and generalizations, *Bull. Amer. Math. Soc.*, **5** (1981), 131-172.

[9] On iterates of hyperbolic transformations of Teichmüller space, *Amer. J. Math.* **105** (1983), 1-11.

Bers L. and Kra I.

[1] (Editors): What is a Kleinian group? A crash course on Kleinian groups, Lecture Notes in Math., No.400, *Springer-Verlag*, 1974.

Besse A.

[1] Manifolds all of whose geodesics are closed, Ergebnisse der Mathematik und ihrer Grenzgebiete, No.93, *Springer-Verlag, Berlin*, 1978.

[2] (Editor): Géométrie Riemannienne en dimension 4, *Cedic/Fernand Nathan, Paris*, 1981.

Best L.A.

[1] On torsion-free discrete subgroups of *PSL*(2, *C*) with compact orbit space, *Canad. J. Math.* **23** (1971), 451-460.

Bianchi L.

[1] Sui gruppi de sustituzioni lineari con coefficienti appurtenenti a corpi quadratici imaginare, *Math. Ann.* **40** (1892), 332-412.

Bing R.H.

[1] The geometric topology of 3-manifolds, *Amer. Math. Soc. Coll. Publ.* No.40, *Providence* (1983).

Boileau M. and Zieschang H.

[1] Heegard genus of closed orientable Seifert 3-manifolds, *Invent. Math.* **76** (1984), 455-468.

Bonahon F.

[1] Bouts des variétés hyperboliques de dimension 3, Prepublication Orsay, 85 T 08 (1985).

[2] Bouts des variétés hyperboliques de dimension trois, *Ann. of Math.*, **124** (1986), 71-158.

Bonahon F. and Otal J.-P.

[1] Variétés hyperboliques à géodésiques arbitrairement courtes, Prepublication Orsay (1987).

Bonahon F. and Siebenmann L.

[1] The classification of Seifert fibred 3-orbifolds, Low-dimensional topology (Chelwood Gate, 1982), London Math. Soc. Lect. Note Ser. 95, Ed. by R. Fenn, *Cambridge Univ. Press*, *Cambridge*, 1985.

Borel A.

[1] Compact Clifford-Klein forms of symmetric space, *Topology* **2** (1963), 111-122.

[2] Introduction aux groupes arithmétiques, *Hermann, Paris* (1969).

Borel A. and Wallach N.

[1] Continuous cohomology, discrete subgroups, and representations of reductive groups, *Ann. of Math. St.*, **94** *Princeton Univ. Press, Univ. of Tokyo Press, Tokyo,* (1980).

Bowditch B.H.

[1] Geometrical finiteness for hyperbolic groups, Preprint,, Univ. of Warwick.

Brin M.G.

[1] Torsion-free actions on 1-acyclic manifolds and the loop theorem, *Topology* **20**:4 (1981), 353-364.

Brin M.G., Johanson K. and Scott P.
[1] Totally peripheral 3-manifolds, *Pacific J. Math.* **118** (1985), 37-51.

Brin M.G. and Thickstun T.L.
[1] On the proper Steenrod homotopy groups and proper embeddings of planes into 3-manifolds, *Trans. Amer. Math. Soc.* **289**:2 (1985), 737-755.
[2] Open irreducible 3-manifolds which are en 1-movable, Preprint (1985).
[3] On properly embedding non-compact surfaces in arbitrary 3-manifolds, Proc. London Math. Soc. (3) **54** (1987) No.2, 350-366

Brooks R.
[1] On the deformation theory of classical Schottky groups, *Duke Math. J.* **52** (1985), 1009-1024.
[2] Circle packings and co-compact extensions of Kleinian groups, *Invent. Math.,* **86**:3 (1986), 461-469.

Brown E.M., Brown M.S. and Feustel C.D.
[1] On properly embedding planes in 3-manifolds, *Proc. Amer. Math. Soc.* **55** (1976), 461-464.

Brown M.
[1] The monotone union of open n-cells in an open n-cell, *Proc. Amer. Math. Soc.* **12** (1961), 812-814.

Brunner A.M., Frame M.L., Lee Y.W and Wielenberg N.J.
[1] Classifying torsion-free subgroups of the Picard group, *Trans. Amer. Math. Soc.* **282** (1984), 205-235.

Bugaenko V.O.
[1] Groups of automorphisms of unimodular hyperbolic quadratic forms over the ring $\mathbb{Z}((\sqrt{5}+1)/2)$, *Vestnik Moskov. Univ., Ser.*1 *Mat. Mekh.* **5** (1984), 6-12. (= Mosc. Univ. Math. Bull. **39**:5 (1984), 6-14).

Burago Yu.D.
[1] Appendix to the Russian translation of the book by J.A. Wolf /1/, *Nauka, Moscow*, 1982, 461-474 (Russian).

Burago Yu.D. and Zalgaller V.A.
[1] Convex sets in Riemannian space of non-negative curvature, *Uspekhi Mat. Nauk* **32**:3 (1977), 3-56. (= Russian Math. Surveys **32**:3 (1977),1-57.)

Canary R.D., Epstein D.B.A. and Green P.
[1] Notes on notes of Thurston, Analytic and geometric aspects of hyperbolic space (Ed. by D.B.A. Epstein), *London Math. Soc. Lect. Note Series* **111**, *Cambridge Univ. Press* (1987), 3-92.

Cannon J.W.

[1] The combinatorial structure of cocompact discrete hyperbolic groups, *Geom. Dedicata* **16** (1984), 123-148.

Cannon J.W., Epstein D.B.A., Holt D.F., Paterson M.S. and Thurston W.P.

[1] Word processing and group theory, Preprint, Geometry Supercomputer Project, 1990.

Caraman P.

[1] n-dimensional quasi-conformal (QCf) mappings, Revised, enlarged and translated from Romanian by author, *Editura Academiei Romane, Bucharest; Abacus Press, Tunbridge Wells, Kent; Haessner Publ. Incl, Newfoundland, N.J.*, 1974.

Cartan E.

[1] La géométrie des groupes simples, *Ann. di Math.* **4** (1927), 209-256.

[2] Complément au mémoire "Sur la géométrie des groups simples", *Ann. di Math.* **5** (1928), 253-260.

Cassels J.W.C.

[1] An introduction to diophantine approximation, *Cambridge Univ. Press, New York* (1957).

Cayley A.

[1] The theory of groups: graphical presentations, *Amer. J. Math.* **1** (1878), 174-176.

[2] On the theory of groups, *Amer. J. Math.* **11** (1889), 139-157.

Chen Su-Shing

[1] A theorem of Ahlfors for hyperbolic spaces, *Trans. Amer. Math. Soc.* **242** (1978), 401-406.

Chen Su-Shing and Greenberg L.

[1] Hyperbolic spaces, In: Contributions to analysis (a collection of papers dedicated to Lipman Bers), edited by L.V. Ahlfors et al., *Academic Press* (1974), 49-87.

Chinburg T.

[1] Volumes of hyperbolic manifolds, *J. Diff. Geom.* **18** (1983), 783-789.

[2] A small arithmetic hyperbolic 3-manifold, *Proc. Amer. Math. Soc.* **100** (1987), 140-144.

Chinburg T. and Friedman E.

[1] The smallest arithmetic hyperbolic 3-orbifold, *Invent. Math.* **86**:3 (1986),507-527.

Chucrow V.

[1] On Schottky groups with applications to Kleinian groups, *Ann. of Math.* **88** (1968), 47-61.

Cohen M.M.
[1] A course in simple-homotopy theory, Graduate texts in Math., No.10, *Springer-Verlag* (1973).

Coxeter H.S.M.
[1] Discrete groups generated by reflections, *Ann. of Math.* **35** (1934), 588-621.
[2] The complete enumeration of finitegroups of the form $R_i^2=(R_iR_j)_j^{k_i}=1$, *J. London Math. Soc.* **10** (1935), 21-25.

Coxeter H.S.M. and Moser W.O.J.
[1] Generators and relations for discrete groups, *Springer-Verlag* (1965).

Crowell R.H. and Fox R.H.
[1] Introduction to knot theory, *Ginn and Co., Boston* (1963).

Culler M, Gordon C.McA., Luecke J. and Shalen P.B.
[1] Dehn surgery on knots, *Ann. of Math.* **125** (1987), 237-300.

Culler M. and Shalen P.B.
[1] Varieties of group representations and splitting of 3-manifolds, *Ann. of Math.* **117** (1983), 109-146.

Damian F.L.
[1] Discrete torsion-free hyperbolic groups generated by translations, Aspects of discrete geometry, *Kishinev, Stiinza Acad. Publ.* (1988), 77-82 (Russian).
[2] To construction of hyperbolic four-dimensional manifolds, Geometry of discrete groups, *Kishinev, Stiinza Acad. Publ.* (1990), 79-84.

Danilov V.I.
[1] The geometry of toric varieties, *Uspekhi Mat. Nauk* **33**:2 (1978), 85-134. (= Russian Math. Surveys **33** :2 (1978), 97-154.)

Davis M.W.
[1] Groups generated by reflections and aspherical manifolds not covered by Euclidean space, *Ann. of Math.* **117** (1983), 293-324.
[2] Coxeter groups and aspherical manifolds. Algebriac topology, Aarhus (1982); Lect. Notes in Math., No.1051, *Spring-Verlag, Berlin-New York*, 1984.

Dehn M.
[1] Über die Topologie des dreidimensionalen Raumes, *Math. Ann.* **69** (1910), 137-168.
[2] Über unendliche discontinuerliche Gruppen, *Math. Ann.* **71** (1912), 116-144.

Deny J. and Lions T.L.
[1] Les espaces du type de Beppo Levi, *Ann. Inst. Fourier* **5** (1955), 305-370.

Derevnin D.A.
[1] On the distance between fixed points of a discrete isometry group in the Lobachevski space, Dep. VINITI N9010-B96 (*Sibirsk. Matem. Zh.)* (1986) (Russian).
[2] The discreteness conditions for hyperbolic isometry groups and automorphisms of hyperbolic manifolds, Cand. Phis.-Math. Sci. Degree Thesis, *Novosibirsk St. Univ., Novosibirsk* (1987).

Dodziuk J.
[1] A lower bound for the first eigenvalue of a finite-volume negatively curved manifold, *Bol. Soc. Bras. Mat.* **18**:2 (1987), 23-34.

Dodziuk J. and Randel B.
[1] Lower bounds for λ_1 on a finite-volume hyperbolic manifold, *J. of Diff. Geom.* **24** (1986), 133-139.

Dold A.
[1] Lectures on algebraic topology, Die Grundlehren der matimatischen Wissenschaften 200, *Springer-Verlag, Berlin*, 1972.

Dolgachev I.
[1] Integral quadratic forms: applications to algebraic geometry (after V. Nikulin), Bourbaki Seminar Vol.1982/83, *Asterique* 105-106, *Soc. Math. France, Paris*, 1983, 1-33.

Dubrovin B.A., Novikov S.P. and Fomenko A.T.
[1] Modern geometry, *Nauka, Moscow*, 1979. (= Modern geometry - methods and applications, Part I. The geometry of surfaces, transformation groups, and fields, Graduate texts in Math. **93**, *Springer-Verlag, New York- Berlin*, 1984.)
[2] Modern geometry: Methods of homology theory, *Nauka, Moscow*, 1984. (English translation by *Springer-Verlag*).

Dunbar W.D.
[1] Fibred orbifolds and crystallographic groups, Ph.D. Thesis, *Princeton University* (1981).

Eberlein P.
[1] Rigidity of lattices of non-positive curvature, *Ergodic Th. and Dynam. Syst.* **3** (1983), 47-85.

Efremovich V.A. and Tikhomirova E.S.
[1] Equimorphisms of hyperbolic spaces, *Izv. Akad. Nauk SSSR, Ser. Mat.* **28** (1964),, 1139-1144 (Russian).

Eilenberg S.P. and MacLane S.
[1] Cohomology theory in abstract groups, Part 1, *Ann. of Math.* **48** (1947), 51-78.

Epstein D.B.A.
[1] Finite presentations of groups and 3-manifolds, *Quart. J. Math.* **12** (1961), 205-212.
[2] Projective planes in 3-manifolds, *Proc. London Math. Soc.* **11** (1961), 469-484.
[3] Ends, topology of 3-manifolds and related topics, edited by M.K. Fort, *Prentice Hall* (1962).
[4] Isomorfismi conformi e geometria iperbolica (Appunti redatti da G. D'Ambra), *Editrice Univ. Levrotto & Bella, Torino* (1984).

Epstein D.B.A. and Marden A.
[1] Convex hulls in hyperbolic space, a theorem of Sullivan, and measured pleated surfaces, Analytic and geometric aspects of hyperbolic space, Warwick-Durham (1984), *London Math. Soc. Lecture Note Ser.* **111** *Cambridge Univ. Press* (1987), 113-254.

Epstein D.B.A. and Penner R.C.
[1] Euclidean decomposition of non-compact hyperbolic manifolds, *J. Diff. Geom.*, **27** (1988), 67-80.

Farrell F.T. and Hsiang W.C.
[1] The topological Euclidean space form problem, *Invent. Math.* **45**:2 (1978), 181-192.
[2] On Novikov's conjecture for non-positive curved manifolds, I, *Ann. of Math.* **113** (1981), 199-209.
[3] Addendum to "On Novikov's conjecture for non-positive curved manifolds".

Farrell F.T. and Jones L.E.
[1] Algebraic *K*-theory of hyperbolic manifolds, *Bull. Amer. Math. Soc.* **14**:1 (1986), 115-119.
[2] Implication of geometrization conjecture for the algebraic *K*-theory of 3-manifolds, Geometry and topology, Ed. C. McCrory and T. Shifrin, Lect. Notes in Pure & Appl Math. **105**, *Marcel Dekker, New York*, 1987, 109-113.
[3] *K*-theory and dynamics, *Ann. of Math.*, **124** (1986), 531-569.

Fathi A. Leudenback F, Poenary V. et al.
[1] Travaux de Thurston sur les surfaces. Les feuilletages mesurés, *Asterique* 66-67, *Soc. Math. de France, Paris*, 1979.

Fenchel W. and Nielsen J.
[1] On discontinuous groups of isometric transformations of the non-Euclidean plane, Studies and essays presented to R. Courant on his 60th birthday, January 1948, *Interscience Publ., New York* 1948, 117-128.

Floyd W.J.
[1] Group completions and limit sets of Kleinian groups, *Invent. Math.* **57** (1980), 205-218.

Floyd W.J., Hoare A.H.M. and Lyndon R.C.
[1] The word problem for geometrically finite groups, *Geom. Dedicata* **20** (1986), 201-207.

Fomenko A.T.
[1] Multidimensional Plateau problem on Riemannian manifolds. On the problem of the algorithmical recognizability of the standard three-dimensional sphere, Proceedings of the International Congress of Math., Vancouver (1974), 515-525.
[2] Topological variational problems, *Mosk. Gos. Univ., Moscow* (1984) (Russian).
[3] The topology of constant energy surfaces in integrable Hamiltonian systems and obstructions to integrability, *Izvestiya Akad. Nauk Ser. Mat. SSSR* **50** (1986), 1276-1307 (= Math. USSR-Izv. **79** (1987), 629-650.)
[4] A Morse theory for integrable Hamiltonian systems, *Dokl. Akad. Nauk SSSR* **287** (1986) 1071-1075. (= Soviet Math. Dokl. **33** (1986), 502-506.)

Fomenko A.T. and Zieschang H.
[1] On the topology of the 3-dimensional manifolds arising in Hamiltonian mechanics, *Dokl. Akad. Nauk SSSR* **294** (1987), 283-287. (= Soviet Math. Dokl. **35** (1987), 529-534.)
[2] On typical topological properties of integrable Hamiltonian systems, *Bochum, Ruhr-Universität,* Preprint 92/1987, 1-47.

Fomenko A.T. and Matveev S.V.
[1] Constant surfaces of Hamiltonian systems, enumeration of three-dimensional manifolds in increasing order of complexity and computation of volumes of closed hyperbolic manifolds, *Uspekhi Matem. Nauk* **43** (1988), 5-22. (= Russian Math. Surveys **43**:1 (1988), 3-24.).

Fomenko A.T. and Trofimov V.V.
[1] Integrable systems on Lie algebras and symmetric spaces, *Gordon and Breach Science Publishers, New York* (1987).

Ford L.R.
[1] Automorphic functions, *Chelsea, New York* (1951).

Fox R.H.
[1] A quick trip through knot theory. Topology of 3-manifolds and related topics, *Prentice-Hall*, 1962, 120-167.

Freedman M.H.

[1] The topology of four-dimensional manifolds, *J. Diff. Geom.* **17** (1982), 357-453.

[2] A conservative Dehn's lemma, Low dimensional topology, Contemp. Math. Vol.20, *Amer. Math. Soc., Providence,* 1983, 121-130.

[3] A geometric reformation of 4-dimensional surgery, *Topol. Appl.* **24** (1986), 133-141.

Freedman M.H., Hass J. and Scott P.

[1] Least area incompressible surfaces in 3-manifolds, *Invent. Math.* **71** (1983), 609-642.

Freedman M.H. and Scora R.

[1] Strange actions of groups on spheres, *J. Diff. Geom.* **25** (1987), 75-98.

[2] Strange actions of groups on spheres, II, Holomorphic functions and moduli II, Proc. Workshop, Berkeley (1986), *Publ. Math. Sci. Res. Inst.* **11** (1988), 41-57.

Freudenthal H.

[1] Uber die Enden topologischer Räume und Gruppen, *Math. Z.* **33** (193?), 692-713.

Fricke R.

[1] Uber eine besondere Klass discontinuierlicher Gruppen reeler linearer Substitutionen, *Math. Ann.* **38** (1891), 50-81, 461-476.

Fricke R. and Klein F.

[1] Vorlesungen über die Theorie der automorphen Funktionen, Vols. 1-2, *Teubner, Leipzig* (1926).

Fried D.

[1] Closed similarity manifolds, *Comm. Math. Helv.* **55** (1980), 576-582.

Fried D. and Goldman W.

[1] Three-dimensional affine crystallographic groups, *Adv. in Math.* **47** (1983), 1-49.

Fried D., Goldman W. and Hirsch M.

[1] Affine manifolds with nilpotent holonomy, *Comment. Math. Helv.* **56** (1981), 487-523.

[2] Affine manifolds and solvable groups, *Bull. Amer. Math. Soc.* **3** (1980), 1045-1047.

Gabai D.

[1] Genera of the arborescent links, *Memoirs Amer. Math. Soc.* **59** (1986) **339**.

Garland H. and Raghunathan M.S.
[1] Fundamental domains for lattices in (*R*)-rank 1 semisimple Lie groups, *Ann. of Math.* **92** (1970), 279-326.

Gehring F.W.
[1] Rings and quasi-conformal mappings in space, *Trans. Amer. Math. Soc.* **103** (1962), 353-393.
[2] Extension theorems for quasi-conformal mappings in *n*-space, *J. Analyse Math.* **19** (1967), 149-167.
[3] Extension theorems for quasi-conformal mappings in *n*-space, Proc. Internat. Congress of Math., Moscow (1966), *Nauka, Moscow*, 1968, 313-318.

Gehring F.W. and Martin G.
[1] Discrete quasi-conformal groups, I, *Proc. London Math. Soc.* **55** (1987), 331-358.
[2] Generalization of Kleinian groups, Preprint, *Math. Sci. Res. Inst., Berkeley* (1986).

Gehring F.W. and Palka B.P.
[1] Quasi-conformally homogeneous domains, *J. Analyse Math.* **30** (1976), 172-199.

Gehring F.W. and Väisälä
[1] The coefficients of quasi-conformality of domains in space, *Acta Math.* **114** (1965), 1-70.

Gerstenhaber M.
[1] On the algebraic structure of discontinuous groups, *Proc. Amer. Math. Soc.* **4** (1953), 745-750.

Gieseking H.
[1] Analytische Untersuchungen über topologische Gruppen, Ph.Dr. Thesis, Münster (1912).

Giffen C.H.
[1] The generalized Smith conjecture, *Amer. J. Math.* **88** (1966), 187-198.

Gilman J.
[1] On the Nielsen type and the classification for the mapping class group, *Adv. in Math.* **40** (1981), 68-96.
[2] Structures of elliptic irreducible subgroups of the modular group, *Proc. London Math. Soc.* **47** (1983), 27-42.
[3] Inequalities in discrete subgroups of $PSL(2, \mathbb{R})$, Preprint *MSRI, Berkeley* (1986).
[4] A geometric approach to the hyperbolic Jørgensen inequality, *Bull. Amer. Math. Soc.* **16**:1 (1987),91-92.

Goldman W.M. and Millson J.J.
[1] Local rigidity of discrete groups acting on complex hyperbolic space, *Inv. Math.* **88** (1987), 495-520.

Golubev V.V.
[1] Investigations in the theory of univalent functions singular points (Russian), *Sci. Notes Saratov St. Univ.* **1**:3 (1924); **2**:1 (1924); **4**:2 (1925); **5**:2 (1926), **6**:3 (1927) (= Univalent analytic functions, automorphic functions, *Fizmatgiz, Moscow* (1961), 197-370).

Gordon C.McA.
[1] Dehn surgery and satellite knots, *Trans. Amer. Math. Soc.* **275** (1983), 687-708.

Gordon C.McA. and Litherland R.A.
[1] Incompressible planar surfaces in 3-manifolds, *Topol. Appl.* **18** (1984), 121-144.

Gottschalk W.H. and Hedlund G.A.
[1] Topological dynamics, *Amer. Math. Soc. Colloquium Publ.* No.26, *Providence* (1955).

Greenberg L.
[1] Discrete subgroups of the Lorentz group, *Math. Scand.* **10** (1962), 85-107.
[2] Fundamental polyhedra for Kleinian groups, Ann of Math. **84** (1966), 433-442.
[3] Commensurable groups of Möbius transformations. Discontinuous groups and Riemann surfaces, edited by L. Greenberg, Ann. of Math. St. No.79, *Princeton Univ. Press, Princeton*, 1974, 227-237.
[4] Finiteness theorem for Kleinian groups. Discrete groups and Automorphic functions, edited by W.J. Harvey, *Academic Press*, 1977, 324-365.

Gromov M.
[1] Manifolds of negative curvature, *J. Diff. Geom.* **13** (1978), 223-230.
[2] Hyperbolic manifolds, groups and actions, Riemann surfaces and related topics: Proceedings of the 1978 Stony Brook Conference, edited by I. Kra and B. Maskit, Ann. of Math. St. No.97, *Princeton Univ. Press, Princeton,* 1981, 183-213.
[3] Curvature, diameter, and Betti numbers, *Comment. Math. Helv.* **56** (1981), 179-195.
[4] Structures métriques pour les variétés Riemanniennes, *Cedic/Fernand Nathan, Paris* (1978).
[5] Hyperbolic manifolds according to Thurston and Jørgensen, Bourbaki Seminar No.546, Lectures Notes in Math. No.842, *Springer-Verlag, Berlin-New-York*, 1981, 40-53.
[6] Hyperbolic groups, Essays in Group theory, Publ. MSRI **8**, Ed. S.M. Gerstein, *Springer-Verlag* (1987), 75-263.

Gromov M., Lawson H.B. and Thurston W.

[1] Hyperbolic 4-manifolds and conformally flat 3-manifolds, Publ. *Math* IHES **68** (1986), 27-45.

Gromov M. and Piatetski-Shapiro I.I.

[1] Non-arithmetic groups in Lobachevski spaces, Publ. *Math. IHES* **66** (1988), 93-103.

Gromov M. and Rokhlin V.A.

[1] Imbeddings and immersions in Riemannian geometry, *Uspekhi Mat. Nauk* **25**:5 (1970), 3-62. (= Russ. Math. Surveys **25**:5 (1970), 1-57.)

Grushko I.A.

[1] The bases of the free product of groups, *Matem. Sb.* **8** (1940), 169-182 (Russian).

Gunning R.C.

[1] Special coordinate coverings of Riemann surfaces, *Math. Ann.* **170** (1967), 67-86.

[2] Affine and projective structures on Riemann surfaces, Riemann surfaces and related topics: Proceedings of the 1978 Stony Brook Conference, edited by I. Kra and B. Maskit, Ann. of Math. St. No.97, *Princeton Univ. Press* (1981), 225-244.

Gusevskii N.A.

[1] A problem in the theory of Kleinian groups, *Dokl. Akad. Nauk SSSR* **234** (1977), 277-279 (= Soviet Math. Dokl. **18** (1977), 629-631).

[2] Topological characterization of a Schottky group, *Sibirsk. Mat. Zh.* **20**:3 (1979), 651-655 (= Siberian Math. J. **20** (1979), 454-457).

[3] Schottky groups in space, *Dokl. Akad. Nauk SSSR* **280** (1985), 30-33. (= Soviet Math. Dokl. **31** (1985), 21-24.)

Gusevskii N.A. and Kapovich M.E.

[1] Conformal structures on three-dimensional manifolds, *Dokl. Akad. Nauk SSSR* **290**:3 (1986), 537-541. (= Soviet Math. Dokl. **34** (1987), 314-318.)

Gutsul I.S.

[1] Compact three-dimensional manifolds of constant negative curvature, Geometry of positive quadratic forms, *Trudy Mat. Inst.Steklov* **152** ((1980), 89-96.

[2] On completeness of incomplete hyperbolic structures, Geometry of discrete groups, *Kishinev, Stiinza Acad. Publ.* (1990), 66-78.

Gutsul I.S. and Makarov V.S.

[1] A property of Fëdorov groups of a Lobachevskii space. Algebra, number theory and their applications, *Trudy Mat. Inst. Steklov* **148** (1978), 106-108.

[2] Incomplete 3-manifolds modelled Lobachevskii space, Aspects of discrete geometry, *Kishinev, Stiinza Acad. Publ.* (1988), 69-76.

Harvey W.

[1] Chabauty spaces of discrete groups. Discontinuous groups and Riemann surfaces, edited by L. Greenberg, Ann. of Math. St. No.79, *Princeton Univ. Press, Princeton*, 1974, 239-246.

[2] (Editor): Discrete groups and automorphic functions, *Academic Press, London*, 1977.

Hass J., Rubinstein H. and Scott P.

[1] Covering spaces of 3-manifolds, *Bull. Amer. Math. Soc.* **16**:1 (1987), 117-119.

Hatcher A.

[1] Hyperbolic structures of arithmetic type on some link complements, *J. London Math. Soc.* **27**:2 (1983), 345-355.

Hausdorff F.

[1] Grundzüge der Mangenlehre, *Veit & Co., Leipzig*, 1914.

Hedlund G.A.

[1] Fuchsian groups and transitive horocycles, *Duke Math. J.* **2** (1936), 530-542.

Hejhal D.A.

[1] Monodromy groups and linearly polymorphic functions, *Acta Math.* **135** (1975), 1-55.

Helgason S.

[1] Differential geometry and symmetric spaces, *Academic Press*, 1962.

Hempel J.

[1] 3-manifolds, *Princeton Univ. Press, Princeton*, 1976.

[2] Orientation reversing involutions and the first Betti number for finite coverings of 3-manifolds, *Invent. Math.* **67** (1982), 133-142.

Hilbert D.

[1] Mathematical problems, Göttinger Nachrichten (1900), 253-297; reprinted *Bull. Amer. Math. Soc.* **8** (1902), 437-473; Mathematical developments arising from Hilbert problems, *Providence* (1976).

Hilden H.M., Losano M.T. and Montesinos J.M.

[1] Universal knots, *Bull. Amer. Nath. Soc.* **8** (1983), 449-450.

[2] The Whitehead link, the Borromean rings, and the knot 9_{46} are universal, *Collect. Math.* **34**:1 (1983), 19-28.

[3] On the universal group of the Borromean rings, *Diff. Topology*, Ed. by U. Kosehorke, Lect. Notes Math. 1350, *Springer-Verlag*, 1989, 1-13.

Hilden H.M., Losano M.T., Montesinos J.M. and Whitten W.
[1] On universal groups and 3-manifolds, *Invent. Math.* **87** (1987), 441-456.

Hirsch M.W. and Thurston W.P.
[1] Foliated bundles, flat manifolds and invariant measures, *Ann. of Math.* **101** (1975), 369-390.

Hopf E.
[1] Fuchsian groups and ergodic theory, *Trans. Amer. Math. Soc.* **39** (1936), 299-314.
[2] Statistik der geodätischen Linien in Manigfaltigkeiten negative Krummung, *Ber. Verh. Sachs. Acad. Wiss. Leipzig* **91** (1939), 261-304.

Hopf H.
[1] Über die Abbildungen der dreidimensionalen Sphäre auf die Kugelfläche, *Math. Ann.* **104** (1931), 637-665.

Hu Sze-Tsen
[1] Homotopy theory, *Academic Press*, 1959.

Husch L.S. and Price T.M.
[1] Finding a boundary for a 3-manifold, *Ann. of Math.* **91** (1970), 223-235.

Inoue H. and Yano K.
[1] The Gromov invariant of negatively curved manifolds, *Topology* **21** (1982), 83-91.

Ishihara S.
[1] Homogeneous Riemann spaces of four dimensions, *J. Math. Soc. Japan* **7** (1955), 345-370.

Ivascu D.
[1] The Schottky space in dimensions greater than two. Complex analysis, Joensuu (1978), Lecture Notes in Math. No.747, *Springer-Verlag, Berlin*, 1979, 167-177.

Jaco W.H.
[1] Finitely presented subgroups of three-manifold groups, *Invent. Math.* **13** (1971), 335-346.
[2] Lectures on three-manifold topology, C.B.M.S. Regional Conference Series in Math., Vol.43, *Amer. Math. Soc., Providence*, 1980.

Jaco W.H. and Shalen P.B.
[1] Seifert fibered spaces in 3-manifolds, *Mem. Amer. Math. Soc.* **21** No.220 (1979).

Johannson K.
[1] Homotopy equivalences of 3-manifolds with boundaries, Lecture Notes in Math. No.761, *Springer-Verlag, Berlin*, 1979.

Johnson D.L.
[1] Topics in the theory of group representations, London Math. Soc. Lecture Notes Series **42**, *Cambridge Univ. Press, Cambridge-New York*, 1980.

Johnson D. and Millson J.J.
[1] Deformation spaces associated to compact hyperbolic manifolds, Discrete groups in geometry and analysis: Papers in honour of G.D. Mostow on his sixtieth birthday, *Birkhauser* (Progress in Math. 67) *Boston*, 1987, 48-106.

Jørgensen T.
[1] Some remarks on Kleinian groups, *Math. Scand.* **34** (1974), 101-108.
[2] On discrete groups of Möbius transformations, *Amer. J. Math.* **98** (1976), 739-749.
[3] Compact 3-manifolds of constant negative curvature fibering over a circle, *Ann. of Math.* **106** (1977), 61-72.
[4] Closed geodesics on Riemann surfaces, *Proc. Amer. Math. Soc.* **72** (1978), 140-142.

Jørgensen T. and Marden A.
[1] Two doubly degenerate groups, *Quart. J. Math.* **30** (1979), 143-156.
[2] Generic fundamental polyhedra for Kleinian groups, Holomorphic Functions and Moduli II: Proc. of Workshop, Berkeley/Calif. 1986, Publ. *MSRI*, 1988, v.11, 69-86.
[3] Algebraic and geometric convergence of Kleinian groups, Preprint *MSRI, Berkeley*, 1986.

Jørgensen T., Marden A. and Maskit B.
[1] The boundary of classical Schottky space, *Duke Math. J.* **46** (1979), 441-446.

Kac V.G. and Vinberg E.B.
[1] Quasi-homogeneous cones, *Mat. Zametki* **1** (1967), 347-354 (= Math. Notes Acad. Sci. USSR **1** (1967), 231-235.

Kagan V.F.
[1] Principles of surface theory, Vol.2, *OGIZ, Moscow-Leningrad*, 1948 (Russian).

Kaplinskaya I.M.
[1] The discrete groups that are generated by reflections in the faces of simplicial prisms in Lobachevskii spaces, *Mat. Zametki* **15** (1974), 159-164. (= Math. Notes **15** (1974), 88-91.)

Kazhdan D.A. and Margulus G.A.
[1] A proof of Selberg's hypothesis, *Mat. Sb.* **75** (1968), 163-168.

Kemeny J.G, Knapp A.W., and Snell J.L.
[1] Denumerable Markov chains, *Van Nostrand*, 1966.

Khovanskii A.F.
[1] Hyperbolic sections of polyhedra, toroidal manifolds, and discrete groups in Lobachevski space, *Funktsional'nyi analiz i ego prilozheniya* **20**:1 (1986), 50-61.

Kimel'fel'd B.N.
[1] Homogeneous domains on the conformal sphere, *Mat. Zamet.* **8** (1970), 321-328. (= Math. Notes **8** (1970), 653-656.)
[2] Quasi-homogeneous domains on a conformal sphere, *Uspekhi Mat. Nauk* **33**:2 (1978), 193-194.

Klimenko E.Ya.
[1] On a class of discrete groups in Lobachevski space, In: Current topics in complex analysis, Abstracts, *Tashkent* (1985), 54 (Russian).

Kobayashi S. and Nomizu K.
[1] Foundations of differential geometry, *Interscience Publ., New York*, 1963.

Koebe P.
[1] Über die Uniformisierung der algebraischen Kurven, Part 2, *Math. Ann.* **69** (1910), 1-81.

Kondo S.
[1] Enriques surfaces with finite automorphism group, *Japan J. Math.* **12** (1986), 192-282.

Kourouniotis C.
[1] Deformations of hyperbolic structures on manifolds of several dimensions, Ph.D. Thesis, *London Univ.* (1984).
[2] Deformations of hyperbolic structures, *Math. Proc. Cambridge Phil. Soc.* **98** (1985), 247-261.
[3] Bendings in the space of quasi-Fuchsian structures, Preprint *Univ. of Crete* (1988).

Kra I.
[1] Deformations of Fuchsian groups, *Duke Math. J.* **36** (1969), 537-546.
[2] On affine and projective structures on Riemann surfaces, *J. Analyse Math.* **22** (1969), 285-298.
[3] Deformations of Fuchsian groups, Part 2, *Duke Math. J.* **38** (1971), 499-508.
[4] On spaces of Kleinian groups, *Comm. Math. Helv.* **47** (1972), 53-69.

[5] Automorphic forms and Kleinian groups, Maths. Lecture Note Series, *Benjamin, Reading*, 1972.
[6] On the Nielsen-Thurston-Bers type of some self-maps of Riemann surfaces, *Acta Math.* **146** (1981), 231-270.

Kra I. and Maskit B.
[1] Remarks on projective structures. Riemann surfaces and related topics: Proceedings of the 1978 Stony Brook Conference, edited by I. Kra and B. Maskit, Ann. of Math. St. No.97, *Princeton Univ. Press, Princeton*, 1981, 343-360.

Krushkal' S.L.
[1] Quasi-conformal mappings and Riemann surfaces, *Nauka, Novosibirsk*, 1975. (= English translation: *John Wiley, New York* (1979)).
[2] A certain property of the limit sets of Kleinian groups, *Dokl. Akad. Nauk SSSR* **225** (1975), 500-502; **237** (1977), 256. (= Soviet Math. Dokl. **16** (1975), 1497-1499.)
[3] Supports of Beltrami differentials for Kleinian groups, *Dokl. Akad. Nauk SSSR* **231** (1976), 799-801 (= Soviet Math. Dokl. **17** (1976), 1642-1644).
[4] Some rigidity theorems for discontinuous groups, Mathematical analysis and related questions, Ed. A.A. Borovkov, *Nauka, Novosibirsk*, 1978, 69-82 (= English translation with changes: *Amer. Math. Soc. Transl.* (2) **122** (1984), 75-83).
[5] To the problem of the supports of Beltrami differentials for Kleinian groups, Romanian-Finnish Seminar on Complex Analysis, Lectures Notes in Math. No.743, *Springer-Verlag, Berlin*, 1979, 132-134.
[6] A remark on the rigidity of quasi-conformal deformations of discrete isometry groups of hyperbolic space, *Sibirsk. Mat. Zh.* **21**:5 (1980), 52-57 (= Siberian Math. J. **21** (1980), 683-687).
[7] Some remarks on Kleinian groups, Riemann surfaces and related topics: Proceedings of the 1978 Stony Brook Conference, edited by I. Kra and B. Maskit, Ann. of Math. St. No.97, *Princeton Univ. Press, Princeton*, 1981, 361-366.

Krushkal' S.L., Apanasov B.N. and Gusevskii N.A.
[1] Kleinian groups in examples and problems, *Izd. Novosibirsk. Gos. Univ., Novosibirsk*, 1978 (Russian).
[2] Uniformization and Kleinian groups, *Izd. Novosibirsk. Gos. Univ., Novosibirsk*, 1979 (Russian).
[3] Kleinian groups and uniformization in examples and problems, *Nauka, Novosibirsk*, 1981. (= Trans. Math. Monographs **62**, *Amer. Math. Soc.* (1986).)

Kuiper N.H.
[1] On conformally flat spaces in the large, *Ann. of Math.* **50** (1949), 916-924.
[2] On compact conformally Euclidean spaces of dimension 2, *Ann. of Math.* **52** (1950), 478-490.

[3] Compact spaces with a local structure determined by the group similarity transformations in $\mathbb{R}^n$, *Indag. Math.* **12** (1950), 411-418.
[4] Hyperbolic 4-manifolds and tessellations, Publ. *Math IHES,* **68** (1988), 47-76.
[5] Fairly symmetric hyperbolic manifolds, Actes du Congrès de Géométrie d'Avignon 1988, to appear (Preprint *IHES*, 1989).

Kulkarni R.S.
[1] Infinite regular coverings, *Duke Math. J.* **45** (1978), 781-796.
[2] On the principle of uniformization, *J. Diff. Geom.* **13** (1978), 109-138.
[3] Some topological aspects of Kleinian groups, *Amer. J. Math.* **100** (1978), 897-911.
[4] Groups with domains of discontinuity, *Math. Ann.* **237** (1978), 253-272.

Kulkarni R.S. and Pinkall U.
[1] Uniformization of geometric structures with applications to conformal geometry, Differential geometry: Proc. Peñiscola, Spain (1985), Ed. A.M. Naveira a.o., (Lecture Notes Math. 1209) *Springer-Verlag, Berlin - New York,* 1986, 190-210.
[2] (Editors), Conformal Geometry, Aspects of Math. V.E. 12, *Friedr. Vieweg und Sohn, Braunschweig/Wiesbaden,* 1988.

Kulkarni R.S. and Shalen P.B.
[1] On Ahlfors's finiteness theorem, Preprint (1984).

Lafontaine J.
[1] Modules de structures conformes plates et cohomologie de groupes discrets, *C.R. Acad. Sci. Paris* **297** (1983), 655-658.

Lanner E.
[1] On complexes with transitive groups of automorphisms, *Comm. Sém. Math. Univ. Lund* **11** (1950), 1-71.

Lavrent'ev M.A. and Belinskii P.P.
[1] Certain problems of geometric function theory, *Trudy Mat. Inst. Akad. Nauk SSSR* **128**:2 (1972), 34-40 (= *Amer. Math. Soc. Trans. Providence* (1973)).

Lehner J.
[1] Discontinuous groups and automorphic functions, *Amer. Math. Soc., Providence*, 1964.

Löbell F.
[1] Beispiele geschlossener dreidimensionaler Clifford-Klenscher Räume negative Krümmung, *Ber. Sächs. Akad. Wiss. Leipzig* **83** (1931), 167-174.

Lok W.L.

[1] Deformations of locally homogeneous spaces and Kleinian groups, Ph.D. Thesis, *Columbia University* (1984).

Lyndon R.C. and Schupp P.E.

[1] Combinatorial group theory, Ergebnisse der Matematik und ihrer Grenzegebiete **89**, *Springer-Verlag, Berlin*, 1977.

Macbeath A.M.

[1] Groups of homeomorphisms of a simply connected space, *Ann. of Math.* **79** (1964), 473-488.

[2] Commensurability of cocompact three-dimensional hyperbolic groups, *Duke Math. J.* **50** (1983), 1245-1253.

[3] Generic Dirichlet polygons and the modular group, *Glasgow Math. J.* **27** (1985), 129-141.

Macbeath A.M. and Hoare A.H.M.

[1] Groups of hyperbolic crystallography, *Math. Proc. Cambridge Phil. Soc.* **79** (1976), 235-249.

MacLane S. and Eilenberg S.P.

[1] Cohomology theory in abstract groups, *Ann. of Math.* **48** (1947), 51-78.

Magnus W.

[1] Non-Euclidean tessellations and their groups, *Academic Press*, 1974.

Magnus W, Karrass A. and Solitar D.

[1] Combinatorial group theory, *Interscience Publ., New York*, 1966.

Mahler K.

[1] On lattice points in n-dimensional starbodies, I. Existence theorems, *Proc. Royal Soc. London* **1008** (1946: A187), 151-187.

Makarov P.V.

[1] On a construction of four-dimensional semiregular polyhedra. Problems of discrete geometry, *Stiinza Acad. Publ., Kishinev*, 1988, 139-150.

Makarov V.S.

[1] A class of decompositions of the Lobachevskii space, *Dokl. Akad. Nauk SSSR* **161** (1965), 277-278. (= *Soviet Math. Dokl.* **6** (1965), 400-401).

[2] On a certain class of discrete groups in Lobachevskii space having an infinite fundamental region of finite measure, *Dokl. Akad. Nauk SSSR* **167** (1966), 30-33 (= *Soviet Math. Dokl.* **7** (1966), 328-331.)

[3] On a class of two-dimensional Fedorov groups, *Izv. Akad. Nauk SSSR Ser. Mat.* **31** (1967), 531-542 (= *Math. USSR-Izv.*).

[4] The Fedorov groups of 4-dimensional and 5-dimensional Lobachevskii spaces, Studies in general algebra No.1, *Kishinev Gos. Univ., Kishinev*, 1968, 120-129 (Russian).
[5] Examples of non-compact decompositions of the Lobachevskii space Λ^n for $n \leq 9$, Studies in discrete geometry, *Stiinza Acad. Publ., Kishinev*, 1974, 120-123 (Russian).
[6] A non-compact partition of 10-dimensional Lobachevskii space. Geometry of positive quadratic forms, *Trudy Mat. Inst. Steklov* **152** (1980), 162-164 (= *Amer. Math. Soc. Trans., Providence* (1981)).
[7] Geometric methods of construction of discrete groups of motions in a Lobachevskii space, Problems in geometry, Vol.15 *VINITI, Moscow*, 1983, 3-59.
[8] Matrix representations of the fundamental groups of some three-dimensional manifolds of constant negative curvature having a finite volume, Problems in discrete geometry, *Stiinze, Kishinev*, 1988, 122-138 (Russian).
[9] On fundamental polyhedron of a discrete group of motions of the Lobachevski space, its combinatoric and deformation, Geometry of discrete groups, *Stiinze, Kishinev*, 1990, 110-121 (Russian).

Makarov V.S. and Gutsul I.S.
[1] Non-compact three-dimensional manifolds of constant negative curvature that have finite measure. Geometry of positive quadratic forms, *Trudy Mat. Inst. Steklov* **15** (1980), 165-169 (= *Amer. Math. Soc. Trans.* (1981)).
[2] An example of an incomplete non-rigid three-dimensional manifold of finite volume having a locally Lobachevski metric, Abstracts for the Leningrad International Topology Conference, *Nauka, Leningrad*, 1982, 95 (Russian).

Maltsev A.I.
[1] The isomorphic matrix representations of infinite groups, *Matem. Sb.* **8** (1940), 405-422 (Russian).
[2] The class of homogeneous spaces, *Izvestiya Akad. Nauk SSSR Ser. Mat.* **13** (1949), 9-32 (Russian).

Marden A.
[1] The geometry of finitely generated Kleinian groups, *Ann. of Math.* **99** (1974), 383-462.
[2] Universal properties of Fuchsian groups in the Poincaré metric, Discontinuous groups and Riemann surfaces, edited by L. Greenberg, Ann. of Math. St. No.79, *Princeton Univ. Press*, 1974, 315-341.
[3] Schottky groups and circles, Contribution to Analysis (a collection of papers dedicated to Lipman Bers), edited by L.V. Ahlfors et al, *Academic Press*, 1974, 273-278.
[4] Isomorphisms between Fuchsian groups, Advances in complex function theory, Lecture Notes in Math. No.505, *Springer-Verlag*, 1976, 56-78.

Marden A. and Maskit B.
[1] On the isomorphism theorem for Kleinian groups, *Invent. Math.* **51** (1979), 9-14.

Margulis G.A.
[1] The isometry of closed manifolds of constant negative curvature with the same fundamental group, *Dokl. Akad. Nauk SSSR* **192** (1970), 736-737. (= *Soviet Math. Dokl.* **11** (1970), 722-723.)
[2] Arithmeticity of non-uniform lattices, *Funckts. Analiz i Pril.* **7**:3 (1973), 88-89. (= *Funct. Anal. Appl.* **7** (1973), 245-246.)
[3] Discrete groups of motions of manifolds of non-positive curvatur Proc. International Congress Math., Vancouver (1974), Vol.2, 21-34. (= *Amer. Math. Soc. Trans.* **109**:2 (1977), 33-45).
[4] Arithmeticity and finite-dimensional representations of uniform lattices, *Funkts. Anal. i Pril.* **8**:3 (1974), 77-78 (= *Funct. Anal. Appl.* **8** (1974), 258-259.)
[5] Arithmetic properties of discrete subgroups, *Uspekhi Mat. Nauk* **29** (1974), 49-98 (= *Russian Math. Surveys* **29** (1974), 107-156).
[6] Arithmeticity of non-uniform lattices in weakly concompact groups, *Funkts. Anal. i Pril.* **9**:1 (1975), 35-44 (= *Funct. Anal. Appl.* **9**:1 (1975), 31-38).
[7] Arithmeticity of irreducible lattices in semisimple groups of rank greater than 1, Supplement to: Discrete subgroups of the Lie groups, *Mir, Moscow*, 1977, 277-313 (Russian).
[8] Factor groups of discrete subgroups and measure theory, *Funkts. Anal. i Pril.* **12**:4 (1978), 64-76 (= *Funct. Anal. Appl.* **12**:4 (1978), 294-305).
[9] Finiteness of quotient groups of discrete subgroups, *Funkts. Anal. i Pril.* **13**:3 (1979), 28-39.
[10] Complete affine locally flat manifolds with free fundamental group, Automorphic functions and number theory, II, Proceedings of the Scientific Seminars of Leningrad Division of the Steklov Mathetmatics Institute, Academy of Sciences of the USSR, Vol.134, *Nauka, Leningrad*, 1984, 190-205 (Russian).

Martin G.J.
[1] Quasi-conformal and bi-Lipschitz homeomorphisms, uniform domains and the quasi-hyperbolic metric, *Trans. Amer. Math. Soc.* **292** (1985), 169-191.
[2] On discrete Möbius groups in all dimensions. A generalization of Jørgensen's inequality, *Acta Math.* **163** (1989), 253-298.

Martio O. and Srebro U.
[1] Automorphic quasi-meromorphic mappings in $\mathbb{R}^n$, *Acta Math.* **135** (1975), 221-247.
[2] On the existence of automorphic quasi-meromorphic mappings in $\mathbb{R}^n$, *Ann. Acad. Sci. Fenn. Ser.* A1 **3** (1977), 123-130.

Maskit B.

[1] A characterization of Schottky groups, *J. Analyse Math.* **19** (1967), 227-230.

[2] On boundaries of Teichmüller spaces and Kleinian groups, Part 2, *Ann. of Math.* **91** (1970), 608-638.

[3] On Klein's combination theorem, *Trans. Amer. Math. Soc.* **120** (1965), 499-509.

[4] On Klein's combination theorem, Part 2, *Trans. Amer. Math. Soc.* **131** (1968), 32-39.

[5] On Klein's combination theorem, Part 3, Advances in the theory of Riemann surfaces, Proc. Conf. Stony Brook 1969, ed. by L.V. Ahlfors et al., Ann. of Math. St. No.66, *Princeton Univ. Press, Princeton*, 1971, 297-310.

[6] On Poincaré's theorem for fundamental polygons, *Adv. in Math.* **7** (1971), 219-230.

[7] A theorem on planar covering surfaces with applications to 3-manifolds, *Ann. of Math.* **81** (1965), 341-355.

[8] On a class of Kleinian groups, *Ann. Acad. Sci. Fenn., Ser.* AI **442** (1969), 1-8.

[9] On the classification of Kleinian groups, Part 2: Signatures, *Acta Math.* **138** (1973), 17-42.

[10] Kleinian groups, *Springer-Verlag, Berlin*, 1988.

[11] Some deformations of codimension two discrete groups, *J. Math. Kyoto Univ.* **26** (1986), 77-80.

Matveev S.V.

[1] Special skeletons of piecewise linear manifolds, *Mat. Sb.* **92**:2 (1973), 282-293. (= *Math. USSR-Sb.* **21** (1973), 279-291 (1974).)

[2] Transformations of special spines and the Zeeman conjecture, *Izv. Akad. Nauk SSSR* **51**:5 (1987), 1104-1116 (= *Mat. USSR Izv.* **31**:2 (1988), 423-434).

[3] Complexity of 3-dimensional manifolds and their enumeration in order of increasing complexity, *Dokl. Akad. Nauk SSSR* **301** (1988), 280-283 (= *Soviet Math. Dokl.* **38**:1 (1989), 75-78).

[4] The complexity theory of 3-manifolds, Preprint 88-13, *Inst. Math. Acad. Nauk UKr.SSR, Kiev* (1988).

Matveev S.V. and Savvateev V.V.

[1] Three-dimensional manifolds having simple special spines, *Coll. Math.* **32** (1974), 83-97.

McCullough D.

[1] Homotopy groups of the space of self-homotopy equivalences, *Trans. Amer. Math. Soc.* **264** (1981), 151-163.

[2] Compact 3-manifolds with infinitely-generated groups of self-homotopy equivalences, *Proc. Amer. Math. Soc.* **91**:4 (1984), 625-629.

[3] Mappings of reducible 3-manifolds, Preprint, *Math. Sci. Res. Inst., Berkeley* (1984).

[4] Twist groups of compact 3-manifolds, Preprint, *Topology* (1984).

McCullough D. and Miller A.
[1] Homeomorphisms of 3-manifolds with compressible boundary, *Mem. of Amer. Math. Soc.* **61** (1986), No.344.

Mednykh A.D.
[1] Semidirect products of discontinuous transformation groups, *Dokl. Akad. Nauk SSSR* **225** (1975), 1016-1017. (= *Soviet Math. Dokl.* **16** (1975), 1578-1580.)
[2] Automorphism groups of three-dimensional hyperbolic manifolds, *Dokl. Akad. Nauk SSSR* **285** (1985), 40-44. (= *Soviet Math. Dokl.* **32** (1985), 633-636.)

Medynkh A.D. and Vesnin A.Yu.
[1] On three-dimensional hyperbolic manifolds of Löbell type, Complex analysis and applications '85 (Varna 1985), *Bulg. Acad. Sci., Sofia* (1986), 341-350.

Mednykh A.D. and Volynets I.A.
[1] Introduction to the theory of Kleinian groups, *Omsk State Univ., Omsk*, 1984 (Russian).

Meeks W.H. III, Simon L. and Yau S.-T.
[1] Embedded minimal surfaces, exotic spheres, and manifolds with positive Ricci curvature, *Ann. of Math.* **116** (1982), 621-659.

Meeks W.H. III and Yau S.-T.
[1] The equivariant loop theorem for three-dimensional manifolds and a review of the existence theorems for minimal surfaces, The Smith conjecture, edited by H. Bass and J. Morgan, *Academic Press*, 1984.

Menasco W.W.
[1] Polyhedra representation of link complements, Low dimensional topology, Contemp. Math. Vol.20, *Amer. Math. Soc.* (1983), 305-326.

Metzger T.A.
[1] On a fundamental set for the limit set of a Fuchsian group, Preprint, *Univ. of Pittsburgh* (1976).
[2] On the exponent of convergence for infinitely generated Fuchsian groups, Complex analysis, Joensuu (1987), Lecture Notes Math. **1351** *Springer-Verlag*, 1988, 243-251.

Meyer J.
[1] Präsentation der Einheitengruppe der quadratischen Form $F(X) = -X_0^2 + X_1^2 + \ldots + X_{18}^2$, *Arch. Math.* **29** (1977), 261-266.

Meyerhoff R.
[1] The cusped hyperbolic 3-orbifold of minimum volume, *Bull. Amer. Math. Soc.* **13** (1985), 154-156.

Millson J.
[1] On the first Betti number of a constant negatively curved manifold, *Ann. of Math.* **104** (1976), 235-247.

Millson J. and Raghunathan M.S.
[1] Geometric construction of cohomology for arithmetic groups, Part 1, Geometry and analysis, Papers dedicated to the memory of V.K. Patodi, Proc. Indian Acad. Sci. Math. Sci. **90**, *Springer-Verlag*, 1981, 103-123.

Milnor J.W.
[1] Hilbert's problem 18: on crystallographic groups, fundamental domains, and on sphere packing, Mathematical developments arising from Hilbert problems, Proc. Symp. Pure Math. Vol.28, *Amer. Math. Soc.* (1976), 491-506.
[2] On fundamental groups of complete affinely flat manifolds, *Adv. Math.* **25** (1977), 178-187.
[3] Computation of volume, Chapter 7 of Thurston's Lecture Notes (1978-79).
[4] Hyperbolic geometry: the first 150 years, *Bull. Amer. Math. Soc.* **6** (1982), 9-25.
[5] A unique decomposition theorem for 3-manifolds, *Amer. J. Math.* **84** (1962), 1-7.

Molnar E.
[1] An infinite series of compact non-orientable 3-dimensional space forms of constant negative curvature, *Ann. Glob. Analysis Geom.* **1**:3 (1983), 37-49; **2**:2 (1984), 253-254.
[2] Twice-punctured compact Euclidean and hyperbolic manifolds and their twofold coverings, Proc. Coll. on Diff. Geometry, Hungary (1984).

Montesinos J.M.
[1] Representing 3-manifolds by a universal branching set, *Math. Proc. Cambridge Phil. Soc.* **94** (1983), 109-123.
[2] Classical tessellations and three-manifolds, *Springer-Verlag, Berlin*, 1987.

Morgan J.W.
[1] On Thurston's uniformization theorem for three-dimensional manifolds, The Smith conjecture, edited by H. Bass and J.W. Morgan, *Academic Press*, 1984, 37-125.
[2] Group action on trees and compactification of the space of classes of $SO(n, 1)$-representations, *Topology*, **25** (1986), 1-33.

Morgan J.W. and Shalen P.B.
[1] Valuations, trees, and degenerations of hyperbolic structures, I, *Ann. of Math.* **120** (1984), 401-476.
[2] Degenerations of hyperbolic structures, II: measured laminations in 3-manifolds, *Ann. of Math.* **127** (1988), 403-456.

Morokuma T.
[1] On discontinuous groups acting on real hyperbolic space I, *Proc. Japan Acad.* **52** (1976), 359-362.

Mostow G.D.
[1] Quasi-conformal mappings in *n*-space and the rigidity of hyperbolic space forms, *Inst. Hautes Etudes Sci. Publ. Math.* **34** (1968), 53-104.
[2] Strong rigidity of locally symmetric spaces, Ann. of Math. St. No.78, *Princeton Univ. Press*, 1973.
[3] On a remarkable class of polyhedra in complex hyperbolic space, *Pacific J. Math.* **86** (1980), 171-276.

Mostow G.D. and Tamagawa T.
[1] On the compactness of arithmetically defined homogeneous spaces, *Ann. of Math.* **76** (1962), 446-463.

Munkholm H.J.
[1] Simplices of maximal volume in hyperbolic space, Gromov's norm and Gromov's proof of Mostow's rigidity theorem (following Thurston), Topology Symposium, Siegen (1979), edited by U. Koschorke and W.D. Neuman, Lecture Notes in Math. Vol.788, *Springer-Verlag, Berlin*, 1980, 109-124.

Myers K.
[1] Free involutions of lens spaces, *Topology* **20** (1981), 313-318.

Myrberg P.J.
[1] Die Kapazität der singularen Menge der linearen Gruppen, *Ann. Acad. Sci. Fenn* **10** (1941), 1-10.

Nevanlinna R.
[1] Uniformisierung, *Springer-Verlag*, 1953.

Newstead P.E.
[1] Introduction to module problems and orbit space, Lectures In Math. & Physics **51**, TATA Institute of Fund. Research,Bombay, *Narosa Pub. House, New Dehli*, 1978.

Nicholls P.J.
[1] On the distribution of orbits in a Kleinian group, *Proc. London Math. Soc.* **29**:3 (1974), 193-215.
[2] Garnett points for Fuchsian groups, *Bull. London Math. Soc.* **12** (1980), 216-218.
[3] The orbital counting functions for discrete groups of infinite volume, *J. London Math. Soc.* **23**:2 (1981), 460-468.
[4] A lattice point problem in hyperbolic space, *Michigan Math. J.* **30** (1983), 273-287.

[5] Discrete groups on the sphere at infinity, *Bull. London Math. Soc.* **15** (1983), 488-492.
[6] Ford and Dirichlet regions for discrete groups of hyperbolic motions, *Trans. Amer. Math. Soc.* **282** (1984), 355-365.

Nielsen J.
[1]-[3] Untersuchungen zur Topologie der geschlossenen zweiseitigen Flächen, *Acta Math.* Part 1: **50** (1927), 189-358; Part 2: **53** (1929), 1-76; Part 3: **58** (1932), 87-167.
[4] Über Gruppen linearer Transformationen, *Mitt. Math. Gesellschaft, Hamburg* **8** (1940), 81-104.

Nikulin V.V.
[1] Quotient groups of groups of automorphisms of hyperbolic forms of subgroups generated by two-reflections, *Dokl. Akad. Nauk SSSR* **248** (1979), 1307-1309. (= *Soviet Math. Dokl.* **20**:5 (1979), 1156-1158.)
[2] Quotient groups of groups of automorphisms of hyperbolic forms by subgroups generated by 2-reflections. Algebraic-geometric applications, Current problems in Mathematics Vol.18, *VINITI, Moscow*, 1981, 3-114 (Russian).
[3] On the arithmetic groups generated by reflections in Lobachevskii spaces, *Izv. Akad. Nauk SSSR Ser. Mat.* **44** (1980), 637-669 (= *Math. USSR-Izv.* **16**:3 (1981). 573-601).
[4] On the classification of arithmetic groups generated by reflections in Lobachevskii spaces, *Izv. Akad. Nauk SSSR Ser. Mat.* **45** (1981), 113-142 (= *Math. USSR-Izv.* **18**:1 (1982), 99-123).
[5] Involutions of integer quadratic-form and their applications to real algebraic geometry, *Izv. Akad. Nauk SSSR Ser. Mat.* **47** (1983), 109-188 (= *Math. USSR-Izv.*).
[6] Surfaces with a finite group of automorphisms and a Picard group of rank 3. Algebraic geometry and its applications, *Trudy Mat. Inst. Steklov* **165** (1984), 119-142. (= *Amer. Soc. Trans.* (1985).)
[7] On a description of the automorphism groups of Enriques surfaces, *Dokl. Akad. Nauk SSSR* **277** (1984), 1324-1327 (= *Soviet Math. Dokl.* **30** (1984), 282-285).
[8] Discrete reflection groups in Lobachevski spaces and algebraic surfaces, Proc. International Congress Math., Berkeley (1986).

Nikulin V.V. and Shafarevich I.R.
[1] Geometries and groups *Nauka, Moscow*, 1983. (= Springer Series in Sov. Math., *Springer-Verlag, Berlin - New York*, 1987.)

Novikov P.S.
[1] On algorithmic unsolvability of word problems in group theory, *Trudy Steklov Inst. Mat. Akad. Nauk SSSR,* **44** (1955) (= *Amer. Math. Soc. Trans.* **9** (1958)).

Novikov S.P.
[1] Homotopically equivalent smooth manifolds, Part 1, *Izv. Akad. Nauk SSSR* **28** (1964), 365-474.

Ohshika K.
[1] Teichmüller spaces of Seifert fibred manifolds with infinite π_1, *Top. Appl.* **27** (1987), 75-93.

Ohtake H.
[1] On discontinuous subgroups with parabolic transformations of the Möbius groups, *J. Math. Kyoto Univ.* **25** (1985), 807-816.

Orlik P.
[1] Seifert manifolds, Lecture Notes in Math. Vol.291, *Springer-Verlag*, 1972.

Papakiriakopulos C.
[1] On Dehn's lemma and the asphericity of knots, *Ann. of Math.* **66** (1957), 1-26.
[2] On solid tori, *Proc. London Math. Soc.* **7** (1957), 281-299.

Patterson S.J.
[1] The limit set of a Fuchsian group, *Acta Math.* **136** (1976), 242-273.

Poincaré H.
[1] Théorie des groupes fuchsiennes, *Acta Math.* **1** (1882), 1-62.
[2] Mémoire sur les groupes kleinéens, *Acta Math.* **3** (1883), 49-92.

Pólya G. and Szegö G.
[1] Aufgaben und Lehrsätze aus der Analysis, Vol.2, *Springer-Verlag*, 1964.

Prasad G.
[1] Strong rigidity of Q-rank 1 lattices, *Invent Math.* **21** (1973), 255-286.

Prokhorov M.N.
[1] Absence of discrete groups of reflections with a non-compact fundamental polyhedra of finite volume in a Lobachevskii space of high dimension, *Izv. Akad. Nauk SSSR* **50**:2 (1986), 413-424 (= *Math. USSR-Izv.* **28**:2 (1987), 401-411).

Przytycki J.H.
[1] Incompressibility of surfaces after Dehn surgery, *Michigan Math. J.* **30** (1983), 289-308.

Pyatetskii-Shapiro I.I. and Shafarevich I.R.
[1] Torelli's theorem for algebraic surfaces of type $K3$, *Izv. Akad. Nauk SSSR Ser. Mat.* **35** (1971), 530-572 (= *Math. USSR-Izv.*).

Raghunathon M.S.
[1] Discrete subgroups of Lie groups, *Springer-Verlag*, 1972.

Reimann H.M.

[1] On the parametric representation of quasi-conformal mappings, Symposia Math. XVIII, Convegno sulle Transformarzioni Quasiconforme e Questioni Connesse, *INDAM, Rome* (1974), 421-428, *Academic Press, London*, 1974.

[2] Ordinary differential equations and quasi-conformal mappings, *Invent. Math.* **33** (1976), 247-270.

[3] Invariant extension of quasi-conformal deformations, Commentationes in honorem olli Lehto LX Annos Nato, Ed. by F.W. Gehring and I.S. Louhivaara, *Ann. Acad. Sci. Fenn., Ser.* AI, Math., **10** (1985), 477-492.

Reshetnyak Yu.G.

[1] Certain geometric properties of functions and mappings with generalized derivatives, *Sibirsk. Matem. Zh.* **7** (1966), 886-919 (= *Siberian J. Math.* **7** (1966)).

[2] Liouville's conformal mapping theorem under minimal assumptions of continuity, *Sibirsk. Mat. Zh.* **8** (1967), 835-840 (= *Siberian J. Math.*).

[3] Spatial mappings with bounded distortion, *Nauka, Novosibirsk*, 1982 (Russian).

[4] Stability theorems in geometry and analysis, *Nauka, Novosibirsk*, 1982 (Russian).

de Rham G.

[1] Differentiable manifolds, *IL, Moscow*, 1956 (Russian translation).

[2] Sur les polygones générateurs de groupes fuchsiens, *Enseignement Math.* **17**:2 (1971), 49-61.

Riley R.

[1] Discrete parabolic representations of link groups, *Matematika* **22** (1975), 141-150.

[2] An elliptical path from parabolic representations to hyperbolic structures, Topology of low-dimensional manifolds, Proc. 2nd Sussex Conf., Chelwood Gate (1977), ed. by R. Fern, Lecture Notes in Math. Vol.722, *Springer-Verlag, Berlin*, 1979, 99-133.

Robinson R.M.

[1] Undecidability and non-periodicity for tilings of the plane, *Invent. Math.* **12** (1971), 117-209.

[2] Undecidable tiling problems in the hyperbolic plane, *Invent. Math.* **44** (1978), 259-264.

Rolfsen D.

[1] Knots and links, *Publish or Perish Inc., Berkeley*, 1976.

Rosenberger G.

[1] All generating pairs of all two-generator Fuchsian groups, *Arch. Math.* **46** (1986), 198-204.

Rosenlicht M.

[1] On quotient varieties and the affine embedding of certain homogeneous spaces, *Trans. Amer. Math. Soc.* **101** (1961), 211-223.

Rourke C.P. and Sanderson B.J.

[1] Introduction to piecewise linear topology, *Springer-Verlag, Berlin-New York*, 1982.

Satake I.

[1] On a generalization of the notion of a manifold, *Proc. Nat. Acad. Sci. USA* **42** (1956), 359-363.

Scharlemann M.

[1] Smooth spheres in $\mathbb{R}^n$ with four critical points are standard, *Invent. Math.* **79** (1985), 125-141.

Schwarzenberger R.L.E.

[1] *n*-dimensional crystallography, Research Notes in Math., Vol.41, *Pitman, Boston, Mass. - London*, 1980.

Scott G.P.

[1] Finitely generated 3-manifold groups are finitely presented, *J. London Math. Soc.* **6** (1973), 437-440.

[2] Compact submanifolds of 3-manifolds, *J. London Math. Soc.* 7 (1973), 246-250.

[3] Fundamental groups of non-compact 3-manifolds, *Proc. London Math. Soc.* **34**:3 (1977), 303-326.

[4] The geometries of 3-manifolds, *Bull. London Math. Soc.* **15** (1983), 407-487.

[5] There are no fake Seifert fibre spaces with infinite π_1, *Ann. of Math.* **117** (1983), 35-70.

Seifert H.

[1] Topologie drei-dimensionalen gefasserts Räume, *Acta Math.* **60** (1932), 147-288.

[2] Komplexe mit Seitenzuordnung, *Nachr. Akad. Wiss., Göttingen* **6** (1975), 49-80.

Seifert H. and Threlfall W.

[1] Topologie, *Deterich'sche Verlagsbuchhandlung, Leipzig* (1948).

Selberg A.

[1] On discontinuous groups in higher-dimensional symmetric spaces. Contributions to Function Theory, *Tata Institute, Bombay* (1960), 147-164.

[2] Recent developments in the theory of discontinuous groups of motions of symmetric spaces, Proc. 15 Scand. Congress, *Springer-Verlag*, 1970, 99-120.

Semenov V.I.
[1] One-parameter groups of quasi-conformal homeomorphisms in Euclidean space, *Sibirsk. Mat. Zh.* **17** (1976), 177-193. (= Siberian Math. J. **17** (1976), 140-152.)

Serre J.P.
[1] Trees, (translated from French), *Springer-Verlag, Berlin-New York*, 1980.

Serrin J.
[1] Mathematical principles of classical fluid mechanics, *Springer-Verlag, Berlin* (1959).

Shafarevich I.R.
[1] (Editor): Algebraic surfaces, *Trudy Mat. Inst. Steklov* **75** (1965). (= *Amer. Math. Soc. Trans.,* **75** (1967)).
[2] Foundations of algebraic geometry, *Nauka, Moscow*, 1972. (= Basic algebraic geometry,*Springer-Verlag, New York and Heidelberg*, 1974.)

Shalen P.B.
[1] Separating incompressible surfaces in 3-manifolds, *Invent. Math.* **52** (1979), 105-127.
[2] Representations of 3-manifold groups and applications in topology, Proc. International Congress of Math., Berkeley (1986), 607-614.

Shimizu H.
[1] On discontinuous groups operating on the product of the upper half planes, *Ann. of Math.* **77** (1963), 33-71.

Siegel C.L.
[1] Discontinuous groups, *Ann. of Math.* **44** (1943), 674-689.
[2] Bemerkung zu einem Satze von Jacob Nielsen, *Math. Tidskr. B.* (1950), 66-70.

Soma T.
[1] The Gromov invariant of links, *Invent. Math.* **64** (1981), 445-454.

Spanier E.H.
[1] Algebraic topology, *McGraw-Hill, New York*, 1966.

Stallings J.R.
[1] On the loop theorem, *Ann. of Math.* **72** (1960), 12-19.
[2] On fibering certain 3-manifolds, Topology of 3-manifolds and related topies, *Prentice-Hall*, 1962, 95-100.
[3] Homology and central series of groups, *J. Algebra* **2** (1965), 170-181.

Stallings J.R. and Wolf A.R.

[1] The Todd-Coxeter process, using graphs, Combinatorial group theory and topology, Ed. S.M. Gersten and J.R. Stallings, *Ann. of Math. St.* **111** *Princeton Univ. Press, Princeton* 1987, 157-162.

Stanley R.P.

[1] The upper bound conjecture and Cohen-Macaulay rings, *Studies in Appl. Math.* **54** (1975), 135-142.

Stein E.M.

[1] Singular integrals and differentiability properties of functions, *Princeton Univ. Press, Princeton*, 1970.

Stein S.K.

[1] Tiling space by congruent polyhedra, *Bull. Amer. Math. Soc.* **80** (1974), 819-820.

Sturm J. and Shinnar M.

[1] The maximal inscribed ball of a Fuchsian group, Discontinuous groups and Riemann surfaces, edited by L. Greenberg, Ann. of Math. St. Vol.79, *Princeton Univ. Press*, 1974, 439-443.

Sulanke R. and Wintgen P.

[1] Differentialgeometrie und Faserbündel, Lehrbücher und Monographien aus dem Gebiete der exakten Wissenschaften, Math. Reihe, Band 48, *Birkhäuser Verlag , Basel-Stuttgart*, 1972.

Sullivan D.P.

[1] On the ergodic theory at infinity of an arbitrary discrete group of hyperbolic motions, Riemann surfaces and related topics: Proceedings of the 1978 Stony Brook Conference, edited by I.Kra and B. Maskit, Ann. of Math. St. Vol.97, *Princeton Univ. Press, Princeton*, 1981, 465-496.

[2] Growth of positive harmonic functions and Kleinian group limit sets of zero planar measure and Hausdorff dimension two, Geometry Symposium, Utrecht (1980) (N. Kiper 60th birthday volume), Lecture Notes in Math. Vo.894, *Springer-Verlag, Berlin-New York*, 1981, 127-144.

[3] A finiteness theorem for cusps, *Acta Math.* **147** (1981), 289-299.

[4] Discrete conformal groups and measurable dynamics, *Bull. Amer. Math. Soc.* **6** (1982), 57-73.

[5] Seminar on conformal and hyperbolic geometry (Notes by M. Baker and J. Seade), Preprint, *IHES* (1982).

[6] Entropy, Hausdorff measures old and new, and limit sets of geometrically finite Kleinian groups, *Acta Math.* **153** (1984), 259-277.

[7] Quasi-conformal homeomorphisms and dynamics, Part 2: Structural stability implies hyperbolicity for Kleinian groups, *Acta Math.* **155** (1985), 243-260.

[8] Travaux de Thurston sur les groupes quasi-fuchsiens et les variétés hyperboliques de dimension 3 fibrés sur S^1, Bourbaki Seminar, Vol.1979/80, Lecture Notes in Math. Vol.842, *Springer-Verlag, Berlin-New York*, 1981.

Sullivan D.P. and Thurston W.P.
[1] Manifolds with canonical coordinate charts, some examples, *Enseign. Math.* (2) **29** (1983), 15-25.

Susskind P.D. and Swarup G.A.
[1] Limit sets of gemetrically finite hyperbolic groups, Preprint (1988).

Teichmüller O.
[1] Extremale quasiconforme Abbildungen und quadratische Differentiale, *Abh. Preuss. Akad. Wiss., Math.-Naturw. Kl.* **22** (1940), 3-197.

Tetenov A.V.
[1] Invariant components of Kleinian groups in space, Proceedings of the 17th All-Union Scientific Students' Conference, Mathematics, *Novosibirsk State Univ.*, 1979, 134-141 (Russian).
[2] Finitely generated Kleinian groups in space, *Sibirsk. Mat. Zh.* **21**:5 (1980), 88-99 (= *Siberian Math. J.* **21** (1980), 709-717.
[3] Locally finite fundamental domains of discrete groups in space, *Sibirsk. Mat. Zh.* **23** (1982), 175-178. (= *Siberian Math. J.* **23** (1982)).
[4] On a number of invariant components of a Kleinian group in space, Preprint No.17, Institute of Mathematics, Siberian Division, *Academy of Sciences of the USSR, Novosibirsk* (1982) (Russian).
[5] Kleinian groups in space and their invariant components, Ph.D. Thesis, Institute of Mathematics, Siberian Division, Academy of Sciences of the USSR, Novosibirsk (1982) (Russian).
[6] The discontinuity set and quotient manifold of a Kleinian group, to appear.

Thurston W.P.
[1] The geometry and topology of 3-manifolds, Lecture Notes, Princeton Univ., Ch.1-9 (1978/79): Ch.11, 13 (1980).
[2] Hyperbolic structures on 3-manifolds, I: Deformations of acylindrical manifolds, *Ann. of Math.* **124**:2 (1986), 203-246.
[3] Hyperbolic structures on 3-manifolds, II: Surface groups and 3-manifolds which fibre over a circle, Preprint, *Princeton Univ.* (1980/1986).
[4] Three-dimensional manifolds, Kleinian groups and hyperbolic geometry, *Bull. Amer. Math. Soc.* **6** (1982), 357-381.
[5] Universal links, Preprint, *Princeton Univ.* (1982).
[6] A norm for the homology of 3-manifolds, *Mem. Amer. Math. Soc.* **59** (1986).
[7] Hyperbolic structures on 3-manifolds, III: Deformations of 3-manifolds with incompressible boundary, Preprint (1980/1986).

[8] Earthquakes in two-dimensional hyperbolic geometry, Low-dimensional topology and Kleinian groups, (Coventry and Durham 1984), London Math Soc. Lect. Note Series **112**, Cambridge Univ. Press, 1986, 91-110.

Tietze H.
[1] Über die topologischen Invarianten mehrdimensionaler Mannigfaltigkeiten, *Monatsh. Math. Phys.* **19** (1906), 1-118.

Tsuji M.
[1] Potential theory in modern function theory, *Maruzen, Tokyo*, 1959.

Tucker T.W.
[1] A correction to a paper by A. Marden, *Ann. of Math.* (2) **102** (1975), 565-566.

Tukia P.
[1] A quasi-conformal group not isomorphic to a Möbius group, *Ann. Acad. Sci. Fenn. Ser.* AI **6** (1981), 149-160.
[2] On two-dimensional quasi-conformal groups, *Ann. Acad. Sci. Fenn. Ser.* AI **5** (1980), 73-78.
[3] The Hausdorff dimension of the limit set of a geometrically finite Kleinian group, *Acta Math.* **152** (1984), 127-140.
[4] Quasi-conformal extension of quasi-symmetric mappings compatible with a Möbius group, *Acta Math.* **154** (1985), 153-193.
[5] On isomorphisms of geometrically finite Möbius groups, *Inst. Hautes Etudes Sci. Publ. Math.* **61** (1985), 171-214.
[6] On limit sets of geometrically finite Kleinian groups, *Math. Scand.* **57** (1985), 29-43.
[7] A rigidity theorem for Möbius groups, *Invent. Math.*, **97** (1989), 405-431.
[8] Differentiability and rigidity of Möbius groups, *Invent. Math.* **82**:3 (1985), 557-578.
[9] On quasi-conformal groups, *J. Analyse Math.* **46** (1986), 318-346.

Väisälä J.
[1] Lectures on n-dimensional quasi-conformal mappings, Lecture Notes in Math. Vol.229, *Springer-Verlag, Berlin - NewYork,* 1971.
[2] Bi-Lipschitz and quasi-symmetric extension properties, *Ann. Acad. Sci. Fenn. Ser* A1 **11** (1986), 239-274.

Venkov B.A.
[1] On arithmetic group of automorphisms of an indefinite quadratic form, *Izv. Akad. Nauk SSSR* **2** (1937), 139-170 (Russian).
[2] On a class of Euclidean polyhedra, *Vestnik of Leningrad State Univ., Ser. Math-Phys.* **2** (1954), 11-31 (Russian).

Vinberg E.B.
[1] Discrete groups generated by reflections in Lobachevskii spaces, *Mat. Sb.* **72** (1967), 471-488.

[2] Some examples of crystallographic groups in Lobachevski spaces, *Mat. Sb.* **78** (1969), 633-639.
[3] Discrete linear groups that are generated by reflections, *Izv. Akad. Nauk SSSR Ser. Mat.* **35** (1971), 1072-1112.
[4] The groups of units of certain quadratic forms, *Mat. Sb.* **87** (1972), 18-36. (= *Math. USSR-Sb.* **87** (1972), 17-35.)
[5] The unimodular integral quadratic forms, *Funkts. Analiz. i Pril.* **6** (1972), 24-31.
[6] Some arithmetical discrete groups in Lobachevskii spaces. Discrete subgroups of Lie groups and applications to moduli (International Coll. Bombay, January 1973), *Oxford University Press, Bombay*, 1975, 323-348.
[7] Absence of crystallographic groups of reflections in Lobachevskii spaces of large dimension, *Trudy Moskov. Mat. Obshch.* **47** (1984), 68-102.
[8] Discrete groups of reflections in Lobachevskii spaces of large dimension, Proc. International Congress Math (Warsaw, August 1983), Warsaw (1984), 585-592.
[9] The two most algebraic *K*3 surfaces, *Math. Ann.* **265** (1983), 1-21.
[10] Hyperbolic groups of reflections *Uspekhi Mat. Nauk* **40** (1985), 29-66. (= *Russian Math. Surveys* **40**:1 (1985), 31-75.)

Vinberg E.B. and Kaplinskaya I.M.
[1] The groups $O_{18,1}(Z)$ and $O_{19,1}(Z)$, *Doklad. Akad. Nauk* **238** (1978), 1273-1275 (= *Soviet Math. Dokl.* **19** (1978), 194-197).

Van Vleck E.B.
[1] On the combination of non-loxodromic substitutions, *Trans. Amer. Math. Soc.* **20** (1919), 299-312.

Volynets I.A.
[1] Fuchsian groups in even-dimensional spaces, *Sibirsk. Matem. Zh.* **27**:3 (1986), 200-203 (= *Siberian Math. J.* **27** (1986)).

Vuorinen M.
[1] Conformal geometry and quasi-regular mappings, Lecture Notes in Math. **1319**, *Springer-Verlag*, 1988.

Wada M.
[1] Conjugacy invariants and normal forms of isometries of hyperbolic space, Ph.D. Thesis, Columbia Univ. (1986).

Waldhausen F.
[1] Eine Verallgemeinerung des Schleifensatzes, *Topology* **6** (1967), 501-504.
[2] Gruppen mit Zentrum und 3-dimensionale Manigfaltigkeiten, *Topology* **6** (1967), 505-517.
[3] On irreducible 3-manifolds which are sufficiently large, *Ann. of Math.* **87** (1968), 56-88.

[4] Eine Klass von 3-dimensionalen Manigfaltigkeiten, *Invent. Math.* (1967), 308-333; **4** (1967), 88-117.

Wall C.T.C.
[1] Geometries and geometric structures in real dimension 4 and complex dimension 2, Geometry and topology, Lecture Notes in Math. **1167** *Springer-Verlag, Berlin- New York*, 1986, 286-292.

Wang Hsen Chung
[1] Topics on totally discontinuous groups. Symmetric spaces (Pure Appl. Math. Vol.8), *Dekker, New York*, 1972, 460-485.

Waterman P.L.
[1] An inscribed ball for Kleinian groups, *Bull. London Math. Soc.* **16** (1984), 525-530.
[2] Fuchsian groups and algebraic number fields, Ph.D. Thesis, Univ. of Aberdeen (1982).

Waterman P. and Wolpert S.
[1] Earthquakes and tessellations of Teichmüller space, *Trans. Amer. Math. Soc.* **278** (1983), 157-167.

Weber C. and Seifert H.
[1] Die beiden Dodekaederäume, *Math. Z.* **37** (1983), 237-253.

Weeks J.
[1] Hyperbolic structures on three manifolds, Ph.D. Thesis, Princeton Univ. (1985).

Weil A.
[1] On discrete subgroups of Lie groups, Part 1, *Ann. of Math.* **72** (1960), 369-384.
[2] On discrete subgroups of Lie groups, Part 2, *Ann. of Math.* **75** (1962), 578-602.
[3] Remarks on the cohomology of groups, *Ann. of Math.* **80** (1964), 149-157.

Weyl H.
[1] Theorie der Darstellung kontinuierlicher half-einfacher Gruppen durch lineare Transformationen, *Math. S.* **23** (1925), 271-309; **24** (1926), 328-395, 789-791.

Whitehead J.H.C.
[1] Combinatorial homotopy, *Bull. Amer. Math. Soc.* **55** (1949), 211-245.
[2] On 2-spheres in 3-manifolds, *Bull. Amer. Math. Soc.* **64** (1958), 157-167.

Wielenberg N.J.
[1] Discrete Möbius groups: fundamental polyhedra and convergence, *Amer. J. Math.* **99** (1977), 861-877.

[2] The structure of certain subgroups of the Picard group, *Math. Proc. Cambridge Phil. Soc.* **84** (1978), 427-436.
[3] Hyperbolic 3-manifolds which share a fundamental polyhedron, Riemann surfaces and related topics: Proceedings of the 1978 Stony Brook Conference, edited by I. Kra and B. Maskit, Ann. of Math. St. Vol.97, *Princeton Univ. Press* (1981), 505-513.

Wolf J.A.
[1] Spaces of constant curvature, University of California, Berkeley, 1972.

Wolpert S.
[1] The Fenchel-Nielsen deformation, *Ann. of Math.* **115** (1982), 501-528.

Yamamoto H.
[1] Constructibility and geometric finiteness of Kleinian groups, *Tôhoku Math. J.* **32** (1980), 353-362.

Zieschang H, Vogt E. and Coldewey H.D.
[1] Surfaces and planar discontinuous groups,(translated from German) Lecture Notes in Math. Vol.835, *Springer-Verlag, Berlin*, 1980.

Zimmer R.J.
[1] Ergodic theory and semisimple groups, Monographs in Math. **81**, *Birkhauser Verlag, Basel-Boston Mass.*, 1984.

INDEX

accidental parabolic element 265
affine structure 35
Ahlfors finiteness theorem 141
Ahlfors 0-measure problem 138, 305
Aleksandrov theorem on space tessellation 192
Aleksandrov theorem, generalization 200
Alexander sphere theorem 282
algorithmic problems 231
Andreev hyperbolic polyhedron classification 111
angle regular tessellation 118
approximation point 128, 246
arithmetic co-compact group construction 85
arithmetic group 113
arithmetic hyperbolic volume 328
arithmetic manifold 328
atom 56
automatic structure 235
automorphic function 242, 374
automorphism of a ball 13
automorphism of a half-space 11
averaging operator on functions and measures 244

Beardon-Maskit conditions 132
Beltrami differential 370, 381
bending deformation 383, 387, 397
bending hyperplane 393
bending hypersurface 392
bending homeomorphism 394
bending, infinitesimal deformation 399
bending of manifold 383, 392
bending of Riemann surface 123, 383
bending line of pleated surface 299
Bers-Kra decomposition theorem 379
Bers-Krushkal' problem on deformations 370, 377
B-group 63, 265, 345
Bieberbach decomposition 54
Bieberbach group 54, 94

Bieberbach theorem 54, 94
binary icosahedron group 312
block-building construction 275
Bonahon end theorem 299
Bonahon compactification of manifold 302
Borel series 142
Borel-Wallach problem 383
Borromean rings 217, 389
Bott integral 329
boundary element of a group 285
boundary group 410
boundary homomorphism 23
boundary of deformation space 409
boundary of group 225
boundary of manifold 1
boundary of Nielsen hull 122, 176
branching order 308
Busemann function 31

canonical cutting of manifold 311, 317
Caratheodory convergence 241
Cartan-Hadamard theorem 310
Cayley graph 223
Cayley graph, embedding 226
Cayley graph, geodesics 224
Cayley graph, tree 225
characteristic of quasi-conformal mapping 371, 381
circle bundle over 2-torus 314
classical uniformization 307
C-matrix 110
h-cobordism 268, 271
h-cobordism with boundary 268
combination of groups 204, 207, 212
combination of groups, topological aspects 259
combination, parabolic 345
compact core of manifold 292
compact-open topology 416
compactification of manifold 290, 301
complement of knot 320
complex of a filling 192
complex of a filling, regular 193
complex, singular 23
complexity of manifold 330
component of discontinuity set 55
component of discontinuity set, invariant 55, 154
component of discontinuity set, stabilizer 55, 150
component of discontinuity set, non-maximal stabilizer 286
component of discontinuity set, thick 165, 168
component of discontinuity set, thin 167
composition of knots 420

composition of knots, genus 420
conformal homeomorphism 3, 333
conformal manifold 35, 333
conformal sphere 4
conformal structure 333, 372
conformal structure, almost uniformized 338
conformal structure, marked 355
conformal structure, non-uniformized 338, 339
conformal structure, uniformized 334, 338, 343
conformal uniformization 334, 345
conformal uniformization, conjecture 334, 343
conjugacy class 357
conjugation problem 231
connected sum of manifolds 205, 257, 335
convergence of groups, algebraic 237
convergence of groups, geometric 237
convergence of groups, polyhedral 239
convergence of groups, strong 237
convergence of group representations 237
convergence of manifolds 204, 237, 321
convex core 125
convex Nielsen hull 122, 168
convex Nielsen hull, boundary 122, 176
convex space 310
covering homotopy 20
covering mapping (projection) 20
covering, planar 309
covering of orbifold 48
covering, regular 21, 309
covering space 20
covering, universal 21, 49
coverings, equivalent 21
Coxeter polyhedron 108
Coxeter diagram 109
crossed homomorphism 362
crossed homomorphism for bending 399
cup-product 365
cup square 365
cusp domain 127
cusp domain, hyperbolic 128, 253
cusp end 96, 251
cusp end, standard 143, 253
cusp point 127
cycle of an edge 71
cylinder theorem 255

deck transformation 21
decidability of the word problem 234
Dedekind ζ-function of a field 328
defining subgroup of a covering 22

defining relation 72, 231
deformation of group 361, 387, 416
deformation of group, bending 383, 392, 397
deformation of group, quasi-conformal 361, 416
deformation of group, infinitesimal 367, 398
deformation of group, stamping 400
deformation of group, stamping-with-torsion 404
deformation of group, stretching 416
deformation of structure 360
deformation of structure, infinitesimal 365
deformation space 381, 387
deformation space, singularities 398, 402
deformation space, non-connectedness 412
deformation, trivial 361, 396
deformations, equivalent 361
Dehn algorithm 232
Dehn reduced word 232
Dehn lemma 255
Dehn surgery 258, 324
Dehn surgery, generalized 258
Dehn surgery, parameters 258
Dehn surgery along Borromean rings 323
Dehn surgery along figure-eight knot 322
Dehn surgery along Whitehead link 323
density point 376
developing map (development) 41, 49
deviator of deformation tensor 363
dilatation of quasi-conformal mapping 2
Dirac measure 245
Dirichlet plane 68
Dirichlet polyhedron 68, 133
disc sum of manifolds 257
discontinuity set 52
discontinuum 206
discrete representation of a group 237
dodecahedron Poincaré space 311
double of manifold 263, 346

edge relation 72
Efremovich-Tikhomirova extension 367
Eilenberg-MacLane cohomology 362
elliptic space 26
elliptic transformation 15
elliptic transformation, prime 84
end of manifold 290
end of manifold, almost tame 293
end of manifold, classification 292
end of manifold, geometrically finite 292
end of manifold, geometrically tame 293
end of manifold, isolated 290

end of manifold, parabolic 96, 291
end of manifold, simply degenerate 300
end of manifold, standard cusp 143, 253
end of manifold, topologically tame 297
Enriques surface 115
exact sequence 24
exceptional fibre of Seifert space 315
excision of knots and links 257
excisive triad 25
extreme point of convex hull 170
Euclidean space form 150, 252
Euclidean structure 35

Fenchel-Nielsen theorem 283
fibre bundle over circle 319
fibre S^1-bundle over surface 312
fibred solid torus 315
figure-eight knot complement 35
filling a space 192
filling a space, incomplete 193
filling a space, locally finite 197
filling a space, locally similar 194
filling a space, non-regular 193
filling a space, projection 192
filling a space, regular 193
finiteness theorems for Kleinian groups 141, 253
foliation 297
Fox-Artin wild arc 272
Fox-Artin sphere 273
free Abelian group of rank k 40
free group 225
free product of groups 204, 260
free amalgamated product of groups 207, 261
Fuchsian equivalent 379
Fuchsian representation of a group 360
fundamental group 20
fundamental group of an orbifold 48
fundamental group, automorphism 290, 366
fundamental domain 64
fundamental domain in set of geodesics 247
fundamental domain in the limit set 75
fundamental domain, homotopy type 412
fundamental domain, local finiteness 118, 121
fundamental polyhedron 117
fundamental polyhedron, generic 240
fundamental polyhedron theorem 72, 200
fundamental polyhedron, Dirichlet 68
fundamental polyhedron, Ford (isometric) 65, 67
fundamental polyhedron, pseudo-Dirichlet 69
fundamental polyhedron, with convex ends 144

Garland-Raghunathan rigidity theorem 203
Garnet point 249
genus of a knot 420
genus of a Riemann surface 206
geodesic 2
geodesic, asymptotic 31
geodesic, bending 170, 176
geodesic, limit bending 176, 179
geodesic currents 300
geodesic, glancing 235
geodesic lamination 170, 300
geometric isomorphism of groups 283
geometric structure 34, 355
geometric structure on link/knot complement 35, 211, 320
geometric uniformization 310
geometries, homogeneous 311
$\mathcal{G}$-manifold 35
Golubev non-rigid groups 372
good orbifold 48
good representation of a group 358
$\mathcal{G}$-orbifold 44
Gram matrix 109
graph-manifold 330
Green's measure 245
Green's function 243
Gromov invariant 324
Gromov hyperbolic group 237
Gromov norm 324
Gromov theorem on volumes 324
Gromov concept of uniformization 310
group, Abelian 53
group action, conservative 242, 249
group action, dissipative 242, 249
group action, ergodic 242
group action, free 54
group action in horoballs 101, 142
group action on a tangent space 107
group action on a vector field 364
group, co-compact 85
group, co-finite 113
group of compact type 139
group, crystallographic 108
group, crystallographic, classification 111
group, deck transformation 21
group decomposition 389
group, degenerate 63, 410
group, discontinuous 51
group, discrete 51
group, elementary 52
group, free 83, 225

group, free Abelian 40, 205
group, Fuchsian 61, 78
group, function 63, 263
group, fundamental 20, 48
group, generated by reflections 108
group, geometrically finite 126, 132
group, geometrically infinite 141, 153
group, geometrically non-tame 297
group, geometrically tame 293
group, Heisenberg 314
group, homology 23
group, Kleinian 51
group, Lorentz 32
group, Möbius 6
group, modular 358
group, nilpotent 314
group, non-degenerate B-group 265, 345
group, non-rigid 368, 370
group, orthogonal 32, 360
group, partially degenerate 410
group, pea-pod 180
group, Picard 51
group, Picard, subgroups 211
group, quasi-Fuchsian 61, 180, 383
group, rigid 367, 371
group, Schottky 61, 265
group, Schottky-type 265
group, solvable 314
group, unimodular 311
group, web 61, 185
group with limit topological circle 419
(G, X)-homeomorphism 41, 333
(G, X)-manifold 35
(G, X)-manifold, complete 42, 49
(G, X)-orbifold 45
(G, X)-structure 35, 355
(G, X)-structure, almost uniformized 338
(G, X)-structure, uniformized 338

Haken manifold 315
Hamiltonian system 329
Hamiltonian field 329
Hamiltonian field, integrability obstruction 329
handlebody 261
harmonic function 306
Hausdorff dimension 139
Hausdorff dimension of the limit set 139
Hausdorff measure 139
Hausdorff metric 140
Hausdorff topology 140

Heegard splitting 259
Heisenberg group 314
Hermit-Mahler compactness criterion 86
h-cobordism 268
h-cobordism with boundary 268
HNN-extension 212
holonomy homomorphism 41, 49, 334
holomony theorem 358
homogeneous domain 16
homogeneous space 311, 355
homology 22
homology cobordism 267
homology cobordism with boundary 268
homotopy class 19
homotopy equivalence 19
homotopy type 20
homotopy of mappings 19
homotopy of paths 20
Hopf manifold 203
horoball 31, 101
horoball, radius 108
horosphere 30, 101
horosphere, critical 249
horospherical limit point 249
hyperbolic cobordism 267
hyperbolic cobordism, non-trivial 271
hyperbolic metric 13, 27-30
hyperbolic retraction 123
hyperbolic (Lobachevskii) space: ball model 29
hyperbolic (Lobachevskii): half-space Poincaré model 30
hyperbolic (Lobachevskii): hyperboloid model 27
hyperbolic (Lobachevskii): projective model 28
hyperbolic (Lobachevskii): pseudo-Euclidean model 27
hyperbolic (Lobachevskii): lower hemisphere model 29
hyperbolic structure 35, 321
hyperbolic structure, complete 198, 200
hyperbolic structure, geometrically finite 144
hyperbolic structure, marked 355
hyperbolic surface 356
hyperbolic simplex 36
hyperbolic ideal simplex 324
hyperbolic ideal simplex, complex parameters 196
hyperbolic ideal simplex of maximal volume 324
hyperbolic regular simplex 37, 324
hyperbolic transformation 15
hyperbolic volumes, set 322
hyperbolic volume, lowest 327
hyperbolic volumes, ordinals 317
hyperbolic 2-plane bundle over surface 421

icosahedral point 84
icosahedron group 312
incompressible torus 219, 316
index of fibred solid torus 315
index of subgroup 34
invariant component 54, 154
inexact element 286
irrational parabolic group 69
irrational parabolic point 69
irrational screw translation 69, 96
isometric sphere 9, 67
isometry, Euclidean 3, 6
isometry, hyperbolic 13, 27-30
isometry, spherical 27
isomorphism of groups 283
isomorphism of groups, geometrical 283
isomorphism of groups, intersection-preserving 287
isomorphism of groups, orientation consistent 288
isomorphism of groups, orientation-preserving 288
isomorphism of groups, orientation-reversing 288
isomorphism of groups, strongly type-preserving 287
isomorphism of groups, type-preserving 283
isoenergy surface 330

Jacobi matrix 2
Jacobian, generalized 66
Jaco-Shalen mapping theorem 316
Jaco-Shelen splitting theorem 316
Jørgensen convergence effect 238
Jørgensen group 404
Jørgensen inequality 83
Jørgensen hyperbolic volume theorem 322

Khovanskii-Prokhorov finiteness theorem 115
Klein bottle 40
Klein combination 204
Klein-Poincaré uniformization theorem 308
Kneser theorem 255
knot complement structures 320
knot, polygonal 419
knot, satellite 320
knot, tame 420
knot, torus 320
knot, wild 419
Koebe theorem on cuts 309
Koebe uniformization principle 308
Kolmogorov-Alexander cup product 365
Krein-Milman theorem 170
Krushkal'-Mostow problem on deformations 367

*K*3-type surface 115
Kulkarnbi-Pinkall uniformization theorems 335, 338

lamination 299
Lanner simplexes 110
lattice 86
leaf of lamination 299
length function 310
length space 310
length space, convex 310
lens space 331
Lie algebra of a group 362
Lie bracket 365
Lie subgroups of the Möbius group 17
lifting paths 21
limit set 52
limit set as the boundary of group 229
limit set, horospherical 248
limit set, wildly knotted 281, 419
Liouville theorem 8
Liouville torus 329
Liouville conformality condition 363
Lobachevskii (hyperbolic) space 27
locally convex space 310
locally homogeneous structure 355
locally homogeneous structure, marked 355
loop 20
loop theorem 255
loops product 20
Lorentz group 32
Lorentz transformation 32

manifold, affine 35
manifold, aspherical 256
manifold, Bieberbach 150, 252, 312
manifold, closed 1
manifold, conformal 35, 333
manifold, conformally flat 334
manifold, elliptic 35
manifold, Euclidean 35
manifold, geometrically atoroidal 316
manifold, geometrically non-tame 297
manifold, geometrically tame 293, 300
manifold, Haken 315
manifold, homotopically atoroidal 316
manifold, Hopf 203
manifold, hyperbolic 35
manifold, hyperbolic rigid 367
manifold, hyperbolic, of finite volume 321
manifold, hyperbolic, of minimal volume 327

manifold, hyperbolic, topology 318
manifold, irreducible 315
manifold, Kleinian 251
manifold, locally convex 310
manifold, non-Haken 318
manifold of class $\mathbb{H}$ 330
manifold, piecewise linear 35
manifold, Poincaré 256
manifold, prime 315, 317
manifold, Riemannian 1
manifold, Riemannian, conformally flat 334
manifold, smooth 1, 35
manifold, sufficiently large 315
manifold, topological 1
manifold, universal 257
mapping, conformal 3, 333
mapping, G-equivariant 186
mapping, proper 198
mapping, quasi-conformal 2
mapping, quasi-regular 386
mapping torus hyperbolic theorem 319
Marden isomorphism theorem 383
Marden theorems on hyperbolic 3-manifold 259
Margulis constant 93, 291
Margulis cusp 97, 291
Margulis lemma 93
Margulis manifold decomposition 92, 143, 291
Margulis tube 97, 291
marked surface 356
Maskit combination theorems 207, 213
Maskit theorems on planar coverings 309
matrix representation of Möbius group 32
maximal ball 169
maximal ball covering 170, 173
maximal disc 170
maximal conical domain 172
maximal cylinder-domain 171
Mayer-Vietoris sequence 24
Mayer-Vietoris sequence, absolute 25
Mayer-Vietoris sequence, reduced 26
Mayer-Vietoris sequence, relative 25
measured lamination 301
metrics on Cayley graph 223, 225
Micky Mouse example 122
Möbius group 6
Möbius homogeneous domain 16
model polyhedron 192
moduli problem 355
modulus of sector box 374
Mostow's rigidity theorem 366

Nielsen convex hull 122, 168
Nielsen theorem 354
Nikulin finiteness theorems 113
Nikulin upper bound for polyhedron side complexity 114
nil-geometry 313
non-maximal stabilizer 286
non-measurable set 75
non-peripheral subgroup 215, 218
normal subgroup spananed by loops 309
Novikov conjecture 271

octahedral space filling 195
orbifold 43
orbifold, barber shop 45
orbifold, billiard table 45
orbifold, conformal 45
orbifold, covering 48
orbifold, good 49
orbifold, hyperbolic 45, 198
orbifold, hyperbolic, minimal volume 327
orbifold, singular set 45
orbifold, totally geodesic 392, 398
orbifold with boundary 49
orbifold, universal covering 49
orbit 52
order relation on bending plane set 393
oricycle stretching 417

parabolic group 54
parabolic group of irrational type 69
parabolic group, complex 359
parabolic Ω-peak 107
parabolic transformation 15
parabolic cusp 127
parabolic point, rank 127
path metric 169
path, quasi-geodesic 233
path, rectifiable 169
pea-pod group 180
P-harmonic function 244
Picard group 51, 212
planar surface 309
pleated surface 292
Poincaré conjecture 256, 258, 317
point of approximation 128, 247
Poisson formula 244
Poisson measure 241
polyhedron 117, 193
polyhedron side 117, 193

precisely invariant domain 101
precisely invariant horoball theorem 98, 101
projective complex space 115
projective space 26
projective structure 35, 337
pseudo-Dirichlet plane 69
pseudo-Dirichlet polyhedron 69
pseudo-isometry 226, 366
punctured solid cylinder 253
punctured solid Euclidean space form 252
punctured solid torus 150, 253

quadratic form 85, 113
quadratic form, admissible 113
quadratic form, automorphism group 27, 85, 113
quadratic form, reflective 113
quadratic form, 2-reflective 115
quasi-algebraic set 360
quasi-conformal mapping 2
quasi-conformal deformation of group 361, 416
quasi-conformal vector field 363
quasi-conformality coefficient 2
quasi-Fuchsian group 61, 383, 409
quasi-Fuchsian representation 360, 383
quasi-Fuchsian representation, space 360
quasi-Fuchsian representation, space boundary 409
quasi-Fuchsian representation, space singularities 400
quasi-hyperbolic metric 190
quotient manifold 54
quotient variety of representations 358

recursive function 234
reflection 5
regular fibre of Seifert space 315
Reimann-Thurston extension theorem 285
Riemannian manifold 1
Riemannian metric 2
Riemann space 26
Riemann surface, marked 356
rigid group 367, 380
rigidity of hyperbolic manifold 365
rigidity of hyperbolic tessellation 202
rigidity theorems 202, 365, 371, 378
R-reduced word 233

sector box 375
Seifert fibre space 312, 315
Seifert fibre space, almost trivial 312
Seifert-van Kampen theorem 255
side pairing 259

similarity 9
similarity structure 336
simple polyhedron 111
singular set of orbifold 45
singularity of Nielsen hull boundary 185
size of hyperbolic manifold 8
slide reflection 347
sol-geometry 314
solid angle of orbit 243
solid torus 321
space of constant curvature 26
space of developments 355
space of quasi-Fuchsian representations 360, 383
special spine 330
special spine, vertex 330
spine of manifold 330
spine of manifold, almost special 330
sphere theorem 256
spherical layer 373
spherical metric 5, 242, 373
spherical structure 35
stable representation of group 358
standard simplex 23
stereographic projection 4
Stolz domain 129
suborbifold 48
Sullivan convex hull theorem 186
Sullivan cusp finiteness theorem 142
Sullivan ergodic theorems 248, 249
Sullivan rigidity theorem 371
Sullivan stability theorem 416
surface layer 267, 344
surface uniformation 308

tangent fibre space 1
thick submanifold 92, 291
thin submanifold 92, 143, 291
Thurston hyperbolization theorem 320
Thurston non-rigidity theorem 204
Thurston structural conjecture 310, 317, 318
Thurston structural theorem for knot complements 320
Teichmüller metric 357, 382
Teichmüller space of Fuchsian group 381
Teichmüller space of Riemann surfaces 355
Teichmüller space of Seifert manifold 422
tessellation of a space 193
tessellation of a space, regular 193, 200
tessellation of a space, Penrose 193
Tetenov group with k invariant component 56
Tetenov hyperbolic cobordism theorem 268

Tetenov finiteness theorem on invariant component 154, 162
Tietze transformation 280
Todd-Coxeter algorithm 234
topology of algebraic convergence 237, 355
topology, compact-open 416
topology of geometric convergence 237, 321
topology of development convergence 355
topology on Möbius group 51
torus 40
torus bundle over circle 315, 337
torus sum of manifolds 344
torus layer 40
totally geodesic submanifold 392, 396
totally geodesic surface 401
trace function 83
transcendentals 83
Tukia compactness principle 139
Tukia isomorphism theorems 285, 289
Tukia limit set homeomorphism 285
Tukia theorem on limit set dimension 139
two-reflection 115

uniformization, conformal 333, 344
uniformization, conformal, conjecture 334
uniformization, criteria 339
uniformization, geometrical 310
uniformization principle, general 307
uniformized (G, X)-structure 338
unipotent element 107
universal group 258
universal link 216, 257

variety of representations 358
variety of representations, quotient 358
vector field, left-invariant 364, 372, 399
Venkov's arithmetical space forms 85
Vinberg's finiteness theorem 114
virtually Abelian group 53, 336
virtually discrete group 339
virtually nilpotent group 336
virtually solvable group 336
visible dimension 78
visual boundary of a group 224

Waldhausen theorem 284
wandering point 339
Wang theorem 321
Web group 61, 185
well-ordering in hyperbolic volumes 325
Whitehead group 272

Whitehead link 38
wild limit sphere 281
wild limit knot 419
wildly quasi-symmetric embedding 282
word metric 223
word problem 232

Zassenhaus neighbourhood 84

9 780792 302162

Batch number: 10824416

Printed by Printforce, the Netherlands